Textbook of Veterinary Physiological Chemistry

▼

Third Edition

Larry R Engelking

Professor of Biomedical Sciences, Emeritus
Cummings School of Veterinary Medicine
Tufts University

AMSTERDAM • BOSTON • HEIDELBERG • LONDON
NEW YORK • OXFORD • PARIS • SAN DIEGO
SAN FRANCISCO • SINGAPORE • SYDNEY • TOKYO
Academic Press in an imprint of Elsevier

ELSEVIER

Academic Press is an imprint of Elsevier
32 Jamestown Road, London NW1 7BY, UK
225 Wyman Street, Waltham, MA 02451, USA
525 B Street, Suite 1800, San Diego, CA 92101-4495, USA

Copyright © 2015, 2011 Elsevier Inc. All rights reserved.

No part of this publication may be reproduced, stored in a retrieval system or transmitted in any form or by any means electronic, mechanical, photocopying, recording or otherwise without the prior written permission of the publisher. Permissions may be sought directly from Elsevier's Science & Technology Rights Department in Oxford, UK: phone (+ 44) (0) 1865 843830; fax (+44) (0) 1865 853333; email: permissions@elsevier.com. Alternatively, visit the Science and Technology Books website at www.elsevierdirect.com/rights for further information.

This book and the individual contributions contained in it are protected under copyright by the Publisher (other than as may be noted herein).

First Edition. 2004 Teton NewMedia.

Notices
Knowledge and best practices in this field are constantly changing. As new research and experience broaden our understanding, changes in research methods, professional practices, or medical treatment may become necessary.

Practitioners and researchers must always rely on their own experience and knowledge in evaluating and using any information, methods, compounds, or experiments described herein. In using such information or methods they should be mindful of their own safety and the safety of others, including parties for whom they have a professional responsibility.

To the fullest extent of the law, neither the Publisher nor the author, contributors, or editors assume any liability for any injury and/or damage to persons or property as a matter of product liability, negligence or otherwise, or from any use or operation of any methods, products, instructions, or ideas contained in the material herein.

British Library Cataloguing-in-Publication Data
A catalogue record for this book is available from the British Library.

Library of Congress Cataloging-in-Publication Data
A catalog record for this book is available from the Library of Congress.

ISBN : 978-0-12-391909-0

For information on all Academic Press publications visit our website at store.elsevier.com.

14 15 16 17 10 9 8 7 6 5 4 3 2 1

Table of Contents

▼

Section III Examination Questions 238

Section III Addendum / Section IV Introduction 252

Section IV: Vitamins and Trace Elements
  Chapter

Section IV Examination Questions 330

Acknowledgments

▼

Gratitude is expressed for major contributions by **Dr. Neal Brown**, from the Department of Pharmacology and Molecular Toxicology at the University of Massachusetts Medical School, who spent considerable time working with *Section II* of this text. The extensive knowledge he brought to this project along with his excellent organizational approach and writing skills are deeply appreciated. Gratitude is also expressed for assistance provided by **Drs. James Baleja** and **Gavin Schnitzler** from the Biochemistry Department of Tufts Medical School, who reviewed several sections of this text.

A number of years ago **Dr. David Leith** from Kansas State Veterinary School provided this author with much of the basic content and rationale for Chapter 92, and I remain indebted to him for his contribution. Dr. Leith also deserves credit for assisting numerous other basic scientists and clinicians to understand and apply the Peter Stewart approach to acid-base chemistry. In addition to Dr. Leith, important contributions were made to this approach by the late **Dr. Vladimir Fencl** from the Departments of Anesthesia and Medicine, Brigham and Women's Hospital and Harvard Medical School. Dr. Fencl will be remembered for his seminal contributions to the discipline of acid-base chemistry, which are well documented in his research articles.

One of my mentors, the late **Professor Rudolf (Rudy) Clarenburg** (1931-1991), will also be remembered by his numerous students, colleagues and friends as an inspirational and effective force in the teaching and study of veterinary physiological chemistry. His text, the ***Physiological Chemistry of Domestic Animals***, Mosby Year Book, 1992, was a guiding light during the developmental years of this text, and he taught many of us the importance of presenting complex biochemical information in a succinct, accurate, practical and relevant manner -- "what is it they need to know, and why do they need to know it." Hopefully readers will find that his teachings had a significant impact on this author.

Thanks also go to **Beth Mellor**, a medical illustrator from our Educational Media Center, who spent many hours working on the Figures and text material for the second and third editions, and to **Nida Intarapanich, V'16**, for her editorial assistance and attention to detail. **Professors John Rush** and **Susan Cotter**, also from Tufts, are acknowledged for their invaluable collegial assistance in helping this author prepare Case Studies for the third edition, and **Elizabeth Gibson**, **Mary Preap** and **Janice Audet** from **Academic Press/Elsevier** deserve credit for helping to bring this third edition to press.

I am also indebted to many past and current students of the **Cummings School of Veterinary Medicine** for their conscientious efforts in detecting minor errors and inconsistencies in this work, and for suggesting avenues for improvement. Highly motivated, bright and engaged students are skilled at detecting ambiguity, vagueness, and lack of clarity, and their constructive comments have been greatly appreciated.

Larry R. Engelking

Preface to the First Edition

▼

This text has been written primarily for veterinary students, interns and residents, and for practicing veterinarians who wish to update their general knowledge of physiological chemistry. Emphasis has been placed on instructional figures and tables, while text material has been held to a minimum. Several multiple choice questions at the end of each chapter will aid in gauging the reader's comprehension of the subject matter, while overviews at the head of each chapter summarize key concepts.

To many veterinary students and clinicians, chemical reactions and pathways are valueless unless they are applied to practical situations, and explained through proper biomedical reasoning. When important biochemical concepts are extracted and resynthesized into rational guidelines for explaining physiological events, then the subject matter becomes relevant, and perhaps more importantly, memorable. Care has been exercised in the preparation of this text to present a clear and concise discussion of the basic biochemistry of mammalian cells, to relate events occurring at the cellular level to physiological processes in the whole animal, and to cite examples of deviant biochemical events where appropriate. Consideration of each major pathway includes a statement regarding the biomedical importance the pathway holds for the cell, tissue or the organism as a whole; a description of key reactions within the pathway; where reactions occur within the cell; the tissue or organ specificity of the pathway; how the pathway is regulated, and how it is coordinated with other important metabolic pathways so that homeostasis is maintained. Clinical examples are frequently used to emphasize connections between common disease states and biochemical abnormalities.

Themes traditionally encountered in cell biology, biochemistry, histology, nutrition, and physiology provide a framework for integration in this text. Emphasis has been placed on **metabolism**, with topics sequenced to permit efficient development of a sound knowledge base in physiological chemistry. *Section I* encounters areas of amino acid, protein, and enzyme chemistry that set the stage for a discussion of nucleotide and nucleic acid metabolism in *Section II*. *Section III* covers carbohydrate and heme metabolism, and *Section IV* discusses important biomedical aspects of vitamin and trace element chemistry. *Section V* presents an in-depth analysis of lipid metabolism, while *Section VI* is devoted to a discussion of sequential metabolic events in starvation and exercise to show where the biochemistry of protein, carbohydrate and lipid metabolism converge. Lastly, *Section VII* is devoted to an in-depth yet practical analysis of acid-base chemistry. While *Sections I* through *VI* provide a foundation for a course in veterinary physiological chemistry, some lecturers may prefer to cover material contained in *Sections VI and VII*, and in *Chapters 7, 11, 38, 60, 64,* and *72* in their physiology courses.

Although care has been taken to include relevant subject matter in a concise, up to date, accurate and reliable fashion, all authors are fallible, with this one being no exception. Therefore, when errors or serious omissions are detected, or if clarity of presentation should be improved, constructive feedback would be genuinely appreciated.

While at times first year veterinary students might find the scope of this text daunting, the messages implicit, interactions among concepts subtle, and the biochemical vocabulary intricate, it is my hope that when the following two questions arise, *"what is the significance of this information,"* and *"why do I need to know it,"* that the answers will become obvious.

Larry R. Engelking

Preface to the Second Edition

▼

Six years have passed since publication of the first edition. Corrections have been instituted and updates added, but the overall length of the text has remained largely unchanged. Learning objectives have been added to each chapter in order to focus student attention upon key concepts presented.

Although the scope and depth of this text should satisfy the basic requirements of most veterinary physiological chemistry courses, when readers desire further information on contemporary aspects of molecular biology (e.g., molecular genetics, regulation of gene expression, recombinant DNA & genetic technology), other educational resources can be consulted.

One objective of this text has been to better prepare students for their organ systems physiology courses. Attempts have been made to convey important biochemical and metabolic concepts in a concise, logically-sequenced, well-illustrated and reliable fashion. However, if readers detect errors, or if the scope, depth or clarity of presentation should be improved, comments and suggestions would be genuinely appreciated.

It is hoped that readers will again find this text to be practical, informative, and relevant to their scholastic needs.

Larry R. Engelking

Preface to the Third Edition

▼

Five years have passed since publication of the second edition. Appropriate revisions and updates have again been instituted, and six **case studies** have been added. The cases were created to help readers realize how an understanding of basic biochemical principles can assist them in appreciating interconnections between metabolic pathways and metabolic disorders.

Additionally, **sectional examinations** have been added to better provide readers with a self-assessment of their understanding. While questions and learning objectives are provided at the end of each chapter, study questions found there are largely limited to material contained in that chapter alone. Sectional exam questions are somewhat more difficult, they possess a broader scope, and they are reasonably good evaluation tools. Students are advised to think through the questions, using them to gage their conceptual understanding of relevant material. (*Merely circling and memorizing the correct answer to a question will largely be non-productive*).

The overall length of this text has increased approximately 175 pages over the previous edition. However, much of this increase is due to the incorporation of sectional exams, case studies, and a more detailed index.

It is again hoped that readers will find this text, its revisions and additions to be practical, informative, and relevant to their educational needs.

Larry R. Engelking

SECTION

I

Amino Acid and Protein Metabolism

Chemical Composition of Living Cells

Overview

- Hydrogen, oxygen, nitrogen, carbon, sulfur and phosphorus normally make up more than 99% of the mass of living cells.
- Ninety-nine percent of the molecules inside living cells are water molecules.
- Cells normally contain more protein than DNA.
- Homogenous polymers are noninformational.
- All non-essential lipids can be generated from acetyl-CoA.
- Like certain amino acids and unsaturated fatty acids, various inorganic elements are dietarily "essential."
- Most all diseases in animals are manifestations of abnormalities in biomolecules, chemical reactions, or biochemical pathways.

All living organisms, from microbes to mammals, are composed of chemical substances from both the inorganic and organic world, that appear in roughly the same proportions, and perform the same general tasks. **Hydrogen, oxygen, nitrogen, carbon, phosphorus**, and **sulfur** normally make up more than 99% of the mass of living cells, and when combined in various ways, form virtually all known organic biomolecules. They are initially utilized in the synthesis of a small number of building blocks that are, in turn, used in the construction of a vast array of vital macromolecules (**Fig 1-1**).

There are four general classes of macromolecules within living cells: **nucleic acids, proteins, polysaccharides**, and **lipids**. These compounds, which have molecular weights ranging from 1×10^3 to 1×10^6, are created through polymerization of building blocks that have molecular weights in the range of 50 to 150. Although subtle differences do exist between cells (e.g., erythrocyte, liver, muscle or fat cell), they all generally contain a greater variety of **proteins** than any other type of macromolecule, with about 50% of the solid matter of the cell being protein (15% on a wet-weight basis). Cells generally contain many more protein molecules than DNA molecules, yet **DNA** is typically the largest biomolecule in the cell. About **99%** of cellular molecules are **water** molecules, with water normally accounting for approximately 70% of the total wet-weight of the cell. Although water is obviously important to the vitality of all living cells, the bulk of our attention is usually focused on the other 1% of biomolecules.

Data in **Table 1-1** regarding the chemical composition of the unicellular *Escherichia coli* (*E. coli*) are not greatly different for multicellular organisms, including mammals. Each *E. coli*, and similar bacterium, contains a single chromosome; therefore, it has only one unique

Copyright © 2015 Elsevier Inc. All rights reserved.

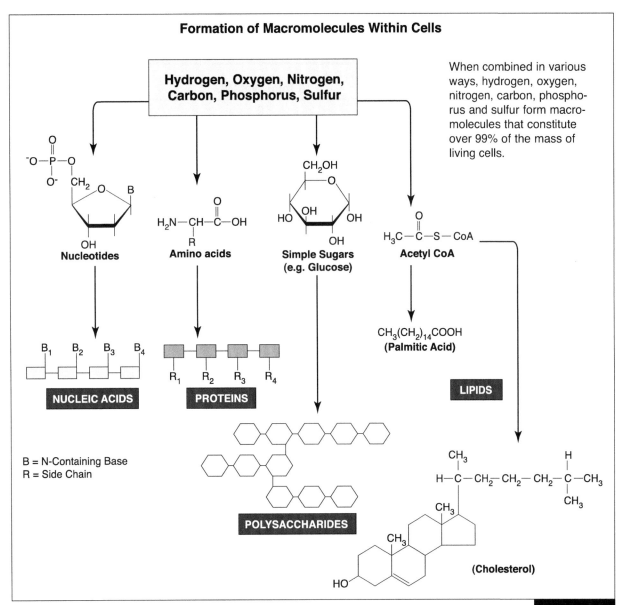

Formation of Macromolecules Within Cells

When combined in various ways, hydrogen, oxygen, nitrogen, carbon, phosphorus and sulfur form macromolecules that constitute over 99% of the mass of living cells.

B = N-Containing Base
R = Side Chain

Figure 1-1

DNA molecule. Mammals, however, contain more chromosomes, and thus have different DNA molecules in their nuclei.

Nucleic Acids

Nucleic acids are nucleotide polymers (from the Greek word **poly**, meaning "several," and **mer**, meaning "unit"), that store and transmit genetic information. Only 4 different nucleotides are used in nucleic acid biosynthesis. Genetic information contained in nucleic acids is stored and replicated in **chromosomes**, which contain

genes (from the Greek word **gennan**, meaning "to produce"). A chromosome is a deoxyribonucleic acid (DNA) molecule, and genes are segments of intact DNA. The total number of genes in any given mammalian cell may total several thousand. When a cell replicates itself, identical copies of DNA molecules are produced; therefore the hereditary line of descent is conserved, and the genetic information carried on DNA is available to direct the occurrence of virtually all chemical reactions within the cell. The bulk of genetic information carried on DNA provides instructions for the

Table 1-1
Approximate Chemical Composition of a Rapidly Dividing Cell (*E. coli*)

Material	% Total Wet Wt.	Different Kinds of Molecules/Cell
Water	70	1
Nucleic acids		
DNA	1	1
RNA	6	
Ribosomal		3
Transfer		40
Messenger		1000
Nucleotides and metabolites	0.8	200
Proteins	15	2000-3000
Amino acids and metabolites	0.8	100
Polysaccharides	3	200
(Carbohydrates and metabolites)		
Lipids and metabolites	2	50
Inorganic ions	1	20
(Major minerals and trace elements)		
Others	0.4	200
	100	

Data from Watson JD: Molecular Biology of the Gene, 2nd ed., Philadelphia, PA: Saunders, 1972.

assembly of every protein molecule within the cell. The flow of information from nucleic acids to protein is commonly represented as **DNA —> messenger ribonucleic acid (mRNA) —> transfer RNA (tRNA) —> ribosomal RNA (rRNA) —> protein**, which indicates that the nucleotide sequence in a gene of DNA specifies the assembly of a nucleotide sequence in an mRNA molecule, which in turn directs the assembly of the amino acid sequence in protein through tRNA and rRNA molecules.

Proteins

Proteins are amino acid polymers responsible for implementing instructions contained within the genetic code. **Twenty different amino acids** are used to synthesize proteins, about half are formed as metabolic intermediates, while the remainder must be provided through the diet. The latter group is referred to as "**essential**" amino acids (see Chapter 3). Each protein formed in the body, unique in its own structure and function, participates in processes that characterize the individuality of cells, tissues, organs, and organ systems. A typical cell contains thousands of different proteins, each with a different function, and many serve as enzymes that catalyze (or speed) reactions. Virtually every reaction in a living cell requires an enzyme. Other proteins transport different compounds either outside or inside cells {e.g., lipoproteins and transferrin (an iron-binding protein) in plasma, or bilirubin-binding proteins in liver cells}; some act as storage proteins (e.g., myoglobin binds and stores O_2 in muscle cells); others as defense proteins in blood or on the surface of cells (e.g., clotting proteins and immunoglobulins); others as contractile proteins (e.g., the actin, myosin and troponin of skeletal muscle fibers); and others are merely structural in nature (e.g., collagen and elastin). Proteins, unlike glycogen and triglyceride, are usually not synthesized and stored as nonfunctional entities.

Polysaccharides

Polysaccharides are polymers of simple sugars (i.e., monosaccharides). (The term "saccharide" is derived from the Greek word **sakchar**, meaning "sugar or sweetness.") Some polysaccharides are **homogenous polymers** that contain only one kind of sugar (e.g., glycogen), while others are complex **heterogenous polymers** that contain 8-10 types of sugar. In contrast to heterogenous polymers (e.g., proteins, nucleic acids, and some polysaccharides), homogenous polymers are considered to be "**noninformational**." Polysaccharides, therefore, can occur as functional and structural components of cells (e.g., glycoproteins and glycolipids), or merely as noninformational storage forms of energy (e.g., glycogen). The 8-10 monosaccharides that become the building blocks for heterogenous polysaccharides can be synthesized from glucose, or formed from other metabolic intermediates (see Chapter 20).

Lipids

Lipids (from the Greek word **lipos**, meaning "fat") are naturally occurring, nonpolar substances that are mostly insoluble in water (with the exceptions being the short-chain volatile fatty acids and ketone bodies), yet soluble in nonpolar solvents (like chloroform and ether). They serve as membrane components (cholesterol, glycolipids and phospholipids), storage forms of energy (triglycerides), precursors to other important biomolecules

(fatty acids), insulation barriers (neutral fat stores), protective coatings to prevent infection and excessive gain or loss of water, and some vitamins (A, D, E, and K) and hormones (steroid hormones). Major classes of lipids are the saturated and unsaturated fatty acids (short, medium, and long-chain), triglycerides, lipoproteins {i.e., chylomicrons (CMs), very low density (VLDL), low density (LDL), intermediate density (IDL), and high density lipoproteins (HDL)}, phospholipids and glycolipids, steroids (cholesterol, progesterone, etc.), and eicosanoids (prostaglandins, thromboxanes, and leukotrienes). All lipids can be synthesized from **acetyl-CoA**, which in turn can be generated from numerous different sources, including carbohydrates, amino acids, short-chain volatile fatty acids (e.g., acetate), ketone bodies, and fatty acids. **Simple lipids** include only those that are esters of fatty acids and an alcohol (e.g., mono-, di- and triglycerides). **Compound lipids** include various materials that contain other substances in addition to an alcohol and fatty acid (e.g., phosphoacylglycerols, sphingomyelins, and cerebrosides), and **derived lipids** include those that cannot be neatly classified into either of the above (e.g., steroids, eicosanoids, and the fat-soluble vitamins).

Although the study of physiological chemistry emphasizes organic molecules, the **inorganic elements** (sometimes subdivided into macro-minerals, trace elements, and ultra trace elements), are also important (see Chapter 48). Several are "**essential**" nutrients, and therefore like certain **amino acids** and **unsaturated fatty acids**, must be supplied in the diet. Inorganic elements are typically present in cells as ionic forms, existing as either free ions or complexed with organic molecules. Many "**trace elements**" are known to be essential for life, health, and reproduction, and have well-established actions (e.g., cofactors for enzymes, sites for binding of oxygen (in transport), and structural components of nonenzymatic macromolecules;

see Chapters 48-52). Some investigators have speculated that perhaps all of the elements on the periodic chart will someday be shown to exhibit physiologic roles in mam-malian life.

Because life depends upon chemical reactions, and because most all diseases in animals are manifestations of abnormalities in biomolecules, chemical reactions, or biochemical pathways, physiological chemistry has become the language of all basic medical sciences. A fundamental understanding of this science is therefore needed not only to help illuminate the origin of disease, but also to help formulate appropriate therapies. The chapters that follow were designed, therefore, to assist the reader in developing a basic rational approach to the practice of veterinary medicine.

OBJECTIVES

- Identify six elements that normally comprise over 99% of the living cell mass.

- Summarize the approximate chemical composition of a living cell.

- Give examples of functionally important intra- and extracellular proteins.

- Distinguish homogenous from heterogenous polymers, and give some examples.

- Understand basic differences between compound, simple and derived lipids.

- Indicate how and why the inorganic elements are essential to life.

- Recognize why a basic understanding of physiological chemistry is fundamental to a clinical understanding of disease processes.

QUESTIONS

1. The most prevalent compound in a living cell is normally:
   a. Protein.
   b. Nucleic acid.
   c. Water.
   d. Lipid.
   e. Polysaccharide.

2. The basic building block for all lipids is:
   a. Water.
   b. Acetyl-CoA.
   c. Phosphorus.
   d. Nucleic acid.
   e. Arginine.

3. The largest biomolecule in a living cell is usually:
   a. Glycogen.
   b. Protein.
   c. Cholesterol.
   d. Deoxyribonucleic acid.
   e. Triglyceride.

4. Which one of the following is a largely homogenous polymer, and therefore "noninformational"?
   a. mRNA
   b. Phospholipid
   c. Protein
   d. Hydrogen
   e. Glycogen

5. Select the FALSE statement below:
   a. Some inorganic elements are considered to be "essential" nutrients.
   b. Triglycerides are considered to be "simple" lipids.
   c. Some polysaccharides are complex polymers in that they contain several different types of sugars.
   d. Virtually every reaction in a living cell requires an enzyme.
   e. Only 10 "essential" amino acids are used in the synthesis of proteins.

6. About 50% of the solid matter in a cell is normally composed of:
   a. Nucleic acids.
   b. Protein.
   c. Carbohydrate.
   d. Lipid.
   e. Inorganic ions.

7. Proteins, unlike glycogen and triglycerides, are usually not synthesized as nonfunctional entities.
   a. True
   b. False

7. a
6. b
5. e
4. e
3. d
2. b
1. c

Chapter 2

Properties of Amino Acids

Overview

- Mammalian proteins are largely composed of 20 standard amino acids.
- At physiologic pH, most amino acids exist as zwitterions.
- Hydrophilic amino acids are found on the surface of proteins.
- Hydrophobic amino acids are located in the interior of proteins.
- Cystine covalently links different regions of polypeptide chains with disulfide (-S-S-) bonding.
- The L form of amino acids is found in most mammalian proteins.
- The BCAAs exhibit similar physical and chemical characteristics.

The thousands of different proteins found in the mammalian organism are composed largely of **20 standard amino acid monomers**, with each conveniently designated by either a one-letter symbol, or a three-letter abbreviation (**Fig. 2-1**). Besides these 20, numerous other biologically active amino acids occur in the mammalian organism as derivatives of one or more of these 20 (see Chapters 3 and 12). Some of the "derived" amino acids occur in protein, yet most occur in biologically free or combined form, and are not found in protein.

The first amino acid isolated from a protein hydrolysate was **glycine** in 1820, and the last of the standard 20, **threonine**, was isolated in 1935. Nineteen of these 20 standard amino acids can be represented by the general structure shown in **Fig. 2-2**.

At physiologic pH, the central, α-carbon atom (C_α, because it is adjacent to the acidic carboxyl group) is normally bonded to a hydrogen atom (**H**), a variable side chain (**R**), a protonated amino group (**-NH$_3^+$**), and a non-protonated carboxyl group (**-COO$^-$**). The exception, proline, contains an imino group (**-NH$_2^+$**) in place of the amino group, and the α-carbon becomes part of a cyclic structure formed between the R group and the imino group. Since side chains are distinct between amino acids, they give each a certain structural and functional identity.

As indicated above, at a **physiologic pH of 7.4**, most amino acids exist as dipolar ions (or **zwitterions**) rather than as unionized molecules. Their overall ionization states, however, vary with pH. In **alkaline** solution (pH >> 7.4), the carboxyl group will usually remain ionized (**-COO$^-$**); however, the amine group may become unionized (**-NH$_2$**). In acidic solution (pH << 7.4), the opposite generally holds true {i.e., the carboxyl group may become unionized (**-COOH**), while the amine group remains ionized (**-NH$_3^+$**)}. For glycine, the **pK** of the carboxyl moiety {i.e., the **pH** at which **half is ionized (-COO$^-$) and half is unionized (-COOH)**} is 2.3, and the pK of the amine moiety is 9.6. The carboxyl and amine pK's for each amino acid are slightly different, and they depend

Copyright © 2015 Elsevier Inc. All rights reserved.

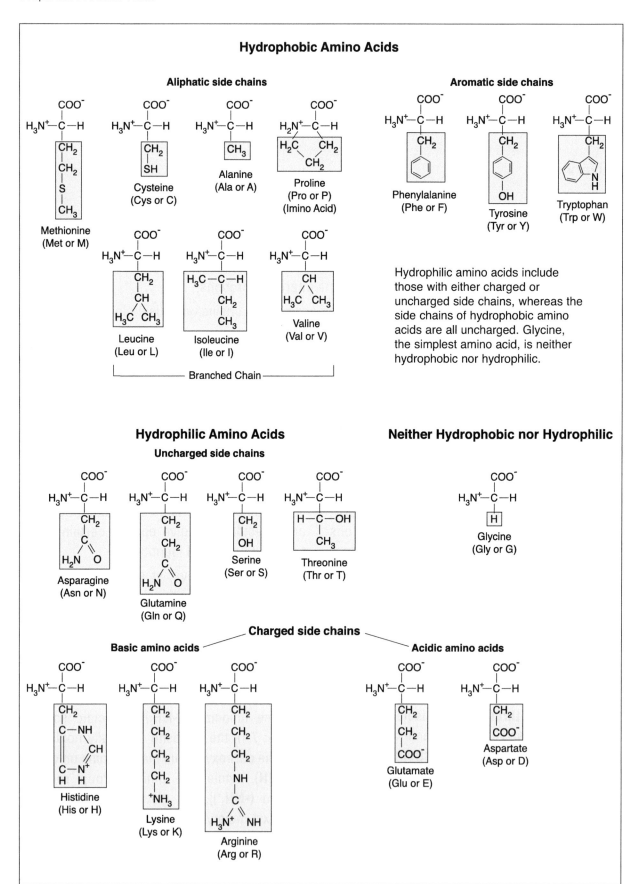

Hydrophilic amino acids include those with either charged or uncharged side chains, whereas the side chains of hydrophobic amino acids are all uncharged. Glycine, the simplest amino acid, is neither hydrophobic nor hydrophilic.

Figure 2-1

largely upon variables such as temperature, ionic strength, and the microenvironment of the ionizable group (see Chapter 84).

Standard Amino Acid Structure

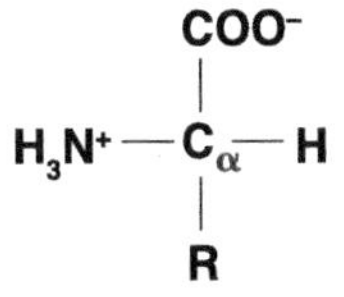

$$H_3N^+ - \underset{\underset{R}{|}}{\overset{\overset{COO^-}{|}}{C_\alpha}} - H$$

Figure 2-2

Hydrophilic Amino Acids

Amino acids with positively charged side chains include the **basic** amino acids **histidine**, **lysine**, and **arginine**. Histidine is only weakly charged at physiologic pH. In contrast, amino acids with negatively charged side chains are **acidic**, and include **glutamic** and **aspartic acids** (which exist as glutamate and aspartate). In some proteins, arginine is found to substitute for lysine, and aspartate for glutamate, with little or no apparent affect on either structure or function. **Serine**, a hydroxylated version of alanine, and **threonine** have uncharged side chains containing an -OH group, and thus interact strongly with water in the formation of hydrogen bonds. They are much more hydrophilic (and thus water-loving) than, for example, their non-hydroxylated cousins, alanine and valine.

The uncharged side chains of **asparagine** and **glutamine** have amide groups with even more hydrogen-bonding capacity. However, asparagine and glutamine are easily hydrolyzed by acid or base to their respective non-amide forms (aspartic acid and glutamic acid). These nine amino acids are all hydrophilic, and thus interact favorably with water. They are frequently found on the surface of proteins where their side chains are exposed to an aqueous medium.

Hydrophobic Amino Acids

The side chains of the hydrophobic amino acids interact poorly with water, and therefore are usually located toward the interior of proteins. They consist largely of hydrocarbons, except for the sulfur atoms of **methionine** and **cysteine**, and the nitrogen atom of **tryptophan**. In the category of **aromatic amino acids** (**AAAs**) are **phenylalanine**, **tyrosine**, and **tryptophan**. The R group in phenylalanine contains a benzene ring, that in tyrosine contains a phenol group, and the R group in tryptophan contains a heterocyclic structure known as an **indole**. In these three amino acids the aromatic moiety is attached to the α-carbon through a methylene ($-CH_2-$) bridge. The AAAs are important hepatic metabolites (see Chapter 8), and although they are hydrophobic, the phenol (**-OH**) group of tyrosine and the **-NH** in the indole group of tryptophan allow them to interact with water, thus making their properties somewhat ambiguous.

Valine, **leucine**, and **isoleucine** are the **branched-chain amino acids** (**BCAAs**), and each bears an aliphatic side chain with a methyl group branch. They are dietarily "essential" in all higher animals, which lack relevant biosynthetic enzymes. Physical and chemical similarities between these amino acids are emphasized by the finding of homologous proteins in various organisms in which these BCAAs replace each other in certain positions without greatly altering the functional properties of the proteins. Although found in many tissues, the BCAAs are extensively catabolized by muscle tissue for energy purposes during the final phase of starvation (see Chapter 76).

Because the **-SH** group of **cysteine** allows it to dimerize through disulfide (**-S-S-**) bonding, this amino acid frequently exists in proteins in its oxidized form, **cystine** (**Fig. 2-3**). Cystine covalently links different regions of polypeptide chains, thus stabilizing proteins and making them more resistant to denaturation. However, when the -SH group of cysteine remains free, it is quite hydrophobic. **Proline**, which is usually considered hydrophobic, is often found in the bends of folded polypeptide chains, and is particularly prevalent in **collagen** (see Chapter 3). Proline can be formed through cyclization of glutamate.

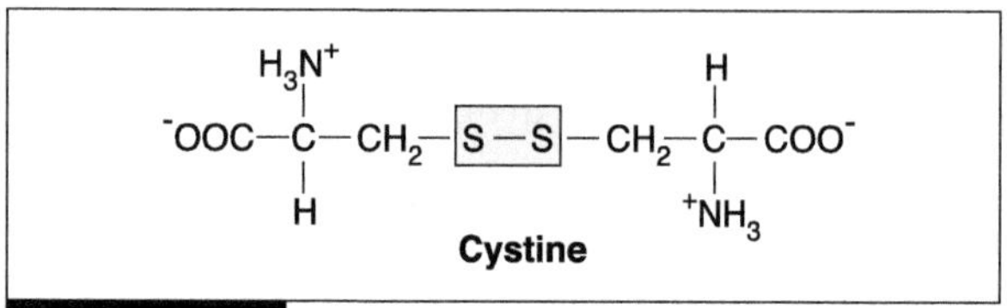

Cystine

Figure 2-3

Neither Hydrophobic nor Hydrophilic

Glycine, which has only one hydrogen atom for a side chain, is the simplest amino acid known, and is neither hydrophobic nor hydrophilic. Because of its simple structure, it can fit into many spaces in polypeptide chains, and is therefore found on both the surface and interior of protein molecules.

Enantiomers

The structures of all amino acids (except glycine) are asymmetrically arranged around the **α-carbon atom**, with their D- and L-isomers being mirror images of each other. Emil Fischer arbitrarily assigned the carboxy group on top, the R group on bottom, and the amine group on the **left** for the **L-isomer**, and on the **right** for the **D-isomer** (**Fig. 2-4**). These **stereoisomers** (also called optical isomers or enantiomers), cannot be interconverted without breaking a chemical bond. With rare exceptions, only the **L forms** of amino acids are found in mammalian proteins; however, D-amino acids are found in bacterial cell walls and in some antibiotics produced by microorganisms.

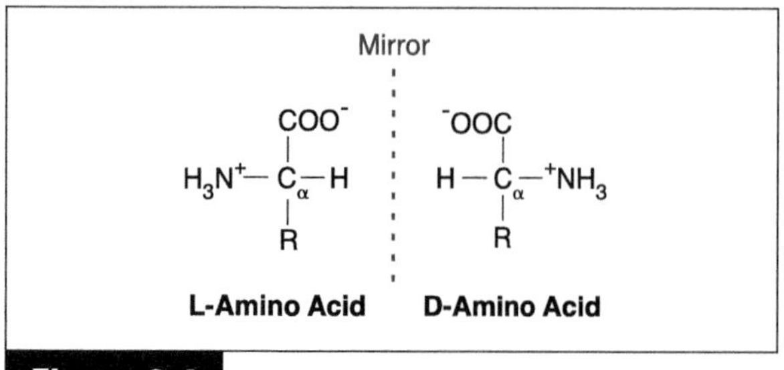

L-Amino Acid   **D-Amino Acid**

Figure 2-4

The phenomenon of stereoisomerism, also called **chirality** (from the Greek word **cheir**, meaning "hand" or "handedness" -- the property of not being superimposable on a mirror image), occurs with all compounds having an asymmetric carbon atom (i.e., one with 4 different substituents). Therefore, the amino acids in **Fig. 2-1** that have two asymmetric carbon atoms, threonine and isoleucine, have 4 enantiomers each.

OBJECTIVES

- Draw the general structure of an amino acid, and indicate how the ionization state changes with pH.

- Classify the 20 standard amino acid monomers into six different families.

- Recognize where the hydrophilic and hydrophobic amino acids are generally   located in proteins.

- Explain why certain amino acids are dietarily "essential."

- Identify the AAAs and the BCAAs, and recognize where they are usually metabolized.

- Understand basic structural and functional differences between cysteine and cystine.

- Although homocysteine is not one of the standard 20 amino acids found in protein, predict which form it usually assumes when it accumulates in blood and urine (see Chapter 16).

- Recognize which amino acid sterioisomer is usually found in mammalian protein.

- Discuss why the normal plasma BCAA:AAA ratio is roughly 3:1, and identify factors that could alter it (see Chapters 8 and 76).

- Explain why Thr has four enantiomers, while Ser has only two.

QUESTIONS

1. **Which one of the following amino acids has 4 enantiomers?**
   a. Glycine
   b. Isoleucine
   c. Leucine

d. Methionine
e. Asparagine

2. **Which amino acid below is considered to be the "simplest" amino acid known?**
   a. Alanine
   b. Proline
   c. Glycine
   d. Glutamic acid
   e. Aspartic acid

3. **At pH 7.4:**
   a. Most amino acids exist as dipolar zwitterions.
   b. The amine group of most amino acids will remain in the unionized form (i.e., -NH₂).
   c. All amino acids will exist in a hydrophilic form.
   d. The carboxy group of most amino acids will remain protonated (i.e., -COOH).
   e. All amino acids will exist in a hydrophobic form.

4. **Which amino acid below is a hydroxylated form of alanine?**
   a. Glutamine
   b. Glycine
   c. Serine
   d. Threonine
   e. Tyrosine

5. **Hydrophilic amino acids:**
   a. Are usually found on the interior of proteins.
   b. Include those with aromatic side chains.
   c. Include those with positively charged side chains which are acidic.
   d. Are usually found on the surface of proteins.
   e. Include those with negatively charged side chains which are basic.

6. **Which one of the following amino acids is generally found in the "bends" of folded polypeptide chains?**
   a. Serine
   b. Proline
   c. Phenylalanine
   d. Glutamine
   e. Cystine

7. **Select the FALSE statement below:**
   a. Most mammalian proteins contain the L-isomers of amino acids.
   b. The structures of all amino acids (except Gly) are asymmetrically arranged around the α-carbon atom.
   c. Cystine is an oxidized form of cysteine.
   d. Branched-chain amino acids are dietarily essential.
   e. Arginine is an acidic amino acid that is also hydrophobic.

8. **Which one of the following is a branched-chain amino acid?**
   a. Lys
   b. Ala
   c. Trp
   d. Leu
   e. Phe

9. **Which one of the following is an aromatic amino acid?**
   a. Phe
   b. Met
   c. Ile
   d. Asp
   e. Gly

10. **Which one of the following amino acids is sulfated?**
    a. Ser
    b. Ile
    c. Asn
    d. Trp
    e. Met

11. **Urea is formed from the terminal side chain of which amino acid (Ch. 10)?**
    a. Val
    b. Asp
    c. Lys
    d. Cys
    e. Arg

12. **Which amino acid is used in catecholamine formation (Ch. 39)?**
    a. Gly
    b. Tyr
    c. Glu
    d. Asn
    e. Trp

12. b
11. e
10. e
9. a
8. d
7. e
6. b
5. d
4. c
3. a
2. c
1. b

Chapter 3

Amino Acid Modifications

Overview

- Hydroxyproline and hydroxylysine are required for collagen formation.
- Desmosine and isodesmosine are needed for elastin formation.
- β-Alanine is required for pantothenic acid formation.
- γ-Carboxyglutamate is involved in blood coagulation.
- Taurine is an essential amino acid for cats.
- GABA is the major inhibitory neurotransmitter in the brain.
- Phenylalanine is needed for hepatic tyrosine formation.
- Most non-essential amino acids can be interconverted with carbohydrate metabolites through aminotransferase reactions.

In addition to the 20 standard amino acid monomers identified in Chapter 2, several others have been isolated from hydrolyzates of mammalian protein, and over 150 others are known to occur biologically, but are not found in protein. Most, however, are thought to be formed through posttranslational modification of one or more of the standard 20. The following deserve special attention.

Modified Amino Acids Found in Protein

Methylhistidine and **methyllysine (Fig. 3-1)** are methyl derivatives of histidine and lysine, respectively, and are known to occur in certain muscle proteins. **Hydroxyproline** is a derivative of **proline** found in the fibrous connective tissue protein known as **collagen**, as well as in some plant proteins. Collagen, being a part of the hide, connective tissue of blood vessels, tendons, cartilage, and bones, is the **most abundant protein** in the mammalian organism (about **30%** of total body protein, and **6%** of the body weight). The hydroxylation of proline to

hydroxyproline requires **ascorbic acid (vitamin C)**, and a deficiency of this vitamin leads to inadequate levels of hydroxyproline, producing weak fibers, fragile blood vessels, weak tendons and loose ligaments (see Chapter 39). Bruises and hematomas, weakness, joint laxity and tooth loss are classic signs of vitamin C deficiency (i.e., **scurvy**).

Although proline and hydroxyproline are important constituents of collagen, **glycine** is frequently found in the third repeating position (-proline-hydroxyproline-glycine-) of its most common tripeptide. Since glycine is small and has no side chain, it allows collagen to twist into a tight and strong 3-stranded helical structure **(Fig. 3-2)**. **Hydroxylysine**, a 5-hydroxy derivative of lysine, is also prevalent in collagen.

Desmosine and **isodesmosine** are formed by the oxidation and crosslinking of four lysine side chains, and are prevalent in the connective tissue protein known as **elastin**. Elastin, unlike other proteins, is capable of undergoing two-way stretch, and is typically found

Copyright © 2015 Elsevier Inc. All rights reserved.

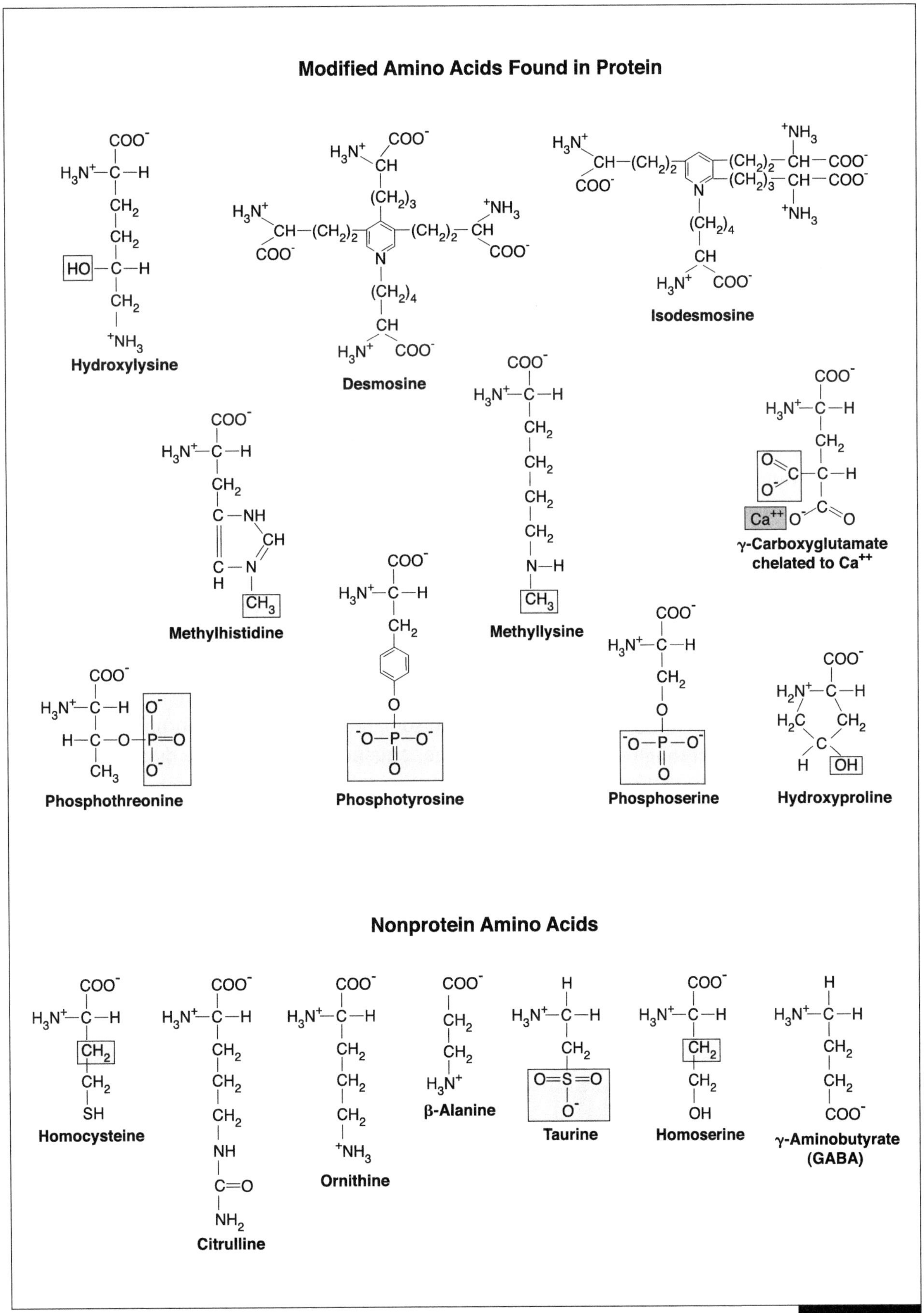
Modified Amino Acids Found in Protein
Hydroxylysine
Desmosine
Isodesmosine
Methylhistidine
Methyllysine
γ-Carboxyglutamate chelated to Ca^{++}
Phosphothreonine
Phosphotyrosine
Phosphoserine
Hydroxyproline
Nonprotein Amino Acids
Homocysteine
Citrulline
Ornithine
β-Alanine
Taurine
Homoserine
γ-Aminobutyrate (GABA)

Figure 3-1

Triple Helix of Collagen

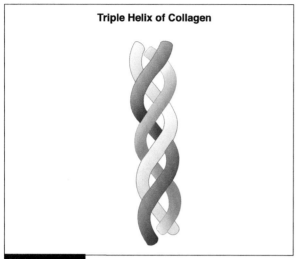

Figure 3-2

with collagen in connective tissue associated with smooth muscle (e.g., blood vessels; **Fig. 3-3**). Arteries typically contain more elastin than veins, and although cardiac output may change over time, a rather continuous flow of blood through the vascular system occurs by distention of the aorta and its branches during ventricular contraction (**systole**), and elastic recoil of the walls of large arteries during ventricular relaxation (**diastole**). Although blood moves rapidly through the aorta and its arterial branches, these branches become narrower, and their walls become thinner toward the periphery. From a predominantly elastic structure, the aorta, the peripheral arteries become more muscular until at the arterioles the smooth muscle layer predominates over the elastic layer.

The hydroxyl groups contained in the side chains of **threonine**, **tyrosine**, and **serine** can be phosphorylated, thus forming **phosphothreonine**, **phosphotyrosine** and **phosphoserine**, respectively. These phosphorylated amino acids are of particular importance to regulatory proteins.

The production of certain biologically active protein clotting factors in the liver involves carboxylation of their glutamic acid residues to γ-**carboxyglutamate**. **Prothrombin (factor II)** contains ten of these residues, which allow chelation of Ca^{++} in a specific protein-Ca^{++}-phospholipid interaction that is essential to blood coagulation (see **Fig. 3-1**). The carboxylase enzyme that catalyzes this reaction is **vitamin K-dependent**, and therefore inhibited by 4-hydroxydicoumarin (**dicumarol**), also known as **warfarin** (see Chapter 47). Nutritional vitamin K deficiency is generally associated with fat malabsorption, which can be caused by pancreatic dysfunction, liver and/or biliary disease, atrophy of the intestinal mucosa, or most anything that causes **steatorrhea** (i.e., bulky, fatty, smelly stools). In addition, loss of microbial flora in the large

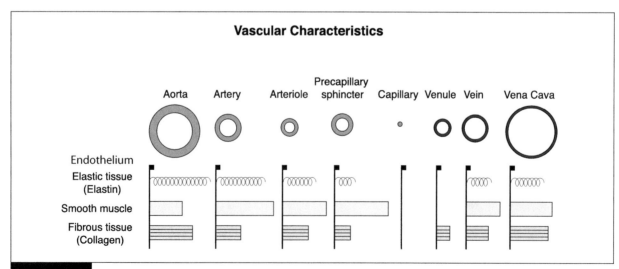

Figure 3-3

Modified from West, JB, Best and Taylor's Physiological Basis of Medical Practice, Williams & Wilkins.

bowel through excessive use of antibiotics can also result in vitamin K deficiency when dietary intake is restricted.

The ten modified amino acids shown in **Fig. 3-1** are genetically distinct since there are no triplet code words for them, and therefore they must arise by enzymatic modification after their parent amino acids have been inserted into respective polypeptide chains. If they are ingested preformed, they will not be inserted into either collagen or elastin.

Nonprotein Amino Acids

Many amino acids not incorporated into proteins are nonetheless important precursors or intermediates in metabolism. **β-Alanine**, for example, is a building block of the B-complex vitamin, **pantothenic acid (vitamin B_5)**, an important constituent of coenzyme A (see Chapter 41). **Homocysteine** and **homoserine** are intermediates in **methionine** metabolism, and **ornithine** and **citrulline** are important intermediates in the synthesis of **arginine** in the hepatic urea cycle (see Chapter 10).

Taurine (2-aminoethane-sulfonic acid) was discovered in ox bile in 1827, and is formed by oxidation of the sulfhydryl group of cysteine to **$-SO_3^-$**, then decarboxylation (see Chapter 62). **Cats**, however, cannot synthesize enough taurine to meet body needs, and therefore require it in their diet. Taurine deficiency in cats is associated with central retinal degeneration and blindness, dilated cardiomyopathy, reproductive failure, retarded body growth, and skeletal deformities in the young. Taurine reportedly participates in maintaining the structure and function of retinal photoreceptors, it appears to be a Ca^{++}-antagonist in heart muscle, it modulates neurotransmitter release in the CNS, and it, along with glycine, is used by the liver to conjugate lipophilic bile acids, thus rendering them soluble in the watery medium of bile (see Chapter 62). Although the mechanisms of taurine actions are still under investigation, it appears that this sulfur-containing amino acid may influence numerous cellular processes, including cell division, osmotic balance, antioxidation, muscle contraction, and even the hepatic conjugation of xenobiotics. It may act by influencing cellular ionic fluxes, particularly those of Ca^{++}, or, through its control of glutamate synthesis, indirectly regulating the excitation threshold of cell membranes.

γ-Aminobutyric acid (GABA, also known as γ-aminobutyrate or 4-aminobutyrate) and **glycine** are important **inhibitory neurotransmitters** in the mammalian central nervous system (CNS). γ-Aminobutyrate is widely distributed in the substantia nigra, globus pallidus, and hypothalamus. Although it is considered to be the major inhibitory neurotransmitter in mammalian brain tissue, it is also known to occur in high concentrations in pancreatic islet tissue, in the enteric nervous system of the gut wall (where it serves as both an excitatory and inhibitory transmitter), and in renal tissue. Some **tranquillizing drugs**, including alcohol, barbiturates, and benzodiazepines (e.g., valium), reportedly act by increasing the effectiveness of GABA at postsynaptic receptor sites. γ-Aminobutyrate is formed in gray matter of the brain through decarboxylation of glutamate. It is normally degraded intraneuronally to succinate by two sequential reactions; the first catalyzed by **γ-aminobutyrate aminotransferase (GABA-AT)**, and the second by **succinate-semialdehyde dehydrogenase**.

Essential and Nonessential Amino Acids

Essential amino acids, which generally have a longer half-life than the nonessential ones, are those that are required in the diet since the body cannot synthesize them in adequate amounts to maintain protein biosynthesis **(Table 3-1)**. If even one essential (or nonessential) amino acid is absent, the remaining 19 cannot be used, and they become catabolized thus leading to a negative nitrogen balance. Essential amino acids vary depending on species and age.

The **branched-chain amino acids**, (**leucine**, **isoleucine** and **valine**) are routinely oxidized in muscle tissue, and **phenylalanine** is needed for hepatic **tyrosine** biosynthesis {which is then used for catecholamine biosynthesis (e.g., dopamine, norepinephrine and epinephrine) in nerve tissue, as well as thyroid hormone biosynthesis}. **Methionine** is needed for cysteine formation, and **tryptophan** is used for serotonin (5-hydroxytryptamine) and melatonin formation. Most nonessential amino acids can be interconverted with carbohydrate metabolites through aminotransferase (i.e., transamination) reactions (see Chapter 9). However, there are no *in vivo* aminotransferase reactions for lysine and threonine, and, in addition, histidine, phenylalanine and methionine are not metabolized to any significant extent by these reactions. Hence, they are all "**essential**" dietary amino acids.

The ordinary diet of domestic animals usually contains more than adequate amounts of both essential and nonessential amino acids. Therefore, these categories are of practical significance only in disease, when specific supplements are administered, or when one is designing an animal diet. If, for any reason, dietary amino acid supply is insufficient, the need to synthesize specific proteins for vital physiologic actions results in a redistribution of amino acids among proteins. For example, **hemoglobin (Hb)** is degraded to the extent of about **1%/day** as erythrocytes die, a loss normally balanced by resynthesis. In amino acid deficiency, relatively less Hb is synthesized because the degree of anemia is more tolerable than a deficiency of certain other proteins. Additionally, there is a definite sequence in which body proteins are lost during starvation in order to maintain the blood glucose concentration (see Chapter 76).

Table 3-1	
Essential Amino Acids	
Adults and Young	**Additional for Young**
Isoleucine	Arginine
Leucine	Glycine (Chickens)
Lysine	Histidine
Methionine	
Phenylalanine	
Taurine (Cats)	
Threonine	
Tryptophan	
Valine	

OBJECTIVES

- Recognize how and why proline and lysine hydroxylations improve the quality of life.

- Identify the basic structure of the vasculature, and how relative proportions of elastin, smooth muscle and collagen change from the aorta to the vena cava.

- Discuss why it is that resistance vessels (arter-ies) contain more desmosine and isodesmosine than capacitance vessels (veins).

- Recognize and discuss the relationship between γ-carboxyglutamate and hydroquinone (see Chapter 47).

- Discuss why dietary modified amino acids cannot be inserted directly into developing polypeptide chains.

- Introduce yourself to the relationship between ornithine and citrulline in the hepatic urea cycle (see Chapter 10).

- Know why taurine is an essential amino acid for cats, and recognize the basic   signs and symptoms of taurine deficiency.

- Show how GABA is formed in CNS neurons.

- Identify the essential amino acids for young and adult animals, and explain why they are essential.

- Understand why relatively less hemoglobin is synthesized in essential amino acid deficiency.

QUESTIONS

1. **The most abundant protein in the mammalian organism is:**
   a. Myosin.
   b. Albumin.
   c. Actin.
   d. Collagen.
   e. Elastin.

2. **Vitamin C is required in the formation of which modified amino acid below?**
   a. Methylhistidine
   b. Hydroxyproline
   c. Desmosine
   d. Homoserine
   e. Ornithine

3. **Which modified amino acid below is a major constituent of elastin?**
   a. Isodesmosine
   b. Methyllysine
   c. Hydroxyproline
   d. Phosphothreonine
   e. β-Alanine

4. **Which modified amino acid below is best associated with blood coagulation?**
   a. Phosphoserine
   b. Homoserine
   c. γ-Carboxyglutamate
   d. Taurine
   e. γ-Aminobutyrate

5. **Which one of the following amino acids is used in the hepatic conjugation of bile acids?**
   a. Glutamate
   b. Serine
   c. Homoserine
   d. Ornithine
   e. Taurine

6. **Which of the following are inhibitory neurotransmitters in the brain?**
   a. Taurine and Epinephrine
   b. γ-Aminobutyrate and Glycine
   c. Glutamate and Tryptophan
   d. Desmosine and Methylhistidine
   e. Hydroxylysine and β-Alanine

7. **All of the following are "essential" amino acids in young cats, EXCEPT:**
   a. Methionine.
   b. Phenylalanine.
   c. Arginine.
   d. Taurine.
   e. Alanine.

8. **Which amino acid below is routinely oxidized in muscle tissue?**
   a. Met
   b. Tyr
   c. Ser
   d. Trp
   e. Ile

9. **Select the FALSE statement below:**
   a. A negative nitrogen balance may follow the absence of an essential amino acid in the diet.
   b. Taurine deficiency in cats is associated with retinal degeneration.
   c. Essential amino acids generally have a longer biologic half-life than nonessential amino acids.
   d. Benzodiazepines (e.g., valium) act by blocking the action of GABA at postsynaptic receptor sites.
   e. Homocysteine is structurally related to methionine.

10. **The normal Hb turnover rate is:**
    a. 1%/day.
    b. 16%/day.
    c. 33%/day.
    d. 66%/day.
    e. 100%/day.

10. a
9. d
8. e
7. e
6. b
5. e
4. c
3. a
2. b
1. d

ANSWERS

Chapter 4

Protein Structure

Overview

- The primary structure of a protein is determined by the amino acid sequence, which in turn determines the secondary, tertiary, and quaternary structures.
- Hydrogen bonds in the secondary structure form between the carboxyl and amino groups of the peptide bond.
- Bonding in the α-helix occurs between one carboxyl and the NH of another amino acid residue further along the chain.
- Hydrogen bonding in the β-pleated sheet occurs between peptide bonds in chains running parallel or antiparallel to each other.
- The tertiary structure of a protein is stabilized by interaction of the amino acid side chains.
- The quaternary structure of a protein consists of the noncovalent interaction of protein chains with other proteins and coenzymes.
- Amyloidosis refers to a group of pathologic conditions caused by the   presence of misfolded and insoluble fibrillar amyloid proteins.
- Plasma proteins consist of a mixture of simple proteins, glycoproteins and lipoproteins, and most are produced by the liver.
- Plasma proteins help to produce the oncotic pressure that assists in keeping water within the vascular system.
- Albumin is normally the most abundant plasma protein.

An animal may contain as many as **100,000** different types of protein. Following synthesis on the ribosome, each protein molecule must fold into the specific conformational state encoded in its amino acid sequence in order to be capable of carrying out its physiologic action. Understanding how this process occurs has proven to be one of the most challenging problems in structural biology, and is a crucial step in the development of strategies to prevent and treat debilitating diseases where proteins fail to fold correctly, or fail to remain in their correctly folded positions (e.g., the **amyloidoses**).

The biologically active form of a protein is a three-dimensional molecule, consisting of a **primary**, **secondary**, and sometimes **tertiary** and **quaternary** structure. It is held together by both **covalent** and **noncovalent** bonds, with the major types of covalent bonds being the **peptide bonds** that link the various amino acids

Copyright © 2015 Elsevier Inc. All rights reserved.

together in polypeptide chains, the **disulfide bonds** that form between and within those chains, and sometimes the **amide bonds** that form between certain carboxyl or amine side chains. Noncovalent interactions (i.e., hydrogen, hydrophobic, and ionic bonds) help to stabilize the three dimensional structure.

Primary Structure

Primary protein structure is defined by the sequence of amino acids held together by rather rigid **peptide bonds**. Amino acids in polypeptide chains are generally numbered from the **N-** to **C-terminus**. These bonds effectively join α-**carbon atoms** of adjacent amino acids by combining the α-carboxyl group of one with the α-amino group of the other (**Fig. 4-1**). The side chains of these amino acids (R groups) are arranged in a trans-configuration (on opposing sides of the peptide bond), which reduces steric interaction.

Amino acids can also be held together by disulfide linkages, and in rare circumstances by amide bonds. Amide bonds can be created between amino acid side chains containing carboxyl (**glutamate** or **aspartate**) and amine (**lysine** or **arginine**) groups (**Fig. 4-2**). Disulfide linkages occur between two **cysteine** residues, thus forming cystine (see Chapter 2). Not all proteins contain disulfide or amide bonds, and those that do are more likely to be found extracellularly than intracellularly. The polypep-

Peptide Bond Formation in the Primary Structure of a Protein

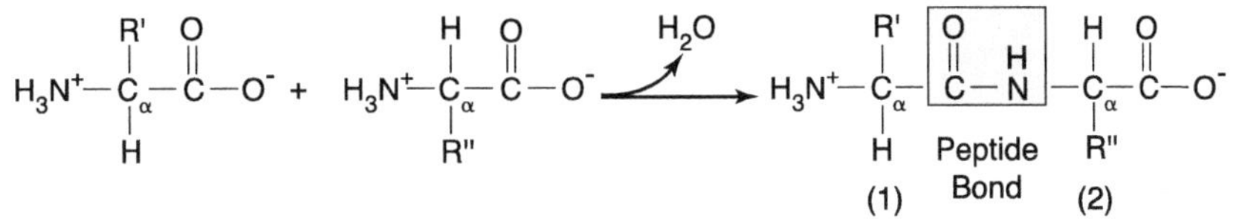

Figure 4-1

Amide Bond Formation in the Primary Structure of a Protein

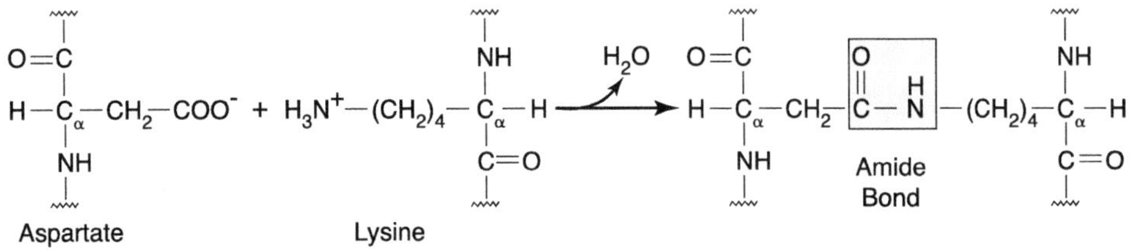

Aspartate　　　Lysine　　　Amide Bond

Figure 4-2

Insulin

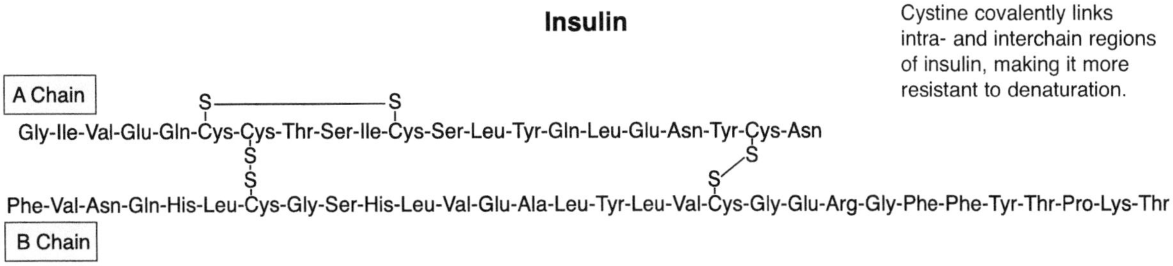

Cystine covalently links intra- and interchain regions of insulin, making it more resistant to denaturation.

A Chain

Gly-Ile-Val-Glu-Gln-Cys-Cys-Thr-Ser-Ile-Cys-Ser-Leu-Tyr-Gln-Leu-Glu-Asn-Tyr-Cys-Asn

Phe-Val-Asn-Gln-His-Leu-Cys-Gly-Ser-His-Leu-Val-Glu-Ala-Leu-Tyr-Leu-Val-Cys-Gly-Glu-Arg-Gly-Phe-Phe-Tyr-Thr-Pro-Lys-Thr

B Chain

Figure 4-3

tide hormone **insulin**, containing 51 amino acid residues, is a good example of a protein having both intra- and interchain disulfide bonds (**Fig. 4-3**). Generally, polypeptides with 50 or more amino acid residues are referred to as proteins.

Secondary Structure

Although multiple possible secondary structures exist in mammalian proteins, only 3 are frequently observed: α-**helix**, β-**pleated sheets**, and β-**bends**.

α-**Helical structures** are stabilized by hydrogen bonding between the amide hydrogen of one peptide bond, and the carbonyl oxygen

of another that is four amino acid residues upstream in the polypeptide chain (**Fig. 4-4**). The right-handed α-helix is the most common secondary structure found in protein, with 3.6 amino acid residues observed per turn of the helix (**Fig. 4-5**). α-**Keratin** (the major protein of skin, hair and nails), **myosin**, the thick filament of muscle, and the **globulins** (i.e., hemoglobin, myoglobin, immunoglobulins, and the many plasma proteins synthesized by the liver), are examples of proteins that have extensive α-helical structures.

An α-helix is generally stabilized by amino acids in the primary structure with uncharged side chains, and, although there is some debate, it may be destabilized by those with bulky or charged side chains (Chapter 2, see **Fig. 2-1**). The only amino acid that does not readily participate in the α-helix is **proline** (due to its bond angles and rigidity of structure). When proline is encountered in the primary structure, the α-helix stops.

β-**Pleated sheets** are formed through interchain hydrogen bonding between the amide hydrogen of a peptide bond in one chain, and the carbonyl oxygen of a peptide bond in another chain (**Fig. 4-6**). Protein chains are extended with amino acid side chains (i.e., R groups) alternating above and below the backbone. Parallel sheets form when the two chains are aligned in the same direction from the amino to carboxyl end, and antiparallel sheets form

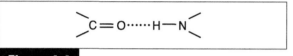

Figure 4-4

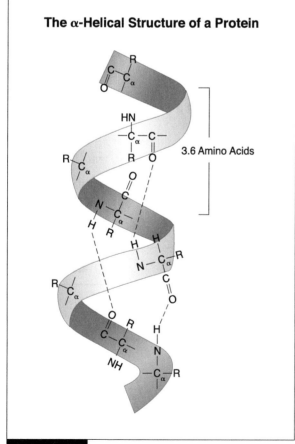

The α-Helical Structure of a Protein

3.6 Amino Acids

Figure 4-5

β–Pleated Sheets in the Secondary Structure of a Protein

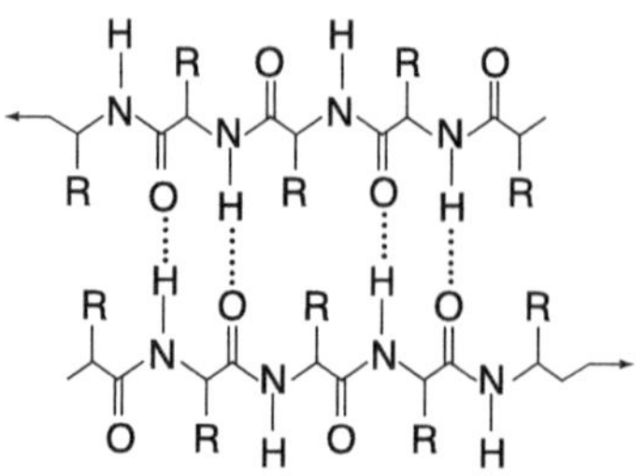

Figure 4-6

when the two chains have opposing polarities. Because certain side chains exhibit steric interactions, the β-structure is preferred when the primary structure contains amino acids with small R groups. **β-Keratin**, found in silk and the beaks of birds, has a strong β-structure with no disulfide bonds.

β-Bends consist of four amino acid residues, and serve to reverse polypeptide chains in tightly folded globular proteins. They are stabilized by hydrogen bonding between the first carbonyl oxygen and the last amide hydrogen in the bend, and contain either glycine or proline, or both.

Tertiary Structure

The tertiary structure of a protein refers to the overall three-dimensional arrangement of its polypeptide chain in space. It is generally stabilized by outside polar hydrophilic hydrogen and ionic bond interactions, and internal hydrophobic interactions between nonpolar amino acid side chains (**Fig. 4-7**). Additional posttranslational covalent bonds in the tertiary structure may be formed with prosthetic groups. Tertiary folding begins while the protein is being molded into its primary polypeptide sequence. It is known to be guided through this process by **chaperones**, which are a family of proteins that bind hydrophobic patches and prevent them from being aggregated prematurely into nonfunctional entities.

Based upon their tertiary structure, proteins are often divided into **globular** or **fibrous** types. Fibrous proteins, like α-keratin, have elongated rope-like structures that are strong and hydrophobic. Globular proteins, like the plasma proteins and the immunoglobulins, are more spherical and hydrophilic.

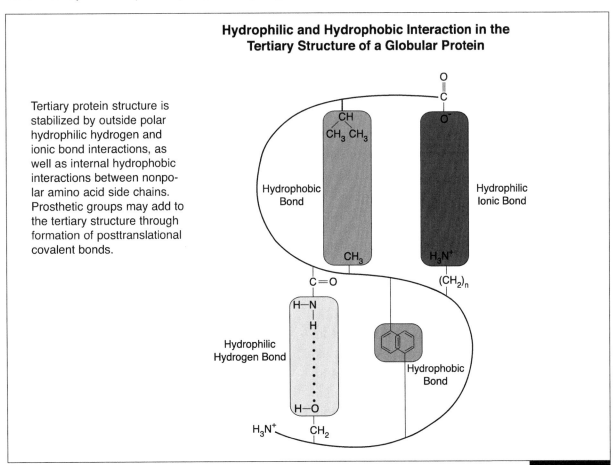

Hydrophilic and Hydrophobic Interaction in the Tertiary Structure of a Globular Protein

Tertiary protein structure is stabilized by outside polar hydrophilic hydrogen and ionic bond interactions, as well as internal hydrophobic interactions between nonpolar amino acid side chains. Prosthetic groups may add to the tertiary structure through formation of posttranslational covalent bonds.

Figure 4-7

Quaternary Structure

Proteins consisting of more than one polypeptide chain will generally have a quaternary structure, stabilized by the same general types of interactions found in the tertiary structure (i.e., both hydrophobic and hydrophilic bonding, occasionally including disulfide bonds). **Hemoglobin**, which combines two α-chains, two β-chains, and four heme groups, is an example of a functional protein having a quaternary structure. Sometimes **enzymes** are combined into a quaternary structure that enables them to act sequentially upon their substrates.

Protein Misfolding

When proteins fail to fold correctly, they lose their ability to carry out their normal physiologic actions, and often become toxic. Cells usually have mechanisms to rid themselves of these aberrant, misfolded entities, but perhaps aging and other factors slow or harm the process. An example of such a protein, in this case with an **enriched (antiparallel) β-pleated sheet**, is **amyloid** (or **amylin**). The term **amyloidosis** refers to a group of pathologic conditions of diverse etiologies, characterized by the accumulation of insoluble fibrillar amyloid proteins in various organs and tissues such that vital functions are compromised (e.g., the heart, kidneys, gastrointestinal, respiratory and nervous systems). Amyloid proteins are manufactured by malfunctioning bone marrow, and the liver of some dogs overproduces **serum amyloid A (SAA)**, which can accumulate in glomerular capillary membranes of the kidney, leading to secondary **renal amyloidosis** (and proteinuria). About 23% of dogs with glomerular disease are thought to have amyloidosis. **Pancreatic amyloidosis** is associated with **diabetes mellitus** in cats and primates, and insulinomas in dogs. Islet amylin initially forms intracellularly in insulin-secreting β-cells, then accumulates extracellularly after exocytosis or cell death. It is apparently stored and co-secreted with insulin. Islet amyloid can surround β-cells, isolating them from adjacent pancreatic tissue and blood capillaries, and it can act as a barrier to the diffusion of glucose and other nutritive substances.

Extracellular β-amyloid deposits in **brain tissue** are associated with the neurodegenerative **transmissible spongiform encephalopathies (TSEs)**. These disorders, which may be genetic, infectious, or sporadic, involve modification of the secondary-tertiary structure of the **prion protein (PrP)**, a self-propagating condition leading to large aggregates of these molecules. Diseases include **scrapie** in sheep, and **bovine spongiform encephalopathy (BSE)** in cattle (mad cow disease). It is not clear how **infective prion isoforms (PrPSc)**, or their potential smaller "infectious agents," make their way from the digestive tract to the brain, but penetration of the oral or intestinal mucosa seems likely to some investigators, as the PrPSc molecules themselves appear to be resistant to protease digestion.

Misfolded proteins in humans are associated with **amyloid cardiomyopathy**, and the neurodegenerative **Lou Gehrig's disease (amyotrophic lateral sclerosis (ALS))**, **Parkinson's disease**, and **Alzheimer's disease** (note: **Tau protein deposits** are also associated with Alzheimer's disease). In the prion diseases above, and perhaps in these conditions as well, toxic proteins appear to spread from cell to cell, inducing healthy proteins to misfold. All of these disorders are intriguing because evidence is accumulating that the formation of highly organized amyloid aggregates is a generic property of polypeptides, and not simply a feature of the few proteins associated with recognized disorders. That such aggregates are not normally found in properly functioning biologic systems is apparently a testament to evolution, in this case to a variety of mechanisms that most likely **inhibit their formation**. Future discoveries in this area will likely be revealing.

Protein Denaturation

Denaturation of a protein generally causes its biological activity to be lost, as well as its solubility. Extremes of **pH** or **temperature**, as well as detergents or other substances that reduce disulfide bonds, can cause protein denaturation. All degrees of structure higher than the primary one are generally destroyed when a protein (or a nucleic acid) is denatured. Particular peptide bonds in the primary structure can also be disrupted with extremes of pH. Additionally, tryptophan is typically destroyed in acid hydrolysis, whereas glutamine and asparagine are deaminated to their respective acids. Amino acids are usually racemized in base hydrolysis (i.e., they lose their optical activity), and serine and threonine are destroyed.

Unlike with nucleic acids, protein renaturation does not readily occur, even when they are placed in a biologically friendly environment. However, since the primary structure directs *in vivo* folding, if this structure remains intact, certain degrees of renaturation become possible.

Plasma Proteins

Blood plasma, which is usually about 4% of the body weight, consists of many constituents including H_2O, nutrients and metabolites, proteins, various hormones and electrolytes (e.g., Na^+, K^+, Ca^{++}, $HPO_4^=/H_2PO_4^-$, Mg^{++}, Cl^-, HCO_3^-; see **Appendix Table II**). The normal total plasma protein concentration ranges from **5.5-8.9 gm/dl** in domestic animal species, and it is a complex mixture including **simple proteins, glycoproteins** (Chapter 20), and various **lipoproteins** (Chapter 63). Thousands of **antibodies (immunoglobulins)** are also present in plasma, although under normal conditions the amount of any one is usually quite low. Comparative dimensions and molecular masses of a few plasma proteins are shown in **Fig. 4-8. Hemoglobin** is normally restricted to erythrocytes, **albumin, β₁-glob-**

ulin, α₁-**lipoprotein (HDL)** and **fibrinogen** are synthesized by the liver, and the γ-**globulins** are produced by lymphoid tissue (i.e., plasma cells). Plasma proteins constitute about 7% of plasma by volume; thus only 93% of the plasma volume is H_2O (a correction that is usually ignored).

Most capillary beds (with the exception of sinusoidal membranes of the liver) are relatively impermeable to plasma proteins, and these molecules exert an osmotic force (an **oncotic pressure** of about **28 mmHg**) across the capillary wall that helps to keep water in plasma. Thus in **hypoproteinemic conditions**, (which can be brought about by many factors including starvation, liver disease, malabsorption syndromes and nephrosis), **edema** (i.e., excess accumulation of fluid in interstitial fluid spaces of the body) can and will develop. In addition to providing this oncotic pressure, plasma proteins exert many other functions ranging from participation in blood clotting,

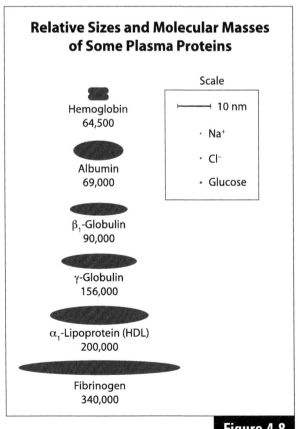

Relative Sizes and Molecular Masses of Some Plasma Proteins

Hemoglobin
64,500

Albumin
69,000

β₁-Globulin
90,000

γ-Globulin
156,000

α₁-Lipoprotein (HDL)
200,000

Fibrinogen
340,000

Scale

├───────┤ 10 nm

· Na⁺

· Cl⁻

· Glucose

Figure 4-8

buffering actions (i.o., they constitute about 7% of the blood buffering capacity; Chapter 85), serving as antibodies, and carriers for various hormones, lipophilic solutes and drugs.

Albumin normally accounts for **60%** of the plasma protein concentration, with the total body albumin pool being about 40% intravascular, and 60% in the skin. Roughly 6-10% of the exchangeable albumin pool, on average, turns over per day. **Globulins** constitute nearly **36%** of total plasma protein, and **fibrinogen**, which participates in blood clotting, accounts for about **4%**. In certain pathophysiologic states, such as the inflammatory bowel diseases, considerable amounts of plasma protein, including albumin, may be lost into the intestinal lumen through an inflamed mucosa. Patients with these conditions have **protein-losing gastroenteropathies**, and the half-life of plasma albumin may be reduced to as little as 1 day.

In other inflammatory states, with various forms of cancer, or secondary to certain types of tissue damage, the plasma protein concentration will **increase**. The proteins involved are referred to as **acute-phase proteins** (or reactants), and include C-reactive protein (CRP; so named because it reacts with the C polysaccharide of pneumococci), α1-antitrypsin, haptoglobin, α1-acid glycoprotein, and fibrinogen. CRP elevations are reported to vary from as little as 50% to as much as 1000-fold. **Cytokines** are proteins made by cells that affect the behavior of other cells. **Interleukin-1 (IL-1)**, a cytokine released from mononuclear phagocytes, is thought to be the principal (but not sole) stimulator of hepatic acute-phase protein biosynthesis.

OBJECTIVES

- Differentiate between amide and peptide bonds in the primary structure of a protein.
- Recognize how intra- and interchain disulfide bonds can contribute to the extracellular structure of a protein.
- Provide some examples of proteins containing α-helical secondary structures, and explain how proline affects those structures.
- Explain how β-pleated sheets are formed in the secondary structure of a protein, and understand how antiparallel β-pleated sheets contribute to amyloid formation.
- Recognize what amyloidosis is, as well as how where it is formed.
- Identify the primary electrolytes and proteins of plasma.
- Explain why hypoproteinemia leads to edema.
- Contrast structural differences between globular and fibrous proteins.
- Give an example of a protein possessing a quaternary structure.
- Recognize the importance of PrPSc molecules, and how they might originate.
- Understand how protein denaturation and renaturation occur.
- Discuss the relationships between TSEs, Peyer's patches, and intestinal protease digestion (see Chapter 7).
- Draw a protein peptide bond, and also an amide bond.
- Understand why highly organized amyloid aggregates are not normally present in properly functioning physiologic systems.
- Identify and be familiar with common constituents reported in a serum profile (see **Appendix Table II**).
- Recognize what percentage of the plasma volume is normally accounted for by protein.
- Understand how and why edema occurs in hypoproteinemic conditions.
- Identify the primary constituent of plasma protein, and discuss its physiologic function.
- Recognize what is meant by a "protein-losing gastroenteropathy."

QUESTIONS

1. **Select the TRUE statement below regarding proteins:**
   a. The biologic action of a protein is dependent only upon its primary structure (i.e., its amino acid sequence).
   b. There are normally only 20 different types of proteins in animals (since there are only 20 different amino acids).
   c. In mammalian organisms, mutations occur when proteins fail to fold correctly.
   d. The formation of highly organized amyloid aggregates may be a generic property of polypeptides.
   e. Proteins with a single polypeptide chain generally possess a quaternary structure.

2. **Select the FALSE statement below regarding protein structure:**
   a. The primary structure of a protein directly influences its secondary, tertiary, and quaternary structures.
   b. The secondary structure of a protein is held together by hydrogen bonds between carboxyl and amino groups of the peptide bonds.
   c. Disulfide bonds are a part of the secondary structure of all mammalian proteins.
   d. The tertiary structure of a protein refers to the overall three-dimensional arrangement of its polypeptide chain in space.
   e. α-Helical structures in proteins are stabilized by hydrogen bonding between amide hydrogens and carbonyl oxygens of different peptide bonds.

3. **The β-pleated sheets that form in the secondary structure of a protein are:**
   a. Stabilized by hydrogen bonding between separate chains that run either parallel or antiparallel to one another.
   b. Composed largely of amino acids with large hydrophobic side chains.
   c. Also known as β-bends.
   d. Stabilized by amide bonds that form between certain carboxyl or amine side chains.
   e. All left-handed.

4. **Chaperones:**
   a. Are peptide bonds in the primary structure of a protein that help to define the secondary structure.
   b. Are polar hydrophilic bonds that normally stabilize the outer regions of the α-helix of proteins.
   c. Help to unite the secondary and tertiary structures of proteins after the primary structure has been formed.
   d. Are carbohydrate derivatives that help to provide the energy needed for protein synthesis.
   e. Guide polypeptide chains into their tertiary structure while the primary sequence is being formed.

5. **Protein denaturation:**
   a. Is readily reversible in mammalian organisms.
   b. Generally increases its water solubility (i.e., hydrophilicity).
   c. Cannot occur if the structure has been stabilized by disulfide bonds.
   d. Can affect all levels of protein structure.
   e. Will occur with extremes of pH, but not of temperature.

6. **Globular proteins:**
   a. Generally do not have a tertiary and/or quaternary structure.
   b. Are usually linear and hydrophobic.
   c. Usually have elongated, strong rope-like structures.
   d. Usually have a primary structure, but not a secondary structure.
   e. Are usually spherical and hydrophilic.

7. **Peptide bonds in proteins join the:**
   a. R-group of one amino acid with the α-carboxyl group of another.
   b. α-Carboxyl group on one amino acid with the α-amino group of another.
   c. Sulfide group of one amino acid with the amide group of another.
   d. Side chains of two adjacent amino acids.
   e. R-Group of one amino acid with the α-amino group of another.

8. **The biologically active form of insulin contains how many amino acid residues?**
   a. 5
   b. 51
   c. 510
   d. 5,100
   e. 51,000

ANSWERS

8. b
7. b
6. e
5. d
4. e
3. a
2. c
1. d

Chapter 5

Properties of Enzymes

Overview

- Most enzymes are proteins.
- Enzymes are used to diagnose and treat disease.
- Some enzymes require coenzymes.
- Allosteric interactions can either facilitate or inhibit enzyme-catalyzed reactions.
- Phosphorylation of an enzyme may either increase or decrease its activity.
- Proteolytic enzymes in the lumen of the digestive tract are activated irreversibly.
- Synthetases stimulate synthesis using ATP.

Enzymes are mediators of metabolism that catalyze reactions responsible for the "dynamic" events of life. Virtually every chemical reaction in a living cell requires an enzyme, and, although the majority of enzymes in the mammalian organism are found within cells, several are secreted by cells and function in circulating blood, in the digestive tract, or in other extracellular fluid spaces. The number of different types of chemical reactions in a mammalian cell is large: an **animal cell**, for example, may contain **1000-4000 different types of enzymes**, each of which catalyzes a single reaction or set of closely related reactions. Certain enzymes are found in the majority of cells because they catalyze common cellular reactions (i.e., the synthesis of proteins, nucleic acids, and phospholipids, and the conversion of glucose and O_2 into CO_2 and H_2O, which produces much of the chemical energy used in animal cells). Other enzymes are found only in a particular type of cell, such as a hepatocyte or a neuron, because they catalyze reactions unique to that cell type. Also, some mature cells, including erythrocytes and epidermal cells, may no longer be capable of making proteins or nucleic acids, yet these cells retain specific enzymes that they synthesized at an earlier stage of differentiation.

Most enzymes are proteins; however, it is now recognized that RNA and other molecules may also exhibit highly specific catalytic activity. As an enzyme, RNA is referred to as a **ribozyme**.

Enzyme concentrations in plasma are frequently used for the **diagnosis of disease**. For example, increased concentrations of **lactate dehydrogenase (LDH)** in plasma may be indicative of liver disease or heart muscle damage, and increased concentrations of exocrine pancreatic enzymes (e.g., **amylase**, **lipase** and **trypsinogen**) in plasma may be indicative of pancreatic injury. Since exocrine pancreatic enzymes are low molecular weight proteins, they also appear in urine when blood levels rise. Enzymes may also be used to **treat disease**. For example, **streptokinase** can be used to dissolve blood clots. Some attempts to control the AIDS virus are also focused on inhibiting enzymatic processes involved in maturation of viral coat proteins.

Copyright © 2015 Elsevier Inc. All rights reserved.

General Properties of Enzymes

Enzymes are substrate-specific catalysts that enhance reaction rates without themselves being used up in the process. Like other catalysts, enzymes increase the rates of reactions that are already energetically favorable, and in many cases they affect both the forward and reverse reactions. Generally, a small amount of enzyme will influence a large amount of reactive substrate:

$$A + B \xrightarrow{\text{Enzyme}} C$$

A reversible reaction is one in which a product (**C**) can be formed from its substrates (**A + B**), and the substrates can be formed from the product. For most mammalian reactions, the balance of this reaction greatly favors one direction over the other (since most products disappear almost as fast as they are formed), so they appear to be **unidirectional** (i.e., physiologically irreversible).

Enzyme Nomenclature

The name of an enzyme usually indicates its function, and the suffix **-ase** is commonly appended to the name of the type of molecule on which the enzyme acts. For example, **aminotransferases** transfer amino groups between substrates, **mutases** shift a phosphate group from one oxygen atom to another within the same molecule, **phosphatases** remove phosphate residues, and **proteases** degrade proteins (**Table 5-1**).

Coenzymes

Some enzymes require, in addition to their substrate, a second organic molecule known as a **coenzyme**, without which they are inactive. Most coenzymes are linked to enzymes by noncovalent forces. Enzymes requiring coenzymes include those which catalyze **oxidoreductions, group transfer** and **isomeriza-**tion reactions, and reactions that form **covalent bonds. Hydrolytic reactions**, on the other hand, such as those catalyzed by digestive enzymes, do **not** require coenzymes. As an example of a coenzyme at work, consider the importance of the ability of skeletal muscle working anaerobically to convert pyruvate to lactate (**Fig. 5-1**). This reaction, catalyzed by the enzyme **lactate dehydrogenase** (**LDH**), serves merely to oxidize the reduced coenzyme **NADH** to **NAD⁺**. Without NAD^+, glycolysis cannot continue and anaerobic ATP synthesis (and hence skeletal muscle contraction) ceases (see Chapter 26). In highly aerobic heart muscle, the lactate that is exported from exercising skeletal muscle tissue can be reconverted to pyruvate (thus regenerating **NADH**; see Chapter 80). Reducing equivalents from **NADH** can be shuttled across mitochondrial membranes to ultimately enter oxidative phosphorylation, and the pyruvate can enter the TCA cycle (see Chapters 34-36).

Control of Enzyme Activity

Enzyme activity is controlled in several important ways to ensure that it is available at the right time, in the right place, and at the right concentration.

Induction of Gene Expression

The presence of a **substrate** can sometimes induce enzyme synthesis, thus resulting in a substantial increase in enzyme concentration. **Steroid** and **thyroid hormones** are also known to increase enzyme synthesis, and then

Lactate Dehydrogenase (LDH)

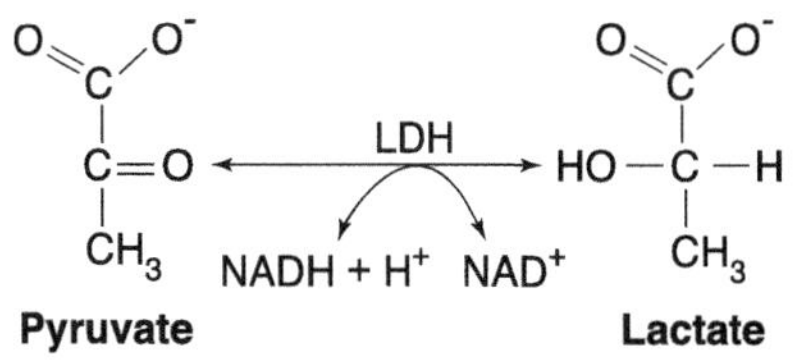

Figure 5-1

Table 5-1		
Enzyme Nomenclature		
Enzymes usually have the suffix-ase attached to either the kind of enzymatic activity they possess, or the substrate they act upon.		
Activity:	Aminotransferases	Transfer amino groups.
	Dehydrogenases	Catalyze oxidation/reduction reactions.
	Hydrolases	Cleavage of bonds between carbon and some other atom by the addition of H_2O across the bond.
	Isomerases	Interconversion of isomers by the transfer of a group from one position to another within the same molecule (e.g., ketoses to aldoses).
	Kinases	Transfer phosphates from ATP to enzymes or metabolic intermediates (thus forming, for example, glycogen phosphorylase a, or glycolytic intermediates like glucose-6-phosphate), or from an intermediate to ADP (GDP, CDP or UDP).
	Ligases	Formation of C-O, C-S, C-C, and C-N bonds by joining two molecules together in reactions requiring energy, which is usually supplied by ATP hydrolysis.
	Lyases	Nonhydrolytic cleavage of C-C, C-S, and some C-N bonds by the addition of a group to a double bond.
	Mutases	Shift a phosphate group from one oxygen atom to another within the same molecule.
	Oxidoreductases	Oxidation/reduction reactions that transfer electrons from one compound to another.
	Peptidases	Degrade peptides.
	Phosphatases	Remove phosphates.
	Phosphorylases	Cleave bonds by orthophosphate (phosphorolysis), and add phosphate (e.g., glucose-1-phosphate formation from glycogen). This is in contrast to hydrolysis, which is cleavage by water.
	Proteases	Degrade proteins.
	Synthases	Stimulate synthesis without using ATP.
	Synthetases	Stimulate synthesis using ATP.
	Transferases	Transfer of a functional group (e.g., amino, acyl, phosphate, or methyl) from a donor to an acceptor molecule.
Substrate:	Arginase	Facilitates hydrolysis of arginine.
	Urease	Facilitates hydrolysis of urea.

the synthesized enzymes can be subsequently activated by **polypeptide hormones** {or those derived from amino acids (e.g., the **catecholamines**)}.

Noncovalent Enzyme-Substrate Interactions

Substrates are bound to enzymes through multiple noncovalent interactions between binding residues at the active sites, and complementary chemical groups on the substrate (sites 1 and 2 in **Fig. 5-2**). This is particularly true of enzymes that have several subunits (i.e., multimeric enzymes). Binding of the substrate to one active site can sometimes increase the affinity of binding at other sites. In the **template model**, the active sites have a rigid structure, and in the **induced-fit model** they are more flexible. In the latter model, which may be more physiologic, substrate binding to an active site induces a conformational change in the enzyme that leads to an exact fit at all active sites between the enzyme and substrate (analogous to the way a sock takes on the shape of a foot).

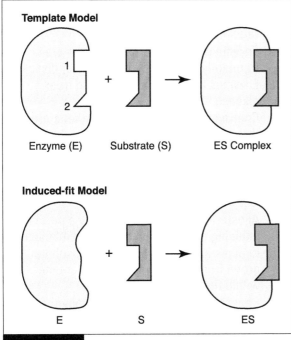

Template Model

Enzyme (E)   Substrate (S)   ES Complex

Induced-fit Model

E   S   ES

Figure 5-2

Feedback Inhibition

Allosteric interactions also play a part in feedback control of enzyme activity. In biosynthetic pathways, groups of enzymes frequently work together in sequential order to carry out metabolic processes. The reaction product of the first reaction quickly becomes the substrate of the second, and so on (**Fig. 5-3**). If the regulatory enzyme is the first enzyme in the pathway, then it may be controlled by the final end-product. When the end-product increases above steady-state concentrations, it may bind to an allosteric site on the regulatory enzyme, thus distorting its active sites and inhibiting its activity (in a non-competitive fashion; see Chapter 6).

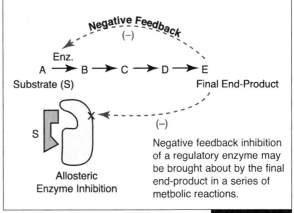

Negative Feedback
(−)

Enz.
A ⟶ B ⟶ C ⟶ D ⟶ E
Substrate (S)   Final End-Product

S

(−)

Negative feedback inhibition of a regulatory enzyme may be brought about by the final end-product in a series of metbolic reactions.

Allosteric
Enzyme Inhibition

Figure 5-3

Reversible Covalent Modification

The catalytic activity of an enzyme can sometimes be controlled by covalent modification. An example is **glycogen phosphorylase**, which exists in either its active, phosphorylated form, or its inactive, dephosphorylated form. This enzyme has two polypeptide chain subunits, each with a serine residue that is phosphorylated at its hydroxyl group by a kinase, or dephosphorylated by a phosphatase (**Table 5-1**). Similarly, glycogen synthase also undergoes reversible covalent modification. When phosphorylated it is inactive, but upon dephosphorylation it becomes active (see Chapter 23).

Control proteins

Enzymatic activity can also be controlled by the binding of specific stimulatory (**s**) or inhibitory (**i**) proteins. Examples are the **G$_s$** and **G$_i$** proteins in plasma membranes. These proteins were so named because their combination with **GTP** activates them. Upon binding of ligands (e.g., certain polypeptide hormones) to outer plasma membrane receptors, these proteins can either activate or inhibit membrane-bound **adenylate cyclase**, and thus regulate intracellular **cAMP** formation. Another example is the intracellular protein **calmodulin**, which becomes activated after binding 4 molecules of **Ca^{++}**. This activated protein then binds many intracellular enzymes, thus modifying their activities (see Chapter 58).

Proteolytic Activation

In contrast to the above modifications, which are all reversible, enzymes can also be activated **irreversibly**, (a process that would naturally occur only once in the life of an enzyme). These enzymes are synthesized as inactive precursors (zymogens or proenzymes), and then activated by cleavage of one or more specific peptide bonds. Good examples are the proteolytic enzymes associated with **digestion** in the gastrointestinal lumen (e.g., **pepsinogen** of the stomach, and the **exo-** and **endopeptidases** of the pancreas; see Chapter 7).

OBJECTIVES

- Explain why most enzymatic cellular reactions are physiologically irreversible.

- Provide examples of coenzyme-assisted reactions, as well as those that do not require coenzymes.

- Identify different ways in which enzyme synthesis can be increased, or synthesized enzymes can be activated.

- Given the name of an enzyme, identify its function.

- Indicate how negative feedback inhibition of enzymatic activity is accomplished.

- Explain how reversible covalent modification can alter the direction of glycogen formation or degradation.

- Indicate how G-stimulatory and G-inhibitory protein activity is controlled.

- Provide examples of irreversible proteolytic enzyme activation.

- Differentiate between the induced-fit and template models of enzyme substrate interaction.

- Distinguish between the activities of phosphatases and phosphorylases, and between those of synthases and synthetases, and provide some examples.

- Provide examples of hydrolytic reactions that do not require coenzymes, and oxidoreductase reactions that do.

- Give examples of mature cells retaining inactive enzymes synthesized at an earlier stage of differentiation.

- Define the word "enzyme," and explain how they function as catalysts.

QUESTIONS

1. **Select the FALSE statement below:**
   a. Enzymes are substrate-specific catalysts that are not used up in the reactions they regulate.
   b. Enzyme concentrations in plasma are frequently used in the diagnosis of disease.
   c. Enzymes are sometimes used to treat disease.
   d. Coenzymes are inorganic prosthetic groups, that help to facilitate enzyme-catalyzed reactions.
   e. Not all enzymes are proteins.

2. **Select the TRUE statement below:**
   a. Allosteric interactions may play a part in inducing enzyme activity, but not inhibiting it.
   b. Kinases are known to shift a phosphate group from one oxygen atom to another within the same molecule.
   c. Phosphatases are known to remove phosphate groups from phosphorylated compounds.

d. Peptidases are required for the formation of peptide bonds in proteins.

e. When pyruvate is converted to lactate, NADH is formed.

3. Which one of the following enzymes is activated irreversibly?

a. Pepsinogen

b. Glycogen phosphorylase

c. Glucokinase

d. Lactate dehydrogenase

e. Hexokinase

4. Which one of following best characterizes enzymatic activity in the digestive lumen?

a. Shifting of phosphate groups from molecule to molecule

b. Hydrolysis

c. Extensive synthetase activity

d. Aerobic

e. Absent following a meal

5. How many different types of enzymes are normally present in a mammalian cell?

a. 20

b. 200

c. 2000

d. 20,000

e. 200,000

6. Hydrolases catalyze:

a. The cleavage of bonds between carbon and some other atom by the addition of H_2O across the bond.

b. Oxidation/reduction reactions.

c. The transfer of phosphate residues from ATP to enzymes or metabolic intermediates.

d. The formation of C-O, C-S, C-C, and C-N bonds by joining two molecules together in reactions requiring energy.

e. Phosphate removal.

7. Which hormone below acts by increasing enzyme synthesis in its target cells?

a. Oxytocin

b. Cortisol

c. Vasopressin

d. Parathyroid hormone

e. ACTH

8. Urease (Ch. 10):

a. Is a urea cycle enzyme found in liver tissue.

b. Is also known as arginase.

c. Is a proteolytic enzyme secreted by the exocrine pancreas.

d. Is an enzyme found in microbes of the rumen.

e. Removes phosphate groups from the urea molecule.

9. Synthases:

a. Transfer amino groups from proteins to phospholipids.

b. Facilitate the hydrolysis of macromolecules.

c. Help to catalyze oxidation/reduction reactions.

d. Catalyze interconversion of amino acid isomers.

e. Stimulate synthesis without using ATP.

10. Which suffix below refers to an enzyme?

a. -ate

b. -ose

c. -ic

d. -ine

e. -ase

11. Enzymes that transfer phosphate from ATP to enzymes (or metabolic intermediates) are referred to as:

a. Mutases.

b. Phosphorylases.

c. Kinases.

d. Phosphatases.

e. Ligases.

12. Coenzymes are NOT required for:

a. Isomerization reactions.

b. Reactions that form covalent bonds.

c. Oxidoreductions.

d. Hydrolytic reactions of digestive enzymes.

e. Reactions catalyzing group transfer.

12. d

11. c

10. e

9. e

8. d

7. b

6. a

5. c

4. b

3. a

2. c

1. d

ANSWERS

Chapter 6

Enzyme Kinetics

Overview

- The effect of substrate concentration on the reaction rate of an enzyme-catalyzed reaction can be described mathematically using the Michaelis-Menten equation.
- The Michaelis-Menten constant (K_m) is the substrate concentration at which the reaction rate is one-half its maximal value ($1/2\ V_{max}$).
- Double reciprocal plots of substrate concentration/reaction velocity (i.e., Lineweaver-Burk plots) will yield linear data.
- Competitive enzyme inhibitors can be overcome by increasing the substrate concentration; noncompetitive inhibitors cannot.
- Noncompetitive inhibition does not alter K_m, but it does reduce V_{max}.
- Competitive inhibition does not alter V_{max}, but it increases K_m.
- An enzyme with a high K_m has a low affinity for its substrate.
- Irreversible inhibitors act kinetically like noncompetitive inhibitors.
- Isozymes are organ specific, and therefore useful diagnostic agents.
- Blood levels of the cardiac isozyme of troponin-I (cTn-I) are commonly used as markers of myocardial cell damage and necrosis (see **Case Study #3**).
- Blood levels of N-terminal-pro-Brain Natriuretic Peptide (NT-proBNP) are used as markers of cardiac hypertrophy or dilation in animals with congestive heart failure.

A good substrate may be cleaved 10,000 times faster than a poor substrate due to enzyme specificity. Enzyme kineticists study factors that influence the rates of enzyme-catalyzed reactions. **Leonor Michaelis** and **Maud Menten** first demonstrated that these reactions occur in two stages (**Fig. 6-1**). The **substrate (S)** first binds to its **enzyme (E)**, thus forming an **enzyme-sub-**strate complex (ES). This binding occurs quickly, but in a reversible fashion (k_1 and k_2 being rate constants for these processes). The enzyme-bound substrate is next transformed into a **product (P)** that dissociates from the enzyme, a process that is also reversible (rate constants k_3 and k_4). The catalytic power of an enzyme on a given substrate involves two basic variables: K_m, which is a measure of the affinity of the enzyme for its substrate, and V_{max}, which is a measure of the maximal velocity of enzymatic catalysis. Although the terms reaction rate and reaction velocity are used interchange-

$$E + S \underset{k_2}{\overset{k_1}{\rightleftharpoons}} ES \underset{k_4}{\overset{k_3}{\rightleftharpoons}} E + P$$

Figure 6-1

Copyright © 2015 Elsevier Inc. All rights reserved.

ably, **velocity** is usually expressed in terms of a change in the concentration of substrate or product per unit time, whereas **rate** refers to a change in total quantity (moles, grams) per unit time.

Substrate Saturation Curves

The effect of substrate concentration on reaction velocity, as shown in **Fig. 6-2**, can be described mathematically using the **Michaelis-Menten equation**:

$$V = V_{max}[S]/(Km + [S])$$

Where:

V = Reaction velocity

[S] = Substrate concentration

V_{max} = Maximal reaction velocity

K_m = [S] at 1/2 V_{max}

Derivation of this equation is based upon **four** assumptions:

1) That [S] is large relative to the enzyme concentration ([E]), therefore formation of the **ES** complex does not significantly alter [S];

2) That [P] at the beginning of the reaction is insignificant, therefore the reaction rate described by k_4 above can be ignored;

3) That the rate at which **ES** is transformed into **E + P** is rate-limiting (i.e., the slowest step in the reaction sequence); and

4) That the formation of the **ES** complex is rapid and reversible, and the rate at which it is formed is equal to the rate at which it disappears.

At low [S], reaction velocities in **Fig. 6-2** can be seen to be low, and increase with increasing [S] until points are reached where higher concentrations show only marginal increases in reaction rate. Theoretically, no matter how high the [S] is raised, the reaction rate will asymptotically approach, but not quite reach, plateau. This point is referred to as the **maximal reaction velocity** (**V_{max}**), where the enzyme becomes saturated with its substrate, and cannot function any faster. Values for **V_{max}** are usual rate units (e.g., **μmol/min**).

The **Michaelis-Menten constant** (**K_m**) is the substrate concentration (e.g., **mM**) at which the reaction velocity is one-half its maximal value (**1/2 V_{max}**), and, as stated above, is a measure of the affinity of the enzyme for its substrate. An enzyme with a **low affinity** for its substrate will have a **high K_m**, and one with a high affinity will have a **low K_m**. Both **V_{max}** and

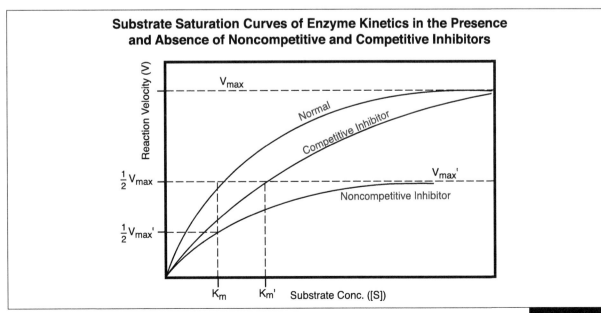

Substrate Saturation Curves of Enzyme Kinetics in the Presence and Absence of Noncompetitive and Competitive Inhibitors

Figure 6-2

K_m are sensitive to changes in **pH, temperature, ionic strength,** and the presence or absence of **enzyme inhibitors**.

Double Reciprocal Plots

Estimates of V_{max} and K_m can be obtained from non-linear substrate **saturation curves** (as in **Fig 6-2**); however, when linear data are desired, **double reciprocal (Lineweaver-Burk)** plots are generally employed (**Fig. 6-3**). The equation for these linear constructs is obtained by taking the reciprocal of the Michaelis-Menten equation:

$$1/V = (K_m/V_{max}) \, (1/[S]) + (1/V_{max})$$

The slope for this equation is K_m/V_{max}, the point at which the linear regression line intersects the **y-axis** is numerically equivalent to $1/V_{max}$, and the point at which it intersects the **x-axis** is $-1/K_m$. Thus, a Lineweaver-Burk plot provides readily identifiable **x-** and **y-intercepts** that in turn define the two kinetic variables, K_m and V_{max}.

Enzyme Inhibitors

Enzymes can be poisoned or inhibited therapeutically by various chemicals. Irreversible inhibitors bind enzymes and dissociate slowly, if at all. An example is the nerve gas diisopropyl-phosphofluoridate (DIPF), which binds acetylcholinesterase (**AChE**), thus prolonging the action of the neurotransmitter, acetylcholine (**ACh**). Death may ensue from nerve gas toxicity due to laryngeal spasm. Less potent reversible anticholinesterases are used clinically to treat diseases such as **Myasthenia Gravis**. Reversible inhibitors are generally classified as being either competitive, or noncompetitive.

Reversible, Competitive Inhibitors

Competitive inhibitors can be considered structural analogues of the substrate, and thus compete for the same active binding sites on the enzyme (**Fig. 6-4**). Because of this competition, if enough substrate is provided, the effect of the competitive inhibitor can be overcome (i.e., the substrate will ultimately occupy all binding sites). This is demonstrated in **Fig. 6-2**, and in **Fig. 6-3**. The V_{max} of the enzyme-catalyzed reaction in the presence of a competitive inhibitor remains unchanged from normal; however, the **apparent K_m (K_m')** for the substrate is increased since a higher concentration of

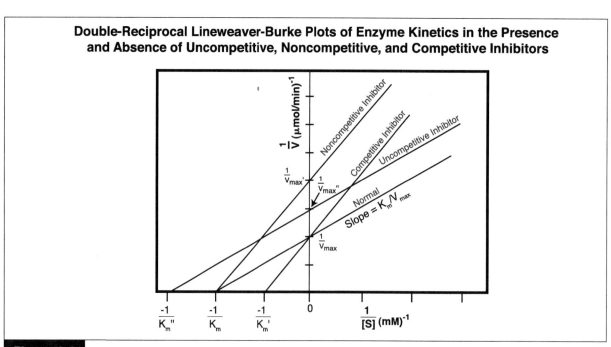

Double-Reciprocal Lineweaver-Burke Plots of Enzyme Kinetics in the Presence and Absence of Uncompetitive, Noncompetitive, and Competitive Inhibitors

Figure 6-3

substrate is required to overcome inhibitory effects of the competitor.

Reversible, Noncompetitive Inhibitors

Noncompetitive inhibitors are usually structurally unrelated to the substrate, and bind to an **allosteric site** on the enzyme (i.e., one other than the active site) (**Fig. 6-4**). Thus, both the substrate and inhibitor theoretically can bind to the enzyme at the same time. The inhibitor, however, usually **distorts the active sites**, thus altering conformation of their catalytic residues (which further reduces effectiveness of the **ES** complex). Since the effect of a noncompetitive inhibitor cannot be reversed by increasing substrate concentrations, K_m remains unchanged; however, there is a reduction in the apparent V_{max} (V_{max}') (**Figs 6-2, 6-3**).

Uncompetitive Inhibitors

A third type of inhibition is uncompetitive, and both K_m and V_{max} are changed (K_m'' and V_{max}''), but the slope remains constant (**Fig. 6-3**). Although these inhibitors do not react with the free enzyme, they can combine with the **ES** complex to slow down or prevent formation of product. This type of inhibition may be seen in reactions involving two substrates.

Irreversible Inhibitors

Irreversible inhibitors generally react covalently with an amino acid side chain on the enzyme to form a stable complex that is permanently inactivated. Some attach to the active sites, and are known as "mechanism based suicide substrates." Kinetic effects of irreversible inhibitors are similar to those of noncompetitive inhibitors above (i.e., V_{max} is decreased, while K_m remains unaffected).

Therapeutic Inhibitors

Most therapeutic inhibitors are either mechanism-based suicide substrates, or competitive inhibitors. Examples are given in **Table 6-1**.

Isozymes

Isozymes (or isoenzymes) catalyze similar reactions, but differ from each other slightly in chemical structure, and therefore kinetic properties. They are usually organ-specific, and therefore exploited for diagnostic purposes. Most enzymes present in plasma are released during normal cell turnover. However, increased amounts of particular isozymes appearing in plasma usually reveal trauma or pathologic processes. They are usually separated from each other in the diagnostic laboratory via electrophoresis.

Creatine phosphokinase (CPK) and **lactate dehydrogenase (LDH)** are two isozymes

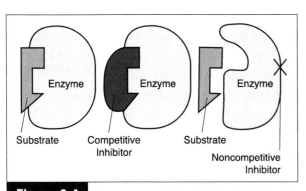

Figure 6-4

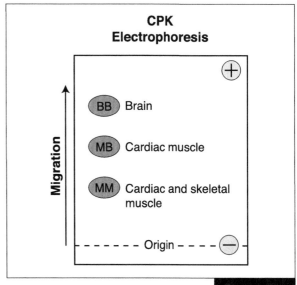

Figure 6-5

Table 6-1			
Examples of Therapeutic Inhibitors			
Drug	**Type**	**Therapeutic Use**	**Target Enzyme**
Allopurinol	Competitive and Suicide	Gout	Xanthine Oxidase
Aspirin	Suicide	Anti-inflammatory	Cyclooxygenase (COX 1,2)
Captopril	Competitive	Hypertension	Angiotensin-Converting Enzyme (ACE)
Coumadin	Competitive	Anticoagulant	γ Glutamylcarboxylase
Edrophonium	Competitive	Myasthenia Gravis	Acetylcholinesterase
5-Fluorouracil	Suicide	Cancer	Thymidylate Synthase
Methotrexate	Competitive	Cancer	Dihydrofolate Reductase
Mevastatin	Competitive	Hypercholesterolemia	HMG-CoA Reductase

HMG = 3-Hydroxy-3-Methylglutaryl; CoA = Coenzyme A

Table 6-2	
Examples of Serum Enzymes Used for Diagnostic Purposes	
Enzyme	**Purpose**
Alanine aminotransferase	Hepatic necrosis in dogs & cats.
Alkaline phosphatase	Bile duct obstruction and bone disease.
Amylase	Acute pancreatitis.
Aspartate aminotransferase	Liver disease in large animals; myopathies, hemolysis, and intestinal disease.
Creatine phosphokinase	Myopathies (& myocardial infarction).
Lactate dehydrogenase	Myocardial infarction; brain, kidney, pancreatic, lung, spleen, liver and erythrocyte disorders.
Lipase	Acute pancreatitis.
Sorbitol dehydrogenase	Liver disease in large animals.

commonly used for diagnostic purposes (**Table 6-2**). The **CPK** enzyme (see Chapter 77) contains two dimers assembled from muscle (**M**) and brain (**B**) subtypes. Skeletal muscle **CPK** is almost entirely the **MM** isozyme, brain **BB**, and cardiac muscle 15% **MB** and 85% **MM** (**Fig. 6-5**). Only cardiac muscle has the **MB** dimer. Lactate dehydrogenase is a tetramer, assembled from **M** and heart (**H**) subunits, with 5 different isozymes appearing via electrophoresis. Following a myocardial ischemic episode, **LDH** release into plasma usually occurs after that of **CPK** and **troponin-I**. The troponins (**Tn-I**, **Tn-T** and **Tn-C**) constitutue a regulatory complex of globular muscle proteins of the I band that normally inhibit contraction by blocking interaction of actin and myosin. Another polypeptide from ventricular myocytes, **N-terminal-pro-Brain**

Natriuretic Peptide (NT-proBNP), is usually elevated in cardiac hypertrophy (or dilation) in animals with congestive heart failure.

Cofactors and Coenzymes

Enzymes are globular proteins, with each having a specific function because of its unique structure. However, optimal enzyme activity usually depends upon the cooperation of **nonprotein substances** known as **cofactors**. The molecular relationship between an enzyme and its cofactor is properly referred to as a **holoenzyme**, while the protein component stripped of its cofactor is an **apoenzyme** (which usually exhibits low activity, or none at all). Not all enzymes, however, have cofactors.

There are **inorganic cofactors**, which include several cations such as Zn^{++}, Mg^{++}, Mn^{++}, Fe^{++}, Cu^{++}, K^+ or Na^+, and **organic cofactors** (i.e., **coenzymes**), which consist of about a dozen substances of diverse structure. The cofactor participation of inorganic cations represents one reason why these minerals are **essential for life**.

Metal ion cofactors bind in a transient, dissociable manner to either the **apoenzyme** or a **co-substrate** (such as ATP), and, as previously stated, they must be present in the medium surrounding an enzyme for catalysis to properly occur. Enzymes requiring an inorganic cofactor are termed **metal-activated enzymes** to distinguish them from the **metalloenzymes** (for which metal ions serve as **prosthetic groups**).

The role of a **cofactor** is either **1)** to alter the three-dimensional structure of the protein and/or bound substrate to maximize interaction, or **2)** to actually participate in the overall reaction as another substrate (i.e., a co-substrate). **Organic cofactors** operate according to role **#2** above, with the chemistry of this participation best described in terms of the coenzyme acting as a donor or acceptor of a particular chemical grouping relative to the other substrate(s). The grouping may be CO_2, a methyl (**-CH$_3$**) group, an amino (**-NH$_2$**) group, or **electrons**, to name a few. Accordingly, **coenzymes** are sometimes referred to as **group transfer agents**, and those that are altered would be **co-substrates**. The **water-soluble B vitamins** supply important components of numerous coenzymes (see Chapters 40-43). Several of these coenzymes contain, in addition, the **adenine**, **ribose** and **phosphoryl moieties** of **AMP** or **ADP**. **Nicotinamide** is a component of the redox coenzymes **NAD$^+$** and **NADP$^+$**, whereas **riboflavin** is a component of the redox coenzymes **FMN** and **FAD**. **Pantothenic acid** is a component of the acyl group carrier **coenzyme A**, as well as **4-phosphopantetheine** (an important prosthetic group in acyl carrier protein (ACP)). As a pyrophosphate, **thiamin** participates in oxidative decarboxylation reactions, and **folic acid**, **biotin** and **cobalamin** are coenzymes functioning in **one-carbon transfer**. Throughout succeeding chapters the activity of each inorganic/organic cofactor will be described and discussed as the need arises.

OBJECTIVES

- Define the K_m (or K_m') and V_{max} of an enzyme catalyzed reaction, and discuss factors that alter these kinetic variables.

- Explain the relationship between K_m and an enzyme's affinity for its substrate.

- Recognize why the V_{max} of an enzyme catalyzed reaction remains unchanged by a competitive inhibitor, yet K_m' increases.

- Discuss similarities and differences between standard non-linear Michaelis-Menten plots, and double reciprocal (Lineweaver-Burk) plots.

- Identify changes (if any) in K_m' and V_{max} during competitive, noncompetitive and uncompetitive enzyme inhibition.

- Give an example of a therapeutic suicide substrate, and discuss its mechanism of action.

- Explain why isozymes have been exploited for diagnostic purposes, and provide some examples.

- Know why noncompetitive inhibitors have no affect on K_m, while competitive inhibitors do.

- Recognize why cardiac troponins are used to assess myocardial damage following an ischemic episode, and understand the relationship between BNP, blood volume, cardiac myocytes, and natriuresis (see **Case Study #3**).

- Discuss how cofactors and coenzymes influence enzymatic reactions.

QUESTIONS

1. The variable that measures the affinity of an enzyme for its substrate is:
   a. k_1
   b. K_m
   c. V
   d. V_{max}
   e. [S]

2. Temperature and pH are known to affect:
   a. [S]
   b. [E]
   c. K_m but not V_{max}
   d. V_{max} but not K_m
   e. K_m and V_{max}

3. The maximal velocity of an enzyme-catalyzed reaction is:
   a. Decreased in the presence of a competitive inhibitor.
   b. Found on the x-axis of a Lineweaver-Burk plot.
   c. Known as the Michaelis-Menten constant.
   d. Increased in the presence of a noncompetitive inhibitor.
   e. Decreased in the presence of an irreversible inhibitor.

4. Which tissue type has the highest concentration of MB dimer for creatine phosphokinase?
   a. Skeletal muscle
   b. Brain
   c. Heart
   d. Kidney
   e. Liver

5. Which one of the following therapeutic inhibitors is considered to be a "suicide substrate"?
   a. Aspirin
   b. Captopril
   c. Coumadin
   d. Edrophonium
   e. Mevastatin

6. The y-intercept on a Lineweaver-Burk plot is:
   a. $1/2\ V_{max}$
   b. $1/2\ K_m$
   c. $1/V_{max}$
   d. $1/K_m$
   e. $-1/K_m$

7. Which one of the following is an accurate representation of the Michaelis-Menten equation?
   a. $V = V_{max} + [S] / (K_m - [S])$
   b. $V = V_{max}\ [S] / K_m\ [S]$
   c. $V = V_{max}\ [S] / (K_m + [S])$
   d. $V = V_{max} + [S] / K_m\ [S]$
   e. $V = (V_{max}/[S])(K_m/[S])$

8. All of the following are TRUE of competitive inhibitors, EXCEPT:
   a. They are usually structural analogues of the substrate.
   b. They reduce the apparent K_m (K_m') of the reactions they inhibit.
   c. The V_{max} of the enzyme-catalyzed reaction in the presence of a competitive inhibitor usually remains unchanged from normal.
   d. They are sometimes used therapeutically to control the activity of target enzymes.
   e. Their effects can usually be reversed by increasing the substrate concentration.

9. Which type of enzyme inhibitor below is known to affect both V_{max} and K_m?
   a. Irreversible
   b. Uncompetitive
   c. Noncompetitive, reversible
   d. Competitive, reversible
   e. Therapeutic

9. b

8. b

7. c

6. c

5. a

4. c

3. e

2. e

1. b

ANSWERS

Chapter 7

Protein Digestion

Overview

- Tissue protein turnover is normally greater than dietary protein assimilation.
- Tissue protein conjugation to ubiquitin targets it for degradation.
- Very little dietary protein escapes intestinal digestion.
- Most protein digestion and absorption occurs in the small intestine.
- Trypsin inhibitor prevents pancreatic autodigestion.
- Small peptides are normally found in the portal circulation.
- Similar transport mechanisms for amino acids and glucose exist in the small intestine and in proximal tubular epithelial cells of the kidney.
- Rennin is a digestive enzyme produced by the abomasum, while renin is a circulating enzyme produced by the kidney.

Proteins targeted for digestion in animals originate from three primary sources:

1) Tissue proteins,

2) Exogenous dietary proteins, and

3) Endogenous proteins in the digestive tract from either sloughed cells or those present in exocrine secretions.

Tissue Protein Turnover

The normal breakdown of tissue protein usually contributes two- to three-fold more available amino acid than does dietary protein. Each day a normal adult mammal degrades about **1-2%** of his/her body protein, principally from **liver** and other **visceral organs**. Protein turnover occurs at different rates in different tissue types. Immunoglobulins and collagen, for example, normally have half-lives measured in years, whereas liver proteins, plasma proteins and regulatory enzymes have half-lives measured in hours to days. In general,

the proteins of visceral organs turn over at a faster rate than those of skeletal muscle and connective tissue. However, during the latter phases of physiologic **starvation** or during **cachexia**, muscle protein is degraded at a faster rate than normal (see Chapter 76).

Tissue protein degradation, like protein synthesis, is a carefully regulated process. Protein conjugation to the 74-amino acid polypeptide known as **ubiquitin** targets it for degradation. This polypeptide is highly conserved, and has been found in species ranging from bacteria to primates. Animals appear to possess complex mechanisms by which abnormal proteins are recognized and degraded more rapidly than normal proteins. These mechanisms, however, may not operate optimally in some disease states.

About 75% of amino acids generated from tissue protein digestion are used for the resynthesis of new tissue protein. The remainder

Copyright © 2015 Elsevier Inc. All rights reserved.

enter hepatic gluconeogenesis (and/or ketogenesis), as well as biosynthetic pathways for a variety of specialized products (see Chapters 8 & 37). Amino acids removed from the tissue pool are replaced through digestion of dietary protein, and from *de novo* biosynthesis. Amino acids in protein, unlike glucose and fatty acids in glycogen and triglyceride, respectively, are stored in the body in a functional capacity, for virtually all proteins are considered to possess anatomic and/or physiologic activity.

Gastrointestinal Protein Digestion

Although **pepsins** can digest up to 15% of peptides and protein in the stomach, by far the bulk of protein digestion (and most all of amino acid absorption), occurs in the small intestine (**Fig. 7-1**). Very little protein escapes intestinal digestion (see Appendix).

Pepsinogen I is synthesized and secreted by chief cells of the oxyntic glandular region of the stomach, and **pepsinogen II** is synthesized and secreted by the pyloric region of the stomach. Both are activated to pepsins in the gastric lumen by **HCl** secreted by fundic parietal cells (**Fig. 7-2**). Maximal acid secretion correlates best with pepsinogen I levels.

Pepsinogen secretion is stimulated by the hormone **secretin** (i.e., "nature's antacid," secreted into blood from specialized S-cells in the duodenum). Unlike secretin, **gastrin** (from G-cells of the gastric antrum) does not directly stimulate pepsinogen secretion by acting on chief cells. Gastrin stimulates acid secretion, and acid in contact with the surface of the gastric mucosa stimulates afferent nerves of the intramural plexuses which, through a cholinergic reflex, in turn stimulate pepsinogen secretion.

The milk clotting enzyme, **rennin** (also known as **chymosin**, but not to be confused with **renin**, produced by juxtaglomerular cells of the kidney), is a hydrolase that catalyzes the cleavage of a single bond in casein to form soluble paracasein, which then reacts with Ca^{++} to form a curd. The commercial preparation of this enzyme, **rennet**, is used for making cheese and custards. Rennin is present in the abomasal juice of the calf, lamb, and kid, and may be secreted in small amounts as well by the stomachs of other young animals. The pepsins themselves, however, have milk-clotting activity, and therefore are thought to perform rennin's function in non-ruminant animals.

Both the pepsins and rennin are endopeptidases with broad specificity and low pH optima, cleaving peptide bonds on the carboxyl side of aromatic and acidic amino acids, leucine, and methionine (see Chapter 2). Proteins are partially degraded by these enzymes to a mixture of oligopeptides that enter the duodenum, and the low pH action of these enzymes is terminated when gastric contents are mixed with alkaline bile and pancreatic juice in the duodenum. Oligopeptides in the duodenum stimulate release of **cholecystokinin (CCK)**, a hormone that in turn stimulates gallbladder contraction and acinar pancreatic enzyme secretion. This hormone also has a mild effect on stimulating pancreatic ductular $NaHCO_3$ secretion.

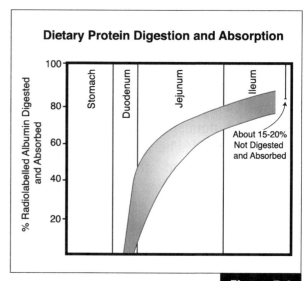

Dietary Protein Digestion and Absorption

About 15-20% Not Digested and Absorbed

Figure 7-1

Control of Gastric Pepsinogen Secretion

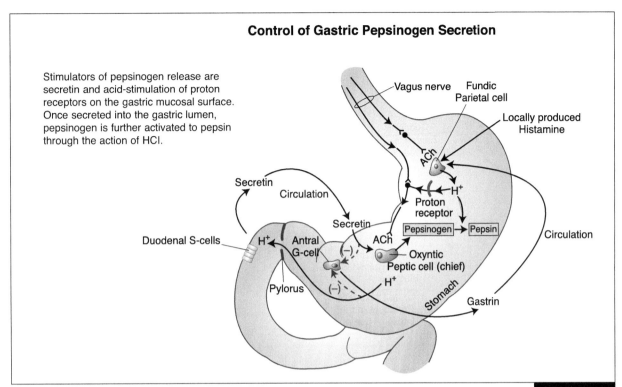

Stimulators of pepsinogen release are secretin and acid-stimulation of proton receptors on the gastric mucosal surface. Once secreted into the gastric lumen, pepsinogen is further activated to pepsin through the action of HCl.

Figure 7-2

Pancreatic proteases (i.e., exo- and endopeptidases), which are zinc-containing enzymes, are secreted in an inactive form (see Chapter 49). Inside the duodenal lumen, **enterokinase** (an enteropeptidase present on the duodenal brush border), initiates a cascade of proteolytic events that lead to the activation of these pancreatic proteases (**Fig. 7-3**). Enterokinase cleaves a hexapeptide from the N-terminal end of trypsinogen, thus activating **trypsin**. Trypsin (an endopeptidase), in turn, catalytically activates more of itself, as well as other endo- (**chymotrypsin, elastase**) and **exopeptidases** (**carboxypeptidase**). A compound known as **trypsin inhibitor** is found in the cytoplasm of pancreatic acinar cells, and not in the contents of secreted zymogen granules. It helps to keep trypsin in its inactive form (**trypsinogen**), such that pancreatic acinar and ductular cells will not be destroyed by protease activity.

Luminal digestion of a protein meal in the small bowel results in small and large peptides, and free amino acids (**Fig. 7-4**). Small and large peptides, which are mostly di- to hexapeptides, and which resist complete hydrolysis by pancreatic proteases, represent about two-thirds of total protein digestion products. As with free amino acids, the small **di-** and **tripeptides** can be absorbed intact by carrier-mediated processes. Once inside the cytoplasmic compartment of intestinal mucosal cells, these peptides can either be further hydrolyzed by intracellular peptidases, or diffuse directly into portal blood (which explains why certain small biologically active peptides exert effects when given orally). Larger **tetra-**, **penta-** and **hexapeptides** are poorly absorbed, and are hydrolyzed further by **brush border peptidases** (which account for about **20%** of total mucosal peptidase activity). Brush border hydrolysis results in free amino acids and small peptides, which can then be absorbed into mucosal cells.

Diffusion, at one time, was presumed to be the primary mechanism for the intestinal absorption of small peptides and amino acids released through protein hydrolysis. Subsequently, it

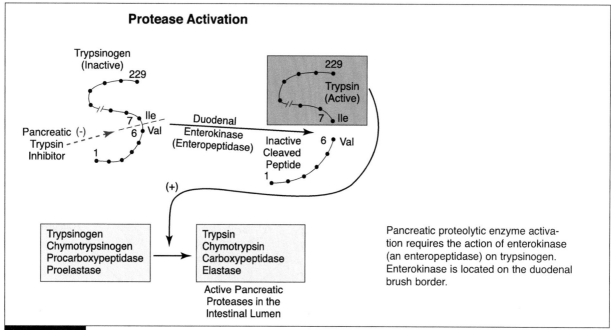

Protease Activation

Pancreatic proteolytic enzyme activation requires the action of enterokinase (an enteropeptidase) on trypsinogen. Enterokinase is located on the duodenal brush border.

Figure 7-3

was shown that the **L-isomers** of amino acids were absorbed by carrier-mediated mechanisms much in the same manner as glucose. **D-Amino** acids are apparently transported only by passive diffusion. In most cases that have been examined, as with glucose, these carrier-mediated processes require sodium (**Na⁺**) as part of a larger complex (see Chapter 38).

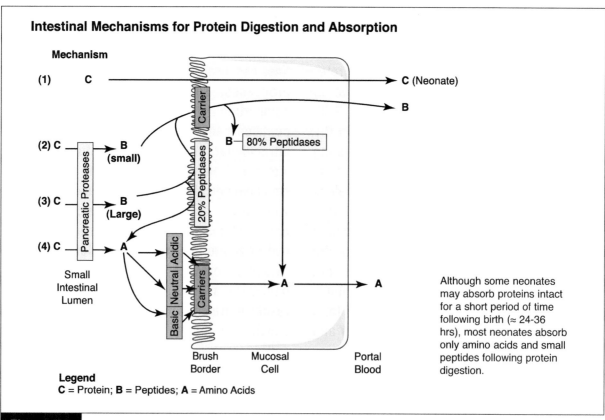

Intestinal Mechanisms for Protein Digestion and Absorption

Although some neonates may absorb proteins intact for a short period of time following birth (≈ 24-36 hrs), most neonates absorb only amino acids and small peptides following protein digestion.

Legend
C = Protein; **B** = Peptides; **A** = Amino Acids

Figure 7-4

Separate **Na$^+$-dependent intestinal transporters** have been identified for the **basic**, **acidic** and **neutral** amino acids, as well as for glycine and the imino acid proline. These Na$^+$-dependent transporters are also found in the brush-border of **proximal renal tubular epithelial cells**, where filtered glucose and amino acids are normally returned to the circulation. The sustained transport of amino acids requires that **Na$^+$** be pumped back out of absorptive cells in exchange for potassium (**K$^+$**), a reaction catalyzed by **Na$^+$/K$^+$-ATPase**. In certain disease states (e.g., **Hartnup syndrome**), transporters for one or more amino acid groups may be defective in both the intestinal mucosa and the proximal tubular cells of the kidneys (see Chapter 41).

In the neonate, **immunoglobulins** (which constitute about one-half of protein found in colostrum), can be absorbed intact for a short period of time (approximately 24 to 36 hours following birth), but most are subsequently digested within mucosal cells. Intraluminal protein is not totally digested by the newborn because the stomach is not yet producing enough **HCl** to activate pepsinogen, and **colostrum** contains **trypsin inhibitor** that prevents luminal activation of trypsin. Although immunoglobulin absorption confers passive immunity to most mammalian neonates (meaning immunoglobulins are endocytosed by mucosal cells and then exocytosed into intestinal lymph), lesser amounts reach the circulation of the primate infant. Primates acquire most of their passive immunity earlier during the fetal stage, when maternal immunoglobulins are transferred across the placenta from dam to fetus.

It has been shown that **prion proteins (PrPs)** can be absorbed by the adult gut to appear in **Peyer's patches** (gut-associated lymphoid tissue and ganglia of the enteric nervous system) prior to detection within the CNS. These infective proteins appear to be resistant to intestinal protease digestion, and give rise to a number of potentially fatal **transmissible spongiform encephalopathies (TSEs**; see Chapter 4).

PrPs are also found endogenously, but when their structures are modified abnormally they become life-threatening. Apparently endogenous PrPs play beneficial roles during embryonic development, where they help cells to communicate with each other.

OBJECTIVES

- Compare the normal rate of tissue protein turnover to gastrointestinal protein digestion.

- Identify the polypeptide known to target abnormal endogenous proteins for degradation.

- Recognize primary sites of dietary protein digestive and absorptive action.

- Name the gastric proteases, recognize what stimulates their secretion, and discuss their digestive actions.

- Outline the manner in which the pancreatic exo- and endopeptidases are activated.

- Identify and discuss the intestinal mechanisms for protein digestion and absorption.

- Compare and contrast the Na$^+$-dependent processes of intestinal and renal amino acid and glucose absorption.

- Compare early to late neonatal protein digestion and absorption.

- Indicate how prion proteins are thought to be absorbed by the gut.

- Discuss the patho-physiologic significance of having different carriers on the mucosal cell surface for basic, neutral and acidic amino acids.

- Explain the physiologic significance of small di- and tripeptides appearing in the hepatic portal circulation following a protein meal.

- Understand why it is important for the pancreas to secrete proteases in their inactive forms, while amylases, ribo- and deoxyribonucleases and lipases are secreted in their active forms.

- Distinguish between renin and rennin, and discuss their physiologic differences.

QUESTIONS

1. **Which one of the following normally contributes the greatest quantity of amino acids to the available body pool per day?**
   a. Tissue protein turnover
   b. Dietary protein
   c. Sloughed cells in the intestinal lumen
   d. Exocrine digestive secretions
   e. Turnover of plasma proteins

2. **All of the following exert positive influences on gastric pepsinogen release, EXCEPT:**
   a. Gastrin.
   b. Secretin.
   c. HCl.
   d. Acetylcholine.
   e. Enterokinase.

3. **Which one of the following is an endopeptidase secreted by the abomasum?**
   a. Gastrin
   b. Renin
   c. Trypsin
   d. Rennin
   e. Carboxypeptidase

4. **Which one of the following converts chymotrypsinogen to chymotrypsin in the intestinal lumen?**
   a. Enteropeptidase (enterokinase)
   b. Elastase
   c. Trypsin
   d. Pepsin
   e. Carboxypeptidase

5. **Trypsin inhibitor is:**
   a. Found in colostrum.
   b. Found on the brush border of mucosal cells in the duodenum.
   c. Secreted by the stomach.
   d. Normally found in saliva.
   e. Normally secreted into the duodenum from the pancreas.

6. **Select the FALSE statement below:**
   a. Approximately two hours following a protein meal, amino acids and small peptides should be found in hepatic portal blood.
   b. The L-isomers of amino acids are absorbed in the small intestine through a Na^+-dependent, carrier-mediated system.
   c. Similar transport mechanisms exist in both the intestine and the kidney for the absorption of amino acids and glucose.
   d. There is no evidence for protein absorption by the intestinal tract.
   e. Protein can be digested in the stomach, but not absorbed.

7. **Tissue protein conjugation to which of the following compounds targets it for degradation?**
   a. Ubiquitin
   b. Chymosin
   c. Enterokinase
   d. Renin
   e. Trypsinogen II

8. **Which protein below normally has the slowest turnover rate?**
   a. Fibrinogen
   b. Phosphofructokinase (PFK)
   c. Sex hormone binding globulin
   d. Albumin
   e. Collagen

9. **Approximately what percentage of amino acids generated from tissue protein digestion are normally used for resynthesis of new tissue protein?**
   a. 1%
   b. 25%
   c. 50%
   d. 75%
   e. 100%

10. **Approximately what percentage of the peptidase activity of the small intestine is located in the brush border?**
    a. 2%
    b. 20%
    c. 40%
    d. 60%
    e. 80%

11. **Pancreatic proteases normally contain:**
    a. Selenium.
    b. Vanadium.
    c. Iron.
    d. Cobalt.
    e. Zinc.

11. e

10. b

9. d

8. e

7. a

6. d

5. a

4. c

3. d

2. e

1. a

ANSWERS

Chapter 8

Amino Acid Catabolism

Overview

- Branched chain amino acids are oxidized by brain and muscle tissue, whereas aromatic amino acids are metabolized in the liver.

- Tetrahydrobiopterin, a relative of folic acid, helps convert Phe to Tyr in the liver.

- Intestinal cells prefer to oxidize Gln and Asn, and release NH_4^+, CO_2, and Ala into the portal circulation.

- Over 50% of amino acid release from skeletal muscle during starvation is accounted for by Ala and Gln.

- The favored gluconeogenic substrate of the liver is Ala, and of the kidneys is Gln.

- Several of the common 20 amino acids can enter the TCA cycle in more than one location.

- While some amino acids are strictly glucogenic or ketogenic, others like Phe can give rise to both glucose and ketone bodies in the liver.

- An inadequate amount of essential amino acids in the diet will lead to a negative nitrogen balance.

Following protein digestion in the intestinal tract, free amino acids are transported via the portal circulation to the liver, which then plays a pivotal role in determining the type and amount of amino acids released into the general (systemic) circulation for supply to non-hepatic tissues. The liver generally prefers to metabolize **aromatic amino acids** {**AAAs**; phenylalanine (**Phe**), tryptophan (**Trp**), and tyrosine (**Tyr**)}, while allowing the **branched chain amino acids** {**BCAAs**; leucine (**Leu**), isoleucine (**Ile**), and valine (**Val**)} to pass on into the systemic circulation (**Fig. 8-1**). The amino acid concentration leaving the liver in the fed state is usually 2-3 times higher than in starvation, and more than half are **BCAAs**. In muscle, BCAA transaminase helps to produce α-keto acid derivatives of these AAs that can be oxidatively decarboxylated for energy purposes.

Hepatic Metabolism of Phenylalanine

In the fed state, approximately three-quarters of the **Phe** entering the liver is converted to **Tyr**, with the other quarter incorporated into proteins. During starvation, Phe can be catabolized in the liver to acetoacetate, a ketone body, and fumarate, an intermediate in the TCA cycle and a gluconeogenic substrate (**Fig. 8-2**).

Conversion of Phe to Tyr requires oxygen and the enzyme **phenylalanine hydroxylase**, which is a **mixed-function oxygenase** since one atom of the O_2 appears in the product (Tyr), and the other in H_2O (**Fig. 8-2**). The cofactor reductant

Copyright © 2015 Elsevier Inc. All rights reserved.

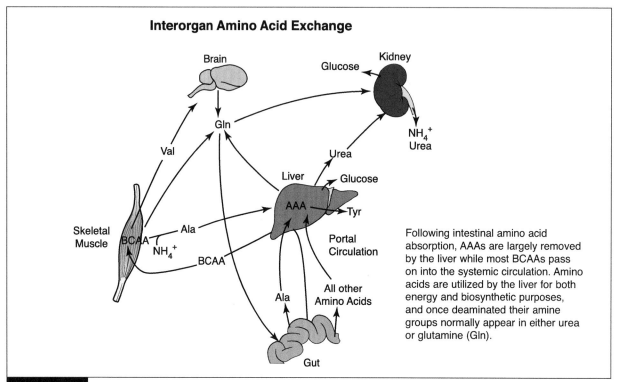

Figure 8-1

---

for this reaction is **tetrahydrobiopterin**, a relative of **folic acid**. Unlike its relative, however, this cofactor can be synthesized in the mammalian organism, and therefore is not considered to be a vitamin. Tetrahydrobiopterin is initially formed by the reduction of **dihydrobiopterin** by NADPH in a reaction catalyzed by **folic acid reductase (FAR)**, which is capable of reducing the basic pteridine nucleus at either positions 5, 6, 7, or 8 (see Chapter 16). NADPH is also used to reduce the **quinonoid** form of **dihydrobiopterin** back to tetrahydrobiopterin in a reaction catalyzed by **dihydropteridine reductase**. The quinonoid form of this cofactor is a normal co-product of the phenylalanine hydroxylase reaction. Hereditary deficiencies of phenylalanine hydroxylase are prominent in the human population, thus leading to accumulation of Phe in blood, tissues, and urine (**phenylketonuria; PKU**). This condition causes severe neurological symptoms, and if present in the animal population may go unrecognized due to early death.

Pteridine is the bicyclic nitrogenous parent compound of the pterins, which are endogenous derivatives of 2-amino-4-hydroxypteridine (**Fig. 8-3**). Some pterins, such as **xanthopterin** (2-amino-4,6-dihydroxypteridine), serve as eye and wing pigments in insects.

The BCAA/AAA Ratio

The **BCAAs** are primarily metabolized by **muscle tissue**, where the concentration of BCAA transaminase is high. This enzyme is only modestly expressed in liver tissue. In normal animals, the plasma **BCAA/AAA ratio** is usually about 3:1. In **liver disease** (or portosystemic shunt), there is decreased hepatic clearance of **AAAs**, owing to hepatocellular damage and/or vascular shunting, and increased utilization of **BCAAs** in muscle tissue. This can lead to a shift in the **BCAA/AAA ratio** from **3:1 to 1.5:1** or less. Since both groups of amino acids compete for entry into the brain by a similar mechanism, this great increase in AAAs serves to augment their entry into the central nervous system (CNS). Normal synthesis of the excitatory catecholamine

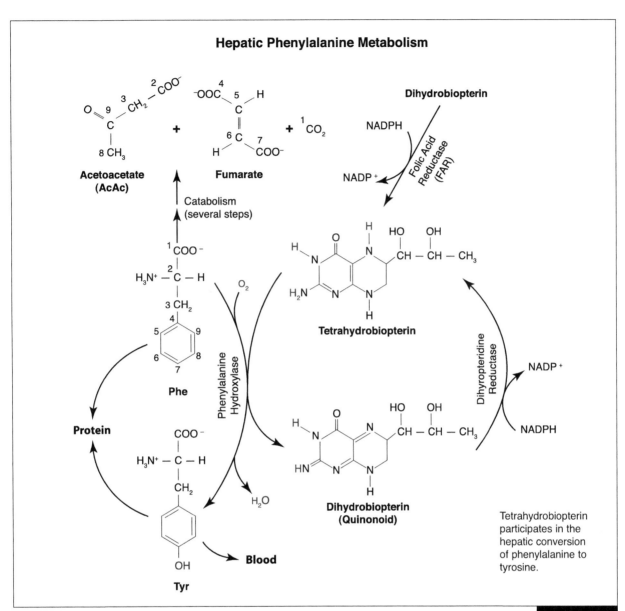

Hepatic Phenylalanine Metabolism

Tetrahydrobiopterin participates in the hepatic conversion of phenylalanine to tyrosine.

Figure 8-2

The Pteridine Nucleus of Tetrahydrobiopterin and Folic Acid

Xanthopterin

Figure 8-3

neurotransmitters, **dopamine (DA)**, **norepi-nephrine (NE)**, and **epinephrine (Epi)**, depends on adequate brain uptake of the **BCAAs**, as well as normal uptake of **Tyr** (see Chapter 12). When brain uptake of BCAAs is impaired and AAA uptake is enhanced, the synthesis of "**false**" **neurotransmitters** exceeds that of normal neurotransmitters. Both **Tyr** and **Phe** serve as precursors of two such "**false**" neuro-transmitters, **octopamine** and β-**phe-nylethan-olamine**. Large quantities of these compounds as well as substances with GABA-like activity (see Chapters 3 & 12) may arise from the bowel following bacterial degradation of protein. Failure of the liver to remove these products allows them access to the brain, where they flood presynaptic nerve endings and displace normal neurotransmitters, preventing the brain from responding normally to stimuli. **Tryptophan** also increases in the brains of patients with hepatic coma. Tryptophan is a precursor to serotonin (**5-hydroxytryptamine, 5-HT**), also a potent inhibitory neurotransmitter (see Chapter 12). Thus, increased production of both weak ("false") and "true" inhibitory neurotransmitters serves to severely depress brain function in **hepatoencephalopathy**. Specially formulated amino acid mixtures that are rich in BCAAs and low in AAAs are sometimes used to normalize the plasma amino acid profile of patients with this condition.

Intestine

Intestinal cells **utilize glutamine (Gln)** and **asparagine (Asn)** as major sources of energy, producing several byproducts, including **citrul-line, ammonium ion (NH$_4^+$), CO$_2$** and **alanine (Ala)**, which are released into the portal circulation (along with absorbed amino acids that are not metabolized). In starvation and/or recuperation following intestinal surgery, **Gln**, released from the liver, skeletal muscles, or provided parenterally, is oxidized by the intestine at a rate higher than normal.

Skeletal Muscle

The **catabolism of BCAAs** in skeletal muscle for energy purposes (e.g., during starvation), usually results in the transfer of amino groups to **pyruvate** (thus forming Ala), and/or **glutamate (Glu)** (thus forming Gln). **More than 50% of amino acid release from skeletal muscle during starvation is accounted for through Ala and Gln** (see Chapter 76). **Alanine** is the favored gluconeogenic substrate of **hepatocytes**, while **Gln** is the favored gluconeogenic substrate of **proximal renal tubular epithelial cells** (see Chapters 11 and 37). During starvation, small amounts of **Val** are also released from skeletal muscle, with this amino acid becoming the preferred substrate for amino acid oxidation in brain tissue.

Brain

When Val is oxidized in brain tissue during starvation, the amino group is normally trans-ferred to Glu, thus forming **Gln**, which is the preferred manner in which the brain rids itself of **ammonia (NH$_3$;** see Chapter 76). Since NH$_3$ is toxic to brain tissue, the ability of supporting glial cells to maintain Gln formation remains integral to neuron survival not only during starvation, but in other physiologic and patho-physiologic states in which **hyperammonemia** develops.

Kidney

The kidney is, like the liver, a **gluconeogenic** and **ammoniagenic** organ (see Chapter 11). Most **NH$_4^+$** produced in the kidney and ulti-mately excreted in urine has its origin in **Gln**, which is removed from blood or the glomer-ular filtrate by proximal tubular epithelial cells of the kidney following release by skeletal muscle, the liver, and brain. The renal release of **NH$_4^+$** from Gln occurs in two sequential steps catalyzed by the mitochondrial enzymes **glutaminase (A)** and **glutamate dehydroge-nase (B)**, respectively (see Chapter 9):

$$\left[\begin{array}{cc} \textbf{(A)} & \textbf{(B)} \\ \textbf{Gln} \longrightarrow \textbf{Glu} + \textbf{NH}_4^+ \longrightarrow \\ \alpha\textbf{-Ketoglutarate } (\alpha\textbf{-KG}^=) + \textbf{2NH}_4^+ \end{array} \right]$$

These reactions are important during the ketoacidosis associated with starvation, since the **divalent anion**, α-**KG**$^=$, can buffer protons during its renal conversion to glucose, and the NH$_4^+$ can pass into the renal filtrate to be excreted in urine (see Chapters 11 and 76).

Liver

With the general exception of Gln and Asn, and the three BCAAs, the catabolism of most other amino acids begins in the liver, where their amino groups are removed and incorporated into urea, or α-KG$^=$ (thus forming Glu and then Gln; see Chapters 10 and 11). The carbon skeletons remaining are either oxidized to CO$_2$ and H$_2$O, or used as substrates for **glucose** formation (gluconeogenesis) and/or **ketone body** formation (ketogenesis; **Fig. 8-4** and Chapter 71). Deaminated amino acids that enter the tricarboxylic acid (TCA) cycle at the level of acetyl-CoA or acetoacetyl-CoA can form ketone bodies (i.e., acetoacetic acid and β-hydroxybutyric acid), but there is no net conversion of their carbon atoms to glucose since the two carbon atoms of acetyl-CoA are lost in the sequential conversion of citrate to α-KG$^=$, and α-KG$^=$ to succinyl-CoA (see Chapter 34).

Several of the common twenty amino acids can enter the TCA cycle in more than one location (**Fig. 8-4**). **Isoleucine** and the three aromatic amino acids (**Phe, Trp,** and **Tyr**), are examples of amino acids that are both **glucogenic** and **ketogenic**. However, since carbon atoms from **Leu** and **Lys** can only be converted to acetyl-CoA or acetoacetyl-CoA, they are considered **ketogenic only**. The key to understanding the ketogenic action of Leu was the discovery that 1 mol of CO$_2$ was "fixed" per mole of isopropyl groups from Leu converted ultimately to acetoacetic acid. This CO$_2$ fixation requires the B-complex vitamin **biotin**, and forms β-**methylglutaconyl-CoA** as an intermediate (see Chapter 42).

In order for carbon atoms from an amino acid to receive net conversion to glucose in either the liver or kidney, they must enter the gluconeogenic pathway at either **pyruvate**, **oxaloacetate** (Asp and Asn), α-**KG**$^=$, **propionyl-CoA** (and then on to succinyl-CoA), or **fumarate** (see Chapter 37).

Nitrogen Balance

Since loss of protein and its derivatives in feces is normally small (see Chapter 7), the amount of nitrogen in urine is usually considered to be a reliable indicator of the amount of irreversible protein and amino acid breakdown. When the amount of nitrogen consumed in the diet is equal to that excreted in urine, the animal is considered to be in nitrogen balance. Most nitrogen is excreted in urine as **urea**, **uric acid**, **allantoin** and/or **NH**$_4^+$ (see Chapter 10). A **positive nitrogen balance** is associated with periods of **growth**, **pregnancy**, **lactation**, and/or **recovery from metabolic stress** (e.g., surgery), where net protein synthesis exceeds protein degradation. Conversely, a **negative nitrogen balance** occurs when there is a net loss of body protein. **Forced immobilization**, **starvation**, **senescence**, **diabetes mellitus**, **glucocorticoid excess**, **infection**, and **traumatic injury** are conditions associated with a net loss of body protein in animals, and therefore a negative nitrogen balance. Additionally, when any one of the nutritionally **essential amino acids** necessary for protein synthesis becomes unavailable (see Chapter 3), protein synthesis is reduced and the other amino acids that would have gone into protein are deaminated, with their nitrogen excreted in urine as either urea or NH$_4^+$. It is possible that some protein is sacrificed from intact cells following injury to provide amino acids (especially **Gln**

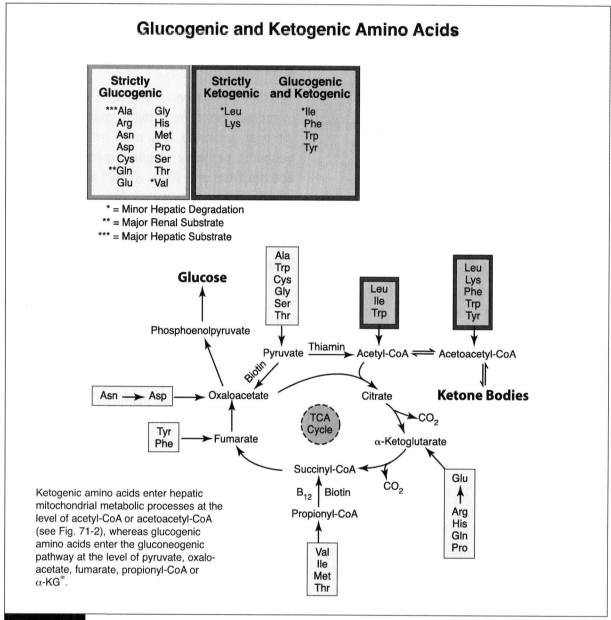

Figure 8-4

needed for purine and pyrimidine nucleotide (and thus DNA and RNA biosynthesis)), during formation of new cells in the healing process (see Chapters 13-15).

OBJECTIVES

- Recognize where most BCAA transaminase is located.

- Outline and discuss the hepatic metabolism of phenylalanine.

- Explain why dihydrobiopterin acts as substrates for FAR.

- Discuss why the plasma BCAA/AAA ratio may be reduced in liver disease, and why this may contribute to CNS depression.

- Provide a reason why glutamine may be added to parenteral fluids following intestinal surgery.

- Explain why more Ala and Gln are released from muscle tissue during starvation than is explicable through their contractile protein content.

- Describe how the brain rids itself of potentially toxic ammonia.

- Identify the amino acid that gives rise to the bulk of glucose produced by the kidney during starvation, and the bulk of ammonium ion excreted in urine.

- Describe the basis for the glucogenic and/or ketogenic amino acid designation, and classify each amino acid accordingly.

- Discuss why urinary nitrogen excretion is a reliable indicator of amino acid turnover, and give examples of physiologic situations leading to a positive or negative nitrogen balance.

QUESTIONS

1. **The adult liver generally prefers to:**
    a. Metabolize branched chain amino acids
    b. Metabolize aromatic amino acids
    c. Catabolize urea
    d. Oxidize ketone bodies
    e. Produce "false" neurotransmitters

2. **Which one of the following amino acids is a favored substrate of the intestine?**
    a. Phe
    b. Trp
    c. Ala
    d. Gln
    e. Val

3. **Which one of the following amino acids, produced from pyruvate and ammonium ion, is released from skeletal muscle tissue at a high rate during starvation?**
    a. Ala
    b. Gln
    c. Val
    d. Ser
    e. Tyr

4. **Which one of the following is the preferred amino acid substrate of brain tissue during starvation?**
    a. Gln
    b. Glu
    c. Val
    d. Leu
    e. Ala

5. **Which one of the following amino acids can be used to form ketone bodies in the liver, but not glucose?**
    a. Leu
    b. Cys
    c. Met
    d. Glu
    e. Phe

6. **Pregnancy is generally associated with:**
    a. A significant decrease in the plasma BCAA:AAA ratio.
    b. A negative nitrogen balance.
    c. Decreased clearance of "false" neurotransmitters from the hepatic portal circulation.
    d. A positive nitrogen balance.
    e. Hepatoencephalopathy

7. **All of the following are associated with a negative nitrogen balance, EXCEPT:**
    a. Insulin excess
    b. Glucocorticoid excess
    c. An essential amino acid deficiency
    d. Traumatic injury
    e. Infection

8. **The cofactor reductant involved in the hepatic conversion of Phe to Tyr is:**
    a. Tetrahydrofolate
    b. NADH
    c. Tetrahydrobiopterin
    d. Folate
    e. Biotin

9. **Which one of the following amino acids is both glucogenic and ketogenic?**
    a. Gly
    b. Lys
    c. Met
    d. Gln
    e. Phe

10. **PKU is best associated with a deficiency in:**
    a. Folic acid reductase.
    b. Phenylalanine hydroxylase.
    c. Dihydropteridine reductase.
    d. Folate.
    e. Phe.

ANSWERS

10. b
9. e
8. c
7. a
6. d
5. a
4. c
3. a
2. d
1. b

Chapter 9

Transamination and Deamination Reactions

Overview

- Hydrocarbons from amino acids can appear in carbohydrates or lipids.
- Essentially all of the α-amino nitrogen from amino acid catabolism can be funneled through glutamate.
- Glutaminase, asparaginase, and glutamate dehydrogenase are all deaminases.
- ALT concentrations are elevated in the cytosol of hepatocytes from small animal species.
- AST is present in both the cytosol and mitochondria of liver cells, and plays a part in the malate shuttle.
- Pyridoxal phosphate, tetrahydrofolate, biotin, and vitamin B_{12} are coenzymes needed for amino acid catabolism.
- Transaminases and the reactions they catalyze are considered to be essential for life.

Amino acids released from dietary or intracellular proteins are catabolized in similar fashions. Their degradation begins with the removal of the α-**amino group**, thus leaving an α-**keto acid** hydrocarbon skeleton (**Fig. 9-1**). Exceptions are threonine and lysine, and the cyclic imino acids, proline and hydroxyproline. Most NH_4^+ released from amino acids is due to the coupled activity of transaminases and glutamate dehydrogenase (see below); however, amino acid **oxidases** in liver and kidney tissue are flavoprotein enzymes. These enzymes oxidize amino acids to α-imino acids, then add H_2O with release of NH_4^+, thus forming the α-keto acid. The reduced flavin ($FADH_2$) is reoxidized by O_2, forming hydrogen peroxide (H_2O_2), which is split by **catalase** in many tissues to H_2O and O_2 (see Chapter 30). Further metabolism of the α-keto acid hydrocarbon skeleton is complex, since trans- or deamination of each amino acid yields a different skeleton, yet all

merge at various points with carbohydrate and lipid metabolism via the **tricarboxylic acid** (**TCA**) cycle (see **Fig. 8-2** and Chapters 12 and 34). Depending on the point of entry, hydrocarbons derived from amino acid catabolism can either be oxidized to CO_2 and H_2O, or can be used for the synthesis of carbohydrates (the glucogenic amino acids), or lipids (the ketogenic amino acids).

Alpha-amino group removal from amino acids occurs through either **transamination** or **deamination** reactions. Transamination reactions are catalyzed by **transaminases** (also called **aminotransferases**). The net effect of transamination is to collect **nitrogen** from different amino acids onto keto acid acceptors such as **pyruvate** {thus forming **alanine (Ala)**}, **oxaloacetate (OAA)** {thus forming **aspartate (Asp)**}, or α-**ketoglutarate (α-KG$^=$)** {thus forming **glutamate (Glu)**} (**Fig. 9-2**). Since Ala is also a substrate for **glutamate pyruvate transaminase** (also called

Copyright © 2015 Elsevier Inc. All rights reserved.

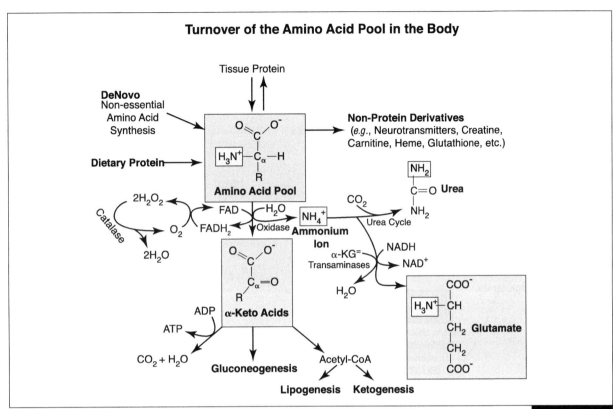

Turnover of the Amino Acid Pool in the Body

Figure 9-1

glutamate-α-ketoglutarate transaminase, or **serum glutamate pyruvate transaminase, SGPT**), essentially all of the α-amino nitrogen from amino acid catabolism can be funneled through **Glu,** thus making it a central source of nitrogen for both biosynthesis and excretion. Transamination reactions, which are found in most cells of the body, are readily reversible.

There are four coenzymes intimately involved in amino acid catabolism: **pyridoxal phosphate** (derived from vitamin B_6), **tetrahydrofolic acid** (**THFA**), **biotin** (a CO_2 shuttler), and **vitamin B_{12}** (cobalamin). Most reactions that involve the α-carbon atom of an amino acid require **pyridoxal phosphate** as a coenzyme (e.g., transaminations, amino acid decarboxylations, and some deaminations; see Chapter 42). **Tetrahydrofolic acid** provides a critical link between amino acid and nucleotide biosynthesis by carrying one-carbon (C_1) fragments from amino acids to intermediates of purine and pyrimidine nucleotides (see Chapters 14-16). Glucogenic amino

acids that are converted to either pyruvate or propionyl-CoA require **biotin-driven carboxylations** to further convert them to oxaloacetate and methylmalonyl-CoA, respectively (see **Fig. 8-4** and Chapter 42). Once methylmalonyl-CoA is formed, **vitamin B_{12}** is required to rearrange it to succinyl-CoA in the pathway of gluconeogenesis (see Chapters 37 and 43). Thus, amino acids such as valine (Val), isoleucine (Ile), methionine (Met), and threonine (Thr) require both biotin and cobalamin in order to become substrates for hepatic glucose production.

Deamination Reactions
Glutamate Dehydrogenase (GLDH)

The oxidative deamination of **Glu,** catalyzed by mitochondrial GLDH, results in the direct release of **ammonium ion** (NH_4^+), which can then be used for mammalian **urea** synthesis in the liver (see **Fig. 9-2** and Chapter 10). Urea is nontoxic, and is the primary end-product of nitrogen metabolism in **mammals**. Many fish

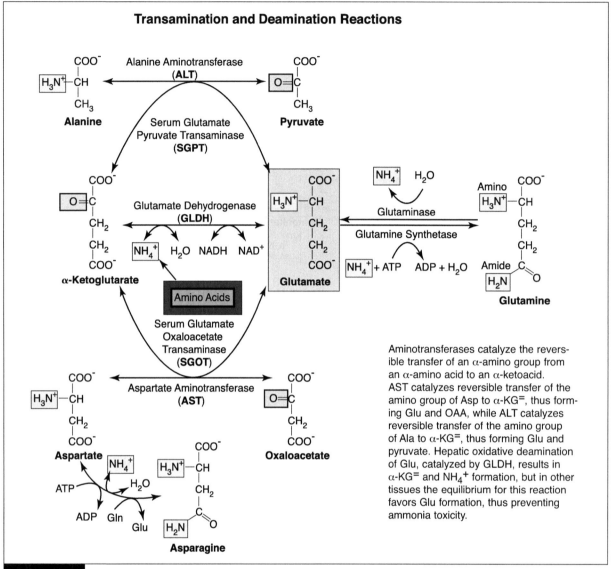

Transamination and Deamination Reactions

Aminotransferases catalyze the reversible transfer of an α-amino group from an α-amino acid to an α-ketoacid. AST catalyzes reversible transfer of the amino group of Asp to α-KG$^=$, thus forming Glu and OAA, while ALT catalyzes reversible transfer of the amino group of Ala to α-KG$^=$, thus forming Glu and pyruvate. Hepatic oxidative deamination of Glu, catalyzed by GLDH, results in α-KG$^=$ and NH$_4^+$ formation, but in other tissues the equilibrium for this reaction favors Glu formation, thus preventing ammonia toxicity.

Figure 9-2

are ammonotelic; they excrete nitrogen primarily as the more toxic ammonia (NH$_3$), and allow water to dilute it. **Insects, land snails, birds,** and most **reptiles** which must conserve water, are **uricotelic** (see Chapter 17). Their excretion of **uric acid** and its salts in urine can be considered a successful terrestrial adaptation since this compound has low water-solubility, and therefore does not draw water with it osmotically when excreted.

Glutamate is the only amino acid in mammalian tissues undergoing oxidative deamination at an appreciable rate. This reaction utilizes either **nicotinamide adenine**

dinucleotide (NAD$^+$), or its phosphorylated derivative **(NADP$^+$)** as an oxidizing agent, producing the reduced forms of these cofactors, **NADH** or **NADPH**. Glutamate dehydrogenase also catalyzes the reverse reaction, allowing NH$_4^+$ to be scavenged in the presence of α-KG$^=$ and NADH (or NADPH). The Glu so formed can then be used for synthesis of biomolecules, or can be transported to the liver for final nitrogen disposal.

GLDH is concentrated in the **livers** of **rodents, ovine, bovine, equine, feline,** and **canine** species. Other tissues containing GLDH in lower concentrations include kidney, brain, muscle, and intestinal mucosal cells. The biological half-life of GLDH in plasma is about 14 hours; therefore, if used for diagnosis of liver disease, activity of this enzyme should be determined soon after sample collection.

Glutaminase and Asparaginase

The side chain amide groups of **glutamine (Gln)** and **asparagine (Asn)** can be released by hydrolysis, also resulting in NH_4^+ formation (**Figs 9-2 and 9-3**). **Glutaminase** and **asparaginase**, respectively, catalyze these reactions. **Glutaminase** activity is high in **GABA** and **Glu-secreting neurons** of the brain. Activity is also high in **proximal renal tubular epithelial cells** where it **increases** in **metabolic acidosis**, and decreases in **metabolic alkalosis** (see Chapter 11). In addition, certain tumors reportedly exhibit abnormally high requirements for glutamine and asparagine, therefore these enzymes are being examined as potential antitumor agents.

Unlike the GLDH reaction, the **Glu <—> Gln** and **Asp <—> Asn** reactions cannot be reversed by a singular enzyme. Conversion of **Glu —> Gln** requires **glutamine synthetase**, and conversion of **Asp —> Asn** requires **asparagine synthetase** (**Fig. 9-3**). The glutamine synthetase reaction, occurring mainly in muscle, liver, kidney, and glial cells of the brain, exhibits similarities to and differences from the GLDH reaction. Both "**fix**" inorganic nitrogen, the former into an **amide**, and the latter into an **amino** linkage (see **Fig. 9-2**); and together they represent a coupled net exergonic process, the former hydrolyzing ATP, and the latter oxidizing NAD(P)H. Formation of Asn from Asp, catalyzed by asparagine synthetase (**Fig. 9-3**), somewhat resembles Gln formation; however, since the mammalian enzyme uses Gln rather than NH_4^+ as a nitrogen source, mammalian asparagine synthetase does not technically "fix" inorganic nitrogen. By contrast, the **bacterial** form does.

In addition to the above oxidative and hydrolytic deamination reactions, threonine, glycine, histidine, and serine are also known to undergo deamination, but by nonoxidative means. They are also known to be transaminated.

Transamination Reactions
Alanine Aminotransferase (ALT)

During starvation, **Ala** is important in transferring amino groups from skeletal muscle to liver, where they can be incorporated into **urea** or **Gln**, and the pyruvate hydrocarbons remaining can be incorporated into **glucose** (see Chapter 76). Although this enzyme is mainly found in the **cytosol of hepatocytes**, it is also present in neural, pancreatic, renal, intestinal, skeletal, and cardiac muscle tissue. Alanine aminotransferase concentrations are comparatively low in the livers of pigs, horses, and ruminant animals, and high in the livers of

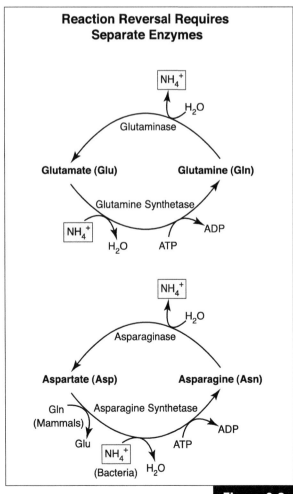

Figure 9-3

primates, dogs, cats, and other small animal species. The serum half-life is short (about 2 to 4 hours), therefore the activity of ALT in plasma should also be determined soon following sample collection.

As indicated above, transamination of **Ala** to **pyruvate** via **ALT** (also called **SGPT**) involves formation of **Glu** from α-**KG**= (Fig. 9-2). Once pyruvate is formed, it can be converted to lactate, decarboxylated to acetyl-CoA, or carboxylated to OAA (see Chapter 27).

Aspartate Aminotransferase (AST)

This enzyme is present in both the cytosol and mitochondria of **liver cells**, and both fractions can be separated electrophoretically. Large amounts of AST are also present in skeletal and cardiac muscle tissue, as well as erythrocytes, the kidneys, and intestinal mucosal cells. Although AST is not tissue specific, it is nonetheless an excellent screen for hepatic necrosis, and it generally remains elevated in the circulation for 2-3 days. It is frequently used to diagnose liver disease, but may not be predictive of hepatobiliary functional capacity.

Cytoplasmic transamination of **Asp** to **OAA** via activation of **AST** also involves formation of **Glu** from α-**KG**= (**Fig. 9-2**). **Glutamate oxaloacetate transaminase** (abbreviated **SGOT**, **serum glutamate oxaloacetate transaminase**) catalyzes incorporation of the amine group from Asp into α-**KG**=, thus forming Glu. The OAA generated from this cytoplasmic reaction can now accept a reducing equivalent from NADH in the formation of malate, thus generating NAD+ needed for anaerobic glycolysis. Malate can now be shuttled into mitochondria where these reactions are reversed (see **Malate Shuttle**, Chapters 35 and 36).

Other Transaminases

Other transaminases also exist in mammalian organisms (e.g., **cysteine transaminase**, **ornithine transaminase**, and **tyrosine α-ketoglutarate transaminase**). Since there are no reported metabolic defects associated with reactions catalyzed by any of the transaminases, like reactions of the tricarboxylic acid cycle, all are considered essential for life.

In summary, frequently the first chemical event occurring in amino acid degradation and the last step in amino acid biosynthesis is a **transamination** reaction. In this reaction an α-amino group is transferred from a donor amino acid to an acceptor α-keto acid to yield the α-keto acid of the donor amino acid, and the amino acid of the original α-keto acceptor. Reversible transamination reactions, which occur in most mammalian cells but are particularly prevalent in hepatocytes, are catalyzed by pyridoxal phosphate-dependent enzymes known as transaminases, or aminotransferases. Essentially all of the α-amino nitrogen from amino acid catabolism can be funneled through **Glu**, thus making it a central source of nitrogen for both biosynthesis and excretion. Although less prevalent than transamination, the conversion of an amino acid to its corresponding α-keto acid also occurs by **oxidative deamination**. Glutamate is the only amino acid in mammalian tissues undergoing oxidative deamination at an apprciable rate, and the reversible enzyme catalyzing this reaction, **GLDH**, uses NAD+ or its phosphorylated derivative (NADP+) as an oxidizing agent. Other amino acids can be deaminated by flavin-dependent oxidases that use either FAD (D-amino acids) or FMN (L-amino acids) as electron acceptors. In each instance the reduced flavoprotein can react directly with O_2 to produce H_2O_2, which is then decomposed to H_2O and O_2 by catalase.

OBJECTIVES

- Identify the reversible reactions catalyzed by the aminotransferases (transaminases) AST (SGOT) and ALT (SGPT), and give the major locations, substrates and products of these reactions.

- Discuss how α-keto acids are derived from amino acids.

- Identify four coenzymes intimately involved with amino acid catabolism, and discuss their functions.

- Describe the reaction catalyzed by GLDH, outline its regulation, and note how cofactors in this reaction link nitrogen metabolism with energy generation.

- Discuss how Glu and Gln are interconverted, and identify where the enzymes catalyzing these conversions are located.

- Explain why the first chemical event occurring in amino acid degradation and the last step in amino acid biosynthesis are usually transamination reactions.

- Understand why most all of the α-amino nitrogen from amino acid catabolism is funneled through glutamate.

- Explain why it is important to have AST located in both mitochondria and the cytoplasm of liver cells.

QUESTIONS

1. Most ammonia generated in the body normally comes from the oxidative deamination of which amino acid?
   a. Ala
   b. Glu
   c. Asn
   d. Asp
   e. Gln

2. Which one of the following coenzymes carries one-carbon fragments from amino acids to intermediates of purine and pyrimidine nucleotides?
   a. Glutamine
   b. α-Ketoglutarate
   c. Tetrahydrofolate
   d. Aspartate
   e. Vitamin B_{12}

3. Which one of the following enzymes is involved with the conversion of aspartate and α-ketoglutarate to oxaloacetate and glutamate, respectively?
   a. ALT
   b. GLDH
   c. SGPT
   d. SGOT
   e. Asparagine synthetase

4. Glutamate dehydrogenase is present in all of the following, EXCEPT:
   a. Erythrocytes.
   b. Liver.
   c. Kidney.
   d. Brain.
   e. Muscle.

5. There are no known metabolic defects associated with the conversion of α-KG$^=$ to Glu, Ala to pyruvate, Asp to OAA, or Gln to Glu. This indicates that:
   a. These reactions are unimportant.
   b. These reactions are essential for life.
   c. Investigators have been largely funded to do other studies.
   d. Investigators have withheld information from us.
   e. Amino acid catabolism is unessential for life.

6. Essentially all of the α-amino nitrogen from amino acid catabolism can be funneled through:
   a. Acetyl-CoA
   b. α-Keto acids
   c. Catalase
   d. Asparagine
   e. Glutamate

7. The oxidative deamination of glutamate yields NH_4^+ and:
   a. NAD^+
   b. Oxaloacetate
   c. Gln
   d. Pyruvate
   e. α-KG$^=$

8. The primary end-product of nitrogen metabolism in mammals is:
   a. NH_3
   b. NH_4^+
   c. Urea
   d. NADH
   e. Uric acid

8. c
7. e
6. e
5. b
4. a
3. d
2. c
1. b

ANSWERS

Chapter 10

Urea Cycle (Krebs-Henseleit Ornithine Cycle)

Overview

- Urea synthesis occurs primarily in the liver.
- Portal-caval shunts and acquired or inherited defects in urea cycle enzymes promote hyperammonemia.
- Aspartate serves as a nitrogen donor in the cytoplasmic phase of hepatic urea formation.
- Mitochondrial carbamoyl phosphate formation is rate-limiting in hepatic urea formation.
- A mitochondrial citrulline-ornithine antiporter exists in the inner mitochondrial membrane of periportal hepatocytes.
- Fumarate, generated in the cytoplasmic portion of the urea cycle, can either be reutilized therein, or it may leak away into the TCA cycle.
- Urea nitrogen is used by rumen microbes for protein biosynthesis.
- Arginase is a liver-specific enzyme.

Although amino acids are metabolized extensively throughout the body, **urea synthesis occurs primarily in the liver**, with minor amounts also formed in astrocytes of the brain. In ureotelic animals, urea is the primary form in which nitrogen appears in urine.

The two primary pathways by which nitrogen is transferred from amino acids to urea involve transamination and deamination reactions. Transaminases channel amino groups from several amino acids into glutamate (Glu), which can then be deaminated by glutamate dehydrogenase (GLDH), or the amino group can be transaminated onto oxaloacetate (OAA), thus forming **aspartate (Asp)** and α-**ketoglutarate** (α-**KG**$^=$) (**Fig. 10-1**). **Carbon dioxide** (**CO$_2$** in the form of bicarbonate, **HCO$_3^-$**), the **ammonium ion** (**NH$_4^+$**) generated from deamination of glutamine (**Gln**) or **Glu**, or from **ammonia (NH$_3$)** entering

directly from portal blood, and **Asp** now become substrates for hepatic urea synthesis (**Fig. 10-2**).

The overall stoichiometry for the **urea cycle** (also known as the Krebs-Henseleit ornithine cycle), can be described by the following equation:

$$\left[\begin{array}{c} \textbf{NH}_4^+ + \textbf{HCO}_3^- + \textbf{Aspartate} + \textbf{3ATP} \longrightarrow \\ \textbf{Urea} + \textbf{Fumarate} + \textbf{2ADP} + \textbf{AMP} + \textbf{4Pi} \end{array} \right]$$

The concentration of **NH$_3$** in portal blood is usually high following a protein meal, and it may be transiently increased by the release of additional **NH$_4^+$** from hepatic **glutaminase** and **GLDH** activity. Ammonia is an allosteric activator of glutaminase. However, by the time hepatic portal blood reaches the systemic circulation, the **NH$_3$** concentration has usually been reduced by about 50-fold.

Copyright © 2015 Elsevier Inc. All rights reserved.

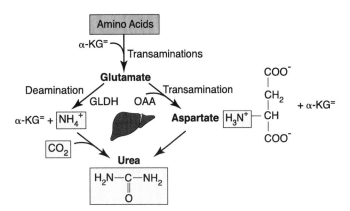

General Nitrogen Pathway from Amino Acids to Urea

Transamination and deamination reactions in periportal hepatocytes help to transfer nitrogen from amino acids ultimately to urea. CO_2, in the form of HCO_3^-, NH_4^+, and the amine group of Asp are substrates for urea biosynthesis.

Figure 10-1

Different hepatocytes lying along the sinusoid have different complements of active enzymes. **Periportal hepatocytes**, located near the portal vein, are the first liver cells to receive blood from the intestine, and they are rich in **carbonic anhydrase (CA)**, **glutaminase**, **GLDH**, and **urea cycle enzyme** activity. Bicarbonate generation in these hepatocytes is dependent upon CA activity, which in turn is influenced by the acid/base status. During periods of **metabolic acidosis**, for example, hepatic portal CO_2 and HCO_3^- concentrations decrease, and the low pH environment inhibits activity of hepatocellular CA and glutaminase, thereby limiting substrate availability for the urea cycle (see Chapter 11).

The five reactions required in periportal hepatocytes for urea formation are depicted in **Figure 10-2**. The first two occur in **mitochondria**, and the last three in **cytoplasm**. Starvation or an increased protein intake can alter concentrations of individual urea cycle enzymes 10- to 20-fold.

Carbamoyl Phosphate Formation

Mitochondrial carbamoyl phosphate formation, the first and also rate-limiting reaction in the urea cycle, is catalyzed by **carbamoyl phosphate synthetase-1 (CPS-1)**. A CPS-2 exists in the cytoplasm; however, it uses a different nitrogen source, and participates in pyrimidine nucleotide rather than urea biosynthesis (see Chapter 14). **N-Acetylglutamate**, whose steady-state level is determined by the rates of its synthesis from acetyl-CoA and Glu, is a cofactor required as an allosteric activator of CPS-1. Under the influence of **CPS-1, HCO_3^-, NH_4^+, 2ATP** and **H_2O** are condensed in the formation of **carbamoyl phosphate, 2ADP,** and an **inorganic phosphate (Pi)**.

Citrulline Formation

The second reaction, catalyzed by **ornithine transcarbamoylase**, involves mitochondrial condensation of **carbamoyl phosphate** with **ornithine**, thus forming **citrulline** and **Pi**. A **citrulline-ornithine antiporter**, which transports ornithine into mitochondria in exchange for citrulline, is located in the inner mitochondrial membrane. Ornithine and citrulline are basic amino acids not found in protein, but they are integral components of the urea cycle. **Mucosal cells** of the digestive tract are also known to produce citrulline and export it to the circulation for uptake by hepatocytes and renal tubular epithelial cells (**Fig. 10-3**). Although **arginine (Arg)** can be synthesized in the liver and kidneys from citrulline, only liver

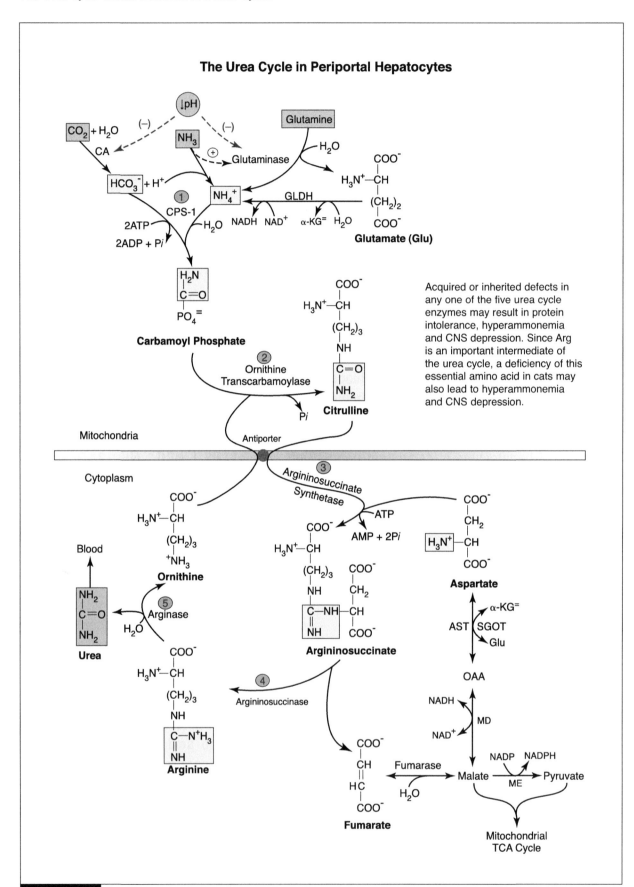

Figure 10-2

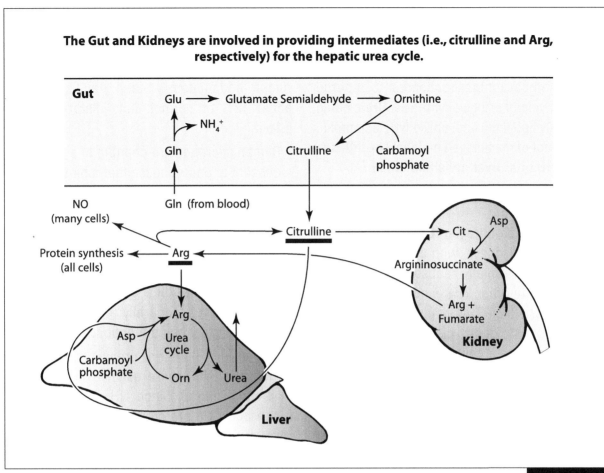

The Gut and Kidneys are involved in providing intermediates (i.e., citrulline and Arg, respectively) for the hepatic urea cycle.

Figure 10-3

cells possess the arginase needed to convert it to urea (see below).

Argininosuccinate Formation

The third reaction, catalyzed by **argininosuccinate synthetase**, involves cytoplasmic condensation of **citrulline** with **aspartate**, thus forming **argininosuccinate**. Linkage occurs via the **α-amino group** of aspartate, which ultimately provides the second nitrogen of urea. Energy required for this condensation is provided by the hydrolysis of **ATP** to **AMP** and **2Pi**, and this reaction can also occur in the kidneys.

Arginine and Fumarate Formation

The fourth reaction (which can also occur in the kidneys), catalyzed by **argininosuccinase**, involves cleavage of **argininosuccinate** into **arginine** and **fumarate** (i.e., the carbon skeleton of aspartate). Fumarate now forms a link with other pathways, including cytoplasmic reformation of aspartate, or entry into the mitochondrial tricarboxylic acid (TCA) cycle. **Fumarate** is hydrated to form **malate** in the presence of **fumarase**, an enzyme found in both mitochondria and the cytoplasm of liver cells. The malate so formed may be shuttled into mitochondria to enter the TCA cycle, or it may be converted to **pyruvate** {by **malic enzyme (ME)**} or **oxaloacetic acid (OAA)** {by **malate dehydrogenase (MD)**}. The OAA may then undergo transamination by accepting an amino group from Glu to reform aspartate (see Chapter 9), thus completing an **entirely cytoplasmic transamination route for entry of nitrogen into the urea cycle**.

Urea Formation

The fifth reaction, catalyzed by cytoplasmic

Mn++-containing **arginase** (see Chapter 51), involves removal of the **urea** side chain from **arginine**, thus forming **ornithine**, which is then transported back into mitochondria to undergo another cycle of urea biosynthesis. **Urea**, being sufficiently lipophilic and, unlike NH_3, a nontoxic end product of mammalian metabolism, **diffuses out of periportal liver cells into blood**.

Disposal of Urea

The **blood urea nitrogen (BUN)** pool is freely filtered by the kidneys, with about 50% of that filtered being normally reabsorbed into blood (**Fig. 10-4**). The other 50% is normally excreted into urine. The high BUN levels that often occur in patients with kidney disease are generally considered to be a consequence, not a cause, of impaired renal function.

Although approximately 75% of the BUN pool is ultimately excreted through urine, about 25% normally moves into the digestive tract where **bacterial urease** activity is high. The NH_3 formed can be funneled into bacterial protein synthesis, or be absorbed directly into the hepatic portal circulation. Urea enters the

reticulorumen through salivation, or directly from the circulation, and it also diffuses across mucosal cells of the colon. That entering the rumen endogenously each day is equal to about 12% of normal daily dietary nitrogen intake.

Rumen microbes are capable of synthesizing all essential and nonessential amino acids for their own protein biosynthetic purposes. This can occur from **nonprotein nitrogen (NPN**, such as **urea**), and appropriate **hydrocarbons** (from **cellulose**). Ruminant animals can grow, reproduce, and lactate, although not at optimal rates, when the diet contains no protein. Dietary protein is generally required, however, for optimal growth and function.

Due to high bacterial urease activity in the rumen, the $[NH_3]$ there usually ranges from 5 to 8 mg/dl (mg%). Since NH_3 is a buffer {i.e., it can accept a proton (H^+)}, excess urea can raise the $[NH_3]$ of rumen fluid, thus causing it to become more alkaline. Therefore, too much dietary protein or urea can lead to a **metabolic alkalosis** in ruminant animals, and make it more difficult for the rumen wall to absorb volatile fatty acids (i.e., butyrate, propionate, and acetate) produced through microbial cellulose digestion. These essential metabolites are more readily absorbed when in the unionized, free acid form. The normal rumen pH range is 5 to 7.

Abnormalities in Urea Biosynthesis

Certain liver diseases that affect the urea cycle, particularly acquired or inherited defects in any one of the five urea cycle enzymes, may have severe consequences for the mammalian organism since there are no alternative pathways for urea biosynthesis. For example, **arginosuccinate synthetase** deficiency, although rare, has been reported in dogs. Patients exhibit protein intolerance,

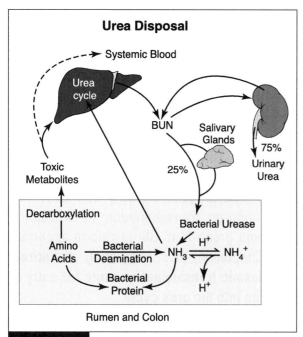

Figure 10-4

hyperammonemia, and a hepatoencephalopathy that leads to CNS depression, coma and death if left untreated. Treatment generally requires measures to control the hepatoencephalopathy, including a low-protein diet and oral antibiotics. Vascular abnormalities that shunt portal blood into the systemic circulation without perfusing hepatic sinusoids (i.e., portal-caval shunts; see **Case Study #4**) compromise the liver's ability to detoxify NH_3 through the urea cycle, and many other liver diseases are known as well to result in decreased urea cycle enzyme activity with hyperammonemia.

Cats with **arginase deficiency**, or those fed diets low in arginine, generally require more dietary ornithine. **Arginase** is liver-specific since it is restricted to the urea cycle, and when it leaks into blood following hepatocellular damage it is usually cleared from the circulation faster than the transaminases (ALT and AST). Therefore, a decrease in the serum activity of this enzyme can sometimes be a useful prognostic indicator of liver cell regeneration following acute liver injury.

The most common urea cycle enzyme deficiency in humans is that of **ornithine transcarbamoylase**. Mental retardation and death often result, but the occasional finding of normal development in treated patients indicates to some that the mental retardation may be caused by excess NH_3 before adequate therapy. The gene for this enzyme resides on the X chromosome, and males are generally more seriously affected than heterozygotic females. In addition to NH_3 and amino acids appearing in blood in increased amounts, **orotic acid (OA)** also increases, presumably because the carbamoyl phosphate that cannot be used to form citrulline diffuses into the cytosol, where it condenses with Asp, ultimately forming OA (see Chapter 14). Hepatic OA accumulation can occur in animals with arginine, ornithine or ornithine transcarbamoylase deficiency, all being associated with fatty liver infiltration (i.e., **steatosis**; see Chapter 72).

OBJECTIVES

- Identify the primary pathways by which nitrogen is transferred from amino acids to urea.

- Know the primary substrates for hepatic urea biosynthesis.

- Explain why periportal hepatic urea biosynthesis is reduced in metabolic acidosis.

- Identify the immediate precursors of the nitrogen and carbon atoms of urea, and name the carrier molecule on which urea is assembled during the urea cycle.

- Account for the ATP requirements of the urea cycle.

- Name five key enzymes of the urea cycle, note their intracellular locations, and indicate the molecular connection between this cycle and the TCA cycle.

- Identify reactions of the urea cycle that also occur in the kidneys and/or mucosal cells of the gut.

- Once produced, describe how urea is normally disposed of by the body.

- Explain why ruminant animals can continue to grow, reproduce and lactate when placed on protein-deficient diets.

- Describe several factors that could increase or decrease the blood urea nitrogen (BUN) concentration.

- Explain why cats fed a diet low in Arg generally required more dietary Orn.

- Indicate how excess dietary protein or urea can lead to metabolic alkalosis in ruminant animals.

- Recognize the rate-limiting mitochondrial reaction of hepatic urea formation, and compare it to a similar reaction occuring in the cytoplasm (see Chapter 14).

- Recognize how microbes in the rumen can synthesize amino acids (and thus protein) from cellulose and urea.

QUESTIONS

1. **Which one of the following amino acids serves as a cytoplasmic nitrogen source in urea formation?**
   a. Ornithine
   b. Aspartate
   c. Alanine
   d. Serine
   e. Tyrosine

2. **Which of the following enzymes in the urea cycle are associated with ATP utilization?**
   a. CPS-1 and ornithine transcarbamoylase
   b. Argininosuccinate synthetase and argininosuccinase
   c. Arginase and ornithine transcarbamoylase
   d. Argininosuccinase and arginase
   e. CPS-1 and argininosuccinate synthetase

3. **Select the FALSE statement below:**
   a. An entirely cytoplasmic transamination route for the entry of nitrogen into the urea cycle can develop from fumarate formation.
   b. Ornithine generally enters mitochondria of periportal hepatocytes in exchange for argininosuccinate.
   c. Hyperammonemia can result from acquired or inherited defects in any one of the urea cycle enzymes.
   d. Urea is a normal constituent of saliva in ruminant animals.
   e. About 25% of the blood urea nitrogen pool normally diffuses into the intestinal tract.

4. **Which one of the following is an allosteric activator of CPS-1?**
   a. H⁺
   b. Aspartate
   c. Citrulline
   d. N-Acetylglutamate
   e. Urea

5. **Urea synthesis occurs primarily in the:**
   a. Liver.
   b. Brain.
   c. Intestine.
   d. Kidney.
   e. Textbooks.

6. **Acquired or inherited defects in any one of the five major hepatic urea cycle enzymes could have devastating consequences for the mammalian organism, because:**
   a. The mitochondrial TCA cycle would be impaired.
   b. There are no alternative pathways for urea biosynthesis.
   c. Bicarbonate would disappear from blood.
   d. Ammonia would disappear from blood.
   e. All transamination reactions would cease.

7. **Substrate availability for the urea cycle is reduced in:**
   a. High protein feeding.
   b. Cachexia.
   c. Metabolic alkalosis.
   d. Metabolic myopathy.
   e. Metabolic acidosis.

8. **Approximately what percentage of the urea filtered by the kidney is reabsorbed into blood?**
   a. 5%
   b. 25%
   c. 50%
   d. 75%
   e. 100%

9. **Excessive dietary intake of urea in ruminant animals will predispose them to:**
   a. Metabolic alkalosis.
   b. Arginine deficiency.
   c. Hepatoencephalopathy.
   d. Cardiomyopathy.
   e. Kidney disease.

10. **Cats with arginase deficiency generally require more dietary:**
    a. Ammonia.
    b. Glutamine.
    c. Ornithine.
    d. Aspartate.
    e. Carbamoyl Phosphate.

11. **Urea cycle intermediates can be synthesized in mucosal cells of the gut, and in the kidneys:**
    a. True
    b. False

11. a
10. c
9. a
8. c
7. e
6. b
5. a
4. d
3. b
2. e
1. b

ANSWERS

Chapter 11

Glutamine and Ammonia

Overview

- Nitrogen and carbon flux from the liver to the kidneys increases in metabolic acidosis.
- Glutamine production by perivenous hepatocytes increases in metabolic acidosis, while urea production by periportal hepatocytes decreases.
- Renal gluconeogenesis increases in prolonged metabolic acidosis.
- Ammonia (and several drugs) move into the tubular filtrate by nonionic diffusion.
- pH control of the urea cycle is overridden in respiratory acidosis.
- The kidney generates glutamine in metabolic alkalosis.
- Hepatic bicarbonate consumption is increased in metabolic alkalosis.

Ammonia Toxicity

Ammonia (NH_3) is normally generated by enteric bacteria and absorbed into portal venous blood, which thus contains higher concentrations of this compound than systemic blood. Since a healthy liver promptly removes NH_3 from portal blood, peripheral blood is nearly NH_3-free. This is essential since NH_3 is toxic to the central nervous system. Should portal blood bypass the liver, the NH_3 concentration may then rise to toxic levels in systemic blood. This follows severely impaired liver function, or development of collateral vascular communications between hepatic portal and systemic veins.

Activity of the hepatic urea cycle normally maintains the peripheral blood concentration of NH_3 at about 0.02 mM in mammals. Any impairment of the cycle increases this, and although there is no precise concentration of NH_3 at which toxicity becomes apparent, it is considered that a blood concentration of about 0.2 mM or above is dangerous (i.e., ten times normal). Ammonia toxicity in animals is usually associated with vomiting and coma, and may be due to **liver damage**, **portosystemic shunt**, **poisons**, **viral infection**, or a **deficiency** of one or more **enzymes** of the **urea cycle** (see Chapter 10).

Nitrogen and Carbon Flux Between Liver and Kidney

Prolonged starvation and **diabetes mellitus** (**DM**) are examples of physiologic situations resulting in **metabolic acidosis**, which in turn imposes demands upon the organism to deal more effectively with potential nitrogenous buffers such as NH_3. Although past concepts of acid-base control focused largely upon the role of the **kidneys** and **lungs** in adjusting the **H⁺ concentration** of extracellular fluid, reconsideration of established pathways and more recent observations led to a conceptual change from a two- to a three-organ regulatory effort (i.e., **lungs**, **kidneys**, and **liver**). Since this development considers an additional regulatory site for systemic pH control, it has broadened our understanding of derangements in acid-base homeostasis. This interorgan team effort not only involves **nitrogen flux**, but also

Copyright © 2015 Elsevier Inc. All rights reserved.

carbon flux, resulting in a partial shift of **gluconeogenesis** from **liver** to **kidney** in **metabolic acidosis** (see Chapter 87).

Although only a few studies have been performed to support this concept under *in vivo* conditions, studies in rats, dogs, and primates indicate that our basic concept of acid-base regulation should be expanded to include hepatic and renal nitrogen and carbon flux.

Hepatic urea and **glutamine (Gln)** biosynthesis represent two sequential pathways for **ammonia (NH₃)** removal from blood. Enzymes for urea biosynthesis are located primarily in **periportal hepatocytes** (see Chapter 10),

while those for **Gln biosynthesis** are located in **perivenous hepatocytes** (i.e., those nearer the central vein) (**Fig. 11-1**). This structural and functional organization of the two pathways for NH_3 detoxification indicates that any flux change through the urea cycle will determine substrate supply for perivenous Gln formation.

Several mechanisms have been described for adjusting the rate of hepatic urea formation (being equivalent to hepatic bicarbonate (HCO_3^-) removal) to the requirements of acid-base homeostasis. Acid-base parameters sensed by the liver for adjusting HCO_3^- disposal via urea synthesis include the H^+ and HCO_3^- concen-

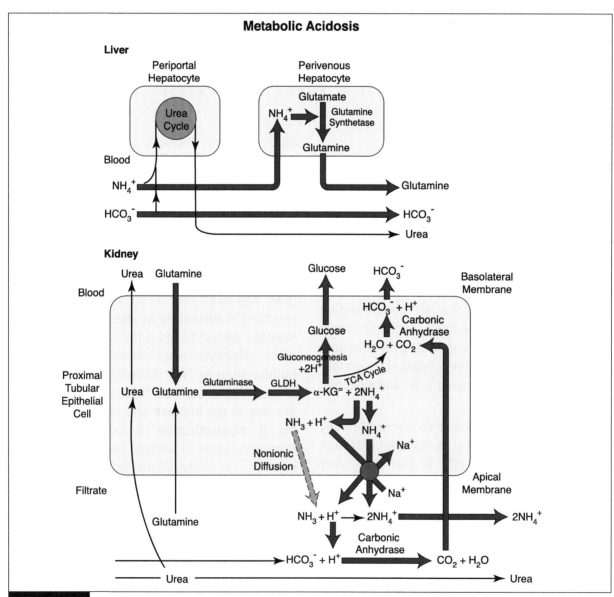

Figure 11-1

trations, as well as the **Pco₂**. A **decrease** in extracellular (and thus intracellular) **pH** results in a **decrease in carbonic anhydrase** and **glutaminase** activity of periportal hepatocytes, and thus a **decrease in urea biosynthesis** (see Chapter 10). The pH-dependence of these substrate-providing reactions is highly sensitive; lowering the extracellular pH from 7.4 to 7.1 is reportedly followed by a 5-fold reduction in hepatic glutaminase activity, and a 3-fold reduction in carbonic anhydrase activity.

In **metabolic acidosis**, extracellular concentrations of CO_2 and HCO_3^- are decreased, and urea biosynthesis is increasingly (up to 90%) dependent upon carbonic anhydrase-catalyzed HCO_3^- formation inside liver mitochondria (which is strongly controlled by pH). Consequently, **urea biosynthesis**, and therefore HCO_3^- removal, **decreases** and **perivenous ammonium ion (NH_4^+)** is increasingly incorporated into **Gln**, the substrate for **renal ammoniagenesis** and **gluconeogenesis** (see **Fig. 11-1**). In proximal renal tubular epithelial cells, 3 out of 5 carbon atoms of Gln can be converted to glucose. This coupling of gluconeogenesis increases in metabolic acidosis in parallel to ammoniagenesis, whereas hepatic gluconeogenesis decreases (see Chapter 75).

Under catabolic conditions leading to a negative nitrogen balance, muscle protein is degraded. The nitrogen liberated from this protein is excreted in urine as either urea or NH_4^+, depending on acid-base status. **Alanine (Ala)** typically becomes a major nitrogen carrier from muscle to liver supporting hepatic, but not renal gluconeogenesis, and consequently, ureagenesis. Under low carbohydrate or ketogenic conditions, however, the contribution of the **Ala cycle decreases**, and **hepatic Gln formation** becomes increasingly significant. In accordance with this concept, metabolic acidosis usually leads to a decrease in hepatic urea formation, and a concomitant fractional increase in urinary NH_4^+ excretion. However, in cases of **diabetic ketoacidosis** the plasma **creatinine** and **blood urea nitrogen (BUN)** concentrations are sometimes increased

due to a combination of increased protein degradation, dehydration, and diminished renal perfusion (i.e., **pre-renal azotemia**; see Chapter 88).

In **metabolic acidosis, Gln** is removed from both the filtrate and peritubular capillary beds by **proximal renal tubular epithelial cells**, where its two nitrogen groups are enzymatically removed; one NH_4^+ is liberated from the amide side group (**deamidation**), the other from the α-amino nitrogen (**deamination**). The remaining structure is the divalent anion, **α-ketoglutarate (α-$KG^=$)**. Once produced, α-$KG^=$ can be metabolized in proximal renal tubular cells to either glucose, or CO_2 and H_2O. **Both pathways consume two protons, and therefore effectively "buffer" excess H^+**.

Dispute exists about the way in which **Gln**, and hence **NH_4^+**, is handled by the kidney. Some argue that the **NH_4^+** generated in proximal renal tubular cells merely competes with H^+ for **Na^+/NH_4^+** and/or **Na^+/H^+** exchange (see **Fig. 11-1**). Others argue that because the pK′ of the ammonia buffer system is **9.0** {the pK′ being the pH at which the protonated (NH_4^+) and unprotonated (NH_3) species of a weak acid (NH_4^+) are present at equal concentrations}, an intracellular pH range of 7.4 to 9.0 would favor dissociation of **NH_4^+** into NH_3 and H^+ (see Chapter 85). Ammonia, being uncharged and lipid-soluble, diffuses easily across cell membranes down its concentration gradient (into tubular fluid). In the proximal tubular filtrate it reassociates with H^+ because of the low pH environment, thus reforming **NH_4^+** (which exhibits only limited permeability to apical tubular membranes since it is charged). The **pH** of the proximal tubular lumen, being normally about **6.9** and even lower in metabolic acidosis, clearly **favors NH_4^+** formation. This process by which NH_3 is passively secreted and then changed to NH_4^+ in the tubular lumen, thus maintaining a concentration gradient for further NH_3 secretion, is called "**nonionic diffusion**." Salicylates and a number of other drugs that are weak bases or weak acids are also secreted by nonionic diffusion into the renal filtrate. They diffuse into tubular fluid at a rate that is dependent upon the pH of the filtrate. Although this second hypothesis regarding NH_3 diffusion seems logical, it nonetheless does not provide direct buffering capacity since every H^+ bound by NH_3 in the tubular filtrate was effectively generated from NH_4^+ in proximal renal tubular cells (i.e., no net gain in buffering capacity).

In **respiratory acidosis**, extracellular concentrations of HCO_3^- and CO_2 are usually elevated, and therefore uncatalyzed HCO_3^- formation inside liver mitochondria provides sufficient substrate for carbamoyl phosphate, and thus urea formation (see Chapter 10). Therefore, pH control in periportal hepatocytes via carbonic anhydrase and glutaminase is overridden in respiratory acidosis. In chronic respiratory acidosis, there is no decrease in urea excretion at the expense of renal NH_4^+ excretion the way there is in chronic metabolic acidosis (see Chapter 90).

In **metabolic alkalosis**, carbonic anhydrase and glutaminase activity in periportal hepatocytes is increased, thus stimulating urea biosynthesis and, therefore, HCO_3^- consumption (**Fig. 11-2**). Glutamine synthetase activity in perivenous hepatocytes is decreased, thus resulting in a net decrease in hepatic Gln production. However, **renal Gln production**, HCO_3^- and **urea excretion** are all **increased** in

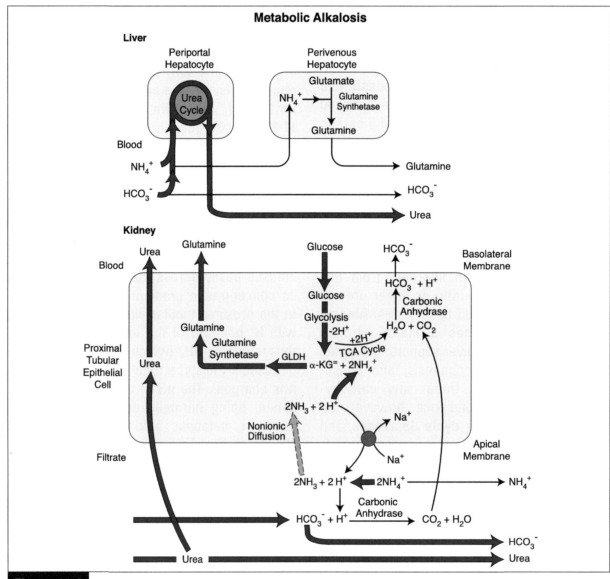

Figure 11-2

metabolic alkalosis, while **renal ammoniagenesis** and **H⁺ secretion** are **decreased** (see Chapter 89).

OBJECTIVES

- Explain why the degree of NH_4^+ and/or HCO_3^- incorporation into urea determines substrate supply for hepatic Gln formation.

- Discuss why there is an increase in renal gluconeogenesis and ammoniagenesis during a ketoacidosis.

- Recognize why renal Gln production as well as urinary HCO_3^- and urea excretion increase during metabolic alkalosis.

- Identify the manner in which the kidneys buffer protons in metabolic acidosis.

- Explain why there is a decrease in hepatic urea production during metabolic acidosis, but not during respiratory acidosis.

- Understand why the BUN concentration may increase during metabolic alkalosis.

- Identify several factors that could lead to NH_3 toxicity.

- Explain how and why systemic pH control involves the lungs, kidneys and liver (see Chapters 82-93).

- Outline the hepatic and renal handling of nitrogenous compounds during each of the four primary acid-base disturbances.

QUESTIONS

1. **Select the FALSE statement below regarding metabolic acidosis:**
    a. Blood concentrations of HCO_3^- and CO_2 (i.e., the Pco_2) are typically decreased in metabolic acidosis.
    b. Glutaminase activity in periportal hepatocytes is usually reduced in metabolic acidosis.
    c. Hepatic urea synthesis is usually reduced in metabolic acidosis.
    d. Renal gluconeogenesis may be increased in diabetic ketoacidosis.
    e. Urinary NH_4^+ excretion is usually decreased in metabolic acidosis.

2. **In metabolic acidosis, movement of NH_3 from proximal renal tubular epithelial cells into the tubular filtrate, where it is then "trapped" as**

NH_4^+**, is an example of:**
    a. Coupled active transport.
    b. Facilitate diffusion.
    c. Nonionic diffusion.
    d. Na^+/NH_3 antiport.
    e. NH_3/NH_4^+ antiport.

3. **Select the TRUE statement below regarding respiratory acidosis:**
    a. Extracellular concentrations of HCO_3^- are typically decreased.
    b. Urinary urea excretion usually does not change.
    c. Renal HCO_3^- reabsorption usually decreases.
    d. Hepatic urea formation is usually increased.
    e. The Pco_2 of blood usually decreases.

4. **Which one of the following usually increases in metabolic alkalosis?**
    a. Hepatic glutamine production
    b. Renal glutamine production
    c. Renal ammoniagenesis
    d. Renal H^+ secretion
    e. Urinary NH_4^+ excretion

5. **Select the FALSE statement below:**
    a. Alanine is the amino acid most likely to undergo gluconeogenesis in the liver.
    b. Nitrogen groups are removed from glutamine in the kidney via both deamidation and deamination.
    c. Salicylates are secreted into the renal filtrate via the process of nonionic diffusion.
    d. Glutamine is the amino acid most likely to under go gluconeogenesis in the kidney.
    e. The NH_3 concentration of systemic blood is normally greater than that of hepatic portal blood.

6. **The normal mammalian plasma ammonia concentration is about:**
    a. 0.02 mM.
    b. 0.20 mM.
    c. 2.0 mM.
    d. 20 mM.
    e. 200 mM.

7. **The pK' of the ammonia buffer system is:**
    a. 5
    b. 6
    c. 7
    d. 8
    e. 9

ANSWERS

7. d
6. a
5. e
4. b
3. b
2. c
1. e

Chapter 12

Nonprotein Derivatives of Amino Acids

Overview

- Tyrosine is used to synthesize melanin, the catecholamines, and the thyroid hormones.
- Tryptophan gives rise to serotonin, melatonin, and niacin.
- Glutathione, derived from glutamate, glycine and cysteine, is an abundant sulfur-containing compound in cells that helps prevent oxidative stress.
- Heme is a porphyrin derived from glycine and succinyl-CoA.
- Carnitine is derived from lysine and methionine, while creatine is derived from glycine, methionine and arginine.
- Nitric oxide, a vasodilator, is derived from arginine.
- Histidine is decarboxylated to histamine in many cell types.
- Glycine is an inhibitory neurotransmitter.
- Lysine, like leucine, is ketogenic, but not glucogenic.
- Aspartate is an important component of the malate and dicarboxylic acid shuttles.
- Serine is important in both phospholipid and glycolipid metabolism.

Amino acids are converted into a variety of important nonprotein derivatives in the body, including **glucose** and **glycogen, triglycerides** and **ketone bodies, purines** and **pyrimidines,** several **biogenic amines** that act as **hormones** and **neurotransmitters, heme, nicotinamide, creatine, glutathione,** and **carnitine (Figs. 12-1** and **12-2).** In addition, many proteins contain amino acids that have been modified for a specific function (see Chapter 3). Nonprotein derivatives of the following nine amino acids have particular physiologic significance.

Tyrosine (Tyr)

Cells of neural origin (e.g., chromaffin cells), convert Tyr to **catecholamines** (i.e., **dopamine, norepinephrine** and **epinephrine).** The catecholamines function as neurotransmitters in the sympathetic (SNS) and central nervous systems (CNS), and they also serve as circulating hormones. **Tyrosine hydroxylase** is considered to be the rate-limiting enzyme in catecholamine biosynthesis (see Chapter 39), and its dysfunction in the CNS is associated with Parkinson's disease.

In melanocytes, Tyr can be converted to **melanin.** While dopamine is an intermediate in the formation of both melanin and norepinephrine, different enzymes hydroxylate Tyr in melanocytes than in neural cells.

Other than dietary input, Tyr is formed from the essential aromatic amino acid, **phenylalanine (Phe)** in the liver (see Chapter 8). Thus, some forms of liver disease lead to a Tyr, and thus catecholamine, deficiency.

The **thyroid hormones** {tetra- (T_4) and triiodo-

Copyright © 2015 Elsevier Inc. All rights reserved.

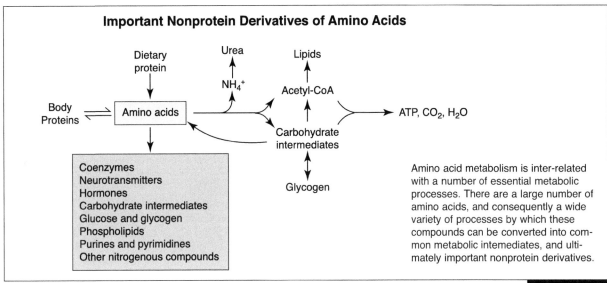

Important Nonprotein Derivatives of Amino Acids

Coenzymes
Neurotransmitters
Hormones
Carbohydrate intermediates
Glucose and glycogen
Phospholipids
Purines and pyrimidines
Other nitrogenous compounds

Amino acid metabolism is inter-related with a number of essential metabolic processes. There are a large number of amino acids, and consequently a wide variety of processes by which these compounds can be converted into common metabolic intemediates, and ultimately important nonprotein derivatives.

Figure 12-1

thyronine (**T₃**), and reverse T₃ (**rT₃**)} are synthesized from **Tyr** and **iodide** (**I⁻**) in follicular cells of the thyroid gland (see Chapter 52). Like Phe, part of the Tyr molecule, when catabolized, enters hepatic gluconeogenesis at the level of **fumarate**, while another part enters ketogenesis at the level of **acetoacetyl-CoA** (see Chapter 8). Therefore, this amino acid is considered to be both **glucogenic** and **ketogenic**.

Tryptophan (Trp)

This essential amino acid is required for **serotonin (5-hydroxytryptamine, 5-HT)** and **melatonin** biosynthesis. Serotonin is found in the pineal gland, intestine, and platelets, as well as in neurons of the CNS. It is synthesized and stored in the pineal gland, CNS, and intestine, and platelets take up 5-HT from the circulation. Serotonin (and melatonin) are thought to help regulate several physiologic activities, including sleep, body temperature, and blood pressure (by stimulating vascular smooth muscle contraction). The conversion of 5-HT to melatonin in the pineal gland is regulated by the **light-dark cycle**. Melatonin is thought to help regulate circadian rhythms, as well as the release of pituitary hormones involved in reproductive functions. Tryptophan is also used in the biosynthesis of the B vitamin-derivative,

nicotinate mononucleotide (NMN), which is a precursor to two important coenzymes used in many cytoplasmic and mitochondrial oxidation/reduction reactions, NAD⁺ and NADP⁺ (see Chapter 41). For every 60 mg of Trp, 1 mg of NMN is usually generated. Thus, animals can become niacin deficient, and exhibit the classic signs of "**pellagra**," through consumption of a diet that is poor in either niacin, the B₃ vitamin that gives rise to NMN, or Trp. Tryptophan, like Tyr, is both glucogenic and ketogenic.

Histidine (His)

Histidine can be decarboxylated to **histamine** in several cell types. Histamine is stored in **mast cells**, and when released plays a role in mediating various allergic and inflammatory reactions. In many vascular beds histamine causes vasodilation through H₁ receptor stimulation, yet in others vasoconstriction through H₂ receptor stimulation. Histamine also causes constriction of bronchiolar smooth muscle, and in the stomach participates in the stimulation of **HCl release** from parietal cells. In the liver His is glucogenic, but not ketogenic, and in erythrocytes the His residues of hemoglobin participate in oxyhemoglobin binding.

Histamine is present in neurons of the posterior hypothalamus, where it plays a role as a neurotransmitter in arousal, the pain threshold, sexual behavior, anterior pituitary hormone release, blood pressure and thirst regulation. H_3 receptors in the brain are largely presynaptic, and thus participate in the negative feedback of further histamine release. Histamine is a biogenic amine (like serotonin and the catecholamines), and it is degraded by **histamine-N-methyltransferase** and **monoamine oxidase** to methylimidazole-acetic acid.

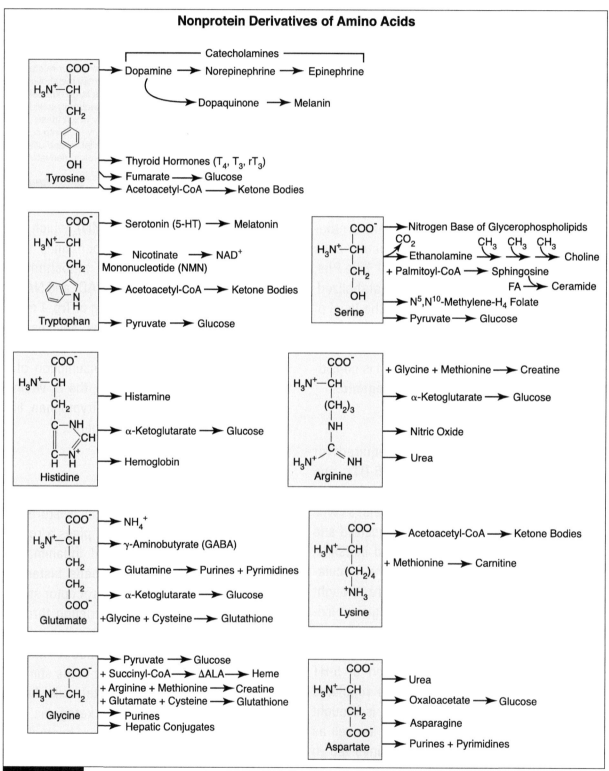

Figure 12-2

Glutamate (Glu)

As discussed in Chapter 9, Glu is the only amino acid in mammalian tissues undergoing oxidative deamination at an appreciable rate, and therefore is a major contributor to the body's **ammonium ion (NH_4^+) pool**. Glutamate and Asp are the most abundant excitatory neurotransmitters in the CNS, and removal of the α-carboxyl group of Glu forms **GABA**, the most abundant inhibitory neurotransmitter in the CNS (see Chapter 3). Glutamate can also be amidated to form **glutamine (Gln)** in astrocytes of the brain and perivenous hepatocytes, as well as other tissues (see Chapters 8-11).

Glutamate, when combined with **glycine** and **cysteine**, forms **glutathione (GSH)**, which is an abundant sulfur-containing compound in cells. Glutathione maintains protein sulfhydryl groups in their reduced state, and in **erythrocytes** is essential in **reducing hydrogen peroxide (H_2O_2) levels**, thus preventing oxidative damage and hemolysis (see Chapters 30 and 31). Glutathione also plays an important role in **eicosanoid biosynthesis**, where it serves as a reductant in the cyclooxygenase reaction, and as a substrate in peptidyl-leukotriene biosynthesis (see Chapters 68 and 69).

Glutamate is a glucogenic, but not ketogenic amino acid, entering the TCA cycle at the level of α-ketoglutarate (α-KG$^=$). This amino acid, through conversion to Gln, also contributes along with glycine and aspartate to biosynthesis of the **purine ring** (see Chapter 15).

Glycine (Gly)

The entire Gly molecule is included in the **purine ring** structure, and, as discussed above, this amino acid is also used in the biosynthesis of **GSH**. The first and rate-limiting step in the biosynthesis of **heme**, which occurs in many body tissues, but particularly in **liver** and **bone marrow**, involves the condensation of Gly with succinyl-CoA, thus forming

Δ-aminolevulinic acid (Δ-ALA, see Chapter 32). All **porphyrin** hydrocarbons are derived from Gly and succinyl-CoA, with porphyrins being cyclic compounds that have an affinity for binding metal ions, particularly iron. Porphyrins contain four **pyrrole rings**, which characterize the structures of the heme moieties of **hemoglobin** and **myoglobin**, as well as the **cytochromes**. Heme degradation results in **bilirubin** formation, an end product of metabolism that is normally excreted in bile. Abnormalities in heme biosynthesis result in a group of diseases known as **porphyrias** (see Chapter 32).

Glycine, methionine, and arginine are used in the biosynthesis of **creatine**, which serves to bind high-energy phosphate groups in muscle and brain tissue (see Chapter 77). Creatine phosphate is unstable, and normally undergoes a rather steady, spontaneous, and nonenzymatic degradation to **creatinine**. Creatinine, being an inactive end product of metabolism, enters blood and is freely filtered by the kidney. Its appearance in urine is usually proportional to muscle mass, and its concentration in plasma is generally considered to be a reflection of the adequacy of glomerular filtration. The biosynthesis of creatine begins in the kidneys, and is completed by the liver.

Glycine is also an **inhibitory neurotransmitter** in the CNS, and is released from **Renshaw cells** (i.e., inhibitory interneurons) in the spinal cord.

Several endogenous metabolites and exogenous xenobiotics are excreted in bile and urine as water-soluble glycine conjugates. Examples include the primary **bile acids** (i.e., glycocholic acid and glycochenodeoxycholic acid; see Chapter 62), and **hippuric acid** formed in the liver from the food additive, benzoate.

Arginine (Arg)

In addition to serving as a substrate for urea and creatine biosynthesis, Arg is used in the biosynthesis of **nitric oxide (NO)**, a powerful

vasodilator and **neurotransmitter**. Nitric oxide biosynthesis is prevalent in the **CNS**, in **vascular endothelial cells**, and in **macrophages**. Different isozymes are found in each of these locations, thereby allowing NO to serve more than one function. Endothelial-derived NO acts through **cyclic guanosine monophosphate (cGMP)**, and generally relaxes vascular smooth muscle. Brain-derived NO also acts through cGMP, but its overall effects are less well characterized. Macrophage-derived NO is thought to participate in tumor and bacterial cell destruction. Arginine, through conversion to α-KG$^{=}$, is glucogenic, but not ketogenic in the liver. In *Streptomyces*, this amino acid donates to the structure of the antibiotic, **streptomycin**.

Lysine (Lys)

Along with methionine, Lys is used in the hepatic biosynthesis of **carnitine**, a compound needed to shuttle **long-chain fatty acids** across mitochondrial membranes (see Chapter 55). Although largely derived from the liver (and diet), carnitine is abundant in tissues capable of oxidizing long-chain fatty acids (e.g., muscle, liver, and kidney tissue). Like leucine (Leu), carbon atoms from Lys can be found in acetyl-CoA and ketone bodies, but there is no net conversion of these amino acids to glucose. Therefore, Leu and Lys are ketogenic, but not glucogenic (see Chapter 8).

OBJECTIVES

- Identify eight important nonprotein derivatives of Tyr, and explain why liver disease could lead to a deficiency of these derivatives.

- Explain why a Trp deficiency could lead to a melatonin deficiency as well as signs of pellagra.

- Discuss different roles for histamine (obtained via decarboxylation of His) in the body.

- Identify and discuss the major roles of Glu and GABA in the CNS, and describe structural differences between these molecules.

Aspartate (Asp)

In addition to serving as a substrate for asparagine, urea, purine, and pyrimidine biosynthesis (see Chapters 9, 10, and 15), Asp transamination is an integral part of both the cytoplasmic and mitochondrial portions of the **malate** and **dicarboxylic acid shuttles** (see Chapters 35 and 36). Therefore, Asp plays a key role in both hepatic and renal gluconeogenesis. Since Asp is easily interconvertible with Glu in transamination reactions (Fig 9-2), it has been difficult to distinguish neurons in the brain that use Glu as an excitatory neurotransmitter from those that use Asp. This difficulty is further compounded by the fact that both amino acids stimulate common receptors.

Serine (Ser)

Serine can be decarboxylated to **ethanolamine**, which can further be methylated in **choline** formation. All three compounds (serine, ethanolamine and choline), can be used as nitrogen bases in **glycerophospholipid** formation (see Chapter 57). Serine also participates in the synthesis of **sphingosine**, and therefore becomes a part of all **ceramides** (see Chapter 59). Serine, a gluconeogenic amino acid that can enter the dicarboxylic acid shuttle at the level of pyruvate (see Chapters 8 and 37), also participates in N^5,N^{10}-methylene-H_4 folate biosynthesis (see Chapter 16).

- Recognize the significance of the TCA intermediate formed following Glu deamination.

- Explain the biological significance of glycochenodeoxycholate, and its relationship to both triglyceride and amino acid metabolism (see Chapter 62).

- Identify six different physiologic compounds partially synthesized from Gly.

- Discuss the relationship of Arg to the urea cycle, as well as to creatine and NO biosynthesis.

- Indicate the relationship between Lys and

ketone bodies, as well as between Lys and tissues capable of long chain fatty acid oxidation.

- Identify the antiporter responsible for moving Asp across inner mitochondrial membranes (see Chapter 36).

- The methylene group donated by Ser to H_4 folate in N^5, N^{10}-methylene-H_4 folate formation eventually ends up on which sulfated amino acid (in the presence of vitamin B_{12})?

- Know which amino acids contribute carbon and/or nitrogen atoms to the pyrimidine and purine rings (see Chapter 14 & 15).

QUESTIONS

1. Dopamine is synthesized from which amino acid?
a. Trp
b. Glu
c. His
d. Gly
e. Tyr

2. Melatonin is a derivative of which amino acid?
a. Trp
b. Glu
c. His
d. Gly
e. Tyr

3. Which one of the following is known to stimulate vascular smooth muscle contraction, thus elevating blood pressure?
a. GABA
b. GSH
c. 5-HT
d. NO
e. His

4. Glutamate, when combined with glycine and cysteine, forms:
a. Nitric oxide.
b. Carnitine.
c. GABA.
d. Glutathione.
e. Creatine.

5. The pyrrole rings of heme are formed from succinyl-CoA, and which amino acid?
a. Glutamate
b. Glycine
c. Histidine
d. Lysine
e. Arginine

6. Which one of the following compounds, synthesized from lysine and methionine, acts to shuttle long-chain fatty acids across mitochondrial membranes?
a. Malate
b. Δ-Aminolevulinic acid
c. Creatinine
d. Creatine
e. Carnitine

7. Melanin is derived from which amino acid?
a. Trp
b. Tyr
c. Gly
d. Lys
e. Asp

8. Which amino acid is used in the biosynthesis of NAD^+?
a. Asp
b. Ser
c. Arg
d. His
e. Trp

9. Which amino acid is used in both sphingosine and N^5, N^{10}-methylene-H_4 folate biosynthesis?
a. Gly
b. Try
c. Arg
d. Ser
e. Glu

10. Removal of the α-carboxyl group of glutamate forms:
a. An inhibitory neurotransmitter in the brain.
b. Glutamine.
c. Sphinogosine, the precursor to creamide.
d. Histamine, which is a biogenic amine.
e. Fumarate, which is an intermediate in the TCA cycle.

11. Which amino acid below is used for urea, creatine and nitric oxide biosynthesis?
a. Arg
b. Lys
c. Trp
d. Ser

ANSWERS

11. a
10. a
9. d
8. e
7. b
6. e
5. b
4. d
3. c
2. a
1. e

Addendum to Section I

Although amino acid and protein metabolism at first glance may be a seemingly more complex discipline than nucleotide, carbohydrate and lipid metabolism, an understanding of general principles involved is necessary for veterinarians, since it is important to many physiological processes, and has significant implications in widely differing clinical situations.

Although molecular nitrogen, N_2, is abundant in the atmosphere, it must first be "fixed," that is, reduced from N_2 to NH_3 by microbes and plants, and then incorporated into protein before it can be utilized by animals. **Essential amino acids**, which animals cannot synthesize, must be supplied in the diet, but the remaining **nonessential amino acids** can be formed endogenously from metabolic intermediates via transamination reactions using amino nitrogen from other amino acids. Following deamination, amino nitrogen can then be incorporated into **urea** or **glutamine** by the liver, and the hydrocarbon skeletons that remain can be oxidized to CO_2 and H_2O via the TCA cycle (Chapter 34), used to synthesize glucose (Chapter 37), or form ketone bodies (Chapter 71). Several amino acids are also precursors to other important biomolecules (e.g., purines (Chapter 15), pyrimidines (Chapter 14), neurotransmitters (Chapter 3 & 9), and hormones such as epinephrine (Chapter 39) and thyroxine (Chapter 52)).

Section I has dealt with the general properties of amino acids, those that are incorporated into protein or non-protein derivatives, basic protein structure, the important properties of enzymes and their kinetics, intestinal protein digestion, amino acid catabolism, transamination and deamination processes, interconversions of amino acids, as well as the removal and excretion of NH_3. As a part of overall NH_3 metabolism, the synthesis and degradation of Glu, Gln, Asp, Ala and Arg have been given particular emphasis, the importance of hepatic urea biosynthesis has been presented, and nitrogen and carbon flux between the liver and kidneys has been introduced. As readers encounter the details of acid-base balance in **Section VII**, re-examination of Chapters 10 and 11 may be helpful.

Introduction to Section II

Section II now deals with nucleotide and nucleic acid metabolism. Following a brief introductory chapter on the general characteristics of nucleotides, pyrimidine and purine biosynthetic pathways are detailed, followed by a chapter on eukaryotic and prokaryotic folic acid metabolism. Relationships of this important water-soluble vitamin to vitamin B_{12}, megaloblastic anemia, and cancer chemotherapy are also discussed. This section concludes with a chapter on nucleic acid and nucleotide turnover, with particular emphasis on purine and pyrimidine salvage and degradative pathways, and their relevance to animal health and disease. Although nucleotides are the building blocks of nucleic acids, and are thus integral to protein synthesis, they also serve multiple other functions, including being parts of several coenzymes (e.g., NAD^+ and $NADP^+$), and donors of phosphoryl groups (e.g., ATP or GTP). They also play a part in carbohydrate (e.g., UDP-glucose; Chapter 23, and UDP-glucuronide; Chapter 29) and lipid metabolism (CDP-diacylglycerol; Chapter 57, and UDP-Gal & UDP-Glu; Chapter 59), and certain second messengers (e.g., cAMP & cGMP) are regulatory nucleotides.

Nucleotide and Nucleic Acid Metabolism

SECTION

II

Chapter 13

Nucleotides

Overview

- Nucleotides are phosphorylated nucleosides.
- A nucleoside is a combination of a nucleic acid base and a sugar.
- ATP is a nucleotide that participates in numerous energy transduction reactions.
- NTPs are the ultimate building blocks of nucleic acids.
- Nucleic acid polymerases are referred to as pols.
- In mammalian cells, ribonucleotides are converted to deoxyribonucleotides at the NDP level.
- DNA and RNA are structurally similar polymers.
- Nucleoside monophosphates are converted to their respective di- and triphosphate derivatives via phosphoryl transfer from ATP.
- Nucleotides can be synthesized from small organic molecules, and they are conserved through salvage pathways.

Nucleotides furnish the building blocks of ribo- and deoxyribonucleic acid (**RNA and DNA**). DNA, in turn, is the permanent genetic material of animal and plant cells, as well as bacteria and some viruses. It contains structural information for the synthesis of some 50,000 proteins needed for general cellular metabolism and differentiated function in mammals, and it confers uniqueness to each organism. Most DNA is found in the nucleus; however, small amounts are also present in mitochondria. Unlike nuclear DNA, which is inherited from both parents, mitochondrial DNA is inherited exclusively from the mother.

In addition to forming DNA and RNA, nucleotides perform a variety of intracellular functions in **energy transduction reactions**. For example, hydrolysis of the nucleotide adenosine triphosphate (**ATP**) supplies needed energy for many cell reactions, and

in several metabolic pathways nucleotides act as activated carriers of carbohydrates, amino acids, lipids, sulfate, and methyl groups. They are structural components of several coenzymes, including coenzyme-A (**CoA.SH**) and nicotinamide adenine dinucleotide (**NAD⁺**; see Chapter 41), and flavin adenine dinucleotide (**FAD**), and they are important allosteric regulators of key intracellular enzymes. For example, cyclic adenosine- and guanosine monophosphate (**cAMP** and **cGMP**, respectively) are second messengers that mediate the effects of several ligands that bind to plasma membrane receptors.

Nucleotides can be synthesized from small organic molecules available in cells, or they can be generated through **salvage pathways** that recycle their nitrogen bases (see Chapter 17). In rapidly dividing cells that are synthesizing large quantities of RNA and DNA (e.g.,

Copyright © 2015 Elsevier Inc. All rights reserved.

bone marrow, intestinal epithelial cells, and many cells of young animals), the need for nucleotides is generally greater than in nondividing cells. It is in these rapidly dividing cells of adult animals that the salvage pathways predominate. Several chemotherapeutic agents exert their effects by interfering with nucleotide biosynthesis (see Chapters 14-17).

Nucleotide Structure

Nucleotides are composed of a nitrogen base (i.e., a **purine** or **pyrimidine**), a cyclic **pentose**, and one or more **phosphate** groups (**Fig. 13-1**). The nitrogen base plus the pentose (ribose or deoxyribose) is known as a **nucleoside**, with addition of phosphate forming a **nucleotide**.

Two **purines**, **adenine** and **guanine**, and three **pyrimidines**, **cytosine**, **uracil**, and its methylated derivative, **thymine**, are found in DNA and RNA (**Fig. 13-2**). **Uracil** is found only in RNA, **thymine** only in DNA, and the others are present in both. Carbon and nitrogen atoms in the purine ring are usually numbered 1 through 9 (see structure of adenine), whereas those in the pyrimidine ring are numbered 1 through 6 (see structure of uracil). Carbon atoms in the nucleoside's pentose moiety are assigned prime numbers (1' through 5') to distinguish them from those of the base to which they are attached. **Ribose** has a **hydroxyl (OH)** group attached to the 2'-carbon atom, which is replaced by a **hydrogen (H)** atom in **deoxyribose (Figs. 13-1, 13-3, and 13-4)**.

Generically, **nucleotides** are referred to as **nucleoside phosphates (NPs)**, which are further distinguished on the basis of their pentose structure as either **ribonucleoside- or 2'-deoxyribonucleoside phosphates**. The specific state of their phosphorylation is defined by notation of the number of phosphates (mono, M; di, D; or tri, T), and their position (e.g., 2', 3', 4', or 5') on the pentose sugar. The most frequently encountered nucleoside phosphates found in nature have their respective phosphate(s) esterified to the 5'pentose carbon of the nucleoside, and, as summarized in **Fig. 13-1**, they assume three major forms--a **nucleoside 5'-<u>mono</u>phosphate** (abbreviated conventionally as **NMP** for the ribose form, or **dNMP** for the 2'-deoxyribose form), a **nucleoside 5'-diphosphate** (i.e., **NDP** or **dNDP**), and a **nucleoside 5'-triphosphate** (i.e., **NTP** or **dNTP**). The NTP has *three* phosphates tandemly esterified to the hydroxyl of the 5' carbon of the pentose sugar, while the NDP and NMP forms each have, respectively, two and one phosphate(s) esterified at the same position. Definition of the position of a specific phosphate atom in an NDP or NTP

A Nucleotide is a Nucleoside Esterified to one or more Phosphate Groups

Cyclic 3', 5' AMP (cAMP)

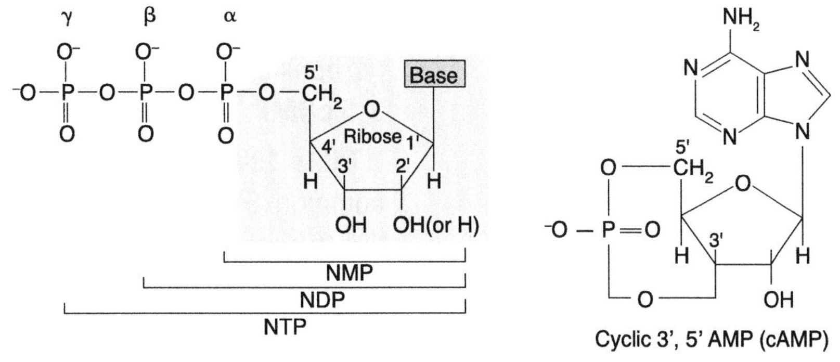

Figure 13-1

Purines and Pyrimidines, The Nitrogenous Bases of Nucleotides

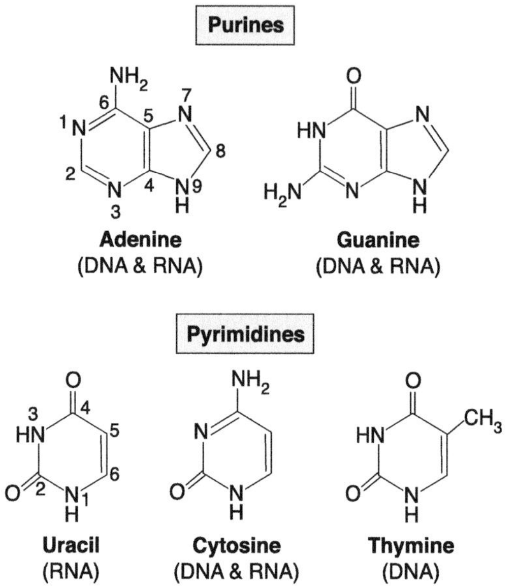

Purines

Adenine
(DNA & RNA)

Guanine
(DNA & RNA)

Pyrimidines

Uracil
(RNA)

Cytosine
(DNA & RNA)

Thymine
(DNA)

Figure 13-2

Nucleic Acid Polymerase Mechanism

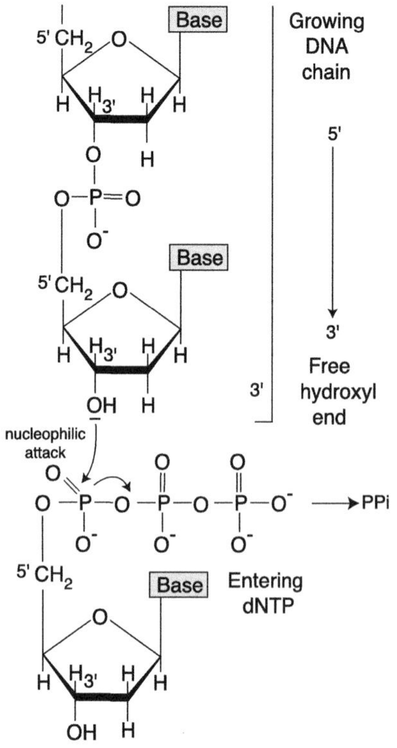

Growing DNA chain

5'

3'

Free hydroxyl end

nucleophilic attack

Entering dNTP

→ PPi

Figure 13-4

Phosphodiester Bonds of DNA and RNA

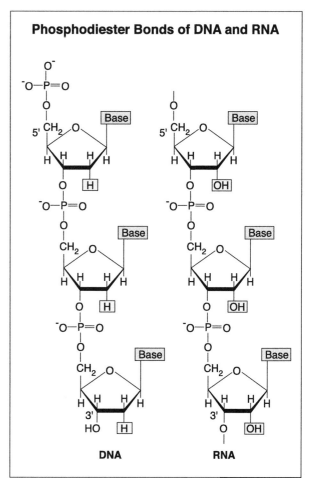

DNA

RNA

Figure 13-3

is based upon its distance from the pentose carbon to which it is ultimately tethered. As shown in **Fig. 13-1**, the phosphate nearest this carbon is defined as the **α-phosphate**, the next is the **β-phosphate**, and the third (in the case of a triphosphate) is the **γ-phosphate**. A less common, but no less important NMP form, is the cyclic NMP (cNMP), in which the phosphate is esterified not to a <u>single</u> pentose carbon atom, but to <u>two</u> carbons - usually carbons 3' and 5'. The cNMPs, **3',5'cAMP** and **3',5'cGMP**, exemplify such cNMPs.

Table 13-1 summarizes abbreviations of the common 5'-mono-, di-, and triphosphates of the **ribonucleosides adenosine**, **guanosine**, **cytidine**, and **uridine**, and the corresponding **2'-deoxyribonucleosides**, **2'-deoxyadenosine (dA)**, **2'-deoxyguanosine (dG)**, **2'-deoxycytidine (dC)**, and **2'-deoxythymidine (dT)**.

Table 13-1
Nucleotide Nomenclature

Abbreviations of ribonucleoside 5'-phosphates

Base	Mono-	Di-	Tri-
Adenine	AMP	ADP	ATP
Guanine	GMP	GDP	GTP
Cytosine	CMP	CDP	CTP
Uracil	UMP	UDP	UTP

Abbreviations of deoxyribonucleoside 5'-phosphates

Base	Mono-	Di-	Tri-
Adenine	dAMP	dADP	dATP
Guanine	dGMP	dGDP	dGTP
Cytosine	dCMP	dCDP	dCTP
Thymine	dTMP	dTDP	dTTP

Note: Nucleoside phosphates using pyrimidines as the nitrogen base are, by convention, referred to as "idines" (e.g., uridine, cytidine and thymidine), whereas those using purines are referred to as "osines" (e.g., adenosine and guanosine).

Nucleoside monophosphates are converted to their respective diphosphate derivatives via phosphoryl transfer from ATP, catalyzed by **nucleoside monophosphate kinase**. The nucleoside diphosphates are then converted to their triphosphate derivatives by **nucleoside diphosphate kinase** at the expense of another ATP. The ATPs used in these phosphorylations are derived primarily through mitochondrial oxidative phosphorylation, and secondarily through reactions of the Embden-Meyerhoff pathway and the tricarboxylic acid cycle (see Chapters 24-27, 34, and 36).

Polynucleotide Structure and Synthesis

The major polymeric forms of nucleotides found in prokaryotes and eukaryotes are the nucleic acids, RNA and DNA. As shown schematically in **Fig. 13-3**, **RNA** and **DNA** are similar polymers, differing in composition only in the 2' substituent of their respective pentoses (i.e., **OH** vs **H**). Both consist of tandem nucleoside units strung together by phosphodiester bonds between their **3'** and **5'** carbons.

Nucleic acid polymerization is catalyzed by nucleic acid polymerases (pols) - **DNA polymerases** in the case of DNA, and **RNA polymer-ases** in the case of RNA. All known polymerases use 5'NTPs (e.g., riboNTPs for RNA, or dNTPs for DNA) as their nucleotide substrates, and they operate by the mechanism exemplified by the polymerization of a dNTP into DNA, as shown schematically in **Fig. 13-4**. Specifically, the polymerase catalyzes polymerization of the 5' α-phosphate of an NTP unit to the "free" OH of the NMP on the 3' "primer" end. The β- and γ-phosphates are eliminated as **pyrophosphate (PPi)**.

In summary, nucleotides are the building blocks of RNA and DNA, and they also participate in energy transduction reactions in a variety of metabolic pathways. Nucleotides possess a nucleoside moiety, consisting of a cyclic pentose esterified to a ring nitrogen of a heterocyclic pyrimidine or purine base, and a mono-, di- or triphosphate group attached to one or more of the pentose hydroxyl groups. Nucleotides are generically referred to as nucleoside phosphates (NPs), and are further distinguished on the basis of their pentose structure as either ribonucleoside- or 2'-deoxyribonucleoside phosphates. They can be synthesized from small organic molecules available in cells, or they can be generated through salvage pathways that recycle their nitrogen bases. The most common NPs in nature have their respective phosphate(s) esterified to the 5' pentose carbon of the nucleoside. The phosphate nearest this carbon is the α-phosphate, the next is the β-phosphate, and the third (in the case of a triphosphate) is the γ-phosphate. Nucleoside monophosphates are converted to their respective di- and triphosphate derivatives via phosphoryl transfer from ATP.

DNA and RNA are similar polymers, differing in composition only in the 2'substituent of their respective pentoses (i.e., H vs OH). Both consist of tandem nucleoside units strung together by phosphodiester bonds between their 3' and 5' carbons. Nucleic acid polymerization is catalyzed by nucleic acid polymerases (pols), which use 5'-NTPs as their nucleotide substrates.

OBJECTIVES

- Recognize several ways in which nucleotides contribute to energy transduction reactions.

- Identify cell types in the adult organism that continually synthesize large quantities of RNA and DNA.

- Outline the ways in which nucleotides can be salvaged by the organism (see Chapter 17).

- Know the difference between a nucleoside and a nucleotide.

- Identify structural differences between and among the purines and pyrimidines present in DNA and RNA.

- Identify the purine and pyrimidine nucleoside triphosphates involved in RNA biosynthesis.

- Contrast primary structural differences between RNA and DNA, and between AMP and cAMP.

- Discuss how nucleoside monophosphates are converted to their respective triphosphate derivatives.

- Identify and explain the steps involved in nucleic acid polymerization.

QUESTIONS

1. **Which one of the following is a nucleoside?**
   a. ATP
   b. Adenosine
   c. GDP
   d. AMP
   e. Adenine

2. **Select the FALSE statement below regarding nucleotides:**
   a. They are products of the reaction catalyzed by nucleic acid polymerases.
   b. They are the building blocks of both DNA and RNA.
   c. They are structural components of several coenzymes.
   d. In several metabolic pathways, they act as carriers of carbohydrates, amino acids, and lipids.
   e. They are composed of a nitrogen base, a pentose, and one or more phosphate groups.

3. **The hydroxyl group (OH) attached to the 2'-carbon atom of ribose is replaced by what in deoxyribose?**
   a. A sulfhydryl group
   b. Oxygen
   c. Hydrogen
   d. Phosphate
   e. Ammonia

4. **Which one of the following is a purine found in both DNA and RNA?**
   a. Cytosine
   b. Thymine
   c. Uracil
   d. Adenosine
   e. Guanine

5. **Which one of the following is NOT characteristic of a nucleic acid polymerase?**
   a. It may use dNTPs as substrates.
   b. It may use NTPs as substrates.
   c. It catalyzes growth of the polymer from the 5' end.
   d. It catalyzes attack of a hydroxyl group on the α-phosphate of an incoming NTP.
   e. It catalyzes a synthetic reaction in which PPi is released as one of the products.

6. **Coenzyme A possesses a nucleotide:**
   a. True
   b. False

6. a
5. c
4. e
3. c
2. a
1. b

Chapter 14

Pyrimidine Biosynthesis

Overview

- Glutamine, aspartate, ATP, and CO_2 are used in the biosynthesis of pyrimidine NTPs.
- Mitochondrial carbamoyl phosphate is not used in pyrimidine biosynthesis.
- Pyrimidine biosynthesis is regulated differently in prokaryotes than in eukaryotes.
- Cytoplasmic CPS-2 involved in pyrimidine biosynthesis is distinct from mitochondrial CPS-1 involved in urea biosynthesis.
- In pyrimidine NTP biosynthesis, the ring is formed before the sugar of PRPP is added.
- UTP inhibits CPS-2 in mammals, while both PRPP and ATP stimulate its activity.
- Orotate accumulation in liver cells is associated with steatosis.
- UDP can be used to form UTP (and then CTP), or it can be utilized in dTMP biosynthesis.

Purine and pyrimidine nucleosides are derived from common precursors: **glutamine (Gln)**, **carbon dioxide (CO_2)**, **aspartate (Asp)**, **5-phosphoribosyl-1-pyrophosphate (PRPP)**, and, for thymine nucleotides, **tetrahydrofolate derivatives**. The following discussion will emphasize various phases of the pyrimidine biosynthetic pathway, and also show how this pathway is regulated in both mammals and prokaryotes who require optimal supplies of uridine triphosphate (UTP) and cytidine triphosphate (CTP) for ribonucleic acid (RNA) formation.

Pathway Summary

The biosynthesis of RNA requires the two **purine nucleoside triphosphates (NTPs** or **rNTPs)**, **adenosine triphosphate (ATP)** and **guanosine triphosphate (GTP)**, and the two **pyrimidine NTPs**, **UTP** and **CTP** (see Chapter 13). In mammals and most prokaryotes, the *de novo* pathway for the biosynthesis of UTP and CTP begins with the assembly of the pyrimidine

ring from simple precursors (**Fig. 14-1**), and proceeds in five phases (**Fig. 14-2**). In **phase one**, CO_2, an **amide (or amido) group** of the amino acid **Gln**, and the **gamma phosphate of ATP** are assembled by the cytoplasmic enzyme **carbamoyl phosphate synthetase-2 (CPS-2)** to generate the carbamic acid derivative, **carbamoyl phosphate (CAP**; note that this compound also participates in urea biosynthesis (see Chapter 10)). In **phase two**, CAP is joined with **Asp** by the enzyme **aspartate transcarbamoylase (ATCase**, a zinc-containing enzyme; see Chapter 49), to form **carbamoyl aspartic acid (CAA**; or carbamoyl aspartate). In **phase three**, CAA sheds a molecule of H_2O and cyclizes to form the ring of the first pyrim-

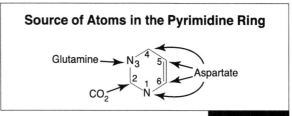

Source of Atoms in the Pyrimidine Ring

Figure 14-1

Copyright © 2015 Elsevier Inc. All rights reserved.

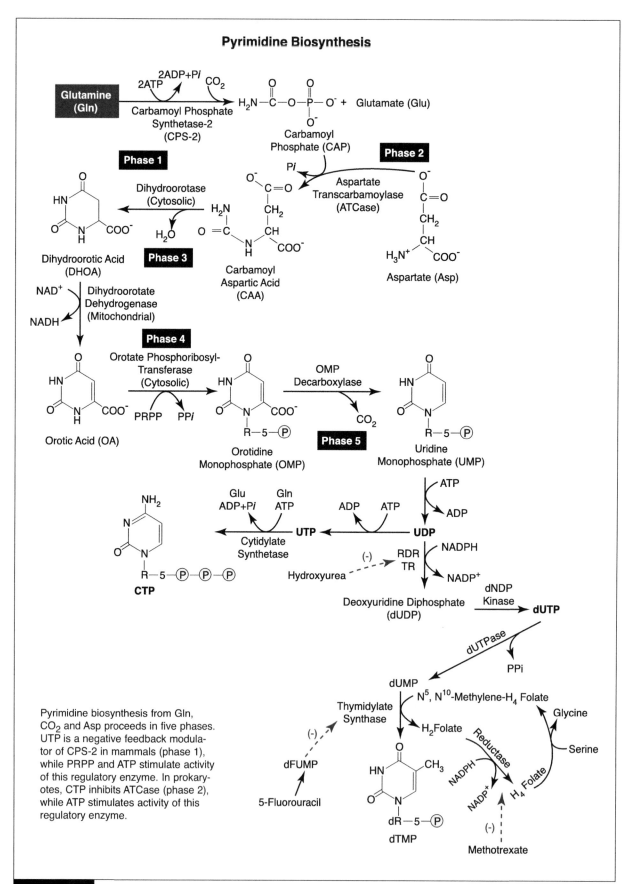

Pyrimidine Biosynthesis

Pyrimidine biosynthesis from Gln, CO_2 and Asp proceeds in five phases. UTP is a negative feedback modulator of CPS-2 in mammals (phase 1), while PRPP and ATP stimulate activity of this regulatory enzyme. In prokaryotes, CTP inhibits ATCase (phase 2), while ATP stimulates activity of this regulatory enzyme.

Figure 14-2

idine, **dihydroorotic acid (DHOA)**. In <u>phase four</u>, DHO is oxidized to **orotic acid (OA)** by **dihydrooratate dehydrogenase** (the only mitochondrial enzyme in pyrimidine biosynthesis), and OA, in turn, receives a **ribose 5-phosphate** from **PRPP** (see step one of purine biosynthesis, Chapter 15), to directly generate its **ribonucleoside 5′-monophosphate** form, OMP.

From a pathophysiologic perspective, OA accumulation in liver cells tends to facilitate formation of a defective, nonglycosylated apoprotein component of very low-density lipoprotein (VLDL), that apparently cannot function adequately in lipid transport. Orotic acid hepatic accumulation, which can occur in animals with arginine deficiency, is thus associated with **fatty liver syndrome** (see Chapter 72).

In <u>phase five</u>, OMP is converted by simple decarboxylation to **uridine 5'-monophosphate (UMP)**, which, in turn, is serially phosphorylated to its **5'-<u>diphosphate form</u>, UDP**, and then to its **5'-<u>triphosphate form</u>, UTP**. In the last step of phase five, with the expenditure of a molecule of ATP, an amide group derived from **Gln** is added to the 4 position of the uracil ring, yielding **CTP**.

As an alternate route, UDP can be sequentially converted to deoxy-UMP (**dUMP**), and N^5,N^{10}-methylenetetrahydrofolate (N^5,N^{10}-methylene-H_4 folate) can be used as a methyl donor to form **dTMP**. Dihydrofolate (H_2 folate) is the other reaction product, and the enzyme cata-lyzing this reaction is **thymidylate synthase**. Thymidine synthesis is the only reaction in which a one-carbon transfer and folate oxidation occur simultaneously. For the coenzyme to function again in one-carbon transfer, H_2 folate must be reduced to H_4 folate by the enzyme **dihydrofolate reductase**. **Metho-trexate**, an analog of folate, inhibits this conversion, and **5-fluorouracil**, once converted to **dFUMP**, inhibits thymidylate synthase. Both are anticancer agents (see Chapter 16).

Pathway Regulation

In order to ensure an appropriate cellular supply of CTP and UTP, mammals and prokaryotes evolved "**feedback**" mechanisms for controlling activity of this *de novo* biosynthetic pathway. In both systems, an early, rate-limiting step of the pathway is subject to either inhibitory or stimulatory regulation by one or more molecules generated in a later phase of the pathway. The target for this regulation in mammals is the phase one enzyme, **CPS-2**, and the major regulatory molecules are **UTP**, **PRPP**, and **ATP**. Each of these molecules works by binding CPS-2 **allosterically**, at a site distinct from its catalytic domain (see Chapter 6). Specifically, **UTP** is a **negative** allosteric effector of **CPS-2**, thus inhibiting its activity. In contrast, **PRPP** and **ATP** are both **positive** allosteric effectors of this enzyme, thus stimulating its activity.

CTP and **ATP** are the specific allosteric regulators of this pathway in prokaryotes, and their rate-limiting target enzyme is the phase two-specific **ATCase**. In this system **CTP** and **ATP** are, respectively, **inhibitory** and **stimulatory**.

Unusual Physical Properties of Relevant Early Stage Mammalian Enzymes

The respective activities of **CPS-2**, **ATCase**, and **dihydroorotase** in mammalian cells are not found in separate enzymes. Rather, each resides, respectively, in one of three independent catalytic domains within a single multifunctional cytoplasmic protein named **CAD** (acronym derived from the first letters of the names of the three respective enzymes). Such **multifunctionality** is thought to allow adjacent catalytic sites to facilitate rapid and complete transfer of intermediates, and also to assure output of the various products in equal amounts. The two enzymatic activities that convert, respectively, orotate (OA) to orotodylate (OMP), and OMP to UMP, also reside as two distinct catalytic domains of a single, bifunctional protein.

It is also important to note significant differences between the **CPS-2** activity of the CAD protein, and that of an identical activity found in a second enzyme, **CPS-1**. Although both CPS activities are readily found in cells such as **periportal hepatocytes**, the CAD-associated **CPS-2** activity is located in the **cytoplasm**, while that for **CPS-1** is specific for **mitochondria**, where it functions primarily in the urea cycle (see Chapter 10 and **Table 14-1**). In contrast to CPS-2, CPS-1 is not subject to allosteric regulation by UTP, PRPP, and ATP, and the amount of mitochondrial CPS-1 is normally about 10-times greater than that of cytosolic CPS-2, thus reflecting a much greater emphasis on hepatic periportal **urea** than **pyrimidine** biosynthesis.

In summary, pyrimidine biosynthesis begins with the formation of CAP from Gln, ATP, and CO_2. This **phase one** reaction is catalyzed by cytoplasmic CPS-2, an enzyme distinct from mitochondrial CPS-1 that is functional in urea biosynthesis. Compartmentalization thus provides independent pools of CAP for each process. Next, the **phase two** condensation of CAP with Asp forms CAA, a reaction catalyzed by ATCase. Ring closure and the formation of DHOA occurs in **phase three** via loss of H_2O, a process facilitated by the enzyme dihydroorotase. Abstraction of hydrogens from C-5 and C-6 by NAD^+ introduces a double bond into the ring structure, thus forming OA. This **phase four** reaction is catalyzed

by the mitochondrial enzyme, dihydroorotate dehydrogenase. All other enzymes of pyrimidine biosynthesis are cytosolic. Phase four transfer of a ribose phosphate moiety from PRPP occurs next, forming OMP, a reaction catalyzed by orotate phosphoribosyltransferase. In **phase five**, decarboxylation of OMP forms UMP, the first true pyrimidine ribonucleotide. Phosphate transfer from ATP yields UDP and UTP in sequential reactions analogous to those for phosphorylation of purine nucleoside mono-phosphates (see Chapter 13). Phase 5 is basically complete when UTP is aminated to CTP by Gln and ATP.

An alternate route for UDP involves formation of **dTMP**. Reduction of ribonucleoside diphosphates (NDPs) to their corresponding dNDPs involves reactions analogous to those for purine nucleosides. NADPH serves in the reduction of UDP to dUDP, which is then dephosphorylated to dUMP. This compound may further accept sequential phosphates from ATP in the formation of dUTP (not shown in **Fig. 14-2**), or it can be methylated at C-5 by N^5,N^{10}-methylene-H_4 folate, catalyzed by thymidylate synthase, thus forming dTMP.

Feedback modulation of this pathway occurs at the level of CPS-2 in mammals, and at the level of ATCase in prokaryotes. UTP inhibits CPS-2 in mammals, whereas both PRPP and ATP stimulate its activity. In prokaryotes, CTP and ATP, respectively, inhibit and stimulate ATCase activity.

Table 14-1		
Comparison of CPS-1 and CPS-2		
	CPS-1	**CPS-2**
Location	Liver Mitochondria	Cytosol of Many Cells
Pathway	Urea Cycle	Pyrimidine Biosynthesis
Nitrogen Source	NH_4^+	Glutamine
Allosteric Inhibitor	----	UTP

OBJECTIVES

- Identify the sources of all carbon and nitrogen atoms in the pyrimidine ring.

- Summarize the five phases of pyrimidine biosynthesis.

- Identify the only mitochondrial enzyme that participates in pyrimidine biosynthesis.

- Explain why OA accumulation is associated with fatty liver syndrome.

- Show how UDP can be converted to either UTP or dUDP.

- Identify the intermediate common to both pyrimidine and urea biosynthesis.

- Indicate how the pyrimidine biosynthetic pathway is regulated in both mammals and prokaryotes.

- Discuss the significance of CAD to the pyrimidine biosynthetic pathway, and draw similarities in this regard to the purine biosynthetic pathway.

- Contrast anatomic and functional differences between CPS-1 and CPS-2.

- Describe the route by which dTMP can be derived from UDP.

QUESTIONS

1. **Carbon atoms 4, 5, and 6 of the pyrimidine ring are normally derived from which amino acid?**
    a. Alanine
    b. Glutamine
    c. Tyrosine
    d. Methionine
    e. Aspartate

2. **Which one of the following is TRUE of carbamoyl phosphate synthetase-2 (CPS-2)?**
    a. It catalyzes the conversion of glutamine to carbamoyl phosphate in liver mitochondria.
    b. It uses aspartate as a nitrogen source.
    c. It is the rate-controlling reaction in mammalian pyrimidine biosynthesis.
    d. It is the rate-controlling reaction in microbial pyrimidine biosynthesis.
    e. It is subject to allosteric regulation by UMP and AMP.

3. **Which one of the following compounds is considered to be the "parent" pyrimidine nucleotide?**
    a. Phosphoribosylpyrophosphate (PRPP)
    b. Carbamoyl aspartic acid (CAA)
    c. Uridine monophosphate (UMP)
    d. Dihydrofolate (H2 folate)
    e. Orotidine monophosphate (OMP)

4. **Which one of the following is NOT an allosteric regulator of the de novo pyrimidine biosynthetic pathway?**
    a. OMP
    b. CTP
    c. PRPP
    d. ATP
    e. UTP

5. **Select the FALSE statement below regarding pyrimidine biosynthesis:**
    a. Orotate is formed before orotidylate.
    b. Feedback modulation of this pathway occurs at the level of CPS-2 in mammals, and at the level of ATCase in prokaryotes.
    c. The pathway for pyrimidine biosynthesis involves both cytoplasmic and mitochondrial reactions.
    d. UTP is formed by amination of CTP.
    e. Glutamine serves as a nitrogen donor in this pathway.

6. **Which one of the following enzymes facilitates ring closure in pyrimidine biosynthesis?**
    a. OMP decarboxylase
    b. Dihydroorotase
    c. CPS-2
    d. Thymidylate synthase
    e. ATCase

7. **5-Fluorouracil inhibits which enzyme in pyrimidine biosynthesis?**
    a. CPS-2
    b. ATCase
    c. Cytidylate synthetase
    d. Thymidylate synthase
    e. Dihydroorotase

8. **The only mitochondrial enzyme in pyrimidine biosynthesis is:**
    a. Cytidylate synthetase
    b. ATCase
    c. Dihydroratate dehydrogenase
    d. OMP decarboxylase
    e. Carbamoyl phosphate synthetase-2

ANSWERS

1. e
2. c
3. e
4. a
5. d
6. b
7. d
8. c

Chapter 15

Purine Biosynthesis

Overview

- All organisms, with the exception of protozoa, synthesize purines.
- Our knowledge of the purine biosynthetic reaction sequence largely evolved from studies in birds.
- PRPP availability determines the rate of purine biosynthesis in cells.
- Gly, Asp, and Gln are amino acids used in purine biosynthesis.
- All intermediates involved in purine biosynthesis are constructed on a ribose 5-phosphate "scaffold."
- AMP and GMP are synthesized from IMP, the "parent" nucleotide.
- Regulation of purine biosynthesis occurs at the beginning and end of the pathway.

The multistep biosynthetic pathway for purine nucleoside triphosphates (NTPs) has much in common with that for pyrimidine NTPs. The **ribose** sugar and **α-phosphate** are generated from **5'-phosphoribosyl-1-pyrophosphate (PRPP)**, in which the β- and γ-phosphates are sequentially donated from "**pre-existing**" **ATP (peATP)**, arising primarily through oxidative phosphorylation of ADP, and secondarily through generative reactions of the Embden-Meyerhoff pathway and the tricarboxylic acid (TCA) cycle (see Chapters 24-27, 34, and 36).

The purine ring structure is formed stepwise from a variety of amino acids and simple one-carbon units (**Fig. 15-1**). For convenience, the biosynthetic pathway that gives rise to this ring structure can be separated into three major phases; **1)** **PRPP** biosynthesis, **2)** formation of **inosine 5'-monophosphate** (**IMP**; the "**parent**" nucleotide), and **3)** formation of **AMP**, **GMP**, and their respective 5'-triphosphates from **IMP**. Specific steps are summarized in **Fig. 15-2**, and a discussion concerning the regulation of purine biosynthesis follows.

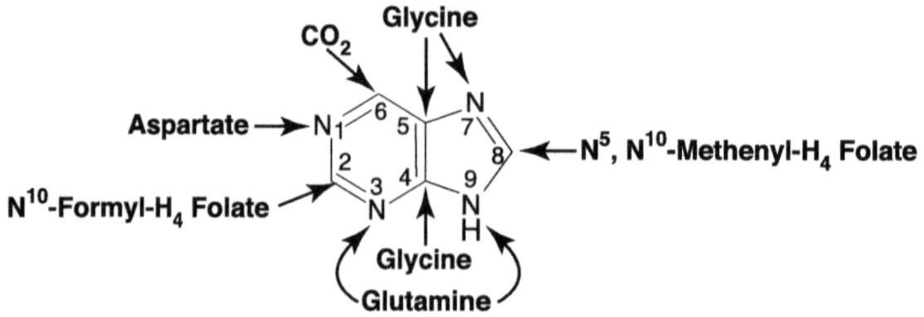

Figure 15-1

Copyright © 2015 Elsevier Inc. All rights reserved.

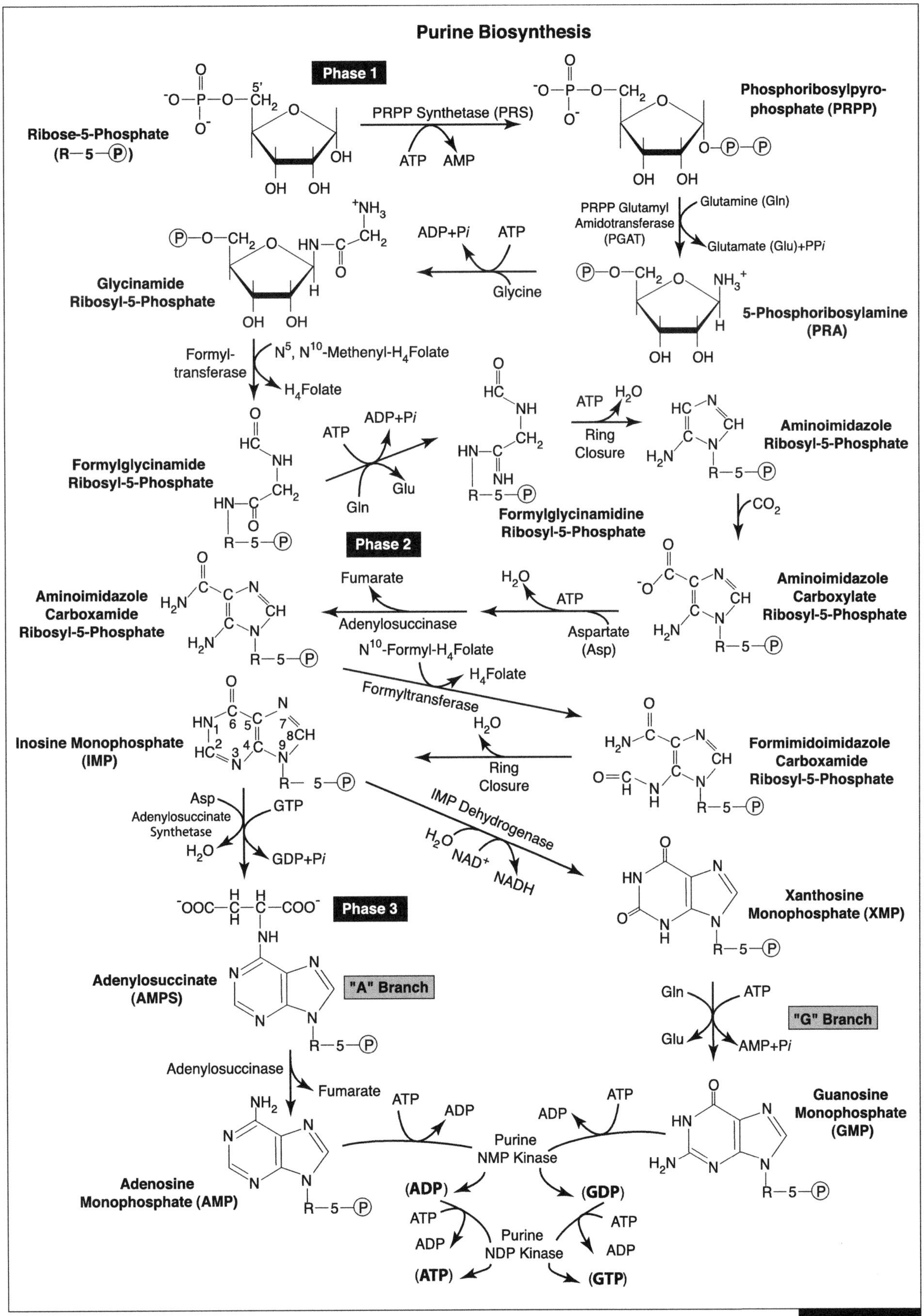

Purine Biosynthesis

Figure 15-2

Phase One - PRPP Biosynthesis

Although this high-energy ribose donor is exploited elsewhere in nucleotide salvage (see Chapter 17), and pyrimidine biosynthesis (see Chapter 14), it is especially important as an early substrate in the *de novo* purine NTP pathway (**Fig. 15-2**). PRPP is synthesized from ribose 5-phosphate (R-5-P) and peATP by **PRPP synthetase** (**PRS**, step 1; **Fig. 15-2**). As summarized at the end of this chapter, the availability of PRPP in the cell is tightly controlled via allosteric "negative feedback" regulation of PRS activity by various purine nucleotides produced in later stages of the pathway, namely IMP, AMP, GMP, and ADP.

Phase Two - Formation of IMP (the parent NMP)

While the pyrimidine ring is assembled *before* it receives a ribose phosphate from PRPP (see Chapter 14), the purine ring is assembled in the *opposite* order, exploiting the ribose phosphate moiety as a construction "**scaffold**". The first addition to the ribose scaffold is an amino group, thus forming **phosphoribosylamine (PRA)**. This compound is formed by reaction of the amido (or amide) group of **glutamine (Gln)** with **PRPP** in the presence of **PRPP glutamyl amidotransferase (PGAT)**. **PRA formation** is a **physiologically irreversible, rate-limiting reaction**, and therefore PGAT, like PRS, is subject to extensive negative feedback regulation by some of the same ribonucleotides that control PRS activity (i.e., IMP, AMP, and GMP; see below).

The next step in phase two is the condensation of **PRA** with **glycine (Gly)** to form **glycinamide ribosyl-5-phosphate**. In this reaction, Gly provides what will ultimately become ring carbons 4 and 5, and ring nitrogen 7 (**Fig. 15-1**). The formyl group of N^5,N^{10}-methenyl-tetrahydrofolate (N^5,N^{10}-methenyl-H$_4$ folate) becomes carbon 8 with the formation of **formylglycinamide ribosyl-5-phosphate** in a reaction catalyzed by **formyl-transferase**. The amide nitrogen from another Gln molecule is next transferred onto the scaffold, thus becoming ring nitrogen 3 of **formylglycinamidine ribosyl-5-phosphate**. In the next reaction, dehydration is accompanied by **ring closure** to form **amino-imidazole ribosyl-5-phosphate**. The initial event in this ring closure is phosphoryl transfer from peATP, followed by nucleophilic attack of the resulting phosphate by the adjacent amino nitrogen. The next step, an **ATP-** and **biotin-independent carboxylation** reaction, yields **aminoimidazole carboxylate ribosyl-5-phosphate**, creating ring carbon atom 6. Condensation of **Asp** with aminoimidazole carboxylate ribosyl-5-phosphate followed by **dehydration** and release of **fumarate** then forms **aminoimidazole carboxamide ribosyl-5-phosphate**. The second step in this two-stage process requires the enzyme **adenylosuccinase**, and yields ring nitrogen 1. This two-step reaction sequence is analogous to that found in the hepatic urea cycle, where citrulline and Asp are converted to arginine (Arg; see Chapter 10).

Ring carbon 2 is next added in a **formyltransferase**-catalyzed reaction involving a second H$_4$ folate derivative, N^{10}**-formyl-H$_4$ folate**. The product, **formimidoimidazole carboxamide ribosyl-5-phosphate** undergoes dehydration and ring closure to complete phase two, thus forming the **parent** purine nucleotide, **inosine monophosphate (IMP)**.

Phase Three - Formation of AMP, GMP, and the Respective 5'-triphosphates

Phase three of the purine biosynthetic pathway forks into an "**A**" **branch** that yields **AMP**, and a "**G**" **branch** that generates **GMP**. Thereafter, AMP and GMP follow a common pathway to their respective NDP and NTP forms.

The "A" branch: Conversion of IMP to AMP involves the simple replacement of the 6-keto group on the ring structure with an amino group. This replacement is accomplished in two enzymatic steps. In the first, which is catalyzed by **adenylosuccinate synthetase**, the keto group of IMP reacts with the amino group of Asp to form **adenylosuccinate (AMPS)** (Note: this reaction requires peGTP rather than peATP as an energy source, and thus provides a potential target for feedback regulation by GTP). In the second step, **adenylosuccinase** catalyzes the release of **fumarate** from AMPS, thus yielding **AMP**.

The "G" branch: Formation of GMP, like the formation of AMP, requires the addition of a single amino group to a ring carbon of IMP - in this case carbon 2. As in AMP formation from IMP, GMP formation is a two-step process. The first step is an **NAD⁺**-dependent oxidation of the 2 carbon by **IMP dehydrogenase**, which generates the 2,6-diketo derivative, **xanthosine monophosphate (XMP)**, and **NADH**. The second step involves amination of the 2-keto group of XMP to form **GMP**, exploiting the amido group of **Gln** and pyrophosphorolysis of **peATP** as a source of energy.

Formation of NDP and NTP Forms of Adenine and Guanine

AMP and **GMP** are reversibly phosphorylated to their respective NDP forms (i.e., **ADP** and **GDP**) by a **purine NMP kinase**. ADP and GDP, in turn, are phosphorylated to their respective NTP forms by a **purine NDP kinase**. Neither of these kinases shows high specificity for adenine or guanine as the purine constituent, and both exploit gamma phosphoryl transfer from peATP.

Regulation of Purine Biosynthesis
Efficacy of Polycatalytic Proteins

Each reaction in phase one and phase two of the purine biosynthetic pathway is catalyzed by a different enzyme in **prokaryotes**. However, in **eukaryotes** gene fusion has given rise to

single polypeptides with multiple catalytic functions like those of the early enzymes in the pyrimidine biosynthetic pathway. For example, the sequential, multistep conversion of **5-phosphoribosylamine to aminoimidazole carboxamide ribosyl-5-phosphate** is catalyzed by a single protein.

Regulation of PRPP Production and its Utilization

As noted above, **PRS** and **PGAT**, the first two enzymes of the purine biosynthetic pathway, are subject to allosteric inhibition by nucleotides formed in latter steps of the pathway. **PRS** is sensitive to high concentrations of **IMP, AMP, GMP**, and **ADP** (see Chapters 14 & 16), while **PGAT** is inhibited by **IMP, AMP**, and **GMP**. **PGAT** is even more tightly regulated by the availability of its PRPP substrate. The cellular **PRPP** concentration of cells is usually about 10-100 times lower than the Michaelis constant (K_m) for the PGAT-catalyzed reaction. Therefore, small increases in the concentration of PRPP result in proportional increases in the rate of 5-phosphoribosylamine formation.

Regulatory Sites in Phase Three

As noted above, there are also at least two targets for "late" ribonucleotide-specific feedback regulation among the enzymes that convert IMP to AMP and GMP. One is **adenylosuccinate synthase**, and the other is **IMP dehydrogenase**.

In summary, IMP is the parent nucleotide from which both **AMP** and **GMP** are generated. The biosynthesis of IMP from the amphibolic intermediate **ribose-5-phosphate** is accomplished in 2 phases. The first phase involves synthesis of PRPP, in which transfer of pyrophosphate from ATP to ribose-5-phosphate is catalyzed by **PRS**, and the second phase involves **10 reaction sequences**, resulting in ring closure of IMP. The pathway then branches, with the "**A**"

branch leading from **IMP** to **AMP**, and the "**G**" **branch** leading from **IMP** to **GMP**. Regulation occurs at the beginning (at **PRS** and **PGAT**), and end of the pathway (at **IMP dehydrogenase** and **adenylosuccinate synthase**).

OBJECTIVES

- Identify the sources of all carbon and nitrogen atoms in the purine ring.

- Summarize the three phases of purine biosynthesis.

- Identify the high-energy ribose donor in both purine and pyrimidine biosynthesis, and show how it is used in nucleotide slavage (see Chapter 17).

- Describe how cellular availability of the ribose donor above is controlled.

- Name the ATP- and biotin-independent carboxylation reaction of purine biosynthesis.

- Indicate how peATP is used in the phosphorylation of AMP to ATP, and GMP to GTP.

- Discuss the regulation of PRPP formation and its utilization, and identify all nucleotides involved.

- Identify regulatory sites in phase three of purine biosynthesis, and list the nucleotides involved in this regulation.

- Describe similarities and differences between pyrimidine and purine biosynthesis.

- Explain how steps involved with ring closure in purine biosynthesis differs from those in pyrimidine biosynthesis.

QUESTIONS

1. All of the following contribute carbon atoms to purines, EXCEPT:
a. CO_2
b. Glycine
c. N^5,N^{10}-Methenyl-H_4 folate
d. Aspartate & glutamine
e. N^{10}-Formyl-H_4 folate

2. Formation of which one of the following intermediates in purine biosynthesis is considered rate-limiting?
a. 5-Phosphoribosylamine
b. Glycinamide ribosyl-5-phosphate
c. Aminoimidazole carboxylate ribosyl-5-phosphate
d. Formimidoimidazole carboxamide ribosyl-5-phosphate
e. Aminoimidazole carboxamide ribosyl-5-phosphate

3. The "parent" purine nucleotide is:
a. AMP.
b. GMP.
c. IMP.
d. CMP.
e. UMP.

4. Nucleotide diphosphate kinase is required for the conversion of:
a. AMP to GMP.
b. ADP to dADP.
c. ADP to AMP.
d. GDP to GTP.
e. AMP to ADP.

5. The intracellular concentration of which one of the following is thought to be the most important factor regulating activity of the rate-controlling enzyme in purine biosynthesis?
a. IMP
b. AMPS
c. XMP
d. GMP
e. PRPP

6. PRPP, a ribose donor in purine biosynthesis, is also used in:
a. The hepatic urea cycle.
b. Nucleotide salvage.
c. The Embden Meyerhoff pathway.
d. Heme biosynthesis.
e. The uronic acid pathway.

7. The ATP- and biotin-independent carboxylation reaction in purine biosynthesis:
a. Converts XMP to PRPP.
b. Is a Phase 3 reaction.
c. Converts IMP to AMPS.
d. Is a Phase 2 reaction.
e. Converts GMP to AMP.

7. d
6. b
5. e
4. d
3. c
2. a
1. d

ANSWERS

Chapter 16

Folic Acid

Overview

- Thymidylate synthase is required for the synthesis of DNA.
- Folic acid can be synthesized by bacteria, but not animals.
- Drugs that inhibit H_2 folate reductase, or thymidylate synthase, are effective anticancer agents.
- The megaloblastic anemia associated with vitamin B_{12} deficiency can be partially alleviated by extra folate in the diet.
- Serine becomes a methyl donor in dTMP, and thus contributes positively to DNA formation.
- Thioredoxin and NADPH aid in the conversion of NDPs to their corresponding dNDPs for DNA synthesis.
- Antimetabolites that inhibit ribonucleoside diphosphate reductase (RDR) reduce the number of dNTPs available for DNA synthesis.

Folic Acid and its Active Form, Tetrahydrofolate

The water-soluble vitamin, **folic acid** (**FA, folate**, or **pteroylglutamate**), consists of a molecule of **para-aminobenzoic acid** (**PABA**), with its amino end attached to a **pteridine**, and its carboxyl group attached to the α-amino group of **glutamic acid** (**Glu**) (**Fig. 16-1** and Chapter 8). **Tetrahydrofolate** (**THFA** or **H₄ folate**), the active **cofactor** form of FA, is formed by reduction of the pteridine ring at positions 5, 6, 7, and 8 by the enzyme, **folic acid reductase** (**FAR**; **Fig. 16-2**). THFA is an essential molecule in both prokaryotic and eukaryotic cell metabolism, donating one-carbon units and other small molecular building blocks to the biosynthesis of **purines** (see Chapter 15, phase two of the pathway), **dTMP** (see Chapter 14 and Fig. 16-2), and selected amino acids, including **methionine** (Met; see Chapter 43).

Folate Metabolism in Animals vs Bacteria

Neither animals nor bacteria can survive without a source of FA. Animals cannot synthesize FA and, therefore, they evolved a means

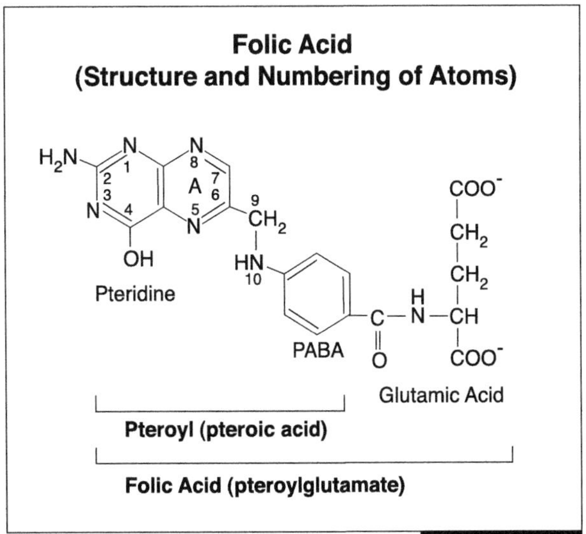

Folic Acid (Structure and Numbering of Atoms)

Pteridine

PABA

Glutamic Acid

Pteroyl (pteroic acid)

Folic Acid (pteroylglutamate)

Figure 16-1

Copyright © 2015 Elsevier Inc. All rights reserved.

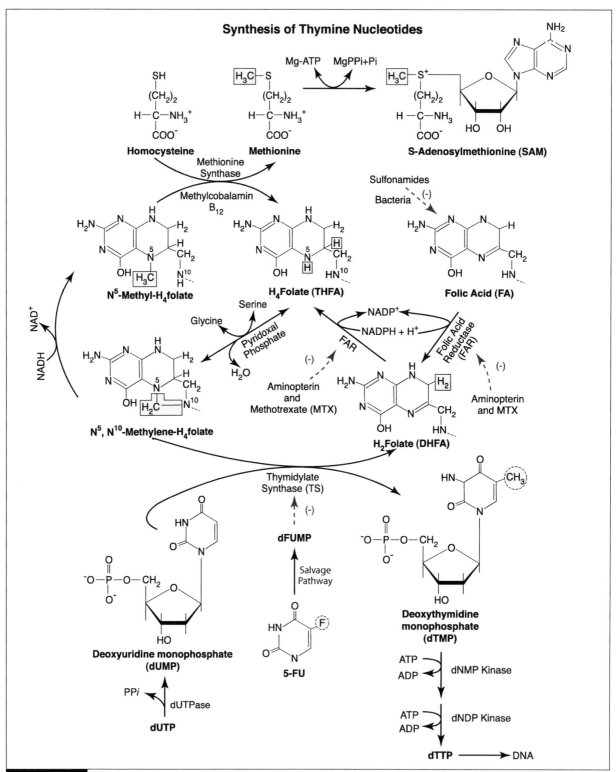

Figure 16-2

of assimilating preformed FA from their diet. Yeast, liver, and leafy plants are major sources of this water-soluble vitamin. In contrast to animals, **bacteria** cannot assimilate FA from their environment, and therefore they evolved enzymatic machinery to synthesize it from an appropriate pteridine, PABA, and Glu. The difference between animals and bacteria in their capacity to assimilate and synthesize FA has been exploited to design bacteria-selective

sulfonamide antibiotics ("sulfa" drugs). Sulfonamides are analogs of PABA, and thus act as competitive inhibitors of PABA's participation in bacterial *de novo* FA biosynthesis. Since bacteria cannot assimilate preformed FA from their environment, the competitive inhibition of PABA incorporation severely reduces their FA levels, thus inhibiting growth and replication. In contrast, **animals**, which **assimilate FA** rather than synthesize it from PABA, generally suffer no ill effects from sulfonamides.

THFA-mediated One-carbon Metabolism

Dietary FA is frequently polyglutamated. Extra glutamates are removed enzymatically in the small intestine prior to FA absorption, and most of the FA is then reduced in mucosal cells of the gut. Reduction requires the presence of NADPH and FAR, and proceeds in two steps - first to **dihydrofolate** (**DHFA** or **H$_2$ folate**), and then to **THFA** (see **Fig. 16-2**). Once formed, THFA picks up the one-carbon methylene group from **serine (Ser)** to form **N^5,N^{10}-methylene-H$_4$ folate**. This compound plays a central role in one-carbon unit metabolism. For example, it can be used as a source of the methyl carbon of **thymine** (see Chapter 14 and **Fig. 16-2**), or it can be enzymatically reduced with NADH to **N^5-methyl-H$_4$ folate**, an important cofactor in methylcobalamin (vitamin B$_{12}$)-dependent methylation of homocysteine to Met (see Chapter 43 and **Fig. 16-2**). The methyl group of **Met** can become highly reactive through addition of an adenosyl group from ATP, thus forming **S-adenosylmethionine (SAM)**. The high transfer potential of this methyl group allows it to move to a number of acceptors (see Chapter 57).

Folate Plasma Concentrations

As indicated above, dietary **FA** is normally absorbed in the proximal part of the small intestine (**jejunum**), with mucosal cells reducing most of it to **DHFA**, and then to **THFA**. Therefore, **low plasma folate** concentrations can be indicators of proximal bowel dysfunction. On the other hand, intestinal bacterial overgrowth will sometimes promote excessive folate production, thus creating **high plasma folate** concentrations.

Megaloblastic Anemia (MA)

Vitamin B$_{12}$ is the largest compound absorbed intact from the terminal ileum, and its absorption requires **intrinsic factor (IF)**, which is synthesized by parietal cells of the stomach or abomasum, and by pancreatic ductular cells of dogs and cats. Gastrectomy or pancreatic insufficiency can therefore compromise intestinal B$_{12}$ absorption if IF is not provided, thus leading to inefficient maturation of erythrocytes (i.e., anemia; see Chapter 43). In the absence of B$_{12}$, homocysteine is not methylated and therefore, most cellular THFA becomes "trapped" as N^5-methyl-H$_4$ folate, creating a shortage of the forms of THFA required for other essential biosynthetic processes - in particular the synthesis of the essential DNA precursor, **dTMP** (see **Fig. 16-2**). With dTMP deficiency, a deficiency of dTTP ensues, and DNA synthesis is inhibited. As a consequence, erythroblast division and maturation in bone marrow is compromised, promoting formation of **megaloblasts** with characteristic large, polymorphic nuclei. The N^5-methyl-H$_4$ folate trap resulting from B$_{12}$ deficiency also leads to impairment of the two THFA-dependent steps in purine biosynthesis (see Chapter 15), and to the accumulation of homocysteine with attendant **homocystinuria**. (Note: Since extracellular sulfhydryl compounds are quickly oxidized to disulfides, homocysteine in blood and urine exists mainly as homocystine.) Although not well understood, the molecular basis for the neurologic signs associated with MA is thought to be secondary to a relative Met deficiency.

The explanation for the development of MA in FA deficiency is similar to that advanced for B$_{12}$ deficiency above. Since B$_{12}$ and an activated form of THFA are common participants in the Met synthase reaction, the MA caused by B$_{12}$ deficiency can be partially alleviated by provision of extra FA in the diet. However, dietary FA supplementation cannot cure the homocystinuria, nor can it alleviate the neurologic signs. Therefore, provision of an assimilable form of B$_{12}$ is required.

Formation of Deoxyribonucleotides

The synthesis of DNA by DNA polymerases requires an adequate supply of the 4 deoxynucleoside triphosphate (dNTP) substrates - dATP, dCTP, dGTP, and dTTP. The *de novo* biosynthetic pathway for these dNTPs involves direct reduction of an **NDP** to the corresponding **dNDP**, followed by ATP-dependent **phosphorylation** of the dNDP product to the dNTP form with an appropriate **dNDP kinase**. NDP reduction is catalyzed by the iron-containing **ribonucleoside diphosphate reductase (RDR, or ribonucleotide reductase)**, an enzyme ubiquitous in animals and prokaryotes (**Fig. 16-3** and Chapter 48). RDR reduces the 2'carbon of the NDP, replacing the **OH** with **H**, and can reduce any of the four common NDPs (CDP,

UDP, ADP, and GDP; see Chapter 16), to their corresponding dNDPs. Reduction of the 2'NDP carbon is accompanied by oxidation of 2 active site sulfhydryl (**SH**) groups to a disulfide (**-S-S-**). The oxidized RDR is then re-reduced by a small dithioprotein, **thioredoxin**, which in turn is re-reduced with **NADPH** by the enzyme **thioredoxin reductase (TR)**. The RDR is subject to a complex pattern of allosteric feedback regulation by dNTPs to ensure that the dNTP pool maintains the balanced composition necessary for efficient DNA biosynthesis.

Conversion of dUTP to its 5-methyl Form, dTTP

As noted above, phosphorylation of the four dNDP products of RDR yields dATP, dCTP, dGTP, and dUTP. Of these, the first 3 may be used directly for DNA synthesis. However, the 4th, dUTP, presents a problem for most prokaryotes and animals because their **DNA** does not contain **uracil**; rather, it contains its 5-methylated form, **thymine**. To ensure that their DNA polymerases receive the requisite supply of dTTP, organisms and cells with thymine-containing genomes evolved a "protective" enzyme, **dUTPase**, to dephosphorylate **dUTP** to **dUMP**, and a second enzyme, **thymidylate synthase (TS)**, to methylate **dUMP** to **dTMP** (**Fig. 16-2**).

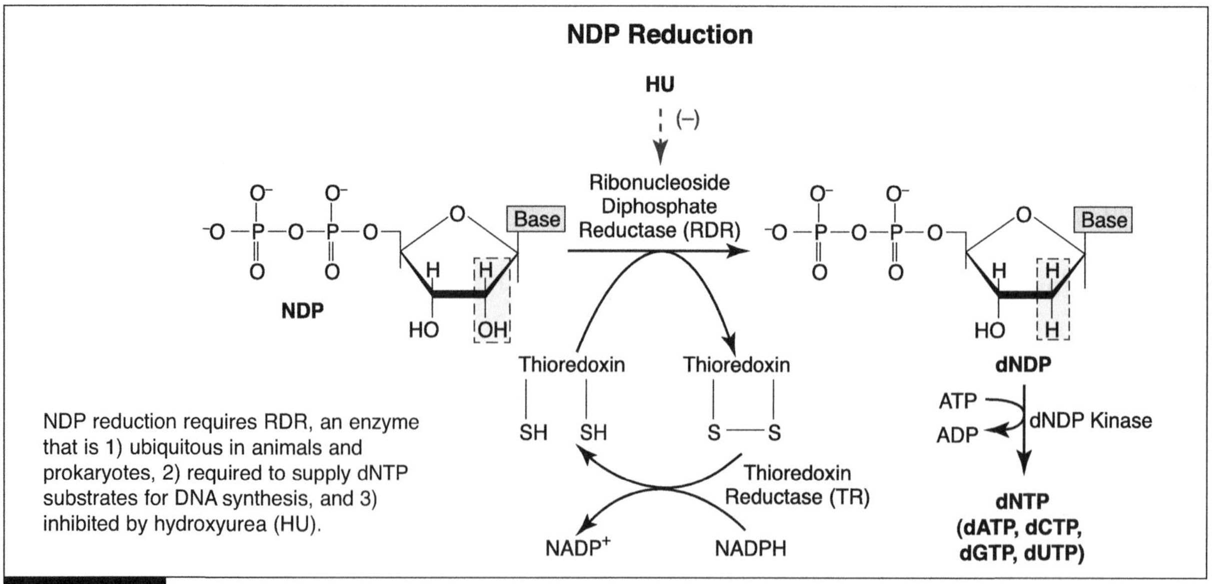

NDP Reduction

NDP reduction requires RDR, an enzyme that is 1) ubiquitous in animals and prokaryotes, 2) required to supply dNTP substrates for DNA synthesis, and 3) inhibited by hydroxyurea (HU).

Figure 16-3

Once formed by TS, dTMP is then sequentially phosphorylated to dTTP by appropriate ATP-dependent dNMP and dNDP kinases.

Chemotherapeutic Drug Targets in dNTP and Folate Metabolism

Many types of cancer cells grow and divide more rapidly than their normal counterparts, and therefore their growth and viability are highly sensitive to **antimetabolites** which inhibit nucleic acid biosynthesis - in particular DNA biosynthesis. There are several examples of antimetabolites that kill cancer and other proliferating cells by interfering with their ability to produce an adequate supply of dNTPs for DNA synthesis. One of the most direct ways to selectively reduce this supply is to **inhibit RDR**. To this end, **hydroxyurea (HU)**, a relatively selective, irreversible inhibitor of RDR, has been found to be a clinically effective suppressor of cell proliferation in both cancer and autoimmune disease. A second important antimetabolite target is TS. Inhibition of TS cuts off the supply of essential dTTP, which, in turn, inhibits DNA synthesis and cell division. One of the best known and most potent TS inhibitors is **dFUMP**, the 5-fluoro derivative of dUMP. Given its charged phosphate group, dFUMP cannot readily penetrate cell membranes. Therefore, a cell-permeable, "prodrug" form - either the base, **5-fluorouracil (5-FU)**, or the deoxyribonucleoside, **5-fluoro 2'-deoxyuridine (5-FUdR)** - can be administered (see Chapter 6). These prodrugs are metabolized to the active dFUMP form intracellularly, via the appropriate salvage pathway (see Chapters 17). A third effective point for attack of the cell's dNTP (and purine NTP) supply is **FAR**. FAR inhibition reduces supply of THFA, which in turn leads to inhibition of the THFA-dependent methylation of dTMP (i.e., the TS reaction; **Fig. 16-2**), and the two THFA-dependent steps in purine biosynthesis (see Chapter 15). The suitability of FAR as an antimetabolite target is illustrated by the considerable success of the drug, **methotrexate (MTX**; see Chapters 6 and 14, and **Fig. 16-2**). MTX is a FA structural analog that acts as a competitive inhibitor of FAR, with an affinity for the enzyme at least 100 times greater than that of FA or DHFA. MTX is widely used in the treatment of leukemias and solid tumors.

OBJECTIVES

- Contrast differences between bacteria and animals in their capacity to assimilate and synthesize FA, and provide rationale for sulfonamide antibiotic therapy.

- Understand how and why the "folate trap" is created, and how it can lead to anemia and homocystinuria (see Chapter 43).

- Explain how Ser, N^5,N^{10}-methylene-H_4 folate, dUMP, methylcobalamin, homocysteine, SAM and biotin participate in one-carbon metabolism.

- Describe the roles of RDR and TR in NDP reduction.

- Explain how dTTP can be formed (sequentially) from dUTP.

- Discuss the manner in which HU, 5-FU, 5-FUdR, MTX and aminopterin act as chemotherapeutic agents.

- Identify factors that could increase or decrease the plasma folate concentration.

QUESTIONS

1. **Which one of the following reduces DNA formation by specifically inhibiting thymidylate synthase?**
   a. Aminopterin
   b. PABA
   c. Sulfonamides
   d. dFUMP
   e. Methotrexate

2. **Sulfonamides are effective:**
   a. Anticancer agents, since they inhibit mammalian H_2 folate formation.
   b. Inhibitors of folate biosynthesis in bacteria.
   c. Drugs for treating the anemia that develops from either folate or vitamin B_{12} deficiency.
   d. Inhibitors of folate biosynthesis in animals.
   e. Anticancer agents, since they inhibit thymidylate synthase.

2. b

1. d

Answers

Chapter 17

Nucleic Acid and Nucleotide Turnover

Overview

- Primates and Dalmatian dogs degrade their purines to uric acid, whereas other domestic animals convert urates to the more soluble allantoin.

- Pyrimidines are catabolized to simpler, water-soluble forms that re-enter metabolic pathways.

- Chronic overproduction or underutilization of PRPP can lead to excess purine nucleotide biosynthesis.

- Allopurinol is a suicide inhibitor of xanthine oxidase.

- Purine bases are usually salvaged through a one-step reaction with PRPP, while pyrimidine bases are usually salvaged through a two-step process involving the sequential action of a nucleoside phosphorylase and a nucleoside kinase.

- Although animals ingest polynucleotides, survival does not require their digestion, absorption, and utilization since ample amounts of purine and pyrimidine nucleotides can be synthesized *de novo*.

As discussed in Chapter 13, a nucleoside triphosphate (NTP) may have several possible fates. It may serve as a polymerase substrate and end up with its NMP moiety incorporated into nucleic acid, it may become a vehicle for transfer and transport of sugars and other small molecules, or, like ATP, it may be cyclized to a 3'-5'NMP. It may also circulate unchanged to serve as a donor of high energy phosphate and pyrophosphate. Whatever the function of a given nucleotide, its **base**, **carbohydrate**, and **phosphate** moieties are constantly "turning over" in response to two dynamic processes: **salvage** and **degradation**.

Release of Bases from Nucleic Acids

Degradation of DNA and RNA in mammals occurs at two major sites - **intracellularly**, during nucleic acid turnover and apoptosis

(programmed cell death), and in the **intestinal lumen**, as part of the process involved in the digestion of dietary nucleic acid (see Appendix).

Whether of cellular or dietary origin, DNA and RNA are degraded to their component parts in a series of sequential enzymatic steps summarized in **Fig. 17-1**. The first step is catalyzed by RNA- or DNA-specific **endonucleases** called, respectively, **DNAases** and **RNAases** (pancreatic endonucleases in the case of intestinal degradation). As the *endo* prefix indicates, these enzymes cut their specific nucleic acid targets **internally**, breaking them into short **oligonucleotides** (**oligos**). The oligos are subsequently degraded from their ends by exonucleases, yielding **5'-** and **3'-ribo-** and **deoxyribo-NMPs**.

Copyright © 2015 Elsevier Inc. All rights reserved.

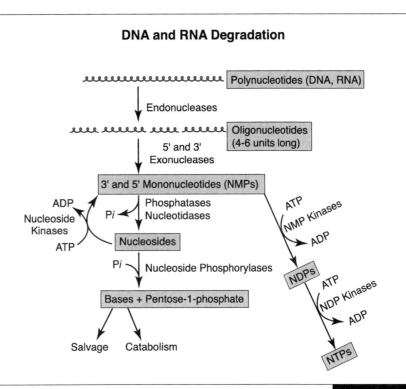

DNA and RNA Degradation

Intestinal and intracellular DNA and RNA degradation takes place in a series of sequential enzymatic steps. The NMPs so generated can either be phosphorylated into useful NTPs, or dephosphorylated into nucleosides. The nucleosides may then be rephosphorylated into NMPs, or be further degraded into their constituent bases and pentose-1-phosphates. The free bases are then either salvaged or catabolized.

Figure 17-1

Nucleotides and Nucleosides

The NMPs released from nucleic acid have two possible fates: **1) ATP-dependent phosphorylation** by **NMP and NDP kinases** to re-form useful NTPs, or **2) dephosphorylation** by **nucleotidases** and non-specific **phosphatases** thus forming **ribo-** or **deoxyribonucleosides**. The nucleosides formed in the latter step also may have two possible fates, depending upon their origin and the context of their formation. They may be re-phosphorylated to their NMP forms by various **nucleoside kinases**, or they may undergo phosphorolytic cleavage by **nucleoside phosphorylases** to liberate their constituent **base** and **pentose (i.e., ribose- or deoxyribose)-1-phosphate**. The free **bases** join bases liberated from the turnover of "free" cellular ribo- and deoxyribonucleotides (i.e., those not previously polymerized into nucleic acid), and finally, these bases are either **a) salvaged** for reconstitution into useful NTP forms, or **b) catabolized** to simpler "dead-end" components (e.g., uric acid in the case of purines).

Salvage of Purine and Pyrimidine Bases

These bases are energetically expensive to synthesize *de novo*, and therefore cells have devised ways to salvage, or recycle, most of them. The salvage pathways are important in **rapidly dividing cells** (e.g., bone marrow, intestinal epithelial cells, and embryonic or fetal cells) where they serve to supplement *de novo* biosynthetic pathways in support of the considerable RNA and DNA biosynthesis needed for cell division, as well as in many non-dividing cells. The liver provides purine and pyrimidine bases as well as their nucleosides for salvage and utilization by tissues incapable of their biosynthesis. For example, the **brain** has a low level of **PRPP glutamyl amidotransferase (PGAT)**, and hence depends in part on exogenous purines plus base salvage. Similarly, **erythrocytes** cannot synthesize **5-phosphoribosylamine (PRA)**, and therefore utilize exogenous purines to form nucleotides (e.g., AMP), and they also make use of the salvage pathways (see Chapter 31). The salvage pathways are also critical for the conversion of several important chemotherapeutic base analogs to their active nucleotide forms (see **dFUMP** formation, Chapter 16).

The "strategy" of base salvage is to return it to a riboNMP form which, in turn, can be re-phosphorylated by ATP-dependent NMP and NDP kinases to yield the ultimate riboNTP form. Base salvage reactions are summarized in **Fig. 17-2**. As indicated, a base may become a riboNMP by two different routes: **1**) a direct, single-step route involving **phosphoribosyl transferase (PRT)**-catalyzed reaction with **PRPP**, or **2**) a two-step, indirect route involving **a)** formation of the **ribonucleoside** via the reversible **ribonucleoside phosphorylase**-catalyzed reaction with **ribose-1-phosphate**, followed by **b) ATP-dependent phosphorylation** of the ribonucleoside by an appropriate **ribonucleoside kinase** to form the **riboNMP**.

Although purines can use the indirect route through the nucleoside, they prefer the direct route. This preference results from the availability of two **purine-specific PRTs**, one specific for **adenine (APRT)**, and the other for **hypoxanthine (HX)** and **guanine (HGPRT)**. In contrast to purine-specific PRTs, there is only one PRT specific for **pyrimidines**, the orotic acid-specific enzyme used in *de novo* pyrimidine biosynthesis (see Chapter 14). Therefore, the major pyrimidine bases, **uracil** and **cytosine**, are primarily salvaged indirectly, using the ribonucleoside phosphorylase/kinase route.

Degradation of Pyrimidine Bases

Major products of pyrimidine degradation are relatively water-soluble, and are usually reused in other areas of metabolism rather than being excreted in urine (**Fig 17-3**). β-**Alanine**,

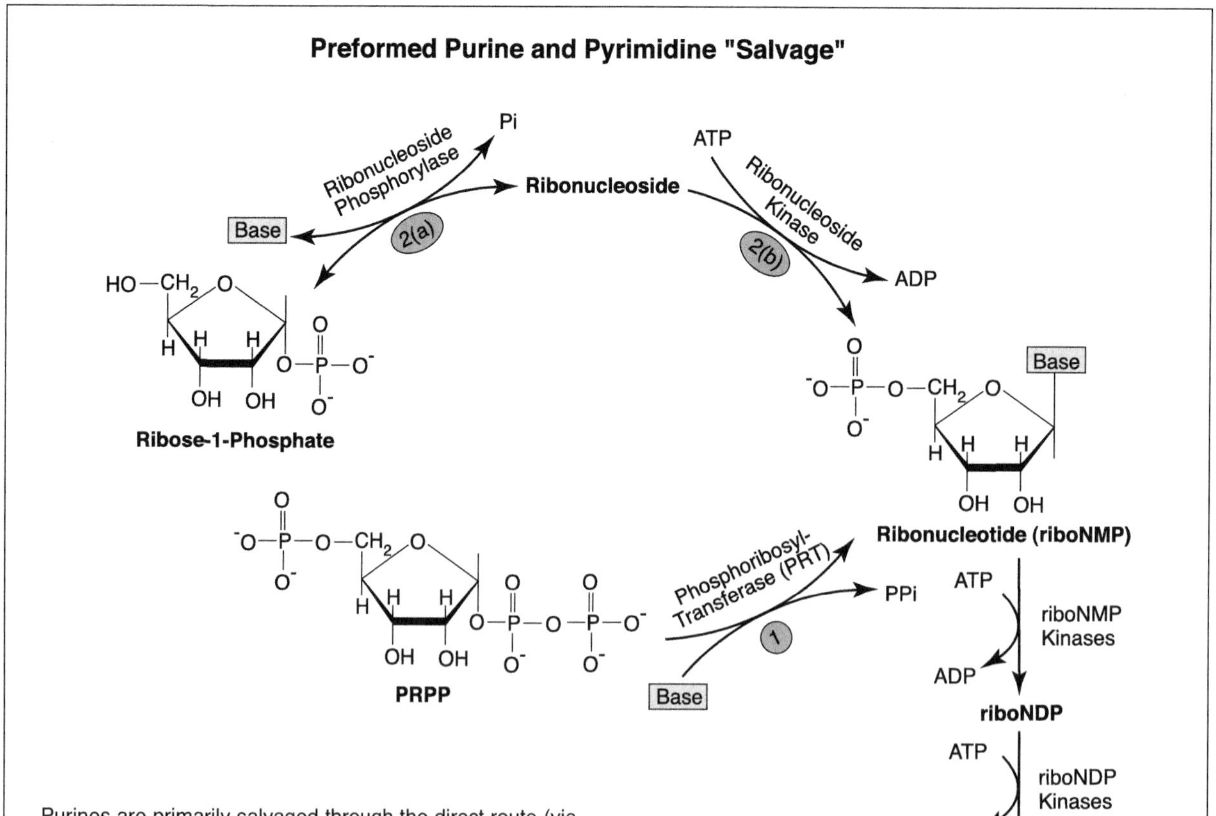

Preformed Purine and Pyrimidine "Salvage"

Purines are primarily salvaged through the direct route (via reaction (1)), whereas pyrimidine base salvage largely involves the indirect route (through reactions 2(a) and 2(b)).

Figure 17-2

from cytosine and uracil degradation, can be used in the biosynthesis of pantothenic acid (vitamin B$_5$ which forms CoA.SH (see Chapter 41)), or in the formation of **acetate**; and β-**aminoisobutyrate**, formed through thymine degradation, can ultimately enter the TCA cycle at the level of **succinyl-CoA** (see Chapters 37, 42 and 43)..

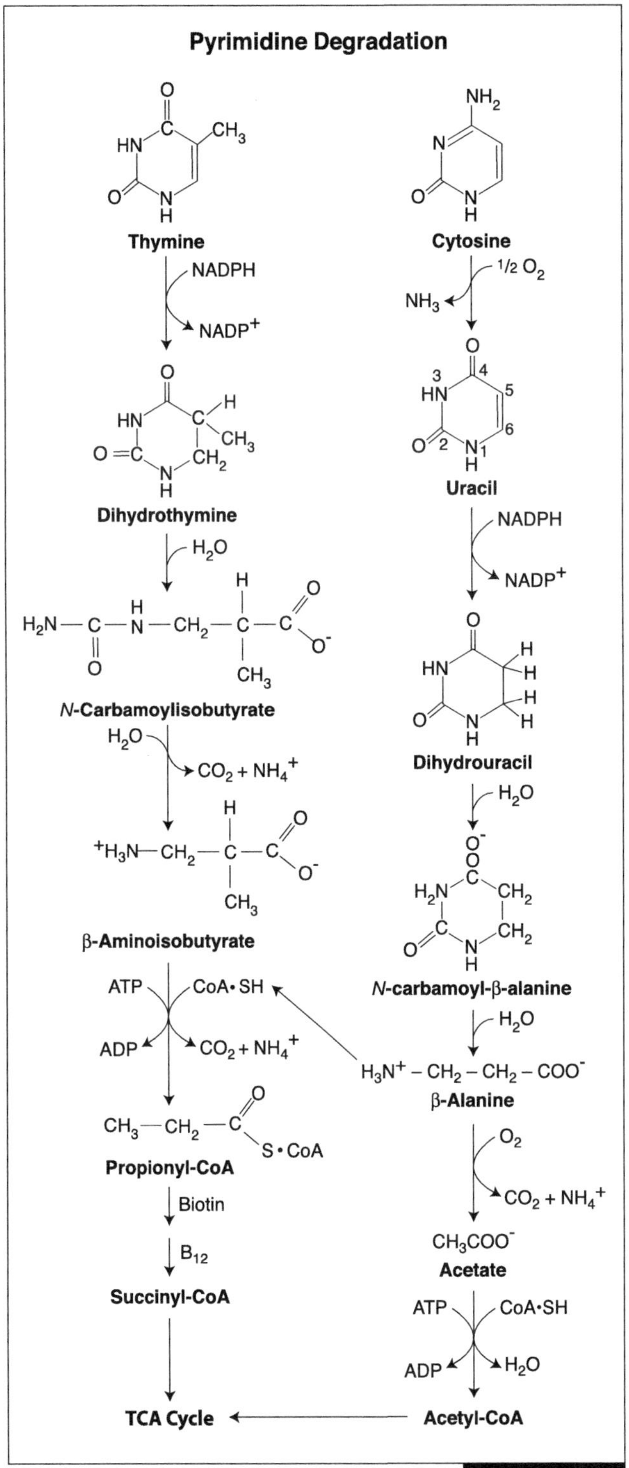

Degradation of Purine Bases

The path of purine catabolism is more complex than that of pyrimidines, and, as noted below, its final step varies significantly (and importantly) among animal species. As shown in **Fig. 17-4**, catabolism of **HX**, **guanine**, and **adenine**, the three major purines, funnels through a single derivative, the 2,6-diketo-purine, **xanthine** (**X**). HX is oxidized directly to X by the molybdenum-dependent flavo-protein enzyme (see Chapter 40), **xanthine oxidase** (**XO**). This enzyme has a wide distri-bution, occurring in milk, the small intestine, muscle (see Chapter 77), liver, and kidney. Guanine also proceeds directly to X through oxidative deamination of its 2 position by the enzyme **guanase**. The ring of adenine may be oxidatively deaminated to HX at the base (adenine) level by **adenosine deaminase**, or it may be deaminated by this enzyme at the **AMP** or **adenosine** level to **IMP** and **inosine**, respectively. IMP is then dephosphorylated by **nucleotidase** to **inosine**, which is then cleaved by **purine nucleoside phosphorylase**, freeing the **HX**. HX then proceeds through two XO-catalyzed oxidations to generate the triketo purine end-product, **uric acid** (**UA**). The XO-catalyzed oxidations produce the toxic **superoxide anion** ($O_2^{-\cdot}$) as a by-product, which under normal circumstances is quickly converted to **hydrogen peroxide** (H_2O_2) by the rather ubiquitous **superoxide dismutase** (**SOD**; see Chapter 30).

Cancer patients with large tumor burdens

Pyrimidine Degradation

Figure 17-3

who undergo radiation or chemotherapy show increased serum and urine concentrations of UA. This increased UA is not due to increased purine nucleotide biosynthesis, but rather to increased destruction of tumor cells that in turn release

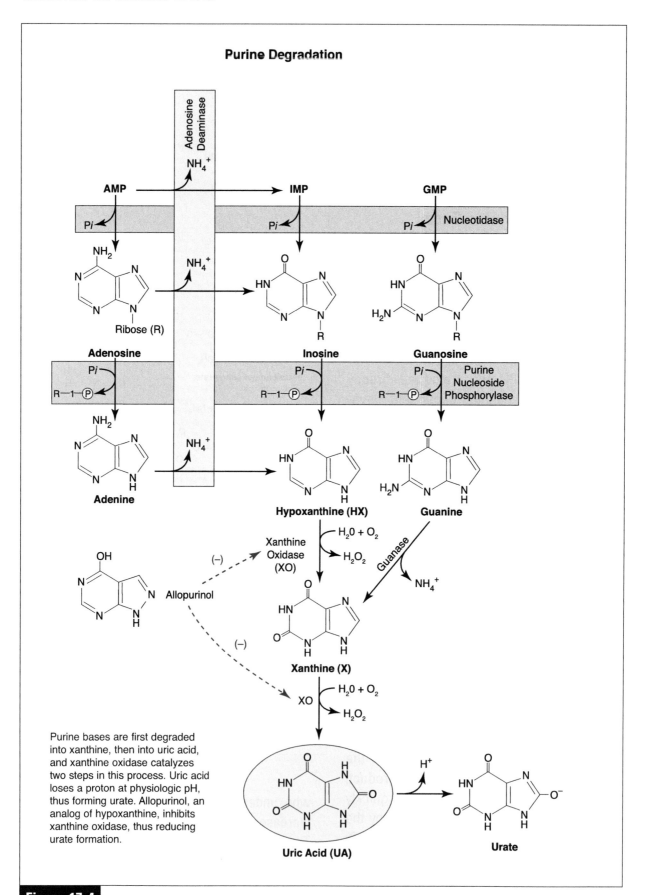

Purine Degradation

Purine bases are first degraded into xanthine, then into uric acid, and xanthine oxidase catalyzes two steps in this process. Uric acid loses a proton at physiologic pH, thus forming urate. Allopurinol, an analog of hypoxanthine, inhibits xanthine oxidase, thus reducing urate formation.

Figure 17-4

degraded nucleic acids and cellular nucleotides that are further metabolized to UA. Some cancer treatments include allopurinol (see below), for the sole purpose of limiting UA buildup.

Excretion of Purine Degradation Products

Uric acid has poor aqueous solubility, and thus, most of it circulates as the slightly more soluble sodium salt, sodium urate (**NaUA**). As indicated in **Fig. 17-5**, animals differ in how they deal with this molecule. **Uricoteles** (i.e., birds, insects, and some reptiles) use UA as the main vehicle for disposal of their nitrogenous waste. In these animals, NaUA reverts to the highly insoluble UA in the low pH of the urinary tract, and passes into urine in a highly concentrated, precipitated form. In **ureotelic** mammals, which dispose of nitrogenous waste primarily as **urea**, NaUA may be handled in two different ways. In most animals, the UA molecule is opened by hepatic uricase and excreted in urine as the more soluble **allantoin**. In fish allantoin is further degraded to **allantoic acid**, and in many microorganisms this compound can be disassembled into **glyoxylate** and two molecules of **urea**. In **Dalmatian dogs** and **primates**, where

hepatic allantoin formation is reduced, NaUA must be excreted largely unchanged in urine.

Uric Acid and Health

Purine excess, whether caused by overproduction, inadequate salvage, inefficient excretion, massive cell destruction, or high dietary purine intake, can lead to **hyperuricemia** - especially in primates and Dalmatian dogs where UA uptake into hepatocytes is reduced. Because of the limited solubility of UA and NaUA, the hyperuricemia may lead to precipitation of UA in the urinary tract and joints, causing **gout** with its attendant urinary tract obstruction (**urolithiasis**), and painful swelling of the joints (**tophi**). Hyperuricemia and gout in humans can also result from two types of genetic defects - one causing loss of feedback control of PRPP synthetase (see Chapter 15), and the other leading to deficiency of the purine salvage enzymes, APRT and HGPRT (**Fig. 17-2**).

The mainstay of **gout therapy** is **allopurinol**, an analog of hypoxanthine. As noted in **Fig. 17-4**, this agent competitively inhibits XO, reducing flow of the more soluble HX and X to the relatively insoluble UA. **Probenecid**

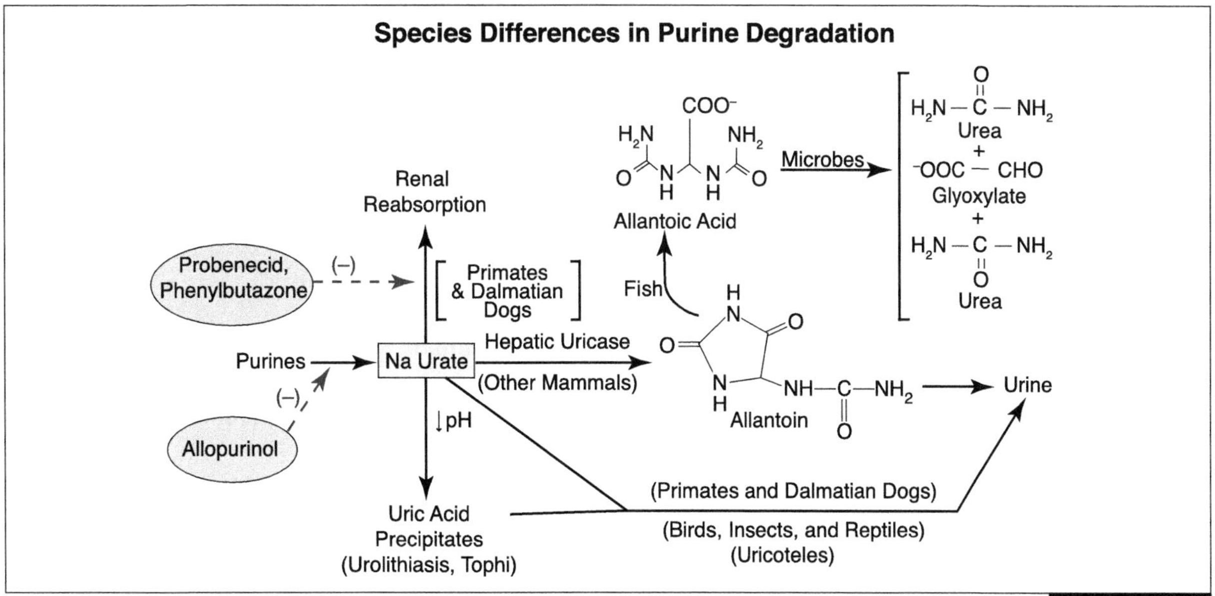

Species Differences in Purine Degradation

Figure 17-5

and **phenylbutazone** are also used to reduce the active proximal tubular reabsorption of UA in primates and Dalmatian dogs, and thus, increase the rate of urinary NaUA excretion. UA appears to be secreted by the nephron of most mammals. Why UA is actively reabsorbed by the kidney of some mammalian species and not others has long been an intriguing question. Studies indicate that **UA** may have **antioxidant activity**, and that normal blood levels in humans are about 6-times greater than those of vitamin C (Hediger MA, 2002).

OBJECTIVES

- Recognize how, where and when nucleic acids are degraded in the body.

- Give two possible fates of the NMPs relased from nucleic acids during DNA and RNA degradation.

- Identify the final end product of adenine and guanine catabolism in mucosal cells of the gut.

- Explain why nitrogenous base salvage is important to rapidly dividing cells, and to certain non-dividing cells (e.g., mature RBCs).

- Identify the two routes by which a nitrogenous base may become a riboNMP during base salvage.

- Contrast primary differences between pyrimidine and purine degradation.

- Discuss the reasoning for adding allopurinol to some cancer treatment regimens.

- Recognize potential factors leading to purine excess, and hyperuricemia.

QUESTIONS

1. **Which one of the following best describes differences and/or similarities between purine and pyrimidine degradation?**
   a. Both purines and pyrimidines are normally degraded to uric acid by all mammals except primates and Dalmatian dogs.
   b. Pyrimidine degradation requires xanthine oxidase, whereas purine degradation requires adenosine kinase.
   c. Purines are typically degraded to AMP, IMP, and GMP, whereas pyrimidines are degraded to CMP, UMP, TMP, and OMP.
   d. The purine ring, unlike the pyrimidine ring, is subjected to oxidation by xanthine oxidase.
   e. Purines and pyrimidines are degraded through similar pathways, and thus share the same degradative enzymes.

2. **Xanthine oxidase, the enzyme that converts hypoxanthine to uric acid, requires which of the following as a cofactor?**
   a. Copper
   b. Iron
   c. Molybdenum
   d. Lead
   e. Magnesium

3. **In most domestic animals, hepatic uricase converts urates primarily to which compound?**
   a. Allopurinol
   b. PRPP
   c. Alloxanthine
   d. Allantoin
   e. Sodium urate

4. **All of the following would favor uric acid crystal formation in the urine of Dalmatian dogs, EXCEPT:**
   a. Probenecid administration following a high dietary purine intake.
   b. Ketoacidosis associated with starvation.
   c. Decreased activity of hypoxanthine guanine phosphoribosyl transferase (HGPRT).
   d. Chronic overproduction of PRPP.
   e. Allopurinol administration.

5. **Adenine, hypoxanthine, and guanine can each be salvaged through a one-step reaction with:**
   a. PRPP.
   b. Uric acid.
   c. Ribose-1-phosphate.
   d. Ammonia.
   e. Allantoin.

6. **β-Alanine can be generated through degradation of:**
   a. Adenine.
   b. Uracil.
   c. Guanine.
   d. Thymine.

6. b
5. a
4. e
3. d
2. c
1. d

ANSWERS

Sections I and II Examination Questions

1. The K_m for Enzyme A is 200 mg/dl, while that for Enzyme B is only 10 mg/dl. Although both enzymes catalyze the same reaction, they are generally found in different tissues. Which statement(s) below is/are TRUE?

   a. The activity of enzyme A cannot be physiologically controlled since it has a high K_m.
   b. All of the active sites on Enzyme B must be filled at a substrate concentration of 10 mg/dl.
   c. The V_{max} for Enzyme A is obtained at a substrate concentration of 200 mg/dl.
   d. At a low substrate concentration, enzyme B is far more capable than enzyme A of producing a common product from this reaction.
   e. All of the above are TRUE.

2. Select the FALSE statement below:

   a. dUTP is used to convert dUTP to dUMP, and dUMP can then be used in the sequential formation of dTTP.
   b. Arginase catalyzes removal of the urea side chain from Arg thus resulting in Asp formation.
   c. The bacterial form of asparagine synthetase can "fix" inorganic nitrogen.
   d. Nonessential amino acids generally have shorter biological half-lives than essential ones.
   e. Cystine normally links different regions of polypeptide chains, thus stabilizing proteins and making them more resistant to denaturation.

3. Which two folic acid derivatives below participate directly in purine biosynthesis?

   a. N^{10}-Formyl-H_4 Folate & N^5,N^{10}-Methenyl-H_4 Folate
   b. N^5-Methyl-H_4 Folate & N^5,N^{10}-Methenyl-H_4 Folate
   c. H_2 Folate & S-Adenosylmethionine
   d. N^5-Methyl-H_4 Folate & N^5,N^{10}-Methylene-H_4 Folate
   e. H_2 Folate & H_2 Folate

4. Which one of the following is an allosteric negative feedback modulator of pyrimidine biosynthesis in mammals?

   a. ATP
   b. IMP
   c. UTP
   d. CTP
   e. PRPP

Copyright © 2015 Elsevier Inc. All rights reserved.

5. If ornithine becomes unavailable to exchange across mitochondrial membranes with citrulline in the urea cycle, the following compound accumulates, and may ultimately lead to excessive orotic acid formation in pyrimidine biosynthesis:
   a. Inosine monophosphate
   b. PRPP
   c. Argininosuccinate
   d. Uridine monophosphate
   e. Carbamoyl phosphate

6. Select the FALSE statement below:

   a. Noncompetitive enzyme inhibitors and the endogenous substrate can bind the enzyme at the same time.
   b. Aspartate aminotransferase is also known as serum glutamate oxaloacetate transaminase.
   c. An argininosuccinate synthetase deficiency would be expected to cause hyperammonemia.
   d. Desmosine and isodesmosine are modified amino acids found in elastin.
   e. The exchangeable plasma albumin pool, which is maintained by the liver, normally turns over at the rate of about 90%/day.

7. Which of the following statements regarding enzyme kinetics is CORRECT?

   a. The V_{max} of an enzyme is a measure of the affinity of that enzyme for its substrate.
   b. A low K_m is associated with low affinity of the enzyme for its substrate.
   c. K_m is the substrate concentration required to achieve the full V_{max} of an enzyme-catalyzed reaction.
   d. The V_{max} of an enzyme-catalyzed reaction is usually reduced by the presence of a non-competitive inhibitor.
   e. K_m is usually expressed in terms of reaction velocity (e.g., moles/sec).

8. Select the TRUE statement(s) below:

   a. Urease is alsoknown as arginase.
   b. Creatinuria is excessive urinary excretion of creatinine.
   c. Tyr, Phe and Trp are ketogenic, but not glucogenic amino acids.
   d. Some cancer patients undergoing chemotherapy may require allopurinol treatment because of increased UA production.
   e. All of the above

9. Place the following structures in proper order (from left to right):

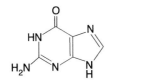

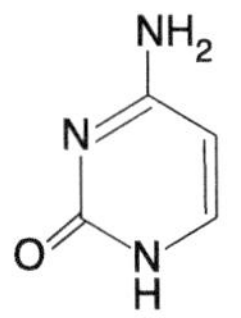

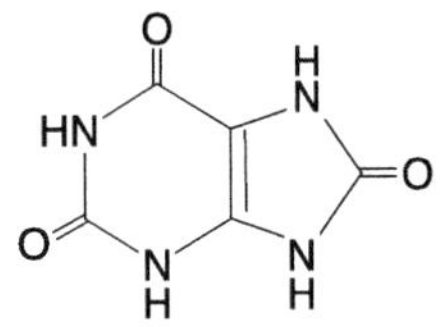

a. Uric Acid, Guanine, Cytosine
b. Guanine, Cytosine, Uric Acid
c. Cytosine, Guanine, Uric Acid
d. Guanine, Uric Acid, Cytosine
e. None of the above, for all three are aromatic amino acids.

10. Select the TRUE statement below:

a. The polypeptide hormone insulin contains both intra- and interchain disulfide bonds between serine residues.
b. Hydrocarbons from dietary amino acids typically appear in animal lipids, but not those from dietary carbohydrates.
c. Base salvage involves the return of purines and pyrimidines to riboNMP forms, which can then be re-phosphorylated to ultimate riboNTP forms.
d. Tryptophan is a nonessential amino acid used for both serotonin and catecholamine biosynthesis.
e. Phosphorylation of an enzyme always renders it active.

11. Hypoxanthine is:

a. A substrate in ALT-catalyzed reactions.
b. An intermediate in UA biosynthesis by several different cell types.
c. Formed from citrulline in the urea cycle.
d. An intermediate in the cytosine degradation pathway.
e. An essential amino acid with an aromatic side chain.

12. Which one of the following is an allosteric activator of CPS-2 in pyrimidine biosynthesis, and the product of a PHASE I reaction in purine biosynthesis?

a. IMP
b. AMPS
c. PRPP
d. XMP
e. GMP

Consider the reaction sequence below:

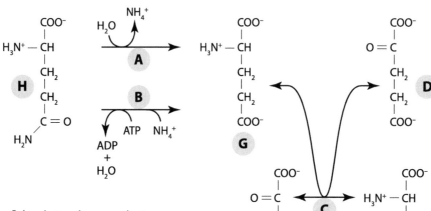

13.
   a. Enzyme **A** is glutamine synthetase.
   b. Enzyme **B** is glutamate dehydrogenase.
   c. Enzyme **C** is aspartate aminotransferase.
   d. Compound **D** is aspartate.
   e. Compound **E** is alanine.

14.
   a. Compound **G** is an α-keto acid.
   b. Compound **H** is formed in neurons, but not in glial cells of the brain.
   c. Compound **F** can be a substrate for an LDH-catalyzed reaction.
   d. Enzyme **A** is SGPT.
   e. Enzyme **C** is found in muscle cells, but not in liver cells.

15.
   a. Compound **D** is an intermediate in the TCA cycle.
   b. Enzyme **B** is glutaminase, an enzyme prevalent in proximal renal tubular epithelial cells.
   c. Compound **E** is a non-protein amino acid.
   d. Compound **H** is an inhibitory neurotransmitter in the brain.
   e. Compound **D** is an amino acid present in many body proteins.

---

16. All of the following are TRUE of the compound to the right, EXCEPT:

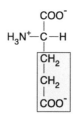

   a. Deaminated to α-ketoglutarate (α-$KG^=$).
   b. Decarboxylated to γ-aminobutyrate (GABA).
   c. A product of an AST-catalyzed reaction.
   d. A substrate for a glutaminase-catalyzed reaction.
   e. A substrate for an ALT-catalyzed reaction.

17. Select the TRUE statement(s) below:

   a. Transaminases are considered to be essential for mammalian life.
   b. Reversible, competitive enzyme inhibitors can be considered structural analogues of the substrate that normally binds with the enzyme.
   c. The liver provides purine and pyrimidine bases for salvage and utilization by tissues incapable of their biosynthesis.
   d. Carbamoyl phosphate can be formed from ATP, CO_2, and the amido group of Gln.
   e. All of the above are TRUE.

18. Which of the following enzyme:inhibitor pairs is INCORRECTLY matched?

   a. Folic acid reductase : Methotrexate
   b. Ribonucleoside diphosphate reductase : Hydroxyurea
   c. Methionine synthase : Methylcobalamin (B_{12})
   d. Folic acid reductase : Aminopterin
   e. Xanthine oxidase : Allopurinol

19. Kinetic data for an enzymatic reaction in the presence (2, 3 & 4) and absence (1) of inhibitors are plotted to the right:

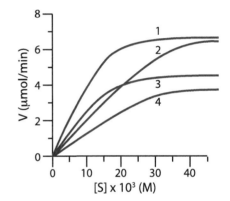

[S] = Substrate Concentration
V = Reaction Velocity

   a. Curve 3 is due to the presence of a competitive inhibitor.
   b. V_{max} in the presence of the non-competitive inhibitor is about 6.5 μmol/min.
   c. The K_m for this enzyme in the absence of inhibitor is less than 1×10^3 molar.
   d. Curve 2 is due to the presence of a non-competitive inhibitor.
   e. All of the above

20. Product negative feedback inhibition of enzyme activity is usually:

   a. Allosteric, and therefore non-competitive.
   b. Mediated through G-stimulatory (G_s) proteins.
   c. Due to a non-reversible covalent modification of the enzyme.
   d. A competitive interaction which reduces V_{max} of the enzyme-catalyzed reaction.
   e. A phenomenon that occurs in amino acid and protein chemistry, but not in nucleotide chemistry.

---

21. Select the FALSE statement regarding the structures below:

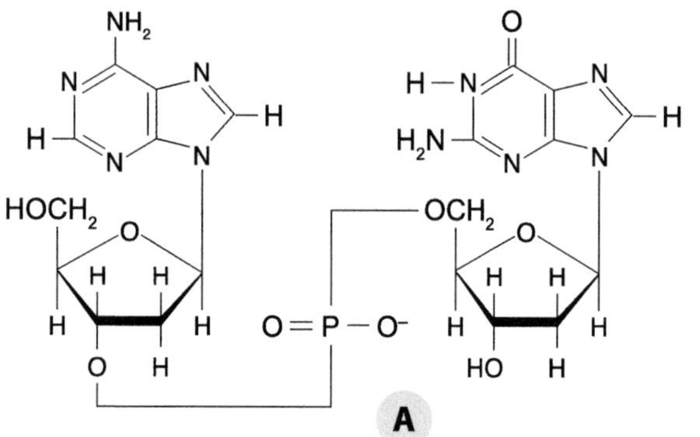

a. **A** would be found in RNA, and **B** in DNA.
b. **E** is an amino acid, and **D** is thymine (a pyrimidine).
c. **B** contains ribose, and **C** contains deoxyribose.
d. **B** is a nucleoside, and **C** contains a phosphate monoester.
e. **A** is a dinucleotide, and **C** contains deoxyribose.

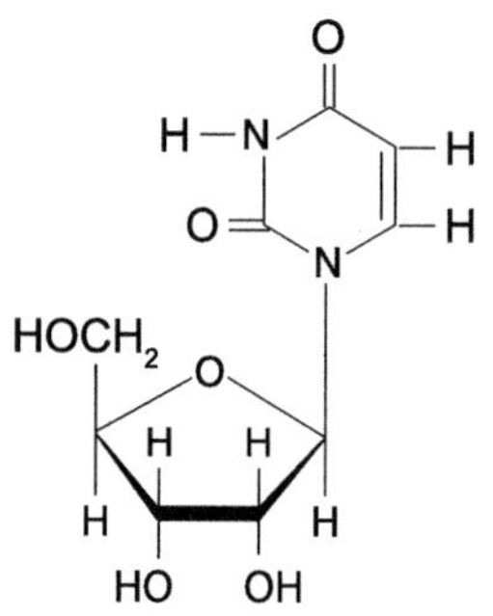

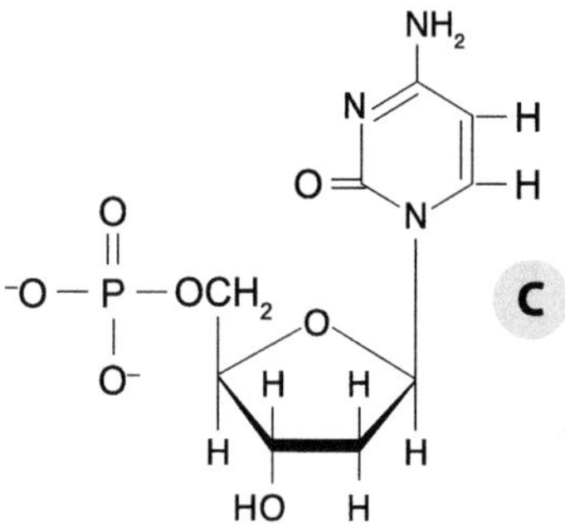

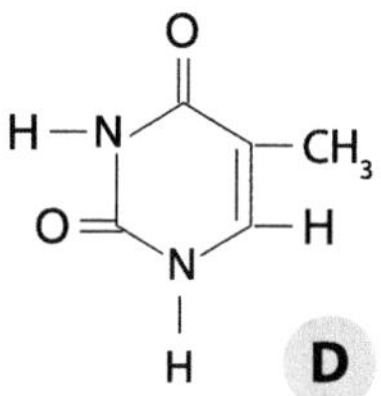

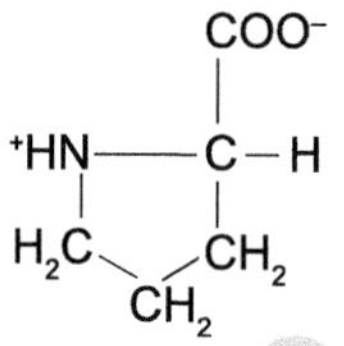

---

22. Select the FALSE statement below:

a. Since extracellular sulfhydryl compounds are quickly oxidized to disulfides, homocysteine in blood and urine exists mainly as homocystine.
b. Isozymes catalyze similar reactions, but differ from each other slightly in chemical structure, and therefore kinetic properties.
c. The plasma BCAA:AAA ratio would be expected to decrease in liver disease.
d. Mammalian cells generally contain many more DNA than protein molecules.
e. Tetrahydrofolate provides a critical link between amino acid and nucleotide biosynthesis by carrying one-carbon fragments from amino acids to nucleotide intermediates.

23. Select the TRUE statement below:

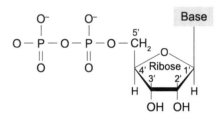

a. The substance to the right is Glu.
b. The most frequently encountered nucleotides in nature
   have their respective phosphate(s) esterified to the
   5′ pentose carbon of the nucleoside.
c. Peptide bonds are created between amino acid side chains containing carboxyl and amide
   groups.
d. Hydrophilic amino acids include the branched-chain and aromatic amino acids.
e. Arginase and glutaminase are found exclusively in renal tubular epithelial cells.

24. Select the FALSE statement below:

a. The substance to the right is a substrate for RDR.
b. There are no "essential nucleotides" that must be
   provided in the diet for the biosynthesis of RNA
   and DNA.
c. Urease-containing bacteria in the rumen and colon are capable of synthesizing urea from
   NH_3.
d. Since the primary structure of a protein
   directs *in vivo* folding, if this structure remains
   intact during denaturation, certain degrees of
   renaturation become possible.
e. The substance to the right is the
   product of a Phe Hydroxylase-catalyzed
   reaction.

25. Which one of the following events occurs in the reaction catalyzed by ribonucleoside diphosphate reductase (RDR):

a. Reduction of ribonucleotides using thiroredoxin and NADPH as reducing agents.
b. Deoxyribonucleoside diphosphates are converted to their corresponding ribonucleotide
   derivatives.
c. Ribonucleoside diphosphates are dephosphorylated.
d. Regeneration of H_4 folate by NADPH.
e. Ribonucleosides are converted to TCA cycle intermediates.

26. Which modified amino acid below is a major constituent of elastin?

a. Isodesmosine
b. Methyllysine
c. Hydroxyproline
d. Phosphothreonine
e. β-Alanine

27. Transaminases or aminotransferases (e.g., AST or ALT) are enzymes that:

a. Are present only in liver tissue.
b. Transfer amino groups from an amino acid to a keto acid.
c. Hydrolyze peptide bonds in protein to yield free amino acids.
d. Transfer methyl groups from one molecule to another.
e. Cleave proteins at the carboxyl end of any arginine/lysine residue.

28. Normally contained within the folic acid molecule are all of the following, EXCEPT:

a. Tryptophan
b. Pteridine
c. Glutamate
d. PABA (para-aminobenzoic acid)
e. Nitrogen

29. Which modified amino acid below is best associated with the chelation of Ca^{++} in feline blood coagulation?

a. Phosphoserine
b. γ-Aminobutyrate
c. Hydroxyproline
d. γ-Carboxyglutamate
e. Methylhistidine

30. All of the following compounds contain nitrogen, EXCEPT:

a. Adenosine
b. Urea
c. Allantoin
d. Fumarate
e. SAM

31. All of the following are amino acid intermediates in the urea cycle, EXCEPT:

a. Citrulline.
b. Arginine.
c. Ornithine.
d. Methionine.
e. Aspartate.

(32-36) Match each phrase below with an appropriate amino acid (options can be used more than once):

        a.   **Asp**
        b.   **β-Ala**
        c.   **Ile**
        d.   **Ser**
        e.   **His**

32. _____ Contributes carbon and nitrogen atoms to the pyrimidine ring.

33. _____ Participates in H_4folate methylation.

34. _____ A component of pantothenic acid (vitamin B_5).

35. _____ Essential branched chain amino acid.

36. _____ Participates in oxyhemoglobin binding as a part of the erythrocytic hemoglobin molecule.

----

37. Sulfonamides are effective antibiotics because they inhibit:

  a. PRPP synthetase.
  b. Bacterial folic acid biosynthesis.
  c. Hepatic uricase.
  d. Bacterial assimilation of folic acid from their external environment.
  e. CPS-2, the rate-limiting enzyme in pyrimidine biosynthesis.

38. Which one of the following is a natural degradative product in cytosine and uracil catabolism?

  a. Propionyl-CoA
  b. Xanthine
  c. H_4Folate
  d. Adenosine Monophosphate
  e. Acetyl-CoA

39. Deficiencies in which of the following are best associated with megaloblastic anemia?

   a. Citrulline and nitric oxide (NO)
   b. Methionine (Met) and urea
   c. Glutamine (Gln) and epinephrine (Epi)
   d. Methylcobalamin (B_{12}) and pteroylglutamate (folic acid)
   e. Histamine and melatonin

40. Which of the following genetic defects would be expected to produce hyperuricemia?

   a. Xanthine oxidase (XO) deficiency
   b. PRPP synthetase (PRS) deficiency
   c. Adenine and guanine phosphoribosyl-transferase (APRAT & GPRT) deficiencies
   d. Glutaminase deficiency
   e. All of the above

41. Which one of the following best describes the substrate concentration at which the rate of a given enzymatic reaction is at one-half maximal velocity?

   a. $K_m \times 2 / 2$
   b. $[S] \times K_m / 4$
   c. $V_{max} / 2$
   d. $K_m / 2$
   e. $1 / V_{max}$

42. Roughly 50% of mammalian cellular solid matter is normally composed of:

   a. Carbohydrate.
   b. Lipid.
   c. Protein.
   d. Nucleic acids.
   e. Inorganic cations and anions.

43. All of the following enzymes catalyze regulatory reactions in purine biosynthesis, EXCEPT:

   a. IMP dehydrogenase.
   b. PRPP glutamyl amidotransferase (PGAT).
   c. Adenylosuccinate synthase.
   d. PRPP synthetase (PRS).
   e. Thymidylate synthase (TS).

Answers

1. d	23. b
2. b	24. c
3. a	25. a
4. c	26. a
5. e	27. b
6. e	28. a
7. d	29. d
8. d	30. d
9. b	31. d
10. c	32. a
11. c	33. d
12. c	34. b
13. e	35. c
14. c	36. e
15. a	37. b
16. d	38. e
17. e	39. d
18. c	40. c
19. c	41. a
20. a	42. c
21. a	43. e
22. d	

Addendum to Section II

The life-roles played by **pyrimidine** and **purine nucleotides** (i.e., nucleoside phosphates) are undeniably important, and go far beyond merely serving as building blocks for RNA and DNA. They participate in metabolic processes as diverse as energy metabolism, protein biosynthesis, signal transduction, and regulation of enzyme activity. When linked to vitamins or vitamin derivatives, nucleotides form a portion of several coenzymes, and as donors and acceptors of phosphoryl groups in metabolism, nucleoside di- and triphosphates (such as ADP and ATP) become principal players in energy transduction processes. Linked to sugars or lipids, nucleotides also constitute key biosynthetic intermediates. The carbohydrate derivatives UDP-glucose and UDP-galactose participate in carbohydrate interconversions, and in the biosynthesis of glycogen. Similarly, nucleoside-lipid derivatives such as CDP-diacylglycerol are intermediates in lipid biosynthesis. The cyclic nucleotides, cAMP and cGMP, serve as second messengers in hormonally regulated events, and GTP and GDP play key roles in the cascade of events that characterize signal transduction pathways. Purine and pyrimidine bases are energetically expensive to synthesize, therefore cells have devised ways to salvage (or recycle) them. When they are finally catabolized, major products of pyrimidine degradation are relatively water-soluble, and are usually reused in other areas of metabolism. Catabolism of purine bases, being far more complex, normally results in **uric acid** formation.

A detailed understanding of nucleotide biochemistry has resulted in the development of various **cancer chemotherapeutic agents** that inhibit nucleic acid biosynthesis. These "antimetabolites" are structural analogs of purine and pyrimidine bases that interfere with specific metabolic reactions. Antifolates (folate analogs) interfere with metabolic steps in which THFA is involved as either a substrate or product, and they have also proven to be effective antitumor agents.

Introduction to Section III

Major pathways of **carbohydrate metabolism** usually begin or end with **glucose**. This section describes the utilization of glucose as a source of energy, its storage in the form of glycogen, release of glucose from glycogen, its association with glycolipids and glycoproteins, and formation of glucose from noncarbohydrate precursors (namely propionate, glycerol, lactate and glucogenic amino acids). Glucose is the only fuel used to any significant extent by certain specialized cells (e.g., erythrocytes), and in the fed state it is a major CNS fuel. Glucose metabolism is defective in some common metabolic diseases (e.g., Cushing's-like syndrome and diabetes mellitus (DM)), with the latter contributing to the development of major medical problems including hypertension, atherosclerosis, microvascular disease, kidney disease, and even blindness.

The manner in which cells protect themselves against **free radicals**, and pathways of **heme metabolism** are also encountered in this section. Topics in erythrocyte, porphyrin and bile pigment biochemistry are closely related. A prevalent clinical condition is **jaundice**, due to plasma bilirubin elevation. This can be due to its overproduction, or failure of its excretion in bile as seen in diseases ranging from hemolytic anemias to hepatobiliary failure.

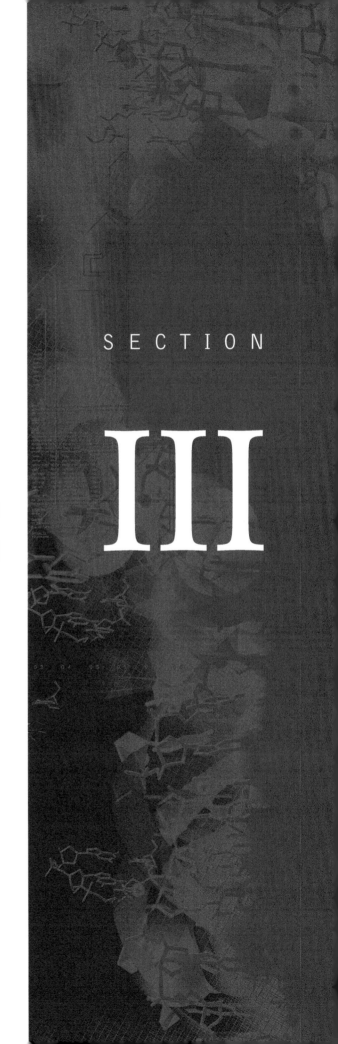

SECTION

III

Carbohydrate and Heme Metabolism

Chapter 18

Carbohydrate Structure

Overview

- Although carbohydrates are the most abundant compounds on earth, they normally constitute only about 1% of the mammalian organism.
- At rest, the central nervous system consumes more carbohydrate than any other organ system.
- Mammalian enzymes involved in carbohydrate metabolism recognize only the D-isomers.
- Fructose is a 6-carbon ketose found in sucrose.
- Maltose is an intermediate in the intestinal digestion of glycogen and starch.
- Monosaccharides such as galactose, glucose, ribose, and fructose usually exist in ring rather than linear forms.
- Reduction of NAD^+ and $NADP^+$ by dehydrogenases changes the character of the nicotinamide ring, with resultant changes in absorption spectra.

Carbohydrates are the most abundant organic compounds on earth. The amount of **cellulose** in plants is estimated at about 10^{11} tons, and the amount of **chitin** in arthropods at about 10^9 tons. However, carbohydrates constitute only about **1%** of the mammalian body.

Complex Carbohydrates

Complex storage carbohydrates include **starch** in plants, and **glycogen** in animals. A normal large dog, for example, possesses about 100 gm of glucose stored as liver glycogen, 250 gm as muscle glycogen, 25 gm as adipose tissue glycogen, and 10 gm as free glucose in the extracellular fluid compartments. At rest, between 160-200 gm of glucose are metabolized per day, with the **brain** being the major consumer ($\approx$120 gm/day). Under normal circumstances, muscle and adipose tissue glycogen are of little use for general metabolism in other organs since these tissues cannot contribute glucose directly to the circulation, and the brain can neither synthesize nor store significant amounts of glycogen. This leaves only 100 gm of liver glycogen, about 3 potatoes in starch equivalents, as the body's main glucose reserve that can be exported to blood. In light of these limited stores of ready-made glucose, it is not surprising that, under conditions other than those immediately after a meal, glucose utilization is reserved mostly for vital functions. In addition, with increasing time after a meal, glucose synthesis from noncarbohydrate sources, predominantly **glucogenic amino acids**, **glycerol**, **pyruvate**, **lactate**, and **propionate** (in herbivores), gains steadily in importance.

Carbohydrates are also present in cell membranes as the polysaccharide portion of **glycoproteins** and **glycolipids**, and they are

Copyright © 2015 Elsevier Inc. All rights reserved.

present in intercellular materials. They are important for the structure of **cartilage** and **bone**, and **plasma proteins**, for example, are mostly glycoproteins. The carbohydrate moieties of glycoproteins are associated with several physiologic actions (see Chapter 20).

Monosaccharides

Monosaccharides are simple sugars that may become basic units of more complex molecules. They are single chains of carbon atoms bearing multiple adjacent hydroxyl groups. The general overall structure is $(HCOH)_n$; hence the name **carbo- (C) hydrate** (H_2O). Those with 3-7 carbon atoms **are the most important for mammalian metabolism.** **Glyceraldehyde** and **dihydroxyacetone** are trioses (3-carbon atoms), **ribose** is a pentose (5-carbon atoms), while **glucose, fructose**, and **galactose** are hexoses (6-carbon atoms) (**Fig. 18-1**). Tetroses are 4-carbon sugars, and heptoses 7-carbon. One of the hydroxyl groups is typically oxidized to either an aldehyde or keto group, creating aldose or ketose sugars. Glyceraldehyde is the smallest **aldose**, and dihydroxyacetone the smallest **ketose**. The convention used to name D- or L-carbohydrates is based on the orientation (right or left) of the hydroxyl group on the highest-numbered asymmetric carbon. Most all carbohydrates involved in mammalian physiology are of the **D-series** (with the exception of L-rhamnose and L-fucose; see Chapter 20). L-carbohydrates are poorly utilized because mammalian enzymes involved in carbohydrate metabolism recognize only the D-isomers. Monosaccharides such as galactose, glucose, ribose, and fructose do not usually exist in linear form. Instead they condense into rings of five or six carbon atoms called **furanose** or **pyranose** rings, respectively (**Fig. 18-1**).

Triose derivatives are formed during catabolism of glucose by **glycolysis** (see Chapters 24-27), while derivatives of trioses, tetroses, pentoses, and the 7-carbon sugar (sedoheptulose), are formed in the breakdown of glucose via the **hexose monophosphate shunt** (see Chapter 28).

Pentoses, NAD+ and NADP+, NADH and NADPH

Pentose polymers such as hemicelluloses, gums, and mucilages constitute part of the simidigestable, "fibrous" material of plants, as exemplified by the **xylose** and **arabinose** polysaccharides. Pentose sugars are also an important part of animal metabolism, as seen with some intermediates of the **hexose monophophate shunt**, and the **uronic acid pathway** (see Chapters 28 & 29).

Ribose is an important constituient of **nucleotides** and **nucleic acids** (see Chapters 13-17). It is also part of the structure of two important coenzymes, **NAD+** and **NADP+** (**Fig. 18-2**). **Nicotinamide adenine dinucleotide (NAD+)** is made up of the base **nicotinamide, ribose**, and **ADP**. An additional **phosphate** may be attached to NAD+, thus creating NADP+. These coenzymes serve as mobile electron acceptors for a number of oxidation-reduction reactions in both the cytosol and in mitochondria (see Chapter 41), and in their reduced forms, **NADH** and **NADPH**, the nicotinamide rings of these dinucleotides have the structures shown in **Fig. 18-3**. Although **NAD+** and **NADP+** are involved in two-electron transfers, only one hydrogen atom (the hydride ion (H^-)) is accepted by the dinucleotide, while the other appears as a proton (H^+), having effectively donated its electron to neutralize the positive charge on the nitrogen atom in the nicotinamide ring. For this reason, the designations **NAD(P)+** and **NAD(P)H** for the oxidized and reduced coenzymes, respectively, are preferred over **NAD(P)** and **NAD(P)H₂**.

Reduction of NAD+ and NADP+ by dehydrogenases changes the character of the nicotinamide ring from **aromatic** to **quinonoid**,

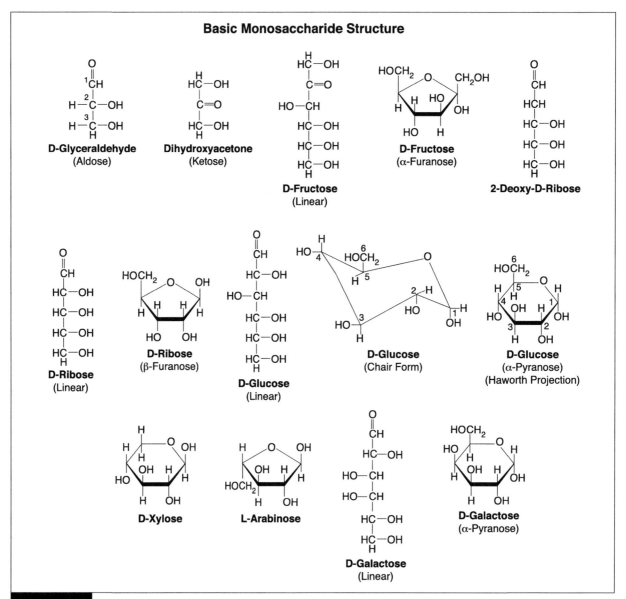

Figure 18-1

with resultant changes in absorption spectra. Since many enzymatic assays rely on differences in spectrophotometric light absorption as sub-strates are converted to products, clinical chemistry labs make substantial use of differences in light absorption between the oxidized (NAD$^+$ and NADP$^+$) and reduced pyridine nucleotides (NADH and NADPH), 260 nm and 340 nm, respectively.

Hexoses

Of the **hexoses**, glucose, fructose, and galactose are physiologically the most important. **Glucose** is a major mammalian fuel, found widely

in fruits and vegetables as a monosaccharide, in disaccharides such as sucrose, maltose, and lactose (**Fig. 18-4**), and in polysaccharides such as glycogen and starch (see Chapter 19). It is converted to other carbohydrates having highly specific functions (e.g., glycogen for storage, ribose in nucleic acids, galactose in lactose of milk, in certain complex lipids, and in combination with protein in glycoproteins). Diseases associated with glucose metabolism include (among others), **diabetes mellitus**, **galactosemia**, and **glycogen storage diseases**.

Fructose is a 6-carbon ketose found in

fruit and honey as a monosaccharide, and in sucrose (a disaccharide of fructose and glucose). Diets high in sucrose can lead to large amounts of fructose entering the hepatic portal vein. Although fructose is not absorbed as rapidly as glucose in the jejunum, it is more rapidly glycolyzed by the liver since it bypasses the step in glycolysis catalyzed by phosphofructokinase, the point where metabolic control is exerted on the rate of glucose catabolism (see Chapter 25). In some animals a significant amount of fructose resulting from the digestion of sucrose is converted to glucose in the intestinal wall prior to passage into the portal circulation. Free fructose is found in seminal plasma, and is secreted in quantity into the fetal circulation of ungulates and whales, where it accumulates in amniotic and allantoic fluids. In all of these situations, free fructose represents a potential source of fuel.

Galactose is a component of lactose (milk sugar), which is synthesized in mammary glands. Galactose is found in some vegetable polysaccharides, and is convertible to glucose in the body (mainly in the liver where it is almost entirely removed from the portal circulation following absorption from the intestine). Galactose also forms part of the polysaccharide chain of many glycoproteins.

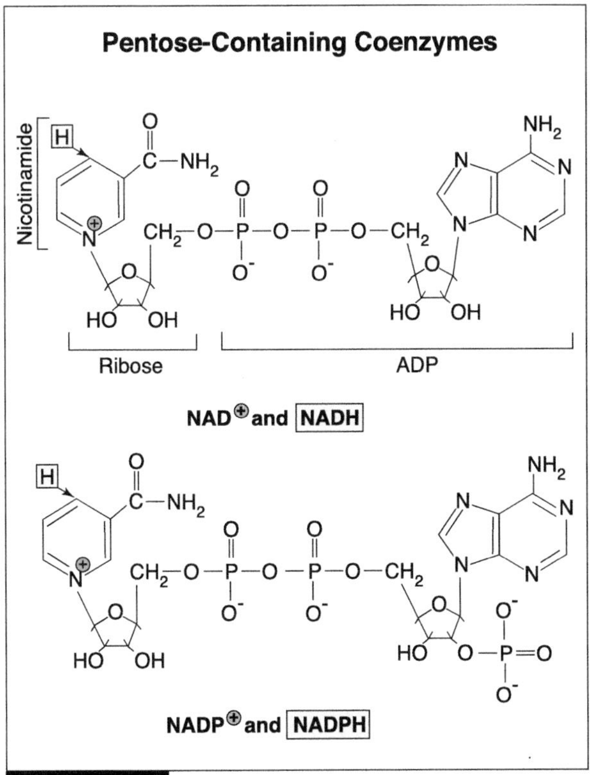

Disaccharides and Trisaccharides

Disaccharides consist of two monosaccharides joined by a glycosidic bond. Structures for the most common disaccharides are shown in **Fig. 18-4**.

Maltose (or malt sugar) is an intermediate in the intestinal digestion (i.e., hydrolysis) of glycogen and starch, and is found in germinating grains (and other plants and vegetables). It consists of two molecules of glucose in an α-(1,4) glycosidic linkage. **Trehalose**, which contains two molecules of glucose linked together somewhat differently from maltose, is a major carbohydrate found in the hemolymph of many insects. It is also found in young mushrooms, where it accounts for about 1.5% of their weight. Cellobiose, the repeating disaccharide unit of cellulose, has β-(1,4) glycosidic linkages which are broken by bacterial cellulases, but not by mammalian constitutive digestive enzymes.

Lactose is found in milk, but otherwise does not occur in nature. It consists of galactose and glucose in a β-(1,4) glycosidic linkage.

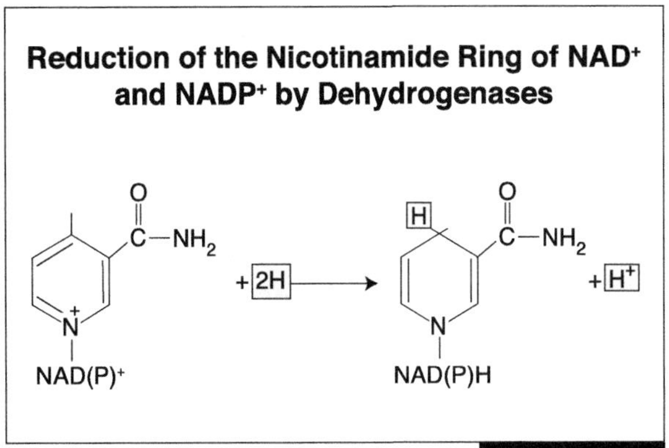

Figure 18-2

Figure 18-3

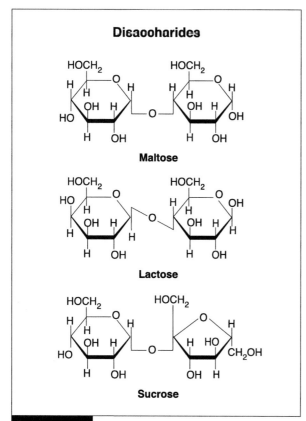

Figure 18-4

Sucrose, or cane sugar, consists of glucose and fructose linked in an α-(1,2) glycosidic bond. It is abundant in the plant world, and is familiar as table sugar. Sucrose and maltose are readily hydrolyzed by disaccharidases found in the brush border of the small intestine (see Chapter 38). Hydrolysis of sucrose to glucose and fructose is sometimes called inversion, since it is accompanied by a net change in optical rotation from *dextro* to *levo* as the equimolar mixture of glucose and fructose is formed on the mucosal surface. Therefore, the intestinal brush border enzyme that hydrolyzes sucrose (i.e., **sucrase**), is sometimes called **invertase** (see Chapter 38).

A number of **trisaccharides** also occur free in nature, and are consumed by animals. **Raffinose** contains glucose, fructose, and galactose held together by both α- and β-glycosidic bonds. This trisaccharide is found in abundance in sugar beets, and several other higher plants. **Melezitose** contains two mole-

cules of glucose and one of fructose, and is found in the sap of some coniferous trees.

OBJECTIVES

- Identify the organ that normally consumes the most glucose at rest.

- Recognize why L-carbohydrates are poorly utilized by mammals.

- Examine structural differences between NAD⁺ and NADPH.

- Understand how clinical chemistry laboratories can distinguish between the oxidized and reduced forms of NAD⁺.

- Recognize the linear and/or ring structures of the most common mammalian monosaccharides (Fig. 18-1).

- Contrast general metabolic characteristics of glucose, fructose and galactose.

- Recognize how brush border oligosaccharidases of the small intestine act on luminal disaccharides (see Chapter 38).

- Know the structures of the common disaccharides (maltose, lactose and sucrose).

- Explain why disaccharides do not appear in the hepatic portal circulation following a meal, yet dipeptides do (see Chapters 7 & 38).

- Identify and discuss the source of the most common trisaccharides present in nature.

- Explain why glycogen present in muscle and adipose tissue is of little use for general metabolism in other organs and tissues of the body.

- Identify the noncarbohydrates used for glucose synthesis in the liver, and discuss why gluconeogenesis is important, particularly to herbivores and carnivores (see Chapter 37).

Questions

1. **Which of the following can contribute glucose directly to the circulation following glycogen breakdown (i.e., glycogenolysis)?**
   a. Muscle
   b. Liver
   c. Adipose tissue
   d. Brain
   e. All of the above

2. The repeating disaccharide unit of cellulose, which contains β-(1,4) glycosidic linkages:
   a. Cannot be hydrolyzed by mammalian constitutive enzymes.
   b. Is referred to as trehalose.
   c. Cannot be hydrolyzed by bacteria.
   d. Is also found in glycogen and starch.
   e. All of the above

3. Which organ of the body generally consumes the most glucose at rest?
   a. Heart
   b. Liver
   c. Brain
   d. Kidney
   e. Lung

4. Which one of the following monosaccharides is a triose and an aldose?
   a. Ribose
   b. Galactose
   c. Fructose
   d. Glyceraldehyde
   e. Sedoheptulose

5. Which one of the following monosaccharides is found in NAD⁺?
   a. Dihydroxyacetone
   b. Fructose
   c. Ribose
   d. Glucose
   e. Galactose

6. Which one of the following monosaccharides, which is a part of the lactose molecule, is synthesized from glucose in the mammary gland?
   a. Galactose
   b. Fructose
   c. Maltose
   d. Ribose
   e. Dihydroxyacetone

7. When NAD⁺ is reduced:
   a. The nicotinamide ring changes from a quinonoid to an aromatic structure.
   b. It continues to absorb UV light at the same optimal wavelength (i.e., 260 nm).
   c. The nitrogen atom in the nicotinamide ring I loses its charge.
   d. The ribose moiety becomes protonated.
   e. A nitrogen atom in adenosine becomes charged.

8. Which one of the following is an intermediate in the intestinal digestion of starch?
   a. Maltose
   b. Galactose
   c. Sucrose
   d. Lactose
   e. Sedoheptulose

9. Approximately what percentage of the mammalian organism is normally comprised of carbohydrate?
   a. 1%
   b. 10%
   c. 25%
   d. 55%
   e. 90%

10. Conversion of galactose to glucose occurs to the greatest extent in which of the following?
   a. The small intestine
   b. The liver
   c. Skeletal muscle
   d. The brain
   e. The mammary gland

11. The 7-carbon sugar, sedoheptulose, is formed in the:
   a. Tricarboxylic acid cycle.
   b. Uronic acid pathway.
   c. Malate shuttle.
   d. Urea cycle.
   e. Hexose monophosphate shunt.

12. Diets high in sucrose can lead to large amounts of which monosaccharide entering the hepatic portal vein?
   a. Galactose
   b. Xylose
   c. Ribose
   d. Arabinose
   e. Fructose

13. Invertase is:
   a. Also known as galactase.
   b. A disaccharide also known as sucrose.
   c. A liver enzyme known to hydrolyze lactose.
   d. An intestinal brush border disaccharidase.
   e. An intestinal enzyme known to convert carbohydrates from their ring to linear forms.

ANSWERS
13. d
12. e
11. e
10. b
9. a
8. a
7. c
6. a
5. c
4. d
3. c
2. a
1. b

Chapter 19

Polysaccharides and Carbohydrate Derivatives

Overview

- Important glucose polymers to animals are glycogen, starch, and cellulose.
- The β-1,4 glycosidic linkages of cellulose prevent it from being digested by constitutive enzymes.
- Glucuronic acid, which is derived from UDP-glucose in the liver, is used to conjugate bilirubin, steroid hormones, and several xenobiotics.
- Vitamin C is derived from carbohydrate.
- Glycosaminoglycans consist of chains of complex carbohydrates, and are found in the extracellular matrix.
- Proteoglycans are glycosaminoglycans attached to protein.
- Inulin, a fructose polymer of plants, can be used to determine the glomerular filtration rate.

Polysaccharides

These macromolecules occur in nature with molecular weights exceeding one million. Their biologic activity is determined largely by the sequence and three-dimensional structure in which their various monosaccharides are arranged. Glucose is the most prevalent monosaccharide unit in polysaccharides, but polysaccharides of mannose, fructose, galactose, xylose, and arabinose are also found in nature. Some polysaccharides are present in **glycoproteins** and in **glycolipids** (see Chapters 20 and 59).

Three important polysaccharides to animals that consist entirely of glucose are **glycogen**, **starch**, and **cellulose**. **Glycosaminoglycans** are important mucopolysaccharides, chitin is an important structural polysaccharide of invertebrates, and **inulin**, a fructose polymer of plants, is sometimes used to evaluate kidney function.

Glycogen, the form in which animal tissues store glucose, has a basic structure that consists of numerous **α-1,4** linked D-glucose molecules, and is sometimes referred to as "**animal starch**". Once in every 11 to 18 linkages, a branching point occurs via an **α-1,6** linkage, which yields a tree-like structure (**Fig. 19-1**).

Glycogen can account for up to 6-8% of the weight of the canine liver; however, this amount may decrease to less than 1% between meals (e.g., overnight; see Chapters 73 and 74). The glycogen content of muscle is normally around 1% by weight, a value that can be similarly decreased following prolonged exercise (see Chapter 79). Then, if a normal diet is consumed, glycogen contents of both liver and muscle can be replenished.

Glycogen is stored in both liver and muscle in granules of varying size. The enzymes required for both the synthesis and degradation of

Copyright © 2015 Elsevier Inc. All rights reserved.

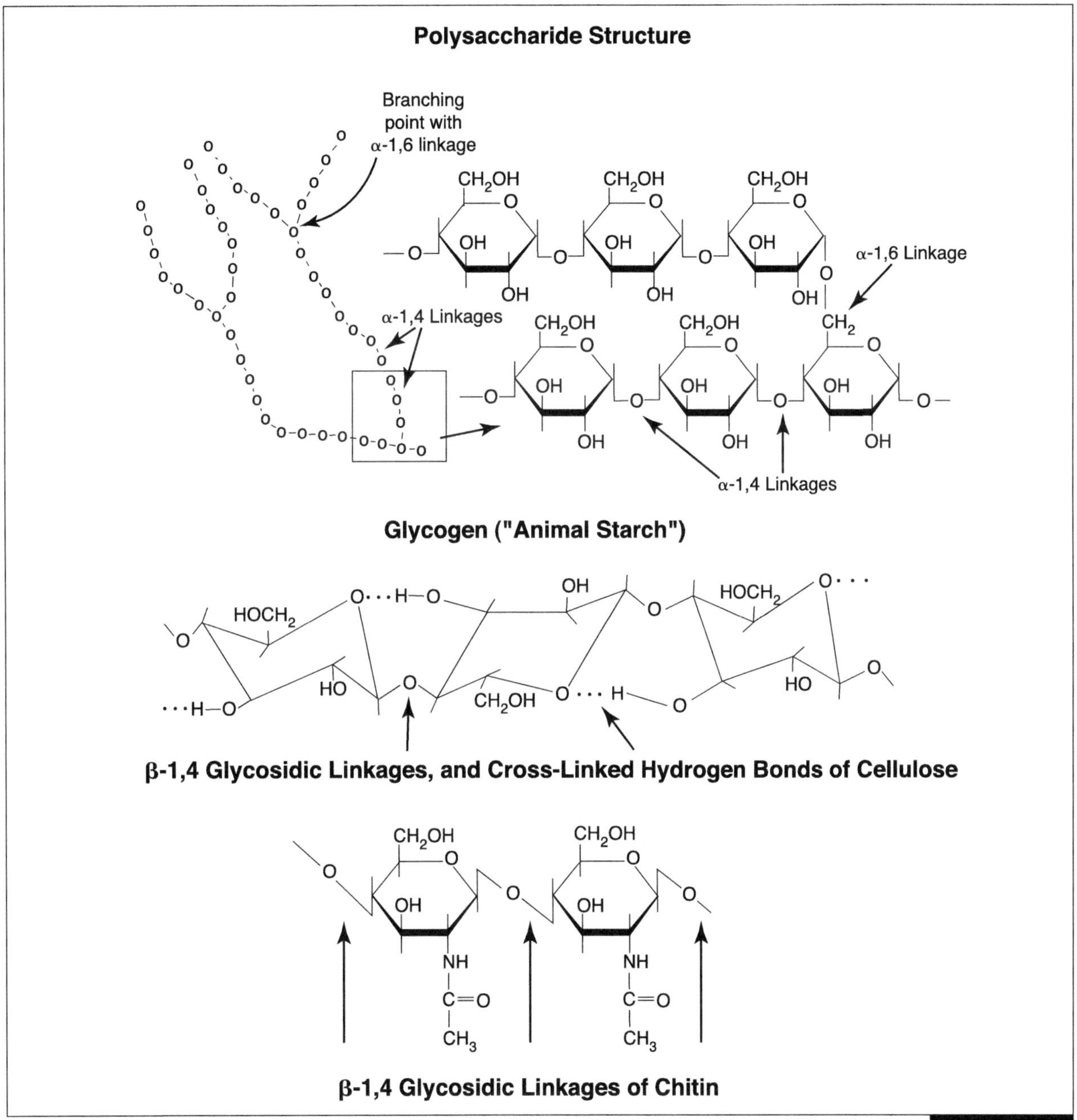

Polysaccharide Structure

Branching point with α-1,6 linkage

α-1,6 Linkage

α-1,4 Linkages

α-1,4 Linkages

Glycogen ("Animal Starch")

β-1,4 Glycosidic Linkages, and Cross-Linked Hydrogen Bonds of Cellulose

β-1,4 Glycosidic Linkages of Chitin

Figure 19-1

glycogen, as well as several factors required for control of these processes, are also found in glycogen granules (see Chapter 23).

<u>Starch</u>, the glucose store of plants, is an important food source of carbohydrate, and is found in cereals, potatoes, legumes, and other vegetables. Starch contains 15 to 20% **amylose**, and 80-85% **amylopectin**. While both glycogen and starch consist totally of D-glucose, amylose is nonbranched, and amylopectin contains about one α-1,6 branch per 30 α-1,4 linkages. **Dextrins** are substances formed in the course of hydrolytic breakdown of starch and glycogen in the digestive tract, by the action of **salivary** and **pancreatic amylase** (see Chapter 38). **Limit dextrins** are the first formed products as hydrolysis reaches a certain degree of branching.

Storage polysaccharides such as starch and glycogen are usually deposited in the form of large granules in the cytoplasm of cells. These granules can sometimes be isolated from cell extracts through differential centrifugation. In times of glucose surplus, glucose units are stored by undergoing enzymatic linkage to the ends of starch or glycogen chains; in times of metabolic need, these glucose units are released enzymatically for use as fuel.

Inulin is a fructose polymer found in tubers and roots of dahlias, artichokes, and dandelions. Since it is hydrolyzable to individual fructose units, it is properly referred to as a **fructosan**. This polysaccharide, unlike plant starch, is easily soluble in warm water, and has been used in clinical evaluation of kidney function. Inulin is freely filtered by the kidneys, it is not reabsorbed or secreted by the nephron, it is not metabolized following injection into blood, it is nontoxic, and therefore its clearance from the circulation closely approximates the **glomerular filtration rate (GFR)**.

Cellulose is insoluble in water, and consists of long, non-branched β-1,4 linked glucose units strengthened by cross-linked hydrogen bonds (see **Fig. 19-1**). Due to β-linkages, cellulose is **nondigestible** by enzymes of mammalian origin (i.e., constitutive enzymes). However, in the digestive tracts of herbivores, cellulase-containing microbes attack the β-linkages, thus making the glucose units of cellulose available as major metabolic entities for the microbes. In return, the microbes make **volatile fatty acids** available for the host (see Chapter 54). This process also takes place (to a limited extent) by microbes in the colons of omnivores and carnivores. Fungi and protozoa secrete cellulases, and the digestion of wood by termites depends on protozoa in their digestive tract. Cellulose, containing over 50% of the carbon in plant material, is the **most abundant organic compound on earth**.

Chitin is a key structural polysaccharide of invertebrates found, for example, in the exoskeletons of crustaceans and insects. It consists of **N-acetyl-D-glucosamine** units joined by the same (undigestible) **β-1,4** glyco-

sidic linkages found in cellulose. Thus, chitin is similar to cellulose except that the C-2 moiety is an acetylated amino group rather than a hydroxyl group (see **Fig. 19-1**).

Carbohydrate Derivatives

Several classes of compounds derived from carbohydrates are metabolically significant. These include **sugar alcohols**, where the carbonyl group is reduced to a hydroxyl group such as glycerol (**Fig. 19-2**). **Glycerol** is a sugar alcohol and a gluconeogenic precursor in the liver (see Chapter 37). It also serves as the backbone of many lipids (to include tri-, di-, and monoglycerides, as well as phospholipids; see Chapter 59). Another important sugar alcohol is **inositol** (or myoinositol), which is formed from glucose 6-phosphate by many different organisms. Inositol is a six-membered polyhydroxy alcohol, that exists in animals primarily as the common triphosphoester isomer, **inositol triphosphate (IP$_3$)**, and in plants as a hexophosphoester. Most IP$_3$ in animals is generated from membrane-bound phospholipid (phosphatidyl inositol), and participates in the intracellular Ca^{++} second messenger system (see Chapter 58).

Sugar acids are another important group of compounds involved with both structural polysaccharides (e.g., hyaluronate), and with clearing metabolites from the body. **Uronic acid anions (Fig. 19-2)**, such as **glucuronate**, result from the oxidation of the non-reducing terminal carbon of UDP-glucose from CH$_2$OH to COO$^-$. Glucuronates are **conjugated** (i.e., linked covalently) to certain metabolites (such as **bilirubin** and **steroids**), and **drugs** by **liver enzymes** to form hepatic glucuronides (**Fig. 19-3**; see Chapter 29). These conjugates have increased water solubility for excretion in either **bile** or **urine**. Glucuronides are also used in the formation of certain glycoproteins, heparin, and vitamin C (see Chapters 20 and 39).

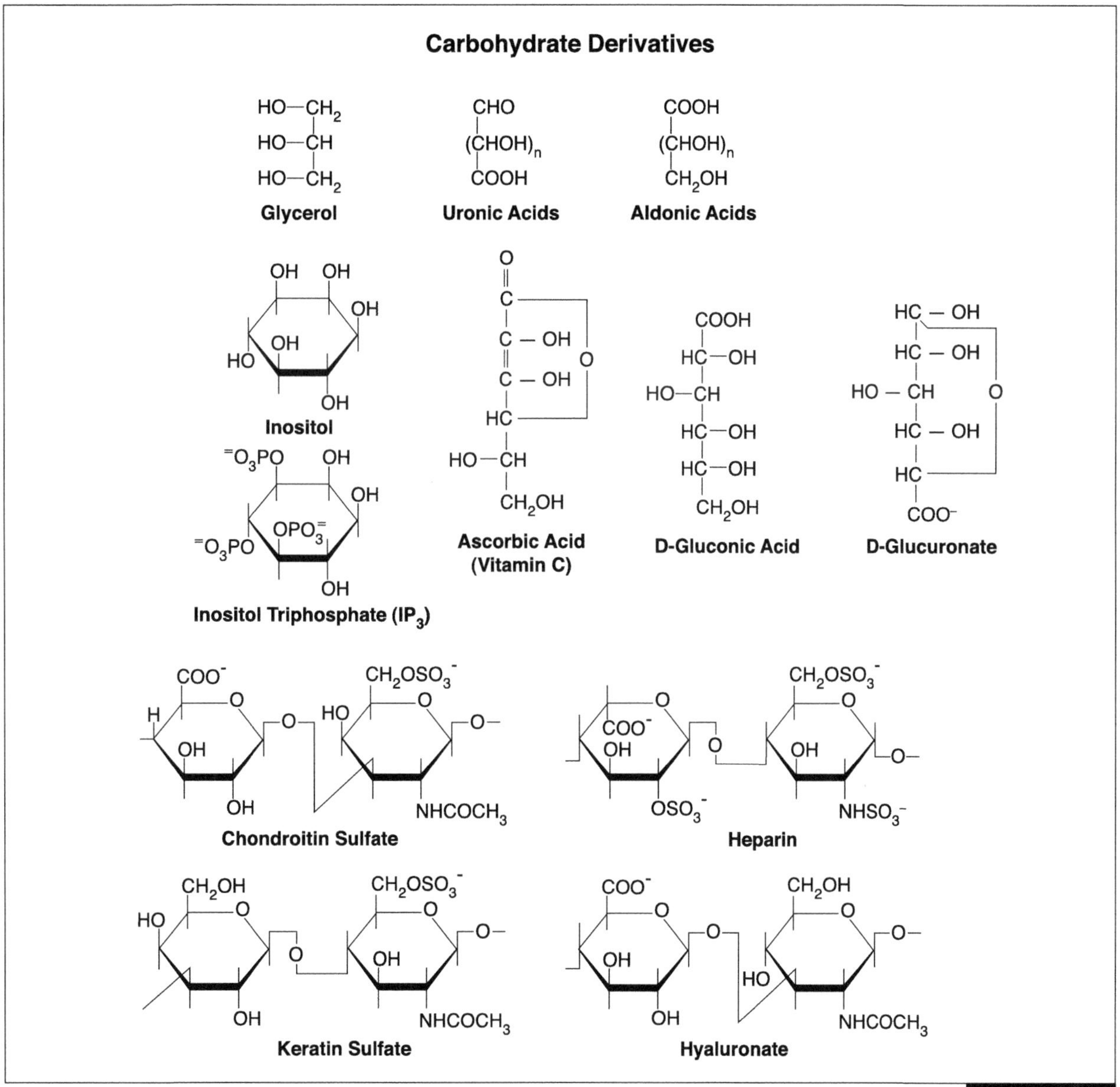

Carbohydrate Derivatives

Figure 19-2

Aldonic acids result from the oxidation of the aldehyde (-CHO) of aldoses, as in the formation of **D-gluconic acid** (see **Fig. 19-2**). The phosphate ester of this acid is an important intermediate in the metabolism of glucose via the hexose monophosphate shunt (HMS) (see Chapter 28). **Ascorbic acid (vitamin C)** is a lactone of a hexonic acid, and an essential nutrient for primates, guinea pigs, flying mammals, fish, and songbirds (see Chapter 39). This vitamin is an **antioxidant** participating in oxidation-reduction reactions, and in the addition of **hydroxyl groups** to proline and lysine in the formation of **collagen** (see Chapter 3). It also assists in the hepatic formation of **bile acids** from cholesterol (see Chapter 62), and in **dopamine** hydroxylation during **catecholamine** biosynthesis.

Deoxysugars lack one hydroxyl group, with perhaps the most important being **2-deoxyribose**, the pentose moiety of **DNA** (see Chapter 13). **Aminosugars** are important components of glycoproteins, membranes, and bacterial cell walls (see Chapter 20).

Glycosaminoglycans (mucopolysaccharides) consist of chains of complex carbohydrates

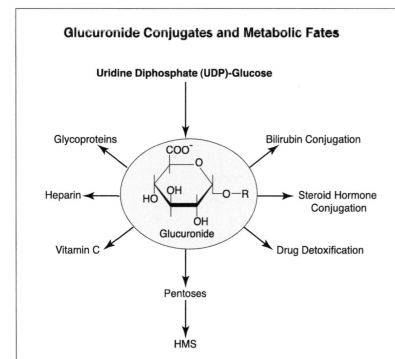

Glucuronide Conjugates and Metabolic Fates

Uridine Diphosphate (UDP)-Glucose

Glycoproteins

Bilirubin Conjugation

Heparin

Steroid Hormone Conjugation

Vitamin C

Drug Detoxification

Pentoses

HMS

COO^-
O
OH
HO
OH
O—R
Glucuronide

Glucuronate, produced from UDP-glucose, is used to form hepatic glucuronide conjugates (e.g., bilirubin, steroid and various drug conjugates), and for the formation of glycoproteins, heparin and vitamin C. Hepatic conjugation increases molecular weight and water solubility for excretion in either bile or urine.

Figure 19-3

characterized by their content of amino sugars and uronic acids. When these chains are attached to a protein, the compound is known as a **proteoglycan**. The amount of carbohydrate in a proteoglycan is usually much greater than that found in a glycoprotein, and may comprise up to 95% of its weight. As ground substance, proteoglycans are associated with structural elements of tissues such as bone, elastin, and collagen. Their property of holding large quantities of water and occupying space, thus cushioning or lubricating other structures, is assisted by the large number of negatively charged groups on their molecules, which, by repulsion, keep the carbohydrate chains apart. Examples are hyaluronate, chondroitin sulfate, keratan sulfate, and heparin (see Fig. 19-2). Keratan sulfate, hyaluronate, and chondroitin sulfate are found in the extracellular matrix of cartilage, where they assist in helping it to cushion compressive forces. Hyaluronate is also found in synovial fluid, the vitreous body of the eye, and in bacteria.

Heparin is an anticoagulant found in the granules of mast cells, which are found most often in the skin, but also in the liver, lungs, and stomach. Mast cell tumors are common in dogs and cats, and affected animals develop bleeding disorders and secondary infections. Since mast cell granules also contain **histamine**, these tumors are also associated with gastric ulceration secondary to increased HCl secretion. Heparin can also bind to lipoprotein lipase (LPL; also called clearing-factor lipase) present in capillary walls, causing release of this enzyme into the circulation (see Chapter 64). The protein molecule of the heparin proteoglycan is unique, consisting exclusively of serine and glycine residues.

Heparan sulfate, which lacks the anticoagulant properties of **heparin**, has fewer N- and O-sulfated groups yet more N-acetyl groups than heparin, and is found on many outer cell surfaces as a proteoglycan. Heparan sulfate is negatively charged, and participates in cell growth and cell-cell communication. It is also found in the basement membrane of the glomerulus, where it plays a major role in determining the charge selectiveness of the renal glomerular filtration barrier.

OBJECTIVES

- List the important polysaccharides of animals that consist entirely of glucose.

- Identify primary structural differences between glycogen, amylose, and amylopectin.

- Explain why glycogen is referred to as animal starch.

- Discuss the reasoning for using inulin clearance as an approximation of the glomerular filtration rate.

- Explain why cellulose and chitin are nondigestible by constitutive enzymes of the mammalian digestive tract.

- Identify some physiologically significant sugar alcohols and sugar acids, and discuss their modes of action.

- Discuss the physiologic significance of D-gluconic acid.

- Describe differences between proteoglycans and glycosaminoglycans, and provide examples of each.

- Distinguish between heparin and heparan.

QUESTIONS

1. **Glycogen is structurally most similar to which of the following?**
   a. Cellulose
   b. Inulin
   c. Chitin
   d. Starch
   e. Glycosaminoglycans

2. **The most abundant organic compound on earth is:**
   a. Starch.
   b. Glycogen.
   c. Cellulose.
   d. Fat.
   e. Protein.

3. **Which one of the following glycosidic link ages in cellulose cannot be attacked by constitutive enzymes?**
   a. α-1,4
   b. β-1,4
   c. α-1,6
   d. β-1,6
   e. γ-1,5

4. **Which one of the following is a sugar alcohol?**
   a. Glucuronic acid
   b. Glycerol
   c. Heparin
   d. Chitin
   e. Vitamin C

5. **Glycosaminoglycans:**
   a. Are found in collagen, but not bone.
   b. Consist of chains of complex carbohydrates.
   c. When attached to protein, are called lipoproteins.
   d. Typically sulfate groups found in most poly saccharides.
   e. Are usually positively charged.

6. **All of the following are true of heparin, EXCEPT:**
   a. It is synthesized as a nonsulfated proteoglycan, which is then deacetylated and sulfated.
   b. It is an anticoagulant found in mast cells.
   c. It can also be correctly referred to as heparan.
   d. It is a glycosaminoglycan.
   e. The protein molecule of the heparin proteoglycan consists exclusively of serine and glycine residues.

7. **Which one of the following is a fructose polymer?**
   a. Chitin
   b. Chondroitin Sulfate
   c. Hyaluronate
   d. Amylose
   e. Inulin

8. **Which sugar alcohol, below, is integral to the calcium messenger system?**
   a. Inositol
   b. Glycerol
   c. Sorbitol
   d. Mannitol
   e. Glucitol

9. **Glucuronide is used for all the following, EXCEPT:**
   a. Bilirubin conjugation.
   b. Vitamin C synthesis.
   c. Glycoprotein formation.
   d. Drug detoxification.
   e. Steroid formation.

ANSWERS

9. e
8. a
7. e
6. c
5. b
4. b
3. b
2. c
1. d

Chapter 20

Glycoproteins and Glycolipids

Overview

- Glycoproteins are found in most organisms.
- Only 8 monosaccharides are associated with the structure of glycoproteins.
- The oligosaccharide chains of glycoproteins are chemical markers used to tag proteins destined to be used outside the cell.
- Oligosaccharide chains are attached to protein through either N- or O-glycosidic bonds.
- Lectins, mucins, and several polypeptide hormones are glycoproteins.
- Glycolipids are carbohydrate-containing lipids with a sphingosine backbone.
- Certain diseases of animals are characterized by abnormal quantities of glycolipids in the central nervous system.
- Carbohydrate residues of membrane-bound glycolipids and glycoproteins are located on the exterior surface of cells.

Glycoproteins

These macromolecules are found in most organisms, from bacteria and viruses to mammals. They are present, for example, in lysosomes, blood, mucus, the extracellular matrix, cell membranes, the nucleus and cytoplasm, and they are associated with several important physiologic processes (**Table 20-1**). Many enzymes, structural proteins, polypeptide hormones, immunoglobulins, antigens, receptors, transport proteins, and blood group substances have covalently linked **oligosaccharide (glycan) chains**, thus establishing them as **glycoproteins**. In many cases, precise roles for their carbohydrate moieties are not clearly established; however, some functions attributed to these oligosaccharides are listed in **Table 20-2**. For example, a major problem in **cancer** is **metastasis**, where problematic cells leave their tissue of origin (e.g., the liver), and migrate through blood to some distant

site (e.g., the lung or brain), growing there in an unregulated fashion. Reports indicate that alterations in the oligosaccharide chains of glycoproteins on the surface of metastatic cancer cells may contribute to this phenomenon. Another proposal regarding the significance of carbohydrate attachment to protein is that the oligosaccharide chains become the identifying chemical markers used to tag proteins that are destined to be utilized **outside** of the cell, or in the membranous network of the cell. Thus, most proteins that are retained for use in the cytoplasm of the cell are **nonglycosylated**.

Although about two-hundred monosaccharides are found in nature, only **eight** are commonly found in the oligosaccharide chains of glycoproteins (**Fig. 20-1**). A widely held belief is that these chains encode considerable biologic information, depending upon their sequences. The carbohydrate content of glycoproteins can

Copyright © 2015 Elsevier Inc. All rights reserved.

Table 20-1
Glycoprotein Functions

Function	Example
• Structural molecules	Collagens
• Lubricate and protect	Mucin
• Transport	Transferrin, Ceruloplasmin
• Immunologic	Immunoglobulins, histocompatibility antigens
• Hormones	TSH, FSH, LH, GH, PRL, ACTH
• Enzymes	Alkaline phosphatase
• Cell attachment-recognition sites	Various proteins involved in cell-cell, virus-cell, bacterium-cell, & hormone-cell interactions.
• Antifreeze	Certain plasma proteins of cold water fish.

Table 20-2
Some Proposed Roles for the Oligosaccharide Chains of Glycoproteins

- Modulate physicochemical properties (e.g., solubility, viscosity).
- Protect against proteolysis (from inside and outside of the cell).
- Affect proteolytic processing of precursor proteins to smaller products.
- Regulate biologic activity (e.g., hormones).
- Affect transport activities of cells (e.g., membrane insertion, intracellular migration, sorting, & secretory activity).
- Affect binding of sperm to the ovulated egg (and subsequent dissolution of the zona pellucida).
- Affect embryonic development and differentiation.
- Target macromolecules to specific destinations.
- Affect homing of lymphocytes to the endothelial lining of lymph nodes.
- Mediate intercellular interactions.
- Blood group determinants.
- Affect sites of metastases selected by some cancer cells.

range from 1% to over 80% by weight, however, less than 100 monosaccharide residues are generally found associated with any one protein residue. This seemingly limited number of glycosidic bonds does not necessarily limit diversity, for these eight monosaccharides may be linked to each other to form several thousand different oligosaccharide chains.

Oligosaccharide chains are attached to protein through either N-glycosidic or O-glycosidic bonds (**Fig. 20-2**). **N-Glycosidic bonds** are formed between the amide group of an asparagine (Asn) chain, and the anomeric carbon of N-acetylglucosamine (GlcNAc). The enzyme catalyzing formation of this bond recognizes Asn residues in the protein having the sequence **Asn-X-Thr** (threonine), or **Asn-X-Ser** (serine), where **X** represents any amino acid. Although some Asn residues with this sequence do not become glycosylated, this sequence is necessary for defining glycosylation sites. **Fucose** (Fuc), which is sometimes found

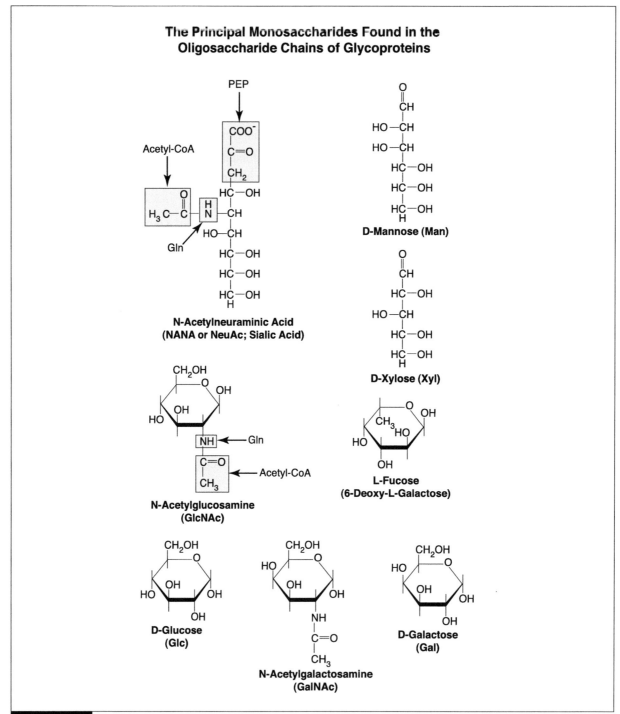

The Principal Monosaccharides Found in the
Oligosaccharide Chains of Glycoproteins

N-Acetylneuraminic Acid
(NANA or NeuAc; Sialic Acid)

D-Mannose (Man)

D-Xylose (Xyl)

N-Acetylglucosamine
(GlcNAc)

L-Fucose
(6-Deoxy-L-Galactose)

D-Glucose
(Glc)

N-Acetylgalactosamine
(GalNAc)

D-Galactose
(Gal)

Figure 20-1

attached to the GlcNAc residues of glycoproteins, is one of the few monosaccharides of the L-configuration found in plants and animals.

O-Glycosidic bonds are formed between the side chain hydroxyl group of either Ser or Thr, and the anomeric carbon of either GalNAc or xylose (Xyl) (**Fig. 20-2**). There is a special type of O-glycosidic linkage found in **collagen**, where Gal is attached to the hydroxyl group of **hydroxylysine** (see Chapter 3). The carbohydrate attached to collagen is always the Gal-Glc disaccharide.

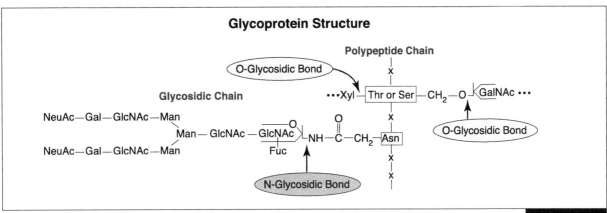

Figure 20-2

Lectins, which were first discovered in plants and microbes, are carbohydrate-binding proteins that agglutinate cells or precipitate glycoconjugates, and several are known to be glycoproteins. Some enzymes, toxins, and transport proteins can be classified as lectins if they have multiple binding sites. **Animal lectins** (or **selectins** such as the C-, S-, P-, and I-types), participate in cell-cell interactions that occur in pathologic conditions such as **inflammation** and **cancer metastasis**.

Mucins (or **mucoproteins**) are glycoproteins that are distributed on the surfaces of epithelial cells of the **respiratory**, **gastrointestinal**, and **reproductive tracts**, and help to lubricate and protect these surfaces. Those with 4% or more carbohydrate are sometimes termed mucoproteins because they exhibit a high viscosity.

Ovulated eggs are surrounded by an extracellular coat called the **zona pellucida (ZP)**. **O-linked oligosaccharides** containing a terminal Gal unit are attached to ZP3, a glycoprotein in this coat, and are recognized by a receptor on the **sperm** surface. Binding triggers release of proteases and hyaluronidases from the sperm, thus dissolving the ZP and allowing fertilization. Another example of the importance of the carbohydrate moiety of glycoproteins can be found in the interaction of **lymphocytes** and **lymph nodes**. Circulating lymphocytes tend to migrate toward their sites of origin. This return, or "homing", is known to be mediated by specific interactions between receptors on the surface of lymphocytes, and carbohydrates on the endothelial lining of lymph nodes. When sialic acid residues from the endothelial surface are enzymatically removed, adhesion of lymphocytes is prevented.

Several **diseases** involving abnormalities in the synthesis and/or degradation of glycoproteins are recognized. As previously indicated, cell surface glycoproteins of metastatic cancer cells can exhibit differences in the structures of their oligosaccharide chains. In one **lysosomal disease**, for example, lysosomal membrane proteins are not properly targeted to the lysosome because of an absence of their Man-6-P recognition signal owing to a genetically determined deficiency of GlcNAc phosphotransferase. Other relatively rare diseases are due to genetic deficiencies in the activities of specific glycoprotein lysosomal hydrolases. These include α- **and β-mannosidosis, fucosidosis, sialidosis, aspartylglycosamminuria**, and **Schindler's disease**. Glycoprotein abnormalities are also thought to be involved in several other disease processes, including **influenza**, **AIDS**, and **rheumatoid arthritis**, (see Chapter 59). Additionally, the evidence that **retinoic acid** is involved in **glycoprotein synthesis** is compelling, since **vitamin A deficiency** results in the accumulation of abnormally low molecular weight intermediates in glycoprotein formation (see Chapter 44).

Glycolipids

Glycolipids are carbohydrate-containing lipids with a **sphingosine** backbone (see Chapter 59). Also attached to the sphingosine moiety is a long-chain fatty acid unit (**Fig. 20-3**). Glycosphingolipids typically account for about 5-10% of lipid in plasma membranes, and are important in intercellular communication.

Cerebrosides are perhaps the simplest glycolipids, in which a singular carbohydrate unit, either glucose (Glc) or galactose (Gal), appears. Without the carbohydrate unit, cerebrosides become **ceramides**. More complex glycolipids, such as **gangliosides**, are found to contain up to seven carbohydrate residues in a branched chain, and include **NeuAc** and **GalNAc** in addition to **Glc** and **Gal** (see Chapter 59).

Studies have shown that the **carbohydrate residues** of membrane **glycolipids** and **glycoproteins** are normally located on the exterior surface of cell membranes (**Fig. 20-4**). This occurs because carbohydrates are hydrophilic, thus preferring the aqueous outside surface of plasma membranes over the more lipid-rich, hydrocarbon core. This orientation assists in maintaining the asymmetric nature of cell membranes, and also allows carbohydrate moieties of these macromolecules to function as cell attachment-recognition sites (**Table 20-1**).

Certain **diseases** of animals are characterized by abnormal quantities of glycolipids in tissues, often the **central nervous system**. They may be generally classified into two groups: 1) true **demyelinating diseases**, and 2) **lysosomal storage disorders** (e.g., gangliosidoses, sphingolipidoses, and leukodystrophies).

In demyelinating disorders, there may be loss of both phospholipids and sphingolipids from neurons of the brain and spinal cord. Affected animals reportedly demonstrate progressive paraparesis and pelvic limb ataxia. The etiology of these conditions is not well established; however, a hereditary basis is suspected.

Ganglioside storage diseases are inherited disorders of lysosomal hydrolase enzymes, that result in accumulation of gangliosides and glycolipid substrates of these hydrolases within lysosomes of neurons and glia throughout the nervous system (see Chapter 59). These diseases are particularly evident in the brain and spinal cord, but also sometimes manifest themselves in peripheral nerves as well. Leukodystrophies are caused by a deficiency of β-galactocerebrosidase, and result in accumulation of galactocerebroside within the nervous system. Effective treatments for demyelinating disorders and lysosomal storage diseases in animals have not been developed, and therefore long-term prognosis is poor.

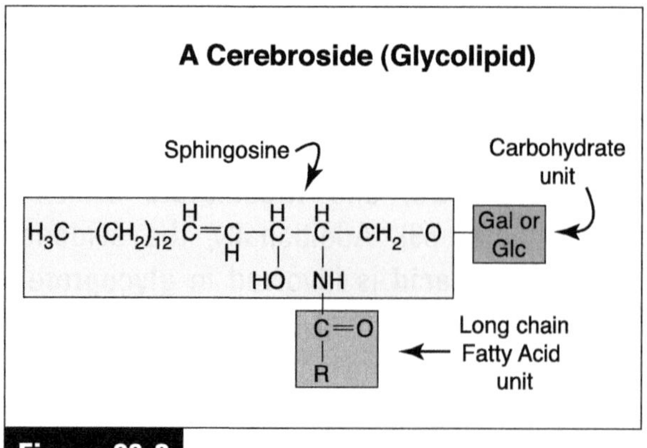

A Cerebroside (Glycolipid)

Sphingosine

Carbohydrate unit

$H_3C-(CH_2)_{\overline{12}}$ C=C—C—C—CH₂—O   Gal or Glc

Long chain Fatty Acid unit

C=O | R

Figure 20-3

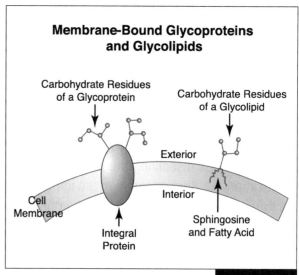

Membrane-Bound Glycoproteins and Glycolipids

Carbohydrate Residues of a Glycoprotein

Carbohydrate Residues of a Glycolipid

Exterior

Interior

Cell Membrane

Integral Protein

Sphingosine and Fatty Acid

Figure 20-4

OBJECTIVES

- Give examples of important glycoproteins and glycolipids, discuss their functions and identify the most abundant glycoprotein in the body.

- Identify extracellular functions for the carbohydrate moieties of glycoproteins and glycolipids, and recognize why most intracellular proteins are nonglycosylated.

- Describe the tripeptide sequence necessary for defining glycoprotein glycosylation sites.

- Name the monosaccharides found in the oligosaccharide chains of glycoproteins, as well as those found in the oligosaccharide chains of glycolipids.

- Explain the proposed mechanism for the "homing" of lymphocytes.

- Provide physiologic roles for lectins and mucoproteins.

- Note the general structure of a cerebroside.

- Recognize causes for and the nature of lysosomal storage diseases (see Chapter 59).

- Know the differences between sphingosine, a ceramide and a cerebroside (see Chapter 59).

- Explain differences between the causes and effects of animal demyelinating diseases, and lysosomal storage disorders.

QUESTIONS

1. **How many different carbohydrate molecules are normally associated with the oligosaccharide chains of glycoproteins?**
   a. 8
   b. 80
   c. 800
   d. 8,000
   e. 80,000

2. **Which type of bond is formed between the side chain hydroxyl group of either Ser or Thr, and the anomeric carbon of either GalNAc or Xyl in glycoproteins?**
   a. Ester
   b. N-Glycosidic
   c. Peptide
   d. O-Glycosidic
   e. Ceramide

3. **Carbohydrate-binding proteins that agglutinate cells or precipitate glycoconjugates are called:**
   a. Cerebrosides.
   b. Mucins.
   c. Lectins.
   d. Glycosphingolipids.
   e. Hormones.

4. **Which one of the following monosaccharides is NOT normally found in glycoproteins?**
   a. Glucose
   b. Fructose
   c. Galactose
   d. Xylose
   e. Mannose

5. **The backbone of animal glycolipids is:**
   a. Glycerol.
   b. Choline.
   c. Sphingosine.
   d. Ethanolamine.
   e. Inositol.

6. **The carbohydrate residues of cell membrane-bound glycolipids and glycoproteins are normally:**
   a. Located on the exterior surface.
   b. Located in the lipophilic core.
   c. Located on the interior surface.
   d. Bound to collagen.
   e. Sucrose polymers.

7. **Which one of the following is NOT a proposed role for the oligosaccharide chains of glycoproteins?**
   a. Affect binding of steroid hormones to receptors on the plasma membrane of their target cells
   b. Mediate intercellular interactions
   c. Affect transport activities of cells
   d. Affect binding of sperm to the ovulated egg
   e. Affect homing of lymphocytes to the endothelial lining of lymph nodes

8. **All of the following are glycoproteins, EXCEPT:**
   a. Alkaline phosphatase
   b. Collagen
   c. Hexokinase
   d. Transferrin

ANSWERS

8. c
7. a
6. a
5. c
4. b
3. c
2. d
1. a

Chapter 21

Overview of Carbohydrate Metabolism

Overview

- In ruminant animals, and to a lesser degree in other herbivores, dietary cellulose is digested by symbiotic microbes to volatile fatty acids.

- Glucose is normally metabolized to pyruvate and lactate in virtually all living mammalian cells.

- The HMS is a source of NADPH for lipid biosynthesis, and ribose for nucleotide formation.

- The uronic acid pathway provides UDP-glucuronate, which can further be utilized for vitamin C formation, for glycoprotein formation, or for the conjugation of endogenous and exogenous lipophilic compounds.

- Glucose and amino acids share a common route of absorption via the hepatic portal circulation.

- Erythrocytes have an absolute requirement for glucose.

- The fetus uses glucose derived from the maternal circulation for both catabolic and anabolic purposes.

- Milk production requires a considerable amount of glucose.

Animals need to process absorbed products of digestion, and dietary content (or lack thereof) generally sets the basic pattern of metabolism in the varied organs and tissues of the body. Important nutritional substrates absorbed by the digestive tract include monosaccharides (mainly glucose), fatty acids (FAs), and amino acids (**Fig. 21-1**). In ruminant animals, and to a lesser degree in other herbivores, dietary cellulose is digested by symbiotic microbes to **short chain (or volatile) fatty acids** (i.e., acetic, propionic, and butyric acid), and therefore tissue metabolism in these animals is adapted to utilize these compounds as major anabolic and catabolic substrates. Most all of the basic products of digestion are processed by their respective cellular metabolic pathways to a common product, **acetyl-CoA**, which can then be coupled with **oxaloacetate** and oxidized in mitochondria to CO_2 and water (with the production of energy in the form of ATP).

Most dietary carbohydrate is used as fuel. It can be degraded immediately as a source of energy, it can be stored as glycogen, or it can be converted to fat for storage.

Glucose is normally metabolized in virtually all living mammalian cells by the pathway of **glycolysis** (see Chapters 24-27). This monosaccharide is a special substrate in mammalian metabolism since the pathway of glycolysis can occur in the absence of oxygen (anaerobic), when the end products are **pyruvate** and **lactate**. However, tissues that can utilize available oxygen (aerobic) are capable of metabolizing

Copyright © 2015 Elsevier Inc. All rights reserved.

Summary

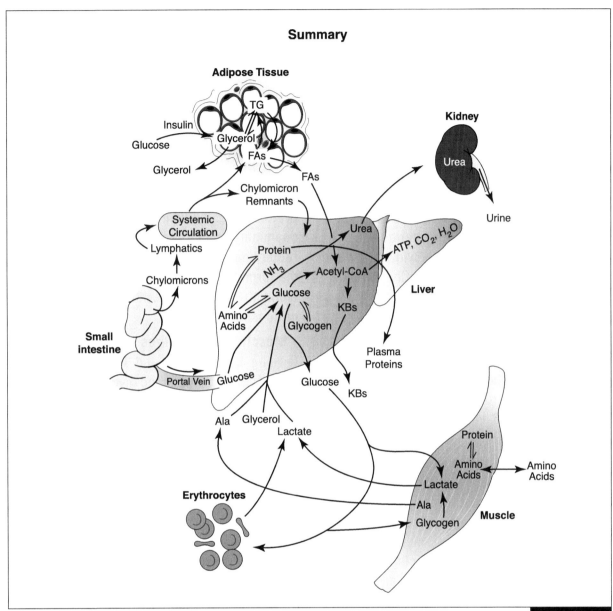

Figure 21-1

pyruvate to **acetyl-CoA**, which can then enter the mitochondrial **citric acid cycle** (also called the Krebs or tricarboxylic acid (TCA) cycle) for complete oxidation to CO_2 and water, with liberation of free energy as ATP in the process of oxidative phosphorylation (see Chapters 34-36). Glucose also takes part in other cellular metabolic pathways as well:

1) Conversion to its storage polymer, **glycogen**, particularly in liver, kidneys, adipocytes, skeletal, and cardiac muscle (see Chapter 23).

2) The **hexose monophosphate shunt (HMS)**, which arises from intermediates of glycolysis (see Chapter 28). The HMS is a source of reducing equivalents (**NADPH**) for lipid biosynthesis, and also the source of **ribose**, which is important for nucleotide and nucleic acid formation. In **mammary tissue**, HMS activity is high during lactation for the production of milk fats, and it is also important to **erythrocytes** who need NADPH for reduced glutathione production (which in turn protects them from oxidative damage; see Chapters 30 and 31).

3) The **uronic acid pathway**, which also arises from glycolytic intermediates, gives rise to uridine diphosphate glucuronate (UDP-glucuronate), which can be used in the biosynthesis of vitamin C in some animals, in glycoprotein formation, or in hepatic conjugation of endogenous and exogenous lipophilic compounds (see Chapter 29).

4) Triose phosphates, which arise from glucose degradation, give rise to **glycerol 3-phosphate**, which in turn is used to form the glycerol backbone of triglyceride (TG; i.e., neutral fat also known as acylglycerol) and most phospholipids (PLs; see Chapter 57).

5) Pyruvate and intermediates of the TCA cycle provide carbon atoms for the synthesis of **amino acids**, and **acetyl-CoA** is the building block for **long-chain fatty acids** and **cholesterol**, the latter being the precursor steroid of all the other steroids synthesized in the body (see Chapters 56 and 61).

Amino acids resulting from the digestion of dietary protein, and glucose resulting from the digestion of dietary carbohydrate, share a common route of absorption via the **hepatic portal vein** (see Chapters 7 and 38). This ensures that both types of metabolites and other small water-soluble products of digestion are initially directed toward the liver. The liver has the primary metabolic function of regulating the blood concentration of most metabolites, particularly glucose and amino acids. In the case of glucose, this is achieved by removing excess glucose from blood and converting it to glycogen (glycogenesis), or to fat (lipogenesis). Since ruminant animals absorb very little glucose from their digestive tract, conversion of dietary glucose to glycogen and fat in the liver is minimal. Most neutral fat in these animals is synthesized from **acetate** (one of the volatile fatty acids, see Chapter 56) in adipocytes.

Between meals, the liver can draw upon its glycogen stores to replenish blood glucose (glycogenolysis), or, in company with the kidneys, to convert noncarbohydrate metabolites such as lactate, glycerol, propionate (herbivores), and glucogenic amino acids, (particularly alanine (Ala) in the liver and glutamine (Gln) in the kidneys), to glucose (gluconeogenesis, see Chapter 37). The maintenance of an adequate concentration of blood glucose is vital for certain tissues in which it is an obligatory fuel (e.g., brain and erythrocytes). The liver also synthesizes the major plasma proteins (e.g., albumin, prothrombin, etc.), and deaminates amino acids that are in excess of requirements. This latter process results in the formation of urea, which is transported via blood to the kidneys where it is excreted in urine (see Chapter 10).

Muscle tissue utilizes glucose as a fuel, forming both lactate and CO_2. It also stores glycogen as a fuel, and synthesizes muscle protein from circulating amino acids. Muscle accounts for approximately 50% of the body mass, and consequently represents a considerable store of protein that can be drawn upon to supply plasma amino acids, particularly during dietary shortage. These, in turn, can be extracted from blood by the liver, with some converted to glucose when needed, and others converted to ketone bodies (KB⁻s, see Chapters 8 and 71).

Cells of the body are specialized in their functions, and tend to emphasize certain metabolic pathways over others. The central role of the **mitochondrion** should be kept in mind while studying intermediary metabolism, since it acts as the focal point and crossroad of carbohydrate, lipid, and amino acid metabolism. In particular, it houses enzymes of the **TCA cycle**, of the respiratory chain and ATP synthesis (i.e., **oxidative phosphorylation**), of **β-oxidation** of fatty acids, and in adult liver cells of **ketone body (KB⁻) production**. In addition, it is the collecting point for carbon skeletons of amino acids after transamination, and for providing these skeletons for the synthesis of nonessential amino acids (see Chapter 9).

Anaerobic glycolysis, **glycogen synthesis**, the **HMS**, and **lipid biosynthesis** all occur in the cytosol. It should be noted that in **gluco-neogenesis**, substances such as lactate and pyruvate that are formed in the cytosol, must enter the mitochondrion and form **oxaloac-etate** before further conversion to glucose (see Chapters 35 and 37).

Membranes of the **endoplasmic reticulum** contain the enzyme system for triglyceride (triacylglycerol) formation, and **ribosomes** are involved with protein biosynthesis.

The use of glucose by peripheral tissues naturally results in a constant drain of glucose from the circulation. Some tissues have an absolute requirement for glucose, while others can use other fuels when glucose is in short supply. **Table 21-1** presents a "short list" of tissues having either a relative or absolute requirement for glucose.

The brain can neither synthesize nor store significant quantities of glucose. Therefore, normal cerebral function requires a continuous supply of energy, largely in the form of glucose. Studies indicate that in normal resting, non-pregnant, non-lactating animals, **over 60%** of glucose consumption per day may occur in the central nervous system.

The persistence of **elevated blood glucose concentrations**, however, can pose a significant threat to diabetic animals. In many tissues, such as nerve, retina, lens, kidney, erythrocytes and small blood vessels, the uptake of glucose is insulin-independent. Therefore, these tissues are most susceptible to chronic complications that evolve from too much available glucose. For example, glucose is highly reactive with proteins, and forms glycosylated proteins

---

Table 21-1
"Short List" of Tissue Glucose Requirements

- The central nervous system (CNS) has an absolute but somewhat adaptable requirement for glucose, which is used to supply energy and also to synthesize lipids. Some portion (but not all) of the energy demands of the CNS may also be supplied by ketone bodies obtained from fatty acid oxidation (i.e., β-oxidation) in the liver.

- Adipose tissue (in non-ruminant animals) may use some glucose for fatty acid synthesis. In all mammals, glucose is used in adipocytes for glycogen deposition, NADPH generation (via the HMS), and to produce glycerol for the reformation of triglycerides.

- Muscle uses glucose for energy metabolism, and to build up glycogen stores. Anaerobic, white muscle tissues rely almost exclusively upon anaerobic glycolysis (largely via stored glycogen), while aerobic, red muscle fibers used in more sustained activity rely mainly on fatty acids and glucose derived from the circulation (see Chapter 80).

- The fetus uses glucose derived from the maternal circulation to build up its energy stores, and to support rapid growth.

- Lactation also results in a heavy glucose demand (a high-producing dairy cow, for example, may use 1,700 gm of glucose per day in milk production).

- Erythrocytes have an absolute requirement for glucose since they lack mitochondria. Rapid changes in the blood glucose concentration (either *in vivo* or that stored in a blood bank) can therefore compromise cell function.

(e.g., **fructosamines**) in nonenzymatic reactions driven solely by the concentration of glucose. Glycosylation of proteins in the lens of the eye, peripheral nerves, and basement membrane of the glomerulus is related to several pathologic changes that develop in diabetic patients. **Glycosylated hemoglobin (HbA₁c)** is a minor form of Hb with carbohydrate covalently linked to the globin chain. Although glycosylation appears to have only a minor effect on the functional properties of Hb, it is useful in monitoring long-term control of blood glucose concentrations in diabetic patients (see Chapter 32). Serum fructosamine concentrations are used similarly.

OBJECTIVES

- Identify and summarize the various locations and routes by which dietary glucose can be metabolized in the body.

- Identify tissues possessing either a "relative" or "absolute" requirement for glucose, as well as those that are insulin-sensitive (see Chapter 22).

- Recognize how serum fructosamines are formed, and why they are clinically significant.

- Explain why the brain exhibits an "adaptable" requirement for glucose, yet erythrocytes maintain an "absolute" requirement.

- Discuss the relative importance of the HMS and the uronic acid pathway to hepatocytes (see Chapters 28 & 29).

- Differentiate the ways in which hepatocytes can metabolize incoming dietary amino acids and glucose from the portal circulation.

- Provide biochemical reasoning for the mitochondrion being the crossroad of carbohydrate, lipid and amino acid metabolism.

QUESTIONS

1. **Which one of the following is NOT typically absorbed directly from the digestive tract into the hepatic portal circulation?**
   a. Lysine
   b. Short- and medium-chain fatty acids
   c. Glucose
   d. Fructose
   e. Chylomicrons

2. **Which one of the following organs is most involved in regulating the plasma glucose concentration?**
   a. Liver
   b. Stomach
   c. Brain
   d. Kidney
   e. Lung

3. **Most neutral fat in ruminant animals is synthesized from:**
   a. Glucose.
   b. Acetate.
   c. Propionate.
   d. Fucose.
   e. Mannose.

4. **Which one of the following has an absolute requirement for glucose since it lacks mitochondria?**
   a. Liver
   b. Brain
   c. Heart
   d. Erythrocyte
   e. Renal tubular cell

5. **All of the following are cytoplasmic processes, EXCEPT:**
   a. Glycogen synthesis.
   b. Hexose monophosphate shunt.
   c. Lipid biosynthesis.
   d. β-Oxidation of fatty acids.
   e. Conversion of glycerol to glucose-6-phosphate.

6. **Select the FALSE statement below:**
   a. In non-ruminant animals, adipose tissue may use some glucose for fatty acid synthesis.
   b. Muscle tissue is known to export glucose into the circulation from stored glycogen.
   c. Adipocytes can store glycogen.
   d. Ruminant animals absorb very little glucose into portal blood from their small intestine.
   e. Glucose is normally metabolized to pyruvate and lactate in virtually all living mammalian cells.

ANSWERS

6. b
5. d
4. d
3. b
2. a
1. e

Chapter 22

Glucose Trapping

Overview

- Only one GLUT transporter (i.e., GLUT 4) has been found to be insulin-dependent.
- GLUT 4 transporters are translocated from the Golgi apparatus to the plasma membrane and back again following insulin receptor binding and dislocation.
- GLUT 3 and GLUT 1, respectively, may be involved with the transport of vitamin C metabolites out of neurons of the CNS and into astrocytes.
- Glucose trapping involves a phosphorylation reaction, which reduces lipophilicity of the product (Glc-6-P).
- Glucokinase and glucose-6-phosphatase are primarily liver enzymes.
- Hexokinase is inhibited by Glc-6-P, yet glucokinase is not.
- Muscle cannot contribute glucose directly to the blood glucose pool.
- Strict carnivores and ruminant animals possess low hepatic glycogen reserves.

Entry of glucose into cells is thought to be carrier-mediated by **integral membrane glucose transporters (GLUTs),** that help to move this monosaccharide down its concentration gradient. Two Na^+-dependent glucose transporter (**SGLT**) isoforms, and five **GLUT** transporter isoforms have been identified in various tissues (**Fig. 22-1**).

The **SGLT 1** and **2** transporters are found in luminal (apical) membranes of proximal renal tubular epithelial cells, while both the **SGLT 1** and **GLUT 5** transporters are found in luminal (apical) membranes of mucosal cells in the small intestine (see Chapter 38). The **SGLT 1** transporter requires 2 Na^+/glucose molecule, while the **SGLT 2** transporter requires only 1, and both are insulin-independent. The **SGLT 1** and/or **2** transporters are responsible for secondary active transport (i.e., symport) of glucose (and galactose) out of the intestine and renal tubular filtrate, against their concen-

tration gradients, and therefore are associated with aerobic processes (i.e., generation of ATP for Na^+/K^+-ATPase utilization). The **GLUT 5** transporter is associated with lesser amounts of anaerobic glucose and fructose (facilitated) diffusion into duodenal and jejunal mucosal cells. Three of the remaining 4 transporters are also **insulin-independent: GLUT 1** is responsible for glucose reabsorption across the basolateral membrane of proximal renal tubular epithelial cells (S-3 segment), and also glucose uptake into erythrocytes, the colon, and across the placenta and blood-brain-barrier, etc.; **GLUT 2** is the pancreatic insulin-secreting β-cell sensor, and also functions in transporting glucose out of intestinal and proximal renal tubular epithelial cells (S-2 segment), and also into and out of the liver; and **GLUT 3** is responsible for basal glucose uptake in neurons, the placenta, and other organs. **GLUT 3** and **GLUT 1**, respectively, may also be

Copyright © 2015 Elsevier Inc. All rights reserved.

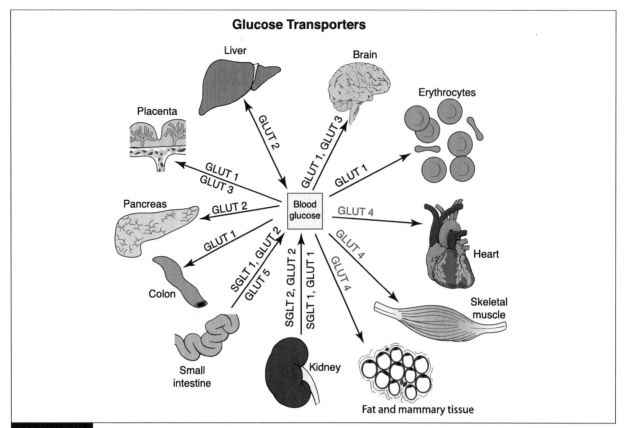

Figure 22-1

involved in the transport of dehydroascorbate (DHC) out of neurons and into astrocytes of the brain (see Chapter 39). **The GLUT 4** transporter is the only one that has been identified as being **insulin-dependent**, and is found primarily on plasma membranes of **muscle and fat tissue**. Transport activity of **GLUT 4** appears to be increased through activation of **protein kinase C (PKC)**.

Studies indicate that in adipocytes the **GLUT 4** transporters are translocated from an inactive pool in the Golgi apparatus to active sites on the plasma membrane following the binding of insulin to its receptors. Following the diffusion of glucose into the cell and the removal of insulin from its binding sites on plasma membrane receptors, the **GLUT 4** transporters are translocated back to the Golgi. This bidirectional translocation process has been found to be both temperature and energy dependent.

Once inside the cell, glucose is rapidly phosphorylated to glucose-6-phosphate (Glc-6-P), a "**trapping**" reaction that serves to keep this important hexose in the cell (since the charged phosphate group precludes diffusion of glucose back out of the cell, **Fig. 22-2**). The Glc-6-P thus formed now becomes a substrate from which multiple pathways branch out. It may be used in the Embden-Meyerhoff pathway (EMP), in glycogen synthesis in hepatocytes and muscle cells, in the hexose monophosphate shunt (HMS), and in the uronic acid pathway of hepatocytes (see Chapters 28 and 29). Enzymes that catalyze glucose phosphorylation, **glucokinase** and **hexokinase**, require energy input from ATP. Other necessary cofactors for these kinases include K^+ and Mg^{++} (the two most prevalent intracellular cations).

The properties of hexokinase and glucokinase are shown in **Table 22-1**, with the most important differences being their comparative locations, kinetic, and regulatory properties.

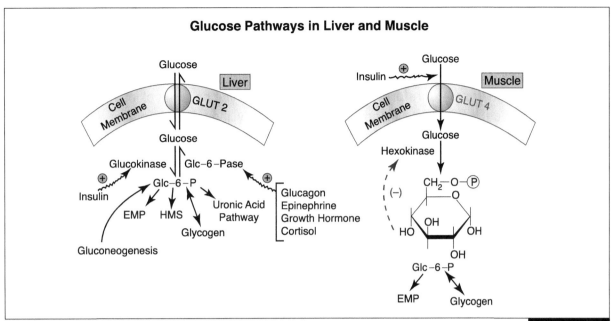

Glucose Pathways in Liver and Muscle

Figure 22-2

Table 22-1
Properties of Glucokinase and Hexokinase

	Glucokinase	Hexokinase
Kinetic parameters		
K_m	High (180 mg%)	Low (0.9 mg%)
Glc-6-P neg. feedback	No	Yes
Inducible	Yes	No
Synthesis/activity	Increased by insulin, decreased by diabetogenic hormones	Not affected by insulin
Tissue distribution	Liver, pancreatic β-cells	Most tissues
Species (relative activity)		
Rat	230	40
Pig	190	38
Dog	110	30
Cat	<2	25
Sheep	<2	8
Cow	<2	23

Hexokinase is present in most all mammalian cells in low concentrations. The low Km for hexokinase (10^{-5}M or 0.9 mg%) means that the enzyme is working at maximal velocity (V_{max}) at extremely low glucose concentrations (see Chapter 6). But this enzyme is powerfully inhibited by its product, Glc-6-P. Hence, when Glc-6-P is not rapidly utilized, that is to say, no longer needed by the cell, hexokinase activity is reduced. This enzyme, also, is not inducible, and although insulin increases **GLUT 4** transporter activity in plasma membranes of muscle fibers and adipocytes, neither its presence nor absence has any direct effect on hexokinase activity. Insulin increases cell membrane permeability of muscle cells and adipocytes to glucose, therefore, glucose entry into these cells becomes limited by the presence or absence of insulin (**Fig. 22-3**), as well as by the cell's ability to metabolize Glc-6-P. When muscle contraction stops, for example, the build-up of intracellular Glc-6-P will limit further phosphorylation of glucose, which in turn should limit further "trapping" of glucose inside inactive cells. Hexokinase will also phosphorylate fructose, but at a much slower rate than glucose (see Chapter 25).

Hexokinase is thought to be the rate-limiting erythrocytic enzyme whose activity declines with age. Thus, erythrocyte senility and turnover may be keyed to this enzyme. As hexokinase activity fades, the ability of the erythrocyte to utilize glucose, its major substrate, declines. Next, intracellular ATP levels drop, electrolyte pumps expire, the cytoskeleton becomes rigid, and the (senile) erythrocyte is destroyed by phagocytic cells (see Chapter 31).

Glucokinase, on the other hand, is primarily a liver enzyme, but is also found on the inner plasma membrane of pancreatic β-cells (cells that synthesize and secrete insulin). It has a high K_m (10^{-2} or 180 mg%), meaning that high concentrations of glucose are required for maximal activity, and it contains Mn^{++} (**Fig. 22-4**). Glucokinase is inducible, so that enzyme activation (and enzyme synthesis) can meet the acute (and chronic) needs for more glucose phosphorylation. Glucokinase activity is not generally inducible, however, in the livers of adult ruminant animals because not enough glucose normally escapes degradation by ruminal microflora. Therefore, very little free glucose is normally absorbed by the ruminant intestinal tract to induce this hepatic enzyme. Unlike hexokinase, glucokinase is not inhibited by its product, Glc-6-P, yet its activity is stimulated by insulin and inhibited by diabetogenic hormones (e.g., epinephrine, cortisol, glucagon, and growth hormone). Glucose phosphorylation (and not cell membrane permeability to glucose) usually becomes rate-limiting for glucose entry into liver tissue.

Net influx of glucose across the plasma membranes of muscle, adipose, and liver cells is enhanced by insulin. Although insulin increases **GLUT 4** transporter activity in plasma membranes of adipose tissue and muscle fibers, it has no direct effect on hexokinase activity. In liver, insulin increases synthesis of glucokinase, but **GLUT 2** transporter activity remains unaffected.

Glucose-6-phosphatase (Glc-6-Pase) catalyzes the breakdown of Glc-6-P to free glucose and inorganic phosphate (Pi). This

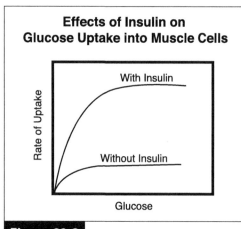

Effects of Insulin on Glucose Uptake into Muscle Cells

With Insulin

Without Insulin

Rate of Uptake

Glucose

Figure 22-3

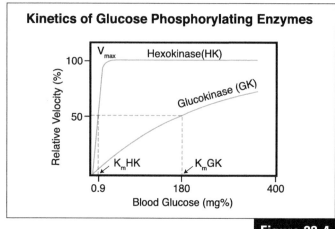

Kinetics of Glucose Phosphorylating Enzymes

V_{max}

Hexokinase(HK)

Glucokinase (GK)

Relative Velocity (%)

100

50

K_mHK K_mGK

0.9   180   400

Blood Glucose (mg%)

Figure 22-4

enzyme is present in liver, kidneys, and gut (to a limited extent), but is absent from other tissues, including muscle and adipose tissue. This effectively means that only those tissues containing Glc-6-Pase can export glucose to the circulation (see Chapter 37).

Hepatic **Glc-6-Pase activity** is stimulated by the **diabetogenic hormones** (i.e., **epinephrine**, **growth hormone**, **glucagon**, and **cortisol**), yet inhibited by insulin; hence, when glucose concentrations in blood are high and, thus, insulin is released by the pancreas, hepatic Glc-6-Pase activity declines while glucokinase activity increases. When blood glucose concentrations decline, and insulin levels drop, hepatic Glc-6-Pase becomes active which helps to supply glucose to the bloodstream. Of course, in diabetes mellitus, where high blood glucose concentrations are present because of a relative insulin deficiency, unbridled high Glc-6-Pase activity aggravates the condition. Consequently, the level of Glc-6-P in liver cells is chronically low, which can lead to impaired hepatic glycogenesis.

The significance of the absence of Glc-6-Pase in muscle cells is that muscle glycogen cannot contribute glucose directly to the blood glucose pool. Instead, muscle cells use their hexose resources exclusively for their own purposes, albeit that anaerobic glycolysis in muscle produces lactate, that can then enter blood to be extracted and converted to glucose by the liver (see Chapter 37).

Given the above discussion, it should be clear why the livers of ruminant animals and strict carnivores (e.g., cats) tend to function mainly as continuous glucose producers (from gluconeogenic substrates). In omnivores, the liver has to deal both with high hepatic portal blood glucose concentrations (after meals), and with periods when it has to produce glucose via glycogen breakdown and gluconeogenesis (between meals). Therefore, their livers act to buffer large changes in blood glucose. Glucose production by the omnivore liver between meals is provided by the break-down of stored glycogen (glycogenolysis, which can provide glucose for about 12-28 hours of starvation), and gluconeogenesis (see Chapters 37 and 74). However, **ruminant animals** and **strict carnivores** have **insignificant liver glycogen reserves** (i.e., low glycogen synthase activity), therefore they must rely exclusively on gluconeogenesis to provide them with adequate amounts of plasma glucose to supply tissue needs.

OBJECTIVES

- Identify the integral membrane glucose transporters (GLUTs & SGLTs), their locations and functions, and differentiate the insulin-insensitive transporters from   the insulin-sensitive transporter.

- Recognize which glucose transporters are most likely involved with vitamin C transport (see Chapter 39).

- Explain how "glucose trapping" occurs in liver and extra-hepatic sites, and how it is regulated.

- Contrast the different kinetic properties of glucokinase (GK) and hexokinase (HK), and provide physiologic explanations for the differences.

- Recognize why hepatic GK activity is low in car-

nivores and ruminants.

- Understand why there is an inverse relationship between the diabetogenic hormones and insulin in the stimulation/inhibition of hepatic Glc-6-Pase and GK.

- Explain the significance of Glc-6-P negative feedback on the HK activity of muscle cells.

- Know why it is important for insulin to remain independent of GLUT 2 activity.

- Provide reasoning for the low hepatic glycogen reserves of carnivores and ruminants.

- Outline the bidirectional movements of adipocyte GLUT 4 transporters in response to insulin binding and removal from its membrane receptors.

- Understand the correlation between erythrocyte senility and HK activity.

- Inidcate how sodium-dependent glucose transporters (SGLTs) function in moving hexoses against their concentration gradients (see Chapter 38).

- Discuss reasons for characterizing glucagon, epinephrine, cortisol and growth hormone as "diabetogenic hormones."

- Contrast the control of GK activity to that of HK.

QUESTIONS

1. **Glucose phosphorylation "traps" this monosaccharide inside cells, because:**
   a. It is conjugated to protein.
   b. Glucose 6-phosphate becomes a substrate for fructose 6-phosphate formation.
   c. The glucose molecule remains bound to hexokinase.
   d. Glucose 6-phosphatase is inhibited by glucose 6-phosphate.
   e. The charged phosphate group reduces its lipophilicity.

2. **Erythrocytic senility appears to be tied to the declining activity of:**
   a. GLUT-1.
   b. Glucose 6-phosphatase.
   c. Glucokinase.
   d. SGLT-2.
   e. Hexokinase.

3. **Which one of the following glucose transporters is insulin-dependent?**
   a. SGLT 1
   b. GLUT 1
   c. GLUT 2
   d. GLUT 3
   e. GLUT 4

4. **Which one of the following is a property of hexokinase, but not of glucokinase?**
   a. Responds to Glc-6-P negative feedback
   b. Possesses a high K_m
   c. Is inducible
   d. Is present in liver, but not muscle tissue
   e. Activity is increased by insulin

5. **Which one of the following cell types can export glucose to the circulation?**
   a. Skeletal muscle
   b. Adipocyte
   c. Neuron
   d. Hepatocyte
   e. Heart muscle

6. **Transport activity of GLUT 4 appears to be enhanced through activation of:**
   a. PLA_2.
   b. PKC.
   c. PLC.
   d. LPL.
   e. cAMP.

7. **The livers of ruminant animals and cats:**
   a. Continually produce glucose, and export it to the circulation.
   b. Cannot respond to insulin.
   c. Possess hexokinase, but not glucose-6-phosphatase activity.
   d. Normally contain more glycogen (per gram of tissue) than the livers of dogs.
   e. Are incapable of gluconeogenesis.

8. **The liver of which animal below normally possesses the greatest glucokinase activity (per gram of tissue)?**
   a. Cat
   b. Sheep
   c. Rat
   d. Cow
   e. Goat

9. **Glucokinase and hexokinase require all of the following co-factors to achieve full activity, EXCEPT:**
   a. Mg^{++}.
   b. ATP.
   c. NAD^+.
   d. K^+.

10. **The GLUT 2 transporter is found in plasma membranes of:**
    a. The placenta and blood-brain-barrier.
    b. Hepatocytes and pancreatic β-cells.
    c. Skeletal muscle and adipose tissue.
    d. Erythrocytes and cardiac myocytes.
    e. Renal glomeruli and neural tissue.

10. b
9. c
8. c
7. a
6. b
5. d
4. a
3. e
2. e
1. e

ANSWERS

Chapter 23

Glycogen

Overview

- Glycogen is the major storage form of carbohydrate in animals.
- Glycogen is stored primarily in muscle and liver tissue (and to a lesser extent in adipose tissue).
- Glycogenolysis in muscle will not result in glucose being added directly to blood.
- Glycogenolysis in liver will result in glucose being added directly to blood.
- Glycogen synthase is activated through dephosphorylation.
- Both the cAMP and Ca^{++} messenger systems are involved in glycogen breakdown.
- Muscle contraction is normally synchronized with glycogenolysis.

Glycogen is the major carbohydrate storage form in animals, and corresponds to **starch** in plants. It occurs mainly in **liver** (up to 6-8% wet weight), and **muscle** (where it rarely exceeds 1% of wet weight). However, because of its greater mass, the whole body muscle glycogen pool is some 3 to 4 times greater than the liver pool. Glycogen is found in the cytosol of cells, and each molecule can contain up to 60,000 glucose residues. It is a **hydrophilic** molecule that exists *in vivo* in highly hydrated glycogen granules. Approximately 65% of glycogen is water. Conversely, **triglyceride**, the major storage form of fat, is anhydrous and **hydro-phobic**, thus making it lighter than glycogen (see Chapters 53 and 57). If a dog stored the same amount of glycogen as it does fat, it would be nearly twice as heavy, and its mobility would be severely reduced. For this and several other reasons, migrating birds also store potential energy primarily as fat.

Glycogen breakdown (glycogenolysis) occurs in liver during both exercise and starvation, whereas muscle glycogenolysis generally requires exercise (see Chapters 74 and 79). Free glucose can be generated from glycogenolysis in the liver, and thus made available to the rest of the body via the bloodstream. However, this does not occur in muscle (since **muscle lacks the enzyme glucose-6-phosphatase (Glc-6-Pase)**). Glycogen broken down in muscle cells must be oxidized therein.

Glycogenesis involves the biosynthesis of glycogen from glucose, glucose metabolites, or metabolic precursors of glucose. **Glycogen storage diseases** are a group of inherited metabolic disorders characterized by deficient mobilization of glycogen or deposition of abnormal forms, leading to muscular weakness, exercise intolerance, and sometimes death. Glycogenesis and glycogenolysis occur via **separate metabolic pathways**, thereby allowing each to operate independently of the other. They should be viewed as continuous, dynamic physiologic processes that are regulated by the presence or absence of various hormones and neurotransmitters.

Copyright © 2015 Elsevier Inc. All rights reserved.

Glycogenesis

The structure of glycogen is represented in **Fig. 23-1**. Branching of the glycogen molecule occurs at an average frequency of every ten glucose residues. Branching increases its solubility as well as the rate at which glucose can be stored and retrieved. Each glycogen molecule has a protein, **glycogenin**, covalently linked to the polysaccharide. Linear glycogen chains consist of glucose molecules linked together by **α-1,4 glycosidic bonds**. At each of the branch points, two glucose molecules are linked together by **α-1,6 glycosidic bonds**. The non-reducing ends of the glycogen molecule are the sites where both synthesis and degradation occur.

The pathway by which glucose-6-phosphate (Glc-6-P) is converted to glycogen is shown in **Fig. 23-2**. Following glucose phosphorylation by hexokinase (HK) or glucokinase, Glc-6-P may be converted to glucose-1-phosphate (Glc-1-P) by the reversible enzyme, **phosphoglucomutase**

(PGM). This reaction, like that for the phosphorylation of glucose, requires Mg^{++} as a cofactor. Glc-1-P is next converted to the active nucleotide, **uridine diphosphate-glucose (UDP-Glc, Fig. 23-3)**, by the action of UDPGlc pyrophosphorylase. UDP-glucose now becomes a branch point for entry into the hepatic uronic acid pathway (via UDP-glucuronate, see Chapter 29), lactose synthesis in the mammary gland (via UDP-galactose), or glycogen synthesis in several tissues (via enhanced activity of glycogen synthase).

Glycogen synthase catalyzes the **rate-limiting step** in glycogenesis. Being a key enzyme, its activity can be **inhibited by phosphorylation**, or **activated by dephosphorylation** (see Chapter 58). Postprandial (i.e., after a meal) conditions activate glycogen synthase activity in various ways. The parasympathetic nervous system (PNS) has an indirect effect via autonomic stimulation of **insulin** release from the pancreas. High levels of glucose also stimulate **insulin**

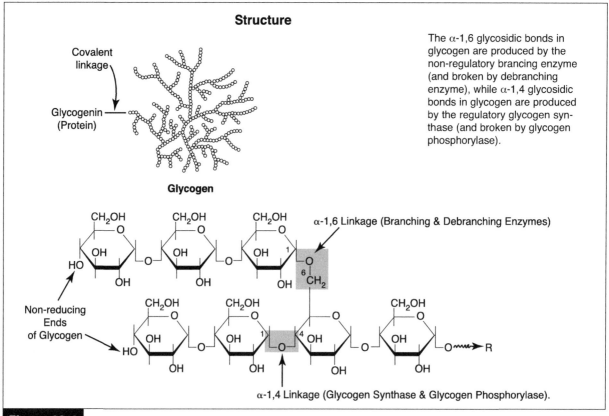

Structure

Covalent linkage

Glycogenin (Protein)

The α-1,6 glycosidic bonds in glycogen are produced by the non-regulatory brancing enzyme (and broken by debranching enzyme), while α-1,4 glycosidic bonds in glycogen are produced by the regulatory glycogen synthase (and broken by glycogen phosphorylase).

Glycogen

α-1,6 Linkage (Branching & Debranching Enzymes)

Non-reducing Ends of Glycogen

α-1,4 Linkage (Glycogen Synthase & Glycogen Phosphorylase).

Figure 23-1

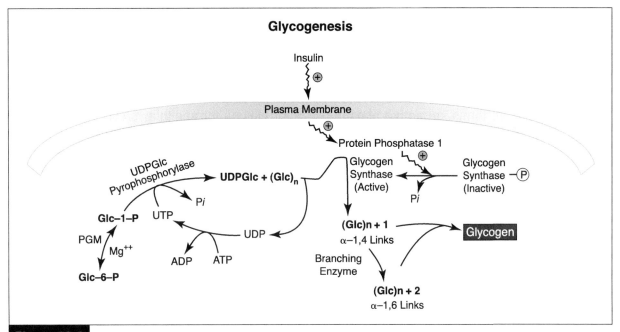

Glycogenesis

Figure 23-2

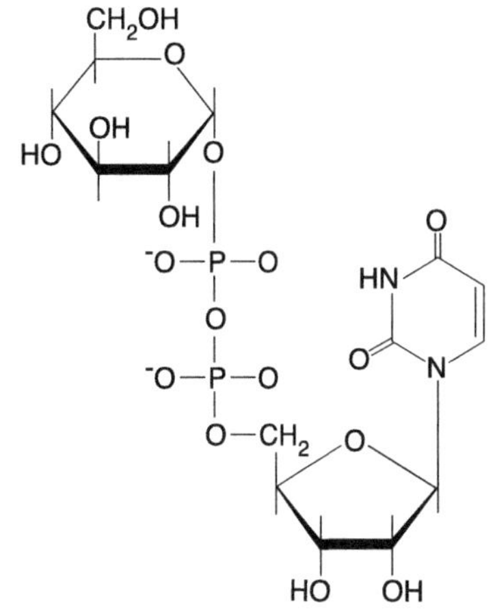

UDP-Glucose (UDPGlc)

Figure 23-3

release. Insulin, the anabolic hormone that promotes storage of dietary bounty, stimulates activity of **protein phosphatase 1**, which in turn stimulates **glycogen synthase** activity by causing its **dephosphorylation**.

When the α-1,4 chain of glycogen extends to 11-15 glucose residues from the nearest branch point, branching occurs. A block of 6-7 glucose residues is moved from the end of one chain to another chain, or to an internal position of the same chain. By catalyzing these α-1,4 —> α-1,6 glucan transfers, the non-regulatory **branching enzyme** helps to create new sites for elongation by glycogen synthase.

Glycogenolysis

Mobilization of glucose from glycogen stores requires two reactions: shortening of the nonreducing ends of α-1,4 glycosidic chains by a regulatory **phosphorylase**, and disassembly of the α-1,6 branch points by a non-regulatory **debranching enzyme**. The first reaction adds **inorganic phosphate (Pi)** rather than H_2O across the glycosidic bond, resulting in the release of **Glc-1-P** which can then be converted to Glc-6-P by PGM. The second reaction removes

a free glucose residue from glycogen which can be quickly phosphorylated to **Glc-6-P** and used in the Embden-Meyerhoff pathway (EMP), or released by liver tissue into blood (**Fig. 23-4**). Additionally, Glc-6-P may also be dephosphorylated by hepatic Glu-6-Pase, and released as free glucose into blood. Approximately 90% of glycogen degradation results in Glc-1-P formation through reaction one above, and 10% in free glucose formation through reaction two.

Glycogen phosphorylase in muscle is immunologically and genetically distinct from that in liver. It is a dimer, with each monomer containing 1 mol of **pyridoxal phosphate (vitamin B₆;** see Chapter 42). Indeed, glycogen phosphorylase may account for as much as 70-80% of this water-soluble vitamin in mammals. This enzyme exists in two basic forms, an **active phosphorylated** form, and an **inactive dephosphorylated** form. Activation of this enzyme involves a cascade of reactions initiated by an increase in intracellular **cyclic-AMP (cAMP)**. The cascade involves an initial **phosphorylation** of **phosphorylase kinase**, which is stimulated by **protein kinase A**. The active phosphorylase kinase next initiates **phosphorylation** of **glycogen phosphorylase** (which becomes active), as well as of **glycogen synthase** (which becomes inactive). Phosphorylase kinase is activated by hormones that elevate intracellular cAMP (e.g., epinephrine acting on **β₂-adrenergic receptors**, and glucagon), whereas **protein phosphatase-1** is activated by insulin. Insulin also activates a cAMP-dependent **phosphodiesterase** in liver cells, which reduces cAMP to its inactive form (5'-AMP), thus favoring glycogen synthesis.

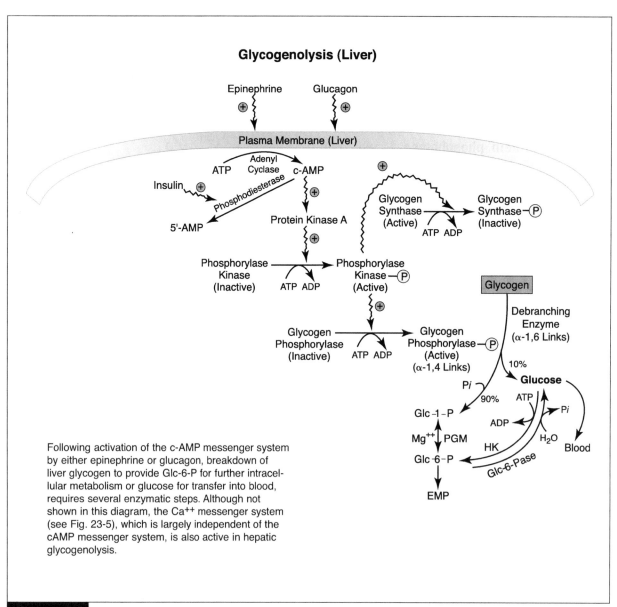

Glycogenolysis (Liver)

Following activation of the c-AMP messenger system by either epinephrine or glucagon, breakdown of liver glycogen to provide Glc-6-P for further intracellular metabolism or glucose for transfer into blood, requires several enzymatic steps. Although not shown in this diagram, the Ca⁺⁺ messenger system (see Fig. 23-5), which is largely independent of the cAMP messenger system, is also active in hepatic glycogenolysis.

Figure 23-4

Phosphorylase kinase can also be allosterically activated by **Ca^{++}**, a mechanism **independent of cAMP** (see Chapter 58). Studies have shown that α_1-**adrenergic receptor** stimulation is also involved in the catecholamine-induced stimulation of glycogenolysis in **both liver and muscle tissue**. This mechanism is particularly important in skeletal muscle, where contraction is initiated by the release of Ca^{++} from the sarcoplasmic reticulum. It allows glycogen degradation, which normally increases several hundred-fold immediately after the onset of contraction, to be synchronized with the contractile process (**Fig. 23-5**). Phosphorylase kinase is a complex enzyme having four different subunits, α, β, γ, and Δ. The Δ subunit is **calmodulin**, a Ca^{++}-binding protein that sensitizes a number of enzymes to small changes in the intracellular Ca^{++} concentration.

Glycogen Storage Diseases

Glycogen storage diseases are known to exist in dogs (usually miniature breed puppies), cats, horses, and primates, and are generally characterized by an inability to form or degrade glycogen in normal metabolic pathways. Examples are a **type I von Gierke's-like disease** (a Glc-6-Pase deficiency), and a **type II Pompe's-like disease** (where glycogen accumulation is minimal due to a glycogen branching enzyme deficiency (GBED)). Clinical signs in type II glycogen storage disease are apparently related to cardiac and skeletal muscle glycogenosis, and include gastric reflux and megaesophagus, systemic muscle weakness, and cardiac abnormalities. A **type III Cori's-like disease** also exists in dogs, which is a debranching enzyme deficiency. As indicated above, these diseases may result

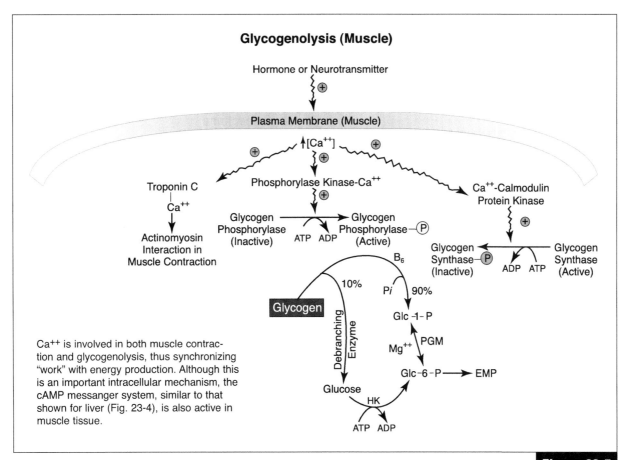

Glycogenolysis (Muscle)

Ca^{++} is involved in both muscle contraction and glycogenolysis, thus synchronizing "work" with energy production. Although this is an important intracellular mechanism, the cAMP messanger system, similar to that shown for liver (Fig. 23-4), is also active in muscle tissue.

Figure 23-5

in either the inability to form glycogen, or in glycogen accumulation in the liver, muscles, kidneys, and nervous tissue. Affected patients become exercise intolerant, blood glucose levels are low, and hyperlipidemia, hepatomegaly, and ketonemia develop. Hormones elaborated in response to critically low blood glucose concentrations in affected animals generally prompt increased rates of gluconeogenesis (and lipolysis), generating a great deal of Glc-6-P which can be subsequently stored as glycogen. Definitive diagnosis of these diseases generally requires enzyme assay of affected tissues (**type I** -- liver, kidneys, and intestinal mucosa; **type II** -- skeletal muscle, white blood cells, and skin fibroblasts: and **type III** -- liver, muscle, and skin fibroblasts). The prognosis for all glycogen storage disorders in animals is poor.

Another glycogen storage abnormality is exhibited in dogs with **steroid** (i.e. glycocorticoid) **hepatopathy**. Affected animals develop hepatomegaly with excessive glycogen accumulation (vacuolar hepatopathy).

OBJECTIVES

- Explain why animals store energy primarily in the form of triglyceride (rather than glycogen).

- Recognize why muscle cannot contribute glucose molecules directly to the circulation from glycogenolysis.

- Provide reasoning for the presence of glycogenesis and glycogenolysis as separate metabolic pathways.

- Outline the pathway for glycogen biosynthesis, and understand its regulation.

- Explain the roles of glycogen phosphorylase, phosphorylase kinase, Ca^{++}, calmodulin, cAMP, protein kinase A, catecholamines, debranching enzyme and vitamin B_6 in glycogen breakdown (see Chapters 42 & 58).

- Characterize the different types of glycogen storage diseases in animals.

QUESTIONS

1. **The intracellular protein normally linked to the polysaccharide chain of glycogen is:**
   a. Glucagon.
   b. Albumin.
   c. Glycogenin.
   d. Calmodulin.
   e. Phosphodiesterase.

2. **Which one of the following enzymes is activated through dephosphorylation by protein phosphatase 1?**
   a. Glycogen Synthase
   b. Glycogen Phosphorylase
   c. Debranching Enzyme
   d. Phosphorylase Kinase
   e. Branching Enzyme

3. **Which one of the following ribonucleotides participates in glycogen formation?**
   a. PRPP
   b. CDP
   c. AMP
   d. GMP
   e. UDP

4. **Which one of the following is associated with the inactivation of phosphorylase kinase in liver cells?**
   a. cAMP
   b. Epinephrine
   c. Protein Kinase A
   d. Glucagon
   e. Phosphodiesterase

5. **A deficiency in which one of the following enzymes would not necessarily be associated with glycogen storage disease (i.e., the inability to mobilize and/or utilize glycogen)?**
   a. Glucose-6-Phosphatase (Liver)
   b. Debranching Enzyme (Muscle)
   c. Hexokinase (Liver)
   d. Glycogen Phosphorylase (Muscle)
   e. Glycogen Phosphorylase (Liver)

6. **Pompe's-like disease is best associated with:**
   a. Steroid hepatopathy.
   b. Branching enzyme deficiency.
   c. Glc-6-Pase deficiency.
   d. Debranching enzyme deficiency.
   e. Hepatic lipidosis.

ANSWERS

1. c
2. a
3. e
4. e
5. c
6. b

Chapter 24

Introduction to Glycolysis (The Embden-Meyerhoff Pathway (EMP))

Overview

- Glycolysis is the major route of catabolism for glucose, fructose, and galactose.
- Anaerobic glycolysis proceeds at a fast pace in fast growing cancer cells, thus resulting in lactic acid production.
- Anaerobic glycolysis is nearly universal among all cell types, although the end products may vary.
- Anaerobic glycolysis is used for "quick energy" in type IIB skeletal muscle fibers.
- The liver, kidneys, brain, and heart normally account for about 7% of the body mass, yet receive almost 70% of the cardiac output at rest. These organs normally consume about 58% of the O_2 utilized in the resting state.
- The cornea and lens rely heavily on anaerobic glycolysis.
- Mature red blood cells rely exclusively on anaerobic glycolysis.

Glycolysis, via the pathway named after its co-discoverers, is the primary route for glucose catabolism in mammalian cells leading to the cytoplasmic production of **pyruvate**, and its subsequent oxidation in the mitochondrial tricarboxylic acid (TCA) cycle through production of either **acetyl-CoA**, or **oxaloacetate** (see **Fig. 24-1**). Glycolysis is also a major pathway for the catabolism of **fructose** and **galactose** derived from dietary **sucrose** and **lactose**, respectively.

Of crucial biomedical importance is the ability of glycolysis to provide **ATP** for cells in the absence of oxygen (O_2), thus allowing skeletal muscle, for example, to continue contracting when aerobic oxidation becomes insufficient. Anaerobic glycolysis also allows poorly perfused tissues to survive, and also those that lack mitochondria (e.g., mature erythrocytes). Conversely, heart muscle, which is adapted for sustained aerobic performance, has limited anaerobic glycolytic potential, and therefore does not perform well under conditions of ischemia (see Chapter 80). A small number of inherited hemolytic anemias occur among domestic animals, in which enzymes of glycolysis (e.g., pyruvate kinase or phosphofructokinase (PFK)), have reduced activity.

Glycolysis is a **highly regulated process**, with just enough glucose being metabolized at any one time to meet the cell's need for ATP. Metabolic intermediates between glucose and pyruvate are phosphorylated compounds, which promote their retention within the cytoplasm (see **Fig. 24-1**). Four molecules of **ATP** are generated from ADP in anaerobic glycolysis: two in the step catalyzed by **phosphoglycerate kinase**, and two in the step catalyzed by **pyruvate kinase**. However, two ATP molecules are consumed during earlier steps of this pathway: the first by the addition of a phosphate residue to glucose in the reaction catalyzed by **hexokinase**, and the second by the addition of

Copyright © 2015 Elsevier Inc. All rights reserved.

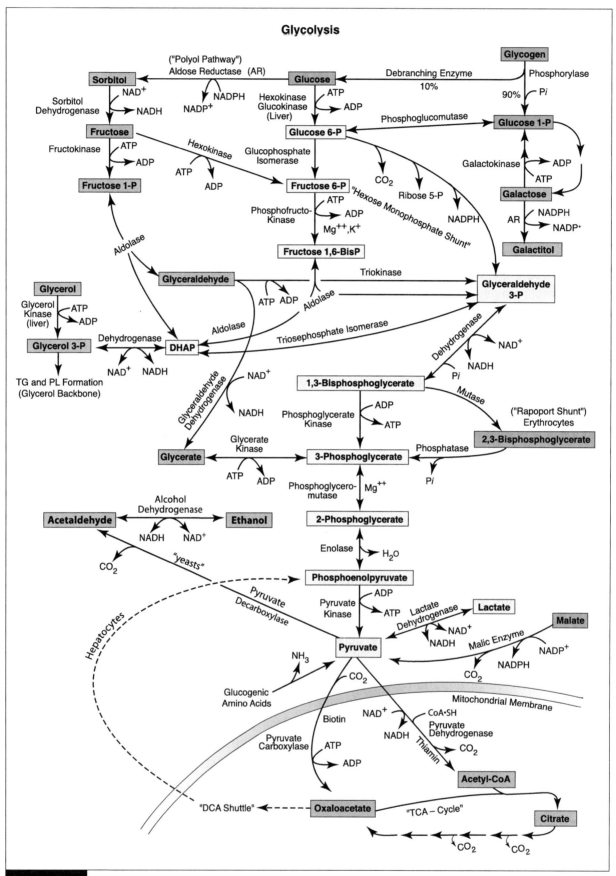

Figure 24-1

a second phosphate to fructose 6-phosphate in the reaction catalyzed by **PFK**. Thus, there is a net gain of **two ATP** molecules in **anaerobic glycolysis** (when starting with glucose). Since the breakdown of glycogen through glucose 1-phosphate does not require ATP, the cytoplasmic phase of **glycogenolysis** can provide a net gain of **3 ATP** molecules.

When oxygen becomes limited to cells, glucose cannot be oxidized completely to CO_2 and H_2O. Cells will thus "ferment" each glucose molecule to two moles of **lactic acid** -- again, with the net production of only two molecules of ATP from each glucose molecule. In fast-growing **cancer cells**, for example, glycolysis frequently proceeds at a much higher rate than can be accommodated by the mitochondrial TCA cycle. Many tumors are also poorly vascularized, thus reducing O_2 availability. Thus, more pyruvate is produced than can be metabolized inside mitochondria. This, in turn, results in excessive production of lactic acid, which produces an acid environment in and around the tumor, a situation that may have implications for certain types of cancer therapy. Lactic acidosis results from other causes as well, including exercise, pyruvate dehydrogenase deficiency, or, for example, grain overload in ruminant animals.

The overall equation for anaerobic glycolysis (from glucose to lactate⁻) is:

$$\text{Glucose} + 2\,\text{ADP} + 2\,\text{Pi} \longrightarrow 2\,\text{Lactate}^- + 2\,\text{ATP} + 2\,H_2O + 2\,H^+$$

The term lactic acid (CH_3-CHOH-COOH) is often used interchangeably with lactate (CH_3-CHOH-COO⁻), which is an anion. Since lactic acid, like many organic acids, is largely dissociated in body fluids (**CH_3-CHOH-COOH ⟶ CH_3-CHOH-COO⁻ + H⁺**), use of the term "lactate" is more appropriate. It should be noted that when lactate accumulates in plasma, it displaces other important anions (e.g., **HCO_3^-** and **Cl⁻**) from extracellular fluids, thus having important implications in acid/base chemistry (see Chapters

86 and 87). When protons (**H⁺**) from lactic acid accumulate in muscle tissue, fatigue ensues because the V_{max} of PFK is lowered, the release of Ca⁺⁺ from the sarco-plasmic reticulum is compromised, actomyosin ATPase activity is reduced, and the conformation of muscle contractile proteins is affected, thus causing pain (see Chapter 81). Much of the lactate in blood normally passes into liver cells (where it can be converted to pyruvate, and then oxidized or used for glucose formation; see Chapter 37), or into cardiac muscle cells which can also convert it to pyruvate, and then oxidize it in mitochondria for energy purposes (see Chapter 80).

Anaerobic glycolysis is nearly universal among all cell types, although the end products may vary. That is, **lactate** (of mammals) may be replaced by a variety of different substances such as **propionate** in bacteria, or **ethanol** in yeast. As pyruvate is converted to lactate (or ethanol), the **NADH** produced in the initial stages of glycolysis is reoxidized to **NAD⁺**, thus allowing anaerobic glycolysis to proceed (**note: conversion of glyceraldehyde 3-phosphate to 1,3-bisphosphoglycerate requires NAD⁺**). This anaerobic fermentation of carbohydrate in yeast forms the basis of the beer and wine industry.

On the other hand, when **ethanol is consumed**, the alcohol dehydrogenase reaction is reversed, and it is converted to acetaldehyde in the cytosol of hepatocytes (**Fig. 24-1**). The acetaldehyde next enters mitochondria to be converted to **acetate** by an acetaldehyde dehydrogenase (see **Case Study #1**). The acetate is easily converted to **acetyl-CoA**, then usually oxidized to CO_2 and H_2O by the electron transfer chain (ETC; see Chapter 36). Reduced NADH is produced by both of these dehydrogenase reactions, therefore the intracellular **NADH:NAD⁺ ratio** can become appreciably increased through large amounts of ethanol ingestion, which can affect a number of important metabolic reactions that use these coenzymes. For example, enzymes involved in **gluconeogenesis** (e.g., lactate dehydrogenase and malate dehydrogenase; Chapter 37) and **fatty acid β-oxidation** (acyl-CoA dehydrogenase; Chapter 55) require NAD^+. Thus, these pathways can be inhibited by excessive alcohol intake, and fasting **hypoglycemia** and the accumulation of excessive hepatic triglyceride (**fatty liver infiltration**) can develop. As previously stated, high levels of NADH also favor lactate formation from pyruvate, accounting for the **lactic acidosis** seen with alcohol over-consumption. This diminishes the concentration of pyruvate required for the pyruvate carboxylase reaction, thus further inhibiting gluconeogenesis. Also, when liver glycogen is depleted in severe cases, and thus no longer available for glycogenolysis, hypoglycemia is further exacerbated.

Ethanol also inhibits **antidiuretic hormone (ADH)** release from the posterior pituitary, thus promoting **diuresis** and **hypertonic dehydration** (see Chapter 94). **Acetaldehyde** is a highly reactive molecule that can form adducts with proteins, nucleic acids, and other molecules, and it appears likely that its ability to react in this way explains some of its toxic effects. Ethanol also appears to interpolate into biologic membranes, expanding them and increasing

their fluidity. When membranes affected are excitable, such as neurons, this results in alterations in their action potentials, it impairs active transport across them, and also affects neurotransmitter release. All of these depress cerebral function and, if severe enough, can produce coma and death from respiratory paralysis.

Why is Anaerobic Glycolysis Necessary?

Although anaerobic glycolysis produces only about 5% of the ATP provided during the catabolism of glucose, there are a number of reasons why it is necessary:

1) There are several instances where animals need **quick energy**. In moving from rest to full flight, for example, aerobic oxidation would require a rapid increase in the O_2 supply, which could only be achieved by increasing the blood supply (which usually takes a number of seconds). Thus, an animal who initiates a sprint from the resting position relies heavily on anaerobic glycolysis.

2) A rapid increase in the O_2 supply to tissues requires a well-developed vascular network. In some instances it may prove inefficient to supply a large body mass (i.e., big muscles),

Table 24-1			
Regional blood perfusion and oxygen consumption in the resting state.			
Region	Body Mass (% Total)	Cardiac Output (% Total)	O_2 Consumption (%Total)
Liver	4.1	27.8	20.4
Kidneys	0.48	23.3	7.2
Brain	2.2	13.9	18.4
Heart M.	0.48	4.7	11.6
Subtotal	**7.26**	**69.7**	**57.6**
Skin	5.7	8.6	4.8
Skeletal M.	49.0	15.6	20.0
Remainder	37.8	6.2	17.6
Whole body	**100.0**	**100.0**	**100.0**

Data from various sources.

with a well-developed blood supply. This is certainly the case for the pectoral muscles of game birds (e.g., pheasants), which are frequently used for escape purposes. In others, the blood supply may be limited because of pathology (e.g., tumors), or physiology (the kidney medulla). In these examples, anaerobic glycolysis may be the major, or only, source of energy.

3) The two major groups of skeletal muscle fibers are **red**, slow-twitch oxidative fibers (**type I**), and **white**, fast-twitch glycolytic fibers (**type IIB**) (see Chapter 80). The type I fibers have high aerobic capacity, and therefore are reasonably fatigue resistant; whereas the type IIB fibers are largely anaerobic. Many fish possess mainly type IIB fibers, with only a thin section along the lateral line being of type I. The lateral line fibers are used during normal periods of swimming, while the large white muscle mass is used for short bursts of rapid activity. Bluefish, however, contain many type I fibers which provide them with far more aerobic capacity.

When resting skeletal muscle is compared to more highly perfused, oxygen-dependent areas of the body (e.g., **liver**, **kidneys**, **brain**, and **heart**), a key distinction becomes apparent: The liver, kidneys, brain, and heart normally account for only about **7%** of the body mass, yet receive almost **70%** of the cardiac output (CO), and consume **58%** of the O_2 utilized in the resting state (**Table 24-1**). **Skeletal muscle** accounts for nearly **50%** of the normal body mass, yet receives only **16%** of the CO at rest, and consumes only **20%** of the O_2 utilized in the resting state. It is no wonder that anaerobic glycolysis is so important in skeletal muscle, since O_2 is being utilized by more "vital" organs in the resting state, even though these organs occupy a rather small fraction of the total body mass. If exercise were to commence quickly from the resting state, anaerobic glycolysis would be mandatory.

4) Aerobic oxidation of carbohydrates, fats, and amino acids is carried out in mitochon-

dria, rather bulky cell organelles. In some cases it may be desirable to reduce the number of mitochondria (because of their bulk) and, in these instances, the cell would be more dependent on anaerobic glycolysis. For example, the **eye** (namely the cornea and lens) needs to transmit light signals with high efficiency. Optically dense structures such as mitochondria and capillaries would reduce this efficiency (and, if they were present in large amounts, animals might literally "see" those extra mitochondria, as well as the blood flowing by in capillaries). Therefore, most of the glucose (over 80%) used by the cornea and lens is normally metabolized anaerobically.

Mature **red blood cells** have no mitochondria, so all of their energy needs are supplied by anaerobic glycolysis (see Chapters 30 and 31). The space is needed for other molecules, in this case hemoglobin, which occupies about 33% of the cell interior. Also, red blood cells are located in a medium (blood plasma), that always has glucose available. On the other hand, heart muscle is an example of a tissue that has retained its aerobic capacity (many mitochondria), but lacks the ability to exhibit powerful contractile forces (like type IIB anaerobic skeletal muscle fibers that have many more actin and myosin filaments (and fewer mitochondria) per unit area).

Historical Perspective

The **EMP** has historical significance since it was the first complete biochemical pathway unraveled. The work spanned a period of nearly 50 years (1890s-1940s), involving numerous "pioneers" of biochemistry. The phosphoesters, ATP and NAD^+ were discovered, their structures determined, and their participations defined. The sequence of conversions was identified, and many of the enzymes involved were isolated and characterized (so students of biochemistry today can enjoy learning about them).

OBJECTIVES

- Define glycolysis and explain its role in the generation of metabolic energy.

- Identify alternative end-points of the glycolytic degradation of glucose, fructose and/or galactose.

- Write the net reaction for the transformation of glucose into pyruvate, and identify locations where ATP and NADH are formed or utilized.

- Account for the net amount of ATP gained through oxidation of one glucose molecule through the EMP.

- Discuss why the LDH reaction is important to both anaerobic and aerobic muscle fibers.

- Recognize why lactic acid accumulation leads to muscle fatigue.

- Explain the pathophysiologic effects of alcohol over-consumption.

- Explain why anaerobic glycolysis is necessary.

- Identify organs receiving the bulk of the cardiac output at rest.

- Explain why mature erythrocytes, the cornea and lens are largely dependent upon anaerobic glycolysis.

- Explain how the intracellular NADH:NAD$^+$ ratio can be altered through alcohol over-consumption (see Case Study #1).

- Explain causes and relationships between alcohol over-consumption, hypoglycemia, fatty liver infiltration, and metabolic acidosis.

QUESTIONS

1. **Metabolic accumulation of lactate in plasma is generally associated with all of the following, EXCEPT:**
   a. Acidosis.
   b. A decrease in the plasma HCO_3^- concentration.
   c. Anaerobic glycolysis in heart muscle.
   d. A decrease in the plasma Cl^- concentration.
   e. Anaerobic glycolysis in type IIB skeletal muscle fibers.

2. **Anaerobic glycolytic enzymes are normally found in which cellular compartment?**
   a. Cytoplasm
   b. Mitochondria
   c. Nucleus
   d. Endoplasmic reticulum
   e. Lysosomes

3. **Which one of the following structures is best associated with anaerobic glycolysis?**
   a. Cardiac muscle fibers
   b. Type I skeletal muscle fibers
   c. The medullary portion of the loop of Henle
   d. The mitochondrion
   e. Smooth muscle cells of blood vessels

4. **During periods of rest, which one of the following normally receives the least amount of blood per unit area?**
   a. Heart muscle
   b. Skeletal muscle
   c. Kidney tissue
   d. The brain
   e. The liver

5. **Anaerobic conversion of pyruvate to lactate:**
   a. Produces NAD$^+$, which is required for further anaerobic oxidation of glucose.
   b. Is the basis of the wine and beer industry.
   c. Occurs normally in cardiac muscle tissue.
   d. Occurs normally in liver tissue (during exercise).
   e. Produces NADH, which is required for further anaerobic oxidation of glucose.

6. **Alcohol over-consumption can promote:**
   a. Hypoglycemia and fatty liver infiltration.
   b. Excessive hepatic NADH production.
   c. Metabolic acidosis.
   d. Hypertonic dehydration.
   e. All of the above

ANSWERS

6. e
5. a
4. b
3. c
2. a
1. c

Chapter 25

Initial Reactions in Anaerobic Glycolysis

Overview

- The "Polyol Pathway" is responsible for conversion of glucose to fructose in diabetic cataract, and in diabetic nerve cells.

- Sorbitol dehydrogenase is present and aldose reductase absent from liver tissue.

- Phosphofructokinase (PFK) catalyzes a key regulatory reaction in glycolysis.

- ATP and citrate powerfully inhibit PFK, while ADP, insulin and NH_3 stimulate its activity.

- Glucagon inhibits hepatic PFK.

- Epinephrine stimulates PFK activity in muscle tissue.

We will now begin our examination of each reaction in the glycolytic pathway, realizing that most all living eukaryotic and prokaryotic cells are dependent upon them. The initial reactions convert **glucose** to **fructose 1,6-bisphosphate (Frc-1,6-bisP)** (**Fig. 25-1**; **note:** glucose and intermediates in glycolysis have been drawn in their linear forms so that structural changes can be more easily recognized). In most cells this conversion occurs in three steps, first a phosphorylation (reaction **#1**), then an isomerization (reaction **#2**), and then another phosphorylation (reaction **#7**). However, in some cell types glucose can take an alternate route to Frc-1,6-bisP through sorbitol and fructose. The initial steps in glycolysis trap glucose in the cell (see Chapter 22), and ultimately form a compound (**Frc-1,6-bisP**) that can be easily cleaved into phosphorylated three carbon units (i.e., **glyceraldehyde 3-phosphate (Gl-3-P)**, and **dihydroxyacetone phosphate (DHAP)**). Since the first reaction in glycolysis (from glucose), i.e., glucose phosphorylation, was covered in Chapter 22, and glycogen degradation was covered in Chapter 23, we will begin with the metabolism of galactose, and its entry into the glycolytic scheme.

Galactose, derived from dietary sources, is largely removed from the portal circulation by the liver and converted in a series of reactions to **glucose 1-phosphate (Glc-1-P)**, and then to **glucose 6-phosphate (Glc-6-P)**. The ability of the liver to accomplish this conversion forms the basis for a test of hepatic function (i.e., the galactose tolerance test). Since Glc-1-P can also be used to synthesize galactose, glucose can be converted to galactose in several different cell types, so that preformed galactose is not essential in the diet. Galactose is synthesized in the body for the formation of lactose (in lactating mammary glands), and it is also a constituent of glycolipids (cerebrosides), and glycoproteins (see Chapters 20 and 59).

The conversion of **Glc-6-P** to **fructose 6-phosphate (Frc-6-P)** is catalyzed by the enzyme **phosphohexose (or glucosephosphate) isomerase** (reaction **#2**). This reaction is readily reversible, and changes an aldopyranose (glucose) to a ketofuranose (fructose). Unlike

Copyright © 2015 Elsevier Inc. All rights reserved.

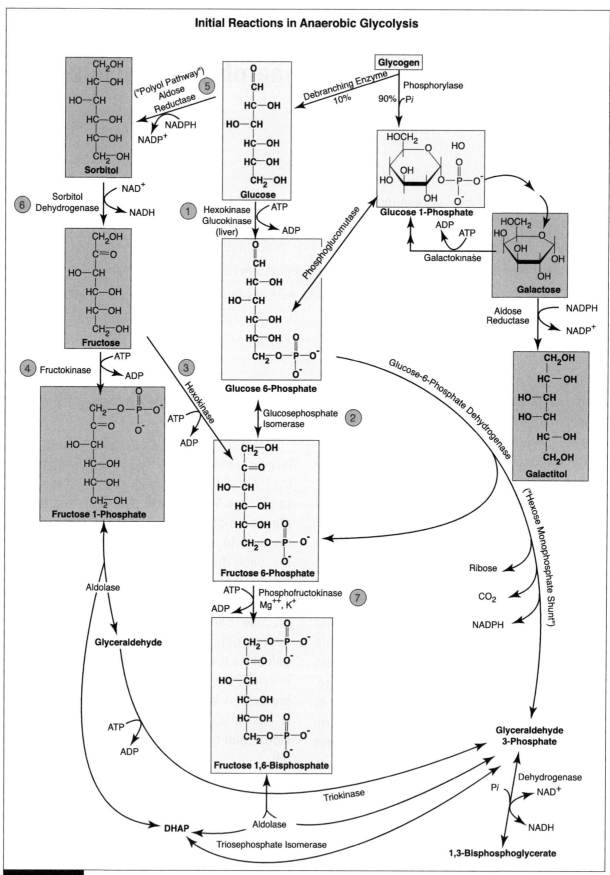

Figure 25-1

the next step in glycolysis (**#7**, i.e., conversion of Frc-6-P to Frc-1,6-bisP), conversion of Glc-6-P to Frc-6-P is not stringently regulated. It should, however, be noted at this point that when Glc-6-P enters the hexose monophosphate shunt (HMS), its products return to the EMP at the level of Frc-6-P and Gl-3-P (see Chapter 28).

Fructose may also be converted in part to **Frc-6-P** (reaction **#3**), and then metabolized via Frc-1,6-bisP. The enzyme catalyzing formation of Frc-6-P is **hexokinase**, the same enzyme that catalyzes conversion of glucose to Glc-6-P. However, much more fructose is converted to **fructose 1-phosphate (Frc-1-P)** in a reaction catalyzed by a highly specific liver enzyme, **fructokinase** (reaction **#4**), which has a **low K_m** (like hexokinase), and has also been demonstrated in kidney and intestine. This enzyme will not phosphorylate glucose, and, unlike hepatic glucokinase, its activity is not affected by starvation or by insulin, which helps to explain why fructose disappears from the blood of diabetic patients at a normal rate. Although it has been recommended that fructose be given to diabetic patients to replenish their carbohydrate stores, most, unfortunately, is metabolized in the intestine and liver, so its value in replenishing carbohydrate stores elsewhere in the body is limited.

It should be noted at this point that both **fructose** and **sorbitol** (a polyol also known as **glucitol**), are found in the **lens**, where they increase in concentration in hyperglycemic diabetic patients, and are involved in the pathogenesis of **diabetic cataract**. The sorbitol pathway from glucose (called the "**Polyol Pathway**"), is responsible for fructose formation, and increases in activity as the glucose concentration rises in **diabetes mellitus** (in those tissues that are not insulin sensitive (i.e., the lens, nerve tissue, intestinal mucosa, erythrocytes, renal tubules, and glomeruli)). Glucose undergoes reduction by NADPH to sorbitol, catalyzed by **aldose reductase** (reaction **#5**), followed by oxidation of sorbitol to fructose in the presence of NAD$^+$ and **sorbitol dehydrogenase (SDH,** reaction **#6**). Although sorbitol and fructose can be metabolized to glycolytic intermediates, this process is slow. Additionally, sorbitol does not diffuse through cell membranes easily, and its accumulation causes osmotic damage by allowing ingress of water with consequent swelling, and eventually cataract formation (clouding of the liquid contents of the lens, probably due to a change in protein solubility). A related polyol, **galactitol** (also known as **dulcitol**), can be formed when **galactose** is reduced by **aldose reductase** and **NADPH**, and it may also accumulate in the lens and participate in the formation of cataracts, particularly in cases of defective hepatic galactose catabolism (i.e., **galactosemia**). This, unfortunately, can be even more serious than sorbitol accumulation since galactitol cannot be further metabolized in the lens.

Cataracts have been prevented in diabetic rats by the use of **aldose reductase inhibitors**. Aldose reductase is also found in the placenta of the ewe, and is responsible for the secretion of sorbitol into fetal blood. The presence of **SDH** in the livers of mammals, including the fetal liver, is responsible for the conversion of sorbitol to fructose. Its elevation in plasma is usually indicative of liver cell damage, particularly in large animal species. The Polyol Pathway, present in the seminal vesicles, is also reported to be responsible for the occurrence of fructose in seminal fluid.

When sorbitol is administered intravenously, it is converted to fructose rather than to glucose, although if given by mouth much apparently escapes intestinal absorption, and is fermented in the colon by bacteria to products such as acetate and histamine (thus sometimes causing abdominal pain). If the liver and intestine of an experimental animal are removed, for example, conversion of injected fructose to glucose does not take place, and the animal succumbs to hypoglycemia unless glucose is administered. Now, back to the EMP.

Reactions **#2** and **#3** are followed by another phosphorylation reaction with ATP (like that forming Glc-6-P), thus converting **Frc-6-P to Frc-1,6-bisP** (reaction **#7**). This reaction, catalyzed by **phosphofructokinase (PFK)**, and also using Mg^{++} and K^+ as cofactors, is a **key regulatory step in glycolysis**. In order to reverse this reaction in hepatic glyconeogenesis (see Chapter 37), PFK must be inhibited and **fructose 1,6-bisphosphatase** must be activated (by glucagon and the other diabetogenic hormones.)

Inhibitors of PFK
ATP
Phosphocreatine
Citrate
Glucagon (liver)
H^+
Stimulators of PFK
AMP and ADP
Frc-6-P
Inorganic phosphate (Pi)
Ammonium ion (NH_4^+)
Epinephrine (muscle)
Insulin

A fall in intracellular pH ($\uparrow[H^+]$) inhibits PFK activity. This inhibition helps to prevent excessive lactic acid formation, and a further precipitous drop in blood pH (acidemia). The concentration of **ATP** is typically about 50 times that of AMP in the cell. Consequently, conversion of only small amounts of ATP to ADP can produce a significant percentage increase in the **AMP** concentration (see Chapter 77). This fact, combined with allosteric PFK activation by AMP, makes PFK activity sensitive to small changes in a cell's energy status, so that the amount of precious carbohydrate spent on glycolysis is accurately controlled. On the other hand, **ATP** and **citrate** powerfully

inhibit PFK. Breakdown of fat (in mitochondria) yields high levels of ATP and citrate (see Chapter 55). Under this condition in muscle tissue, energy is plentiful and PFK inhibition partially spares glucose from further glycolytic breakdown. However, a certain amount of glucose oxidation in muscle is still required during aerobic β-oxidation of fatty acids in order to assure sufficient oxaloacetate availability to keep the TCA cycle functioning (see Chapters 35 and 79).

Phosphocreatine is a short-term buffer for ATP (see Chapter 77), and helps to maintain ATP at normal concentrations, even at the start of a sudden energy demand:

ADP + Phosphocreatine <—> ATP + Creatine

Phosphocreatine levels in resting muscle are typically two to three times greater than ATP levels, and usually fall following the onset of exercise. Therefore, when PFK needs to be activated, the potentiation of ATP inhibition by phosphocreatine is removed. Increased degradation of AMP during exercise leads to **ammonium ion (NH_4^+)** and **Pi** formation (see Chapters 17 and 77), which stimulate PFK activity. **Epinephrine**, a hormone secreted from the adrenal medulla during exercise due to sympathetic nervous system stimulation, activates PFK in muscle tissue, but not in the liver, for during exercise the liver is in a gluconeogenic state.

In summary, the initial cytoplasmic reactions in anaerobic glycolysis generally use either galactose, fructose, glycogen, or more often glucose as substrates. In hyperglycemic patients, the "polyol pathway" becomes important in insulin-independent tissues. Phosphofructokinase is a key regulatory enzyme in anaerobic glycolysis, and the activity of this enzyme is regulated by various factors including the ATP/ADP ratio, the H^+ concentration, phosphocreatine, citrate, glucagon, Frc-6-P, Pi, NH_4^+, epinephrine, and insulin.

OBJECTIVES

- Describe steps in the conversion of glucose to Frc-1,6-bisP, including all intermediates and enzymes, and locations where ATP is consumed.

- Outline two pathways for the conversion of fructose to Gl-3-P.

- Explain how galactose is incorporated into the hepatic EMP, and why preformed galactose is not essential in the diet.

- Recognize the importance of the "Polyol Pathway" to diabetic animals, and discuss potential causes and effects of galactosemia.

- Note where hydrocarbons entering the HMS return to the EMP.

- Describe the numerous factors involved in allosteric regulation of PFK activity, and understand why PFK is considered to be a key regulatory enzyme of the EMP.

- Identify tissues containing aldose reductase and/or SDH.

- Discuss why it is important for hepatic conversion of Frc-1,6-bisP to Frc-6-P to be regulated by an enzyme other than PFK.

- Explain how glycolytic rate control occurs in exercising muscle fibers during a sprint (see Chapter 77).

QUESTIONS

1. **Which one of the following enzymes is known to convert glucose to sorbitol in nerve cells as well as in the lens of the eye?**
   a. Sorbitol dehydrogenase
   b. Fructokinase
   c. Hexokinase
   d. Phosphofructokinase
   e. Aldose reductase

2. **Which one of the following enzymes is reasonably "liver specific" in large animals?**
   a. Hexokinase
   b. Aldose reductase
   c. Phosphofructokinase
   d. Sorbitol dehydrogenase
   e. Glucosephosphate isomerase

3. **Which one of the following compounds is known to stimulate phosphofructokinase activity in muscle cells?**
   a. ATP
   b. Citrate
   c. Ammonium ion
   d. Glucagon
   e. Phosphocreatine

4. **Which one of the following is required in the conversion of glucose to sorbitol?**
   a. Pi
   b. NADPH
   c. NAD^+
   d. NADH
   e. $NADP^+$

5. **Which one of the following compounds cannot be converted to glycolytic intermediates in the lens of the eye?**
   a. Glucose
   b. Sorbitol
   c. Fructose
   d. Galactitol
   e. Fructose 1-phosphate

6. **Which one of the following inhibits PFK activity in muscle cells?**
   a. Glucose
   b. Epinephrine
   c. ADP
   d. ATP
   e. Lactate anion

7. **Which one of the following reversible enzymes is active in both hepatic glycolysis and gluconeogenesis?**
   a. Phosphofructokinase
   b. Aldose reductase
   c. Hexokinase
   d. Fructokinase
   e. Glucosephosphate isomerase

8. **The galactose tolerance test is used to assess the ability of the:**
   a. Liver to catabolize this compound.
   b. Mammary gland to synthesize lactose.
   c. CNS to synthesize cerebrosides.
   d. Lens to synthesize sorbitol and galactitol.
   e. Liver to convert fructose 6-phosphate to fructose 1,6-bisphosphate.

8. a
7. e
6. d
5. d
4. b
3. c
2. d
1. e

ANSWERS

Chapter 26

Intermediate Reactions in Anaerobic Glycolysis

Overview

- The intermediate reactions in anaerobic glycolysis involve the cleavage of fructose 1,6-bisphosphate into two triose phosphates, which are ultimately converted to pyruvate in some ATP-yielding reactions.

- NAD^+ is required for glycolysis to continue, and is used in the conversion of glyceraldehyde 3-phosphate to 1,3-bisphosphoglycerate.

- Carbon atoms from glyceraldehyde, derived through the action of aldolase on fructose 1-phosphate, can enter the glycolytic scheme at the level of glyceraldehyde 3-phosphate, or at the level of 3-phosphoglycerate.

- The first site of ATP production in the EMP is from 1,3-bisphosphoglycerate to 3-phosphoglycerate.

- Conversion of 2-phosphoglycerate to phosphoenolpyruvate in erythrocytes can be prevented with fluoride, thus keeping the plasma glucose concentration from changing in stored blood.

- Conversion of phosphoenolpyruvate to pyruvate is "physiologically irreversible".

- Diphosphoglyceromutase catalyzes formation of an important glycolytic intermediate in erythrocytes.

- The anaerobic phase of glycolysis does not yield as much ATP as the aerobic phase.

The intermediate reactions in anaerobic glycolysis begin by cleavage of the hexose **fructose 1,6-bisphosphate (Frc-1,6-bisP)** into two triose phosphates (i.e., **dihydroxyacetone phosphate (DHAP)** and **glyceraldehyde 3-phosphate (Gl-3-P)**). These phosphorylated intermediates are ultimately used to harvest ATP as they proceed through a series of reactions that oxidize them to **pyruvate (Fig. 26-1)**.

The conversion of **Frc-1,6-bisP** to **Gl-3-P** and **DHAP** is catalyzed by **aldolase** (reaction **#8**), a zinc-containing enzyme (see Chapter 49). This is the only degradative step in the EMP involving a C-C bond. The DHAP and Gl-3-P

generated can also be readily interconverted by the enzyme **triosephosphate isomerase (#9)**. The equilibrium of this reaction strongly favors Gl-3-P formation since this compound is continually being phosphorylated to **1,3-bisphosphoglycerate (1,3-bisPG**; reaction **#15)**. **Fructose 1-phosphate (Frc-1-P)**, derived from dietary fructose or from the **Polyol Pathway** (see Chapter 25), can also be split into glyceraldehyde and DHAP by **aldolase (#10)**.

Next, **glyceraldehyde** can gain access to the EMP via another enzyme present in liver, **triokinase (#11)**, which catalyzes the phosphorylation of **glyceraldehyde** to **glyceraldehyde**

Copyright © 2015 Elsevier Inc. All rights reserved.

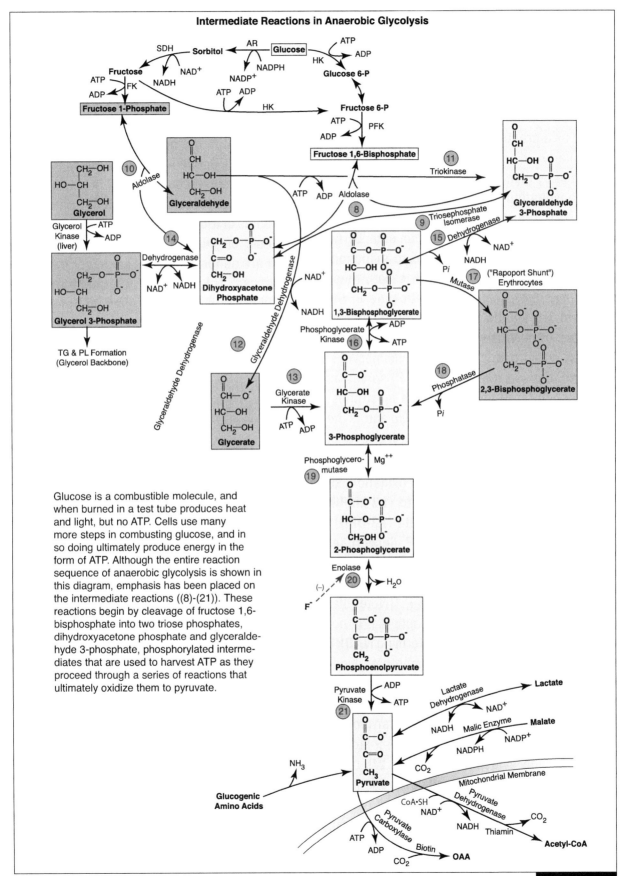

Intermediate Reactions in Anaerobic Glycolysis

Glucose is a combustible molecule, and when burned in a test tube produces heat and light, but no ATP. Cells use many more steps in combusting glucose, and in so doing ultimately produce energy in the form of ATP. Although the entire reaction sequence of anaerobic glycolysis is shown in this diagram, emphasis has been placed on the intermediate reactions ((8)-(21)). These reactions begin by cleavage of fructose 1,6-bisphosphate into two triose phosphates, dihydroxyacetone phosphate and glyceraldehyde 3-phosphate, phosphorylated intermediates that are used to harvest ATP as they proceed through a series of reactions that ultimately oxidize them to pyruvate.

Figure 26-1

3-phosphate (Gl-3-P). Additionally, **glyceraldehyde** can be acted upon by **glyceraldehyde dehydrogenase (#12)** to form **glycerate**, and then **glycerate kinase (#13)** will catalyze formation of **3-phosphoglycerate (3-PG)** from glycerate. Dihydroxyacetone phosphate also serves as a precursor to **glycerol 3-phosphate**, particularly in adipose tissue, a reaction catalyzed by **glycerol 3-P dehydrogenase (#14,** see "**Phosphatidic Acid Pathway**", Chapter 57). Since **NAD$^+$** must be regenerated in order for glycolysis to continue, and since both DHAP and glycerol 3-P are permeable to mitochondrial membranes (where NADH is not), these two compounds sometimes participate in the "**glycerol 3-P shuttle**" (see Chapter 36), where electrons from NADH (rather than NADH itself) are carried across mitochondrial membranes. **Glycerol** (derived from glycerol 3-P) serves as the backbone for triglycerides and most phospholipids (see Chapter 57), and it also enters the hepatic gluconeogenic pathway at the level of glycerol 3-P (see Chapter 37).

Glycolysis proceeds by the oxidation of **Gl-3-P** to **1,3-bisphosphoglycerate (1,3-bisPG;** also called **1,3-diphosphoglycerate (1,3-DPG))**, catalyzed by the enzyme **Gl-3-P dehydrogenase (#15)**; and, because of the activity of **triosephosphate isomerase, DHAP** can be oxidized to **1,3-bisPG** via **Gl-3-P**. (Note that reaction **#15** uses inorganic phosphate (**Pi**), not ATP, and **NAD$^+$** is required, thus forming **NADH**). As stated previously, during anaerobic glycolysis, **NAD$^+$ must be continuously generated for this reaction in order to permit the EMP to continue**. Oxidation of NADH is carried out by either coupling to the pyruvate —> lactate reaction in the cytosol (see Chapter 27), or by coupling to the DHAP —> glycerol 3-P (#14), or oxaloacetate —> malate reactions, where electrons from NADH, rather than NADH itself, are carried across mitochondrial membranes to enter oxidative phosphorylation (see Chapter 36). Also note that **two molecules of 1,3-bisPG** have been generated at this point from **1 molecule of glucose**. This reaction is freely reversible in the liver, and is used in both glycolysis and gluconeogenesis (see Chapter 37).

The high energy phosphate bond formed during the previous reaction (**#15**) is next transferred to ADP during the formation of **3-PG** from **1,3-bisPG**, catalyzed by the enzyme **phosphoglycerate kinase (#16)**. **This is the first site of ATP production in the EMP**. Since two molecules of triose phosphate were formed per molecule of glucose undergoing glycolysis, 2 ATP's are generated at this stage per molecule of glucose. This reaction is a **substrate-level phosphorylation**, a term used to refer to a process in which a substrate participates in an enzyme-catalyzed reaction that yields ATP (or GTP). A phosphate is transferred in this reaction from a high-energy compound (i.e., 1,3-bisPG), that is not a nucleotide. This stands in contrast to oxidative phosphorylation in which electron transport by the respiratory chain of the mitochondrial inner membrane is used to provide the energy necessary for ATP synthesis (see Chapter 36). Although this reaction provides a means for the generation of ATP in anaerobic glycolysis, it can also be used in the reverse direction for the synthesis of 1,3-bisPG at the expense of ATP when hepatic gluconeogenesis is being stimulated (see Chapter 37).

In **erythrocytes** of several mammalian species, **diphosphoglyceromutase (#17)** catalyzes formation of an important intermediate product in this reaction (**2,3-bisphosphoglycerate (2,3-bisPG**, also called **2,3-diphosphoglycerate (2,3-DPG))**, which reduces the hemoglobin binding affinity for oxygen (see Chapter 31). Note that this step effectively bypasses reaction **#16**, and dissipates as heat the free energy associated with the high-energy phosphate of 1,3-bisPG. This loss of high-energy phosphate, which means that there is no net production of ATP when glycolysis takes this "**Rapoport Shunt**", may be of advantage to the economy of the erythrocyte since it would allow glycolysis to proceed when the need for ATP was minimal. The 2,3-DPG formed in erythrocytes can return to the EMP following a dephosphorylation reaction (**#18**).

The next reaction in anaerobic glycolysis involves **phosphoglyceromutase (#19)**, an enzyme that transfers the phosphate from position 3 of **3-PG** to position 2, thus forming **2-phosphoglycerate (2-PG)**. This sets up the production of another high-energy phosphate, as well as more ATP two reactions later.

Enolase (**#20**) catalyzes the dehydration of **2-PG** to **phosphoenolpyruvate (PEP)**, another high-energy compound. This is a remarkable reaction from the standpoint that a high-energy phosphate compound is generated from one with a markedly lower energy level. Although this reaction is freely reversible, a large change in the distribution of energy occurs as a consequence of the action of enolase upon 2-PG. Enolase is inhibited by **fluoride (F⁻)**, a property that is made use of by adding F⁻ to collected blood to inhibit erythrocytic glycolysis prior to estimation of the plasma glucose concentration.

As anaerobic glycolysis continues, the high-energy phosphate of **PEP** is transferred to ADP by the enzyme **pyruvate kinase (#21)**, to generate 2 more moles of ATP (per mole of glucose oxidized), and **pyruvate**. This is another **substrate level phosphorylation** that is accompanied by considerable loss of free energy as heat. However, unlike reaction **#16**, conversion of PEP to pyruvate is **physiologically irreversible**. **Hepatic** pyruvate kinase is **activated by Frc-1,6-bisP** (feed-forward activation), and **inhibited by alanine** and **ATP**. Pyruvate kinase in **muscle** is **not affected by Frc-1,6-bisP**, but it is **inhibited** by **phosphocreatine** and **activated** by a **drop in the ATP/ADP ratio**. The difference in the kinetics and regulation of these two isoenzymes reflects the fact that liver is a gluconeogenic organ, whereas muscle is not.

Anaerobic glycolysis gives a net yield of **two ATP** in the conversion of one glucose to two pyruvate molecules. One ATP is used in the hexokinase reaction, one is used in the PFK reaction, and two ATPs per glucose molecule are generated at the phosphoglycerate kinase and two at the pyruvate kinase reactions (a total of four ATPs generated and two utilized). Anaerobic glycolysis can give a net yield of **three ATP** when the glucose 6-phosphate residue comes from intracellular glycogen (since the hexokinase reaction, which uses one ATP, may not be utilized (see Chapter 23)). If aerobic conditions prevail and pyruvate is allowed to move into mitochondria for acetyl-CoA or oxaloacetate formation, then reducing equivalents from the two NADH generated in the conversion of GI-3-P to 1,3-bisPG (reaction **#15**) can be shuttled into mitochondria for ATP formation (see Chapter 36). Since three ATP are generated in the mitochondrial respiratory chain for each molecule of NADH which enters, a total of six ATP can be generated from cytoplasmic production of NADH during the oxidation of one glucose molecule. This brings the total to **8 ATP** molecules derived (both directly and indirectly) from the cytoplasmic portion of glucose oxidation, or **9** from each molecule of glucose 6-phosphate derived directly from glycogen. The mitochondrial portion of aerobic glucose combustion, however, can yield **30 ATP** equivalents from the complete oxidation of 2 pyruvate molecules through acetyl-CoA, thus bringing the total possible ATP equivalents derived from the complete aerobic oxidation of 1 molecule of glucose to **38** (see Chapter 34). If pyruvate, however, is being converted to oxaloacetate rather than acetyl-CoA (e.g., during fat oxidation in muscle tissue), then this number would be reduced to **30 ATP equivalents**.

With one mol of glucose approximating 686

Kcal worth of energy, and one ATP high-energy phosphate bond being equal to about 7.6 Kcal, it follows that approximately **42% of the energy of glucose is captured in the form of ATP during complete aerobic combustion**. The remaining energy in glucose escapes as heat, which aids in the regulation of body temperature. Of course, eventually all energy derived from glucose oxidation is released as **heat** after ATP is used up while serving its numerous purposes.

OBJECTIVES

- Outline all steps in glycolysis between Frc-1,6-bisP and pyruvate, and recognize the intermediates, enzymes and cofactors that participate in ATP generation.

- Explain the dynamic relationship that exists between glycolysis and the "Glycerol 3-P Shuttle" (see Chapter 36).

- Recognize why glycerol kinase activity is high in liver tissue, and how glycerol 3-P is used in adipose tissue.

- Identify the key reaction in anaerobic glycolysis requiring a steady supply of NAD$^+$, and the reactions that provide this cofactor.

- Discuss the importance of aldolase and triosephosphate isomerase to the EMP.

- Know what the "Rapoport Shunt" is, and why it is important to erythrocytes.

- Explain why fluoride is sometimes added to blood collected from animals.

- Describe how pyruvate kinase activity is regulated in liver and muscle tissue.

- Compare the ATP equivalents derived from anaerobic glycolysis to those derived from aerobic glycolysis.

- Identify reactions in glycolysis that are considered to be "physiologically irreversible," and show how these reactions are circumnavigated in gluconeogenesis (see Chapter 37).

QUESTIONS

1. **Which one of the following enzymes catalyzes the only degradative step in the EMP involving a C-C bond?**
   a. Aldolase
   b. Triokinase
   c. Triosephosphate isomerase
   d. Phosphofructokinase
   e. Glucosephosphate isomerase

2. **Which one of the following phosphorylation reactions uses inorganic phosphate (Pi) rather than ATP?**
   a. Glucose —> Glucose 6-phosphate
   b. Fructose —> Fructose 6-phosphate
   c. Galactose —> Glucose 1-phosphate
   d. Glycerol —> Glycerol 3-phosphate
   e. Glyceraldehyde 3-phosphate —> 1,3-Bisphosphoglycerate

3. **The first reaction that generates ATP in the EMP is:**
   a. Glucose —> Glucose 6-P.
   b. 1,3-Bisphosphoglycerate —> 3-Phosphoglycerate.
   c. Phosphoenolpyruvate —> Pyruvate.
   d. Fructose 1,6-bisphosphate —> DHAP + Glyceraldehyde 3-P.
   e. 2-Phosphoglycerate —> Phosphoenolpyruvate.

4. **Which one of the following glycolytic intermediates is a hexose?**
   a. 3-Phosphoglycerate
   b. Fructose 1,6-bisphosphate
   c. Dihydroxyacetone phosphate
   d. Phosphoenolpyruvate
   e. Glyceraldehyde 3-phosphate

5. **Identify the two enzymes that catalyze substrate-level phosphorylation reactions in the EMP:**
   a. Glucosephosphate isomerase and Triosephosphate isomerase
   b. Phosphofructokinase and Hexokinase
   c. Pyruvate kinase and Phosphoglycerate kinase
   d. Phosphoglyceromutase and Enolase
   e. Aldolase and Glyceraldehyde 3-phosphate dehydrogenase

ANSWERS

1. a
2. e
3. b
4. b
5. c

Chapter 27

Metabolic Fates of Pyruvate

Overview

- Pyruvate occupies a unique position at a major crossroad of carbohydrate, protein, and lipid metabolism.
- Heart muscle removes lactate from the circulation during exercise.
- Primary functions for alanine are incorporation into protein and participation in transamination.
- Cytoplasmic malate and malic enzyme are a source of NADPH for lipogenesis, as well as a source of pyruvate.
- Acetyl-CoA, and increases in mitochondrial ATP/ADP and NADH/NAD$^+$ ratios inhibit pyruvate dehydrogenase.
- Acetyl-CoA is an allosteric activator of pyruvate carboxylase.
- A dietary deficiency of thiamin or niacin will reduce pyruvate dehydrogenase activity, and a biotin or Zn^{++} deficiency will reduce pyruvate carboxylase activity.

Approximately 90% of available metabolic energy that was present in the original glucose molecule remains in the **two moles of pyruvate** produced from it through anaerobic glycolysis. If anaerobic conditions prevail and pyruvate cannot be oxidized to CO_2 and H_2O in mitochondria, it may be converted to **lactate** so that **NAD$^+$ is regenerated** for continued operation of glycolysis and ATP formation in the cytoplasm (**Fig. 27-1**). Under anaerobic conditions, this is quantitatively the most important process for reoxidation of NADH in the cytoplasm of vertebrate cells, and the enzyme catalyzing this reaction is **lactate dehydrogenase (LDH, #22)**. Other effective means of regenerating NAD$^+$ in the cytoplasm are to convert pyruvate (through an amino transferase reaction) to **alanine**, and, under aerobic conditions, to operate the **glycerol 3-phosphate** and **malate shuttles** (see Chapter 36).

Tissues damaged by disease or injury typically release their intracellular enzymes into blood, where subsequent detection and measurement of their activities can provide meaningful information regarding which tissues are most affected (see Chapters 5 and 6). Lactate dehydrogenase exists in the body as several separate isozymes, and the pattern of **LDH isozymes** present in blood is particularly useful in distinguishing, for example, between **myocardial infarction** and **liver disease** (such as infective hepatitis). Two different genes are known to code for LDH. One codes for the **M** or **muscle form**, and the other for the **H** or **heart form**. There are four subunits in LDH, thus giving five possible isozymes: **M$_4$, M$_3$H, M$_2$H$_2$, MH$_3$**, and **H$_4$**. Skeletal muscle contains some of all five isozymes (but predominantly **M$_4$**), and heart has predominantly **H$_4$**. The **H$_4$** isozyme is strongly inhibited by pyruvate; however, the

Copyright © 2015 Elsevier Inc. All rights reserved.

Pyruvate-Branching Reactions

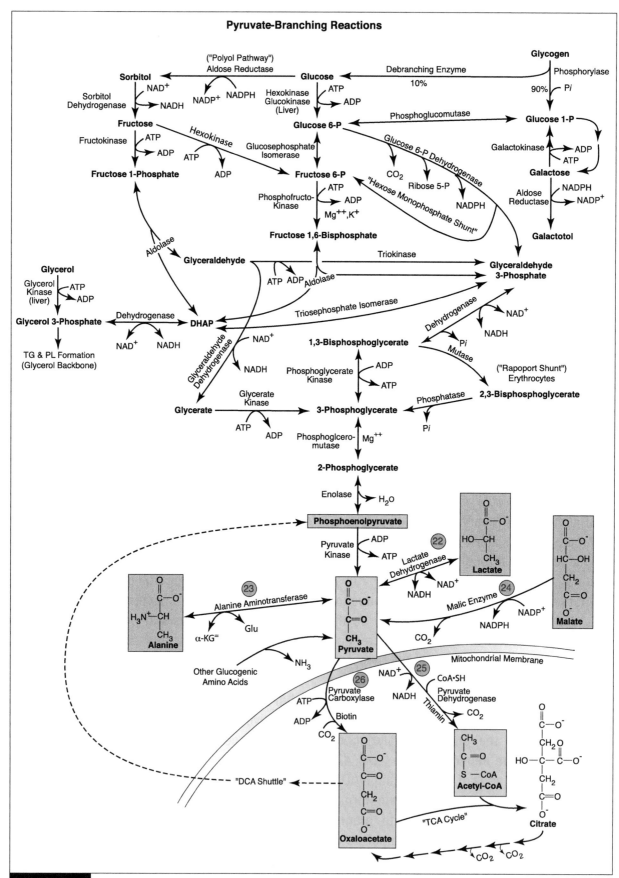

Figure 27-1

M₄ species is not. During periods of exercise when skeletal muscle is producing lactate, heart muscle removes it from the circulation forming pyruvate, which can then enter its numerous mitochondria (see Chapter 80). LDH isozymes are also present in the **liver, kidneys** and **RBCs**, and relatively little tissue injury or hemolysis can result in pronounced circulatory activity since tissue activity is generally high.

The reaction that interconverts **pyruvate** with **alanine** (**Ala**, reaction **#23**), is catalyzed by **alanine aminotransferase** (**ALT**, see Chapter 9). The carbon skeletons of pyruvate and alanine differ by only the substitution of an α-amino group for the carbonyl oxygen (see **Fig. 27-1**). This reaction is reversible, and usually operates in the direction of **alanine —> pyruvate** in **liver** (during gluconeogenesis), and in the direction of **pyruvate —> alanine** in **muscle** (particularly during exercise). Primary functions for alanine are incorporation into proteins and participation in transamination. A large amount of ammonia (NH₃) is transported from muscle and other peripheral tissues to liver in the form of alanine. This process uses pyruvate produced by glycolysis to accept nitrogen from (branched-chain) amino acids in the formation of alanine, which is converted back to pyruvate in the liver where it can then participate in gluconeogenesis. Since the liver supplies glucose to other tissues, pyruvate and alanine constitute a **shuttle mechanism** for carrying nitrogen to the liver to be reutilized or converted to urea (see **Fig. 37-2**). Other glucogenic amino acids that can be converted to pyruvate in the liver include **tryptophan, glycine, serine, cysteine**, and **threonine** (see Chapter 8).

Another enzyme that can give rise to pyruvate in the cytoplasm is **malic enzyme** (**#24**). This enzyme is a decarboxylating enzyme that serves as an additional source of **NADPH** for **lipogenesis**. Although there is little malic enzyme activity in ruminant animals, other mammals that utilize glucose for lipogenesis (by way of citrate), are thought to rely heavily on transferring reducing equivalents from extramitochondrial **NADH** to **NADP** through the combined actions of **malate dehydrogenase** and **malic enzyme**, particularly in liver cells (see Chapter 56). The pyruvate that results from this reaction sequence is then available to reenter mitochondria for conversion to oxaloacetate or acetyl-CoA, or be converted to alanine via ALT (above). In addition to cytoplasmic conversion of malate to pyruvate, it should be noted that malate can also be shuttled back into mitochondria (usually in exchange with either citrate or α-ketoglutarate), or be converted to oxaloacetate (see Chapters 34, 35, and 36).

Mitochondrial formation of **acetyl-CoA** from **pyruvate** is an important irreversible step in animal cells, because they are unable to further convert the acetyl-CoA to glucose. The enzyme that catalyzes this oxidative decarboxylation reaction is **pyruvate dehydrogenase** (**PDH, #25**). The acetyl-CoA generated is now committed to two principal fates:

1) Oxidation to CO₂ and H₂O via the TCA cycle (with generation of ATP through oxidative phosphorylation, see Chapters 34-36), or

2) Incorporation into other compounds (e.g., acetylcholine (an important neurotransmitter), ketone bodies, or citrate.

Once citrate is formed it can either be oxidized in the TCA cycle, or it can diffuse into the cytoplasm where it becomes available for incorporation into various lipids (e.g., fatty acids and steroids).

The activity of PDH is stringently regulated. An increase in the **mitochondrial NADH/NAD⁺, ATP/ADP**, or **GTP/GDP concentration ratios** will **inhibit** this enzyme, as will a buildup of **acetyl-CoA**, whereas presence of **insulin** and an increased **pyruvate** concentration has a **stimulatory** effect. Reduced PDH activity occurs through phosphorylation of a serine residue, whereas activation occurs secondary to an increase in phosphatase activity (and consequent dephosphorylation of PDH). Thus, PDH activity is reduced when the mitochondrial energy level is high and biosynthetic intermediates are abundant (for example during fatty acid β-oxidation), and it is increased when glucose is being funneled into pyruvate and the insulin levels are elevated. **Arsenite or mercuric ions inhibit PDH**, as does a dietary deficiency of the B-complex vitamins, **thiamin** or **niacin** (a source of NAD⁺; see Chapters 40 and 41). This reaction, its cofactor requirements and metabolic inhibitors, is similar to the oxidative decarboxylation reaction of the TCA cycle that is catalyzed by **α-ketoglutarate dehydrogenase** (see Chapter 34).

A PDH deficiency could have serious consequences in those tissues that depend largely on complete aerobic glucose oxidation for ATP generation. This is the case for muscle, kidney, brain and peripheral nervous tissue, and there is evidence that some cases of spino-cerebellar ataxia may be due to PDH deficiency.

Pyruvate carboxylase, a **zinc-containing enzyme** (see Chapter 49), tags a **CO_2** onto **pyruvate**, so that **oxaloacetic acid (OAA)** is formed (reaction **#26**). To drive this reaction energetically, one **ATP** is needed. Furthermore, **Mg⁺⁺** and **Mn⁺⁺** are required, and the B vitamin **biotin** is involved as a shuttler for CO_2 (see Chapter 42). Pyruvate carboxylase activity is especially high in the liver and kidneys, main sites for gluconeogenesis. By producing OAA, this enzyme helps to replenish an important TCA intermediate, and also opens the way for gluconeogenesis from pyruvate and compounds that are converted to pyruvate (such as lactate and various amino acids, see the **dicarboxylic acid (DCA) shuttle**, Chapter 37). A high level of acetyl-CoA inside mitochondria is required for optimal activity of pyruvate carboxylase. **Acetyl-CoA** serves as an allosteric activator

of this enzyme, even though acetyl-CoA itself does not partake in the reaction. The mitochondrial level of acetyl-CoA starts to rise when fatty acids are broken down to fill a demand for energy, and this rise causes pyruvate carboxylase to replenish the TCA cycle (by generating OAA from pyruvate), so that acetyl-CoA can be combusted through citrate formation (**note:** a **six carbon citrate** is formed through the condensation of a **four carbon OAA**, and a **two carbon acetyl-CoA**). This is particularly important in exercising aerobic muscle tissue, where about 66% of the energy demand is met through fatty acid oxidation (and thus acetyl-CoA generation), and 33% through glucose oxidation (and thus OAA generation, see Chapters 35 and 79). Thus, not only is pyruvate carboxylase important in hepatic and renal gluconeogenesis, but it also plays an important role in maintaining appropriate TCA cycle intermediates in exercising muscle and other tissue. These intermediates need to be replenished because they are consumed in some biosynthetic reactions, such as heme biosynthesis (see Chapter 32). This action of pyruvate carboxylase is commonly referred to **anaplerotic**, meaning "to fill up."

Objectives

- Outline the three primary stages of glycolysis, and discuss regulatory reactions controlling each stage.

- Identify four means of regenerating NAD⁺ from NADH so that anaerobic glycolysis can continue.

- Describe regulation of the hepatic, skeletal and cardiac muscle isozymes of LDH.

- Review the importance of the ALT and AST-catalyzed transaminase reactions (see Chapter 9).

- Indicate how amino nitrogen from BCAA oxidation in muscle tissue is normally transported to the liver.

- Explain the actions of malate dehydrogenase and malic enzyme in transferring reducing equivalents from cytoplasmic NADH to NADP in hepatic lipogenesis (see Chapter 56).

- Discuss why there is no net conversion of hydrocarbons from acetyl-CoA to glucose in hepatic gluconeogenesis (see Chapter 37).

- Show how the PDH and pyruvate carboxylase reactions are controlled, and identify the vitamin cofactors involved.

- Understand the differing physiologic roles pyruvate carboxylase plays in providing hydro-carbons for the hepatic "DCA Shuttle," and hydrocarbons for citrate formation in exercising muscle tissue.

- Explain why pyruvate and insulin stimulate PDH activity, while acetyl-CoA stimulates pyruvate carboxylase activity.

Questions

1. **Which one of the following enzymes requires biotin as a cofactor?**
   a. Malic enzyme
   b. Pyruvate dehydrogenase
   c. Alanine dehydrogenase
   d. Pyruvate carboxylase
   e. Lactate dehydrogenase

2. **Which one of the following enzymes has an anaplerotic form of action?**
   a. Pyruvate dehydrogenase
   b. Alanine dehydrogenase
   c. Lactate dehydrogenase
   d. Pyruvate carboxylase
   e. Malic enzyme

3. **Which one of the following LDH isozymes is most prevalent in liver tissue?**
   a. LDH_1
   b. LDH_2
   c. LDH_3
   d. LDH_4
   e. LDH_5

4. **Select the TRUE statement below regarding LDH in heart muscle:**
   a. It is strongly stimulated by pyruvate.
   b. It converts lactate to pyruvate during exercise.

   c. The LDH isozyme in heart muscle is similar to that found in liver tissue.
   d. The LDH isozyme in heart muscle is similar to that found in skeletal muscle tissue.
   e. It is also known to convert alanine to pyruvate.

5. **Which one of the following cytoplasmic enzymes is associated with NADPH generation?**
   a. Lactate dehydrogenase
   b. Malic enzyme
   c. Alanine aminotransferase
   d. Pyruvate dehydrogenase
   e. Pyruvate carboxylase

6. **Under anaerobic conditions the most important glycolytic reaction for generating NAD^+ in the cytoplasm is:**
   a. Glucose —> Glucose 6-P
   b. 3-Phosphoglycerate —> 2-Phosphoglycerate
   c. Alanine —> Pyruvate
   d. Fructose 6-P —> Fructose 1,6-bisphosphate
   e. Pyruvate —> Lactate

7. **Which one of the following enzymes is involved with oxidative decarboxylation of pyruvate?**
   a. Malic enzyme
   b. Lactate dehydrogenase
   c. Alanine aminotransferase
   d. Pyruvate carboxylase
   e. Pyruvate dehydrogenase

8. **During exercise nitrogen is transported from muscle to the liver primarily in the form of:**
   a. Urea.
   b. Uric acid.
   c. NH_4^+.
   d. Lactate.
   e. Alanine.

9. **Acetyl-CoA generated from fat oxidation, is an allosteric activator of:**
   a. Pyruvate carboxylase.
   b. Lactate dehydrogenase.
   c. Pyruvate dehydrogenase.
   d. Malic enzyme.
   e. Alanine aminotransferase.

ANSWERS

1. d
2. d
3. e
4. b
5. b
6. e
7. e
8. e
9. a

Chapter 28

Hexose Monophosphate Shunt (HMS)

Overview

- The HMS generates NADPH for reductive biosynthesis of lipids, and ribose for nucleotide and nucleic acid biosynthesis.

- The HMS has both nonreversible oxidative and reversible nonoxidative phases.

- Muscle tissue is deficient in glucose 6-phosphate dehydrogenase, the rate-limiting enzyme in the oxidative phase of the HMS.

- Liver, adipose, and endocrine tissues possess an active HMS, as does lactating mammary tissue and mature erythrocytes.

- Thiamin deficiency affects the nonoxidative portion of the HMS.

- Sedoheptulose 7-phosphate and erythrose 4-phosphate are intermediates in the HMS.

- In order to completely oxidize glucose in the HMS, GI-3-P must be converted to Glc-6-P, which involves enzymes of the glycolytic pathway working in the reverse direction.

The **HMS** (variously known as the **pentose phosphate**, **phosphogluconate**, or **hexose monophosphate pathway**, **cycle**, or **shunt**), is an alternate cytoplasmic route for the metabolism of **glucose 6-phosphate (Glc-6-P)**. In most tissues, 80-90% of glucose oxidation occurs directly through the Embden-Meyerhoff Pathway (EMP); the other 10-20%, however, may occur in the HMS.

This pathway does not generate **ATP** through substrate-level phosphorylation, but it does generate CO_2 (unlike anaerobic glycolysis). This "alternative" pathway has two main functions:

1) Generation of **NADPH** for reductive biosynthesis of lipids (e.g., fatty acids, cholesterol and other steroids), and

2) Provision of **ribose** residues for nucleotide and nucleic acid biosynthesis (e.g., ATP, NAD^+, FAD, RNA, and DNA).

The HMS is one of three cytoplasmic routes for the production of NADPH (the other two being cytoplasmic isocitrate dehydrogenase and malic enzyme-catalyzed reactions (see Chapter 56)). Since intestinal absorption of ribose is generally too limited to meet metabolic demands, the HMS and the hepatic uronic acid pathway (see Chapter 29) become important sources of pentoses for nucleotide and nucleic acid biosynthesis.

The HMS of animals has both **oxidative (nonreversible)** and **nonoxidative (reversible)** phases, and both can give rise to ribose 5-phosphate (see **Fig. 28-1**). Only the **oxidative phase**, however, gives rise to CO_2 and **NADPH**. In plants, part of the HMS participates in the formation of hexoses from CO_2 in photosynthesis.

The **nonoxidative phase** interconverts three,

Copyright © 2015 Elsevier Inc. All rights reserved.

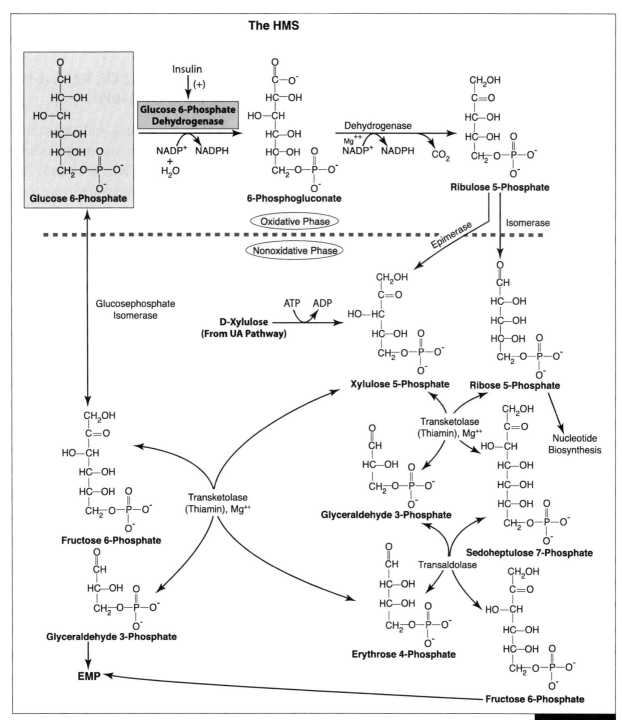

Figure 28-1

four, five, six, and seven-carbon sugars. Most tissues are capable of catabolizing Glc-6-P through both phases, however, **muscle tissue** is deficient in the **rate-limiting enzyme** of this shunt that converts **Glc-6-P** to **6-phosphoglu-conate** (i.e., **glucose 6-phosphate dehydro-genase (Glc-6-PD)**). Therefore, muscle tissue produces only minor amounts of NADPH via the

HMS, and has limited lipid biosynthetic capacity. Muscle tissue does, however, retain the ability to generate ribose 5-phosphate through the non-oxidative phase of the shunt, thereby providing sufficient quantities to meet demands for DNA, and thus protein biosynthesis.

Activity of the HMS is high in **liver** and **adipose tissue**, which require NADPH for the reductive biosynthesis of fatty acids from acetyl-CoA. High HMS activity is also found in **endocrine tissues** which need NADPH for synthesis of cholesterol and steroid hormones (e.g., testes, ovaries, and adrenal cortex), or use ribose for nucleic acids involved in the production of proteinaceous hormones (e.g., insulin, PTH, and the anterior pituitary hormones). In the adrenal medulla and in some nervous tissue, the HMS is needed to provide NADPH for hydroxylation reactions involved in dopamine, norepinephrine, and epinephrine biosynthesis. In the **mammary gland**, HMS activity is high during lactation for the production of milk fats and proteins, but low in the nonlactating state.

Hexose monophosphate shunt activity is also high in mature erythrocytes, the lens and cornea, all of which need **NADPH** for **reduced glutathione** production (which in turn protects them from oxidative damage; see Chapter 30). **Glucose 6-phosphate dehydrogenase deficiency**, with subsequent impaired erythrocytic NADPH generation, has been reported in Weimaraner dogs, and it is the most common genetic enzymopathy known in humans (particularly prevalent among individuals of Mediterranean, Asian, and African descent). Several hundred variants of this enzyme have been identified in erythrocytes. An erythrocytic deficiency in Glc-6-PD can cause an increase in the concentration of methemoglobin, a decrease in the amount of reduced glutathione, an increase in hydrogen peroxide (H_2O_2), and increased fragility of red blood cell membranes. The net result may be hemolysis, which can be exacerbated when subjects are given excessive amounts of oxidizing agents such as aspirin or sulfonamide antibiotics. On the other hand, a relative deficiency of Glc-6-PD in erythrocytes may protect some animals and people from certain parasitic infestations (e.g., falciparum malaria), since the parasites that cause this disease require the HMS and reduced glutathione for optimal growth.

The overall equation for the HMS can be expressed as follows:

$$\boxed{\begin{array}{c} 3 \text{ Glc-6-P} + 6 \text{ NADP}^+ \longrightarrow 3 \text{ CO}_2 + 2 \text{ Glc-6-P} \\ + \text{ Gl-3-P} + 6 \text{ NADPH} \end{array}}$$

This equation erroneously indicates that a single glucose molecule is progressively degraded. This is not, however, the case. The HMS is clearly a more complex pathway than glycolysis. It is multicyclic, in that 3 molecules of Glc-6-P can give rise to 3 molecules of CO_2 and 3 five-carbon residues. The latter can be rearranged to regenerate 2 molecules of Glc-6-P and 1 molecule of the glycolytic intermediate, glyceraldehyde 3-phosphate (Gl-3-P). Since two molecules of Gl-3-P can regenerate Glc-6-P, this pathway can account for the complete oxidation of glucose.

In the first irreversible reaction of the HMS, the **NADP⁺/NADPH** concentration **ratio** exerts primary control over **Glc-6-PD** activity, with both NADP⁺ and NADPH competing for binding to this **rate-limiting enzyme**. As the ratio declines, so does the activity of Glc-6-PD. Glucose 6-phosphate dehydrogenase can also be induced by **insulin**.

The nonoxidative phase of the HMS is controlled primarily by substrate availability. The **NADP⁺/NADPH** concentration ratio is normally about 0.014 in the cytosol of rat liver cells, several orders of magnitude below that for NAD⁺/NADH (i.e., about 700). Mitochondrial ratios for both are about 10:1. A **transhydrogenase** is present on the **inner mitochondrial membrane** that passes electrons from NADH to NADP⁺. Mitochondrial NADPH can be used by glutamate dehydrogenase (see Chapter 9), by hydroxylases involved in the mitochondrial phase of steroid biosynthesis (see Chapter 61), or by mitochondrial enzymes involved in fatty acid chain elongation (see Chapter 56).

In the next reaction, **6-phosphogluconate** is oxidized again by NADP⁺ and decarboxylated to produce **ribulose 5-phosphate**. This compound now serves as a substrate for two different enzymes. It can be **isomerized** to **ribose 5-phosphate**, the compound needed for **nucleotide** biosynthesis (see Chapter 13), or in a separate reaction it can be epimerized at carbon 3 to yield **xylulose 5-phosphate**. These 2 five-carbon sugar phosphates form the starting point for the next series of reactions catalyzed by transketolase and transaldolase. **Transketolase** transfers a two-carbon unit from a ketose to an aldose, a reaction requiring as a coenzyme the B_1 vitamin, **thiamin**, in addition to Mg^{++}. This reaction is severely limited in **thiamin deficiency**, and measurement of erythrocytic transketolase activity is sometimes used as a measure of such (see Chapter 40). **Transaldolase** transfers a three-carbon unit from an aldose to a ketose.

From **Fig. 28-1** it can also be seen that a two-carbon unit is transferred by **transketolase** from **xylulose 5-phosphate** to **ribose 5-phosphate**, thus yielding a three-carbon sugar phosphate, **Gl-3-P**, and a seven-carbon sugar phosphate, **sedoheptulose 7-phosphate**. Using these two sugar phosphates, **transaldolase** allows the transfer of a three-carbon dihydroxyacetone moiety from the ketose **sedoheptulose 7-phosphate**, to the aldose **Gl-3-P**, thus forming the **ketose fructose 6-phosphate (Frc-6-P)**, and the four-carbon aldose, **erythrose 4-phosphate**. Fructose 6-phosphate can now enter the EMP for further oxidation. In another reaction catalyzed by **transketolase**, **xylulose 5-phosphate** and **erythrose 4-phosphate** exchange a two-carbon unit, yielding **Frc-6-P** and **Gl-3-P**.

In order to completely oxidize glucose to CO_2 via the HMS, it is necessary that enzymes be present to convert Gl-3-P to Glc-6-P. This involves enzymes of the glycolytic pathway working in the reverse direction and, in addition, the gluconeogenic enzyme **Frc-1,6-bisphosphatase**, which converts Frc-1,6-bisP

(produced from condensation of Gl-3-P and dihydroxyacetone phosphate), to Frc-6-P (see Chapter 37). This is unlikely, however, in the liver, since **insulin** activates **Glc-6-PD** and inhibits **Frc-1,6-bisphosphatase**.

In summary, the cytoplasmic HMS provides a means for degrading the hexose carbon chain one unit at a time. However, in contrast to the mitochondrial TCA cycle, this shunt does not constitute a consecutive set of reactions that lead directly from Glc-6-P to six molecules of CO_2. Rather, in the nonreversible oxidative phase hexose is decarboxylated to pentose via two NADPH-forming oxidative reactions, and in the reversible non-oxidative phase 3, 4, 5, 6 and 7-carbon sugars are interconverted. NADPH can be used in many different tissues for the reductive biosynthesis of lipids, while ribose 5-phosphate can be used for RNA, DNA, ATP, NAD^+ and FAD biosynthesis. As sugars are rearranged in the HMS, hexoses formed can re-enter the glycolytic sequence, and xyluose 5-phosphate can be accepted from the uronic acid pathway (see Chapter 29).

OBJECTIVES

- Describe the primary functions of the HMS, and identify the tissues in which this shunt is most prevalent.

- Explain how the HMS of muscle tissue produces ribose 5-phosphate for DNA biosynthesis, and why Glc-6-PD is usually absent in this tissue type.

- Know the primary function of the HMS in red blood cells, the lens and cornea.

- Give the overall equation for the HMS.

- Even though it is possible, explain why it is unlikely that the hepatic HMS would account for the complete oxidation of glucose.

- Outline sugar interconversions in the nonoxidative phase of HMS.

- Recognize how the activity of Glc-6-PD is controlled, and explain the symptoms of Glc-6-PD deficiency.

- Explain why RBCs are used to assess thiamin deficiency.

- Identify the glycolytic intermediate siphoned off into the HMS, as well as the glycolytic intermediates where hydrocarbons from this shunt are returned.

- Identify the product from the uronic acid pathway that normally enters the HMS.

QUESTIONS

1. CO_2 is derived from which part of the hexose monophosphate shunt?
   a. The nonreversible oxidative phase.
   b. From the conversion of glucose 6-phosphate to 6-phosphogluconate.
   c. From the conversion of ribose 5-phosphate + xylulose 5-phosphate to sedoheptulose 7-phosphate + glyceraldehyde 3-phosphate.
   d. The reversible nonoxidative phase.
   e. From the conversion of 2 moles of ribulose 5-phosphate to xylulose 5-phosphate and ribose 5-phosphate.

2. The oxidative phase of the hexose monophosphate shunt is least active in which one of the following tissue types?
   a. Liver
   b. Adipose tissue
   c. Endocrine tissue
   d. Lactating mammary gland
   e. Muscle tissue

3. The hexose monophosphate shunt is important to mature erythrocytes because it generates:
   a. Ribose for protein synthesis.
   b. NADPH for reduced glutathione production.
   c. Sedoheptulose 7-phosphate for heme biosynthesis.
   d. Glucose 6-phosphate for glycogen deposition.
   e. 2,3-Diphosphoglycerate, which reduces the affinity of hemoglobin for oxygen.

4. The activity of erythrocytic glucose 6-phosphate dehydrogenase is controlled primarily by:
   a. Substrate availability.
   b. Hormones.
   c. The cytoplasmic NADP⁺/NADPH concentration ratio.

   d. The mitochondrial NADP⁺/NADPH concentration ratio.
   e. Xylulose 5-phosphate negative feedback inhibition.

5. The activity of which enzyme below in the hexose monophosphate shunt would be most affected by thiamin deficiency?
   a. Glucose 6-phosphate dehydrogenase
   b. 6-Phosphogluconate dehydrogenase
   c. Epimerase
   d. Transketolase
   e. Transaldolase

6. Glucose 6-phosphate dehydrogenase deficiency is best associated with:
   a. Uremia.
   b. Malaria.
   c. Mental retardation.
   d. Osteoporosis.
   e. Hemolysis.

7. The seven-carbon intermediate of the HMS is:
   a. Fructose 6-phosphate.
   b. Xylulose 5-phosphate.
   c. Ribulose 6-phosphate.
   d. Sedoheptulose 7-phosphate.
   e. Erythrose 4-phosphate.

8. Intermediates of the hexose monophosphate shunt typically reenter glycolysis at the level of:
   a. Fructose 6-phosphate and glyceraldehyde 3-phosphate.
   b. Dihydroxyacetone phosphate and fructose 1-phosphate.
   c. 1,3-Bisphosphoglycerate and 2-phosphoglycerate.
   d. Phosphoenolpyruvate and pyruvate.
   e. Glucose 1-phosphate and galactose.

9. In order to completely oxidize glucose to CO_2 via the hexose monophosphate shunt, it is necessary that enzymes be present to convert:
   a. Glycogen to glucose 6-phosphate.
   b. Glyceraldehyde 3-phosphate to glucose 6-phosphate.
   c. Fructose 6-phosphate to fructose 1,6-bisphosphate.
   d. Ribulose 5-phosphate to 6-phosphogluconate.
   e. 6-Phosphogluconate to glucose 6-phosphate.

9. b
8. a
7. d
6. e
5. d
4. c
3. b
2. e
1. a

Chapter 29

Uronic Acid Pathway

Overview

- The uronic acid pathway is another cytoplasmic route for the metabolism of Glc-6-P.
- UDP-Glucuronate, formed through the uronic acid pathway, is used for numerous hepatic conjugation reactions.
- Cats have difficulty conjugating, and therefore excreting various drugs from the body that are typically metabolized through the hepatic glucuronide conjugation mechanism.
- Vitamin C is synthesized via the uronic acid pathway in most animals except primates, fish, flying mammals, songbirds, and guinea pigs.
- L-Gulonate serves as a branch point between the uronic acid pathway and the hexose monophosphate shunt through formation of D-Xylulose.
- The sugar moieties of various classes of glycoproteins are derived through the uronic acid pathway.
- Heparin and heparan sulfate formation involve the uronic acid pathway.

The uronic acid pathway is another alternate cytoplasmic route for the metabolism of glucose 6-phosphate (Glc-6-P), which can convert it to **uridine diphosphate glucuronate (UDP-glucuronate)** for use in the biosynthesis of vitamin C and glycoproteins, or hepatic detoxification of endogenous and exogenous compounds (**Fig. 29-1**). Like the hexose mono-phosphate shunt (HMS), the uronic acid pathway does not produce ATP via substrate-level phosphorylation. However, it does generate reducing equivalents in the form of **NADH**, which can subsequently give rise to **ATP** through mitochondrial oxidative phosphorylation (see Chapter 36).

Glucose 6-phosphate is converted to **glucose 1-phosphate (Glc-1-P)** in the first reaction of the uronic acid pathway, which then reacts with uridine triphosphate (UTP) to form the active nucleotide, **UDP-glucose**. These initial reactions, which are also involved in glycogen formation, were previously described in Chapter 23. Next, UDP-glucose is oxidized by an **NAD$^+$-dependent dehydrogenase** to yield **UDP-glucuronate**, using two moles of NAD$^+$. (**Note:** At pH 7.4, most acids are dissociated; hence, the use of the name for the anionic (UDP-glucuronate) rather than the acid form (UDP-glucuronic acid) of this compound is more appropriate). The liver and, to a lesser extent, the kidneys use UDP-glucuronate for conjugation reactions with non-polar, lipophilic endogenous compounds (e.g., steroid hormones and bilirubin, a natural break-down product of heme), and with some lipophilic drugs and toxic substances entering from the diet. These two organs may then excrete these more polar, water-soluble conjugates into bile and/

Copyright © 2015 Elsevier Inc. All rights reserved.

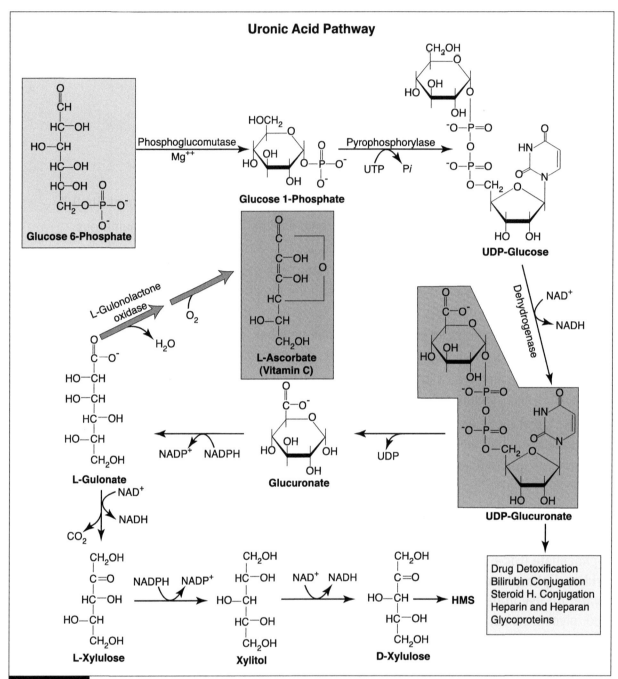

Uronic Acid Pathway

Figure 29-1

or urine. Since glucuronide conjugates are also more soluble in the watery medium of blood plasma, they are less bound to plasma proteins, and therefore are more likely to be filtered by the kidneys and excreted into urine following release from liver cells into blood.

The liver contains several isoforms of **UDP-glucuronosyltransferase** (**UGT**), the enzyme used in these conjugation reactions. Some isoforms appear to be selective for endogenous compounds (e.g., **bilirubin**), while others target exogenous compounds. Hydrophobic compounds (like bilirubin, steroid hormones, xenobiotics, phenobarbital, zidovudine (AZT), and other drugs) typically have one or more hydroxyl groups that become linked to glucuronate through **O-glycosidic bonds**. A typical glucuronide conjugation reaction is shown in

Fig. 29-2. Drug exposure is known to induce synthesis of the glucuronosyltransferase isoform specific for that drug, thereby, over time, enhancing excretion and promoting tolerance.

Cats appear to have a less diverse pattern of UGT isoform expression for exogenous compounds than do other species. This helps to explain why cats have difficulty conjugating, and therefore excreting various drugs from the body (e.g., **aspirin**, **acetaminophen**, **diazepam**, and **morphine**) that are typically metabolized through the hepatic glucuronide conjugation mechanism. Such differences may indeed reflect the evolutionary influence of the carnivorous diet of Feliform species, and resultant minimal exposure to plant-derived toxins which may otherwise keep these glucuronide-conjugating isoforms functional.

Carbohydrate-protein complexes (i.e., **glyco-saminoglycans**, **mucopolysaccharides**, and **glycoproteins** such as **chondroitin sulfate**, **dermatan sulfate**, and **hyaluronic acid**) can also be derived from UDP-glucuronate, and therefore become important in various body structures. Cartilage, bone, skin, umbilical cord, heart valves, arterial walls, cornea, and tendons are representative examples. Hyaluronic acid also occurs in synovial fluid and the vitreous humor of the eye; it serves to lubricate joints and to hold water in interstitial spaces.

In **heparin**, as in dermatan sulfate, the UDP-glucuronate is epimerized to iduronate, and sulfate is added. This compound contains repeating disaccharide units of **glucosamine** and **iduronate**. It is produced and stored in mast cells, and it is also found in the liver, lung, and skin. Heparin is known for its **anticoagulant** and lipid-clearing properties. **Heparan sulfate** (not to be confused with heparin), is present on many cell surfaces, particularly endothelial cells in the walls of blood vessels (where it is produced). It contains repeating disaccharide units of **glucosamine** and **glucuronate**. Heparan sulfate is a negatively charged proteoglycan that participates in cell growth and cell-cell communication. It is also found in the basement membrane of the glomerulus, where it plays a major role in determining the charge selectiveness of the renal glomerular barrier. When this barrier is compromised, plasma proteins begin to appear in urine.

Since further reactions of **UDP-glucuronate** involve the sugar rather than its UDP form, the nucleotide must be removed. Next, glucuronate can be reduced by **NADPH** to yield **L-gulonate**. In most animals L-gulonate can be converted to **L-ascorbate (vitamin C)**. However, in **primates**, **fish**, **flying mammals (Chiroptera)**, **songbirds (Passeriformes)**, and **guinea pigs**, this conversion does not occur because of the absence of the enzyme **L-gulonolactone oxidase**. Therefore, these organisms **require vitamin C in their diets**. Those species that synthesize ascorbate may also benefit from dietary supplementation at critical times (e.g., during rapid growth or stress; see Chapter 39).

A prolonged **vitamin C deficiency** can lead to scurvy, decreased resistance to some infections, and alterations in connective tissue structure. Signs of scurvy include small, subcutaneous hemorrhages, muscle weakness, soft swollen gums, and loose teeth. Vitamin C is a powerful reducing agent and antioxidant, thus influencing several oxidation-reduction reactions in the organism. It functions in the hydroxylation of proline, which is used for

---

Typical Glucuronide Conjugation Reaction

$$R-OH + UDP-Glucuronate \xrightarrow{\text{UGT}} R-O-Glucuronide + UDP$$
(Insoluble)   (Soluble)

Figure 29-2

collagen formation, thus influencing connective tissue formation (see Chapter 3). Vitamin C helps to facilitate bile acid formation in the liver (see Chapter 62), and it also assists in the synthesis of catecholamines from tyrosine (see Chapter 39).

L-Gulonate also serves as a branch point between the **uronic acid pathway** and the **HMS**. Loss of CO_2 from L-gulonate yields **L-xylulose**, which can then be converted to its D-isomer through a preliminary NADPH-dependent reduction to **xylitol**. This product is oxidized via an NAD^+-dependent reaction to **D-xylulose**, which can then enter the HMS following phosphorylation (as **xylulose 5-phosphate**, see Chapter 28).

The uronic acid pathway is known to be affected in several ways. For example, lack of the enzyme needed to convert L-xylulose to xylitol can produce essential or idiopathic **pentosuria**. In this disease considerable quantities of L-xylulose may appear in urine. Administration of aminopyrine or antipyrine have been reported to increase the excretion of L-xylulose in pentosuric subjects.

Additionally, **xylitol** is an artificial sweetner, and in small amounts can cause **canine** hypoglycemia, hypokalemia, and hypophosphatemia (through an increase in insulin release), and in some cases liver dysfunction.

Some drugs are known to increase the rate at which glucose enters the uronic acid pathway. For example, administration of barbiturates such as phenobarbital, or of chlorobutanol to rats has been reported to result in a significant increase in the conversion of glucose to glucuronate, L-gulonate, and L-ascorbate.

In summary, the uronic acid pathway has five major biosynthetic functions:

1) Synthesis of the **sugar moieties** of various classes of **glycoproteins**.

2) Participation in **heparin** and **heparan sulfate** formation.

3) Production of **UDP-glucuronate** for various conjugation reactions.

4) L-ascorbate (**vitamin C**) formation.

5) To serve as a minor route for the formation of **pentoses** (e.g., D-xylulose), which can enter the HMS.

Like the HMS, the uronic acid pathway is an alternate cytoplasmic route for the metabolism of Glc-6-P (**Fig. 29-3**). It does not produce ATP via substrate-level phosphorylation; however, it does generate reducing equivalents which can give rise to ATP through mitochondrial oxidative phosphorylation.

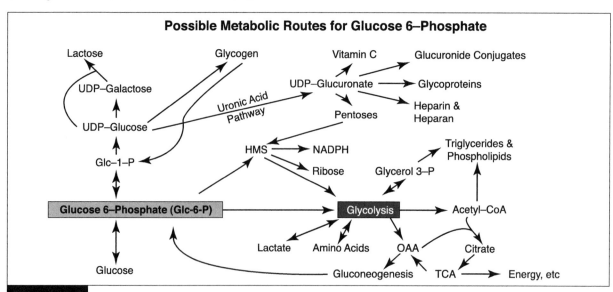

Figure 29-3

OBJECTIVES

- Identify the various routes in which Glc-6-P and UDP-glucose can be metabolized in animal cells.

- Name the five major biosynthetic functions of the uronic acid pathway.

- Explain the physiologic advantage of hepatic glucuronide conjugation.

- Understand why some drugs (e.g., aspirin, acetaminophen, diazepam and morphine) exhibit longer biologic half-lives in cats than in other animal species.

- Distinguish the carbohydrate-protein complexes that can be partially derived from UDP-glucuronate.

- Identify animals that lack L-gulonolactone oxidase, and therefore cannot synthesize L-ascorbate via the uronic acid pathway.

- Contrast and compare CO_2 generation (if present) in anaerobic/aerobic glycolysis, the HMS and the uronic acid pathway.

- Explain why phenobarbital administration may increase vitamin C formation.

QUESTIONS

1. **Which one of the following is true of cats, but not horses?**
   a. They are capable of synthesizing vitamin C from L-gulonate.
   b. They are incapable of conjugating bilirubin with glucuronate.
   c. They are capable of synthesizing the pentose D-xylulose through a series of reactions from UDP-glucuronate.
   d. The have difficulty conjugating aspirin and acetaminophen with glucuronate.
   e. They are incapable of converting glucose 6-phosphate to glucose 1-phosphate.

2. **Which one of the following contains repeating disaccharide units of glucosamine and glucuronate, and plays a role in determining the charge selectiveness of the glomerular barrier?**
   a. Heparan sulfate
   b. Hyaluronic acid
   c. UDP-Glucuronosyltransferase
   d. Heparin
   e. Glycogen

3. **The uronic acid pathway produces:**
   a. ATP via substrate-level phosphorylation.
   b. L-Gulonate from vitamin C.
   c. Pentoses which can enter the HMS.
   d. 6-Phosphogluconate from glucose 6-phosphate.
   e. Reducing equivalents in the form of NADPH.

4. **Absence of which one of the following enzymes prevents primates, bats, guinea pigs, and some aquatic organisms from completing the biosynthesis of ascorbate?**
   a. Phosphoglucomutase
   b. Phosphorylase
   c. Glycogen synthase
   d. L-Gulonolactone oxidase
   e. UDP-Glucuronosyltransferase

5. **The uronic acid pathway:**
   a. Does not produce ATP via substrate-level phosphorylation.
   b. Unlike the hexose monophosphate shunt, occurs in mitochondria.
   c. Uses glucose 6-phosphate dehydrogenase as its rate-controlling enzyme.
   d. Transfers reducing equivalents from NADH to NADPH.
   e. Is found in ruminant animals, but not in carnivores.

6. **Identify the nitrogen-containing intermediate of the uronic acid pathway:**
   a. L-Ascorbate
   b. Xylitol
   c. L-Gulonate
   d. Glucuronate
   e. UDP-Glucose

7. **The active nucleotide, UDP-glucose, can be used as a branch point for entry into:**
   a. Lactose biosynthesis in the mammary gland.
   b. Glycogen biosynthesis in many tissues.
   c. Hepatic L-ascorbate biosynthesis.
   d. All of the above
   e. None of the above

7. d
6. e
5. a
4. d
3. c
2. a
1. d

ANSWERS

Chapter 30

Erythrocytic Protection from O_2 Toxicity

Overview

- Breathing 100% oxygen can be toxic.
- Superoxide anions (O_2^-, i.e., free radicals) are produced normally in the body, and are usually quickly removed through several different cellular processes.
- Hydroxyl radicals ($OH^.$) are considered to be more toxic than superoxide anions or hydrogen peroxide (H_2O_2).
- Hydroperoxidation of membrane lipids adversely affects membrane structure and function.
- Catalase, superoxide dismutase (SOD), glutathione peroxidase, and methemoglobin reductase protect erythrocytes from free radicals.
- Excessive oxidative stress may result in the formation of Heinz bodies in circulating erythrocytes.
- Vitamin E, vitamin C, and perhaps uric acid also act as antioxidants.

Erythrocytes, or **red blood cells (RBCs)**, carry **oxygen (O_2)** from the lungs to tissues, and are involved in the transport of **carbon dioxide (CO_2)** from tissues to the lungs (see Chapter 84). The mature erythrocyte is a highly specialized cell. **Hemoglobin (Hb)**, which constitutes about one-third of its weight, contributes red color and gas-carrying capacity (see Chapter 31). Although RBCs of **primates** may have a 120-day life-span (traveling some 175 miles through the circulation), the life-span of RBCs among domestic animal species varies considerably (e.g., 65 days in the **pig** and 150 days in the **horse**). In smaller animals with a high metabolic rate, the survival time is generally shorter than in larger animals, and red cell survival time generally increases with hibernation. On average, approximately **1%** of circulating erythrocytes are replaced **daily**.

As erythrocytes age, they are removed from the circulation by cells of the **reticuloendothelial (RE) system** (predominantly macrophages in the **spleen**). To balance this degradation, there is a constant need for **hematopoiesis** (production of new RBCs in **bone marrow**). The primary stimulatory factor for red blood cell production is **erythropoietin (EPO)**, a protein hormone produced largely by interstitial cells of the inner renal cortex.

Erythrocytes carry on their surface antigenic determinants commonly known as blood groups. These determinants are commonly, but not always, identified with specific carbohydrate elements of glycoproteins on the cell surface.

Oxygen Toxicity

Although O_2 is essential for mammalian life, breathing 100% oxygen for several hours can

Copyright © 2015 Elsevier Inc. All rights reserved.

be toxic, with damage to the lungs, brain, and retina occurring. At one time this was a major cause of blindness among premature human infants with respiratory distress syndrome. Now, such infants are administered air containing no more than 40% oxygen.

The toxicity of O_2 is due to the production of highly reactive products derived from it. Even under normal circumstances, small amounts of these reduced products are produced; however, RBCs and other tissues are specialized for their removal. These reactive products are the **superoxide anion ($O_2^{-\cdot}$, a free radical)**, **hydrogen peroxide (H_2O_2)**, and the **hydroxyl radical ($OH\cdot$, Fig. 30-1)**. The $O_2^{-\cdot}$ normally has a short half-life (milliseconds), and arises from the acquisition of a single electron (e^-) by a molecule of O_2. It can be generated by the action of ionizing radiation on O_2 (thus causing tissue radiation damage), from xanthine oxidase (an enzyme which catalyzes the oxidation of xanthine to uric acid, see Chapter 17), from flavoprotein oxidases (e.g., aldehyde oxidase), from auto-oxidation of reduced quinones, catecholamines and thiols (i.e., oxidations in the absence of an enzyme), and when O_2 combines with Hb or myoglobin. In contrast to the recent appreciation of the role of superoxide anions, generation of **H_2O_2** in cellular oxidations (and the need to reduce the concentration of this toxic metabolite) has been known for many years. Enzymes known to generate H_2O_2 include D-amino acid oxidase and amine oxidase, as well as **superoxide dismutase (SOD, see below)**.

The **hydroxyl radical ($OH\cdot$)** can be generated in erythrocytes from **H_2O_2** in both the **Haber-Weiss** and **Fenton reactions** (see **Fig. 30-1**). It is considerably more reactive, and therefore more toxic than either **H_2O_2** or **$O_2^{-\cdot}$**. It affects proteins and DNA, as well as **unsaturated fatty acids** in cell membrane phospholipids (**Fig. 30-2** and Chapter 46). **Hydroperoxidation** of membrane lipids has two consequences:

1) It increases the hydrophilic nature of the lipid, which changes membrane structure so that normal function is disturbed.

2) It inhibits some enzymes, thus compromising metabolic processes within the membrane and within the cell.

For example, when red blood cell membranes are sufficiently damaged by hydroperoxidation, cells are more rapidly degraded with resultant anemia. A further problem is that iron atoms in Hb are readily oxidized (even by H_2O_2), and the resulting methemoglobin is unable to properly transport O_2.

Cellular Protection Against Free Radicals

At least four cellular mechanisms appear to play a role in reducing harmful effects of these oxidants (**Fig. 30-3**). Indeed, at normal oxygen tensions they usually eliminate the problem entirely.

The enzyme **SOD** serves to lower normal intracellular $O_2^{-\cdot}$ concentrations to extremely low levels ($<10^{-11}$ M). The H_2O_2 produced is then removed by the action of **catalase**, or

---

Free Radical Formation

A. $O_2 + e^- \longrightarrow O_2^{-\cdot}$

B. $2O_2^{-\cdot} + 2H^+ \longrightarrow H_2O_2 + O_2$

C. $O_2^{-\cdot} + H_2O_2 \longrightarrow O_2 + OH^- + OH\cdot$ (Haber-Weiss Rx.)

D. $Fe^{++} + H_2O_2 \longrightarrow Fe^{+++} + OH^- + OH\cdot$ (Fenton Rx.)

Figure 30-1

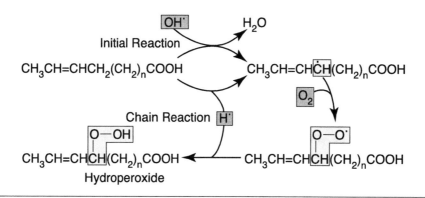

Figure 30-2

Mechanisms for Removing Free Radicals

1. Superoxide Dismutase (SOD) Reaction

$$2O_2^{\cdot\cdot} + 2H^+ \xrightarrow{\text{SOD}} H_2O_2 + O_2$$

2. Catalase (C) Reaction

$$2H_2O_2 \xrightarrow{\text{C}} O_2 + 2H_2O$$

3. Glutathione Peroxidase (GP) and Glutathione Reductase (GR) Reactions

$$H_2O_2 + 2GSH \xrightarrow{\text{GP}} 2H_2O + GSSG$$

$$\text{Lipid-OOH} + 2GSH \xrightarrow{\text{GP}} \text{Lipid-OH} + GSSG + H_2O$$
(Hydroperoxide)

$$GSSG + H^+ + NADPH \xrightarrow[\text{FAD}]{\text{GR}} 2GSH + NADP^+$$

4. Methemoglobin Reductase (MR) Reaction

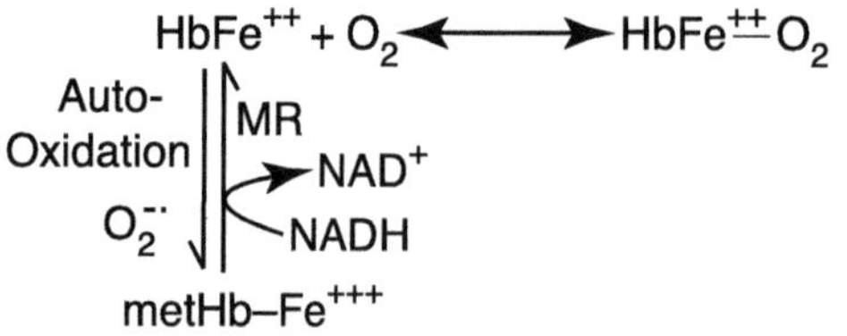

$$HbFe^{++} + O_2 \longleftrightarrow HbFe^{\pm\pm}O_2$$

Auto-Oxidation | MR
→ NAD$^+$
O$_2^{\cdot\cdot}$ ← NADH
metHb–Fe^{+++}

Figure 30-3

---

glutathione peroxidase. **SOD**, a **zinc-** and **copper-containing enzyme**, is present in all aerobic tissues, as well as RBCs (see Chapters 49 and 50).

Catalase is a rather ubiquitous enzyme, that selectively converts H_2O_2 to O_2 and H_2O. Patients deficient in this enzyme, however, show few toxic symptoms, presumably because of the redundant actions of **glutathione peroxi-**

dase.

The tripeptide **glutathione** is present at high concentrations in RBCs (see Chapter 31). The reduced form of glutathione is represented by the abbreviation **GSH**, thus highlighting the importance of the sulfhydryl (-SH) group contributed by cysteine. This group is highly reactive, and may also act non-enzymatically as a free radical scavenger (as well as

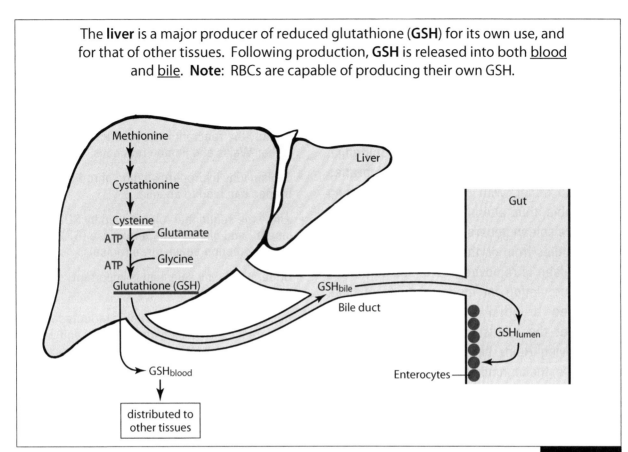

The **liver** is a major producer of reduced glutathione (**GSH**) for its own use, and for that of other tissues. Following production, **GSH** is released into both <u>blood</u> and <u>bile</u>. **Note:** RBCs are capable of producing their own GSH.

Figure 30-4

protecting against H_2O_2 via conversion to H_2O). This vitally important enzyme contains **selenium (Se)**, an essential mineral in the diet (see Chapter 51). In normal RBCs, **GSH** is constantly being oxidized to GSSG, but **GSSG** accounts for less than 1% of total erythrocyte glutathione. Continual reduction of GSSG to GSH is accomplished by the FAD-containing enzyme, **glutathione reductase**, and **NADPH** (produced in the HMS); reduced glutathione (GSH) again becomes available to protect against oxidizing agents. The liver also produces GSH, and exports it into blood and bile for use by other tissues (**Fig. 30-4**).

Methemoglobin reductase does not use NADPH, instead it makes use of **NADH** generated during anaerobic glycolysis in the RBC (see Chapter 31). **Methemoglobin (metHb-Fe^{+++})** is an inactive form of Hb in which the iron has been auto-oxidized from the **ferrous (Fe^{++})** to **ferric (Fe^{+++})** state by superoxides. Methemoglobin is dark-colored, and when present in large quantities in the circulation causes a dusky discoloration of the skin resembling cyanosis. Only heme iron in the ferrous state can carry oxygen reversibly (HbFe^{++} + O_2 <—> HbFe^{++}-O_2), but small amounts of Hb-Fe^{++} still undergo slow auto-oxidation (even in the absence of oxidative stress) to metHb-Fe^{+++}. Methemoglobin reductase converts it back to its functional form. Methemoglobinemia due to methemoglobin reductase deficiency may be inherited or acquired (by ingestion of certain drugs and/or other chemicals).

Excessive oxidative stress may result in the formation of **Heinz bodies** (large rigid structures that distort the RBC membrane). Heinz bodies are formed when the SH groups of Hb become oxidized, and the globin precipitates. **Horses** that feed on **red maple leaves**, for example, sometimes develop an acute hemolytic anemia that is characterized by the formation of Heinz bodies and methemoglobin in erythrocytes. The oxidizing agent in red maple leaves has not been clearly elucidated.

Heinz body formation will generally occur within 24 hours of toxin exposure, and may occur with or without methemoglobinemia. Oxidative injury to erythrocytic membranes may also occur, with changes usually being irreversible, thus altering membrane deformability. The spleen generally functions to remove Heinz bodies from erythrocytes; consequently, if the spleen is removed, Heinz bodies persist in the circulation for longer periods of time. Compared to other domestic animals, the spleen of the **cat** is generally less capable of removing Heinz bodies from erythrocytes. Therefore, these inclusions may be seen in the blood of normal cats. Acetaminophen toxicity is a common cause of Heinz body anemia in cats. Because acetaminophen is normally detoxified in the liver via glucuronidation, and cats are less capable of glucuronidating xenobiotics, the circulating half-life of this drug increases, thus exacerbating the toxicity (see Chapter 29).

Onion ingestion is known to precipitate Heinz body formation in **cattle, horses, dogs,** and **cats**. The toxic agent is reported to be **n-propyl disulfide**, a compound which decreases erythrocytic glucose-6-phosphate dehydrogenase activity, thus decreasing NADPH generation and GSH production.

Vitamin antioxidants available to the body include **tocopherol** (**vitamin E**, see Chapter 46), and **ascorbate** (**vitamin C**, see Chapter 39). Several problems associated with red cell glucose 6-phosphate dehydrogenase deficiency (see Chapter 31) have reportedly improved following vitamin E administration. **Uric acid** is also reported to possess antioxidant activity, and may contribute to longevity in certain vertebrate species.

OBJECTIVES

- Describe the primary functions of erythrocytes, and discuss their turnover rate.

- Show how superoxide anions, H_2O_2 and hydroxyl radicals are normally generated in the body.

- Identify the reactants and products of the Haber-Weiss and Fenton reactions.

- Explain why hydroperoxidation of red cell membranes can lead to anemia.

- Know the sequential roles played by SOD, catalase (C) and glutathione peroxidase (GP) in cellular protection against free radicals.

- Understand why selenium is important to the health of animals.

- Recognize how methemoglobin reductase helps to increase oxy-hemoglobin binding, and why it is needed for NAD^+ generation in erythrocytes.

- Explain how Heinz bodies can be formed in animals.

- Indicate how vitamin E or vitamin C ingestion might improve the condition of an animal with Glc-6-PD deficiency.

- Explain how enterocytes of the digestive tract receive reduced glutathione.

- Know how and why methemoglobinemia develops.

- Recognize why catalase and glutathione peroxidase are somewhat redundant enzymes.

- Recognize the association of reperfusion injury with free radical accumulation (see **Case Study #3**).

QUESTIONS

1. **Which one of the following reactive products derived from oxygen is considered to be the most toxic?**
   a. The superoxide anion ($O_2^-\cdot$)
   b. Hydrogen peroxide (H_2O_2)
   c. The hydroxyl radical ($OH\cdot$)
   d. Water (H_2O)
   e. Superoxide dismutase (SOD)

2. **Hydroperoxidation of membrane lipids has which one of the following consequences for a cell?**
   a. It increases the lipophilic nature of the lipid, thus increasing transport of fat-soluble compounds into and out of the cell.
   b. It increases enzyme activity within the cell membrane.
   c. It causes fatty acids contained in membrane-bound phospholipids to become more saturated.
   d. It increases the hydrophilic nature of the lipids, thus changing membrane structure and decreasing function.
   e. It decreases free radical formation within the membrane.

3. **Which one of the following is an erythrocytic zinc- and copper-containing enzyme?**
   a. Hexokinase
   b. Catalase
   c. Methemoglobin reductase
   d. Glutathione peroxidase
   e. Superoxide dismutase

4. **Which one of the following is a selenium containing enzyme in erythrocytes?**
   a. Hexokinase
   b. Phosphofructokinase
   c. Methemoglobin reductase
   d. Glutathione peroxidase
   e. Superoxide dismutase

5. **Which one of the following erythrocytic enzymes uses NADH?**
   a. Catalase
   b. Phosphofructokinase
   c. Methemoglobin reductase
   d. Glutathione peroxidase
   e. Superoxide dismutase

6. **Methemoglobin is:**
   a. The active form of hemoglobin in which the iron has been auto-oxidized from the ferric (Fe^{+++}) to ferrous (Fe^{++}) state.
   b. Dark-colored (compared with hemoglobin).
   c. Not normally formed in erythrocytes.
   d. Formed by the action of methemoglobin reductase on oxygenated hemoglobin.
   e. Normally found in blood plasma, but not in red blood cells.

7. **H_2O_2 is normally detoxified by:**
   a. Catalase and glutathione peroxidase.
   b. SOD and glutathione reductase.
   c. Methemoglobin reductase and catalase.
   d. The Haber-Weiss and Fenton reactions.
   e. Glutathione peroxidase and SOD.

8. **The usual CBC of a patient on glucocorticoid therapy would be expected to show a:**
   a. Neutrophilia and lymphopenia.
   b. Erythrocytosis.
   c. Hyperglycemia.
   d. Thrombocytosis.
   e. All of the above

9. **Reperfusion injury following myocardial ischemic injury is associated with increased levels of which of the following:**
   a. Superoxide dismutase.
   b. Catalase.
   c. Superoxide and hydroxyl radicals.
   d. Glutathione peroxidase.
   e. All of the above

10. **Heinz bodies are:**
    a. Free radicals.
    b. Precipitated hemoglobin molecules.
    c. Oxidized glutathione molecules.
    d. Methemoglobin molecules.
    e. None of the above

11. **Methemoglobin reductase is needed to:**
    a. Continually return the oxidized form of iron in hemoglobin to the ferrous state.
    b. Convert erythrocytic hydrogen peroxide to water.
    c. Peroxidize unsaturated fatty acids in red cell membranes.
    d. Continually auto-oxidize the iron in erythrocytic hemoglobin.
    e. Form GSH and $NADP^+$ from GSSG, H^+ and NADPH in erythrocytes.

12. **The half-life of the superoxide anion is normally:**
    a. Milliseconds.
    b. Seconds.
    c. Minutes.
    d. Hours.
    e. Days

12. a
11. a
10. b
9. c
8. e
7. a
6. b
5. c
4. d
3. e
2. d
1. c

Chapter 31

Carbohydrate Metabolism in Erythrocytes

Overview

- An insulin-independent GLUT 1 transporter helps to facilitate glucose entry into erythrocytes.

- Ribose 5-phosphate participates in nucleotide salvage inside mature erythrocytes.

- 2,3-BPG, which is generated through the Rapoport-Luebering shunt, may account for up to 50% of available phosphate inside erythrocytes.

- Cl^-, rather than 2,3-BPG, may help to unload O_2 from hemoglobin in the erythrocytes of cats and cattle.

- Mature erythrocytes require purine ribonucleotides, but not pyrimidine ribonucleotides.

- Hexokinase is believed to be the key erythrocytic enzyme whose activity fades with age.

- Respiratory alkalosis promotes an increase in the 2,3-BPG levels of erythrocytes.

The entry rate of **glucose** into red blood cells (RBCs) is normally greater than expected from simple diffusion. It is an example of facilitated diffusion, with the membrane transporter protein involved called a glucose transporter (GLUT), specifically **GLUT 1**. Other GLUT 1 transporters are found in the placenta, kidneys, intestine, and blood-brain-barrier, and all are insulin-independent (see Chapter 22).

The process of glucose entry into erythrocytes is of major importance for their survival. Although reticulocytes (i.e., developing RBCs) are nucleated with considerable quantities of ribosomal and mitochondrial RNA, mature RBCs of mammals have disposed of their intracellular organelles (i.e., nucleus, mitochondria, lysosomes, endoplasmic reticulum, and Golgi). The absence of these structures allows for membrane deformability, and it also allows more room in the cytoplasm for hemoglobin,

the O_2 carrier of blood. Since mature mammalian erythrocytes are missing several important intracellular organelles, they cannot synthesize nucleic acids or proteins, nor can they combust fat for energy. They must thus rely on **glucose** as their sole source of energy.

Birds, herptiles (amphibians and reptiles), and fish, retain a nucleus within their mature red cells; however, it is nonfunctional, and therefore these cells cannot divide. Mammals may have an advantage over these animals since enucleated erythrocytes are thought to be more flexible, and therefore more capable of navigating small capillaries. However, species with nucleated erythrocytes seem to function satisfactorily, despite the presence of these "inferior" structures.

Once glucose enters erythrocytes, it is immediately phosphorylated by **hexokinase (HK)**, an enzyme with a **low K_m** (see Chapter 22). Since

Copyright © 2015 Elsevier Inc. All rights reserved.

glucose 6-phosphate (Glc-6-P) feeds back negatively on HK, decreased metabolic need will generally limit further phosphorylation. However, since glucose cannot be kept out of erythrocytes during hyperglycemia, glucose molecules may still saturate the cell. **Hexokinase is believed to be the key erythrocytic enzyme whose activity fades with age**. A natural result of HK senility is loss of erythrocytic capacity to utilize glucose. Next, the Na^+/K^+- and Ca^{++}-ATPase pumps expire; then, the cytoskeleton becomes rigid and, thus,

RBCs get stuck in narrow vessels (especially the spleen and liver), and are phagocytosed by reticuloendothelial (RE) cells. Other age-related changes occur in the composition of glycoproteins and glycolipids, which attract senile RBCs to lectins on membranes of RE cells.

As previously discussed, the **hexose monophosphate shunt** (**HMS**) of the mature erythrocyte produces **NADPH** and **ribose 5-phosphate** from Glc-6-P (see Chapters 28 and 30, and **Fig. 31-1**). Erythrocytes are quite dependent upon

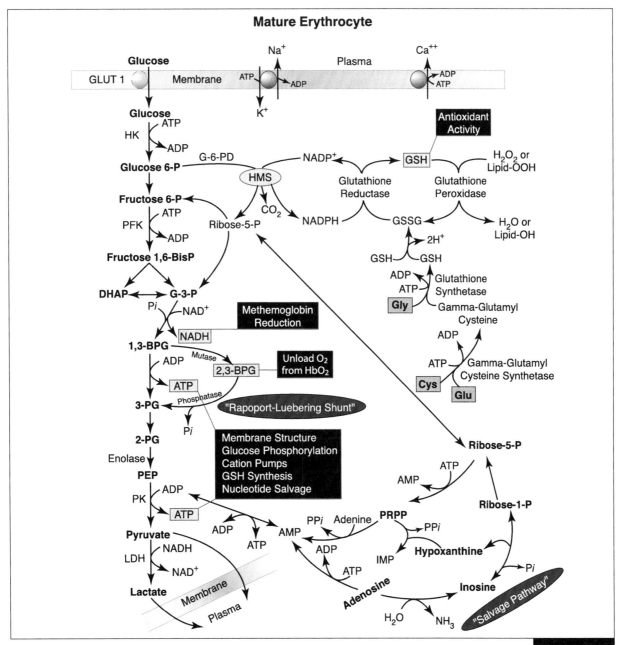

Figure 31-1

NADPH for **glutathione reduction**, and the **ribose 5-P** is utilized in **nucleotide salvage** (see below). Although **glucose 6-phosphate dehydrogenase (Glc-6-PD)** deficiency has not been clearly demonstrated in animals, several million humans demonstrate this hereditary trait, and therefore are susceptible to hemolysis, lipid peroxidation, and other oxidative stresses that may be produced by drugs (such as sulfonamides, primaquine, components of fava beans, or other oxidizing agents).

Approximately 90% of glucose is oxidized anaerobically in RBCs via the Embden-Meyerhoff pathway (EMP), and the end products of metabolism (i.e., **pyruvate** and **lactate)** diffuse into plasma. Lesser amounts of Glc-6-P may pass through the HMS when needed. A second shunt, the **2,3-bisphosphoglycerate (2,3-BPG)** or **Rapoport-Luebering shunt**, is present in the main glycolytic pathway at the step of processing **1,3-bisphosphoglycerate (1,3-BPG)**. Formation of **2,3-BPG** (or **2,3-diphosphoglycerate (2,3-DPG))** by a **mutase** enzyme competes with the formation of **3-phosphoglycerate (3-PG)** by a **kinase**, as both enzymes work on the same substrate. Once 2,3-BPG is formed, it re-enters the EMP at the level of **3-PG**. Up to **15%** of glucose taken up by erythrocytes may pass through this shunt, and since it effectively bypasses the ATP-producing step

directly from 1,3-BPG to 3-PG, this may be economical by allowing glycolysis to proceed when the need for ATP is minimal. In primate erythrocytes, 2,3-BPG may account for over 50% of available phosphorus, and it is about equimolar in quantity with Hb. However, in several animal species studied, erythrocytic 2,3-BPG levels are known to vary considerably (**Table 31-1**).

The most important function of **2,3-BPG** is to **combine with Hb**, causing a **decrease in its affinity for O₂** (and displacement of the oxyhemoglobin dissociation curve to the right). This effectively unloads O_2 from Hb, allowing O_2 to diffuse into tissues where it is needed. This is important in adaptation to high altitude, and in some species the transport of O_2 from maternal to fetal blood (i.e., 2,3-BPG binds more avidly to adult Hb than it does to fetal Hb in most mammals studied, except the **horse** and **pig**, where fetal Hb has been found to be indistinguishable from adult Hb). In animals like **cattle** and **cats**, that have low erythrocytic 2,3-BPG levels, the **chloride anion (Cl⁻)** may be functioning in a similar capacity to that of 2,3-BPG.

Upon moving from sea level to high altitude, many mammalian organisms suffer from acute O_2 shortage, and animals begin to hyperventilate. This leads to **respiratory alkalosis**, which in turn **increases 2,3-BPG levels** in erythrocytes

Table 31-1	
Representative 2,3-BPG Levels in Erythrocytes of Different Animal Species	
Species	**2,3-BPG (mM)**
Rabbit	5.56
Horse	5.28
Pig	5.20
Dog	4.27
Primate	4.00
Dolphin	3.46
Cat	0.50
Cattle	0.46

Data from various sources.

by activating the mutase enzyme. This increase usually takes about 1 hour to occur, and the half-life of 2,3-BPG is about 6 hours. Hormones that increase 2,3-BPG formation, largely through increased erythropoiesis, include **thyroxine**, **growth hormone**, **testosterone**, and **erythropoietin**.

Hemolytic anemias in animals are known to result from genetically determined deficiencies in one or more of the glycolytic enzymes (e.g., pyruvate kinase (PK) or phosphofructokinase (PFK)). Since pyruvate kinase is "downstream" from 2,3-BPG in the EMP, a deficiency in this enzyme would be expected to result in a net **increase** in erythrocytic 2,3-BPG levels, whereas PFK deficiency would be expected to have the opposite effect.

The developing RBC contains **nucleotides** in DNA and RNA. However, those in DNA are lost when the nucleus is extruded, and ribosomal RNA is normally degraded within about 24 hrs of the RBC entering the circulation. The cell, however, can use the resulting purine ribonucleotides from RNA (i.e., adenine and guanine), but pyrimidine ribonucleotides (i.e., cytosine, uracil and thymine) are unwelcome because they compete with ATP and ADP in crucial reactions, and thus interfere with cell function. Erythrocytes must maintain an adequate pool of adenine nucleotides (ATP, ADP, and AMP) to survive. Since the mature erythrocyte cannot synthesize them *de novo*, they are conserved and replenished through a **nucleotide salvage pathway** (see **Fig. 31-1** and Chapter 17). When blood is preserved for transfusion in an adenosine and/or adenine-containing medium, erythrocytes maintain higher levels of ATP, and therefore have increased viability. A continual supply of ATP is needed to maintain membrane deformability, glucose phosphorylation, glutathione (GSH) biosynthesis, and nucleotide salvage. ATP-requiring cation pumps in the plasma membrane independently extrude Ca^{++}, and extrude Na^+ in exchange for K^+ (i.e.,

Na^+/K^+-ATPase). Species differences exist, however, in the relative amounts of Na^+ and K^+ found in erythrocytes. Although most animals maintain high intracellular concentrations of K^+ (and low concentrations of Na^+), cats and most dogs possess low concentrations of K^+ in their mature erythrocytes, due most likely to ever-diminishing ATP supplies. Their reticulocytes, however, contain high concentrations of K^+.

The **white cells of blood**, although normally fewer in number, are concerned with the body's response to injury and infection. Red blood cell metabolism accounts for about 90% of glucose used by blood cells each day, while 10% is normally accounted for by white cell metabolism. It is generally believed that a high rate of glycolysis in white blood cells is associated with their ability to grow and divide rapidly, a property they have in common with cancer cells. The rate of glycolysis in white blood cells is therefore markedly increased when the immune response is activated (e.g., due to infection).

OBJECTIVES

- Explain differences between developing mammalian RBCs are mature RBCs. Also indicate what affects these difference present to overall RBC metabolism.

- Know why glucose floods the erythrocytes of hyperglycemic animals.

- Explain why erythrocytic HK aging correlates directly with RBC turnover.

- Show how NADPH and ribose-5-P are formed in mature erythrocytes, and how they are utilized.

- Indicate where ATP is produced and where it is utilized in mature erythrocytes.

- Describe the two routes 1,3-BPG can take in erythrocytes, and how metabolic alkalosis can affect the direction 1,3-BPG moves. Identify the functional significance of this relationship.

- Indicate how 2,3-BPG (or 2,3-DPG) levels differ among animal species.

- Distinguish why an animal with PK deficiency

may exhibit a difference in RBC O_2 carrying capacity from an animal with PFK deficiency.

• Explain why whole blood is usually preserved in an adenosine and glucose-containing medium, and why older RBCs possess less K^+ than younger RBCs.

QUESTIONS

1. **Mature feline erythrocytes contain:**
   a. Mitochondria.
   b. A high concentration of K^+ (like nerve cells).
   c. Hexokinase with a high K_m.
   d. A nucleus.
   e. Antioxidant activity.

2. **Erythrocytic ATP is needed for all of the following processes, EXCEPT:**
   a. Glutathione reduction.
   b. Cation pumps in the cell membrane.
   c. Nucleotide salvage.
   d. Conversion of fructose 6-phosphate to fructose 1,6-bisphosphate.
   e. Glucose phosphorylation.

3. **Which one of the following erythrocytic enzymes is thought to be the key enzyme whose activity fades with age, and therefore governs the RBC life-span?**
   a. Hexokinase (HK)
   b. Phosphofructokinase (PFK)
   c. Glucose 6-phosphate dehydrogenase (G-6-PD)
   d. Enolase
   e. Pyruvate kinase (PK)

4. **2,3-Bisphosphoglycerate (2,3-BPG) is normally produced by which route in erythrocytes?**
   a. The hexose monophosphate shunt.
   b. The nucleotide salvage pathway.
   c. The Rapoport-Luebering shunt.
   d. The Embden-Meyerhof pathway.
   e. The tricarboxylic acid cycle.

5. **In addition to glucose, which one of the following compounds added to stored blood would most likely help to maintain energy levels within erythrocytes?**

   a. Adenosine
   b. Lactate
   c. Na^+/K^+-ATPase
   d. Glutathione
   e. Ca^{++}

6. **Which one of the following increases activity of 1,3-BPG mutase in erythrocytes, thus increasing production of 2,3-BPG?**
   a. Acidemia
   b. Moving from a high to low altitude.
   c. An increased intracellular pH.
   d. Glutathione
   e. Methemoglobin

7. **Methemoglobin reduction in erythrocytes requires:**
   a. Glutathione
   b. Catalase
   c. ATP
   d. H_2O_2
   e. NADH

8. **Glutathione production requires all of the following, EXCEPT:**
   a. Cys
   b. 2,3-BPG
   c. ATP
   d. Glu
   e. Gly

9. **PFK deficiency would be expected to result in:**
   a. A net increase in erythrocytic 2,3-BPG.
   b. An increase in erythrocytic Na^+/K^+-ATPase activity.
   c. A decreased hemoglobin-O_2 affinity.
   d. Enhanced erythrocyte turnover.
   e. Pyruvate kinase deficiency.

10. **Erythrocytic glutathione reduction requires:**
    a. Glutathione peroxidase.
    b. NADPH.
    c. A mutase enzyme.
    d. Ribose 5-P.
    e. The insulin-independent GLUT-1 transporter.

ANSWERS

10. b
9. d
8. b
7. e
6. c
5. a
4. c
3. a
2. a
1. e

Copyright © 2015 Elsevier Inc. All rights reserved.

Heme Biosynthesis

Chapter 32

Overview

- With the exception of Fe^{++}, all atoms in heme are derived from succinyl-CoA and glycine.
- Part of heme biosynthesis occurs in mitochondria.
- Δ-ALA synthase controls the rate-limiting step in heme biosynthesis.
- Lead inhibits heme biosynthesis.
- Porphobilinogen contains the basic pyrrole ring system used to assemble other mammalian porphyrins.
- Some porphyrins cause photosensitization, and thus skin lesions in animals.
- Harderian glands secrete protoporphyrin IX.
- Δ-ALA can be used in cancer photodynamic therapy.
- Carbon monoxide competes with O_2 for binding to the Fe^{++} of hemoglobin.
- Nitric oxide binds to hemoglobin, which may be of significance in cases of hemolysis.
- Fetal hemoglobin (HbF) and adult hemoglobin (HbA) do not bind 2,3-BPG with equal affinity.

Heme biosynthesis occurs primarily in the bone marrow and liver, although all nucleated cells have the capacity to synthesize this compound. It is derived from **porphyrins**, which are cyclic compounds that have a high affinity for binding metal ions, usually **ferrous (Fe^{++})** or **ferric (Fe^{+++})** iron (**Fig. 32-1**). They contain four **pyrrole rings** linked together by methylene ($-CH_2$-) bridges.

Metalloporphyrins in nature are conjugated to proteins, and form many important biologic compounds. **Hemoglobin (Hb)**, for example, is an erythrocytic iron porphyrin attached to the protein, globin (**Fig. 32-2**). **Myoglobin**, which imparts red color to aerobic muscle fibers, has a structure similar to Hb, and both compounds possess the ability to combine reversibly with O_2. **Cytochromes** are iron porphyrin proteins that act as electron transfer agents in cellular oxidation-reduction reactions. Important examples are the mitochondrial cytochromes associated with oxidative phosphorylation (see Chapter 36), the hepatic cytochrome P-450 system that participates in drug detoxification (i.e., hydroxylation), and the mitochondrial cytochromes involved with steroid hormone biosynthesis. Tissues rich in cytochromes include muscle, liver, kidney, and the steroidogenic tissues (e.g., adrenal cortices, testes, ovaries, and placenta). **Catalases** (see Chapter 30) are ubiquitous iron porphyrin enzymes that degrade H_2O_2 in mammals (similarly to peroxidases in

Heme Biosynthesis

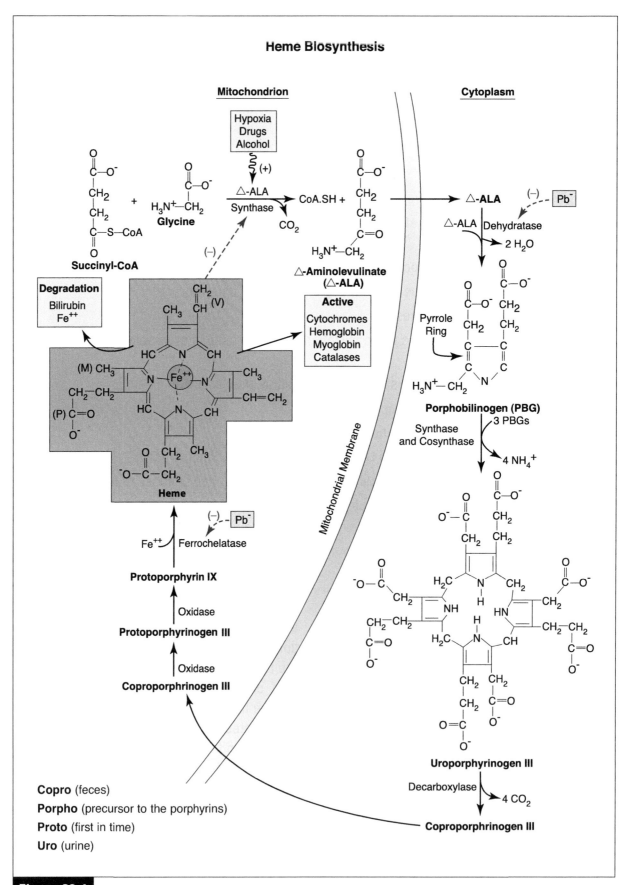

Figure 32-1

plants), and **tryptophan pyrrolase**, another iron porphyrin enzyme, catalyzes oxidation of tryptophan to formyl kynurenine. **Chlorophyll**, a magnesium (Mg^{++})-containing porphyrin, is an important photosynthetic pigment in plants, and also a source of porphyrins in some animal diets.

With the exception of Fe^{++}, all atoms in **heme** come from **succinyl-CoA**, derived from the mitochondrial tricarboxylic acid (TCA) cycle (see Chapter 34), and **glycine (Gly)**. The initial and last three reactions in heme biosynthesis occur in mitochondria, with the intermediate steps being extramitochondrial (see **Fig. 32-1**). The first step is catalyzed by **Δ-aminolevulinate (Δ-ALA) synthase**, a **vitamin B₆-dependent** enzyme (see Chapter 42), and involves condensation of succinyl-CoA and Gly to form Δ-ALA. This enzyme is located in the outer mitochondrial membrane, and its turnover rate is rapid (about 1 hr in mammalian liver), a common feature of an enzyme catalyzing a rate-limiting reaction. The activity of Δ-ALA synthase is subject to **feedback inhibition by heme**, and it is **induced** by **hypoxia** and a variety of compounds including **alcohol (i.e., ethanol)**, **steroids**, and drugs such as the **barbiturates** (phenobarbital and pentobarbital). Once Δ-ALA reaches the cytoplasm, two molecules condense to form **porphobilinogen (PBG)**, the basic pyrrole ring system used to assemble other porphyrins. Formation of PBG is catalyzed by **Δ-ALA dehydratase**, a **zinc-containing** enzyme that is strongly inhibited by **lead (Pb⁻)**. Lead toxicity results in anemia due to decreased heme biosynthesis, and increased amounts of Δ-ALA appear in urine.

Four molecules of PBG next combine to form **uroporphyrinogen III** through the action of two enzymes, **uroporphyrinogen I synthase** and **uroporphyrinogen III cosynthase**. Uroporphyrins were first identified in urine, but they are also found elsewhere in the organism.

Uroporphyrinogen III is next converted to **coproporphyrinogen III** by **decarboxylation** of all its acetate units to methyl groups. Coproporphyrins were first isolated from feces, but they are also found in urine. Coproporphyrinogen III then enters mitochondria, where it is oxidized to **protoporphyrinogen III**, and then to **protoporphyrin IX**. These combined reactions alter the side chains attached to the pyrrole rings, giving them their characteristic one carbon **methyl (M)**, two carbon **vinyl (V)**, and three carbon **propionyl (P)** constituents.

The final step in heme biosynthesis involves incorporation of **Fe⁺⁺** into **protoporphyrin IX** in a reaction catalyzed by **ferrochelatase** (also called **heme synthase**). This enzyme is also inhibited by **Pb⁻**.

The various porphyrin intermediates in heme biosynthesis are **colorless**, but they can be oxidized to red or brown porphyrins. In abnormal porphyrin metabolism, oxidized products can accumulate and stain tissues. These products also absorb light energy, which can lead to **photosensitization** (if animals stay out in UV light), and damage to the skin. The appearance of persons with congenital erythropoietic porphyria, who have deformed faces and venture out only at night to avoid further skin damage, may have been the beginning of the **werewolf** legend.

There are three principal types of photosensitization recognized in domestic animals. One type is caused by ingestion of significant amounts of photodynamic substances not normally present in the diet (e.g., certain poisonous plants), another is due to aberrant pigment production from endogenous sources (e.g., **pink tooth** of cattle or Siamese cats due to excessive presence of porphyrin intermediates), and a third type is due to liver disease in herbivores. In the latter instance, the photosensitizing agent is **phylloerythrin**, a natural porphyrin derivative of chlorophyll produced by intestinal microbes. This photodynamic substance is normally removed from portal blood by the liver, and secreted into bile.

Harderian Glands

These lacrimal (tear) glands are found in amphibians, reptiles, birds and some mammals (i.e., those with nictitating membranes (a third eyelid)), but not in primates. Although the Harderian gland is a deep lacrimal gland, there is a more superficial lacrimal gland also present. Harderian secretions are prominent in rodents, and they drain into the nasolacrimal duct.

Harderian glands were first described in 1694 by **Johan Jacob Harder**, and today are known to produce pheromones, melatonin and protoporphyrin IX. **Protophorphyrin IX** secretions from these glands increase during stress, and stain **red** (or **pink**) under UV light.

Photodynamic Therapy (PDT)

This form of phototherapy uses otherwise nontoxic light-sensitive compounds (e.g., Δ-**ALA**) placed on the skin or injected into the bloodstream, then exposed selectively to light, whereupon they become toxic to targeted malignant or other diseased cells. PDT has also been shown to kill microbial cells, including bacteria, fungi and viruses, and it is sometimes used to treat acne. Although PDT is not widely used in cancer treatment today, it generally has few side-effects. However, it can only be used to treat areas that can be reached by a light source.

Over 60 years ago investigators demonstrated in laboratory animals that **porphyrins** exhibit a preferential affinity for **rapidly dividing cells** including malignant, embryonic, and regenerative cells over quiescent tissues. Because of this they proposed that these agents could be used to treat certain forms of cancer. Once malignant cells have taken up the photodynamic substance, a light source wavelength needs to be appropriate for exciting the photo-sensitizer to produce **reactive oxygen species** (i.e., **free radicals**). Although O_2 is relatively non-reactive under normal conditions, photo-sensitizing chemicals promoted to an excited state upon light absorption cause O_2 to produce **singlet oxygen** (i.e., the **superoxide anion; O^{--}**), which can rapidly attack organic compounds, thus becoming highly cytotoxic. Under normal conditions superoxide anions are eliminated from cells in milliseconds through the action of **superoxide dismutase** (**SOD**; see Chapter 30).

Hemoglobin (Hb)

Myoglobin and Hb are evolutionarily conserved globular proteins. Adult Hb (HbA) has 4 subunits, each containing a heme moiety conjugated to a polypeptide chain, and a 64,450 MW. There are normally two pairs of polypeptide chains in each Hb molecule (designated α- and β-**chains**). Conversely, myoglobin contains only one polypeptide chain, and a single heme group.

In uncontrolled diabetes mellitus, the erythrocytic glucose concentration rises, and glucose molecules attach to the terminal **valine (Val)** in each β-chain of Hb (commonly referred to as **glycosylated Hb**, or **HbA$_{1C}$**, see **Fig. 32-2**). This interaction is thought to have little effect upon the O_2 carrying capacity of Hb; however, HbA$_{1C}$ levels in blood are a useful long-term index of how well the blood glucose concentration of the diabetic animal is being managed.

When Hb binds O_2 (i.e., **oxyhemoglobin**), it attaches to the **Fe^{++}** of heme which is contained in a pocket of hydrophobic amino acids. This is the same site where **carbon monoxide (CO)** or **nitric oxide (NO)** attach, thus forming **carboxyhemoglobin** or **nitrosylhemoglobin**, respectively. The binding affinity of Hb for CO is some **230 times greater** than it is for O_2. Usually, less than 5% of Hb molecules contain bound CO. However, if the percentage is increased through excessive CO exposure, O₂ transport can be severely compromised. Nitric oxide is an important vasodilator, and normally little is bound to Hb. However, in cases of hemolysis, where Hb floats freely in plasma, significant quantities of NO may become Hb-bound, thus leading to hypertension.

Each heme group of Hb is attached to the globin chain by linkage of Fe⁺⁺ to a histidine residue (**Fig. 32-2**). The histidines are arranged so that O₂ can bind with Hb reversibly.

Carbon dioxide (CO₂) does not compete with O₂ for binding to Hb. Rather, small amounts (≈ 20%) of capillary CO₂ bind nonenzymatically to the **amine (NH₂)** terminal of the **globin polypeptide chain**, thus forming **carbamino-hemoglobin**. Most of the CO₂ that enters the RBC in capillaries of the body is converted to H⁺ and HCO₃⁻ through the enzymatic action of **carbonic anhydrase**, and then HCO₃⁻ diffuses down its concentration gradient into plasma

(see Chapter 84). Because deoxyhemoglobin has a higher pK than oxyhemoglobin, the extra H⁺ generated can be immediately bound (i.e., **buffered**) by the **Hb** molecule itself, a process which results in O₂ release in capillary blood. Generally, one H⁺ tends to be bound for each two O₂ molecules released, and this buffering action contributes significantly to the overall buffering capacity of blood. This change in O₂ binding with a slight decrease in pH is called the **Bohr effect**, a phenomenon that is accompanied by a shift in the O₂ dissociation curve to the right (i.e., Hb becomes less saturated at any given partial pressure of O₂). Two other factors that contribute to the Bohr effect in erythrocytes are **2,3-bisphosphoglycerate (2,3-BPG)**, and **carbaminohemoglobin formation** (see Chapter 31).

With the exception of **horses** and **pigs**, the blood of mammalian fetuses contains **fetal Hb (HbF)**. The structure of HbF is similar to that of

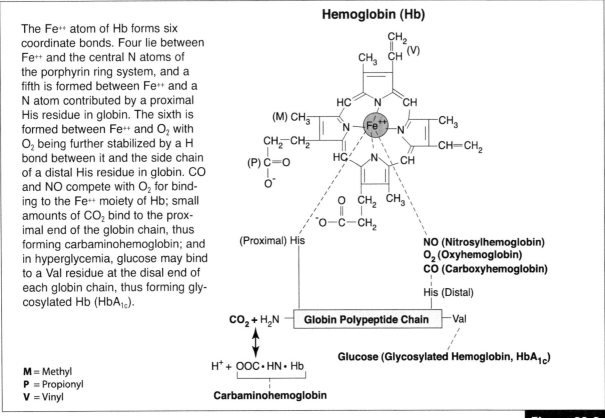

Figure 32-2

HbA, except that the β-chains are replaced by γ-chains (differing amino acid sequences). Fetal Hb binds 2,3-BPG less avidly than does HbA, which helps *in utero* to assure that proper O_2 exchange occurs between the dam and fetus. Soon after birth, however, HbF is normally replaced by HbA.

Anemias and Polycythemia

Common **anemias** (i.e., reductions in the amount of red blood cells or of hemoglobin in the circulation) result from excessive intra- or extravascular hemolysis (see Chapter 33), from impaired biosynthesis of hemoglobin (e.g., iron deficiency; see Chapter 48), or impaired production of erythrocytes (e.g., erythropoietin (EPO), folic acid, or vitamin B_{12} deficiency; see Chapters 16 and 43). In humans the **thalassemias** and **sickle cell anemia** are a group of hereditary hemolytic anemias resulting from variations in the amino acids comprising the α- or β-polypeptide chains of hemoglobin.

Anemia classification is generally undertaken by determining whether evidence of a bone marrow response to the condition is present in blood. With the exception of horses, this is routinely accomplished by determining whether circulating reticulocyte numbers are increased. **Regenerative anemia** will usually show an increase, whereas **nonregenerative anemia** will not. Examples of conditions leading to regenerative anemia include blood loss and increased erythrocyte destruction (hemolysis). Hemolytic anemia usually results in a more dramatic regenerative response than hemorrhagic anemia due to greater iron availability. Nonregenerative anemia may result from several chronic disease conditions, from drug or toxin exposure, or from inflammation. For example, chronic renal failure, hypothyroidism, or liver failure can precipitate a nonregenerative anemia.

Polycythemia (i.e., the presence of excessive amounts of red blood cells in the circulation) may be either relative, or absolute. A **relative polycythemia** may result in an increased hematocrit (Hct; the percentage of the blood volume made up of erythrocytes), but perhaps not an increase in the total red blood cell mass. Dehydration with a loss of plasma volume would be an example. An **absolute polycythemia** may result from increased renal EPO secretion as a response to hypoxemia, or because of oversecretion from malignancy. Polycythemia *rubra vera* results from a malignancy in red blood cell production despite normal or low concentrations of EPO. Any one of these polycythemic conditions causes an increase in blood viscosity, decreased cerebral perfusion, and potentially heart failure.

OBJECTIVES

- Give examples of porphyrins present in the body, and briefly discuss the physiologic roles of each.
- Identify the amino acid, the metal and the TCA intermediate that are needed for hemoglobin biosynthesis.
- Discuss the rate-control of Δ-ALA synthase activity in porphyrin biosynthesis.
- Name two enzymes in heme biosynthesis that are inhibited by lead.
- Identify the principal types of photosensitization recognized in domestic animals, and explain the origin of each.
- Explain what causes HbA_{1c} formation in animals, and what elevated levels signify.
- Know the differences between carboxyhemoglobin, oxyhemoglobin, deoxyhemoglobin, carbaminohemoglobin and nitrosylhemoglobin, and how each can affect blood oxygenation.
- Recognize what the "Bohr Effect" is, and how it pertains to tissue oxygenation.
- Summarize differences between regenerative and nonregenerative anemia, and give examples of conditions leading to each.
- Give examples of conditions leading to relative and absolute polycythemias.

- Understand the association between protoporphyrin IX and Harderian glands.

- Explain how photodynamic therapy works, and why it might be used to treat   certain forms of cancer.

- Recognize the association between PDT and the superoxide anion.

QUESTIONS

1. **Porphyrins are derived from which of the following compounds in animals?**
   a. Oxaloacetic acid and Tyrosine
   b. Succinyl-CoA and Glycine
   c. α-Ketoglutarate and Glutamine
   d. Acetyl-CoA and Alanine
   e. Citrate and Arginine

2. **Which of the following is considered to be the rate-limiting enzyme in heme biosynthesis?**
   a. Catalase
   b. Δ-ALA dehydratase
   c. Ferrochelatase
   d Uroporphyrinogen III cosynthase
   e. Δ-ALA synthase

3. **Lead toxicity would be expected to increase urinary levels of which one of the following intermediates in heme biosynthesis?**
   a. Succinyl-CoA
   b. Δ-Aminolevulinic acid
   c. Porphobilinogen
   d. Coproporphyrinogen III
   e. Uroporphyrinogen III

4. **Which one of the following compounds would most likely cause photosensitivity in ruminant animals with liver disease?**
   a. Phylloerythrin
   b. Hemoglobin
   c. Myoglobin
   d. Carbaminohemoglobin
   e. Δ-Aminolevulinic acid

5. **Carbon monoxide bound to hemoglobin is referred to as:**
   a. Carbaminohemoglobin.
   b. Nitrosylhemoglobin.
   c. Glycosylated hemoglobin (HbA_{1c}).
   d. Carboxyhemoglobin.
   e. Methemoglobin.

6. **The initial and last three reactions in heme biosynthesis occur in:**
   a. The nucleus.
   b. The cytoplasm.
   c. Mitochondria.
   d. The endoplasmic reticulum.
   e. Peroxisomes.

7. **Which of the following is associated with regenerative anemia?**
   a. A decrease in circulating reticulocytes
   b. Chronic renal failure
   c. hypothyroidism
   d. Liver failure
   e. Hemolytic anemia

8. **Uncontrolled diabetes mellitus is best associated with an increase in circulating:**
   a. Erythrocytes.
   b. Fe^{++}.
   c. Coproporphrinogen III.
   d. Carbon monoxide.
   e. HbA_{1c}.

9. **Hemolysis:**
   a. Can lead to hypertension because Hb in plasma binds nitric oxide, thus keeping it from promoting vasodilation.
   b. Can lead to a regenerative anemia if it is extensive.
   c. Can lead to excretion of Hb in urine.
   d. All of the above

10. **PDT has been used to:**
    a. Kill bacteria, fungi and viruses.
    b. Kill rapidly dividing cells.
    c. Treat acne.
    d. All of the above

11. **Hardarian glands are known to secrete:**
    a. Succinyl-CoA.
    b. Protoporphyrin IX.
    c. Hemoglobin.
    d. Coproporphrinogen III.
    e. Porphobilinogen.

11. b
10. d
9. d
8. e
7. e
6. c
5. d
4. a
3. b
2. e
1. b

ANSWERS

Chapter 33

Heme Degradation

Overview

- Heme is normally degraded to bilirubin, an end-product of mammalian metabolism.
- "Shunt bilirubin" represents bilirubin which does not originate from heme of normal senescent erythrocytes.
- Starvation results in an unconjugated hyperbilirubinemia in horses.
- Some dogs are capable of producing bilirubin from heme in their kidneys.
- Urobilinogens are produced from bilirubin by intestinal microbes.
- Indirect bilirubin refers to the unconjugated bilirubin fraction, while direct bilirubin is conjugated.
- Unconjugated bilirubin rarely appears in urine, whereas conjugated bilirubin is more likely to be filtered by the kidney and excreted into urine.
- Unconjugated bilirubin is lipophilic, and has a high affinity for brain tissue.

Hemoglobin (Hb) is catabolized in reticuloendothelial (RE) cells (also known as fixed macrophages or mononuclear phagocytes) of the spleen and bone marrow, where the protein portion, **globin**, is cleaved off, hydrolyzed, and exported as amino acids for reutilization by the organism, while the **heme** portion is digested. **Iron (Fe^{++})** is exported (bound to transferrin in plasma) toward bone marrow (for new erythropoiesis), and other tissues (e.g., liver) for storage. Heme is oxidized to **biliverdin** (a green pigment found, for instance, in high concentration in the bile of snakes and birds), whereby the porphyrin ring is opened up and small amounts of **carbon monoxide (CO)** are formed (**Fig. 33-1**). Carbon monoxide, which is highly toxic, normally competes with O_2 for Hb binding, and it is slowly expired through the lungs (see Chapter 30). Biliverdin is next reduced to **bilirubin** by RE cells, which then diffuses into blood where it is bound to **albumin** and transported to the liver for ultimate removal from blood, conjugation with a glucuronide, and excretion into bile.

The average **production of bilirubin** from heme in adult mammals is about 3-5 mg/kg BW/day. Under normal circumstances, **75-85% of total bilirubin production occurs in the RE system** from degradation of Hb heme derived from senescent erythrocytes. Although all mammalian cells contain heme in the form of hemoproteins, with the exception of the liver the tissue concentrations of these hemoproteins are so low, or their turnover rates so slow (e.g., myoglobin), that this contribution to bilirubin production is normally considered insignificant.

Studies using radiolabeled heme precursors in primates (glycine-^{14}C and Δ-ALA-^{3}H) demonstrated the existence of **two early-labeled plasma bilirubin peaks**, and **one late-labeled**

Copyright © 2015 Elsevier Inc. All rights reserved.

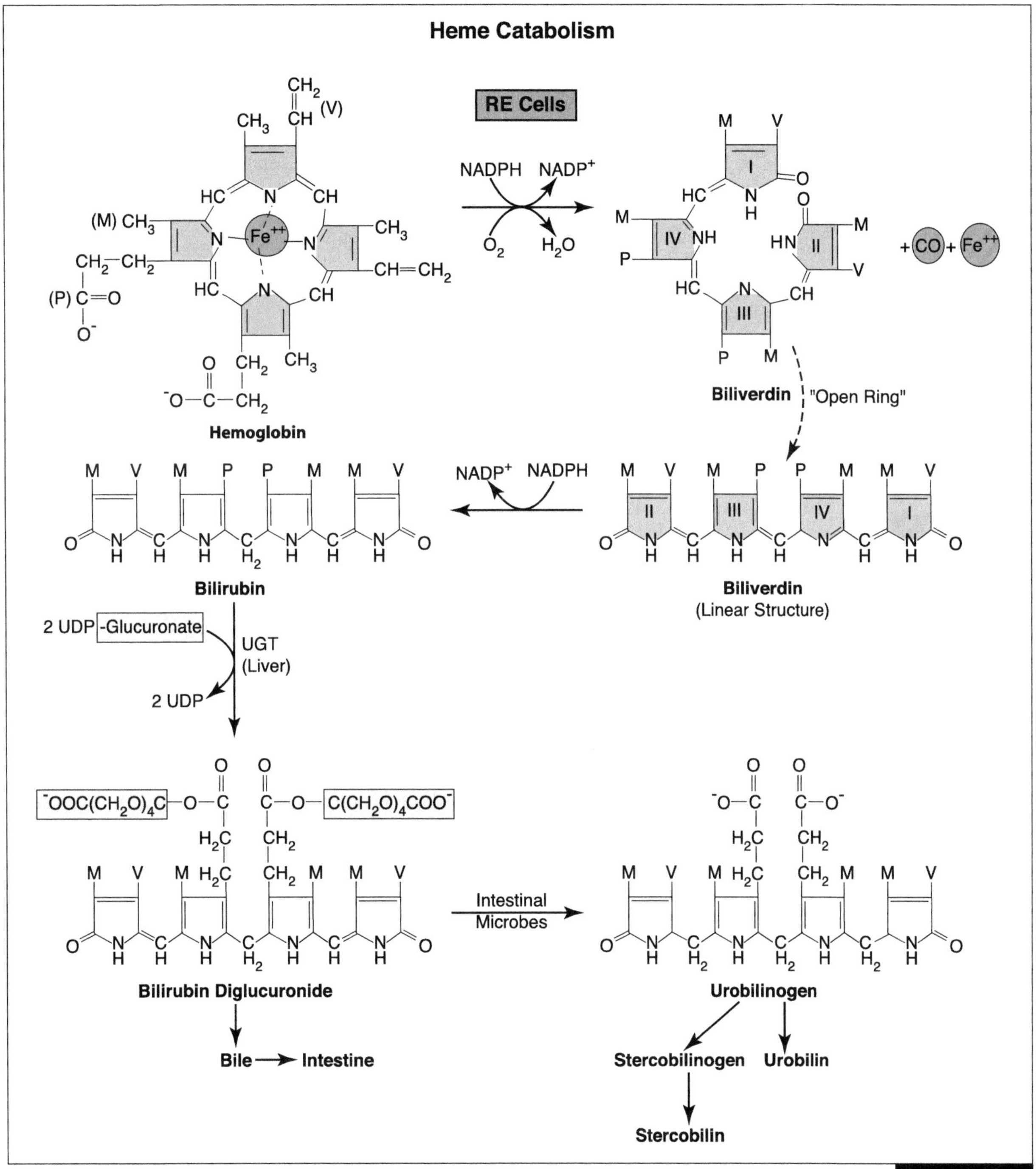

Heme Catabolism

Figure 33-1

peak (Fig. 33-2). The **first peak** occurring 1-6 hours post injection is derived from rapidly turning over substances such as **hepatic hemoproteins** (cytochromes, catalase, and tryptophan pyrrolase), **free hepatic heme**, and **hepatic porphyrins** (uroporphyrin I and III, coproporphyrin I and III, etc.) involved in heme biosynthesis (see Chapter 32). The **second peak**

occurring at 1-3 days is bilirubin derived from erythrocytic bilirubin produced by **ineffective or increased erythropoiesis**, or **intramedullary hemolysis**. Some degree of ineffective erythropoiesis is normal accounting for this fraction. Peaks one and two are termed "**shunt bilirubin**", since they represent bilirubin which does not come from normally senescent erythrocytes.

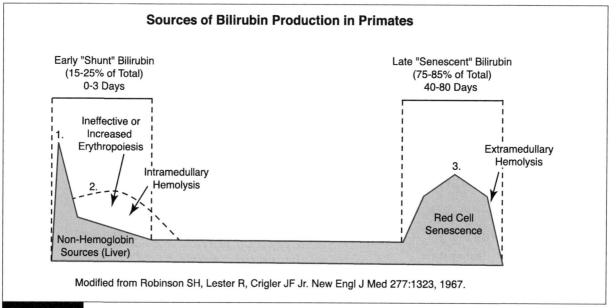

Sources of Bilirubin Production in Primates

Early "Shunt" Bilirubin
(15-25% of Total)
0-3 Days

Ineffective or
Increased
Erythropoiesis

Intramedullary
Hemolysis

Non-Hemoglobin
Sources (Liver)

Late "Senescent" Bilirubin
(75-85% of Total)
40-80 Days

Extramedullary
Hemolysis

Red Cell
Senescence

Modified from Robinson SH, Lester R, Crigler JF Jr. New Engl J Med 277:1323, 1967.

Figure 33-2

Disorders of peak two bilirubin are found with either thalassemia or pernicious anemia (e.g., folate or vitamin B$_{12}$ deficiency; see Chapters 16 and 43). **Peak three** bilirubin usually originates from **normal senescent hemolysis**. In many hemolytic anemias, however, jaundice (i.e., bilirubin accumulation in tissues) is largely due to **extramedullary hemolysis** (peak three). In addition, extramedullary hemolysis causes bone marrow to produce erythrocytes at an increased rate, therefore an even greater amount of ineffective erythropoiesis or intramedullary hemolysis will be occurring, and peak two bilirubin will also be increased.

Hepatic Bilirubin Uptake, Conjugation, and Excretion

The liver takes up bilirubin from sinusoidal blood by way of a specific transport system, moves it to the smooth endoplasmic reticulum, conjugates it, and excretes it into bile (causing bile's **orange-green** color). Transfer across the hepatocyte plasma membrane is bidirectional, and some investigators believe that as much as 30% of bilirubin initially taken up by liver cells may normally reflux back into plasma. Although hereditary defects in the

hepatic uptake mechanism for bilirubin occur, a common cause of unconjugated hyperbilirubinemia in animals (due to the reduced ability of the liver to remove bilirubin from the circulation), occurs with **starvation**. This effect is seen most often with **equine** species.

Studies have shown that bilirubin is conjugated in the liver by esterification with glucuronides in most mammalian species. In the **horse**, **dog, cat, mouse,** and **rabbit,** glucose and xylose are also conjugated to bilirubin. In pony bile, for example, bilirubin diglucuronide, bilirubin diglucoside, bilirubin glucuronide-glucoside, bilirubin monoglucuronide, bilirubin monoglucoside, and bilirubin monoxyloside conjugates are present. Hepatic bilirubin mono- and diglucuronide formation is catalyzed by **UDP-glucuronosyltransferase** (**UGT**, see Chapter 29).

If hemolysis is excessive in **dogs**, free hemoglobin may sometimes be extracted by the **kidney tubules**, where small amounts can be converted to bilirubin, conjugated, and excreted into urine. Although this has not been reported in other animal species, the observation that the canine kidney can produce bilirubin from heme, conjugate it and excrete it into urine may explain the presence of bilirubinuria during periods when the plasma conjugated bilirubin concentration is minimal.

Active secretion of conjugated bilirubin into bile occurs against a large concentration gradient. As conjugated bilirubin reaches the terminal ileum and large intestine, some deconjugation by bacterial enzymes (glucuro-nidases) occurs, and the pigment is subsequently reduced by fecal flora to a group of colorless tetrapyrrolic compounds known as **urobilinogens**. Urobilinogen can be further reduced to **stercobilinogen** which, upon oxidation yields **stercobilin** (which imparts a dark color to feces, **Fig. 33-1**). Only about 15% of urobilinogen formed in the intestine is normally reabsorbed, and the liver normally extracts urobilinogen efficiently from portal blood (**Fig. 33-3**). Urobilinogen is excreted into bile unchanged, returning to the intestine (thereby completing an enterohepatic urobili-nogen cycle). Since the liver extracts most of the urobilinogen from portal blood, normally there are mere traces which pass on into the peripheral circulation. Although urobilinogen is 80% bound to plasma protein, what little escapes the liver quickly reaches the kidneys, where it can undergo glomerular filtration. Urobilinogen excretion into urine correlates well with urinary pH, with higher values being recorded in alkaline than in acidic urine. The colorless urobilinogen is oxidized by light to the highly colored **urobilin**, which can impart color to urine.

Characterization of Plasma Bilirubin

Assay of plasma bilirubin can be diagnostically useful, depending upon the total amount present, its source, and its ratio of unconjugated:conjugated fractions. According to the most commonly used **Van den Burgh** reaction, bilirubin is coupled with diazotized sulfanilic acid to form azobilirubin. The color of this derivative is pH dependent, becoming **blue** under alkaline conditions. The color, thus, is proportional to the concentration. When the reaction occurs in the presence of methanol, both the **unconjugated (UCB)** and **conjugated (CB) bilirubin** pigments react, therefore the concentration is termed **total bilirubin (Table 33-1)**. When water is used in place of methanol, only the water-soluble conjugated pigment reacts, the concentration then is termed **direct-reacting bilirubin**. The unconjugated bilirubin thus can only be resolved as the difference between the two, and is referred to as the **indirect** fraction.

Under normal conditions, most of the bilirubin in plasma is in the unconjugated form (bound to albumin), for biliary excretion of conjugated bilirubin is the preferred route. Therefore, elevations in plasma conjugated bilirubin would be anticipated during pathophysiologic

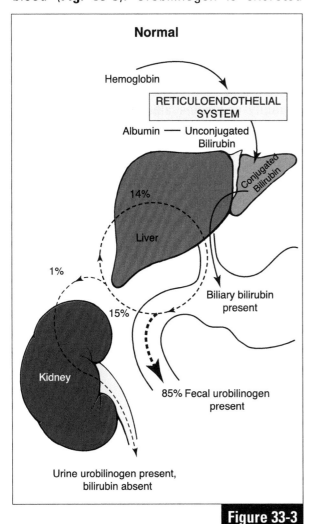

Normal

Hemoglobin

RETICULOENDOTHELIAL SYSTEM

Albumin —— Unconjugated Bilirubin

Conjugated Bilirubin

14%

Liver

1%

15%

Biliary bilirubin present

Kidney

85% Fecal urobilinogen present

Urine urobilinogen present, bilirubin absent

Figure 33-3

Table 33-1
Plasma Bilirubin
UCB = Unconjugated Bilirubin (lipid soluble)
CB = Conjugated Bilirubin (water soluble)
Total = UCB + CB
Direct = CB
Indirect = Total - Direct = UCB

processes associated with biliary obstruction, or in other conditions where liver cells are unable to transport conjugated bilirubin across canalicular membranes into bile, therefore regurgitating it into plasma. It should be noted in this regard that **canalicular excretion is usually rate-limiting for hepatic bilirubin transport**. Therefore, in situations where bilirubin production is increased (e.g., hemolysis), the normal liver continues to conjugate it at a faster rate than it can be excreted into bile; therefore, the residual conjugated bilirubin may be refluxed back into the circulation, adding a conjugated component to an existing unconjugated hyperbilirubinemia.

General **physicochemical properties** of **UCB** and **CB** are shown in **Table 33-2**. Unconjugated bilirubin is relatively insoluble in water, urine and bile, but it is lipophilic and thus has a high affinity for brain tissue. Unconjugated bilirubin is also tightly bound to plasma albumin, and therefore is rarely found in urine. Conversely, CB is more hydrophilic, it is loosely bound to albumin in the circulation, and therefore is more likely to be filtered by the kidneys and excreted into urine. Conjugated bilirubin is also less likely to cross the blood-brain-barrier.

Bilirubin encephalopathy (kernicterus) is an acquired condition caused by extreme unconjugated hyperbilirubinemia which exceeds the capacity of plasma albumin to bind bilirubin, and thus keep it from crossing the blood-brain-barrier. Hemolytic diseases or conditions where the liver is deficient in conjugating

Table 33-2		
Physicochemical Properties of Unconjugated (UCB) and Conjugated Bilirubin (CB)		
Property	**UCB**	**CB**
Van den Bergh reaction	Indirect	Direct
Solubility in aqueous solution	−	+
Lipid solubility	+	−
Attachment to plasma albumin	++	+
Presence in icteric urine	−	+
Presence in bile	−	+
Affinity for brain tissue	+	−
Association with hemolytic jaundice	++	+/−
Association with obstructive jaundice	+	++
Association with starvation	+	−

+ = Present; − = Absent; ++ = Definitely Present; +/− = Equivocally Present

bilirubin are best associated with kernicterus. Since unconjugated bilirubin is lipid soluble, it penetrates neuronal and glial membranes easily and subsequently disrupts oxidative phosphorylation. Patients surviving kernicterus may exhibit a range of neurologic symptoms, including spasticity, muscular rigidity, ataxia and mental retardation.

OBJECTIVES

- Explain how conjugated bilirubin (CB) can be formed from hemoglobin and Glc-6-P, and how it is normally handled by the body.

- Discuss how non-hemoglobin sources of bilirubin may be formed in the liver.

- Identify the meaning of "Shunt Bilirubin," and describe its origin.

- Explain why an unconjugated hyperbilirubinemia in hospitalized equine species may not indicate hepatic dysfunction.

- Discuss how urobilinogen and stercobilinogen are formed, and what increased or decreased amounts of urinary urobilinogen may signify.

- Describe differences between direct, indirect and total bilirubin.

- Recognize the physicochemical properties of UCB and CB, and know what an increase in the blood levels of each potentially represents.

- Distinguish between jaundice and kernicterus, and identify conditions leading to each.

QUESTIONS

1. **"Shunt Bilirubin":**
   a. Comes from late senescent hemolysis of red blood cells.
   b. May include bilirubin derived from ineffective or increased erythropoiesis.
   c. Is also referred to as conjugated bilirubin.
   d. Is indirect bilirubin determined from the Van den Burgh reaction.
   e. Is only seen in abnormal situations.

2. **All of the following are normal derivatives of bilirubin, EXCEPT:**
   a. Biliverdin.
   b. Urobilinogen.
   c. Urobilin.
   d. Stercobilinogen.
   e. Stercobilin.

3. **When a horse quits eating, the concentration of which one of the following would be expected to rise in plasma?**
   a. Urobilinogen
   b. Hemoglobin
   c. Unconjugated bilirubin
   d. Glucose
   e. Stercobilin

4. **Under normal circumstances, the majority of bilirubin production in the body occurs in:**
   a. Hepatocytes.
   b. The lungs.
   c. Bone marrow.
   d. Reticuloendothelial tissue.
   e. The kidneys.

5. **In addition to the liver, in which one of the following locations in dogs can bilirubin be produced from heme, and be conjugated to a glucuronide?**
   a. Spleen
   b. Lung
   c. Brain
   d. Kidney
   e. Bone marrow

6. **Direct-reacting bilirubin is:**
   a. Unconjugated bilirubin.
   b. Conjugated bilirubin.
   c. Total bilirubin.
   d. Biliverdin.
   e. Urobilinogen.

7. **Which one of the following physico-chemical characteristics is true of unconjugated bilirubin, but not conjugated bilirubin?**
   a. High affinity for brain tissue.
   b. Soluble in aqueous solutions.
   c. Loosely bound to plasma albumin.
   d. Normally present in bile.
   e. Is direct-reacting in the Van den Bergh reaction.

7. a
6. b
5. d
4. d
3. c
2. a
1. b

ANSWERS

Chapter 34

Tricarboxylic Acid (TCA) Cycle

Overview

- Genetic abnormalities in TCA cycle enzymes are generally incompatible with life.
- Isocitrate dehydrogenase catalyzes the rate-limiting reaction in the TCA cycle.
- Oxidation of 1 pyruvate molecule through the TCA cycle can yield 15 ATP equivalents.
- Although oxaloacetate and acetyl-CoA are impermeable to mitochondrial membranes, citrate and other intermediates of the cycle are permeable.
- Certain rodenticides can interrupt the TCA cycle.
- Several B vitamins are required as cofactors in the TCA cycle.
- GTP generated in the TCA cycle can be used to drive the conversion of OAA to PEP in hepatic gluconeogenesis.

The **TCA cycle** (also called the **Citric Acid** or **Krebs Cycle**), occurs in the mitochondrial matrix, and is a major integration center for coordinating various pathways of carbohydrate, lipid, and protein metabolism. It plays a major role in transamination and deamination reactions (see Chapter 9), gluconeogenesis from lactate, amino acids, and propionate (see Chapter 37), and indirectly, lipogenesis (see Chapter 56). Although several of these processes are carried out in many different tissues, the liver is the major organ in which all occur to a significant extent. The repercussions, therefore, are profound when, for example, large numbers of hepatocytes are damaged or replaced by connective tissue (as in acute hepatitis and cirrhosis, respectively). Few, if any, genetic abnormalities of TCA cycle enzymes have been reported, thus indicating that such abnormalities are undoubtedly incompatible with life.

The overall reaction catalyzed by the TCA cycle is a sum of the individual reactions (see **Fig. 34-1**):

$$\left[\begin{array}{c} \textbf{Acetyl-CoA + 3 NAD}^+ \textbf{ + FAD + GDP +} \\ \textbf{Pi + 2 H}_2\textbf{O} \longrightarrow \textbf{2 CO}_2 \textbf{ + 3 NADH + FADH}_2 \textbf{ +} \\ \textbf{GTP + CoA.SH} \end{array} \right]$$

The TCA cycle begins by combining a two carbon molecule of **acetyl-CoA** with a four carbon dicarboxylate, **oxaloacetate** (**OAA**), resulting in the formation of a six carbon tricarboxylate, **citrate**. The source of acetyl-CoA for this reaction can be pyruvate generated through anaerobic glycolysis (see Chapter 27), ketogenic amino acids (see Chapter 8), fatty acids (see Chapter 55), or ketone bodies (see Chapter 71). The TCA cycle becomes a terminal furnace for the oxidation of acetyl-CoA if citrate is converted to isocitrate, since subsequent reactions that generate succinyl-CoA release two carbon atoms as CO_2. However, unlike

Copyright © 2015 Elsevier Inc. All rights reserved.

The Mitochondrial TCA Cycle

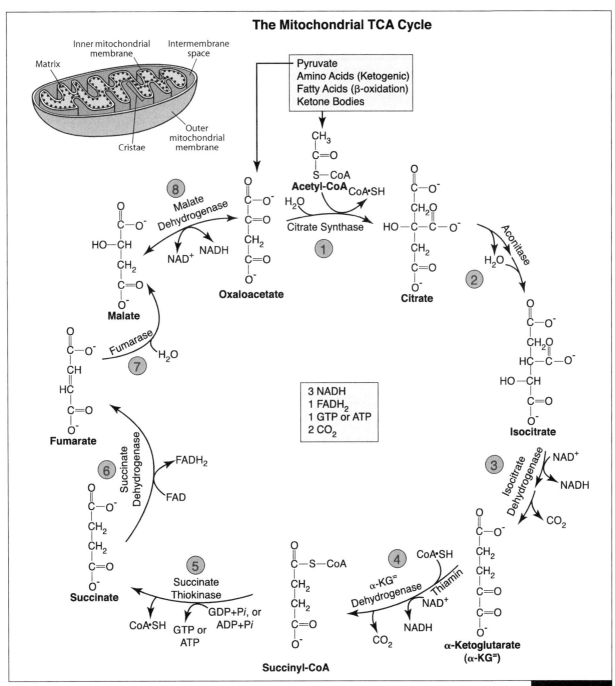

Figure 34-1

acetyl-CoA, OAA can be regenerated by the TCA cycle.

The enzyme catalyzing reaction **#1** of the TCA cycle is **citrate synthase** (also called condensing enzyme; see **Fig. 34-1**). This reaction is exo-thermic, and **physiologically irreversible** in mitochondria. Citrate synthase is inhibited by high mitochondrial concentrations of **ATP**, **citrate**, and **long-chain fatty acyl-CoA** (see

Table 34-1). Since both acetyl-CoA and OAA are impermeable to mitochondrial membranes (like NAD⁺ and NADH), these molecules cannot diffuse out of mitochondria (directly) to serve as substrates for cytoplasmic reactions, or enzyme regulators. However, citrate is permeable to mitochondrial membranes, and when it diffuses into the cytoplasm it inhibits phosphofructokinase (PFK) activity (see Chapter 25),

Table 34-1
Regulation of TCA Cycle Enzymes

Enzyme	Prosthetic Group	Coenzyme	Vitamin	Inhibitor
#1	–	–	–	ATP, Citrate
				Fatty Acyl-CoA
				(long-chain)
#2	Fe^{++}	–	–	Fluoroacetate
				Fluoroacetamide
#3	Mn^{++} or Mg^{++}	NAD^+	Niacin	NADH, ATP
#4 As	Thiamin diphosphate	NAD^+	Thiamin, Niacin	NADH, ATP,
				Hg, Succinyl-CoA
#5	Mn^{++} or Mg^{++}	–	Pantothenic acid	–
#6	Fe^{++}	FAD	Riboflavin	Malonate, OAA
#7	–	–	–	–

and can be converted back to acetyl-CoA and OAA through the activity of citrate cleavage enzyme (or citrate lyase). The acetyl-CoA so generated can now serve as a substrate for lipid biosynthesis (see Chapter 56).

The second reaction sequence in the TCA cycle requires the **aconitase** (aconitate hydratase) enzyme, which converts **citrate** to **isocitrate**. This enzyme contains ferrous iron (Fe^{++}) in the form of an iron-sulfur (**Fe:S**) protein complex. The conversion takes place in two steps, dehydration to **cis-aconitate** (not shown), then rehydration to isocitrate. This reaction sequence is **inhibited** by **fluoroacetate** and **fluoroacetamide**, plant poisons and rodenticides which, in the form of fluoroacetyl-CoA, condense with OAA to form fluorocitrate. This compound further inhibits aconitase, causing citrate accumulation. Fluoroacetate and fluoroacetamide are thus suicide inhibitors since the cell converts them from their inactive states to an active inhibitory unit (fluorocitrate). These poisonous compounds are odorless, tasteless, and water-soluble. Dogs, cats, pigs and other animals that eat deceased fluoroacetate- or fluoroacetamide-poisoned rats or mice may

meet a similar fate.

Isocitrate next undergoes dehydrogenation (reaction **#3**) in the presence of **isocitrate dehydrogenase** (**ICD**) to form **oxalosuccinate** (structure not shown). Two different ICDs have been described. One, which is **NAD^+-specific**, is found only in mitochondria where it is associated with the TCA cycle, and the other, which is **$NADP^+$-specific**, is found in the cytoplasm associated with the outer mitochondrial membrane. **Cytoplasmic ICD** is important in ruminant adipocytes, where it assists in providing NADPH for lipogenesis (see Chapter 56).

Following oxalosuccinate formation, there is a second reaction catalyzed by ICD which involves decarboxylation, and thus formation of **α-ketoglutarate** (**α-KG⁼**). Either **Mn^{++}** or **Mg^{++}** is an important prosthetic component of this **decarboxylation** process (see **Table 34-1**). This rate-limiting reaction sequence in the TCA cycle, catalyzed by ICD, is considered to be **physiologically irreversible**, and is allosterically **inhibited** by **ATP** and **NADH**, and **activated** by **ADP** and **NAD^+**. These compounds affect the K_m of ICD for its primary substrate, isocitrate. **Niacin**, water-soluble vitamin B_3 in the form of NAD⁺ (see Chapter 41), is required in three dehydrogenase-catalyzed reactions of the TCA cycle (reactions **#3**, **#4**, and **#8**).

Next, α-**KG**= undergoes **oxidative decarboxylation** (reaction **#4**) in a manner analogous to the conversion of pyruvate to acetyl-CoA (see Chapter 27). The α-KG= oxidative decarboxylation reaction, catalyzed by α-**KG**= **dehydrogenase**, also requires identical cofactors to those present in the pyruvate dehydrogenase-catalyzed reaction (e.g., **thiamin** (**vitamin B₁** which forms a thiamin diphosphate prosthetic group; see Chapter 40), **niacin** (a source of **NAD⁺**), and **coenzyme A.SH** (**CoA.SH; vitamin B₅**); see Chapter 41). The equilibrium of this reaction is so much in favor of **succinyl-CoA** formation, that it is also considered to be **physiologically irreversible** (like reactions **#1** and **#3**). As in the case of pyruvate decarboxylation, **arsenite** (**As**) and **mercuric** (**Hg**) ions **inhibit** α-**KG**= **decarboxylation** by binding to an -SH⁻ group in the dehydrogenase complex, thus preventing its reoxidation to a disulfide. Physiologic **inhibitors** of this reaction include high mitochondrial levels of **NADH**, **ATP**, and the reaction product, **succinyl-CoA**.

To continue with the TCA cycle, **succinyl-CoA** is next converted to **succinate** by the enzyme **succinate thiokinase** (also called **succinyl-CoA synthetase**, reaction **#5**). This is the only example in the TCA cycle of the generation of a high-energy phosphate (either **ATP** or **GTP**) at the substrate level. In liver cells, for example, where the gluconeogenic pathway is active, **GTP** produced in mitochondria through this reaction can be used by **PEP carboxykinase** to convert OAA to PEP on its way to becoming glucose (see Chapter 37). However, in most cells **ATP** will be the high energy phosphate carrier derived from this reaction. Either **Mn⁺⁺** or **Mg⁺⁺** are prosthetic groups for succinate thiokinase, and the water-soluble **vitamin B₅** (**pantothenic acid**), which is a part of **CoA.SH**, is the cofactor attached to "active" carboxylic

acid residues like succinyl-CoA (see **Table 34-1**).

Succinate is next dehydrogenated to **fumarate**, catalyzed by the enzyme **succinate dehydrogenase** (reaction **#6**). This is the only reaction in the TCA cycle involving transfer of hydrogen from a substrate to a **flavoprotein** (**FAD**), without the participation of NAD⁺. In addition to FAD, this enzyme, like aconitase above, contains iron-sulfur (Fe:S) protein. The water-soluble **vitamin B₂** (**riboflavin**), is used to form **flavin adenine dinucleotide** (**FAD**), and the succinate dehydrogenase complex is physiologically **inhibited** by high mitochondrial concentrations of **OAA**, as well as by **malonate** (*in vitro*). Malonate is structurally similar to succinate.

Fumarase (fumarate hydratase) next catalyzes addition of **H₂O** to **fumarate**, thus yielding **malate** (reaction **#7**). Malate is then converted to **OAA** by **malate dehydrogenase**, a reversible reaction requiring **vitamin B₃** (**niacin**) as **NAD⁺** (reaction **#8**). Although the equilibrium of this reaction strongly favors malate formation, net flux is usually toward OAA formation since this compound, together with the other product of the reaction (NADH), is removed continuously in further reactions. However, during gluconeogenesis in liver tissue, OAA synthesis from pyruvate favors malate formation, which then exits mitochondria and reforms OAA in the cytoplasm. Thus, reversal of this reaction becomes important in the hepatic **dicarboxylic acid cycle** (see Chapter 37).

Three molecules of **NADH** and one of **FADH₂** are produced for each molecule of **acetyl-CoA** catabolized in one revolution of the TCA cycle. These reducing equivalents are next passed on to the respiratory chain in the inner mitochondrial membrane, providing sufficient O₂ is available (see Chapter 36). During passage along the chain, each **NADH** molecule generates **three** high-energy phosphate bonds in the form of **ATP**, and each **FADH₂** generates **two**. A further high energy bond in the form of

GTP or **ATP** is produced during conversion of succinyl-CoA to succinate, thus bringing the total to **12 ATPs** generated for each turn of the cycle. Conversion of pyruvate to acetyl-CoA generates another NADH, and considering that 1 mol of **glucose** yields 2 mol of pyruvate, the mitochondrial (aerobic) portion of glucose oxidation (from 2 molecules of pyruvate through acetyl-CoA) yields **30 ATP** equivalents. This, added to the **8 ATP** equivalents derived (both directly and indirectly) from the cytoplasmic phase, brings the total potential aerobic high-energy phosphate bond production from glucose oxidation to **38 ATP** equivalents (assuming that pyruvate is passed through acetyl-CoA, and not OAA; see Chapter 26).

Exchange Transporters of the Inner Mitochondrial Membrane

While the outer mitochondrial membrane presents little or no permeability barrier to substrates, intermediates or nucleotides of interest, the **inner membrane** limits the types of molecules that can move into or out of mitochondria. Several transport systems have been described for the inner mitochondrial membrane, and some have been rather well-characterized (**Fig. 34-2**). These transport systems are necessary for the uptake and output of **ionized metabolites** while preserving electrical and osmotic balance. In the **antiport** mechanisms shown, the concentration gradient of one solute drives the movement of the exchange solute in the opposite direction.

The inner mitochondrial membrane is freely permeable to uncharged small molecules, such of O_2, H_2O, CO_2 and NH_3, and to monocarboxylate anions, such as acetate, propionate, acetoacetate, and β-OH-butyrate (i.e., the ketone bodies and volatile fatty acids). However, monocarboxylic acids penetrate more freely in their undissociated, more lipid-soluble form. Long-chain fatty acids are transported into mitochondria via the carnitine system (Chapter 55), and there is a special antiporter that exchanges pyruvate with OH^-. However, dicar-

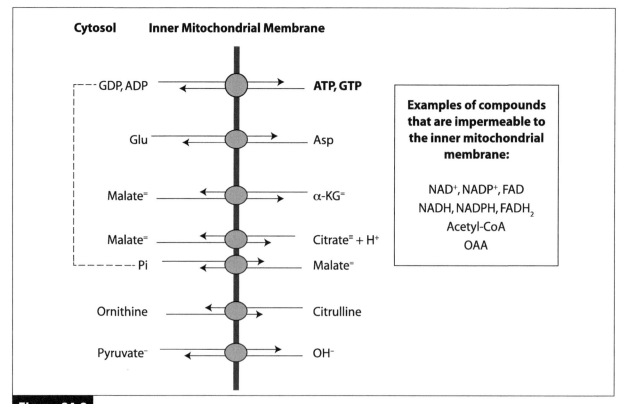

Figure 34-2

boxylate and tricarboxylate anions and amino acids require specific transporters to facilitate their passage across the inner mitochondrial membrane. In the hepatic urea cycle, citrulline exchanges with ornithine (Chapter 10), and Ca^{++} release from mitochondria is thought to be facilitated by exchange with Na^+ and/or H^+ (Chapter 58). The adenine nucleotide transporter allows exchange of ATP with ADP (but not AMP), while inorganic phosphate (Pi, which is needed for ATP formation) gains access to mitochondria via exchange with malate.

OBJECTIVES

- Provide an explanation for the apparent non-existence of genetic TCA cycle enzyme abnormalities.

- Identify the intracellular location of the Krebs Cycle.

- Explain why the Citric Acid Cycle is considered to be an integration center for coordinating various pathways of carbohydrate, lipid and protein metabolism.

- Identify the origins of OAA and acetyl-CoA for entry into the TCA cycle.

- Know the reactions of the TCA cycle (in their appropriate sequence), and name all enzymes involved.

- Indicate steps in the TCA cycle that yield CO_2, NADH, $FADH_2$ and/or GTP.

- Give examples of condensation, dehydration, hydration, decarboxylation, oxidation and substrate-level phosphorylation reactions in the TCA cycle.

- Calculate the ATP yield from the complete oxidation of pyruvate or of acetyl-CoA.

- Indicate control points in the TCA cycle, note the activators and inhibitors, and identify important prosthetic groups and vitamin cofactors involved.

- List the intermediates of the TCA cycle that are impermeable to the inner mitochondrial membrane.

- Recognize how inner mitochondrial membrane transporters operate.

QUESTIONS

1. **Which one of the following is known to inhibit mitochondrial citrate synthase?**
   a. ADP
   b. Oxaloacetate
   c. ATP
   d. Acetyl-CoA
   e. Carnitine

2. **Which one of the following is permeable to mitochondrial membranes?**
   a. Malate
   b. Oxaloacetate
   c. Acetyl-CoA
   d. NAD^+
   e. NADH

3. **Which compound below is a plant poison and rodenticide, which acts as a suicide inhibitor of aconitase, a TCA cycle enzyme?**
   a. Malonate
   b. Arsenite
   c. Succinyl-CoA
   d. Mercury
   e. Fluoroacetate

4. **Which enzyme below catalyzes the rate-limiting reaction in the TCA cycle?**
   a. Citrate synthase
   b. Isocitrate dehydrogenase
   c. α-Ketoglutarate dehydrogenase
   d. Succinate thiokinase
   e. Malate dehydrogenase

5. **All of the following reactions are considered to be "physiologically irreversible", EXCEPT:**
   a. Cytoplasmic (Phosphoenolpyruvate —> Pyruvate).
   b. Mitochondrial (Citrate —> Isocitrate).
   c. Mitochondrial (Isocitrate —> α-Ketoglutarate).
   d. Mitochondrial (α-Ketoglutarate —> Succinyl-CoA).
   e. Mitochondrial (Malate —> Oxaloacetate).

5. e
4. b
3. e
2. a
1. c

Chapter 35

Leaks in the Tricarboxylic Acid (TCA) Cycle

Overview

- The TCA cycle is an amphibolic pathway.
- Anaplerotic reactions replenish TCA cycle intermediates when they leak away from the cycle.
- Oxaloacetate leaks away from the TCA cycle to form pyrimidines and glucose.
- Succinyl-CoA leaks away from the TCA cycle to form the porphyrins (including heme).
- Several intermediates may leak away from the TCA cycle to form proteins.
- Citrate may leak away from the TCA cycle to reform acetyl-CoA and OAA in the cytoplasm, with the former entering into lipid biosynthesis.
- Leaks in the TCA cycle of skeletal muscle fibers are largely replenished through glucose oxidation.
- Amino acids may be used to replenish leaks in the TCA cycle.

The TCA cycle is considered to be an **amphibolic pathway**, having both oxidative and biosynthetic functions. Its oxidative functions involve the complete combustion of acetyl-CoA, whereupon ATP is generated through oxidative phosphorylation (see Chapters 34 and 36). Several intermediates in the TCA cycle, however, can serve as substrates for other biosynthetic pathways, and therefore constitute important "**leaks**" in the cycle. **Citrate**, for example, can be used for fatty acid and/or steroid biosynthesis in the cytoplasm, and **oxaloacetate (OAA)** for glucose biosynthesis. Therefore, it is necessary to replenish these intermediates when they leak away into other pathways so that the TCA cycle can continue to operate. Reactions that fulfill this purpose are known as **anaplerotic (i.e., to "fill-up") reactions**. A major anaplerotic reaction in muscle, for example, is the **carboxylation of pyruvate** to **OAA**.

TCA Cycle Intermediates are Converted to Other Essential Compounds

Citrate is permeable to mitochondrial membranes, and hence, may leave mitochondria to enter the cytoplasm. There, under the influence of **citrate lyase** (also called **citrate cleavage enzyme**), citrate may be split into OAA and acetyl-CoA, with carbon atoms from the latter being used for cytoplasmic lipid biosynthesis (**Fig. 35-1** and Chapter 56). This process is used by liver cells in converting **carbohydrate to fat**. Since the conversion of pyruvate to acetyl-CoA occurs in mitochondria, and since acetyl-CoA is largely impermeable to mitochondrial membranes, citrate serves as a carrier of carbon atoms between the standard pathway of carbohydrate oxidation, and the pathway of lipid (particularly fatty acid) biosynthesis.

Copyright © 2015 Elsevier Inc. All rights reserved.

Leaks in the TCA Cycle

Because of the anabolic involvements for which TCA cycle intermediates leak away, cycle activity would decline if there were no processes whereby these intermediates could be replenished.

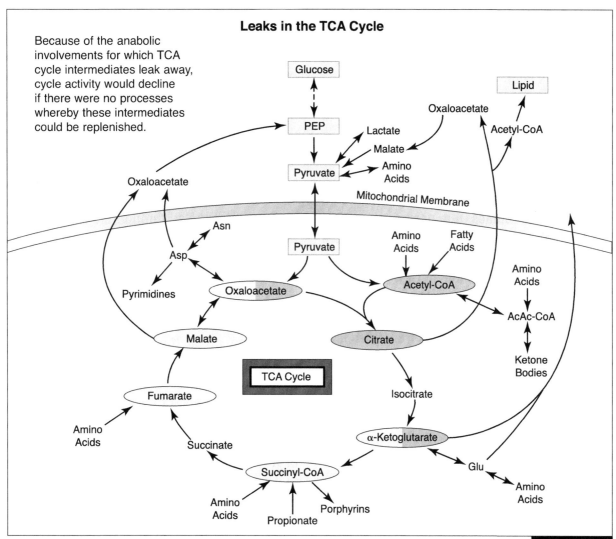

Figure 35-1

Within mitochondria of liver cells, if there is insufficient OAA available to couple with acetyl-CoA in the formation of citrate, a separate pathway of **ketone body formation** from acetyl-CoA (through HMG-CoA) may be activated (see Chapter 71).

Virtually all organisms are capable of synthesizing **amino acids**, although there may be differences in the types and number of amino acids synthesized by any one organism. For example, most plants and bacteria can synthesize all 20 amino acids required for protein biosynthesis (see Chapter 2), but animal cells are capable of synthesizing only certain ones, relying on dietary intake to account for the remainder (see Chapter 3).   Whatever the organism,

some intermediates of the TCA cycle serve as important precursors for amino acid formation. Two particularly important compounds are **OAA** and **α-ketoglutarate (α-KG$^=$)**, while **3-phosphoglycerate** and **pyruvate** from the Embden-Meyerhoff pathway also serve as amphibolic intermediates that can give rise to amino acids. Fumarate, succinyl-CoA, acetyl-CoA, and acetoacetyl-CoA are used in plant and bacterial amino acid biosynthesis, but not in animal. Generally speaking, the anabolic utilization of acetyl-CoA for biosynthetic reactions does not create as much of an operational strain on the TCA cycle as does the removal of one or more of the internal intermediates.

Nearly all mammalian organisms are capable of synthesizing **pyrimidine** and **purine** nucleotides in the same manner, from elementary substances such as CO_2, NH_3, **aspartate (Asp)**, **glycine (Gly)**, **glutamine (Gln)**, **tetrahydrofolate (H_4-folate)**, and **ribose** (see Chapters 13, 14 and 15). Two separate pathways are involved, one for pyrimidines, and another for purines. The first nucleotide product of the pyrimidine pathway is **uridine-5'-monophosphate (UMP)**, which then becomes the parent compound from which other pyrimidine nucleotides are produced (see Chapter 14). **Aspartate** gives rise to carbon atoms 4, 5, and 6 of UMP, as well as to nitrogen atom number 1. Therefore, as the cellular requirement for pyrimidine biosynthesis increases, carbon atoms from **OAA** have a tendency to leak away from the TCA cycle in the form of Asp (see Chapter 36). **Aspartate** and **Gln** also participate in **purine** biosynthesis; however, each donates nitrogen atoms only (see Chapter 15). Therefore, their participation in purine biosynthesis does not remove carbon atoms from the TCA cycle. Purine carbon atoms are derived from Gly, CO_2, N^{10}-formyl-H_4 folate, and N^5,N^{10}-methenyl-H_4 folate (see Chapter 15).

Succinyl-CoA also leaks away from the TCA cycle into **porphyrin** metabolism. Porphyrins are cyclic compounds that have a high affinity for binding metal ions, usually ferrous (**Fe^{++}**) iron, and they give rise to **heme**. Although heme is produced in virtually all mammalian tissues, its synthesis in **bone marrow** and **liver** is most pronounced because of requirements for incorporation into **hemoglobin** and the **cytochromes**, respectively (see Chapter 32). Therefore, porphyrin biosynthesis in these tissues represents another formidable leak in the TCA cycle.

Connections between the TCA cycle to the biosynthesis of **glucose** are made through **OAA**, **malate**, **Asp**, and **phosphoenolpyruvate (PEP)** (see Chapter 37). **Oxaloacetate** itself cannot cross mitochondrial membranes, but, via reversible conversions to either **malate** or

Asp, cytoplasmic and mitochondrial **OAA/Asp** and **OAA/malate** end up functioning as single pools that serve both cell compartments (see Chapter 9). As a gluconeogenic precursor, OAA may be formed from amino acids, lactate, or propionate via either the pyruvate carboxylase or malate dehydrogenase steps, or by deamination of Asp. Phosphoenolpyruvate carboxykinase then converts OAA to PEP on its way to forming glucose. If, however, OAA is completely recycled through the TCA cycle, it does not become a drain on the overall OAA/Asp pool.

Replenishment of TCA Cycle Intermediates

Because of all the anabolic involvements for which TCA cycle intermediates leak away, cycle activity would become deficient (so that not enough acetyl-CoA could be oxidized) if there were no processes whereby these intermediates are replenished. One of the functions of **glucose** is to replenish the TCA machinery, so that **acetyl-CoA** (derived from fat oxidation) can be completely oxidized. **Pyruvate**, formed from glucose via the Embden-Meyerhoff pathway, can be carboxylated by **pyruvate carboxylase** to OAA when there is a need to couple this intermediate with rising titers of acetyl-CoA in the formation of **citrate** (see Chapter 27). The saying goes that "**fats are burned in a carbohydrate flame**." This is certainly true of aerobic muscle fibers, which during sustained exercise may be oxidizing **glucose** (for **OAA** formation), and **fatty acids** (for **acetyl-CoA** formation) in the ratio of about **3:6** (see Chapter 79).

Proteins yield **amino acids** that can also be converted to TCA cycle intermediates through **transamination** and **transdeamination** reactions, and thus, serve a similar purpose as that of glucose oxidation. Generally, all **nonessential amino acids** are in equilibrium with their respective α-**ketoacids** via transamination (e.g., **Ala/pyruvate**, **Asp/OAA**, and **Glu/α-KG$^=$**;

see Chapter 9). For the net conversion of amino acids to TCA cycle intermediates, however, transamination is not enough, since one amino acid is gained for each one lost. For the net conversion to a TCA cycle intermediate, an amino acid must not merely exchange, but rather get rid of the amino group, and this is usually accomplished through **Glu** formation. Glutamate can then be oxidatively deaminated by **glutamate dehydrogenase**; the **NH₃** thus formed can then be converted to **urea** in the liver (see Chapters 9 and 10). By these combined mechanisms, all nonessential, and some essential amino acids can be converted to their corresponding α-ketoacids. Most of these ketoacids are in equilibrium with TCA cycle intermediates or with pyruvate, and are therefore glucogenic. Some amino acids yield **acetyl-CoA** and/or **acetoacetyl-CoA**, and, thus, are ketogenic (see Chapter 8). In general, dietary proteins contain a greater proportion of glucogenic than ketogenic amino acids.

Propionate is also a source for replenishment of TCA cycle intermediates through **succinyl-CoA** formation (see Chapter 37). Though not of importance in the diets of dogs and cats, it is of the utmost importance for ruminant animals who rely on gluconeogenesis from propionate, amino acids, glycerol, and lactate for their glucose supply. Propionate conversion to succinyl-CoA requires the B vitamins **pantothenate** (a source of **coenzyme A.SH**), **biotin** (a **CO₂** shuttler in combination with **Mn⁺⁺**), and **cobalamin** (**B₁₂**; a **cobalt-containing** vitamin which rearranges **methylmalonyl-CoA to succinyl-CoA**) (see Chapters 41, 42, 43, 51, and 52). The B-complex vitamins are usually formed in adequate amounts by bacterial flora present in the digestive tracts of animals, or they are received through the diet.

In summary, it is well established that some intermediates in the TCA cycle are also members of other metabolic pathways. For example, in the liver OAA can be converted

(via PEP) to glucose, but this can only occur if carbon atoms are fed into the cycle in addition to those from acetyl-CoA (e.g., as α-KG⁼ formed from the deamination of Glu, or as succinyl-CoA formed from propionate). In other words, some of the reactions of the TCA cycle, but not the cycle itself, can be used biosynthetically. This "multiple use" of TCA cycle reactions in some tissues will influence the way in which flux of intermediates through the cycle is regulated.

In some respects the TCA cycle can be considered to consist of two subsequent steps: **1)** the span from **acetyl-CoA** and **OAA** to α-KG⁼, and **2)** the span from α-KG⁼ or **succinyl-CoA** to **malate** and/or **OAA**. In some tissues and under certain metabolic conditions this division of the cycle into two separate steps allows us to better understand how it is possible for intermediates to feed into the cycle at, or after, the level of α-KG⁼, and be withdrawn from the cycle at the level of malate or **OAA**. This is accomplished without necessarily interfering with the normal operation of energy generation via the conventional cycle. This division is particularly relevant when amino acid metabolism is being considered. For example, when hepatocytes are conducting gluconeogenesis during sustained aerobic exercise (see Chapters 77-81), malate and OAA are being siphoned away from the cycle to generate PEP in the cytoplasm. Conversely, in exercising muscle cells, the primary substrate entry point is at the level of **acetyl-CoA** (from fatty acid β-oxidation), and **OAA** (from glucose oxidation). However, in both tissue types **net flux** through the TCA cycle remains the same.

Although the TCA cycle functions within mitochondria, several reactions may also occur in the cytoplasm. For example, extramitochondrial malate dehydrogenase plays a role in the malate shuttle for the reoxidation of cytosolic NADH to NAD⁺ (see Chapter 36), and extramitochondrial isocitrate dehydrogenase is a substantial, if not main source of NADPH for fatty acid biosynthesis in adipocytes of ruminant animals (see Chapter 56). Other TCA cycle enzymes have also been found in the cytoplasm (e.g., aconitase), however their roles in this compartment are less well understood.

OBJECTIVES

- Discuss why the TCA cycle is considered by be an amphibolic pathway, and describe the role of anaplerotic reactions assocaited with this cycle.

- Indicate how pyrimidine and porphyrin biosyntheses can represent formidable leaks in the TCA cycle.

- Show how the mitochondrial Asp/OAA pool can be compromised in hepatic gluconeogenesis.

- Indicate how glucose is used to repenish the TCA machinery of exercising aerobic muscle fibers.

- Recognize how proteins and propionate can be used to replenish TCA cycle intermediates.

- Explain why and how the TCA cycle can be divided into two separate but interconnected parts.

- Know why certain enzymes of the mitochondrial TCA cycle are found in the cytoplasm.

- Show how hydrocarbons from acetyl-CoA and OAA can cross mitochondrial membranes.

QUESTIONS

1. **Which one of the following intermediates leaks away from the TCA cycle to enter porphyrin biosynthesis?**
   a. Oxaloacetate
   b. Acetyl-CoA
   c. Fumarate
   d. α-Ketoglutarate
   e. Succinyl-CoA

2. **Which one of the following intermediates leaks away from the TCA cycle to enter pyrimidine biosynthesis?**
   a. Oxaloacetate
   b. Acetyl-CoA
   c. Malate
   d. Succinate
   e. Isocitrate

3. **Which one of the following amino acids can undergo deamination as well as transamination?**
   a. Glu
   b. Asp
   c. Ala
   d. Tyr
   e. Val

4. **Which one of the following is considered to be an anaplerotic reaction for the TCA cycle?**
   a. Succinyl-CoA —> Pyrimidines
   b. Fumarate —> Purines
   c. Pyruvate —> Oxaloacetate
   d. Citrate —> Isocitrate
   e. Acetyl-CoA —> Ketone Bodies

5. **Carbon atoms from all of the following are known to replenish TCA cycle intermediates, EXCEPT:**
   a. Propionate.
   b. Amino acids.
   c. Glucose.
   d. Vitamin B_{12}.
   e. Pyruvate.

6. **Identify the TCA cycle intermediate(s) that is/are impermeable to mitochondrial membranes:**
   a. Citrate
   b. α-KG$^=$
   c. Malate
   d. OAA
   e. All of the above

7. **Carbon atoms are removed from the TCA cycle to form all of the following, EXCEPT:**
   a. Glutamine.
   b. Heme.
   c. Cytosine.
   d. Glucose.
   e. Guanine.

7. e

6. d

5. d

4. c

3. a

2. a

1. e

ANSWERS

Chapter 36

Oxidative Phosphorylation

Overview

- The electron transport chain (ETC) is the major consumer of O_2 in mammalian cells.
- The ETC passes electrons from NADH and $FADH_2$ to protein complexes and mobile electron carriers.
- Coenzyme Q (CoQ) and cytochrome c (Cyt c) are mobile electron carriers in the ETC, and O_2 is the final electron recipient.
- The malate and glycerol 3-P shuttles regenerate cytoplasmic NAD^+ for glycolysis, and deliver reducing equivalents to the mitochondrial ETC.
- Inhibitors of oxidative phosphorylation arrest cellular respiration.
- Uncouplers dissociate oxidation from phosphorylation, and help to generate heat as animals adapt to the cold.

Mitochondrial **NADH** and **FADH₂** are energy-rich molecules because each contains a pair of electrons that has high transfer potential. When these electrons are transferred in the inner mitochondrial membrane between protein carriers to molecular oxygen (O_2), considerable energy is liberated which can be used to generate ATP. This process of **oxidative phosphorylation** utilizes the **electron transport chain (ETC)**, also known as the **respiratory chain**, which is the major consumer of O_2 in mammalian cells.

Movement of Electrons from Cytoplasmic NADH to the Mitochondrial ETC

Intact mitochondrial membranes are **impermeable to NADH and NAD⁺**, and in order for glycolysis to continue, **NAD⁺** must be continually regenerated in the cytoplasm (see Chapter 26). Therefore, reducing equivalents (i.e., electrons) from NADH, rather than NADH itself, are carried across mitochondrial membranes by either **malate (Mal)** or **glycerol 3-phosphate (Fig. 36-1)**, thus allowing for cytoplasmic NAD^+ reformation, and for NADH and/or FADH₂ utilization in the mitochondrial ETC. In the **Mal shuttle**, reducing equivalents from **NADH** are accepted by **oxaloacetate (OAA)**, thus forming **Mal** which crosses mitochondrial membranes via an **α-ketoglutarate (α-KG⁼)-Mal antiporter**. Inside mitochondria, **NADH** is regenerated from **Mal**, and **OAA**, which also cannot cross mitochondrial membranes, is returned to the cytoplasm via reversible conversion to **aspartate (Asp**; see Chapters 9 and 35). The **amine group** from **Asp** is transferred to **α-KG⁼** in the cytoplasm, thus forming **glutamate (Glu)**, which is returned to mitochondria via an **Asp-Glu antiporter**. Inside mitochondria, **Glu** transfers its **amine group** to **OAA**, thus reforming **Asp** and completing the shuttle.

Another carrier of reducing equivalents is **glycerol 3-P**, which, like Mal and Asp, readily traverses mitochondrial membranes. This **shuttle** transfers electrons from **NADH** to

Copyright © 2015 Elsevier Inc. All rights reserved.

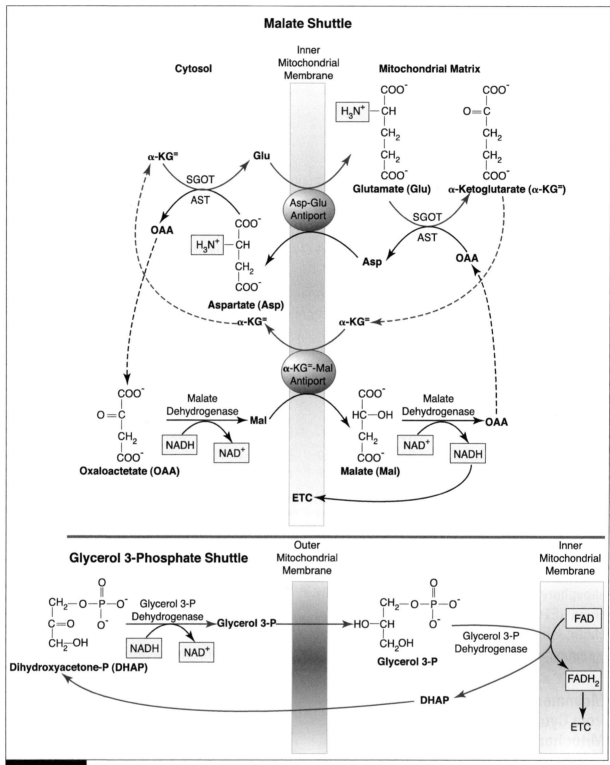

Figure 36-1

dihydroxyacetone phosphate (DHAP), thus forming **glycerol 3-P** (and **NAD⁺**) in the cytoplasm. **Glycerol 3-P** then crosses the outer mitochondrial membrane, and is reoxidized to **DHAP** by the **FAD** prosthetic group of **glycerol** **3-P dehydrogenase**. **FADH₂** is thus formed on the inner mitochondrial membrane, and **DHAP** diffuses back into the cytosol to complete the shuttle. In some species, activity of this shuttle decreases after thyroidectomy. Although it is present in insect flight muscle, the brain, brown adipose tissue, white muscle tissue and the liver of mammals, in other tissues (e.g., heart muscle), mitochondrial **glycerol 3-P dehydrogenase** is deficient. It is therefore believed that the **malate shuttle** is of more universal utility than the **glycerol 3-P shuttle**, particularly since **3** rather than **2 ATP** can be generated per atom of O_2 consumed (see below).

Oxidation and Reduction

Oxidation is a process in which electrons are removed, and **reduction** one in which electrons are gained. When one compound is oxidized, another must be reduced. The ETC functions by passing electrons from compounds with less reductive potential, such as **NADH** and **FADH₂**, to those with more reductive potential, such as **coenzyme Q (CoQ)** and **cytochrome c (Cyt c)**. The final recipient of electrons in the respiratory chain is **molecular oxygen (O_2)**, which together with **hydrogen** forms **water (H_2O)** (**Fig. 36-2**).

The ETC contains four protein complexes and two mobile electron carriers. As membrane-bound complexes, these carriers (**CoQ** and **Cyt c**) accept electrons from the preceding complex, then pass them on to the subsequent complex. **Complex I**, known as **NADH-CoQ reductase**, passes **two electrons** from **NADH** to **CoQ**. It contains 25 different proteins including several nonheme iron proteins, and a covalently-bound **flavin mononucleotide (FMN)**. Like FAD, FMN is derived from water-soluble **vitamin B₂** (**riboflavin**; see Chapter 40). Flavin mononucleotide receives two electrons from **NADH**, and passes them to a set of **iron-sulfur (FeS)** proteins (**Fe₂S₂** and **Fe₄S₄**; nonheme prosthetic groups), which then pass them on to **CoQ**. This compound (**CoQ**), also called **ubiquinone**, is not firmly attached to any protein, and serves as a lipid-soluble **mobile electron carrier**.

The **second complex, succinyl-CoQ reductase**, serves to bind **succinate, FADH₂,** and enzymes containing bound FADH₂ to the rest of the ETC. Two electrons are transferred from **FADH₂** through several **FeS proteins**, and finally to **CoQ**. The **FADH₂** that enters the ETC at this point is either generated from the mitochondrial succinate to fumarate reaction, or from the glycerol 3-phosphate shuttle. Since this route to CoQ bypasses the first ATP-generating step in NADH oxidation, **only 2 (rather than 3) ATP can be generated from FADH₂ oxidation**.

Cytochrome c reductase is the **third complex** of the ETC, and it contains an **FeS protein** as well as **cytochromes b** and **c₁** (which are heme-containing proteins). Like the iron in complex I, the heme iron of complex III is reduced from the ferric (Fe^{+++}) to ferrous (Fe^{++}) state. Complex III transfers two electrons from **CoQ** to **Cyt c**, the second mobile electron carrier. Unlike the large enzyme complexes, Cyt c is a small heme-containing protein with only 104 amino acid residues. It is loosely attached to the inner mitochondrial membrane rather than being embedded in it. As with the other cytochromes, the iron atom of heme in Cyt c changes from the +3 to +2 state on reduction.

The final **complex (IV)** is **Cyt c oxidase**. It transfers a pair of electrons from each of two **Cyt c** molecules and a **copper (Cu)-containing enzyme** to **O_2**. Like complex III, the Cyt c oxidase complex contains heme iron as part of cytochromes **a** and **a₃**. During electron transfer, the copper-containing enzyme changes charge from +2 to +1 on reduction (see Chapter 50).

Phosphorylation

As electrons are passed between **NADH** and **CoQ**, **CoQ** and **Cyt c**, and **Cyt c** and **O_2**, **protons are ejected** from the inner mitochondrial membrane to re-enter the mitochondrial matrix, and flow through protrusions (**F₀ and F₁**) on the inner membrane. The **F₀** component

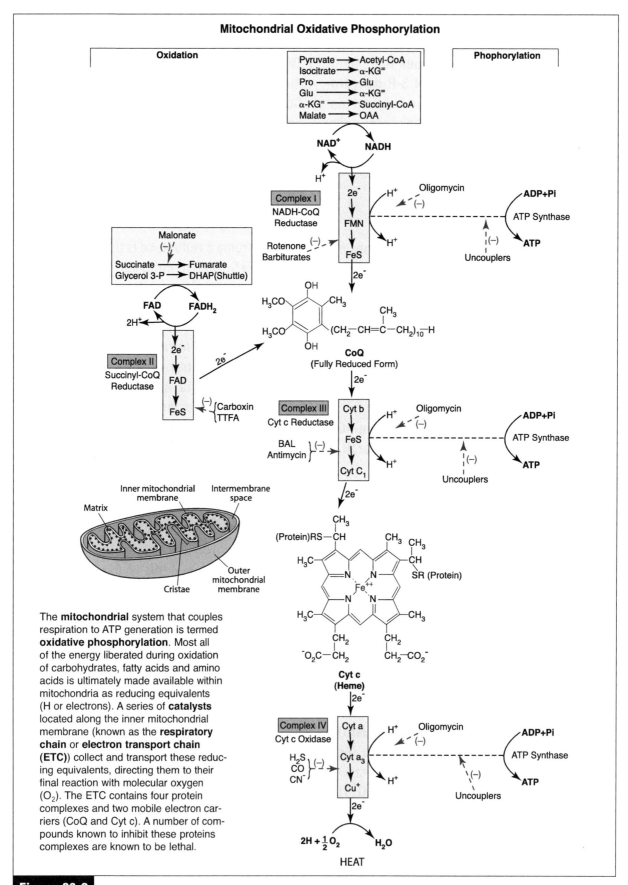

Mitochondrial Oxidative Phosphorylation

The **mitochondrial** system that couples respiration to ATP generation is termed **oxidative phosphorylation**. Most all of the energy liberated during oxidation of carbohydrates, fatty acids and amino acids is ultimately made available within mitochondria as reducing equivalents (H or electrons). A series of **catalysts** located along the inner mitochondrial membrane (known as the **respiratory chain** or **electron transport chain (ETC)**) collect and transport these reducing equivalents, directing them to their final reaction with molecular oxygen (O_2). The ETC contains four protein complexes and two mobile electron carriers (CoQ and Cyt c). A number of compounds known to inhibit these proteins complexes are known to be lethal.

Figure 36-2

spans the inner membrane, forming a H^+ channel, and the F_1 component protrudes into the matrix and contains the active **ATP synthase** site where **ADP** and **Pi** condense to form **ATP**. Once ATP is produced it moves through the inner mitochondrial membrane in exchange for ADP from extramitochondrial sites via a **membrane ADP/ATP antiporter**.

Inhibitors and Uncouplers

Inhibitors that **arrest cellular respiration** may block the ETC at any of **4 sites** (see **Fig. 36-2**). The **first** inhibits Complex I (e.g., barbiturates and the insecticide and fish poison, rotenone); the **second** Complex II (e.g., malonate, carboxin and TTFA (an Fe-chelating agent)); the **third** Complex III (e.g., BAL (dimercaprol), and the antibiotic, antimycin); and the **fourth** Complex IV (e.g., the classic poisons hydrogen sulfide (H_2S), carbon monoxide (CO), and cyanide (CN^-)). Oligomycin, a compound that blocks movement of H^+ through the F_0 channel, inhibits ATP synthesis, as well as oxidation.

The action of **uncouplers** is to **dissociate oxidation from phosphorylation**. When this occurs, NADH and $FADH_2$ are oxidized, **heat** is produced, but none of the energy from oxidation is trapped as ATP. This may be likened unto a car engine in which the clutch has been uncoupled and the engine revved up; a great deal of fuel is burned but, since none of the energy is used for propulsion of the vehicle, all energy escapes as heat. At times this can be metabolically useful, for it is a means of generating heat during hibernation, in the immediate postnatal period, and in animals adapted to the cold.

Examples of exogenous uncouplers are **dinitrocresol, pentachlorophenol, m-chloro-carbonyl cyanide phenylhydrazone (CCCP), valinomycin, gramicidin**, and the compound that has been studied most, **2,4-dinitrophenol**. These compounds allow protons to pass into mitochondria, thereby destroying the proton

(i.e., pH) gradient necessary for ATP production. **Brown adipose tissue** is specialized for heat generation, and contains abundant mitochondria (which impart brown color to the tissue). Large blood vessels of newborns are surrounded by brown adipose tissue, where the oxidation of fatty acids releases heat that helps maintain the temperature of circulating blood.

An inner-membrane protein called **thermogenin** is a natural uncoupler of oxidative phosphorylation, and acts as a transmembrane H^+ transporter. In addition to brown adipose tissue, it is found in muscle-cell mitochondria of seals and other animals adapted to the cold. Like other uncouplers, it short-circuits the proton concentration gradient. **Fatty acids** can also act as endogenous uncouplers in mitochondria containing thermogenin. In turn, **norepinephrine** controls the release of fatty acids from (white) adipocytes, and cold stress leads to **thyroxine** release that also assists in lipolysis, and the uncoupling of oxidation from phosphorylation. Mitochondria containing this protein can thus function as ATP generators, or as miniature furnaces. Additionally, some investigators feel that one function of **peroxisomes** is heat rather than ATP generation, as a product of long-chain fatty acid oxidation (see Chapter 55).

OBJECTIVES

- Explain how the "Malate Shuttle" operates in moving reducing equivalents from the cytoplasm to mitochondria, why it is important, and how Glu, Asp, α-KG$^=$ and Mal are transported across inner mitochondrial membranes.

- Explain why the "Malate Shuttle" is of more universal utility than the simpler "Glycerol 3-P Shuttle."

- Describe the compartments and membranes of mitochondria, and locate the respiratory assemblies they contain.

- Define oxidative phosphorylation and cellular respiration.

- Identify the final recipient of electrons in the ETC.

- Recognize how the four protein complexes and two mobile electron cariers of the ETC operate.

- Compare the processing of reducing equivalents from $FADH_2$ to those from NADH in the ETC.

- Explain how ATP is produced during oxidative phosphorylation, and how it moves into the cytoplasm.

- Show how the effects of an inhibitor differ from those of an uncoupler of oxidative phosphorylation.

- Indicate how thermogenin, fatty acids, norepinephrine and thyroxine can contribute to endogenous mitochondrial heat generation.

- Discuss how peroxisomal β-oxidation can contribute to endogenous heat generation (see Chapter 55).

- Identify inhibitors known to arrest cellular respiration, and recognize the site where each inhbitor operates.

QUESTIONS

1. **The malate shuttle is thought to be of more universal utility than the glycerol 3-phosphate shuttle, because:**
   a. NADH can be transferred intact from the cytoplasm to the mitochondrial matrix using this shuttle.
   b. Mitochondrial $FADH_2$ is produced using the malate shuttle.
   c. Malate uncouples oxidation from phosphorylation inside mitochondria.
   d. Three rather than 2 ATP can be generated per atom of O_2 consumed from mitochondrial NADH generated in the malate shuttle.
   e. $FADH_2$ cannot be oxidized in the electron transport chain.

2. **Which one of the following functions as a lipid-soluble mobile electron carrier in the electron transport chain?**
   a. Cytochrome b
   b. Copper
   c. Flavin mononucleotide
   d. Coenzyme Q
   e. Iron-sulfur protein

3. **Cytochrome c reductase is synonymous with which complex in the electron transport chain?**
   a. I
   b. II
   c. III
   d. IV
   e. V

4. **Which one of the following is known to block movement of H^+ through the mitochondrial F0 channel, thereby inhibiting ATP formation?**
   a. Carbon monoxide
   b. Oligomycin
   c. Carboxin
   d. Barbiturates
   e. Rotenone

5. **Uncouplers of mitochondrial oxidative phosphorylation are associated with all of the following, EXCEPT:**
   a. Hibernation.
   b. A decrease in mitochondrial pH.
   c. An increase in fatty acid oxidation in brown adipose tissue.
   d. Thermogenin.
   e. An increase in ATP generation.

6. **In the malate shuttle, reducing equivalents from NADH are accepted by which compound in the cytoplasm?**
   a. Pyruvate
   b. Malate
   c. Dihydroxyacetone phosphate
   d. Oxaloacetate
   e. Aspartate

7. **Which one of the following compounds associated with the ETC is a heme-containing protein?**
   a. Cytochrome c
   b. Ubiquinone
   c. NADH
   d. Coenzyme Q
   e. Flavin mononucleotide

8. **Which complex serves to bind enzymes containing bound $FADH_2$ to the rest of the ETC?**
   a. NADH-CoQ reductase
   b. Succinyl-CoQ reductase
   c. Cyt c reductase
   d. Cyt c oxidase

8. b
7. a
6. d
5. e
4. b
3. c
2. d
1. d

ANSWERS

Chapter 37

Gluconeogenesis

Overview

- Gluconeogenesis occurs in the liver and kidneys.
- Gluconeogenesis supplies the needs for plasma glucose between meals.
- Gluconeogenesis is stimulated by the diabetogenic hormones (glucagon, growth hormone, epinephrine, and cortisol).
- Gluconeogenic substrates include glycerol, lactate, propionate, and certain amino acids.
- PEP carboxykinase catalyzes the rate-limiting reaction in gluconeogenesis.
- The dicarboxylic acid shuttle moves hydrocarbons from pyruvate to PEP in gluconeogenesis.
- Gluconeogenesis is a continual process in carnivores and ruminant animals, therefore they have little need to store glycogen in their liver cells.
- Of the amino acids transported to liver from muscle during exercise and starvation, Ala predominates.
- β-Aminoisobutyrate, generated from pyrimidine degradation, is a (minor) gluconeogenic substrate.

Gluconeogenesis is a major regulatory process in the **liver** and **kidneys** by which noncarbohydrate substrates; namely **glycerol**, **lactate**, **propionate**, and **glucogenic amino acids**; are converted to **glucose 6-phosphate (Glc-6-P)**, and then to either **free glucose** or **glycogen**. The liver and kidneys are the major organs containing the full complement of gluconeogenic enzymes (i.e., **pyruvate carboxylase, PEP carboxykinase, fructose 1,6-bisphosphatase**, and **glucose 6-phosphatase**, see **Fig. 37-1**).

Gluconeogenesis is needed to meet the demands for **plasma glucose** between meals, which then becomes particularly important as an energy substrate for nerves, erythrocytes, and other largely anaerobic cell types. Failure of this process can lead to brain dysfunction, coma, and death. In **ruminant animals** and **carnivores**, **hepatic gluconeogenesis** is a **continual**, ongoing process that has little or no relation to the frequency of food consumption.

It is clear that even under conditions where **fatty acid oxidation** may be supplying most of the energy requirement of the organism, there is always a certain basal requirement for **glucose**, and glucose is a major fuel providing energy to skeletal muscle under anaerobic conditions (see Chapters 35 and 80). In addition, gluconeogenic mechanisms in the liver are used to clear certain products of metabolism from blood (e.g., **lactate**, produced by erythrocytes, the retina, kidney medulla, and anaerobic muscle fibers; and **glycerol**,

Copyright © 2015 Elsevier Inc. All rights reserved.

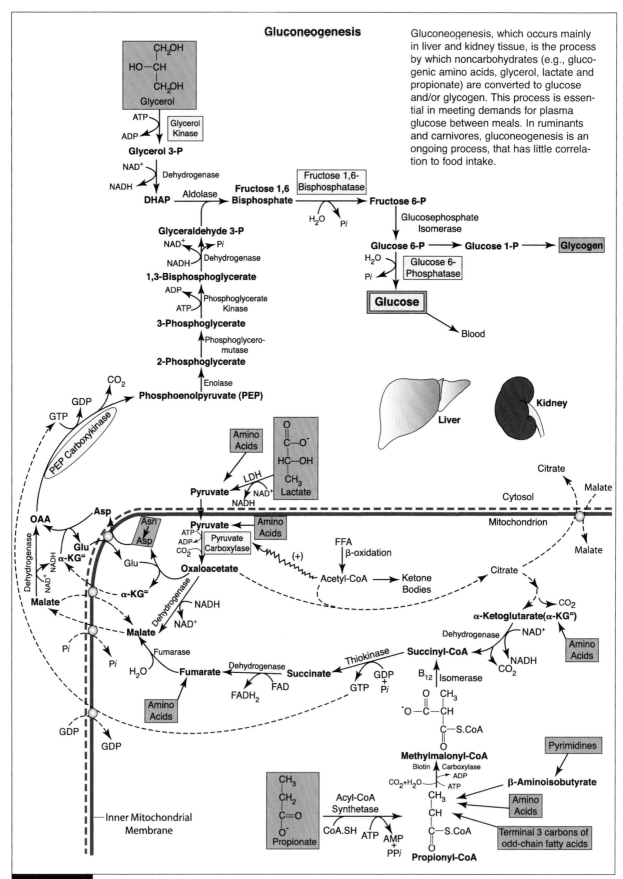

Gluconeogenesis

Gluconeogenesis, which occurs mainly in liver and kidney tissue, is the process by which noncarbohydrates (e.g., glucogenic amino acids, glycerol, lactate and propionate) are converted to glucose and/or glycogen. This process is essential in meeting demands for plasma glucose between meals. In ruminants and carnivores, gluconeogenesis is an ongoing process, that has little correlation to food intake.

Figure 37-1

which is continuously produced by adipose tissue when fat is being mobilized for energy purposes). Of the **amino acids** transported from muscle to liver during starvation, **alanine (Ala)** predominates. This amino acid is part of the **glucose-alanine cycle** which has the effect of cycling glucose from liver to muscle with formation of **pyruvate**, followed by transamination to **Ala**, then transport of **Ala** to **liver**, followed by **gluconeogenesis** back to **glucose** (**Fig. 37-2**). During starvation, **branched-chain amino acid (BCAA) oxidation in muscle** serves as a source of **ammonia (NH₃)**. A net transfer of amino nitrogen from muscle to liver (and then to **urea**), and of potential energy (**glucose**) from liver to muscle is thus effected. The energy needed for hepatic synthesis of glucose from pyruvate (or lactate) is thought to be derived from the β-oxidation of fatty acids.

Gluconeogenic Precursors
Glucogenic Amino Acids

Since amino acids are used for a variety of vital biosynthetic functions in the liver (e.g., protein production), they are generally used conservatively for gluconeogenesis. During their conversion to glucose, their amino groups are irretrievably lost to urea, which then enters blood (i.e., **blood urea nitrogen (BUN)**). The majority of amino acids form TCA intermediates and pyruvate, and are therefore **glucogenic** (see **Fig. 37-1**). Others form acetyl-CoA and, thus, are **ketogenic** (see Chapters 8 and 71). **Acetyl-CoA is not a (net) substrate for gluconeogenesis** since its two carbon atoms are lost as CO_2 in the TCA cycle. Especially important for gluconeogenesis are **Ala** (in the liver), and **glutamine** (in the kidneys, see Chapter 11). It has been estimated that the contribution of amino acids to ruminant gluconeogenesis is about **5-7%** of glucose produced in both the fed and fasted state.

Lactate

A good source of glucose during exercise is lactate (see **Fig. 37-2**, Lactic Acid (Cori) Cycle), and during concentrated carbohydrate feeding (where rumen lactate concentrations are greatly increased), lactate becomes an important hepatic gluconeogenic substrate. Lactate is also readily oxidized in cardiac muscle (which is rich in mitochondria; see Chapter 80).

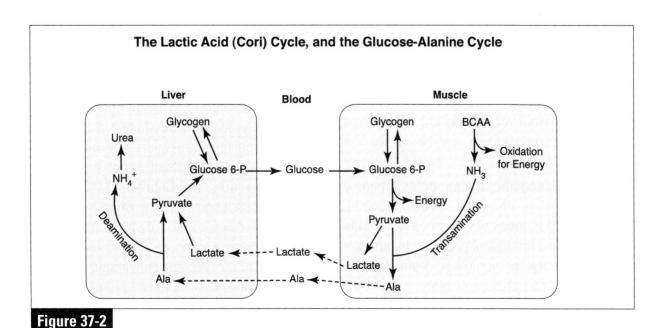

The Lactic Acid (Cori) Cycle, and the Glucose-Alanine Cycle

Figure 37-2

Glycerol

Since **glycerol kinase** activity is absent in white adipose tissue, glycerol becomes a waste product of lipolysis, and is converted to glucose in the liver, and to a lesser extent in the kidney. This substrate becomes a significant source of glucose in **hibernating animals** (e.g., the black bear), where lipolysis becomes necessary for survival (see Chapter 76). It should be noted that the synthesis of glucose from glycerol requires fewer steps (and therefore less energy), than synthesis from other precursors. Glycerol utilization in gluconeogenesis also bypasses the **dicarboxylic acid (DCA) shuttle** (see **pyruvate carboxylase** below), thereby allowing oxaloacetate (OAA) to be reutilized in the TCA cycle.

Propionate

Propionate, a volatile fatty acid (VFA) produced from microbial carbohydrate digestion in ruminants and other herbivores (see Chapter 54), is a major hepatic gluconeogenic substrate. The percentage of glucose derived from propionate in the liver varies with diet (and species), from a maximum of about **70%** under heavy grain feeding in ruminants, to very little during starvation. The importance of propionate as a gluconeogenic substrate is illustrated by the observation that the lactating udder of the goat may utilize **60-85%** of glucose produced by the liver for milk production. In contrast to propionate, **acetate** and **butyrate**, the other two major VFAs produced through microbial carbohydrate digestion, do not contribute carbon atoms directly to the net synthesis of glucose.

Certain glucogenic amino acids (namely isoleucine, valine, threonine, and methionine), the **terminal 3 carbons of odd-chain fatty acids** undergoing mitochondrial β-oxidation, and the **β-aminoisobutyrate** generated from thymine degradation, can also enter hepatic gluconeogenesis at the level of **propionyl-CoA**. While the former may be quantitatively significant to carnivores, and to all animals during starvation, the latter two are not since:

1) Few odd-chain fatty acids exist in mammalian organisms (with the exception of ruminant animals; see Chapter 54), and

2) Only small amounts of β-aminoisobutyrate normally become available to the liver through pyrimidine degradation (see Chapter 17).

Entry of propionate into gluconeogenesis (as well as amino acids that are converted to propionyl-CoA), requires **pantothenate (a source of coenzyme A.SH)**, **vitamin B$_{12}$**, and **biotin** (see **Fig. 37-1**). These vitamins are normally synthesized by microbes inhabiting the digestive tract (see Chapters 41, 42, and 43).

Gluconeogenic Enzymes

The pathway of gluconeogenesis is a partial reversal of the Embden-Meyerhoff pathway. There are four steps, however, in the latter pathway that must be circumnavigated using separate enzymes.

Pyruvate Carboxylase

The metabolic route from pyruvate to OAA and then on to PEP is called the **DCA shuttle**. It allows pyruvate and compounds that can be converted to pyruvate, such as lactate and amino acids, to be metabolized to PEP without having to traverse the physiologically irreversible step from PEP to pyruvate in the opposite direction. Pyruvate carboxylase is a mitochondrial enzyme that is activated by **acetyl-CoA** (although acetyl-CoA itself is not a substrate for this reaction, see Chapter 27). The level of acetyl-CoA begins to rise when fatty acids are broken down to fill a demand for energy. Since OAA itself cannot cross mitochondrial membranes, its reversible conversion to **malate** (and **aspartate (Asp)**) allows carbon atoms to be transported into the cytoplasm where OAA can then be regenerated (see **Fig. 37-1** and Chapter 36).

PEP Carboxykinase

Cytoplasmic conversion of **OAA** to **PEP** is catalyzed by **PEP carboxykinase**, which requires **GTP** to drive this reaction energetically in an uphill direction. This high energy compound (GTP) is normally derived from the conversion of succinyl-CoA to succinate in the TCA cycle (see **Fig. 37-1**). In most mammals the PEP carboxykinase enzyme is located predominantly in the cytosol. However, in the **guinea-pig** liver, 50% of PEP carboxykinase activity is found in mitochondria, and in **birds** this figure approaches 100%.

This reaction is the focal point of gluconeogenesis, and the **rate-limiting step**. PEP carboxykinase, like fructose 1,6-bisphosphatase and glucose 6-phosphatase below, is stimulated by the **diabetogenic hormones** (i.e., **epinephrine**, **growth hormone**, **glucagon**, and the **glucocorticoids**), and **inhibited by insulin**. **Thyroxine** also plays a stimulatory role. Epinephrine and glucagon act by increasing intracellular cyclic-AMP levels, and the glucocorticoids act by increasing the synthesis and activity of the four major gluconeogenic enzymes. Although gluconeogenesis also takes place in proximal renal tubular epithelial cells of the kidney, neither insulin nor glucagon are thought to affect that process.

Fructose 1,6-Bisphosphatase

This enzyme is inducible, and in addition to the diabetogenic hormones above, is activated by high cytoplasmic **ATP** and **citrate** levels. This reaction bypasses the regulatory reaction of glycolysis catalyzed by **phosphofructokinase (PFK)**, which is inhibited by glucagon, ATP, phosphocreatine, H^+ and citrate (see Chapter 25).

Glucose 6-Phosphatase

In liver and kidney tissue (but not in muscle), this enzyme is present to remove phosphate from Glc-6-P, enabling glucose to diffuse into blood. This is usually the final step in gluconeogenesis and in hepatic glycogenolysis, which is reflected by a **rise in the blood glucose concentration** (see Chapter 23). When hepatic Glc-6-P is being overproduced through gluconeogenesis (e.g., during glucocorticoid stimulation), then some may pass on into glycogen formation, thus replenishing intracellular reserves.

Glycerol Kinase

In addition to the four enzymes above, **glycerol kinase**, with the assistance of **ATP**, converts **glycerol** (from white adipose tissue) to **glycerol 3-P** (**Fig 37-1**), and it is also a gluconeogenic enzyme (although it does not participate in reversal of the EMP). This enzyme is found in the liver, kidneys, intestine, brown adipose tissue and the lactating mammary gland.

In summary, gluconeogenesis is the process whereby noncarbohydrate substrates such as glycerol, lactate, pyruvate, propionate, and gluconeogenic amino acids are converted to either free glucose or glycogen. The liver and kidneys are the major organs containing the full complement of gluconeogenic enzymes, and thus are the only organs that can synthesize and export glucose into blood. Gluconeogenesis becomes necessary for meeting the demands for plasma glucose between meals, particularly in tissues that store little glycogen, in those lacking mitochondria, and in those that are insulin-independent. In the liver's of ruminants and carnivores, gluconeogenesis is an ongoing process that has little or no correlation to the frequency of food intake. Failure of this essential, hormonally-mediated hepatorenal process in animals can lead to brain dysfunction, coma, and death.

OBJECTIVES

- Describe the physiologic significance of gluconeogenesis, and list the primary precursors. Show how and why primary precursors in carnivores may differ from those in ruminant animals.

- List the irreversible reactions of glycolysis and identify the enzymatic steps of gluconeogenesis that bypass them.

- Identify the major organs that conduct gluconeogenesis.

- Describe all steps in the conversion of alanine to PEP, and identify the enzymes, intermediates and cofactors involved.

- Calculate the number of high-energy phosphate bonds consumed during gluconeogenesis, and compare it to the number formed during glycolysis.

- Outline the "Lactic Acid (Cori) Cycle" as well as the "Glucose-Alanine Cycle," and discuss the physiologic significance of each.

- Describe all steps in the conversion of propionate to PEP, and identify the enzymes, intermediates and cofactors involved.

- Identify the inner-mitochondrial membrane antiporters that are needed for gluconeogenesis to occur.

- Explain how gluconeogenesis is endocrinologically controlled.

- Recognize how gluconeogenic conversion of Frc-1,6-bisP to Frc-6-P is controlled, and discuss why glycerol kinase is considered to be a gluconeogenic enzyme.

- Explain why the succinate thiokinase reaction in liver mitochondria yields GTP, yet in most other mitochondria, ATP (see Chapter 34).

QUESTIONS

1. **The two major organs containing the full complement of gluconeogenic enzymes are:**
    a. Brain and kidney.
    b. Lung and liver.
    c. Spleen and stomach.
    d. Lung and brain.
    e. Liver and kidney.

2. **The net transfer of carbon atoms from all of the following can be found in glucose, EXCEPT:**
    a. Palmitate.
    b. Lactate.
    c. Propionate.
    d. Glutamine.
    e. Glycerol.

3. **Which compound below normally passes through the hepatic dicarboxylic acid shuttle on its way to forming glucose?**
    a. Lactate
    b. Glycerol
    c. Urea
    d. Phosphoenolpyruvate
    e. Acetyl-CoA

4. **Which one of the following enzymes catalyzes the rate-limiting reaction in gluconeogenesis?**
    a. Pyruvate carboxylase
    b. PEP carboxykinase
    c. Fructose 1,6-bisphosphatase
    d. Enolase
    e. Glucose 6-phosphatase

5. **Which one of the following is best associated with the stimulation of gluconeogenesis?**
    a. Insulin
    b. Parathyroid hormone
    c. Cyclic-AMP
    d. Diacylglycerol
    e. Phosphodiesterase

6. **Entry of propionate into gluconeogenesis requires vitamins:**
    a. E and C.
    b. B_{12} and biotin.
    c. D and K.
    d. Niacin and folate.
    e. B_6 and thiamin.

7. **Which hepatic gluconeogenic enzyme also functions in glycogenolysis?**
    a. PEP carboxykinase
    b. Fructose 1,6-bisphosphatase
    c. Glycerol kinase
    d. Pyruvate carboxylase
    e. Glucose 6-phosphatase

8. **Adipocytes contain all of the following gluconeogenic enzymes, EXCEPT (see Chapter 70):**
    a. Pyruvate carboxylase.
    b. PEP carboxykinase.
    c. Glucose 6-phosphatase.

8. c

7. e

6. b

5. c

4. b

3. a

2. a

1. e

Chapter 38

Carbohydrate Digestion

Overview

- Monosaccharides do not require intestinal digestion prior to absorption.
- Salivary α-amylase and pancreatic α-amylase are similar enzymes.
- Members of the amylase family of enzymes are found in many tissues and organs of the body, however the highest concentrations are in the salivary glands and pancreas.
- Disaccharides cannot be absorbed (intact) by the intestine.
- Intestinal glucose and galactose absorption is Na⁺-dependent, yet insulin-independent.
- Intestinal fructose absorption is normally slower than glucose and galactose absorption.
- Most intestinal glucose absorption is associated with aerobic energy expenditure.
- A β-glucosidase and a β-galactosidase are present in the brush border of the small intestine, which hydrolyze glucose and galactose residues from glucocerebrosides and galactocerebrosides.

Monosaccharides (largely hexoses and pentoses) require no intestinal digestion prior to absorption; however, oligosaccharides must be hydrolyzed to monosaccharides before they can be absorbed. Since mammals lack the enzyme **cellulase**, they are incapable of carrying out the constitutive digestion of cellulose (which contains **β-1,4 glycosidic linkages**). However, they can digest (i.e., hydrolyze) dietary starch and glycogen (which contain α-1,4 and α-1,6 glycosidic linkages). Sites for starch and glycogen digestion are in the mouth and upper small intestine. Most monosaccharide absorption occurs in the duodenum and jejunum.

Salivary α-Amylase (Ptyalin)

With the exception of ruminants and carnivores (who do not secrete salivary α-amylase), this **endo-glycosidase** (or **endo-glucosidase**) initiates hydrolysis of **α-1,4 glycosidic bonds** within dietary glucose polymers, (e.g., glycogen

and starch), but it will not attack **α-1,6** or **terminal α-1,4 glycosidic linkages**. Glycogen is a product of animal metabolism, whereas starch is the comparable plant storage form of glucose (see Chapters 19 and 23). Glucose molecules in glycogen are mostly in long chains held together by **α-1,4 glycosidic bonds**, but there is some chain-branching produced by α-1,6-linkages. **Amylopectin**, which constitutes 80-85% of dietary starch, is similar but less branched, whereas **amylose**, the other component of starch, is a straight chain glucose polymer containing **only α-1,4** linkages.

Oligosaccharides are products of salivary α-amylase digestion. The major oligosaccharides are dextrins (with **α-limit dextrins** being the first formed products), compounds with an average chain length of 8-10 glucose residues containing the α-1,6 branches. Small amounts of **maltose** (an α-1,4 disaccharide), and **maltotriose** (an α-1,4 trisaccharide) are also formed at this point (**Fig. 38-1**). The optimal

Copyright © 2015 Elsevier Inc. All rights reserved.

Hydrolysis of Glycogen and Starch

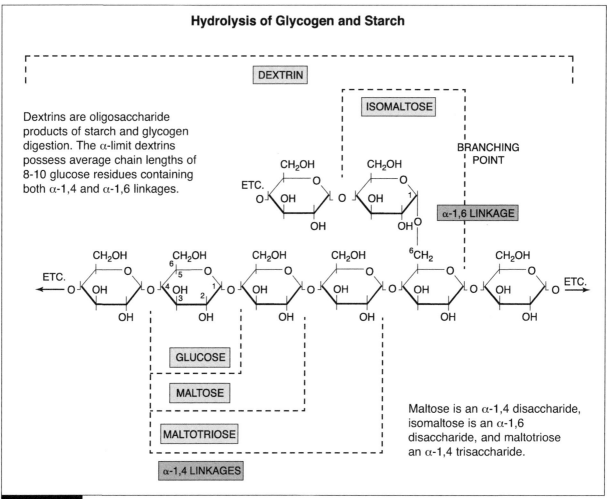

Dextrins are oligosaccharide products of starch and glycogen digestion. The α-limit dextrins possess average chain lengths of 8-10 glucose residues containing both α-1,4 and α-1,6 linkages.

DEXTRIN

ISOMALTOSE

BRANCHING POINT

α-1,6 LINKAGE

GLUCOSE

MALTOSE

MALTOTRIOSE

α-1,4 LINKAGES

Maltose is an α-1,4 disaccharide, isomaltose is an α-1,6 disaccharide, and maltotriose an α-1,4 trisaccharide.

Figure 38-1

pH for salivary α-amylase is 6.7, and its action is inhibited by acidic gastric juice when food enters the stomach. Therefore, little starch and glycogen digestion occurs during the short time food is in the mouth, and even less will occur in the stomach unless it fails to mix its contents fully.

Intestinal Carbohydrate Digestion

There are two phases of intestinal carbohydrate digestion. The first occurs in the **lumen** of the small intestine, and is commonly referred to as the **hormonal** or **pancreatic phase**. The second occurs in **brush border membranes**, and involves the action of a number of integral membrane proteins with saccharidase activity (see **Appendix**). Unlike salivary and pancreatic α-amylase, most of the surface **oligosaccha-**

ridases are exoenzymes which clip off one monosaccharide at a time.

Luminal Phase (Pancreatic α-Amylase)

The entry of partially digested acidic chyme into the duodenum stimulates specialized mucosal cells to release two important polypeptide hormones into blood; **secretin** (from duodenal S cells), and **cholecystokinin** (**CCK**, from duodenal I cells). These hormones then stimulate exocrine pancreatic secretions into the duodenal lumen containing **NaHCO$_3$** (needed to neutralize acidic chyme), and **digestive enzymes** (including α-amylase). Both salivary and pancreatic α-amylase (which are similar enzymes), continue internal starch, glycogen, and dextrin digestion in a favorable neutral duodenal pH environment (i.e., pH 7).

Polysaccharides are digested to a mixture of **dextrins** and **isomaltose** (which contain all of the α-1,6 branch-point linkages), as well as **maltose** and **maltotriose** (**Fig. 38-1**). Most salivary and pancreatic α-amylase is destroyed by trypsin activity in lower portions of the intestinal tract, although some amylase activity may be present in feces.

Brush Border Phase (Oligosaccharidases)

Enzymes responsible for the final phase of carbohydrate digestion are located in brush border membranes of the small intestine, and have their active sites extending into the lumen.

They are generally protected from degradation by a glycocalix mucus coat. However, since they are anaerobic enzymes, like all digestive enzymes, when they are cleaved from the brush border, or mucosal cells are sloughed into the lumen, these enzymes remain active. The reactions they catalyze are indicated in **Fig. 38-2**.

Some of these brush border enzymes have more than one substrate. **Isomaltase** (also called **α-dextrinase** or **α-1,6-glucosidase**), is mainly responsible for hydrolysis of α-1,6-glycosidic linkages in isomaltose and the

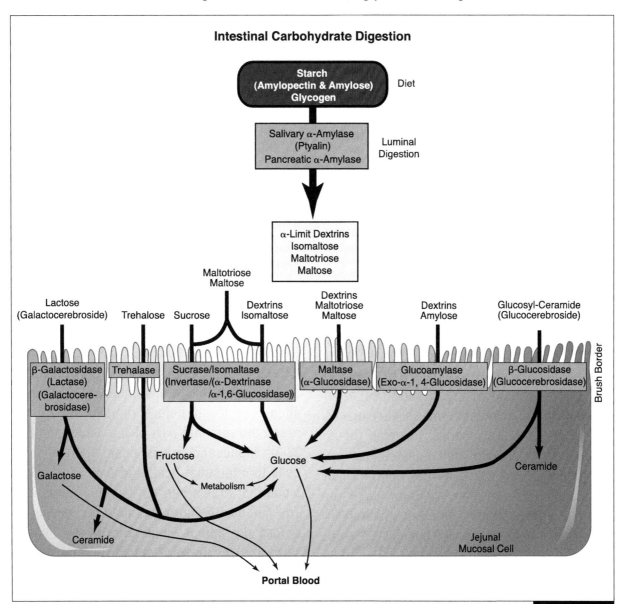

Figure 38-2

dextrins. However, along with **maltase** (also called **α-glucosidase**) and **sucrase** (also called **invertase**), it also breaks down maltotriose and maltose. **Sucrase/isomaltase** is a bifunctional enzyme, having one domain with isomaltase activity, and another with sucrase activity on the same polypeptide. Sucrase and isomaltase are reportedly synthesized as a single polypeptide chain and inserted into the brush border membrane. This protein is then hydrolyzed by pancreatic proteases into sucrase and isomaltase subunits, but the subunits reassociate noncovalently at the intestinal surface.

Dietary disaccharides (sucrose, lactose, and trehalose) are digested by their appropriate **disaccharidases**. **Trehalose** is a rare disaccharide found in young mushrooms, and lactose, which is found in milk, generally disappears from the animal diet following weaning.

A brush border **galactocerebrosidase** (**β-galactosidase** or **lactase**) catalyzes digestion of dietary galactocerebrosides (galactosyl-ceramides; see Chapter 59), as well as lactose (a glucose-galactose disaccharide). Although **lactase** is a labile enzyme, **maltase** and **sucrase** are more adaptive, and therefore inducible by their substrates.

A **glucocerebrosidase** (**β-glucosidase**) is also present in the brush border of the small intestine, which catalyzes digestion of dietary glucocerebrosides. Additionally, another member of the **α-amylase family** (i.e., enzymes found in a number of organs and tissues in addition to the pancreas and salivary glands (e.g., semen, testes, ovaries, fallopian tubes, striated muscle, lung, and adipose tissue)), is present in the brush border of the small intestine. This enzyme, glucoamylase or **exo-α-1,4-glucosidase**, catalyzes hydrolysis of terminal α-1,4-glycosidic bonds in starch, glycogen, and the dextrins.

In **pancreatic insufficiency**, as well in a variety of **bacterial** and **enteric infections**, transient loss of carbohydrate digestive activity can occur. As a result, increased amounts of di-, oligo-, and polysaccharides that are not hydrolyzed by α-amylase and/or intestinal brush border oligosaccharidases, cannot be absorbed: therefore, they reach the distal portion the intestine, which contains bacteria from the lower ileum on down. Bacteria can effectively metabolize carbohydrate polymers because they possess many more types of saccharidases than do mammals. Products of anaerobic bacterial carbohydrate digestion include short-chain **volatile fatty acids** (see Chapter 54), **lactate, hydrogen gas** (H_2), **methane** (CH_4), and **carbon dioxide** (CO_2). These compounds, if not absorbed, can cause fluid secretion, increased intestinal motility, and cramps (i.e., **diarrhea**), either because of increased intraluminal osmotic pressure, distension of the gut, or because of a direct irritant effect of bacterial degradation products on the intestinal mucosa. Some leguminous seeds (e.g., beans, peas, soya) can also be difficult for some animals to digest since they contain a modified sucrose to which one or more galactose moieties are linked. The glycosidic bonds of galactose are in the α-configuration, which can only be split by bacterial enzymes. The simplest sugar of this family is **raffinose** (α Gal(1,6)-α Glc(1,2)-β Fru).

Intestinal Monosaccharide Absorption

Hexoses and pentoses are rapidly absorbed across the wall of the small intestine. Essentially all are removed before remains of a meal reach the terminal part of the ileum (see **Fig. 38-3**). Monosaccharides pass from mucosal cells to interstitial fluid, and then to capillary blood that drains into the hepatic portal vein. The mucosal cell process involves movement across the **apical membrane** on the luminal side, and the **basolateral membrane** on the serosal side of the cell.

Early experiments concerning absorption rates for glucose from solutions perfused through the intestines of guinea pigs,

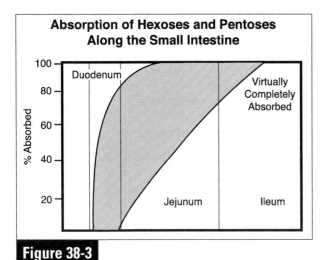

Figure 38-3

showed that the bulk of glucose absorption is **oxygen-dependent (Fig. 38-4)**. Therefore, it was concluded that active transport is involved, and that only a small amount of glucose is normally absorbed via a passive, diffusional mechanism. In order to explain how the active component occurs, a carrier has been identified which binds **both glucose** and **Na⁺** at separate sites, and which transports them both through the apical membrane using sodium's electrochemical gradient (**Fig. 38-5**). Both **glucose** and **galactose** transport are uniquely affected by the amount of **Na⁺** in the intestinal lumen because they share, with Na⁺, the same **cotransporter**, or **symport**, and therefore compete for uptake. This transporter, known as the **sodium-dependent glucose transporter** (**SGLT-1**), is also found in apical membranes

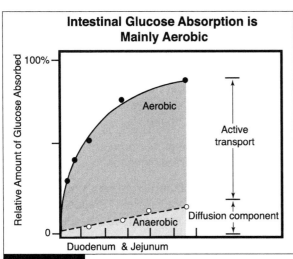

Figure 38-4

of proximal renal tubular epithelial cells (the S-3 segment). In the early part of the proximal tubules (S-1 segment), apical membranes contain an **SGLT-2** transporter, which functions similarly to the **SGLT-1** transporter. Both are **insulin-independent**, they do not require ATP (directly), and they can transport hexoses (e.g., glucose), but not Na⁺, against their concentration gradients. Drugs are being developed for the treatment of **type II (adult-onset) diabetes mellitus** that **inhibit the SGLT-2 transporter** (which is found in the kidneys, but not the intestine), thus promoting glucouria, polyuria, and a decreased blood pressure.

Since the intracellular Na⁺ concentration is low in intestinal and proximal renal tubular epithelial cells, as it is in other cells, Na⁺ moves into these cells, along its concentration gradient. Glucose (and/or galactose in the intestine) moves with Na⁺, with release occurring inside the cell. The Na⁺ is then actively transported into the lateral intercellular space (in a 3:2 exchange with K⁺), and glucose and/or galactose are transported by **facilitated diffusion**, using an insulin-independent **GLUT-2** transporter, into the interstitium and thence into capillary blood. Thus, **Na⁺/glucose/galactose cotransport** is an example of **secondary active transport**, with energy from ATP used to drive **Na⁺/K⁺ ATPase** (see **Fig. 38-5**). Once Na⁺ reaches lateral intercellular spaces, it is free to diffuse down its concentration gradient, either back into the lumen or toward blood. When the apical Na⁺/glucose/galactose cotransporter is congenitally defective, resulting glucose/galactose malabsorption causes severe diarrhea that can be fatal if glucose and galactose are not promptly removed from the diet.

Fructose utilizes a slightly different mechanism for intestinal absorption. The rate of fructose absorption is generally about **3 to 6 times slower** than that of overall glucose absorption, and it is not apparently driven by a secondary-active mechanism. It is, however, saturable. Fructose

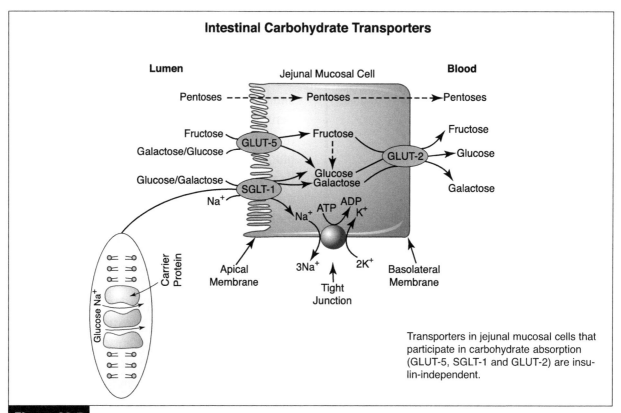

Intestinal Carbohydrate Transporters

Transporters in jejunal mucosal cells that participate in carbohydrate absorption (GLUT-5, SGLT-1 and GLUT-2) are insulin-independent.

Figure 38-5

is transported by Na^+-independent facilitated diffusion into enterocytes (along with some glucose and galactose), using the **insulin-independent GLUT-5 transporter**. It is transported out of enterocytes, along with glucose and galactose, using the **GLUT-2** transporter (see **Fig. 38-5** and Chapter 22). Some fructose is converted to glucose inside mucosal cells. Additionally, glucose (and fructose) can also enter into intracellular metabolism. About 10% of available glucose is thought to enter the hexose monophosphate shunt of enterocytes. Since mucosal cells have a high rate of glycolysis, the lactate so produced anaerobically diffuses into portal blood where it can be extracted and metabolized by the liver.

Intestinal **pentose** absorption is thought to occur by simple passive diffusion, and is not apparently associated with the SGLT or GLUT transporters. The molecular configurations that appear to be necessary for the secondary active transport of monosaccharides by the

SGLT-1 transporter are the following: The OH on carbon 2 should have the same configuration as that in glucose, and an α-pyranose ring must be present. Both of these conditions are present in the **hexoses** glucose and galactose, but not in fructose or the pentoses (see **Fig 18-1**).

OBJECTIVES

- Identify sites where the bulk of constitutive carbohydrate digestion and absorption occur.

- Note similarities and differences between glycogen, starch and cellulose.

- Differentiate dextrin from α-limit dextrin, and the luminal phase of carbohydrate digestion from the brush border phase.

- Identify the polypeptide hormones that participate in carbohydrate digestion.

- Explain why brush border enzymes of the small intestine remain active when mucosal cells possessing them are sloughed into the lumen.

- Identify all of the brush border oligosaccharidases, their substrates and hydrolytic reaction products. Explain why some are considered bifunctional.

- Indicate how and where luminal gluco- and galactocerebrosides are digested and absorbed (see Chapters 59).

- Identify the products of anaerobic bacterial carbohydrate digestion.

- Compare quantitatve aspects of intestinal aerobic and anaerobic carbohydrate absorption, and show how secondary active transport occurs.

- Understand why intestinal carbohydrate absorption is insulin-independent.

- Indicate how the three intestinal hexose transporters operate, and how pentoses are absorbed.

- Explain why intestinal fructose absorption is slower than intestinal glucose absorption.

QUESTIONS

1. **Eighty to 85% of dietary starch consists of:**
   a. Maltose.
   b. Maltotriose.
   c. Galactose.
   d. Amylose.
   e. Amylopectin.

2. **Which enzyme, present in the brush border of mucosal cells in the small intestine, is capable of cleaving α-1,6 glycosidic bonds?**
   a. Isomaltase
   b. Threhalase
   c. Glucoamylase
   d. Lactase
   e. Maltase

3. **The majority of monosaccharide absorption normally occurs in the:**
   a. Stomach.
   b. Jejunum.
   c. Ileum.
   d. Cecum.
   e. Colon.

4. **Which one of the following glucose transporters is found in the basolateral membrane of carbohydrate-absorptive mucosal cells?**
   a. SGLT-1
   b. GLUT-2

   c. GLUT-3
   d. GLUT-4
   e. GLUT-5

5. **Na⁺/glucose/galactose cotransport in the intestine is an example of:**
   a. Passive diffusion.
   b. Na⁺/K⁺ ATPase-driven active transport.
   c. Simple diffusion.
   d. Secondary active transport.
   e. Insulin-dependent transport.

6. **Fructose absorption in the small intestine:**
   a. Is an example of facilitated diffusion.
   b. Is Na⁺-dependent.
   c. Is insulin-dependent.
   d. Shares a common mechanism with pentoses.
   e. Occurs through a secondary active transport mechanism.

7. **Members of the amylase family of enzymes are found in all of the following tissues, EXCEPT:**
   a. Liver.
   b. Testes and ovaries.
   c. Pancreas.
   d. Small intestine.
   e. Lung.

8. **Which one of the following does NOT contain an α-1,6 glycosidic linkage?**
   a. Glycogen
   b. Isomaltose
   c. Starch
   d. Dextrin
   e. Amylose

9. **Which enzyme, present in the brush border of mucosal cells in the small intestine, is capable of catalyzing the digestion of dietary glucocerebrosides?**
   a. Glucoamylase
   b. Maltase
   c. β-Glucosidase
   d. Isomaltase
   e. Invertase

10. **Salivary α-amylase:**
    a. Hydrolyzes small amounts of glycogen and starch in the oral cavity.
    b. Has a pH optimum near that of duodenal juice.
    c. Both of the above amylase.
    d. Is inactive in acidic gastric juice.

ANSWERS

1. e
2. a
3. b
4. b
5. d
6. a
7. a
8. e
9. c
10. c

Section III Examination Questions

1. **Which one of the following is a 6-carbon compound?**

   a. Ribose
   b. Glucokinase
   c. Lactose
   d. 2,3-BPG
   e. Sorbitol

(Questions 2-6)   **Matching (each structure can be used more than once):**

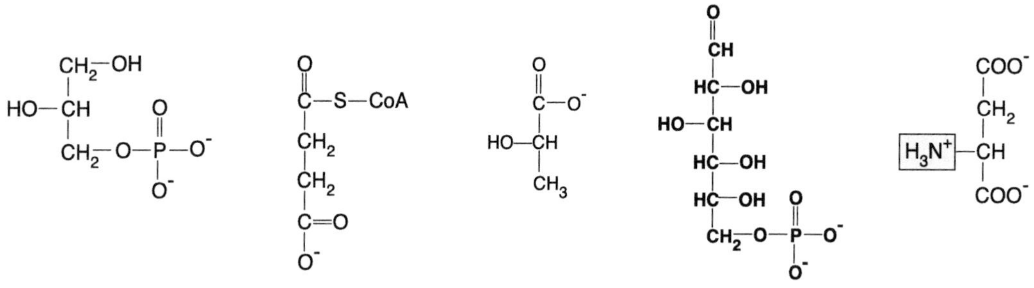

A   B   C   D   E

2. ___ Accumulation of this compound in skeletal muscle fibers inhibits glucose phosphorylation.

3. ___ This gluconeogenic intermediate can be formed from methylmalonyl-CoA in the presence of vitamin B$_{12}$.

4. ___ An end product derived from the anaerobic oxidation of glucose in erythrocytes.

5. ___ This compound becomes either a substrate or a product in AST-catalyzed reactions.

6. ___ This compound is used to form the triglyceride and glycerophospholipid backbone.

---

7. **Which one of the following statements regarding carbohydrates is FALSE?**

   a. Constituents of nucleic acids
   b. Contain carbon, hydrogen, nitrogen and oxygen
   c. Constituents of glycoproteins
   d. Energy stores of plants and animals
   e. Constituents of glycolipids

Copyright © 2015 Elsevier Inc. All rights reserved.

8. Select the FALSE statement below:

a. The rate-limiting step in hepatic bilirubin transport is generally considered to be canalicular excretion into bile.
b. The Cori cycle and the glucose-alanine cycle largely involve contracting muscle fibers, blood and the liver.
c. Lactate dehydrogenase activity is high in cardiac muscle tissue.
d. Thermogenin, a mitochondrial protein, is a natural uncoupler of oxidative phosphorylation, and acts as a trnasmembrane H^+ transporter.
e. Pyruvate kinase catalyzes a reversible reaction that funnels pyruvate toward glucose in gluconeogenesis.

9. Which of the following are INCORRECTLY paired?

a. Uroporphyrinogen : Heme precursor
b. Mitochondrial glycerol 3-P dehydrogenase : T_4-sensitive
c. Triosephosphate isomerase : TCA cycle enzyme
d. Mitochondrial isocitrate dehydrogenase : $NAD^+ \rightarrow NADH$
e. Glycerol kinase : Liver

10. Select the TRUE statement(s) below regarding canine erythrocytes (RBCs):

a. RBCs preserved for transfusion in a cytosine or thymine-containing medium generally possess greater viability than those preserved in a glucose-containing medium.
b. They contain the zinc- and copper-containing enzyme, succinate thiokinase.
c. The plasma membrane GLUT 1 transporter is insulin-independent.
d. They normally export pyruvate, but not lactate into plasma.
e. All of the above

11. Select the TRUE statement(s) below:

a. Conversion of propionyl-CoA to methylmalonyl-CoA is a carboxylation reaction requiring biotin.
b. Δ-Aminolevulinate would have a tendency to increase in the urine of patients with lead toxicity.
c. Kernicterus is an acquired condition caused by extreme unconjugated hyperbilirubinemia.
d. The conversion of 2-phosphoglycerate to phosphoenolpyruvate in hepatocytes is a reversible reaction.
e. All of the above

12. An elevated blood concentration of which one of the following hormones typically reverses gluconeogenesis in the canine liver?

a. Epinephrine

b. Insulin
c. Erythropoietin
d. Cortisol
e. Growth Hormone

13. Select the FALSE statement below:

a. Pyruvate carboxylase is found in liver cells, but not in muscle cells.
b. Gluconeogenesis is used to help replenish hepatic glycogen reserves between meals.
c. The potential energy in glucose is partially lost (as heat and H_2O) as it passes through the aerobic phase of glycolysis and oxidative phosphorylation.
d. Citrate is permeable to mitochondrial membranes, and is known to reduce the activity of cytoplasmic phosphofructokinase.
e. During anaerobic conditions, the glycerol 3-P and malate shuttles would not be working at full capacity, and cytoplasmic NAD^+ would most likely be generated from the LDH-catalyzed reaction.

14. Reduced activity of which one of the following enzymes in mature erythrocytes would lead to increased 2,3-DPG levels?

a. Diphosphoglyceromutase
b. Phosphofructokinase
c. Triosephosphate Isomerase
d. Pyruvate Kinase
e. Hexokinase

15. The amount of precious carbohydrate spent on glycolysis in a muscle cell is normally controlled by all of the following factors, EXCEPT:

a. Allosteric activators and inhibitors of aldose reductase.
b. Cytoplasmic levels of citrate, phosphocreatine, inorganic phosphate and ammonium ion.
c. The plasma epinephrine concentration.
d. The cell's energy status (i.e., cytoplasmic ATP, ADP and AMP levels).
e. Allosteric activators and inhibitors of phosphofructokinase.

16. Select the FALSE statement below:

a. Xylose is a pentose found in the oligosaccharide chains of glycoproteins.
b. Either a glycogen branching enzyme of glycogen debranching enzyme deficiency can cause glycogen storage disease.
c. Cyclic-AMP (c-AMP) stimulates glycogen synthase activity in hepatocytes.
d. Enolase, a glycolytic enzyme required in the conversion of 2-phosphoglycerate to phosphoenolpyruvate, is inhibited by fluoride.
e. Strict carnivores and ruminant animals store little glycogen in their liver cells.

17. Formation of 2,3-bisphosphoglycerate (2.3-BPG) from 1,3-bisphospho-glycerate (1,3-BPG) during erythrocytic glycolysis:

a. Is associated with ATP hydrolysis.
b. Involves NAD^+ reduction.
c. Results in methemoglobin reduction.
d. Requires enolase.
e. Is enhanced by respiratory alkalosis.

18. Glucokinase (compared with hexokinase):

a. Is responsible for the dephosphorylation of glucose 6-P in hepatocytes.
b. Is present on the outer mitochondrial membranes of hepatocytes.
c. Is not inhibited by accumulated glucose 6-P in cells in which it resides.
d. Has a high affinity for its substrate.
e. Synthesis and activity is decreased by insulin.

19. The equine uronic acid pathway is properly associated with all of the following, EXCEPT:

a. Ascorbate production of UDP-glucuronate.
b. A minor route for the formation of pentoses (e.g., xylulose).
c. Synthesis of the sugar moieties of various classes of glycoproteins (e.g., proteoglycans).
d. Hepatic production of UDP-glucuronate for various conjugation reactions.
e. Production of reduced glutathione (GSH) for conversion of H_2O_2 to H_2O.

20. All of the following are cytoplasmic reactions, EXCEPT:

a. $H_2O_2 + 2GSH \rightarrow 2H_2O + GSSG$
b. $Glc\text{-}1\text{-}P + UTP \rightarrow UDPGlc + PPi$
c. $6\text{-Phosphogluconate} + NADP^+ \rightarrow Ribulose\ 5\text{-}P + NADPH + CO_2$
d. $Pyruvate + CO_2 + ATP \rightarrow OAA + ADP$
e. $Glucuronate + NADPH \rightarrow L\text{-Gulonate} + NADP^+ \rightarrow \rightarrow L\text{-Ascorbate}$

21. Select the FALSE statement below:

a. Epinephrine and glucagon-stimulated cyclic-AMP formation in hepatocytes favors glycogenolysis through phosphorylation (i.e., ultimate activation of glycogen phosphorylase and inhibition of glycogen synthase).
b. Since catecholamine levels in blood are elevated during exercise, epinephrine stimulates PFK activity in both exercising muscle cells and hepatocytes.
c. Aldose reductase is an enzyme found in cells of the lens that catalyzes formation of sorbitol from glucose, as well as galactitol from galactose.
d. Phylloerythrin is a porphyrin derivative of chlorophyll formed by intestinal microbes.

e. Methemoglobin is the inactive form of hemoglobin in which the iron has been oxidized from the ferrous (Fe^{++}) to ferric (Fe^{+++}) state by superoxides.

22. **Which of the following enzymes are either absent or largely inactive in red muscle tissue, but are present and active in feline liver tissue?**

    a. Glucose 6-Phosphate Dehydrogenase and PEP Carboxykinase
    b. Glycogen Synthase and Glycogen Phosphorylase
    c. Phosphofructokinase and Isocitrate Dehydrogenase
    d. Pyruvate Dehydrogenase and Pyruvate Carboxylase
    e. Citrate Synthase and Phosphoglucomutase

23. **Mitochondrial conversion of pyruvate to acetyl-CoA in liver tissue is associated with all of the following, EXCEPT:**

    a. Oxidation by a dehydrogenase enzyme.
    b. Stimulation by an elevated NADH:NAD^+ concentration ratio.
    c. Enzymatic allosteric inactivation by acetyl-CoA.
    d. NAD^+ reduction.
    e. Stimulation by insulin.

24. **An animal exhibits the following:**

    - **Elevated direct and indirect plasma bilirubin concentrations**
    - **Elevated urinary bilirubin and urobilinogen excretion**

    Based upon this information alone, the patient most likely has:

    a. Lead toxicity.
    b. Bile duct obstruction.
    c. A glycogen storage disease.
    d. Hemolytic anemia.
    e. A hepatic UDP-glucuonosyltransferase deficiency.

25. **Which one of the following transporters is associated with entry of glucose from the intestinal lumen into mucosal cells of the small intestine, and from the glomerular filtrate into proximal tubular epithelial cells of the kidney?**

    a. Asp-Glu antiporter
    b. Insulin-dependent GLUT 4 transporter
    c. Insulin-independent glucokinase (GK) transporter
    d. Na^+-dependent glucose transporter
    e. Glucose-Malate antiporter

26. The following reaction requires:

$$metHb\text{-}Fe^{+++} \rightarrow Hb\text{-}Fe^{++}$$

a. NADH.
b. Superoxide Dismutase.
c. Selenium.
d. Glutathione.
e. None of the above

27. Select the FALSE statement below regarding glycoproteins:

a. These compounds are found in the membranous network of the cell.
b. Some circulating hormones are glycoproteins.
c. All enzymes of the glycolytic pathway are glycoproteins.
d. Although galactose is found in the oligosaccharide chains of glycoproteins, when present in the portal circulation it is largely cleared and metabolized by the liver.
e. These macromolecules are found in most organisms, from bacteria and viruses to mammals.

28. For questions 28-31, refer to the diagram to the right:

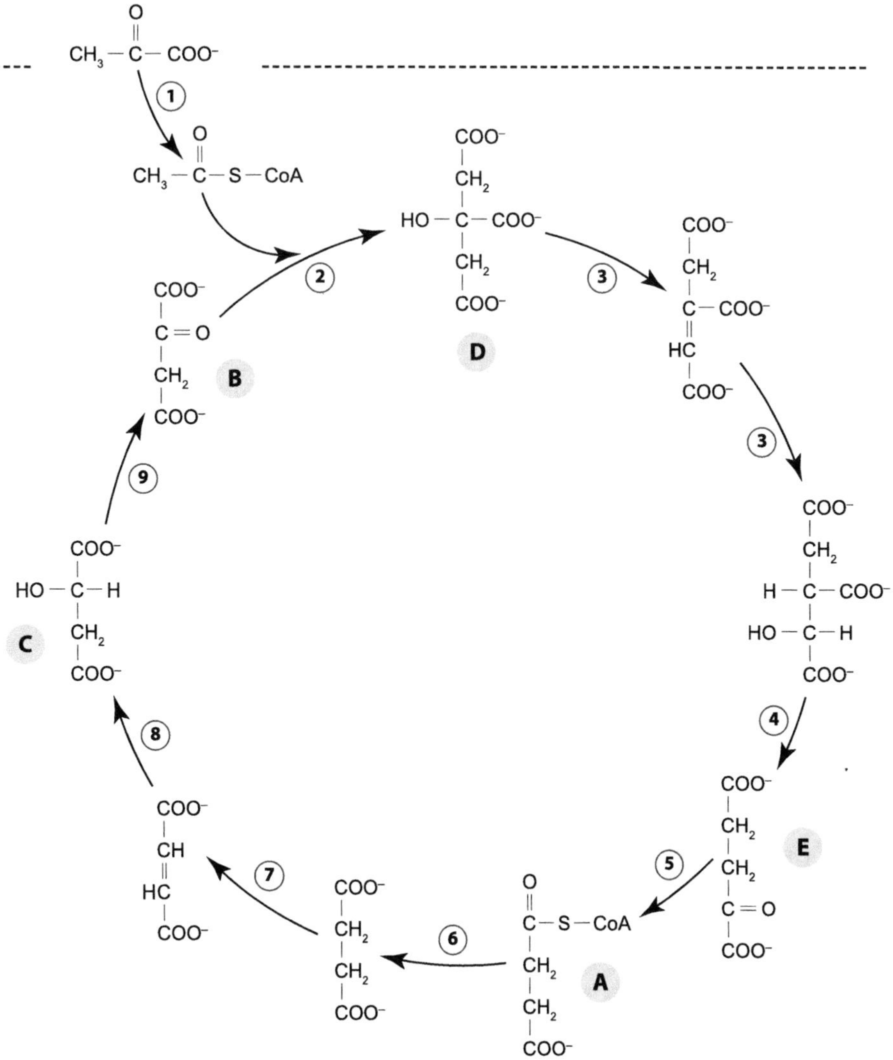

28. CO$_2$ leaks away for this cycle at steps:

    a. 3 and 6.
    b. 4 and 5.
    c. 6 and 8.
    d. 9 and 3.
    e. None of the above

29. Propionate enters the gluconeogenic scheme at point:

    a. A.
    b. B.
    c. C.
    d. D.
    e. E.

30. NADH is produced at point(s):

    a. 1.
    b. 4.
    c. 5.
    d. 9.
    e. All of the above

31. Compound C exchanges across the inner mitochondrial membrane with:

    a. Compound **E**.
    b. Inorganic phosphate.
    c. Compound **D**.
    d. All of the above
    e. None of the above

----

32. Conversion of sorbitol to fructose:

    a. Is the rate-limiting step in gluconeogenesis.
    b. Requires aldose reductase.
    c. Occurs in the equine liver.
    d. Is a mitochondrial process.
    e. Requires AST (SGOT).

33. Gluconeogenic precursors normally include all of the following, EXCEPT:

a. Glycerol.
b. Glutamine.
c. Propionate.
d. Aspartate.
e. Acetate.

34. Select the FALSE statement below:

a. Cytochrome c is a heme-containing protein in the mitochondrial electron transport chain.
b. ATP and ADP permeate hepatic mitochondrial membranes.
c. Xylitol promotes insulin release in dogs, which can lead to hypophosphatemia and hypokalemia.
d. Glycogen synthase is inactive when it is phosphorylated.
e. Transketolase and aldolase function primarily in mitochondrial oxidative phosphorylation.

35. FADH$_2$:

a. Generates 3ATP as electrons from this coenzyme pass through oxidative phosphorylation.
b. Is formed in mitochondria during the conversion of glycerol 3-P to malate.
c. Is typically generated in the cytoplasm, not in mitochondria.
d. Is formed during mitochondrial conversion of succinate to fumarate.
e. None of the above

36. Select the FALSE statement below:

a. DHAP must first be converted to Gl-3-P before passing through the remainder of the EMP.
b. Biologic information is encoded in glycoproteins through the carbohydrate sequences of their oligosaccharide chains.
c. Superoxide dismutase is present in erythrocytes, and virtually all aerobic tissues.
d. The anticoagulant, heparin, is produced and stored in mast cells, and contains repeating disaccharide units.
e. Due to the absence of a complete HMS, muscle tissue cannot form pentoses, and therefore must obtain them from blood.

37. Glucose is metabolized anaerobically by mature erythrocytes for all of the following purposes, EXCEPT:

a. Maintain reduced glutathione formation.
b. Prevent methemoglobin accumulation.
c. Maintain membrane structure, and therefore integrity.
d. Prevent 2,3-BPG formation.
e. Maintain plasma membrane Na$^+$/K$^+$-ATPase and Ca^{++}-ATPase activity.

38. The Uronic Acid Pathway:

a. Involves the conversion of UDP-glucose to UDP-glucuronate.
b. Occurs mainly in mitochondria of aerobic muscle fibers.
c. Utilizes NADH, but generates NADPH for lipid biosynthesis.
d. Serves as a major route for the formation of acetyl-CoA from ascorbate.
e. All of the above

39. Select the TRUE statement(s) below:

a. Serum fructosamine and HbA_{1c} levels are useful in monitoring the long-term control of blood glucose in diabetic patients.
b. Glucuronate is an activated nucleotide found in certain forms of glycoproteins.
c. A rise in the intracellular concentrations of H^+, Frc-6-P and inorganic phosphate (Pi) are known to stimulate PFK activity.
d. N-Acetylneuraminic acid (NeuAc) is found in the oligosaccharide chains of glycoproteins, but not those of glycolipids.
e. All of the above

40. Carbon atoms from which one of the following normally "leak" away from the TCA cycle (following a transamination reaction) to form pyrimidines?

a. Fumarate
b. Isocitrate
c. Acetyl-CoA
d. OAA
e. Succinyl-CoA

41. Examine the reaction below:

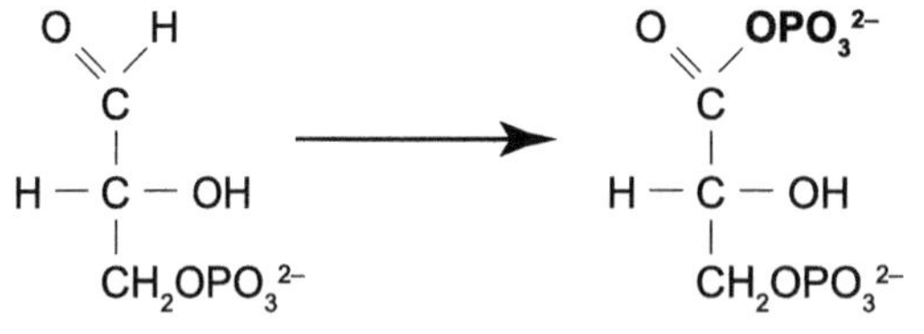

a. This is the rate-limiting reaction in hepatic glycogenesis.
b. This mitochondrial reaction requires Succinate Thiokinase, GDP and inorganic phosphate.
c. This intermediate reaction of the uronic acid pathway requires NADPH.
d. This glycolytic reaction requires inorganic phosphate and NAD^+.
e. This gluconeogenic reaction requires Fructose 1,6-Bisphosphatase and water.

42. **Which of the following are impermeable to the inner mitochondrial membrane?**

    a. ATP, ADP, GTP and GDP
    b. Glycerol 3-P, Malate and Propionate
    c. Oxaloacetate, NAD^+ and Acetyl-CoA
    d. Aspartate, Pyruvate and Glutamate
    e. Citrate, α-Ketoglutarate and Inorganic Phosphate

43. **Bilirubin Diglucuronide:**

    a. Is a conjugated metabolic end-product of mammalian heme catabolism.
    b. Is normally present at a lower concentration in plasma than indirect-reacting bilirubin.
    c. Is normally excreted in bile.
    d. Is more hydrophilic than unconjugated bilirubin.
    e. All of the above

44. **All of the following are correctly paired, EXCEPT:**

    a. Cats : Difficulty conjugating some xenobiotics to glucuronates.
    b. UDP-glucose : Active nucleotide intermediate in heme biosynthesis.
    c. Ubiquinone (CoQ) : Mobile electron carrier in the ETC.
    d. Glucokinase : Inducible hepatic enzyme.
    e. Carbaminohemoglobin : Carbon dioxide bound to Hb.

45. **Under conditions of low blood glucose (hypoglycemia):**

    a. Muscle glucose 6-Phosphatase can dephosphorylate Glc-6-P, thus contributing to the blood glucose pool.
    b. Glucokinase becomes the main intracellular "glucose trapping" enzyme in nerve tissue.
    c. Lactate stores can be converted to glucose in muscle and nerve tissue, thus contributing to the blood glucose concentration.
    d. Liver glycogen can be converted to glucose by a pathway involving glycogen phosphorylase, debranching enzyme, phosphoglucomutase and glucose 6-phosphatase.
    e. All of the above

46. **If oxidation proceeds without phosphorylation in the mitochondrial respiratory chain:**

    a. It is said to be "oxygen deficient," and lactic acid accumulation will occur.
    b. ATP is generated for the sole purpose of heat production.
    c. It is thus "inhibited," and the animal will most likely die.
    d. Electron transport from NADH to O_2 ceases.
    e. It is said to be "uncoupled," and ATP generation will be reduced.

47. Which one of the following stimulates muscle PFK activity?

a. ATP, since a high energy state in the cell necessitates more glucose oxidation.
b. NH_4^+, since the mere intracellular presence of this cation is an indirect indicator of AMP degradation.
c. Citrate, since an increase in the cytoplasmic citrate concentration is reflective of a need for more energy.
d. Phosphocreatine, since the cytoplasmic concentration of this compound usually increases following the onset of exercise.
e. Glucagon, since this hormone, like insulin, stimulates glycolysis.

48. GSH:

a. Is the reduced form of glutathione.
b. Is produced by the liver, and exported to blood for use by other tissues.
c. Is far more prevalent in normal adult erythrocytes than GSSG.
d. Contains cysteine.
e. All of the above

49. All of the following are correctly described decarboxylation reactions, EXCEPT:

a. Pyruvate + NAD^+ + CoA.SH $\rightarrow$ Acetyl-CoA + $\mathbf{CO_2}$ + NADH
b. OAA + GTP $\rightarrow$ Phosphoenolpyruvate + $\mathbf{CO_2}$ + GDP
c. Fructose 6-P + NAD^+ + ATP $\rightarrow$ F-1,6-BisP + $\mathbf{CO_2}$ + NADH + ADP
d. 6-Phosphogluconate + $NADP^+$ $\rightarrow$ Ribulose 5-P + $\mathbf{CO_2}$ + NADPH
e. α-KG$^=$ + NAD^+ + CoA.SH $\rightarrow$ Succinyl-CoA + $\mathbf{CO_2}$ + NADH

50. Select the TRUE statement below:

a. Muscle PFK is inhibited by inorganic phosphate, ADP and citrate, signifying that under these conditions lipid biosynthesis should take precedence over glucose oxidation.
b. Glycerol enters the hepatic gluconeogenic pathway at the level of glycerol 3-P (and subsequently DHAP), and therefore does not traverse the dicarboxylic acid (DCA) shuttle.
c. Anaerobic glycolysis gives a net yield of 3 ATP in the conversion of one glucose to 2 pyruvates, while glycolysis gives a net yield of 2 ATP (when the Glc-6-P residue comes from intracellular glycogen).
d. The liver requires insulin in order to move glucose across its cell membranes.
e. Like the pyruvate kinase and pyruvate dehydrogenase reactions, the lactate dehydrogenase reaction is also "physiologically irreversible."

51. Select the TRUE statement(s) below:

a. Uroporphyrinogen III contains a tetrapyrole ring.
b. Glucogenic amino acids contribute to gluconeogenesis in carnivores, but not in ruminant animals.
c. Conversion of OAA to PEP is known as the Haber-Weiss reaction.
d. The brain can synthesize glucose from non-carbohydrate sources.
e. All of the above

52. A reduction in the red cell count, in the absence of an increase in circulating reticulocytes, is indicative of:

a. A recent hemorrhage.
b. An absolute polycythemia.
c. Hemolytic anemia.
d. A relative polycythemia.
e. Nonregenerative anemia.

53. Which one of the following enzymes is active in hepatic glycogenesis, but not in glycogenolysis?

a. UDP-glucose pyrophosphorylase
b. Glycogen phosphorylase
c. Phosphoglucomutase
d. Debranching enzyme
e. None of the above

54. The following reaction is normally catalyzed by which enzyme?

$$2H_2O_2 \rightarrow O_2 + 2H_2O$$

a. Superoxide Dismutase
b. Glutathione Reductase
c. Mehemoglobin Reductase
d. L-Gulonolactone Oxidase
e. Catalase

55. Select the FALSE statement below:

a. Harderian exocrine secretions are known to contain at least one porphyrin intermediate.
b. Elevations in cellular phosphocreatine have a tendency to reduce the flow of hydrocarbons through glycolysis.
c. A high mitochondrial ATP:ADP ratio has a tendency to reduce the activity of pyruvate dehydrogenase.

d. Nitric oxide and carbon dioxide compete for binding to the Fe'' moiety of hemoglobin.

e. Malate dehydrogenase is found in the cytoplasm as well as in mitochondria.

56. Which of the following are INCORRECTLY matched?

a. Diabetogenic hormones : Glucagon, Growth Hormone, Cortisol, Epinephrine

b. Glucose 6-P dehydrogenase : Inhibited by NADPH

c. Biliverdin : Intermediate in bilirubin formation from heme

d. Glycerol : Exported from adipocytes to blood during lipolysis

e. Propionyl-CoA : End product of purine degradation

57. Select the TRUE statement below:

a. A patient with a glucose 6-phosphatase deficiency would fail to respond to glucagon normally.

b. Erythrocytes possess GLUT 4 transporters.

c. Although glycogen storage diseases are generally characterized by an inability to form or degrade glycogen in normal metabolic pathways, animals possessing these diseases seldom exhibit exercise intolerance.

d. Conjugated bilirubin is normally more tightly bound to albumin in plasma than is unconjugated bilirubin.

e. When epinephrine and glucagon stimulate hepatocytes during exercise, glycogen is broken down in those hepatocytes, with most all of the Glc-6-P formed passing through the EMP.

58. Reducing equivalents in the malate shuttle are carried across mitochondrial membranes as a part of which molecule?

a. Citrate

b. Glu

c. α-KG$^=$

d. Asp

e. Malate

59. A canine patient is diagnosed as having a glucokinase deficiency. Which one of the following consequences would be most likely?

a. Animal exhibits difficulty secreting insulin, and difficulty removing glucose from hepatic portal blood.

b. Animal shows signs of significant exercise intolerance.

c. Animal demonstrates a total inability to metabolize glucose.

d. Animal becomes hypoglycemic between meals.

e. Animal shows signs of megaloblastic anemia.

Answers

1.	e	43.	e
2.	d	44.	b
3.	b	45.	d
4.	c	46.	e
5.	e	47.	b
6.	a	48.	e
7.	b	49.	c
8.	e	50.	b
9.	c	51.	a
10.	c	52.	e
11.	e	53.	a
12.	b	54.	e
13.	a	55.	d
14.	d	56.	e
15.	a	57.	a
16.	c	58.	e
17.	e	59.	a
18.	c		
19.	e		
20.	d		
21.	b		
22.	a		
23.	b		
24.	d		
25.	d		
26.	a		
27.	c		
28.	b		
29.	a		
30.	e		
31.	d		
32.	c		
33.	e		
34.	e		
35.	d		
36.	e		
37.	d		
38.	a		
39.	a		
40.	d		
41.	d		
42.	c		

Addendum to Section III

To present and discuss carbohydrate metabolism thoroughly would require more attention and space than the scope of this book allows. Rather, focus was placed on detailing major carbohydrate anabolic and catabolic pathways (i.e., glycogenesis, glycogenolysis, glycolysis, the HMS & uronic acid pathway, the TCA cycle, oxidative phosphorylation and gluconeogenesis), emphasizing their rate-controls and biomedical significance, identifying and discussing the importance of various carbohydrate derivatives, introducing the glycoproteins and glycolipids, discussing erythrocyte metabolism, heme biosynthesis and degradation, and introducing major physiologic processes involved in carbohydrate digestion.

The many principles of metabolism contained in **Sections I**, **II** and **III** will hopefully become more coherent in subsequent sections of this text. In other words, there is no such thing as an instant or partial appreciation of **metabolism**. On the contrary, it develops gradually; and rote memorization, albeit useful in initially developing a basic language, is a first step only in comprehending and ultimately applying the physiological chemistry subject matter.

Introduction to Section IV

Section IV now deals with the **vitamins** (Chapters 39-47) and **trace elements** (Chapters 48-52). Vitamins are organic compounds required for a number of biochemical functions, they cannot be synthesized by the organism, and thus must be supplied (in small amounts) through the diet. Most vitamin deficiency and toxicity symptoms are well established, and are known to occur throughout the animal kingdom.

Vitamins A (1916) and B_1 (thiamin; 1926) were first discovered, and were found to be fat- and water-soluble, respectively. As additional vitamins were unmasked, they were also shown to be either fat- or water-soluble; therefore, this single property was used as a basis for their classification. Apart from folic acid (Chapter 16) and vitamin C, the water-soluble vitamins were designated members of the B-complex series, and newly discovered fat-soluble vitamins were given D, E and K (koagulation) designations. Besides their solubility characteristics, the water-soluble vitamins share little in common (from a biochemical and physiologic perspective).

Although animals are thought to contain and incorporate most all of the elements known, they are generally divided into macrominerals, microminerals (trace elements), and ultra trace elements. The trace elements are iron, zinc, copper, manganese, selenium, iodine and cobalt. Each is physiologically significant, particularly for reproductive health, with symptoms of toxicity and deficiency occurring. Biochemical and physiologic actions are quite well-defined, and although exceptions exist, most tend to concentrate in the germ of seeds and grains (where the highest concentrations of the B-complex vitamins also reside).

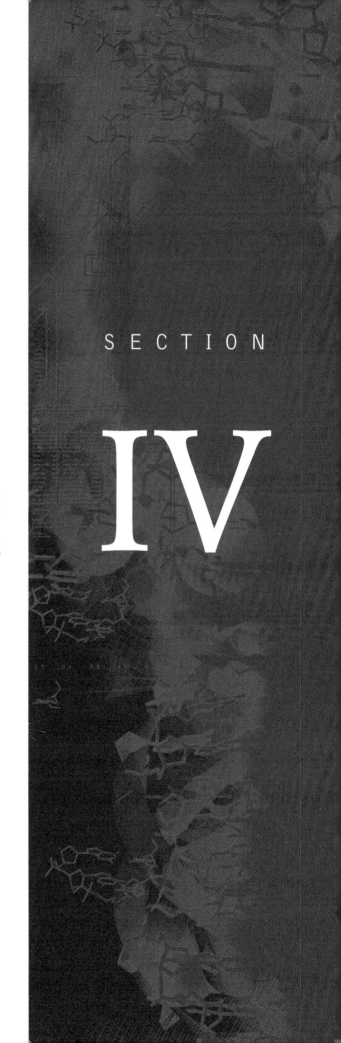

SECTION IV

Vitamins and Trace Elements

Chapter 39

Vitamin C

Overview

- Vitamin C is a powerful reducing agent that participates in several important hydroxylation reactions.
- Na^+-coupled transporters help to facilitate entry of vitamin C into cells.
- Glial cells in the brain regenerate vitamin C from DHA.
- Vitamin C is needed for collagen, carnitine, catecholamine, and bile acid biosynthesis.
- Oxalate is a natural degradation product of vitamin C.
- Vitamin C uses Fe^{++} and Cu^{++} as cofactors, and it enhances intestinal Fe^{++} absorption.
- Vitamin C deficiency can result in "scurvy."
- Although most mammals can synthesize vitamin C from glucose, it cannot be formed in primates, fish, flying mammals, songbirds, or the guinea pig.
- Vitamin C is a natural preservative added to pet food products.

Water-soluble Vitamins

Vitamins are structurally unrelated organic compounds that function in small amounts as metabolic catalysts, usually in the form of coenzymes. **Water-soluble vitamins**, such as **vitamin C (ascorbate)** and the **B-complex vitamins**, are produced by plants and microorganisms; however, most are not synthesized at all or in sufficient amounts by animals to satisfy tissue requirements. Only modest stores of these vitamins are normally available in the mammalian organism, therefore, they need to be routinely supplied from food, fluids, or microorganisms inhabiting the digestive tract.

Vitamin C (Ascorbate)

"Scurvy", a **vitamin C deficiency** disease known since ancient times, was a particular problem for sailors in the 15th-19th centuries, who's diets were often less than adequate on the long voyages they endured. These men would develop swollen legs blotched with capillary hemorrhages, decaying peeling gums with loose teeth, decreased capacity to heal wounds, depression, anemia, and fatigue. **Infantile scurvy** (also known as **Barlow's syndrome or disease**), is associated with similar symptoms.

Although most vertebrates can synthesize vitamin C from glucose (see Chapter 29), it cannot be formed in **primates**, **fish**, **flying mammals**, **songbirds**, or the **guinea pig**. Therefore, these animals require it in their diet. This vitamin is found in both plant and animal foods, but is particularly prevalent in fruits and vegetables. Some bacteria also synthesize ascorbate.

Regardless of whether vitamin C is derived through the diet or from biosynthesis in liver (as in **rodents**), or the kidneys (as in **reptiles**),

Copyright © 2015 Elsevier Inc. All rights reserved.

specific transport mechanisms are required to move it into dependent tissues (**Fig. 39-1**). Ascorbate enters cells via **Na⁺-coupled vitamin C transporters** (**SVCT 1** or **SVCT 2**), and cellular efflux occurs by as yet undescribed mechanisms. The oxidized form of vitamin C (i.e., **dehydroascorbate (DHA)**; **Fig. 39-2**), is thought to exit and enter cells via glucose transporters (**GLUT 3** and **GLUT 1**, respectively; see Chapter 22). In the CNS, **glial cells** (i.e., astrocytic supporting cells) **regenerate vitamin C** from **DHA** via **reduced glutathione (GSH)** oxidation, and then vitamin C is transported back into neurons (see Chapter 90). Neurons exhibit a high level of oxidative metabolism, and thus require protection by this important water-soluble vitamin.

At the molecular level, ascorbate is a **powerful reducing agent**, like the fat-soluble **vitamin E** (Chapter 46), and as such possesses general importance as an antioxidant, thus affecting the body's "redox" potential (i.e., the relative states of oxidation/reduction of other water-soluble substances inside and outside of cells). It is used as a natural **preservative** in pet food products, and is sometimes given to **cats** as a treatment to reverse the **methemoglobinemia** associated with **acetaminophen toxicity** (see Chapters 29 and 30). The physiologic importance of vitamin C as an antioxidant has been documented in **pond turtles**, which possess particularly high concentrations of vitamin C in the **brain** (Rice ME, et al, 1995). These animals exhibit a high tolerance for O_2 depletion during **diving**, and vitamin C may help to prevent oxidative damage to neurons during the reoxygenation period following a hypoxic dive.

Other notable reactions involving ascorbate include **hydroxylations** using molecular oxygen (O_2), that also use either **Fe⁺⁺** or **Cu⁺⁺** as a cofactor (**Table 39-1**). Here, ascorbate is thought to play either of two roles: **1)** as a direct source of electrons for the reduction of

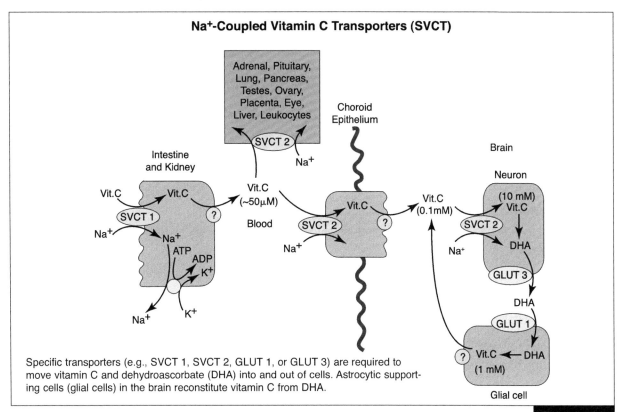

Na⁺-Coupled Vitamin C Transporters (SVCT)

Specific transporters (e.g., SVCT 1, SVCT 2, GLUT 1, or GLUT 3) are required to move vitamin C and dehydroascorbate (DHA) into and out of cells. Astrocytic supporting cells (glial cells) in the brain reconstitute vitamin C from DHA.

Figure 39-1

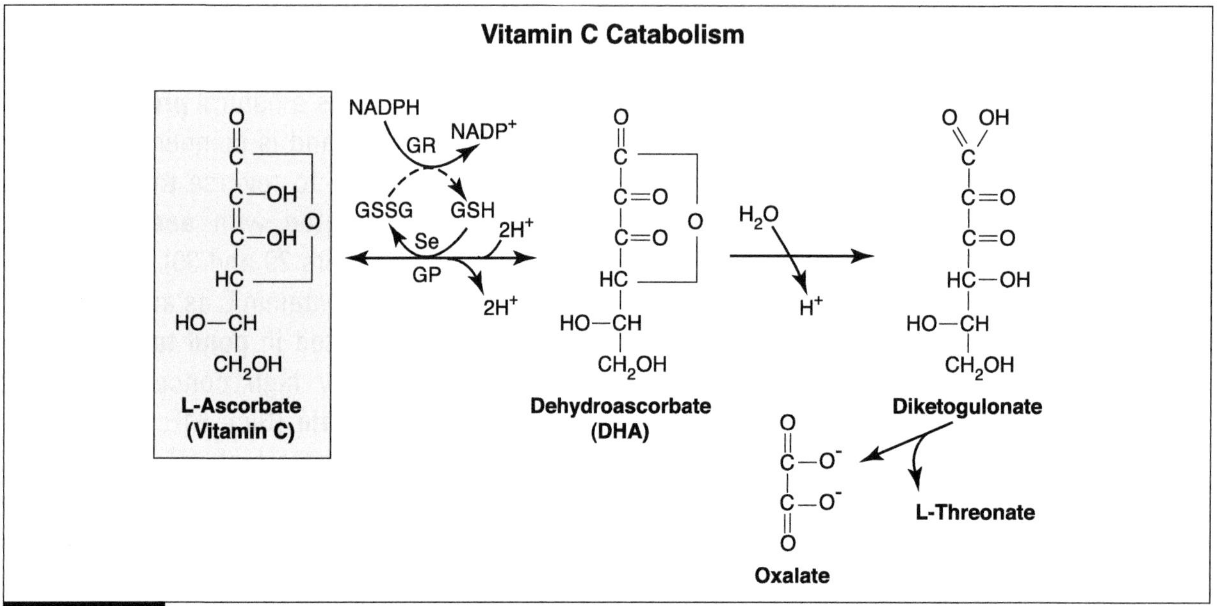

Vitamin C Catabolism

Figure 39-2

O_2 (e.g., as a cosubstrate), or **2**) as a protective agent for maintaining **Fe^{++}** or **Cu^{++}** in their reduced states. Particularly important are hydroxylations involving **hydroxyproline** and **hydroxylysine** formation during collagen biosynthesis in **connective tissue** (primarily in bone and ligaments, **Fig. 39-3** and Chapter 3).

The formation of collagen is important during growth and development, when collagen fibers are constantly being laid down and also removed. Once physical maturity is achieved, there is relatively little collagen turnover (one of the few proteins in the body for which this may be said). An exception is healing from

Table 39-1
Actions of Vitamin C

- Powerful reducing agent ("Antioxidant").
- Enhance intestinal Fe^{++} absorption.
- Reduce cataract formation.
- Enhance leukocyte activity.
- Participate in Cu^{++}-dependent amidation reactions in polypeptide hormone biosynthesis (e.g., GH, CT, and MSH).
- Participate in the amidation of C-terminal glycine residues in the brain by Cu^{++}-dependent enzymes.
- Act as a carrier of sulfate groups in glycosaminoglycan formation (the "ground substance" between cells in all organs).
- Participate in hydroxylation reactions using O_2 (with Fe^{++} or Cu^{++} as cofactors).
  a. Hydroxyproline and hydroxylysine formation during collagen biosynthesis.
  b. Carnitine biosynthesis from lysine and S-adenosylmethionine.
  c. Dopamine hydroxylation during catecholamine biosynthesis.
  d. Hydroxylation of steroid hormones, aromatic drugs, and carcinogens in liver microsomes.

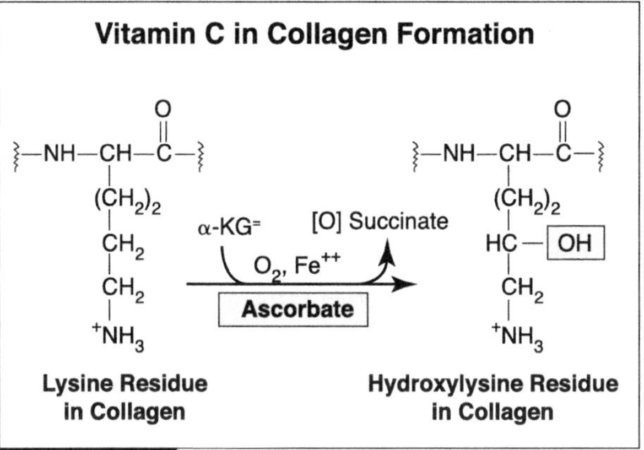

Vitamin C in Collagen Formation

Figure 39-3

tissue injury and scar formation.

Ascorbate has a secondary function in connective tissue metabolism as a carrier for sulfate groups needed in **glycosaminoglycan** formation (chondroitin sulfate, dermatan sulfate, etc.; see Chapter 29). These compounds help to form the gel matrix (or "**ground substance**") between cells in all organs. There would seem to be an obvious connection between these needs for ascorbate in connective tissue metabolism, and the basic symptoms of **scurvy** above.

Although vitamin C supplementation is not considered to be essential in dogs, megadoses of ascorbate fed to the bitch during pregnancy, and provided to the offspring until young adulthood, have been associated with reducing the incidence of **canine hip dysplasia** in animals considered genetically at risk for this condition.

Ascorbate is also used in the biosynthesis of **carnitine** from lysine and S-adenosylmethionine (see Chapters 55 and 57). Carnitine is involved in the transport of long-chain fatty acids across mitochondrial membranes.

Ascorbate is active in **hepatic microsomal drug metabolism**. Both endogenous and exogenous **steroids** are hydroxylated and conjugated in the liver, as are certain nonsteroidal drugs (e.g., barbiturates) and suspected carcinogens. The resulting hydroxylation makes these compounds more water-sol-uble, and thus more likely to be excreted from the body through bile or urine. For example, the first step in hepatic **bile acid** formation from cholesterol (7-α-hydroxycholesterol formation; see Chapter 62) is activated by vitamin C, and **guinea pigs** exhibiting ascorbate deficiency decrease their biliary excretion of bile acids. It has also been shown that rats exposed to toxic polychlorinated biphenyls (PCBs) greatly increase their need for vitamin C.

This vitamin is also concentrated in **leukocytes**. Deficiencies in leukocyte ascorbate concentrations have been reported in some diabetics, leading to a decreased capacity for wound repair and response to infection. Indeed, the enhanced response of neutrophils to chemotactic stimuli, and enhanced proliferation of lymphocytes in response to mitogens, may be related to their vitamin C content. The ascorbate in leukocytes may prevent autoxidation of the oxygen-radical forming system integral to their function, as well as modulate their production of leukotrienes (see Chapters 68 and 69). A propensity for **cataract** formation has also been linked to low ascorbate (and vitamin E) bioavailability, perhaps from a relative lack of reducing equivalents. Ascorbate is normally concentrated in the aqueous humor (see **Fig. 39-1**).

Ascorbate also plays an important role in **biogenic amine** (i.e., **catecholamine**) biosynthesis in the **adrenal medulla**, **central (CNS)**, and **sympathetic (SNS) nervous systems** (**Fig. 39-4**). Indeed, the highest concentrations of this vitamin are usually found in these locations. Vitamin C serves as a cosubstrate in the **hydroxylation of dopamine to norepinephrine**, catalyzed by the enzyme **dopamine β-hydroxylase**. Catecholamines are associated with the ability of animals to deal with stress, and they help to mobilize glycogen and triglyceride for energy purposes (see Chapter 23 and 70).

Vitamin C has also been found to participate in Cu^{++}-dependent amidation reactions in **poly-**

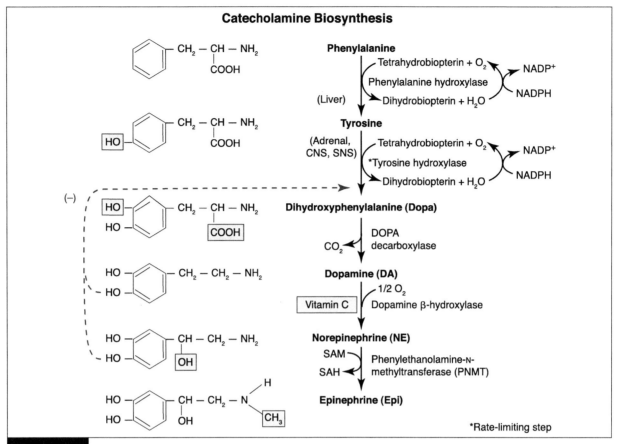

Figure 39-4

peptide hormone biosynthesis (e.g., growth hormone (GH), calcitonin (CT), and melanocyte stimulating hormone (MSH)), and it may be necessary for the evocation of increased numbers of cell surface **acetylcholine receptors** by muscle cells responding to nerve stimuli. This vitamin plays a role in Fe^{++} metabolism, by enhancing conversion of dietary iron from the **ferric (Fe^{+++})** to **ferrous (Fe^{++})** state. Since iron is more readily absorbed from the intestine in the **Fe^{++}** state, vitamin C helps to facilitate its absorption (see Chapter 48). Ascorbate may also be involved in the mobilization of stored Fe^{++}, especially from hemosiderin in the spleen. While a modest intake of vitamin C may be beneficial, excessive intake has a propensity to cause iron overload (i.e., **hemochromatosis**), which can be debilitating.

The capacity of this vitamin to **chelate Ca^{++}** may mean it has a function as well in bone mineral metabolism. In this regard, animal experiments point to a role for this vitamin in **tooth formation**. In dog breeds predisposed to **Cu^{++} hepatotoxicosis** (e.g., Bedlington terriers), discouraging intestinal absorption of Cu^{++} through chelation has been attempted with high oral doses of vitamin C (see Chapter 50). Results of these studies, however, have not been encouraging.

A major pathway of ascorbate catabolism is to **oxalate** (see **Fig. 39-2**). It has been estimated that under normal circumstances, approximately one-third of **urinary oxalate** may be derived from ascorbate catabolism. Large doses of vitamin C (ascorbic acid) will enhance urine acidity, which promotes conversion of urate into **uric acid**, and **oxalate** into **oxalic acid**. Vitamin C overload could also promote formation of **calcium oxalate** kidney stones in susceptible animals.

OBJECTIVES

- Compare the action of intestinal SVCT 1 to that of SGLT 1 (see Chapter 38).

- Explain how and where vitamin C is oxidized and reduced in the CNS.

- Show how vitamin C is transported across the blood-CSF barrier, and how it is transported into and out of neurons and glial cells of the brain (see Chapter 90).

- Summarize the metabolic actions of vitamin C in the body, and recognize the signs and symptoms of scurvy.

- Identify five important hydroxylation reactions where vitamin C acts as a cofactor.

- Describe the role of vitamin C in intestinal iron absorption.

- Explain the proposed relationship between ascorbate overload and urolithiasis.

- Describe all steps in the biosynthesis of epinephrine from phenylalanine, and include the enzymes, intermediates and cofactors involved.

QUESTIONS

1. **Vitamins typically act as _____ in the reactions they participate in?**
   a. Enzymes
   b. Coenzymes
   c. Substrates
   d. Energy sources
   e. Hormones

2. **Which one of the following animals cannot synthesize vitamin C from glucose?**
   a. Horse
   b. Rat
   c. Dog
   d. Guinea pig
   e. Goat

3. **Vitamin C is known to participate in which reaction type?**
   a. Carboxylation
   b. Dehydration
   c. Hydration
   d. Hydroxylation
   e. Decarboxylation

4. **Vitamin C is known to participate in the biosynthesis of all of the following, EXCEPT:**
   a. Bile acids.
   b. Collagen.
   c. Epinephrine.
   d. Carnitine.
   e. Cholesterol.

5. **Absorption of which one of the following from the digestive tract is thought to be enhanced by vitamin C?**
   a. Iron
   b. Glucose
   c. Amino acids
   d. Medium-chain fatty acids
   e. Lactate

6. **Vitamin C is known to be concentrated in all of the following locations, EXCEPT:**
   a. Sympathetic, post-ganglionic neurons.
   b. Skeletal muscle.
   c. Central nervous system.
   d. Leukocytes.
   e. Adrenal medulla.

7. **Which one of the following is a normal degradation product of vitamin C?**
   a. Oxalate
   b. Dopamine
   c. Lactate
   d. Ascorbate
   d. Acetyl-CoA

8. **Select the TRUE statement below regarding vitamin C:**
   a. It cannot be regenerated in the body.
   b. It enters cells via Na⁺-coupled transporters.
   c. It cannot cross the blood-brain-barrier.
   d. It donates reducing equivalents directly to the mitochondrial electron transport chain.
   e. Large doses of this vitamin will alkalinize the urine.

9. **Glutathione (GSH) participates in dehydroascorbate reduction:**
   a. True
   b. False

9. a
8. b
7. a
6. b
5. a
4. e
3. d
2. d
1. b

ANSWERS

Chapter 40

Thiamin (B₁) and Riboflavin (B₂)

Overview

- Thiamin diphosphate is used to facilitate oxidative decarboxylation and transketolase reactions.
- Thiamin deficiency, which limits aerobic metabolism, can be fatal.
- Thiamin is activated to its coenzyme form in brain and liver tissue.
- Several foods exhibit thiamin antagonist activity.
- Riboflavin is used to produce FMN and FAD, and stored forms of this vitamin tend to decompose in the presence of light.
- FAD and FMN are coenzymes containing iron or molybdenum.
- FAD and FMN production is enhanced by thyroid hormones.
- Riboflavin, like bile acids, exhibits an enterohepatic circulation (EHC).
- Erythrocytes can be used to assess both thiamin and riboflavin deficiency.

The **B-complex vitamins** are central to the metabolism of all mammalian cells since they act as **coenzymes** in specific reactions in glycolysis, the hexose monophosphate shunt (HMS), the tricarboxylic acid (TCA) cycle, and in lipid metabolism. They are generally associated with their respective enzymes through covalent bond formation near the active site.

Thiamin (Vitamin B₁)

Thiamin is produced by some microbes, it is present in most plant and animal tissues, and it is involved in several reactions involving **thiamin diphosphate (pyrophosphate)** as coenzyme. An ATP-dependent **thiamin diphosphotransferase** present in **brain** and **liver** tissue is responsible for conversion of this vitamin to its active form. Although some thiamin triphosphate is also formed, and may play a separate role in brain cell viability, approximately 80% of

thiamin in the body is found in the diphosphate form. This vitamin is not stored in the body (to any great extent), except in the sense of being attached to some enzymes. It is heat labile, and therefore easily destroyed. Thiamin is absorbed by both passive and active processes from the digestive tract, and is excreted from the body as multiple metabolites in urine.

There are two general types of reactions in mammals that utilize the activated coenzyme form of this vitamin: **1) oxidative decarboxylation**, and **2) transketolase reactions** (Fig. 40-1). Examples of common oxidative decarboxylation reactions are conversion of **pyruvate to acetyl-CoA** in glycolysis (see Chapter 27), conversion of α-ketoglutarate (α-KG$^=$) to **succinyl-CoA** in the TCA cycle (see Chapter 34), and **decarboxylation of α-ketocarboxylic acid derivatives** of the branched-chain amino acids (BCAAs; leucine, isoleucine, and valine;

Copyright © 2015 Elsevier Inc. All rights reserved.

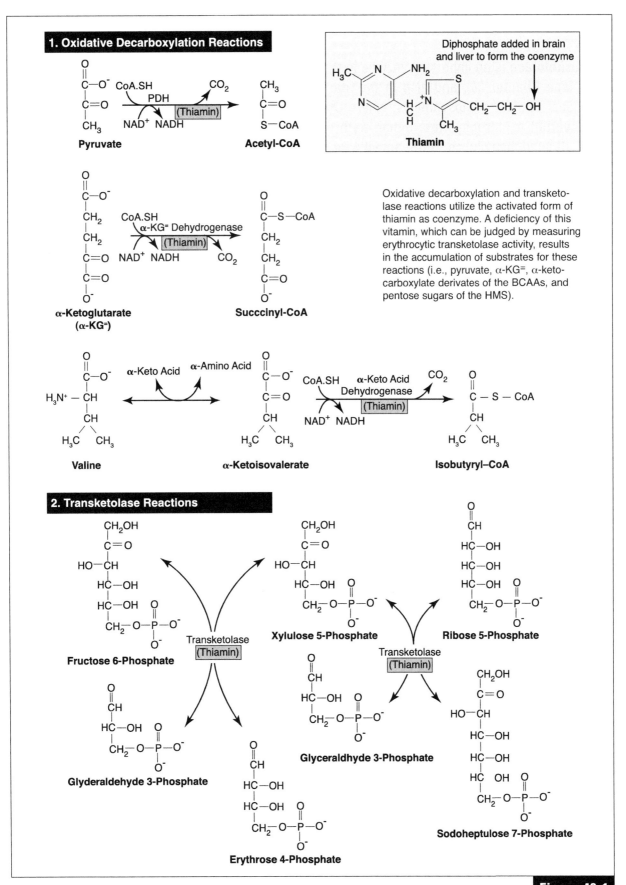

1. Oxidative Decarboxylation Reactions

Pyruvate

Acetyl-CoA

Diphosphate added in brain and liver to form the coenzyme

Thiamin

α-Ketoglutarate
(α-KG⁼)

Succinyl-CoA

Oxidative decarboxylation and transketolase reactions utilize the activated form of thiamin as coenzyme. A deficiency of this vitamin, which can be judged by measuring erythrocytic transketolase activity, results in the accumulation of substrates for these reactions (i.e., pyruvate, α-KG⁼, α-keto-carboxylate derivates of the BCAAs, and pentose sugars of the HMS).

Valine

α-Ketoisovalerate

Isobutyryl–CoA

2. Transketolase Reactions

Fructose 6-Phosphate

Transketolase
(Thiamin)

Xylulose 5-Phosphate

Ribose 5-Phosphate

Glyderaldehyde 3-Phosphate

Glyceraldhyde 3-Phosphate

Transketolase
(Thiamin)

Erythrose 4-Phosphate

Sodoheptulose 7-Phosphate

Figure 40-1

see Chapter 8), particularly in brain and muscle tissue.

Transketolase is an important enzyme in the HMS (see Chapter 28), and also in the "dark reactions" of plant photosynthesis, where CO_2 is converted to carbohydrate. Since erythrocytes depend heavily on HMS activity (and therefore thiamin availability; see Chapters 30 and 31), a deficiency of this vitamin can be judged by measuring erythrocytic transketolase activity. Given the above discussion, thiamin-deficient animals would be expected to exhibit accumulation of substrates involved in the above reactions (e.g., pentose sugars, pyruvate, α-KG$^=$, and the α-ketocarboxylate derivatives of the BCAAs).

Several foods have been found to exhibit **thiaminase** (or thiamin antagonist) activity (e.g., raw fish (e.g., tuna, salmon, and shellfish), rice bran, tannins (coffee and tea), bracken fern (*Pteridium aquilinum*), and horsetail (*Equisetum arvense*)). If **cats**, for example, are fed excessive amounts of raw fish in home-prepared diets, they may develop signs of thiamin deficiency. Additionally, some symbiotic microbes found in the digestive tract also produce thiaminase. **Thiamin deficiency** is associated with "**beriberi**" in primates, "**polioencephalomalacia**" in ruminant animals and horses, and "**Chestak's paralysis**" in the fox, mink, and cat. Generalized symptoms of deficiency include peripheral neuropathy (most marked in the extremities), weakness, tenderness and atrophy of muscles, fatigue, and decreased attention span. Affects on the CNS are largely a result of its importance to aerobic metabolism, and when thiamin is deficient there is conversion to anaerobic glycolysis, local production of lactic acid, and neuronal dysfunction. Affected patients may exhibit vestibular "seizures," fixed and dilated pupils, loss of physiologic nystagmus, plus stupor or coma. The condition may become **fatal** if left untreated. However, the response

to therapy is reported as being dramatic, and the treatment innocuous, so that clinicians have been encouraged to treat whenever the diagnosis is suspected. Routine laboratory tests may be normal, although serum can be tested for vitamin B_1 metabolites, or erythrocytes can be tested for transketolase activity. Within the **human** population, thiamin deficiency is most often associated with **alcoholism**.

Riboflavin (Vitamin B₂)

Riboflavin was first isolated from milk in 1933, and functions as part of two coenzymes, **flavin adenine dinucleotide (FAD)**, and **riboflavin 5'-monophosphate (flavin mononucleotide, FMN; Fig. 40-2)**. Riboflavin 5'-monophosphate is formed by ATP-dependent phosphorylation of riboflavin, whereas FAD is synthesized by a further reaction with ATP in which the AMP moiety of ATP is transferred to FMN. Flavin

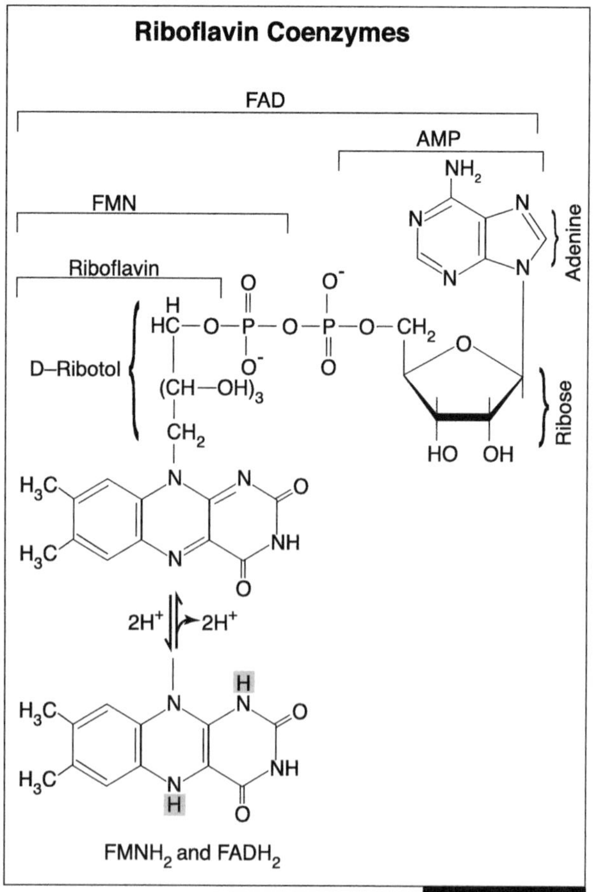

Figure 40-2

adenine dinucleotide and FMN are usually tightly bound to their respective apoenzyme protein, and contain one or more trace elements as essential cofactors (usually **iron** or **molybdenum**; see Chapter 48). **Metalloflavoproteins** are capable of reversible reduction, thus yielding **FADH₂** and **FMNH₂**.

Flavins are useful in physiologic systems in that they are stronger oxidizing agents than NAD^+, thus fitting in further along the electron transport chain (see Chapter 36). They participate in one or two electron processes (and thus reactions with free radicals or metal ions), and in reduced form react directly with O_2 (as in hydroxylation reactions). Riboflavin is thus an enzyme cofactor, or coenzyme, fundamental to many areas of metabolism, and is intimately involved in processes by which the oxidation of glucose and fatty acids are utilized for adenosine triphosphate (ATP) formation, and thus the support of metabolic processes. Riboflavin coenzyme formation (and thus trapping within cells), is initiated through phosphorylation by a **flavokinase**, which is positively regulated by the most active form of thyroid hormone (i.e., **triiodothyronine, T₃**; see Chapter 52).

The metalloflavoproteins are widespread in mammalian organisms, and participate in various oxidation/reduction reactions, exemplified as follows:

- **Ferredoxin reductase** (participates in renal vitamin D activation (see Chapter 45).
- **Succinate dehydrogenase** (which links the mitochondrial TCA cycle to oxidative phosphorylation; see Chapters 34-36).
- **NADH-CoQ reductase** and **succinyl-CoQ reductase** (complexes I and II of the mitochondrial ETC; see Chapter 36).
- Mitochondrial **glycerol 3-phosphate dehydrogenase** in transporting reducing equivalents from the cytosol into mitochondria (i.e., the glycerol 3-phosphate shuttle; see Chapters 35 and 36).

- **Xanthine oxidase** in purine degradation (see Chapter 17).
- **Acyl-CoA dehydrogenase** in mitochondrial fatty acid β-oxidation (see Chapter 55).
- **Glutathione reductase** in erythrocytes for the reduction of oxidized glutathione (see Chapter 30).
- **α-Amino acid oxidase** in hepatic and renal amino acid deamination (see Chapter 9).

All of these enzyme systems are impaired in riboflavin deficiency. Additionally, **glucose oxidase**, a non-mammalian FAD-specific enzyme prepared from **fungi**, is of interest because it is sometimes used *in vitro* for determination of blood glucose concentrations.

Riboflavin is a colored, fluorescent pigment that is synthesized by plants and microbes, but not mammals. In addition to vegetables, yeast, liver, and kidney are usually good sources of riboflavin. In contrast to thiamin, it is relatively heat stable, but **decomposes in the presence of light**. Therefore it is generally stored in dark bottles. Riboflavin absorption occurs primarily in the upper part of the small intestine, and is thought to occur via a carrier-mediated process.

Conversion of riboflavin to its coenzyme derivatives (FAD and FMN) occurs to a great extent in the **liver**. These compounds are thus stored there, and small amounts that enter bile generally return to the liver via the portal circulation following intestinal reabsorption. Thus, riboflavin, like bile acids, exhibits an enterohepatic circulation (EHC; see Chapter 62). It is ultimately degraded by hepatic microsomal mixed function oxidases, with degradative products being excreted in bile and some also entering blood. Degradative products entering blood are generally filtered by the kidneys and appear in urine largely in the form of 7- and 8-hydroxymethyl derivatives.

Riboflavin deficiency causes several non-specific signs and symptoms in animals, including mucus membrane inflammation, alopecia, dermatitis, anemia, photophobia, corneal vascularization, and cataracts. Unlike thiamin, however, deficiency of this vitamin does not usually lead to life-threatening conditions. **Erythrocytic glutathione reductase** activity has been used to assess riboflavin deficiency.

OBJECTIVES

- Recognize the two general types of reactions that utilize the activated coenzyme form of thiamin.

- Explain why erythrocytes are used to assess thiamin deficiency.

- Understand why cats fed excessive amounts of raw fish may develop thiamin deficiency.

- Associate various animal species with either beriberi, polioencephalomalacia or Chestak's paralysis, and discuss the symptoms and causes of these conditions.

- Show how FMN and FAD can can be formed from riboflavin, and explain what the metalloflavoproteins are.

- Show how the flavins (FAD and FMN) participate in oxidative phosphorylation (see Chapter 36).

- Identify and describe eight different biochemical reactions that would be impaired in riboflavin deficiency, and predict the consequences.

- Identify important dietary sources of riboflavin, and explain why it appears in the enterohepatic circulation (EHC).

- Explain why erythrocytic glutathione reductase activity can be used to assess a riboflavin deficiency (see Chapter 30).

QUESTIONS

1. **Which one of the following glycolytic reactions involves thiamin?**
   a. Phosphoenolpyruvate —> Pyruvate
   b. Glucose —> Glucose 6-phosphate
   c. 3-Phosphoglycerate —> 2 Phosphoglycerate
   d. Fructose 6-phosphate —> Fructose 1,6-bisphosphate

   e. Pyruvate —> Acetyl-CoA

2. **Which of the following, if fed to cats in excessive amounts, could cause thiamin deficiency?**
   a. Raw fish (e.g., tuna or salmon)
   b. Fresh vegetables (e.g., squash, corn, or beans)
   c. Raw meat (e.g., beef)
   d. Milk
   e. Poultry (e.g., chicken)

3. **A thiamin deficiency can be judged by measuring the activity of which one of the following enzymes in mature erythrocytes?**
   a. Glutathione reductase
   b. Pyruvate carboxylase
   c. Transketolase
   d. Pyruvate dehydrogenase
   e. α-Ketoglutarate dehydrogenase

4. **Riboflavin 5'-monophosphate (FMN):**
   a. Is formed by ATP-dependent phosphorylation of riboflavin.
   b. Is formed from FAD by an ATP-dependent phosphorylation.
   c. Formation is initiated extracellularly through the activity of riboflavin dehydrogenase.
   d. Formation is inhibited in most cells by triiodothyronine (T$_3$).
   e. Activity is inhibited by molybdenum.

5. **Which one of the following is an FAD-containing enzyme?**
   a. Glycogen synthase
   b. Glutathione reductase
   c. Pyruvate carboxylase
   d. Glucose 6-phosphatase
   e. Glycogen phosphorylase

6. **Conversion of riboflavin to its coenzyme derivatives occurs to the greatest extent in which organ?**
   a. Brain
   b. Liver
   c. Kidney
   d. Lung
   e. Heart

7. **Alcoholics have been known to develop beriberi, steatosis and metabolic acidosis (see Chapter 24):**
   a. True
   b. False

ANSWERS

7. a
6. b
5. b
4. a
3. c
2. a
1. e

Chapter 41

Niacin (B₃) and Pantothenic Acid (B₅)

Overview

- Niacin is a component of NAD^+ and $NADP^+$.

- Dietary nicotinamide, niacin, and Trp can all give rise to NAD^+.

- The Trp content of some food sources may be quantitatively more important in generating NMN, and thus NAD^+, than niacin itself.

- Coenzymes generated from niacin are best associated with dehydrogenase reactions.

- Niacin deficiency can result in pellagra.

- Pantothenic acid gives rise to two coenzymes, 4-phosphopantetheine and coenzyme A (i.e., CoA.SH).

- 4-Phosphopantetheine is a prosthetic group for acyl carrier protein, which participates in fatty acid biosynthesis.

- The reactive thiol (-SH) group of CoA.SH serves as a carrier (and activator) of acyl groups, most notably in degradative energy-yielding pathways.

- Lipoic acid is a B-complex vitamin whose only known function is to participate in the oxidative decarboxylations of a-ketoacids.

Niacin (Vitamin B₃)

This water-soluble vitamin is a component of the most central electron carrier substances in living cells, the **nicotinamide adenine dinucleotides** (i.e., **NAD^+/NADH**, and **$NADP^+$/NADPH**; see Chapter 18), and therefore functions in several important metabolic pathways (e.g., the **Embden Meyerhoff pathway (EMP)**, the **hexose monophosphate shunt (HMS)**, the **tricarboxylic acid (TCA) cycle, oxidative phosphorylation**, and **fatty acid biosynthesis** and **oxidation**). These electron carriers play a widespread role in many **dehydrogenase enzyme reactions** occurring in both the cytosol and within mitochondria. Generally, **NAD^+-linked dehydrogenases** catalyze redox reactions in **oxidative** pathways, whereas **$NADP^+$-linked**

dehydrogenases (or reductases) are found in pathways concerned with **reductive biosynthesis**. The nicotinamide adenine dinucleotides are not tightly bound to their respective enzymes, and are therefore considered by some as true substrates (although they are most commonly referred to as **coenzymes**).

Dietary **nicotinamide, niacin (nicotinate)**, and **tryptophan (Trp)** can all give rise to **nicotinate mononucleotide (NMN)**, a precursor to both **NAD^+** and **$NADP^+$**, by enzymes present in most cells (**Fig. 41-1**). Dietary nicotinamide, which is naturally present in most plant and animal foods, must first undergo deamidation to nicotinate (which is added to many animal pet foods and complete feeds). This compound is then converted to **desamido-NAD^+** by reaction

Copyright © 2015 Elsevier Inc. All rights reserved.

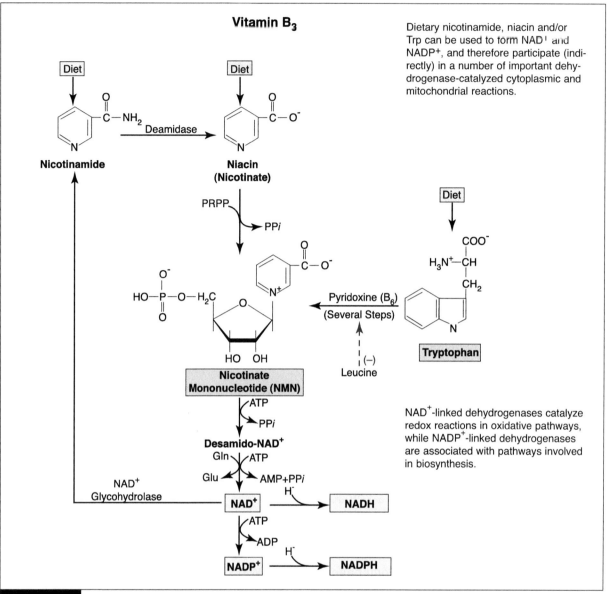

Vitamin B₃

Dietary nicotinamide, niacin and/or Trp can be used to form NAD⁺ and NADP⁺, and therefore participate (indirectly) in a number of important dehydrogenase-catalyzed cytoplasmic and mitochondrial reactions.

NAD⁺-linked dehydrogenases catalyze redox reactions in oxidative pathways, while NADP⁺-linked dehydrogenases are associated with pathways involved in biosynthesis.

Figure 41-1

first with **5-phosphoribosyl-1-pyrophosphate (PRPP)**, forming NMN, and then by adenylylation with ATP. The amido group of glutamine (Gln) then contributes to form the coenzyme **NAD⁺**, which may be further phosphorylated to form **NADP⁺** (see Chapter 18).

As indicated above, NMN is also produced from the **essential** amino acid **Trp**, and although only about 1/60th of dietary Trp is normally utilized in this manner, this amount may increase 3-fold with pregnancy. It should be noted that the Trp content of some food sources may be quantitatively more important

in generating NMN, and thus NAD⁺, than niacin itself. For example, cow's milk normally has about 11 "niacin equivalents" in its Trp content (i.e., amount of Trp/60), yet only one niacin equivalent in its niacin content (an 11:1 ratio). For beef this ratio is about 21:2, eggs 19:1, wheat flour 5:2, yet for corn it is about 2:5. Additionally, **pyridoxal phosphate** (the active form of vitamin B₆; see Chapter 42) is needed to convert **Trp** to **NMN**. Therefore, in some dietary situations, a **vitamin B₆ deficiency** may lead to a **deficiency** of **NMN**.

Excess dietary **leucine** (e.g., from sorghum) has also been reported to contribute to niacin deficiency by inhibiting the key enzyme that converts **Trp** to **NMN**. Other conditions leading to symptoms of niacin deficiency include administration of drugs which divert **Trp** toward **serotonin (5-hydroxy-tryptamine)** formation, and **Hartnup syndrome**, in which Trp absorption is impaired in both the intestinal tract and the kidneys. Proximal renal tubular epithelial cells and mucosal cells in the jejunum contain similar transport proteins that couple the downhill transport of Na+ into cells with the co-transport of amino acids (see Chapter 7). Some of these transporters are specific for the nonpolar, neutral amino acids like Trp, others for the basic and acidic amino acids, and still others for glycine and the imino acid proline (see Chapter 2). Patients with this disorder (as judged from high concentrations of neutral amino acids in urine) are much more likely to have symptoms if they also have poor diets. A patient with a high intake of niacin would be less likely to be symptomatic, and one with a high-protein diet would compensate somewhat for the loss of Trp by increasing absorption of this amino acid in dipeptides and tripeptides (see Chapter 7).

Niacin deficiency results in weakness (lassitude), indigestion and inappetence, and later in the classic signs of "**pellagra**" (i.e., the "**3 Ds**" – **dermatitis, diarrhea, and dementia** –). In addition, some animals reportedly exhibit vomiting with evidence of inflamed mucus membranes. The dementia is said to progress to irritability, sleeplessness, confusion, and eventually delirium and catatonia.

Like most of the other B-complex vitamins, niacin stores in the body are minimal, and it appears to be nontoxic in large doses. Large pharmacologic doses have been used to help lower serum cholesterol. It reportedly reduces the flux of fatty acids from adipose tissue, which leads to less formation of cholesterol-bearing lipoproteins (VLDL —> IDL —> LDL; see Chapters 65 and 67).

Pantothenic Acid (Vitamin B₅)

This water-soluble vitamin, a constituent of pet foods, plants, and animal tissues, has been shown to be an essential **growth factor** for several different species (e.g., chicks, rats, and others). It is found largely as part of two coenzymes, **4-phosphopantetheine** and **coenzyme A** (**CoA** or **CoA.SH**; **Fig. 41-2**). It is customary to abbreviate the structure of the free (i.e., reduced) CoA as CoA.SH, in which the reactive **thiol** (sulphydryl or -SH) group of the coenzyme is designated. It is this thiol group which esterifies with carboxylic acids to form **acyl-CoA** compounds, which are, therefore, **thioesters**:

$$CoA.SH + R\text{-}COOH \longrightarrow R\text{-}CO.SCoA + H_2O$$

An acyl group often linked to CoA is the acetyl unit, thus forming acetyl-CoA. Acyl-CoA compounds are important in both catabolism, as in the oxidation of fatty acids, and in anabolism, as in the synthesis of membrane lipids. The exception to the presence of these coenzyme derivatives in animals appears to be blood plasma, where this vitamin is found primarily as pantothenic acid.

Although little is known about its absorption and metabolism, the fact that plasma contains pantothenic acid in non-coenzyme form indicates that the vitamin B₅ coenzymes in food are broken down before or during intestinal absorption. Pantothenic acid, like other water-soluble vitamins, is freely filtered by the kidneys and excreted into urine.

The important coenzyme actions of intracellular pantothenic acid derivatives are numerous, and quite well documented. Thioesters of **CoA. SH** play a central role in pathways of **energy metabolism**, and **4-phosphopantetheine** is an important prosthetic group in **acyl carrier**

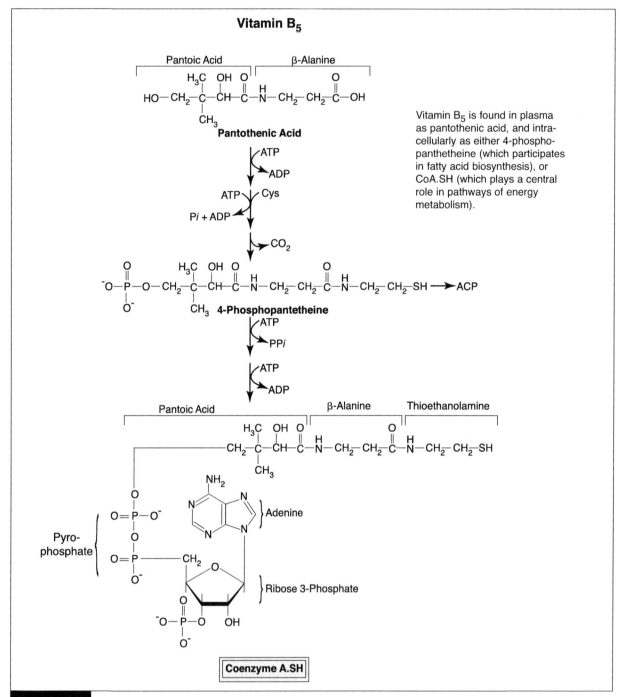

Vitamin B₅

Vitamin B₅ is found in plasma as pantothenic acid, and intra-cellularly as either 4-phospho-pantetheine (which participates in fatty acid biosynthesis), or CoA.SH (which plays a central role in pathways of energy metabolism).

Figure 41-2

protein (**ACP**), which participates in reactions concerned with fatty acid biosynthesis (see Chapter 56). Some examples of reactions involving these pantothenic acid derivatives are as follows:

- The **first step** in the **TCA cycle** (transfer of an acetyl group from acetyl-CoA to oxaloac-etate, thus forming citrate; see Chapter 34).

- The **fifth step** in the **TCA cycle** (conversion of

succinyl-CoA to succinate; see Chapter 34).

- **Activation of long-chain fatty acids** in the cytoplasm and **propionate** in mitochondria (by appropriate acyl-CoA synthetases; see Chapter 55).

- **Mitochondrial β-oxidation of fatty acids** (recipient of acetyl units removed from the fatty acid chain; carrier of the fatty acid moiety; see Chapter 55).

- **Cytoplasmic synthesis of cholesterol** from 3-hydroxy-3-methylglutaryl-CoA (HMG-CoA; see Chapter 61).
- **Mitochondrial synthesis of ketone bodies** from HMG-CoA (see Chapter 71).
- **Acetylcholine biosynthesis** (see Chapter 57).
- **Porphyrin biosynthesis** (conversion of succinyl-CoA + glycine to Δ-aminolevulinate; see Chapter 32).
- **Fatty acid or acetate transfer** to polypeptides, including some enzymes, receptors, and hormones.

Pantothenic acid deficiency is considered to be rare among domestic animal species, except as an accompaniment of general malnutrition. When deliberately induced, deficiency symptoms include vomiting, malaise, and abdominal distress. Another deficiency symptom reported in humans is burning cramps in the extremities (i.e., "burning foot syndrome").

Lipoic acid

Lipoic acid is another B-complex vitamin, whose only known function is to participate in the **oxidative decarboxylations** of α-ketoacids, principally conversion of pyruvate to acetyl-CoA, and α-ketoglutarate to succinyl-CoA, two similar dehydrogenase complexes of the TCA cycle. However, it has not been shown to be essential as a dietary component of animals. As such minute amounts are needed, and because some lower organisms do require it as a growth factor, the question of whether it is produced in the tissues of higher organisms, or is acquired in sufficient quantities through the diet, remains unanswered. Lipoic acid (structure not shown), like thiamin and biotin, contains sulfur, and has been isolated from liver and yeast. It is covalently linked to the enzymes that require it through a peptide bond with the terminal side-chain amino group of lysine.

OBJECTIVES

- Contrast the different types of dehydrogenase reactions using NAD^+ as a coenzyme, to those using $NADP^+$ as a coenzyme.
- Trace biochemical steps in the conversion of dietary nicotinamide and/or nicotinate to NAD^+.
- Explain how dietary pyridoxine (B_6), Leu and/or Trp can affect physiologic supplies of NMN, and thus NAD^+.
- Recognize what is meant by a "niacin equivalent."
- Explain the causes and symptoms of Hartnup syndrome and pellagra.
- Identify the coenzyme forms of pantothenate.
- Describe reactions in the TCA cycle that are dependent upon a coenzyme form of pantothenate.
- Show how vitamin B_5 participates in both fatty acid β-oxidation and fatty acid biosynthesis (see Chapters 55 & 56).
- Indicate how CoA.SH assists in the incorporation of a volatile fatty acid (propionate) into hepatic gluconeogenesis (see Chapter 37).
- Identify vitamin B_5-dependent reactions catalyzed by -ALA synthase, HMG-CoA reductase and HMG-CoA lyase (see Chapters 32, 61 & 71).
- Identify two decarboxylation reactions of the TCA cycle that require lipoic acid.

QUESTIONS

1. **Which one of the following amino acids is used in the biosynthesis of nicotinate mononucleotide (NMN)?**
   a. Alanine
   b. Tyrosine
   c. Leucine
   d. Tryptophan
   e. Glycine

2. **A deficiency of which one of the following B-complex vitamins would make it more difficult to convert Trp to NMN, and therefore potentiate a niacin deficiency?**
   a. Pyridoxine
   b. Folic acid

c. Biotin
d. Riboflavin
e. Thiamin

3. Niacin deficiency is best associated with:

a. Scurvy.
b. Beriberi.
c. Adrenal insufficiency.
d. Pellagra.
e. Megaloblastic anemia.

4. Which of the following enzymes are best associated with niacin (vitamin B₃)?

a. Kinases
b. Carboxylases
c. Dehydrogenases
d. Isomerases
e. Phosphatases

5. 4-Phosphopantetheine is:

a. A prosthetic group for acyl carrier protein.
b. Synthesized from niacin.
c. Used by animals to synthesize pantothenic acid (vitamin B₅).
d. Also known as coenzyme A.
e. A natural product of plant, but not animal metabolism.

6. Coenzyme A.SH is associated with all of the following, EXCEPT:

a. Porphyrin biosynthesis.
b. Ketone body synthesis.
c. Cholesterol synthesis.
d. Mitochondrial β-oxidation of fatty acids.
e. Cytoplasmic glycogen synthesis.

7. Acyl-CoA compounds are properly referred to as:

a. Porphyrins.
b. Thioesters.
c. Fatty acids.
d. Vitamins.
e. Nicotinamides.

8. Lipoic acid is:

a. A fat-soluble vitamin.
b. An essential dietary component of animals.
c. A B-complex vitamin.

d. A short-chain, volatile fatty acid produced by rumen microbes.
e. An unsaturated fatty acid found in the 2-position of numerous membrane-bound phospholipids.

4. Approximately what amount of the dietary Trp intake is used for NMN synthesis?

a. $1/6^{th}$
b. $1/60^{th}$
c. $1/600^{th}$
d. $1/6,000^{th}$
e. $1/60,000^{th}$

10. Which TCA cycle enzymes are associated with pantothenate?

a. Malate dehydrogenase and α-KG⁼ dehydrogenase
b. Succinate dehydrogenase and aconitase
c. Isocitrate dehydrogenase and fumarase
d. Succinate thiokinase and citrate synthase
e. None of the above

11. Pantothenic acid:

a. Functions in the ETC in its coenzyme form (CoQ).
b. Is filtered by the kidneys and excreted in urine.
c. Is found in its active coenzyme form in plasma.
d. Is a fat-soluble vitamin
e. Like carbohydrates, lacks nitrogen.

12. Cholesterol biosynthesis from acetyl-CoA requires pantothenic acid (vitamin B₅), but not niacin (vitamin B₃):

a. True
b. False

13. The oxidative decarboxylation of pyruvate to acetyl-CoA requires:

a. ATP, CO_2, biotin and pyruvate carboxylase.
b. Malic enzyme, NADPH, ATP and lipoic acid.
c. CO_2, vitamin B₆, NADH and biotin.
d. Thiamin, lipoic acid, niacin, pantothenic acid and pyruvate dehydrogenase.

13. d
12. b
11. b
10. d
9. b
8. c
7. b
6. e
5. a
4. c
3. d
2. a
1. d

Chapter 42

Biotin and Pyridoxine (B$_6$)

Overview

- Biotin participates in carboxylation reactions.
- Biotin, pantothenate (B$_5$), and cobalamin (B$_{12}$) are needed by herbivores to move propionate into hepatic gluconeogenesis.
- Biotin-supplemented diets are sometimes fed to young, growing animals.
- Raw egg white reduces intestinal biotin absorption.
- Vitamin B$_6$ (pyridoxine) is used in muscle glycogenolysis, and in erythrocytes it is bound to hemoglobin.
- Pyridoxal phosphate is used in transamination reactions.
- Although rare in animals, a vitamin B$_6$ deficiency can result in increased amounts of amino acid metabolites appearing in urine, and it can reduce conversion of Trp to NAD$^+$.

Biotin

This water-soluble B-complex vitamin is a widely distributed **imidazole derivative** found in plants in the free form, but in animals it is linked to protein. Animal requirements for biotin are generally met through the diet, or through microbial synthesis within the digestive tract.

During digestion, biotin is initially released from protein as a lysine-adduct (**biocytin**), and either further digested to free biotin, or absorbed as such and hydrolyzed within intestinal mucosal cells. The enzyme involved in this process (sometimes referred to as **biotinidase**), may also be the **biotin carrier protein (BCP)** associated with this vitamin both intracellularly and within serum (**Fig. 42-1**).

Intestinal absorption of this vitamin in the jejunum is thought to occur by facilitated diffusion at low concentrations, and by simple diffusion at high concentrations. Although biotin deficiencies are rare, they can

be induced through use of broad-spectrum oral antibiotics over a long period of time (which reduces microbial biosynthesis), or through excessive consumption of raw eggs. Raw egg whites contain the active form of **avidin**, a protein that binds tightly with biotin, thus making it unavailable for absorption from the digestive tract. Therefore, raw eggs should not be fed to young animals who need biotin for proper growth and development. Since heat denatures avidin, cooked egg consumption is not associated with biotin deficiency.

The function of biotin is to participate in **carboxylation** reactions. First, a **carbonic phosphoric anhydride** is formed through combination of a high energy phosphate from **ATP**, CO$_2$ derived from HCO$_3^-$, magnesium (**Mg^{++}**), and potassium (**K$^+$**). Next, **biotin carboxylase** (sometimes referred to as **holoenzyme synthetase**), and manganese (**Mn^{++}**) help to facilitate movement of the activated carboxyl group of carbonic

Copyright © 2015 Elsevier Inc. All rights reserved.

phosphoric anhydride to biotin, thus forming **carboxybiotin**. This compound now becomes available to various carboxylases (or transcarboxylases) for **one carbon transfer**.

There are **four** important biotin-driven "**CO₂-fixing**" reactions that occur in animal cells (**Fig. 42-1**), and they are all catalyzed by substrate-specific carboxylases.

1) Conversion of pyruvate to oxaloacetate (OAA, via **pyruvate carboxylase**; see Chapter 27 and 37).

2) Conversion of propionyl-CoA to methylmalonyl-CoA (via **propionyl-CoA carboxylase**; see Chapter 37).

3) Formation of acetoacetate from leucine (via **β-methylcrotonyl-CoA carboxylase**; see Chapter 8).

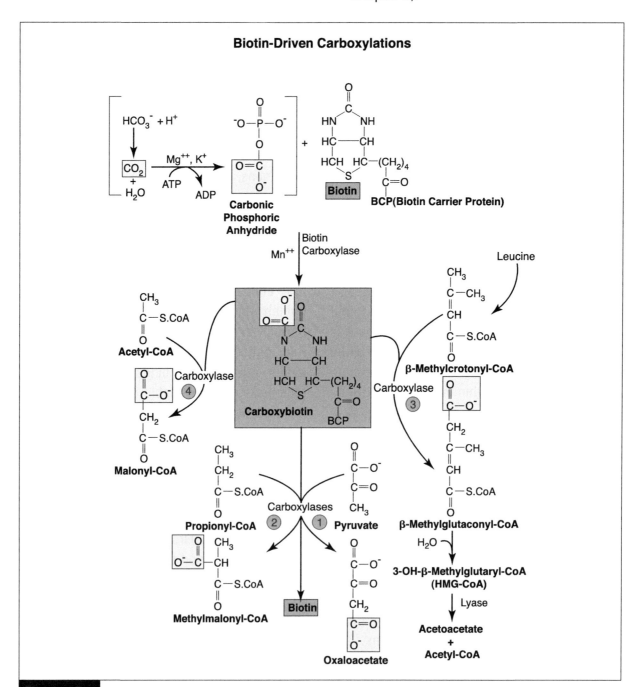

Figure 42-1

4) Formation of malonyl-CoA from acetyl-CoA in fatty acid biosynthesis (via **acetyl-CoA carboxylase**; see Chapter 56).

Hepatic **OAA formation** from pyruvate is allosterically enhanced by acetyl-CoA, a product of mitochondrial fatty acid β-oxidation. This enzyme is particularly important to liver and kidney tissue during gluconeogenesis, and to muscle tissue during exercise.

Propionyl-CoA is formed from the short-chain volatile fatty acid, propionate, from several glucogenic amino acids (i.e., threonine, methionine, isoleucine, and valine), from terminal three carbon segments of odd-chain fatty acids undergoing mitochondrial β-oxidation, and from β-aminoisobutyrate during thymine degradation. Propionate, generated from microbial fermentation of cellulose, is a significant source of carbon atoms for hepatic gluconeogenesis in herbivores, and the glucogenic amino acids listed above can be a significant source of carbon atoms for gluconeogenesis in carnivores, and in starved animals. Since most mammals store few odd-chain fatty acids, their terminal three carbon segments become an insignificant source of propionyl-CoA for gluconeogenesis. Since only modest amounts of β-aminoisobutyrate become available to the liver through pyrimidine degradation, this compound is also considered to be an insignificant gluconeogenic substrate. Following carboxylation of propionyl-CoA to methymalonyl-CoA, further conversion to succinyl-CoA requires cobalamin (vitamin B_{12}; see Chapter 43).

Hepatic catabolism of Leu to **ketone bodies** requires carboxybiotin, and forms β-methylglutaconyl-CoA as intermediate. Leucine and Lys are the only amino acids that are strictly ketogenic (see Chapter 8).

Biotin deficiency is not usually caused by simple dietary deficiency, but by defects in utilization. Symptoms can include dermatitis, alopecia, and muscle weakness. Biotin, like other B-complex vitamins, is not considered to be toxic, and supplementation can reverse these symptoms. In young animals, extra demands on biotin, a vitamin needed for proper

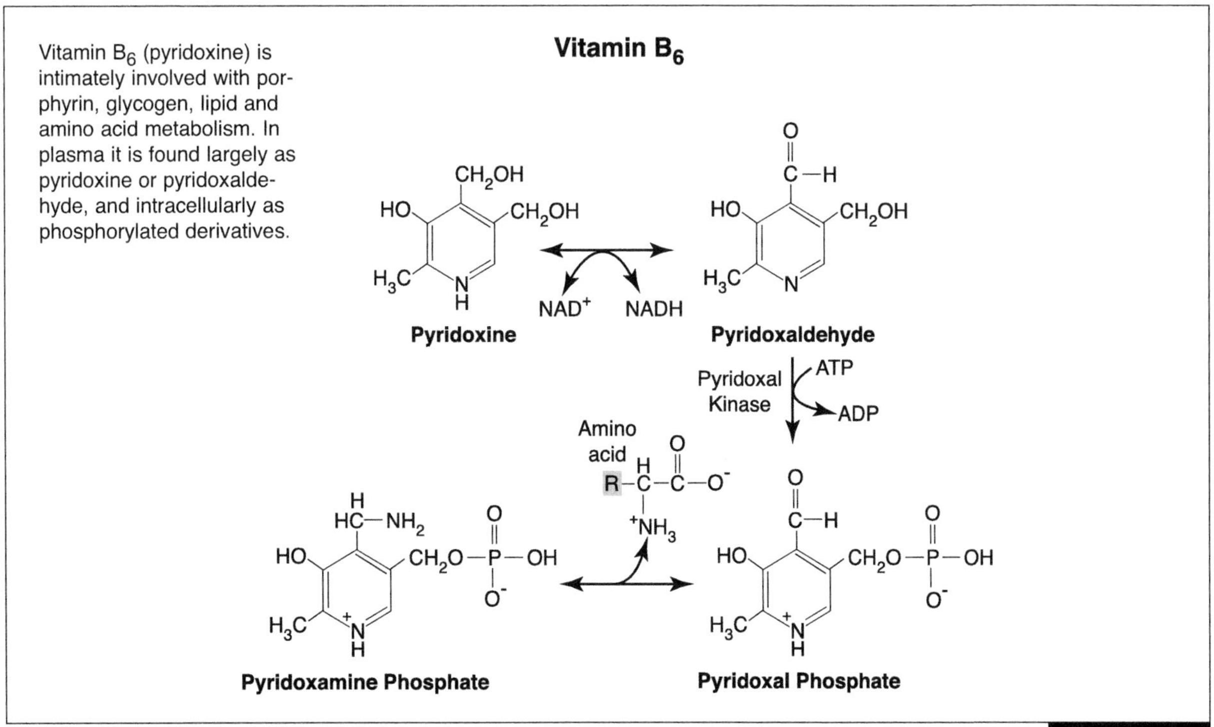
Vitamin B6
Vitamin B6 (pyridoxine) is intimately involved with porphyrin, glycogen, lipid and amino acid metabolism. In plasma it is found largely as pyridoxine or pyridoxaldehyde, and intracellularly as phosphorylated derivatives.
CH2OH
HO
CH2OH
H3C
N
H
NAD+
NADH
Pyridoxine
O
C—H
HO
CH2OH
H3C
N
Pyridoxaldehyde
Pyridoxal Kinase
ATP
ADP
Amino acid
H
R–C–C–O-
+NH3
O
HC—NH2
CH2O—P—OH
O-
Pyridoxamine Phosphate
Pyridoxal Phosphate

Figure 42-2

growth and development, are sometimes met by feeding biotin-supplemented diets.

Pyridoxine (B₆)

Vitamin B_6 is the collective term for **pyridoxine**, the form most prominent in plants, and for the phosphorylated coenzyme derivatives, **pyridoxal** and **pyridoxamine phosphate**, common forms found in animal tissues (**Fig. 42-2**). **Pyradoxine, pyridoxamine and pyridoxaldehyde** are major transport forms available to cells, and account for most of the vitamin in plasma. Once inside cells they become phosphorylated. In erythrocytes, for example, this vitamin becomes concentrated about four- to five-fold, where it is bound mainly to **hemoglobin** (it may enhance O_2 binding).

Vitamin B_6 coenzymes participate in over 100 different enzyme-catalyzed reactions, most of which occur in all living cells, and some of which are only present in liver and kidney cells (main sites of gluconeogenesis). They are involved with muscle **glycogenolysis**, and with **amino acid metabolism**, where they function in numerous transamination, decarboxylation, dehydratase, and side-chain cleavage reactions. Pyridoxal phosphate is an important coenzyme which aids the action of muscle **glycogen phosphorylase**, the enzyme mediating glycogen breakdown (see Chapter 23). **Muscle phosphorylase may account for as much as 70-80% of total body vitamin B₆ in mammals.**

Transamination reactions involving vitamin B_6 include those concerned with the synthesis of nonessential amino acids, and also those concerned with the first step in amino acid catabolism (see Chapter 9). Transaminases are rate-limiting for the catabolism of specific amino acids in the liver. **Decarboxylation** reactions involve the synthesis of several important substances, including neuroactive amines (e.g., serotonin, histamine, and γ-amino butyric acid (GABA)), Δ-aminolevulinic acid

(the first step in heme biosynthesis; see Chapter 32), and intermediates in the synthesis of sphingomyelin, phosphatidyl-choline (lecithin), carnitine, and taurine (important as a bile acid conjugate, and also in brain and eye function). **Dehydratase** reactions include conversion of serine and threonine to their α-keto acids through oxidative removal of the amino group as ammonia. **Side chain cleavage** reactions are exemplified by the formation of glycine from serine, and the splitting of cystathionine in methionine degradation. Thus, vitamin B_6 is intimately involved with **amino acid metabolism**, and also plays an important role in **porphyrin**, **glycogen**, and **lipid metabolism**. In addition to these actions, vitamin B_6 may be involved with growth hormone and insulin release (since deficiency of this vitamin in rats depresses uptake of amino acids by muscle cells). Some studies indicate that it may also be involved in the return (and thus inactivation) of steroid receptor complexes to the cytosol from the nucleus after they have promoted transcription of certain genes (which indicates that it plays a regulatory role in steroid hormone action).

Vitamin B_6 is thought to be absorbed from the digestive tract mainly as **pyridoxine**, and excreted in urine as **pyridoxic acid**. **Deficiencies** due to lack of pyridoxine are rare in animals, and when they occur they are usually part of a general B-complex vitamin deficiency. Due to its importance in amino acid catabolism, a deficiency of this vitamin can result in urinary excretion of increased levels of some **amino acid metabolites** which are normally degraded further (specifically metabolites of tryptophan, methionine, and glycine). Urinary **urea** excretion may also be enhanced, because of a decreased capacity to synthesize nonessential amino acids, resulting in decreased reutilization of NH_3 and amino nitrogen. As indicated in the previous chapter, a lack of B_6 can also **reduce NAD⁺ formation**

from tryptophan. General deficiency symptoms include hyperirritability and convulsive seizures in infants, and inflammation of the oral cavity and oily dermatitis in adults.

In summary, biotin participates in four important "CO_2-fixing" reactions, and pyridoxine (B_6) is intimately associated with porphyrin, glycogen, lipid and amino acid metabolism. The majority of B_6 is complexed with muscle glycogen phosphorylase. Biotin deficiency is generally associated with defects in biotin utilization, while pyridoxine deficiency is usually associated with a broader B-complex vitamin deficiency. Neither vitamin is considered toxic.

OBJECTIVES

- Outline the protein-binding characteristics and requirements of biotin, and indicate how raw egg white consumption might affect biotin availability.

- Identify the substrate, energy, enzyme and cofactor requirements for carboxybiotin formation.

- Contrast and compare the ways in which folic acid, biotin, methylcobalamin and SAM participate in one carbon transfer (see Chapters 16 & 57).

- Identify and summarize the importance of the four key biotin-driven cellular carboxylation reactions.

- Explain the causes of biotin deficiency.

- Identify common extra- and intracellular forms of vitamin B_6.

- Explain the involvement of pyridoxine in glycogen metabolism.

- Recognize why a pyridoxine deficiency could result in an increased BUN concentration.

- Explain why phosphorylated forms of vitamin B_6 are not normally present in plasma.

- Discuss the involvement of vitamin B_6 in nonessential amino acid biosynthesis (see Chapter 9).

- Recognize how (and where) vitamin B_6 participates in heme biosynthesis.

QUESTIONS

1. **Which one of the following compounds, contained in egg white, binds strongly with biotin in the digestive tract, thus preventing its absorption?**
   a. Carnitine
   b. Avidin
   c. Pyridoxine
   d. Ascorbate
   e. Sodium

2. **Biotin is a coenzyme for certain:**
   a. Transamination reactions.
   b. Dehydratase reactions.
   c. Side-chain cleavage reactions.
   d. Carboxylation reactions.
   e. Decarboxylation reactions.

3. **Which one of the following mitochondrial reactions involves biotin?**
   a. Pyruvate —> Acetyl-CoA
   b. α-Ketoglutarate —> Succinyl-CoA
   c. Succinyl-CoA —> Succinate
   d. Fumarate —> Malate
   e. Propionyl-CoA —> Methylmalonyl-CoA

4. **Biotin is involved in ketone body production from which one of the following amino acids?**
   a. Leucine
   b. Glutamine
   c. Alanine
   d. Aspartate
   e. Proline

5. **Approximately 70-80% of vitamin B_6 in mammalian organisms is associated with which enzyme?**
   a. Hepatic glycogen synthase
   b. Renal glucose 6-phosphatase
   c. Muscle glycogen phosphorylase
   d. Adipolytic triglyceride lipase
   e. Cyclooxygenase (COX-1)

6. **Pyridoxine deficiencies are best associated with increased urinary excretion of:**
   a. Glucose metabolites (e.g., lactate).
   b. Ketone bodies.
   c. Amino acid metabolites.
   d. Cholesterol.
   e. Bilirubin diglucuronide.

6. c
5. c
4. a
3. e
2. d
1. b

ANSWERS

Chapter 43

Cobalamin (B₁₂)

Overview

- Vitamin B_{12} is an antipernicious anemia factor.

- Vitamin B_{12} is generally absent from plant and vegetable foods unless they are contaminated by microbes.

- Liver is a good source of the three endogenous forms of vitamin B_{12} (methylcobalamin, 5'-deoxyadenosylcobalamin, and hydroxocobalamin).

- Ileal absorption of B_{12} requires intrinsic factor, which is synthesized by gastric parietal cells, as well as by pancreatic ductular cells in dogs and cats.

- The association of Co^{+++} with B_{12} is the primary recognized action of this trace element in mammalian metabolism.

- Entry of propionate into hepatic gluconeogenesis requires 5'deoxyadenosylcobalamin.

- The metabolism of vitamin B_{12} is intimately entwined with that of folic acid.

- Common symptoms of vitamin B_{12} deficiency include homocystinuria and methylmalonuria.

- A secondary intestinal dysfunction may develop from persistent cobalamin deficiency.

It was recognized in the early 1800's that **pernicious (or megaloblastic) anemia** may (in part) be due to a disorder of the digestive tract and assimilative organs. It was determined in the early 1900's that this condition could be reversed and controlled by eating raw or **mildly-cooked liver**. Therefore, investigators postulated that a gastric "**intrinsic factor (IF)**," combined with an "**extrinsic factor**" from ingested liver, would bring about absorption of an "**antipernicious anemia factor**." The extrinsic factor was later found to be the antipernicious anemia factor, **cobalamin (vitamin B₁₂)**, and the IF was determined to be an important **glycoprotein**, secreted into gastric or abomasal juice by parietal cells, and additionally by pancreatic ductular cells in dogs and cats.

This water-soluble vitamin, which is thought to be the **largest essential nutrient absorbed intact through the distal ileal mucosa**, is generally absent from plant and vegetable foods, unless they are contaminated by microbes. Herbivores generally obtain cobalamin from symbiotic microbes in their digestive tracts (providing sufficient cobalt (Co^{+++}) is available for microbial B_{12} biosynthesis; see Chapter 52). Vitamin B_{12} is conserved by the healthy liver, a particularly good source of this vitamin for ominvores and carnivores.

Cobalamin is a complex molecule, consisting of a "**corrin ring**," which is a more hydrogenated form of the porphyrin ring associated with heme (see Chapter 32), with differences as well in the side chains of the ring. Cobalamin contains **Co⁺⁺⁺** rather than Fe^{++}, and a **5,6-dimethylb-enzimidazole grouping** attached to the corrin

Copyright © 2015 Elsevier Inc. All rights reserved.

ring through a complex linkage, involving an unusual **ribose-phosphate** moiety (**Fig. 43-1**). The 5,6-dimethylbenzimidazole grouping is also chelated to Co^{+++} at the active center of the corrin ring. Vitamin B$_{12}$ exists in four forms that differ in the nature of additional **R groups** attached to Co^{+++}. The cyano derivative, commonly known as **cyanocobalamin**, is the commercially available form. After transport in blood, cobalamin is usually taken-up by target cells as **hydroxocobalamin**, where it can be converted to **methylcobalamin** in the cytoplasm, or **5'-deoxyadenosylcobalamin** in mitochondria. In the liver, it is found in all three forms. The association of Co^{+++} with vitamin B$_{12}$ is the primary recognized function for this trace element in mammalian metabolism (although other secondary actions are also known; see Chapter 52).

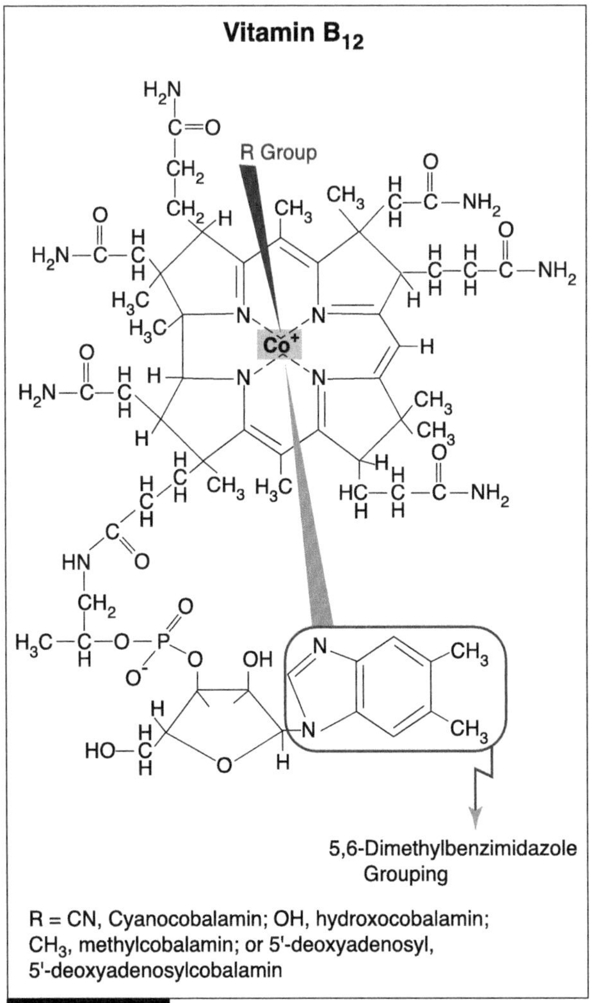

There are **two** important enzymatic reactions in animals that require **vitamin B$_{12}$** (**Fig. 43-2**):

1) Rearrangement of **methylmalonyl-CoA** to **succinyl-CoA** (which requires **5'-deoxyadenosylcobalamin**); and

2) Transfer of a **methyl group** from **N^5-methyltetrahydrofolate** (**N^5-methyl-H$_4$ folate**) to **homocysteine** in the formation of **methionine** (which requires **methylcobalamin**).

The first reaction is important in the sequential conversion of **propionate** to **succinyl-CoA**, an intermediate of the TCA cycle. Propionate is formed from microbial cellulose and starch digestion, from the terminal 3 carbons of odd-chain fatty acids during mitochondrial β-oxidation, from β-aminoisobutyrate during pyrimidine degradation, and from several amino acids during protein degradation (see Chapters 37 and 42). It is of particular significance in the process of **hepatic gluconeogenesis**.

Through the second reaction, which requires methylcobalamin, **H$_4$ folate** is made available to

Vitamin B$_{12}$

R = CN, Cyanocobalamin; OH, hydroxocobalamin; CH$_3$, methylcobalamin; or 5'-deoxyadenosyl, 5'-deoxyadenosylcobalamin

Figure 43-1

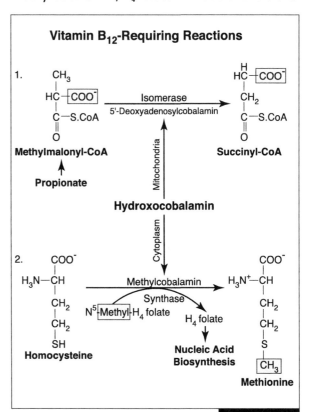

Figure 43-2

participate in **purine, pyrimidine**, and **nucleic acid biosynthesis**. The metabolism of vitamin B$_{12}$ is thus intimately entwined with that of another water-soluble vitamin, **folic acid**, and both are fundamental to one-carbon metabolism (see Chapter 16).

Other reactions involving this vitamin, as a coenzyme or otherwise, cannot be excluded at this time. It has proposed involvement in the synthesis of either the lipid or protein components of **myelin** (see Chapter 59), independent of the methylmalonyl-CoA reaction. This would help to explain the demyelination (or lack of myelination), and nerve degeneration observed in B$_{12}$ deficiency. Investigators have proposed that neurologic disorders associated with B$_{12}$ deficiency may be secondary to a relative deficiency of methionine. A deficiency of this vitamin also appears to result in a loss of tissue **carnitine**, perhaps in the form of a propionic acid adduct entering blood and urine. However, this could occur because of methylmalonate and propionate accumulation due to their lack of conversion to succinyl-CoA via methylmalonyl-CoA isomerase (or mutase). Less carnitine would have consequences for the shuttling of long-chain fatty acids across mitochondrial membranes (see Chapter 55).

Intestinal absorption of B$_{12}$ is far more complex than absorption of other water-soluble vitamins. Dietary B$_{12}$ is frequently bound to protein (B$_{12}$-protein; **Fig. 43-3**). Gastric (or abomasal) HCl and pepsin help to free B$_{12}$ from ingested protein, which allows it to become immediately bound to a number of endogenous **R-proteins**. These proteins are present in saliva and gastric juice, they bind vitamin B$_{12}$ tightly over a wide pH range, and they are both structurally and functionally related to the transcobalamins found in blood and in the liver. These R-proteins have a high affinity for vitamin B$_{12}$.

The rate of **IF** secretion by parietal cells usually parallels their rate of HCl secretion. Intrinsic factor binds B$_{12}$ with less affinity than

do the R-proteins. Thus, most B$_{12}$ released from food in the stomach is bound by R-proteins. Pancreatic **trypsin** and other **proteases** begin degradation of complexes between R-proteins and B$_{12}$ in the duodenum. This degradation greatly lowers the affinities of the R-proteins for B$_{12}$, so that this vitamin can now be appropriately transferred to IF. **Calcium (Ca^{++})** and **bicarbonate (HCO$_3^-$)** from pancreatic and biliary ductular cell secretions, also help to provide conditions necessary for the binding of B$_{12}$ to IF in the duodenum.

When IF binds with B$_{12}$, IF undergoes a conformational change favoring formation of dimers. The brush border plasma membranes of mucosal cells in the distal ileum, a similar site of **bile acid absorption** (see Chapter 62), contain receptor proteins that recognize and bind these **B$_{12}$-IF dimers**. However, they do not recognize and bind the B$_{12}$-R-protein complexes, nor do they bind free B$_{12}$. Consequently, in pancreatic insufficiency, particularly in dogs and cats, the B$_{12}$-R-protein complexes may not be properly degraded, and pancreatic ductular IF secretion may be inadequate, therefore leaving B$_{12}$ tightly bound in these R-protein complexes. Since these complexes are not available for absorption, B$_{12}$ deficiency may ensue.

Following uptake of the B$_{12}$-IF dimer complex by mucosal cells in the distal ileum, B$_{12}$ is slowly released into portal blood. It does not normally appear there until about 4 hours following ingestion, leading some investigators to suggest that the delay may involve mitochondrial incorporation during absorption. Once in portal blood, B$_{12}$ is bound to **transcobalamin II (TC II)**, a β-globulin synthesized largely by the liver, but also by ileal epithelial cells. This B$_{12}$-TC II complex is rapidly cleared from portal blood by the liver through receptor-mediated endocytosis. Within hepatocytes B$_{12}$ becomes bound to a closely related **TC I**, and stored (50-90% of total). Small amounts of free B$_{12}$ are secreted into bile (0.1-0.2% per day), with about

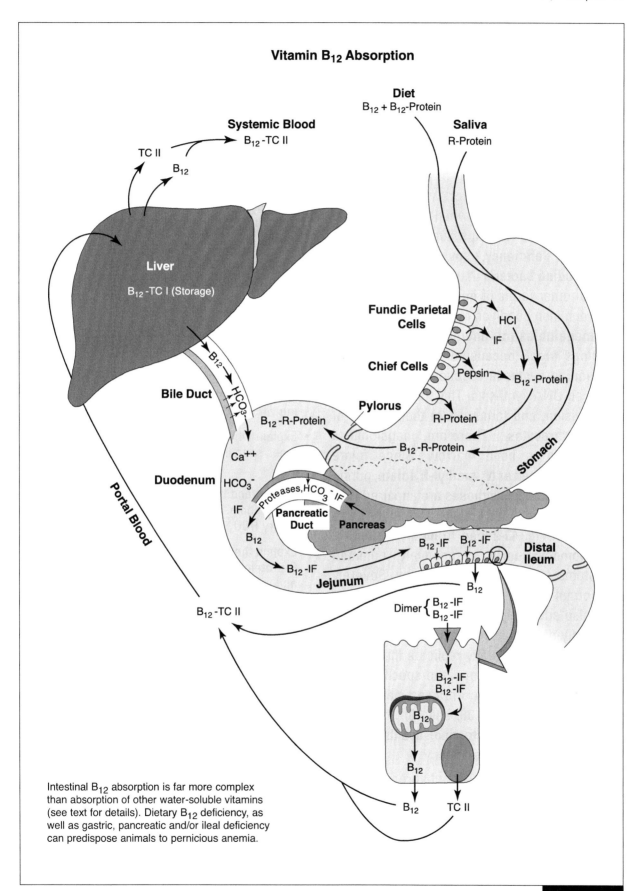

Vitamin B₁₂ Absorption

Intestinal B₁₂ absorption is far more complex than absorption of other water-soluble vitamins (see text for details). Dietary B₁₂ deficiency, as well as gastric, pancreatic and/or ileal deficiency can predispose animals to pernicious anemia.

Figure 43-3

70% of this normally being reabsorbed again by the distal ileum. Because only about 1% of the body store is lost daily in animals, if intestinal B_{12} absorption totally ceases, it generally takes several weeks-months before symptoms of deficiency appear.

In the complete absence of IF, only about 1-2% of an ingested load of B_{12} can be absorbed. When **B_{12} is deficient in the diet**, when **pancreatic insufficiency** ensues, or when **ileal absorption is compromised** (perhaps due to **gastrectomy** and thus **IF deficiency**, to **overgrowth of cobalamin-binding bacteria** in the small bowel, or to several other forms of **intestinal disease**), the most common condition resulting is **pernicious (or megaloblastic) anemia**. This condition develops largely because of impairment of the **methionine synthase** reaction, which causes impaired DNA synthesis. This further prevents cell division, and formation of the nucleus of new erythrocytes. Therefore, megaloblasts accumulate in bone marrow. Since folate becomes trapped as **N^5-methyl-H_4folate**, purine and pyrimidine biosyntheses are impaired (see Chapter 16). Reduced activity of reactions one and two in **Fig. 43-2** lead to **methylmalonuria** and **homocystinuria**, respectively, common symptoms of **vitamin B_{12} deficiency**. Severely subnormal serum cobalamin concentrations may also adversely affect normal proliferation of rapidly-dividing crypt cells in the intestinal mucosa, which generally replicate themselves on a weekly basis. Hence, the specific activities of jejunal digestive enzymes may deteriorate (see Chapters 7, 38, and 60), therefore adding a **secondary intestinal dysfunction** to persistent cobalamin deficiency.

Although vitamin B_{12} is absorbed in the terminal ileum, **folate**, another water-soluble vitamin, is absorbed in the jejunum. Low plasma folate can indicate proximal bowel dysfunction (see Chapter 16), whereas low plasma cobalamin can indicate dysfunction of the terminal ileum. On the other hand, high plasma folate concentrations may be due to intestinal bacterial overgrowth.

OBJECTIVES

- Explain why either folic acid or vitamin B_{12} deficiency could lead to megaloblastic anemia (MA; see Chapter 16).

- Recognize why folic acid supplementation can partially offset the MA caused by vitamin B_{12} deficiency, yet have no effect on the homocystinuria.

- Identify the essential trace element required by microbes in the biosynthesis of hydroxocobalamin, and know why liver is a good dietary source of cobalamin.

- Name the commercial, mitochondrial and cytoplasmic forms of cobalamin.

- Identify and discuss the importance of two important enzymatic reactions in animals requiring vitamin B_{12}.

- Explain why neuronal demyelination is sometimes seen in B_{12} deficiency (see Chapters 57 & 59).

- Understand the relationship between gastric (and abomasal) HCl secretion, and pancreatic/biliary HCO_3^- secretion to vitamin B_{12} absorption.

- Explain the interactions between dietary protein, R-proteins, gastric and pancreatic proteases and intrinsic factor (IF) in B_{12} gastrointestinal transport.

- Indicate why the rate of gastric IF secretion usually parallels the rate of gastric HCl secretion.

- Recognize the anatomic relationship between intestinal bile acid and intestinal B_{12} absorption (see Chapter 62).

- Explain the process of intestinal B_{12} absorption, transport in blood and storage in the liver.

- Identify various causes of vitamin B_{12} deficiency, and explain why homocystinuria, methylmalonuria and intestinal dysfunction may become pathophysiologic signs.

- Explain why high or low circulating levels of vitamin B_{12} or folate can indicate intestinal bacterial overgrowth, or intestinal dysfunction.

QUESTIONS

1. **Vitamin B$_{12}$ is intimately entwined with the action of which water-soluble vitamin?**
   a. Vitamin C
   b. Niacin
   c. Pantothenic acid
   d. Pyridoxine
   e. Folic acid

2. **Vitamin B$_{12}$ helps which of the following substrates gain access to the TCA cycle?**
   a. Lactate
   b. Aspartate
   c. Propionate
   d. Leucine
   e. Medium-chain fatty acids

3. **All of the following are associated with vitamin B$_{12}$ deficiency, EXCEPT:**
   a. Gastrectomy.
   b. Pancreatic insufficiency.
   c. Ileal disease.
   d. Rumen bacterial overgrowth.
   e. Dietary cobalt deficiency.

4. **Cobalamin receptor proteins in the distal ileum normally bind which of the following with the greatest affinity?**
   a. Vitamin B$_{12}$-R-Protein complexes
   b. Vitamin B$_{12}$ in the free form
   c. Vitamin B$_{12}$-IF dimers
   d. Vitamin B$_{12}$-transcobalamin I complexes
   e. Vitamin B$_{12}$-transcobalamin II complexes

5. **Which one of the following is attached to the corrin ring of vitamin B$_{12}$ through a complex linkage involving ribose-phosphate?**
   a. N^5-Methyltetrahydrofolate
   b. 5'-Deoxyadenosylthiamine
   c. Homocysteine
   d. 5,6-Dimethylbenzimidazole
   e. Iron

6. **Which one of the following is the best source of cobalamin?**
   a. Kidney
   b. Liver
   c. Skeletal muscle
   d. Brain
   e. Egg yolk

7. **The largest B-complex vitamin absorbed intact through the adult intestine is thought to be:**
   a. Thiamin.
   b. Riboflavin.
   c. Niacin.
   d. Pantothenic acid.
   e. Cobalamin.

8. **Which one of the following aids in the conversion of homocysteine to methionine?**
   a. Methylcobalamin
   b. 5'-Deoxyadenosylcobalamin
   c. Hydroxocobalamin
   d. Cyanocobalamin
   e. Pyridoxine

9. **Vitamin B$_{12}$ deficiency is best associated with:**
   a. Hyperglycemia.
   b. Hypertension.
   c. Gallstones.
   d. Anemia.
   e. Inflammation.

10. **Vitamin B$_{12}$ deficiency would be expected to lead to which of the following?**
    a. Polycythemia and clotting abnormalities
    b. Amino aciduria
    c. Intrinsic factor deficiency
    d. Uremia
    e. Methylmalonuria and homocystinuria

11. **Transcobalamin II is synthesized by:**
    a. The stomach.
    b. The pancreas.
    c. Hepatocytes.
    d. Biliary ducts.
    e. Salivary glands.

12. **Intrinsic factor (IF) is a:**
    a. Glycoprotein.
    b. Eicosanoid.
    c. Porphyrin.
    d. Lipoprotein.
    e. Nucleotide.

13. **R-proteins that bind B$_{12}$ are produced:**
    a. In salivary glands.
    b. By gastric secretory cells.
    c. By pancreatic ductular cells.
    d. All of the above.
    e. A and B above.

13. e
12. a
11. c
10. e
9. d
8. a
7. e
6. b
5. d
4. c
3. d
2. c
1. e

ANSWERS

Chapter 44

Vitamin A

Overview

- Vitamin A exists as a provitamin in vegetables (i.e., β-carotene).
- Vitamin A exists in three oxidation states; retinal, retinol, and retinoic acid (retin A).
- Retinal plays an important role in vision.
- Retinoic acid plays an important role in reproductive biology, bone remodeling, and epithelial tissue homeostasis.
- Although retinol and retinal are stored in the body, retinoic acid is not.
- Vitamin A is best associated with thyroid hormone action.
- Vitamin A and D excess can be toxic, whereas E and K excess is generally nontoxic.
- Several aspects of vitamin A metabolism are Zn^{++}-dependent.

Fat-Soluble Vitamins

The **fat-soluble vitamins A, D, E**, and **K** are lipophilic, hydrophobic molecules, that are assembled from isoprenoid units, the same building blocks that are used to synthesize cholesterol (see Chapter 61). Because of their hydrophobic nature, they are transported in blood bound either to lipoproteins (e.g., **chylomicrons**; see Chapter 64), or to more specific carrier proteins. They are absorbed from the intestine along with other dietary lipids (see Chapter 60), therefore, abnormalities in lipid absorption resulting in **steatorrhea** can also result in a **fat-soluble vitamin deficiency**.

As a group, the fat-soluble vitamins have diverse biologic actions (e.g., **vitamin A, vision; vitamin D, Ca^{++} and $PO_4^=$ homeostasis; vitamin E, antioxidant activity;** and **vitamin K, blood clotting**). Although at one time **vitamin D** was considered to be only a vitamin, now it is considered to be a **prohormone** as well (a precursor to the 1,25-dihydroxy forms of vitamin D (1,25(OH)₂-D₂ and 1,25(OH)₂-D₃; see Chapter

45). Large quantities of the fat-soluble vitamins can be stored in the **liver** and in adipose tissue, and toxicity can result following excessive intake of **vitamins A** and **D**. Vitamins E and K are generally considered nontoxic.

Vitamin A

This vitamin plays a central role in both **photopic (day)** and **scotopic (night) vision,** in **reproductive biology,** in **bone remodeling,** and in the **maintenance and differentiation of epithelial tissues (Table 44-1).** It exists in three oxidation states (i.e., **retinol,** the alcohol; **retinal,** the aldehyde; and **retinoic acid,** which is not stored in the body; **Fig. 44-1**). These various oxidation states of the natural vitamin, as well as synthetic analogs and metabolites, are called **retinoids,** stemming from their importance to the physiology of the retina. In mammalian organisms, interconversion of retinol with cis- and trans-retinal occurs, but oxidation of trans-retinal to retinoic acid (retin A) is an irreversible process. Thus, retinoic

Copyright © 2015 Elsevier Inc. All rights reserved.

Table 44-1
Actions of Vitamin A

Retinol and Retinal

Vision

Rhodopsin synthesis (rods)

Porphyropsin synthesis (cones)

Retinoic Acid

Growth and differentiation of epithelial cells

Glycoprotein synthesis

Expression/production of growth hormone

Mucus production

Bone remodeling

Reproduction

Spermatogenesis

Placental development

Maintain corpus luteum function

Lung surfactant (phospholipid) production

Stimulate myeloid cell differentiation to granular leukocytes

Induce transglutaminases

Crosslinking of proteins (which is necessary for macrophage function, blood clotting, and cell adhesion)

---

acid can support growth and differentiation of various tissues, but it cannot replace retinol or retinal in their support of the visual system.

Vitamin A exists as a provitamin in vegetables, in the form of the yellow pigment, **β-carotene**. This provitamin consists of two molecules of retinal, joined at the aldehyde ends of their carbon chains. It is only about one-sixth as effective a source of vitamin A as retinol on a weight for weight basis. Carotene-like compounds are known as **carotenoids**, which are important sources of vitamin A to herbivores and omnivores. Ingested β-caro-tenes may be oxidatively cleaved to all-trans retinal (vitamin A_1) by **β-carotene dioxygenase** in the intestine and liver of herbivores and omnivores, but much less so (if at all) in carni-vores. Therefore, carnivores are dependent upon preformed vitamin A (i.e., retinal or retinol) in their diets. β-**Carotene dioxygenase** is induced by thyroxine, therefore hypothyroid herbivores or omnivores may develop β-**caro-tenemia**.

In the intestinal mucosa of all mammals, **retinal** is **reversibly** reduced to **retinol** by a specific **zinc-dependent retinal reductase** (or dehydrogenase), which uses either NADH or NADPH as a coenzyme. This enzyme is also present in rod and cone photoreceptors. The **intestine** also converts some **retinal** to **retinoic acid (retin A)**. Most retinol formed in intestinal mucosal cells is esterified to fatty acids, and packaged into **chylomicrons**. These lipopro-teins are then exocytosed by mucosal cells, where they are incorporated into lymph, and then enter the systemic circulation (see Chapter 64). Once acted upon by lipoprotein lipase on the capillary endothelium servicing adipocytes, they become **chylomicron remnants**, and are subsequently removed from the circulation by the liver (with their content of retinol esters). For transport to tissues, stored retinol esters in the liver are hydrolyzed, and the retinol is bound to **retinol-binding protein (RBP)**. In contrast to retinol, retinoic acid is transported in plasma bound to **albumin**.

Following hepatic secretion of the reti-nol-RBP complex into blood, retinal is taken up into target tissues via cell surface receptors. Once inside extrahepatic target cells (e.g., photoreceptors in the retina), retinol is bound by a **cellular retinol-binding protein (CRBP)**. Retinol, retinal, and retinoic acid bind to nuclear proteins, where they are most likely involved in the control of gene expression. Thus, vitamin A appears to behave in a manner similar to **steroid** and **thyroid hormones**, and certain proteins may be under dual control (by both triiodothyronine, T_3, and vitamin A). It controls protein (enzyme) biosynthesis.

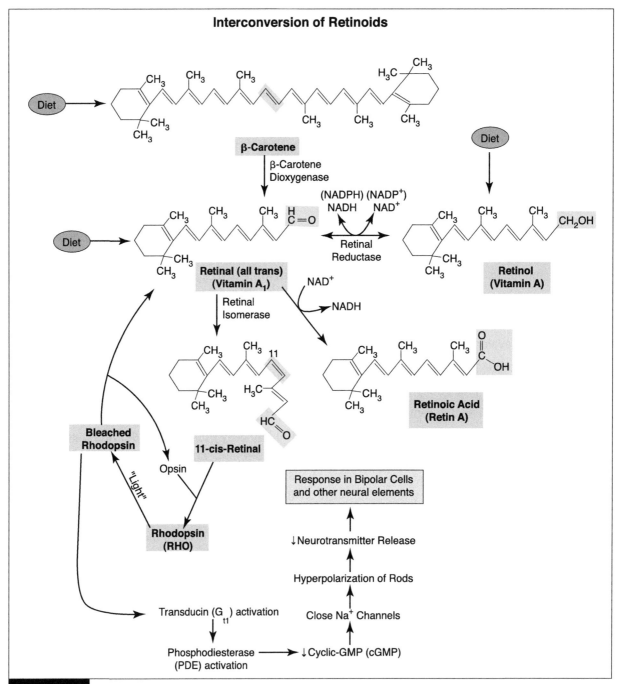

Interconvexsion of Retinoids

Figure 44-1

Vitamin A Toxicity

Hypervitaminosis A will occur after the capacities of these binding proteins have been exceeded. Symptoms generally include skin erythema and desquamation, increased liver size, abdominal pain, nausea, and appetite loss. Excessive quantities of free retinol will damage plasma membranes of the liver and other tissues, as well as hepatic lysosomal membranes. Although retinoic acid is not stored, excessive consumption over time has been reported to cause bone demineralization through increasing the ratio of osteoclasts to osteoblasts. Vitamin A is normally detoxified by the liver, with metabolites being found in both bile and urine. During pregnancy, excessive retinoic acid interferes with segmentation genes, thus causing abnormal fetal development.

Vitamin A and Vision

Photoreceptors of the retina contain discs consisting of lipid embedded with the protein, **opsin**. When this protein combines with 11-cis-retinal in **rod photoreceptors**, it forms **rhodopsin (RHO**, visual purple), the direct recipient of light energy in dim light (non-colored, gray-black vision). As **light bleaches rhodopsin**, opsin and trans-retinal are reformed, and **transducin (G_{t1}, a membrane-bound GTP-binding protein)** is activated. This, in turn, activates a **cyclic-GMP (cGMP)-specific phosphodiesterase** that cleaves cytoplasmic **cGMP** to inactive **5'-GMP** (Fig. 44-2). Cyclic-GMP keeps rod Na^+ channels open, and when **cGMP levels decline, Na^+ channels close**, and the rod becomes **hyperpolarized**. Hyperpolarization of rods leads to a **decrease in transmitter release** from synaptic terminals, which in turn leads to a **response in bipolar cells and other neural elements of the retina that is perceived as light**.

This cascade of reactions generally amplifies light signals, and helps explain the remarkable sensitivity of rod photoreceptors to dim light (which are capable of producing a detectable response to as little as one photon of light). **Cones**, which are used in normal (colored) vision, also contain some retinal, combined with cone iodopsin to form **porphyropsins**, which absorb light of lower energy. Similar steps are thought to be involved in cone activation, however, vitamin A deficiency has a much more marked effect on scotopic than on photopic vision.

In contrast to the semi-structural roles of retinol and retinal in visual excitation, the role of **retinoic acid** is more humoral, particularly in epithelial, osteoid (bone), and gonadal tissues. Indeed, the expression/production of growth hormone itself may be turned-on by a specific retinoic acid hormone receptor complex.

Vitamin A Deficiency

Although the primary signs and symptoms of vitamin A deficiency are usually noted in the visual system, **hypovitaminosis A** also leads to a reduction in mucus-secreting cells and replacement of columnar epithelial cells by thick layers of horny, stratified epithelium in several parts of the organism. This includes

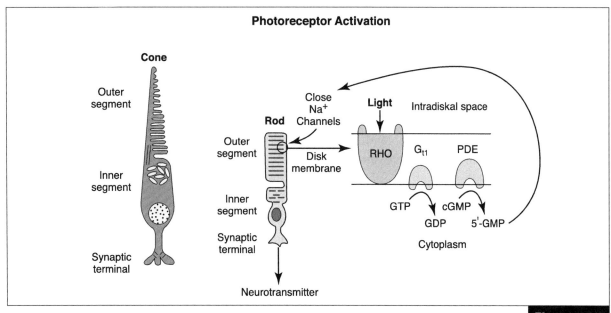

Photoreceptor Activation

Figure 44-2

keratinization of the corneal epithelium, lung, skin, and intestinal mucosa, as well as a drastic reduction in goblet cells within intestinal crypts and villus areas. Although the rates of columnar cell proliferation and migration along the villus are apparently not altered, the synthesis of specific glycoproteins in the intestinal mucosa and liver can be markedly depressed. **Retinoic acid** is needed for gene expression of glycosyl-transferases and fibronectin, as well as trans-glutaminases. Transglutaminases catalyze the crosslinking of proteins by amidating the γ-car-boxyl group of a glutamate residue with an amino group of lysine (which is necessary for macrophage function, blood clotting, and cell adhesion).

Retinoic acid is also essential for the activity of cells in the epiphyseal cartilage, which must undergo a normal cycle of growth, maturation, and degeneration to permit normal bone growth, which is controlled at the epiphyses. **Bone resorption** is retarded in deficiency, although there is apparently no defect in the normal calcification process. Although the mechanism is unclear, it may involve the influence of vitamin A on the osteoclast:osteoblast ratio, since vitamin A up-regulates vitamin D receptors.

Finally, retinoic acid plays a role in **fertility**. In vitamin A deficiency, spermatogenesis is arrested at the spermatid stage in rats, chickens, and cattle, and is reversed with vitamin treatment. Retinoic acid acts on gene transcription in the nucleus of germinal cells of the testes, and CRBP is present in high concentration in the epididymis. Deficiency also interferes with the estrus cycle, placental development, and other aspects of female reproduction in the rat and chicken, causing fetal resorption.   Since several aspects of vitamin A metabolism are **zinc-dependent** (i.e., retinal reductase is a zinc-dependent enzyme), deficiency symptoms of this element are similar to those of vitamin A (see Chapter 49).

In summary, the fat-soluble vitamin A is found in several different forms, including **all-trans-retinal** (vitamin A₁), **11-cis-retinal, rhodopsin** (RHO), **retinol** (the alcohol), and **retin A** (the acid). In animal tissues, and especially the liver, vitamin A is stored as **retinol esters** (with long-chain fatty acids), so that these also become a major dietary source to carnivores. The other major sources of this vitamin are the carotenoid pigments present in green plants. The carotenoids are known as provitamins A, and their conversion to vitamin A involves oxidative fission by **β-carotene dioxygenase** in the intestine of herbivores and omnivores, but not in carnivores. This oxidation forms retinal, the corresponding aldehyde which is the active form of the vitamin in the visual cycles, but which is reduced to the alcohol by a specific zinc-dependent retinal reductase (or dehy-drogenase) for transport and storage in other tissues.

Vitamin A is absorbed from the small intestine with the aid of **bile acids,** and it is incorpo-rated into **chylomicrons** along with several other large lipophilic compounds (see Chapters 62-64). The most readily apparent role of vitamin A is its participation in the detection of **light** by retinal cells. The retina contains two distinct types of photoreceptor cells, cones and rods. In both, the photoreceptor pigments consist of proteins, opsin in rods and iodopsin in cones, linked covalently to retinal. Although similar steps are thought to be involved in both cone and rod activation, vitamin A deficiency has a much more marked effect of scotopic (night) than on photopic (day) vision.

Retinoic acid, which is not stored in the body, plays a central role in bone remodeling, reproductive biology, and the maintenance and differentiation of epithelial tissues. In the absence of this form of the vitamin, there is a reduction in mucus-secreting cells, seen most often in the corneal epithelium, lung, skin, and intestinal mucosa. Retinoic acid is also essential to cells of the epiphyseal cartilage as they undergo normal cycles of growth, maturation, and degeneration. Bone resorption is retarded in vitamin A deficiency, and exacerbated in vitamin A toxicity. Retinoic acid also plays an important role in spermatogenesis, ovarian function, fetal and placental development. Since several aspects of vitamin A metabolism are zinc-dependent, several of the signs and symptoms of zinc-deficiency are also seen with vitamin A deficiency.

OBJECTIVES

- Discuss the cause-effect relationship between steatorrhea and a fat-soluble vitamin deficiency.

- Show why retinoic acid can support growth and differentiation of epithelial tissues, and why it cannot replace retinol or retinal in their support of the visual system.

- Identify multiple actions of the different vitamin A oxidation states, and discuss relationships between zinc and vitamin A deficiency.

- Explain why dietary carotenoids are an unsatisfactory source of vitamin A to carnivores, and why hypothyroidism might induce a state of β-carotenemia in herbivores and omnivores.

- Indicate how vitamin A is absorbed from the intestine, carried in plasma, taken-up by target cells and bound intracellularly. Also discuss the cellular mechanisms of vitamin A action.

- Provide reasoning for each of the signs and symptoms of vitamin A toxicity, as well as those for vitamin A deficiency.

- Explain the participation of vitamin A in the cascade of reactions leading to amplification of light signals in retinal photoreceptors.

- Indicate how vitamin A is thought to be involved in bone resorption (see Chapter 45).

- Understand how retinoic acid toxicity affects the pregnant animal.

- Explain common features of vitamin A, steroid and thyroid hormone action.

QUESTIONS

1. **β-Carotene consists of two molecules of:**
   a. Rhodopsin.
   b. Retinol.
   c. Retin A.
   d. Trans-retinal.
   e. Transducin.

2. **Which one of the following is not stored in the body?**
   a. Retinoic acid
   b. Retinol
   c. Retinaldehyde (retinal)
   d. Vitamin D
   e. Vitamin E

3. **Vitamin A is best associated with the transport and action of which hormone?**
   a. Insulin
   b. Thyroxin
   c. Aldosterone
   d. Calcitonin
   e. Follicle Stimulating Hormone

4. **When rhodopsin is bleached by light in rod photoreceptors, all of the following occur, EXCEPT:**
   a. Transducin becomes activated.
   b. cGMP-dependent phosphodiesterase activity increases.
   c. Na^+ channels close.
   d. Rods depolarize.
   e. Neurotransmitter release from rods declines.

5. **Vitamin A deficiency is associated with all of the following, EXCEPT:**
   a. Keratinization of the corneal epithelium.
   b. Decreased hepatic glycoprotein formation.
   c. Enhanced bone resorption.
   d. Decreased spermatogenesis.
   e. Night blindness.

6. **Vitamin A toxicity has the greatest affect on which organ?**
   a. Brain
   b. Kidney
   c. Liver
   d. Lung
   e. Heart

6. c
5. c
4. d
3. b
2. a
1. d

Chapter 45

Vitamin D

Overview

- Vitamin D can be synthesized in the skin, and it helps to facilitate intestinal Ca^{++} absorption.
- Ergosterol, which occurs in plants, is a vitamin D precursor.
- The active form of vitamin D ($1,25(OH)_2D$) is a steroid hormone.
- Renal activation of vitamin D is regulated by several endocrine factors.
- Liver and/or kidney disease can result in a $1,25(OH)_2D$ deficiency.
- Parathyroid hormone deficiency can be treated through vitamin D supplementation.
- Cholecalciferol rodenticide poisoning can cause symptoms of vitamin D toxicity.
- Glucocorticoids and calcitonin can be used to reverse symptoms of vitamin D toxicity.

Canine **rickets** was induced nearly 100 years ago through dietary manipulation, which could be cured with cod liver oil. The factor responsible was not vitamin A, but **vitamin D**.

Unlike other fat-soluble vitamins, vitamin D can be synthesized by mammalian tissues (i.e., the malpighian layer of the epidermis in herbivores and omnivores); however, this source of the vitamin is not sufficient to meet body needs in dogs and strict carnivores, therefore they depend upon dietary intake. It is generally not produced by microbes, and its mechanism of action is primarily that of a steroid hormone.

Vitamin D_3 (**cholecalciferol**) is produced in the skin from **7-dehydrocholesterol** by a non-enzymatic process, catalyzed by **ultraviolet (UV) light (Fig. 45-1)**. Although sunlight-activated vitamin D_3 formation may be an important source for many animals, this form of the vitamin can also be obtained via the diet. Ultraviolet light is partially filtered out by hair coat, pigmented layers of the skin, and window glass; however, the nose of hairy mammals

may be a particularly important location for D_3 formation.

Ergosterol occurs in plants, milk and yeast, and differs from 7-dehydrocholesterol at its side chain, which contains an extra methyl group at C-24. Ultraviolet (UV) radiation cleaves the B ring of both ergosterol and 7-dehydrocholesterol yielding **ergocalciferol** and **cholecalciferol**, respectively. Irradiation of milk and yeast is a commercial means of producing D_2 from ergosterol, and **dihydrotachysterol (DHT)** is a synthetic analog of D_2. Although the antirachitic potencies of D_2 and D_3 (once properly hydroxylated) may differ between animal species, the following discussion will use vitamin D as a collective term for the two vitamers.

Vitamin D is slowly released from skin into plasma, where it is normally bound by **vitamin D binding protein** (**DBP**; a liver-derived α-globulin known as **transcalciferin**), and subsequently cleared from the circulation by the liver (**Fig. 45-2**).

Absorption of vitamin D in the small intestine

Copyright © 2015 Elsevier Inc. All rights reserved.

Vitamin D Formation

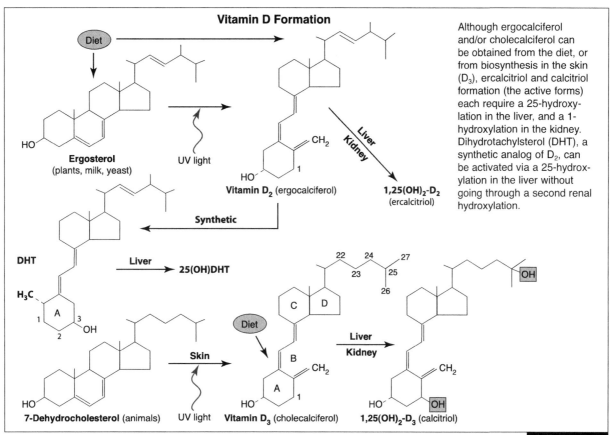

Although ergocalciferol and/or cholecalciferol can be obtained from the diet, or from biosynthesis in the skin (D₃), ercalcitriol and calcitriol formation (the active forms) each require a 25-hydroxylation in the liver, and a 1-hydroxylation in the kidney. Dihydrotachylsterol (DHT), a synthetic analog of D₂, can be activated via a 25-hydroxylation in the liver without going through a second renal hydroxylation.

Figure 45-1

Vitamin D Absorption and Distribution

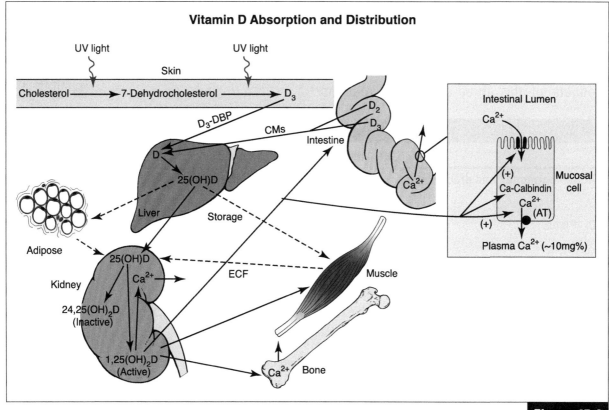

Figure 45-2

requires the presence of **bile acid micelles** (see Chapter 62). Like other fat-soluble vitamins, it is incorporated into **chylomicrons (CMs)** in mucosal cells of the small intestine, and then transferred to lymph (see Chapter 64). Chylomicron remnants containing fat-soluble vitamins are removed by the liver (like the D_3-DBP complex from skin), where D is hydroxylated (in a nonregulated fashion) by a **25-hydroxylase**. This reaction also occurs (at low efficiency) in the intestine. Calcifediol and/or ercalcifediol (25(OH)D) are normally the major forms of this vitamin in blood, and the major storage forms in liver, adipose, and skeletal muscle tissue.

In proximal renal tubular epithelial cells of the **kidney**, (and to a lesser extent in the placenta and in bone), **25(OH)D** is **activated** through **hydroxylation** at **position 1 of the A ring** by another mitochondrial enzyme, **1α-hydroxylase (Fig. 45-3)**. This conversion involves a complex monooxygenase reaction system requiring **NADPH**, **Mg^{++}**, molecular **oxygen (O$_2$)**, and at least three additional components: **1)** a **flavoprotein**, renal ferredoxin reductase (see Chapter 40) ; **2)** an **iron-sulfur protein**, renal ferredoxin; and **3)** **cytochrome P$_{450}$**. The products of this three-component reaction

sequence produce the most potent form of vitamin D, **1,25-dihydroxycholecalciferol (1,25(OH)$_2$D$_3$** or **1,25-DHC; calcitriol;** and **ercalcitriol (1,25(OH)$_2$D$_2$); Table 45-1).**

The activity of **1α-hydroxylase** is positively regulated by several factors, including **low serum ionized Ca^{++} and PO$_4^{\equiv}$** concentrations, **parathyroid hormone (PTH;** needed to increase the serum ionized Ca^{++} concentration), **growth hormone (GH;** Ca^{++} needed for body growth), **placental lactogen (PL;** Ca^{++} needed for fetal

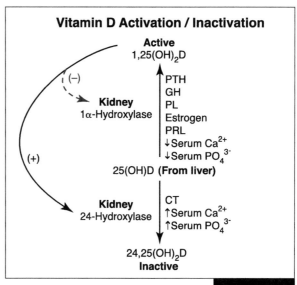

Figure 45-3

Table 45-1	
Various Forms of Vitamin D	
Active hormone* and precursors	**Generic Name**
D_1	(Impure preparation of D_2)
D_2 (Plants)	Ergocalciferol (Calciferol)
25(OH)D$_2$	Ercalcifediol
*1,25(OH)$_2$D$_2$	Ercalcitriol
Reduced D$_2$ (DHT)	Dihydrotachysterol
*25(OH)DHT	25-Hydroxydihydrotachysterol
D_3	Cholecalciferol
25(OH)D$_3$	Calcifediol
*1,25(OH)$_2$D$_3$ (1,25-DHC)	Calcitriol (1,25-Dihydroxycholecalciferol)
Inactive: 24,25(OH)$_2$D (major); 25,26(OH)$_2$D (minor); 1,25,26(OH)$_3$D (minor); Others	

growth), **estrogen** (Ca^{++} needed during pregnancy), and **prolactin** (**PRL**; Ca^{++} needed during lactation). In addition to estrogen, progestins and androgens cause marked increases in the renal 1α-hydroxylase activity of **ovulating birds**.

In general, 1,25(OH)$_2$D serves to enhance serum Ca^{++} levels through facilitating intestinal Ca^{++} absorption, and through promoting the action of PTH on bone and the kidney. With the exception that PTH decreases while 1,25(OH)$_2$D increases renal PO$_4^{\equiv}$ retention, PTH and 1,25(OH)$_2$D exert similar actions, and are generally synergistic with each other. Therefore, this lipid-soluble steroid, which can be absorbed from the intestinal tract, is used to treat patients with PTH deficiency. Since PTH is a protein, if given orally it is destroyed in the digestive tract.

Calcitriol is also an important regulator of its own production. High levels of **1,25(OH)$_2$D inhibit renal 1α-hydroxylase** in a negative-feedback fashion, yet stimulate formation and activity of a renal **24-hydroxylase** (that leads to formation of **24,25-(OH)$_2$D, an inactive** by-product of vitamin D). Other factors known to stimulate activity of 24-hydroxylase include **high** circulating levels of serum ionized **Ca^{++}** and **PO$_4^{\equiv}$**, as well as **thyrocalcitonin** (**CT**). Additional hydroxylations of this molecule are known to occur at positions 23 and 26. Although over 20 metabolites have been found, only **1,25(OH)$_2$D** has been shown to exhibit physiologic activity.

The effects of **1,25(OH)$_2$D** on the transfer of Ca^{++} (and possibly **PO$_4^{\equiv}$** across the **intestinal mucosa** requires special consideration, for it is this location where vitamin D is thought to exert its major control. **It is the only hormone known to effectively promote translocation of Ca^{++} against its concentration gradient in the intestine.**

Although **30-80% of ingested Ca^{++} can be absorbed from the digestive tract**, active absorption occurs primarily in the upper part of the small intestine by a multi-step mechanism

(see **Fig. 45-2**). Calcium is not ionized at neutral pH, thus **gastric acid** (HCl) helps to solubilize calcium salts and free Ca^{++} from bound protein, thus permitting intestinal absorption. **1,25(OH)$_2$D** plays a critical role in duodenal Ca^{++} absorption by opening specific **Ca^{++} channels** in mucosal cell membranes, and stimulating transcription of certain proteins, including **calbindin** and **Ca^{++}-ATPase**. Through these actions 1,25(OH)$_2$D can increase the efficiency of intestinal Ca^{++} influx from the lumen. This phenomenon, known as **transcaltachia**, occurs over a period of seconds to minutes, whereas the effects on transcription take hours. The gradient for Ca^{++} influx is secondarily maintained by the buffer action of calbindin, which as an intracellular Ca^{++}-binding protein exhibits significant homology with calmodulin, and mysoin light chain. It is important for the function of these cells that the concentration of free ionized Ca^{++} in the cytoplasm be maintained at a low concentration of about 10^{-6}M, thus the buffer action of calbindin fulfills this role. The active transport (AT) of Ca^{++} out of mucosal cells is the result of Ca^{++}-ATPase activity, which is also stimulated by **1,25(OH)$_2$D**.

Actions of **1,25(OH)$_2$D** on bone also deserve mention since both PTH and **1,25(OH)$_2$D** have receptors on (PTH) or in (**1,25(OH)$_2$D**) mature **osteoblasts**, but not on/in **osteoclasts**. These two hormones stimulate osteoblasts to produce **cytokines** that accelerate maturation of osteoclasts in a paracrine fashion. Local release of lysosomal enzymes from osteoclasts and end products of glycolysis then create a local environment that favors **bone dissolution**. This system appears to be synergistic, for alone PTH cannot account for its overall action of increasing the serum ionized Ca^{++} concentration. **1,25(OH)$_2$D** also promotes release of a **Gla protein, osteocalcin**, from osteoblasts (see Chapter 47). This protein binds to hydroxyapatite, preventing further mineralization.

Vitamin D and its metabolites are largely excreted into **bile**, with only about 3% normally appearing in urine. Since these are steroids, many are reabsorbed (and thus conserved) by the intestine.

Vitamin D Toxicity

As with vitamin A, vitamin D excess can result in symptoms of **toxicity**. Hypercalcemia and hyperphosphatemia, deposition of Ca^{++} in soft tissues (especially the kidney, heart, lung, and vasculature), hypercalciuria, and kidney stones have been described. Reptile pets seem particularly susceptible. Vitamin D toxicity can also occur, for example, when **cholecalciferol rodenticides** are consumed by dogs and cats, or when too much vitamin D is administered to animals with hypoparathyroidism. Glucocorticoids and CT are sometimes administered to reverse symptoms of toxicity. **Glucocorticoids** interfere with the mechanism of 1,25-DHC action on the intestine and kidney, and **CT** generally exerts opposite actions to both PTH and 1,25-DHC, including activation of renal 24-hydroxylase.

Vitamin D Deficiency

Vitamin D deficiency symptoms include abnormal bone mineralization and deformities (i.e., **rickets** in young animals, and **osteomalacia** in adults), hypocalcemia, and high circulating titers of PTH. Additionally, since vitamin D receptors are present on hair follicles, vitamin D deficiency can promote hair loss. In addition to **hereditary**, **dietary**, and **behavioral causes** (e.g., lack of UV light), **liver** and/or **kidney disease** may also result in a $1,25(OH)_2D$ deficiency. Also, **lead (Pb)** appears to block intestinal $1,25(OH)_2D$-stimulated Ca^{++} absorption, yet vitamin D supplementation enhances Pb uptake. As with vitamin A, normal hepatic stores of vitamin D are generally thought to be capable of supporting animals on vitamin-deficient diets for several months. Therefore, deficiency symptoms usually manifest themselves slowly.

25(OH)DHT, the synthetic analog of D_2, appears to be active in both the intestine and bone of nephrectomized rats. Comparison of the structures of DHT and $1,25(OH)_2D$ (**Fig 45-1**) shows that ring A of DHT is rotated so as to place its 3-hydroxyl group in approximately the same geometrical position as the 1-hydroxyl group of $1,25(OH)_2D$. It seems, therefore, that 25(OH)DHT interacts with receptor sites of $1,25(OH)_2D$ without undergoing 1-hydroxylation, thus bypassing renal mechanisms of metabolic control.

In summary, vitamin D conforms to the definition of both a **hormone** and a **vitamin,** and it plays an important role in control of the Ca^{++} concentration of the extracellular fluid compartment. Whether as a hormone produced in the skin, or as a vitamin provided in the diet, vitamin D has to be chemically modified by two hydroxylations, which occur sequentially in the liver and the kidneys, before it can play an active role in control of Ca^{++} distribution. Discovery of the necessity for hepatic and renal hydroxylations of cholecalciferol led to explanations for a number of clinical problems and their more satisfactory treatments.

Ergocalciferol (D_2) and cholecalciferol (D_3) are seco-steroids in which their B rings have been broken by fission of a carbon-carbon bond. In this case fission caused by UV light on ergosterol of plants producing D_2, and that on 7-dehydrocholesterol in the epidermis of the skin producing D_3.

Once D_2 and/or D_3 reach the liver, the first step in sequential activation occurs, namely hydroxylation by a monooxygenase (hydroxylase) to produce $25(OH)D_2$ and/or $25(OH)D_3$. These compounds, which can be stored in the liver and also in muscle and adipose tissue, are next transported to the kidneys where further hydroxylation takes place in the 1-position of each to produce the active hormones ($1,25(OH)_2D_2$ and/or $1,25(OH)_2D_3$), or in the 24-position of each to produce the inactive hormones ($24,25(OH)_2D_2$ and/or $24,25(OH)_2D_3$). In normal animals the relative plasma concentrations of the three hydroxylated forms (**25-hydroxy-, 24,25-dihydroxy-, and 1,25-dihydroxy-**) are about **100:10:1.**

The active form of vitamin D serves to enhance serum Ca^{++} levels through **facilitating intestinal Ca^{++} absorption,** and through **promoting the action of PTH on bone and the kidneys.** Since most of the actions of 1,25(OH)$_2$D and PTH are similar, vitamin D is frequently used to treat patients with PTH deficiency.

OBJECTIVES

- Identify normal sources of vitamin D$_2$, vitamin D$_3$ and DHT for animals, and discuss the hepatic metabolism of each.

- Discuss the intestinal absorption and plasma transport of vitamin D (see Chapters 62 & 64).

- Outline and explain the reasoning behind the renal endocrine control of vitamin D hydroxylation.

- Identify and discuss steps involved in vitamin D-stimulated intestinal Ca^{2+} absorption.

- Distinguish between the various active and inactive forms of vitamin D.

- Identify and describe primary physiologic roles for vitamin D in bone and in the kidneys.

- Explain why vitamin D is used to treat patients with PTH deficiency.

- Identify the signs and symptoms of vitamin D toxicity, and explain why CT and glucocorticoids can be used to treat this condition.

- Explain the signs and symptoms of vitamin D deficiency, and also explain why 25(OH)DHT is active in the bone and intestine of nephrectomized rats.

QUESTIONS

1. Ergosterol:
   a. Is a plant steroid.
   b. Differs from 7-dehydrocholesterol only in the orientation of its ring structure.
   c. Is a source of vitamin A in animals.
   d. Is vitamin D$_3$.
   e. Is the most active form of vitamin D in animals.

2. Calcifediol:
   a. Is usually the major form of vitamin D found in the blood of animals.
   b. Is 1,25(OH)$_2$D$_3$, the most active form of vitamin D.
   c. Is normally hydroxylated in the kidney through the action of a 25-hydroxylase.
   d. Formation in the liver is closely regulated via the action of several different hormones.
   e. Is a polypeptide.

3. Renal 1α-hydroxylase, the activity of which is needed in the activation of Vitamin D, is thought to be stimulated by all of the following, EXCEPT:
   a. Parathyroid hormone.
   b. Prolactin.
   c. Hypocalcemia.
   d. Calcitonin.
   e. Hypophosphatemia.

4. Vitamin D can be used effectively to treat patients with which one of the following hormone deficiencies?
   a. Insulin
   b. Calcitonin
   c. Cortisol
   d. Thyroid Hormone
   e. Parathyroid Hormone

5. The active form of vitamin D is the only hormone known to "effectively" promote translocation of Ca^{++}:
   a. Against its concentration gradient in the renal tubules.
   b. Against its concentration gradient in the intestine.
   c. Against its concentration gradient, into bone.
   d. Into smooth muscle tissue.
   e. Out of bone.

6. Which one of the following elements is associated with a decline in 1,25(OH)2D-stimulated intestinal Ca^{++} absorption?
   a. Iron
   b. Copper
   c. Lead
   d. Magnesium
   e. Potassium

7. 25(OH)DHT could be used to effectively treat vitamin D-deficient animals with renal dysfunction:
   a.   True
   b.   False

7. a
6. c
5. b
4. e
3. d
2. a
1. a
ANSWERS

Chapter 46

Vitamin E

Overview

- There are several naturally occurring forms of vitamin E in plants.
- Vitamin E is passively absorbed in conjunction with other lipid-soluble vitamins by the intestinal tract, and subsequently packaged into chylomicrons.
- Vitamin E is the first and selenium the second line of defense against peroxidation of lipids contained in cell membranes.
- Tocopherols act as antioxidants by breaking free-radical chain reactions.
- Vitamin C helps to regenerate the active form of vitamin E.
- Vitamin E deficiency can result in erythrocyte fragility, muscular degeneration, steatitis, retinopathy, and reproductive failure.
- Vitamin E excess appears to be non-toxic.

There are several naturally occurring forms of **vitamin E** (**tocopherol**) in plants, and all are isoprenoid-substituted 6-hydroxychromanes (tocols). α-**Tocopherol**, which is methylated in the 5, 7, and 8 position, has the widest natural distribution and greatest physiologic activity; however, the β, γ, and Δ-**tocopherols** are also of dietary significance (**Fig. 46-1**). The β and γ forms are used as natural preservatives in pet foods. With the exception of fat and liver tissue, animal foods are generally considered to be poor sources of this vitamin.

Vitamin E and **selenium** (**Se**) act synergistically in protecting membranes against lipid peroxidation (see below). Vitamin E is absorbed in conjunction with other lipid-soluble vitamins across the intestine, thus requiring the presence of bile acid micelles (see Chapter 62). Malabsorption due to intestinal resection or disease, or that secondary to liver disease, are common causes of vitamin E deficiency. This vitamin is initially distributed from the intestine through the lymphatic circulation as a part of chylomicrons (CMs), from where it is said to "rub off" onto cells, such as erythrocytes. From the liver onward, the distribution of vitamin E follows that of triglyceride and other lipids, via lipoproteins (particularly CMs and liver-derived VLDL; see Chapters 64 and 65) to adipose tissue and cell membranes. Vitamin E is thought to become more evenly distributed throughout the body than the other fat-soluble vitamins, with highest concentrations found in plasma, liver, and adipose tissue. As with the other fat-soluble vitamins, plasma levels of vitamin E are not, however, a good index of total body stores.

Vitamin E appears to be the first and selenium (Se) the second line of defense against peroxidation of unsaturated fatty acids (UFAs) contained in cell membrane phospholipids. Cholesterol and phospholipids in the plasma membrane, as well as those in subcellular membranes (e.g., mitochondria, endoplasmic

Copyright © 2015 Elsevier Inc. All rights reserved.

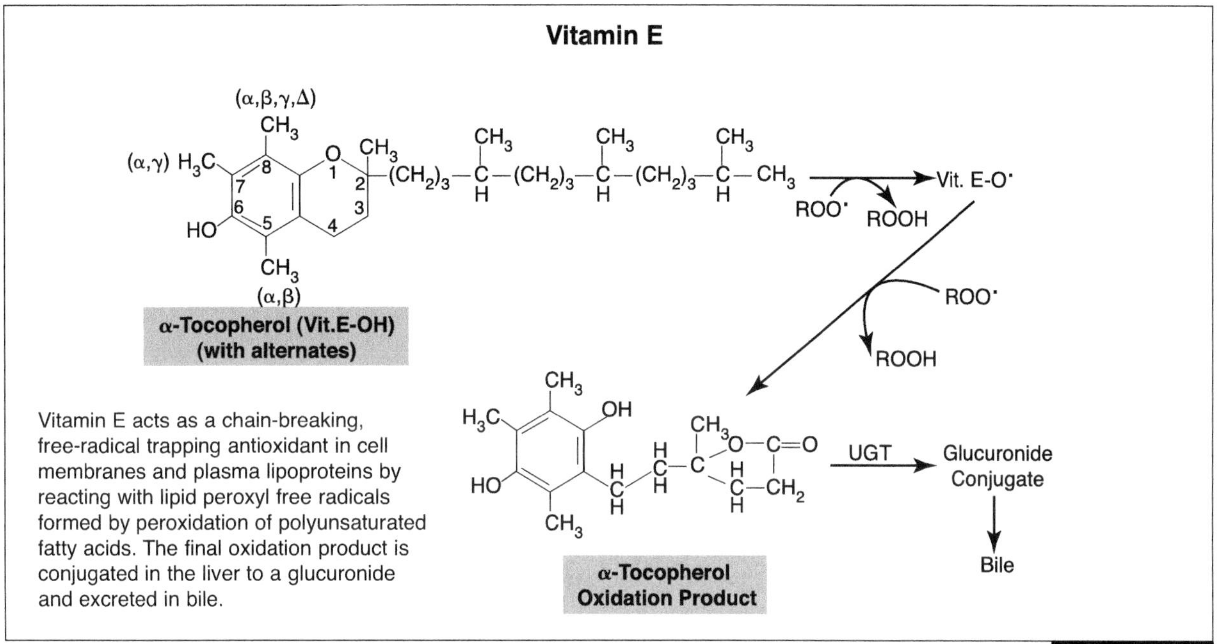

Vitamin E

α-Tocopherol (Vit.E-OH)
(with alternates)

Vitamin E acts as a chain-breaking, free-radical trapping antioxidant in cell membranes and plasma lipoproteins by reacting with lipid peroxyl free radicals formed by peroxidation of polyunsaturated fatty acids. The final oxidation product is conjugated in the liver to a glucuronide and excreted in bile.

α-Tocopherol
Oxidation Product

Figure 46-1

reticulum, etc.), possess high affinities for α-tocopherol. This works against injury to cell membranes, as in red blood cell fragility and the muscular degeneration of animals. It also aids in the normal functioning of the seminiferous epithelium (and therefore sperm production), and assists with implantation (thus sustaining the fetus in the uterus). The **antioxidant** action of tocopherol is effective at high oxygen levels, and thus tends to be concentrated in lipid structures exposed to high partial pressures of O_2 (e.g., membranes of the respiratory tree, the retina, and erythrocytes; see **deficiency symptoms below**).

Tocopherols act as antioxidants by **breaking free-radical chain reactions** through transfer of a phenolic hydrogen (H) to a **peroxyl free radical** of an UFA (**H + ROO· —> ROOH**; **Fig. 46-1**). The need for vitamin E, therefore, partially depends upon the dietary intake of UFAs. Animals consuming large quantities of fish-based diets (which normally contain high amounts of UFAs) without adequate antioxidants such as vitamins C or E, can experience peroxidation of body fat and fat necrosis. Phenoxy free radicals (**Vit. E-O·; Fig. 46-2**) formed may react with the reduced

form of **vitamin C** to regenerate tocopherol, or with other peroxyl free radicals (**ROO·**), thus oxidizing the chromane ring and side chain. This **oxidation product** is typically removed from the circulation by the liver, and through the action of **UDP-glucuronosyltransferase** (**UGT**; see Chapter 29), conjugated to a **glucuronide** and excreted into bile. Elimination of vitamin E in this manner generally requires that new tocopherol be obtained through the diet.

Glutathione peroxidase (**GP**), of which **Se** is an integral component (see Chapters 30 and 51), constitutes the second line of defense against hydroperoxides which could otherwise damage membranes, and inhibit some membrane-bound enzymes (see **Fig. 46-2**). Following the action of **phospholipase A$_2$** on membrane-bound phospholipid, the UFA in position 2 is released into the cytosol where it is now available to enter into eicosanoid biosynthesis (see Chapters 68 and 69). Peroxides generated from these UFAs can now be acted upon by cytoplasmic enzymes such as **GP** and **catalase**, which are capable of converting H_2O_2 to O_2 and H_2O. Catalase is a rather ubiquitous enzyme, present in most aerobic tissues.

Elimination of Phenoxy Free Radicals

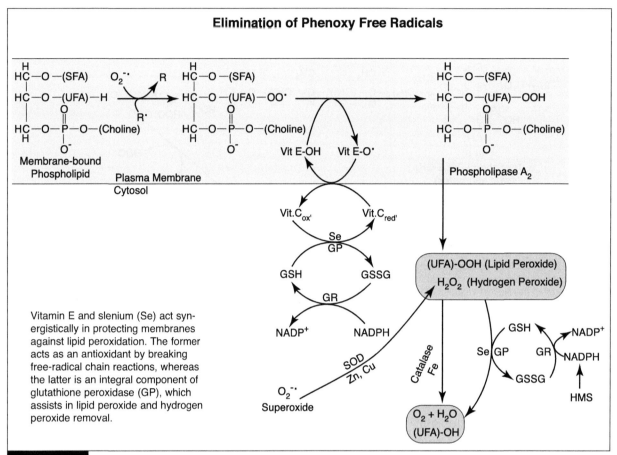

Vitamin E and slenium (Se) act synergistically in protecting membranes against lipid peroxidation. The former acts as an antioxidant by breaking free-radical chain reactions, whereas the latter is an integral component of glutathione peroxidase (GP), which assists in lipid peroxide and hydrogen peroxide removal.

Figure 46-2

However, patients with a catalase deficiency reportedly show few toxic symptoms, presumably because of the redundant actions of GP, particularly in erythrocytes.

For its activity, GP requires the reduced form of **glutathione (GSH)**. Through continual reduction of the oxidized form of glutathione (GSSG) to GSH (accomplished by the FAD-containing enzyme, **glutathione reductase (GR)**, and **NADPH** (produced by the hexose monophosphate shunt (HMS))), GSH again becomes available in the cytosol to protect against a buildup of H_2O_2.

Another source of cytoplasmic H_2O_2 evolves through the action of **superoxide dismutase (SOD)** on oxygen free radicals ($O_2^{-\cdot}$). This zinc- and copper-containing enzyme is present in all aerobic tissues, as well as in erythrocytes (see Chapter 30).

An atom or molecule with one or more unpaired electrons is a free radical, and free radicals are highly reactive because of their tendency to acquire electrons from other substances. Not all reactive oxygen species are, however, free radicals (e.g., singlet oxygen (O_2), and H_2O_2). Three important reactive species that are known to damage tissues are $O_2^{-\cdot}$, H_2O_2, and the **hydroxyl free radical (OH·; Table 46-1**). Although the OH· is highly reactive, it is normally short-lived. Additional sources of reactive species include **xanthine oxidase (XO**; see Chapter 17), which may generate superoxides during reperfusion injury in ischemic organs (see **Case Study #3**), and **cyclooxygenase** and **lipoxygenase** (see Chapters 68 and 69), important control enzymes in eicosanoid pathways that can give rise to OH· and peroxyl radicals. **Uric acid**, the product of XO activity, may itself possess antioxidant activity.

Table 46-1
Reactive Oxygen Species and Their Antioxidants

Abbreviation	Reactive Species	Antioxidants
O_2		Singlet oxygen
Vit A & E, β-carotene		
$O_2^{-\cdot}$	Superoxide free radical	SOD, Vit E, β-carotene
ROO·	Peroxyl free radical	Vit E & C
H_2O_2	Hydrogen peroxide	Catalase, GP

The metabolism of xenobiotics via the cytochrome P_{450} system can also give rise to free radicals, particularly in liver tissue. Because these molecules are so reactive, they generally act close to their point of origin. Therefore, many cell structures become vulnerable to these agents, including structural proteins, enzymes, membranes, and nucleic acids, which can result in mutation and apoptosis.

Vitamin E Deficiency

Vitamin E deficiency can result from severe **fat malabsorption** with consequent steatorrhea, some forms of **cholestatic liver disease**, **abetal-ipoproteinemia**, and **intestinal resection**. In experimental animals vitamin E deficiency has resulted in fetal resorption, premature infants have been born with inadequate reserves, and in males testicular atrophy has occurred.

Excessive lipid peroxidation of membranes and other sites of fat accumulation accounts for most of the symptoms associated with **vitamin E deficiency**. The most clear-cut example is **enhanced erythrocyte fragility**, where RBCs exhibit a marked change in morphology and become easily destroyed. Vitamin E is also essential for the development and maintenance of normal **nerve** and **muscle** cell activity. Either vitamin E or selenium deficiency can result in a massive influx of Ca^{++} into cells; mitochondria become loaded with this element, and reduce their ATP output. This mineral influx results in **muscular degeneration**, and gives muscle a characteristic appearance (i.e., "**white muscle disease**"), which is usually more prominent in young animals. Myocardial involvement with this disease may result in sudden death.

White muscle disease is sometimes confused with the **muscular dystrophies** of animals, which are hereditary degenerative diseases of skeletal muscle. Dystrophic muscles usually contain fewer fibers, an increase in the number and size of nuclei, and myofibrillar degeneration without effective regeneration.

Another outcome of the lack of antioxidant action of vitamin E, and which occurs in its deficiency, is the accumulation of "**lipofuscin**" or "ceroid pigment" granules in many tissues, including the CNS, lungs, kidneys, adipocytes, and muscle. These granules contain oxidized unmetabolizable lipids that have partially crosslinked with protein or peptides to form a hard globule that cannot be disposed of by the organism. These granules normally accumulate with age, and this accumulation is inhibited by a high vitamin E intake, at least in mice. Fatty tissue inflammation (i.e., "**steatitis**"), is also associated with vitamin E deficiency.

Animals with **cholestasis** (e.g., bile duct obstruction) absorb fat-soluble vitamins poorly. In chronic conditions **neuromuscular damage** occurs that can be somewhat alleviated through parenteral administration of vitamin E. Additionally, **retinopathy** can occur in vitamin E deficiency upon exposure to high oxygen tensions. Again, this condition has been reversed with vitamin E administration.

Studies in several animal species have shown that in males, vitamin E deficiency results first in **sperm immotility**, then in degeneration of the seminiferous epithelium, and then **cessation of sperm production**. In females there is a **failure of uterine function** in vitamin E deficiency, with a lack of development of the vasculature that would allow the conceptus to implant in the uterine wall. Although vitamin E supplementation can help to reverse these symptoms, **Se** does not effectively prevent fetal resorption in rats, nor encephalomalacia in chickens, and thus, is not a complete substitute for vitamin E.

Although several health problems have been associated with vitamin E deficiency, as discussed above, high intakes of this vitamin do not appear to be as debilitating as do those for vitamins A, D and K. Vitamin E, however, can act as an **anticoagulant**, therefore hypervitaminosis E may increase the risk of bleeding.

OBJECTIVES

- Identify locations in the body normally containing the highest concentrations of vitamin E.

- Explain why vitamin E is considered to be the first, and Se the second line of defense against membrane lipid peroxidation.

- Understand why animals consuming fish-based diets have an increased need for vitamin E.

- List the reactive oxygen species known to damage tissues and their antioxidants.

- Recognize how vitamin C helps to regenerate the active form of vitamin E.

- Explain how vitamin E is metabolized and excreted by the body.

- Summarize the redundant activities of catalase and glutathione peroxidase, and show how H_2O_2 is produced in aerobic tissues.

- Explain pathophysiologic findings related to vitamin E deficiency.

- Compare vitamin E toxicity symptoms to those of the other fat-soluble vitamins.

QUESTIONS

1. **Which one of the following elements acts synergistically with vitamin E in protecting membranes against lipid peroxidation?**
   a. Cobalt
   b. Zinc
   c. Lead
   d. Selenium
   e. Oxygen

2. **The need for vitamin E partially depends upon the dietary intake of:**
   a. Protein.
   b. Starch.
   c. Sucrose.
   d. Iron.
   e. Unsaturated fatty acids.

3. **Vitamin E is a (an):**
   a. Ergosterol.
   b. Retinoid.
   c. Tocopherol.
   d. Menaquinone.

4. **An atom or molecule with one or more unpaired electrons is a (an):**
   a. Free radical.
   b. Enzyme.
   c. Cytochrome.
   d. Free fatty acid.
   e. Phospholipid.

5. **Vitamin E deficiency is associated with all of the following, EXCEPT:**
   a. Muscular degeneration.
   b. Cholestasis.
   c. Lipofuscin accumulation.
   d. Retinopathy.
   e. Diabetes insipidus.

6. **Which one of the following can regenerate vitamin E?**
   a. SOD
   b. GSH
   c. Vitamin C
   d. Catalase

7. **Hydrogen peroxide is normally acted upon (i.e., detoxified) by:**
   a. Vitamins E and C.
   b. GSH and catalase.
   c. SOD and glutathione reductase.
   d. Superoxides and hydroxyl free radicals.

7. b
6. c
5. e
4. a
3. c
2. e
1. d

ANSWERS

Chapter 47

Vitamin K

Overview

- Vitamin K, which is needed for blood clotting, can be derived from plants, bacteria, and animal tissues.
- A water-soluble form of vitamin K can be converted to the active derivative, hydroquinone.
- The biologic half-life of vitamin K is shorter than the other fat-soluble vitamins.
- The liver is the main repository of vitamin K.
- Vitamin K is required for the hepatic postsynthetic transformation of several protein clotting factors.
- Vitamin K helps to facilitate the γ-carboxylation of glutamate residues, which in turn chelate Ca^{++}.
- A vitamin K cycle exists in the endoplasmic reticulum of liver cells.
- An important therapeutic use of vitamin K is as an antidote in poisoning by dicoumarol or warfarin.
- Vitamin K plays a role in bone metabolism, as well as in the renal reabsorption of Ca^{++}.
- Tissue distributions of this vitamin indicate that its actions are diversified.

Vitamin K ("**koagulation**"), a fat-soluble vitamin required for blood clotting, was discovered in Germany in 1929. The pure **vitamin (K_1)** was obtained from alfalfa in 1939, and later it was realized that bacteria synthesized a second form (**K_2**), with a more unsaturated side chain (**Fig. 47-1**). The **menaquinones** are a series of polyprenoid unsaturated forms of **vitamin K_2** found in animal tissues, and synthesized by bacteria (actinomycete microbes) inhabiting the digestive tract. **Vitamin K_3 (menadione)** is a water-soluble form synthesized commercially, and also capable of being converted to the active form (hydroquinone; HQ).

Vitamins K_1 and **K_2** are naturally occurring polyisoprenoid-substituted **napthoquinones**, that are absorbed from the small intestine with variable efficiency (10-80%). The amount of fat in the diet, as well as bile acid availability through the liver, will influence absorptive efficiency. Vitamins K_1 and K_2 are distributed much like the other fat-soluble vitamins, initially in chylomicrons (CMs), which enter the bloodstream through the lymphatics, then later partially in VLDL and LDL (see Chapters 64 and 65). Vitamin K_3, being water-soluble, can be absorbed in the absence of bile acids, passing directly from intestinal mucosal cells into the hepatic portal circulation.

The **liver** is the main repository of these vitamins, although there appears to be a rapid turnover; consequently, **body pools are thought to be small**. This high rate of turnover, compared to the other fat-soluble vitamins, supports the

Copyright © 2015 Elsevier Inc. All rights reserved.

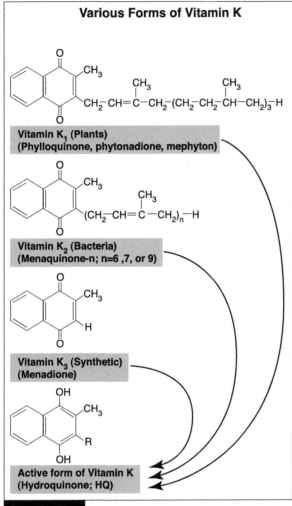

Various Forms of Vitamin K

Vitamin K₁ (Plants)
(Phylloquinone, phytonadione, mephyton)

Vitamin K₂ (Bacteria)
(Menaquinone-n; n=6 ,7, or 9)

Vitamin K₃ (Synthetic)
(Menadione)

Active form of Vitamin K
(Hydroquinone; HQ)

Figure 47-1

need for a continual supply through the diet (**K₁** and/or **K₂**), or from intestinal bacteria (**K₂**).

Although the liver form of this vitamin is **phylloquinone (K₁)** in horses, liver forms in dogs are usually half as **K₁**, and half as the **menaqui-**

nones (**K₂**). Serum, however, contains mostly phylloquinones, indicating that the menaquinones may undergo side-chain saturation in the liver before being released into blood. Vitamin K and its oxidized metabolites are lost from the mammalian organism in both bile and urine.

Vitamin K is required for the **hepatic post-synthetic transformation of clotting factors II** (prothrombin), **VII** (proconvertin), **IX** (plasma thromboplastic component, Christmas factor, or antihemophilic factor B), and **X** (Stuart-Prower factor); as well as anti-coagulation **proteins C**, and **S** (which facilitates the inactivation of factors V_a and VIII_a by facilitating the action of activated protein C). All are initially synthesized in the liver as inactive precursor proteins. In vitamin K deficiency, the liver continues its production of these coagulation proteins, however, they are nonfunctional.

The normal action of these factors in the clotting cascade is dependent upon the **carboxylation** of their terminal **glutamate (Glu)** residues to **γ-carboxyglutamate (Gla; Fig. 47-2)**. Factor II contains ten of these residues, whose negativity allows **chelation of calcium (Ca^{++})** in a specific protein-Ca^{++}-phospholipid interaction essential to their biologic role. Following injury, the binding of Ca^{++} by prothrombin anchors it to phospholipid membranes derived from blood **platelets**. This brings prothrombin

Role of Vitamin K (HQ) in Glutamate Carboxylation

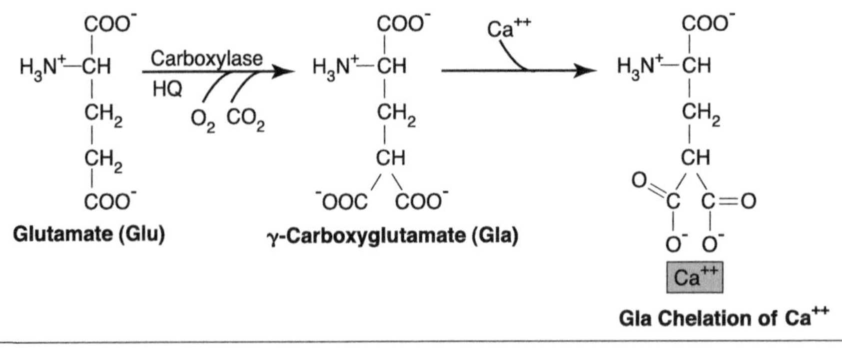

Glutamate (Glu)   γ-Carboxyglutamate (Gla)   Gla Chelation of Ca^{++}

Figure 47-2

into close proximity with two proteins that mediate its conversion into **thrombin** -- factor X_a (a serine protease), and factor V (proaccelerin, a stimulatory globulin). The amino-terminal fragment of prothrombin, which contains the Ca^{++}-binding sites, is released in this activation step. Thrombin, which becomes freed in this way from the phospholipid surface, now becomes available to activate serum **fibrinogen**.

The involvement of vitamin K in the γ-carboxylation of several other proteins has also been demonstrated. These proteins are involved in **bone metabolism** (hydroxyapatite dissolution), and in **connective tissue** and **kidney function**. In all cases the vitamin appears to be required for the γ-carboxylation of specific Glu residues, which in turn allow chelation of Ca^{++}.

A small protein in bone (**osteocalcin** or "**bone Gla protein**"; **BGP**), which contains three Gla residues, is involved in the action of **vitamin D** in bone. The active form of vitamin D, **1,25(OH)₂D** (see Chapter 45), acts on osteoblasts to increase synthesis and secretion of osteocalcin, which then binds to hydroxyapatite, thus preventing further mineralization. Some osteocalcin also escapes into blood, especially if vitamin K status is compromised. Conditions associated with **hyperparathyroidism**, characterized by enhanced bone mineral turnover, thus show increases in blood osteocalcin levels.

A Gla protein found in the **kidney** is thought to be involved in the reabsorption of Ca^{++} by renal tubular epithelial cells (a function also shared by PTH and vitamin D). In addition, Gla proteins have been found in calcium-containing kidney stones.

Tissue distributions of vitamin K and its metabolites tend to indicate that the action of this vitamin is diversified, and not confined to a few tissues. Uptake by the spleen reportedly equals that of the kidney, at least in rats, and is exceeded by that of the lungs, skin, and muscle. Thus, it is likely that we are only beginning to understand the full functional significance of the phyllo- and menaquinones.

A **vitamin K cycle** exists in the endoplasmic reticulum of liver cells, where the **2,3-epoxide** product of the HQ carboxylation reaction is converted by **2,3-epoxide reductase** to the **quinone** form of the vitamin (using an as yet unidentified dithio reductant; **Fig. 47-3**). This reaction is inhibited by the anticoagulants, dicoumarol (**4-hydroxydicoumarin**) and **warfarin** (**Fig. 47-4**), with accumulation of the 2,3-epoxide resulting in the formation of an abnormal prothrombin that does not appropriately bind Ca^{++}. Normal reduction of the **quinone** to the **hydroquinone** form of the vitamin (by **NADPH**) completes the cycle for regenerating the active form. An important therapeutic use of vitamin K is as an antidote to poisoning by dicoumarol or warfarin. The quinone and water-soluble dione forms of the vitamin will bypass the inhibited 2,3-epoxide reductase step, thus providing an important source of the active form.

Dicoumarol is found in **spoiled sweet clover**, which contains a mold that produces this metabolite and can cause fatal hemorrhagic episodes in animals that graze on hay containing this compound. This coumarin derivative is used clinically as an anticoagulant to prevent thromboses in patients prone to clot formation. Both **dicoumarol** and **warfarin**, a structurally related 2,3-epoxide reductase inhibitor, are the active ingredients in rat poisons. Horses, swine, dogs, and cats that consume rodenticides containing these anticoagulants generally require vitamin K treatment. The duration of antagonism depends on the amount consumed, as well as on the physiologic half-life of these compounds (which can extend from days to weeks).

Vitamin K Deficiency

Vitamin K deficiency states are somewhat difficult to establish since the requirement for this vitamin in many species is met by intes-

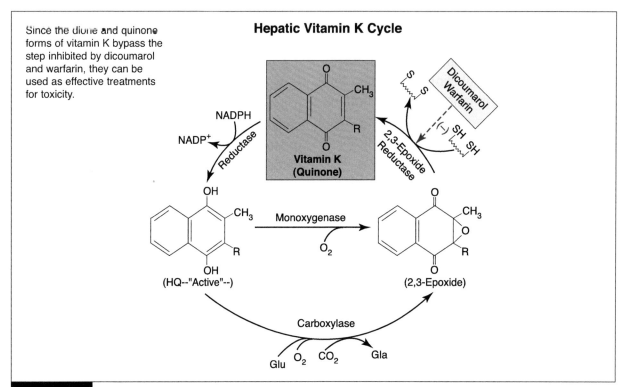

Since the dione and quinone forms of vitamin K bypass the step inhibited by dicoumarol and warfarin, they can be used as effective treatments for toxicity.

Hepatic Vitamin K Cycle

NADPH
NADP⁺
Reductase
Vitamin K (Quinone)
Dicoumarol Warfarin
2,3-Epoxide Reductase
(HQ--"Active"--)
Monoxygenase
O₂
(2,3-Epoxide)
Carboxylase
Glu O₂ CO₂ Gla

Figure 47-3

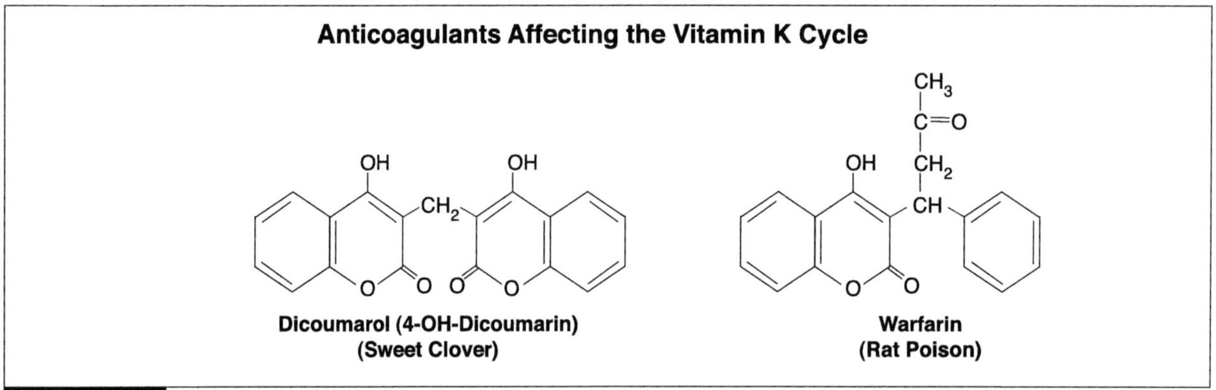

Anticoagulants Affecting the Vitamin K Cycle

Dicoumarol (4-OH-Dicoumarin)
(Sweet Clover)

Warfarin
(Rat Poison)

Figure 47-4

tinal microbial synthesis. Ruminant animals generally meet their vitamin K requirements through rumen **microbial biosynthesis** (with subsequent absorption in the small intestine), and the **coprophagy** practiced by rodents, rabbits, and other species provides vitamin K, produced in the large bowel. Horses generally receive sufficient vitamin K from pasture, hay, and intestinal bacteria to meet their needs. Dogs receive both **K₁** and **K₂** in their diets, and cats derive their quinones from eating meat.

Vitamin K deficiency can be caused by fat malabsorption, which may be associated with severe liver disease and/or biliary obstruction, pancreatic dysfunction, atrophy of the intestinal mucosa, or any cause of steatorrhea. In addition, sterilization of the large bowel by antibiotics can result in deficiency when dietary intake is limited. Vitamin K deficiency can present itself before other fat-soluble vitamin deficiencies, since the turnover of this vitamin is normally rather high. In newborn puppies, a transient form of vitamin K deficiency may occur secondary to malnutrition of the bitch during gestation, or the marginal inadequacy of fetal hepatic protein synthesis.

Vitamin K Toxicity

Although few toxicity symptoms have been reported in animals, vitamin K_3 (menadione), the synthetic water-soluble form of the vitamin used clinically, can apparently react with sulfhydryl groups on proteins, and therefore become toxic. Gastrointestinal disturbances and anemia have been associated with vitamin K excess.

OBJECTIVES

- Identify dietary and structural differences between the phylloquinones and menaquinone, and discuss differences in hepatic forms of vitamin K between dogs and horses.

- Describe how most ruminants, horses, rodents, rabbits, dogs and cats satisfy their vitamin K requirements.

- Explain why orally administered K_3 may appear at higher concentrations in the hepatic portal circulation than orally administered K_1 or K_2.

- Recognize why the liver can continue its production of clotting factors in vitamin K deficiency, yet those factors remain dysfunctional.

- Indicate how hepatic γ-carboxyglutamate is formed, where it is found, and how it functions.

- Discuss relationships that exist between bone Gla protein, $1,25(OH)_2D$, retinoic acid, PTH and Ca^{++} homeostasis.

- Understand why vitamin K is sometimes used as a therapeutic antidote to sweet clover and rat poisoning.

- Explain why vitamin K deficiency can present itself before other fat-soluble vitamin deficiencies, and identify vitamin K deficiency symptoms.

- Understand why newborn puppies might exhibit signs of vitamin K deficiency.

QUESTIONS

1. **The water-soluble form of vitamin K is:**
   a. K_1.
   b. K_2.
   c. K_3.
   d. K_4.
   e. K_5.

2. **Which one of the fat-soluble vitamins normally has the highest turnover rate?**
   a. A
   b. D
   c. E
   d. K

3. **In vitamin K deficiency:**
   a. The liver stops its production of prothrombin.
   b. The liver stops its production of albumin.
   c. The liver continues its production of prothrombin, however, it is nonfunctional.
   d. Dicoumarol can be effectively used as a supplement.
   e. Warfarin administration would have no effect.

4. **Which one of the following enzymes is vitamin K-dependent?**
   a. Glutamate carboxylase
   b. Vitamin K reductase
   c. 2,3-Epoxide reductase
   d. Purine nucleoside phosphorylase
   e. PEP Carboxykinase

5. **Which fat-soluble vitamins are best associated with osteocalcin of bone?**
   a. K and E
   b. D and A
   c. K and A
   d. D and E
   e. D and K

6. **Dicoumarol is best associated with spoiled:**
   a. Milk.
   b. Sweet clover.
   c. Fish.
   d. Meat.
   e. Bread.

7. **Biliary obstruction could cause vitamin K deficiency:**
   a. True
   b. False

ANSWERS

7. a
6. b
5. e
4. a
3. c
2. d
1. c

Chapter 48

Iron

Overview

- Iron is the most abundant trace element in mammalian organisms.
- The principal function of iron in the body involves oxygen transport.
- Only 3-6% of the iron present in the diet is normally absorbed by the intestine.
- Iron absorption is an active process, with most absorption occurring in the upper part of the small intestine.
- Cobalt, zinc, copper, and manganese compete, somewhat, with iron for intestinal absorption.
- Unless bleeding occurs, iron is not easily excreted from the body.
- Iron, which is continually interconverted between the ferric and ferrous state, can cause free radical formation.
- Iron toxicity results in pancreatic and liver damage, which is exacerbated in vitamin E deficiency.
- Young, fast-growing animals are vulnerable to iron deficiency.
- Most metalloflavoproteins contain iron.

Trace Elements

The mammalian organism normally contains and incorporates most all of the elements in the periodic table. The twenty-six most important can be subdivided into "**macrominerals**" (i.e., minerals required in comparatively large amounts in the diet), "**trace elements**" (i.e., microminerals), and "**ultra trace elements**" (those that are almost undetectable in body fluids and tissues; **Table 48-1**). Each trace element is physiologically important, with symptoms of both deficiency and toxicity occurring. Among the ultra trace elements, **lead** and **arsenic** toxicities are most frequently encountered in animals.

The trace elements are known to be essential for good health and reproduction, and each has well-defined actions. For example, they may serve as cofactors in enzymatic reactions, components of body fluids (electrolytes), sites for binding of oxygen (in transport), and structural components of nonenzymatic macromolecules. Although exceptions exist, most tend to concentrate in the germ of seeds and grains, where the highest concentrations of the B-complex vitamins are found. A discussion of the metabolism of each trace element can be found in the following five chapters, with many of their interactions, relations to disease, deficiency, and toxicity symptoms identified.

Iron (Fe)

Iron is the most abundant trace element in mammalian organisms, and one of the two most abundant in nature. Approximately **70%** of that present in the body is normally asso-

Copyright © 2015 Elsevier Inc. All rights reserved.

Table 48-1
Common Elements of the Body

Macrominerals	Trace Elements	Ultra Trace Elements
Calcium (Ca)	Iron (Fe)	Chromium (Cr)
Phosphorus (P)	Zinc (Zn)	Fluorine (F)
Magnesium (Mg)	Copper (Cu)	Nickel (Ni)
Sodium (Na)	Manganese (Mn)	Boron (B)
Chloride (Cl)	Selenium (Se)	Molybdenum (Mo)
Potassium (K)	Iodine (I)	Arsenic (As)
Sulfur (S)	Cobalt (Co)	Vanadium (V)
		Bromine (Br)
		Silicon (Si)
		Lead (Pb)
		Lithium (Li)
		Tin (Sn)

ciated with hemoglobin (Hb; see Chapter 32). Lesser amounts are associated with the heme-containing **cytochromes** and **iron-sulfur proteins** of electron transport (see Chapter 36), and the enzymes of hepatic drug metabolism (i.e., those involving the **cytochrome P$_{450}$ system**). Also, the widely distributed enzyme of DNA synthesis (**ribonucleoside diphosphate reductase (RDR)**) contains iron (see Chapter 16), as do several enzymes involved in the metabolism of biogenic amines (e.g., **tyrosine** and **tryptophan hydroxylases** that initiate formation of DOPA and serotonin, respectively). The **myeloperoxidase** of leukocytes is an iron-containing enzyme involved in bacterial killing, and the liver heme enzymes, **catalase** and **tryptophan oxygenase**, also contain iron. Many flavoprotein enzymes contain either molybdenum or iron as essential cofactors. These **metalloflavoproteins** are widespread, and participate in many important oxidation/reduction reactions (see Chapter 40).

Although **70%** of iron in the body is associated with **Hb**, highly variable amounts are associated with **ferritin**, a multisubunit protein present in all cells, but especially those of the liver, spleen, and bone marrow. Another **3-5%** is bound to muscle **myoglobin**, **hemosiderin** (a granular protein-iron complex and breakdown product of ferritin), and the heme-containing **cytochromes** (**Fig. 48-1**). Most iron in plasma is transported bound to a β$_1$-globulin known as **transferrin**, or **siderophilin**. The movement of

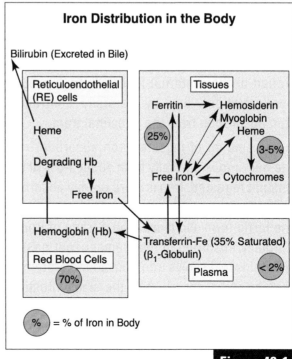

Iron Distribution in the Body

Bilirubin (Excreted in Bile)

Reticuloendothelial (RE) cells

Heme

Degrading Hb

Free Iron

Tissues

Ferritin → Hemosiderin
Myoglobin
Heme

25%

3-5%

Free Iron ← Cytochromes

Hemoglobin (Hb)

Red Blood Cells

70%

Transferrin-Fe (35% Saturated)
(β$_1$-Globulin)

Plasma   < 2%

% = % of Iron in Body

Figure 48-1

iron from hepatic storage sites to transferrin involves **ceruloplasmin**, a protein which carries about 60% of Cu^{++} in plasma (see Chapter 50). Normally, transferrin is about 35% saturated with iron. When heme is degraded to bilirubin by reticuloendothelial cells, the iron is normally recycled (see Chapter 33).

From its association with these specific proteins, it is clear that the principal functions of iron in the body involve **oxygen transport** within blood and muscle, and **electron transfer** in relation to energy metabolism. It is also intimately involved in **cell proliferation**, the production and disposal of **oxygen radicals** (and hydrogen peroxide (H_2O_2); see Chapter 30), systemic **hormone action**, and in some aspects of **immune defense**.

Iron has the capacity to accept and donate electrons readily, interconverting between the **ferric (Fe^{+++})** and **ferrous (Fe^{++})** states. This capability makes it a useful component of cytochromes, oxygen-binding molecules (e.g., Hb and myoglobin), and many enzymes. However, iron can also damage tissues by catalyzing conversion of H_2O_2 to **free-radicals** that attack cell membranes, proteins, and DNA. Organisms generally have difficulty excreting iron from the body, and therefore deal with this problem by tightly regulating the iron concentration of their internal fluids (through protein sequestration), and by carefully controlling iron absorption from the intestinal tract.

The amount of ingested iron absorbed from the gut ranges normally from about **3-6% of the amount ingested**. Iron is more readily absorbed in the ferrous state, but most dietary iron is in the ferric form. No more than a trace of iron is normally absorbed in the stomach, but gastric **HCl** helps to dissolve iron from bound protein so that it can be absorbed by the small intestine. The importance of this function is indicated by the fact that iron deficiency anemia is a troublesome and frequent complication of partial gastrectomy. **Ascorbic acid (vitamin C)** and

other reducing agents in the diet also help to facilitate conversion of iron from the ferric to ferrous state (see Chapter 39).

Heme is also absorbed by the small intestine using an **HCP-1** transporter (not shown in **Fig. 48-2**), and the Fe^{++} that it contains is released inside mucosal cells. Other dietary factors, such as **phytic acid**, **phosphates** and **oxalates**, can reduce iron availability by forming insoluble compounds with it. Pancreatic juice, because of its alkaline nature, also tends to reduce iron absorption.

Iron absorption is an active process, with most absorption occurring in the **upper part of the small intestine**. Other mucosal cells can transport iron, but the duodenum and adjacent jejunum contain most of the carriers. **Cobalt**, **Zn^{++}**, **Cu^{++}**, and **Mn^{++}** appear to **compete**, somewhat, with **Fe^{++}** for intestinal absorption (see Chapters 49-52).

Iron must cross two membranes to be transferred across the absorptive epithelium of the small intestine (**Fig. 48-2**). Each transmembrane transporter is coupled to an enzyme that changes the oxidation state of iron. The **apical transporter** has been identified as **DMT-1**, and acts in concert with a recently identified **ferric reductase**. The **basolateral transporter** requires **hephaestin**, a **ferroxidase-type** protein that converts iron back to the ferric state.

The erythropoietic factor that regulates iron absorption is thought to be **erythropoietin (EPO)**, a glycoprotein hormone of 165 amino acids. Erythropoietin blood levels rise in such conditions as anemia or pregnancy, and its half-life in the circulation is about 5 hours. This hormone also stimulates the bone marrow to produce red cells, however, it takes about 2-3 days to realize this effect since red cell maturation is relatively slow. In adult animals, about 85% of EPO originates from the kidneys (endothelial cells of peritubular capillaries), and 15% from the liver. During fetal life, the major site of both EPO and red cell production is the

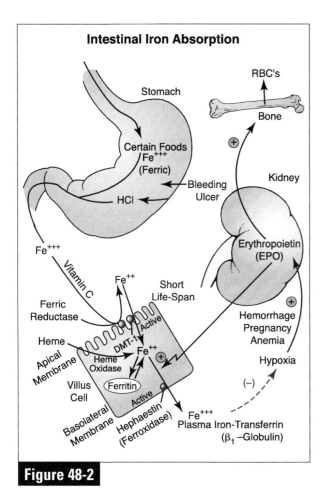

Intestinal Iron Absorption

Figure 48-2

liver. However, in some forms of adult kidney disease, the liver cannot compensate for loss of EPO synthesis, and anemia develops.

Iron Toxicity

If more iron is absorbed than excreted, **iron overload** results. Large ferritin and hemosiderin deposits are associated with **hemochromatosis**, a syndrome characterized by pigmentation of the skin, pancreatic damage with diabetes ("**bronze diabetes**"), cirrhosis, a high incidence of hepatic carcinoma, and gonadal atrophy. Hemochromatosis can be produced by prolonged, excessive iron intake, as well as by a congenital disorder in which the mucosal regulatory mechanism behaves as if iron deficiency were present, and absorbs iron at a high rate ("idiopathic hemochromatosis"). Hereditary hemochromatosis is, unfortunately, a common genetic disorder of humans.

Chelated iron in concentrations of 0.1 to 15% can be found in some commercial plant foods, and iron supplements are frequently present in the home. Although low levels of Fe^{+++} can be found in bile, pancreatic juice, and urine, it should be noted that **a distinctive feature of iron metabolism is the absence of a specific mechanism for iron excretion from the body (unless bleeding occurs)**.

Iron Deficiency

Iron-deficiency anemia is an important syndrome in many animals leading to weakness and lethargy due to decreased O_2 transport to tissues. It can result from dietary deficiency of absorbable iron, or from chronic blood loss through parasitism or hemorrhage. Blood that is reabsorbed from a body cavity or tissue pocket typically recycles the iron.

Young, fast-growing animals are particularly vulnerable to iron deficiency since they have a high demand for iron-containing compounds, and lack reserves. A major contributing factor is the low level of iron in milk, although colostrum is normally high in this mineral. For example, **piglets** housed in buildings that allow no access to earth usually require supplemental iron if anemia is to be avoided, but iron toxicity can result, especially if vitamin E deficiency coexists. Numerous gastrointestinal disturbances are also associated with iron deficiency, where there is an increased incidence of diarrhea and malabsorption.

In summary, iron has the capacity to readily interconvert between the ferric and ferrous states, which makes it a valuable component of the cytochromes and O_2-binding molecules. Seventy percent of that in the body is usually associated with Hb, with the remainder associated with myoglobin, ferritin and transferrin, and with various enzymes and heme-containing molecules. The amount of iron absorbed from the digestive tract is normally about 3-6% of that injested, and following absorption this trace element is highly conserved and reutilized. Excessive iron absorption results in **hemochromatosis**, and iron deficiency results in **anemia**.

OBJECTIVES

- Describe all steps involved in intestinal iron absorption, discuss the control and efficiency of this process, and identify the ways in which iron circulates in blood.

- Discuss the quantitative and qualitative partitioning of iron in various pools of the body.

- Identify relationships between iron and NDP reduction, and between iron and catecholamine biosynthesis (see Chapters 16 & 39).

- Show how the copper-binding protein, ceruloplasmin, is involved with iron transport.

- Understand iron's involvement in the production and disposal of free radicals, and in O_2 transport and electron transfer (see Chapters 30, 32 & 36).

- Recognize the trace elements that compete with iron for intestinal absorption.

- Compare intestinal iron efficiency to that of other trace elements, and understand why hemochromatosis is a common human genetic disorder.

- Contrast iron-deficiency anemia to other causes of anemia.

QUESTIONS

1. **Approximately what percentage of iron ingested is normally absorbed by the intestine?**
   a. 5%
   b. 28%
   c. 50%
   d. 70%
   e. 95%

2. **The apical transporter for iron in the small intestine:**
   a. Is known as hephaestin.
   b. Will transport iron only in the ferric (Fe^{+++}) state.
   c. Is a ferroxidase.
   d. Is known as ferritin.
   e. Acts in concert with a ferric reductase.

3. **Most iron in plasma is normally:**
   a. Bound to albumin.
   b. Bound to myoglobin.
   c. Bound to transferrin.
   d. Filtered by the kidneys.
   e. Contained in the cytochromes.

4. **Select the TRUE statement below regarding iron:**
   a. Hemochromatosis is a form of iron-deficiency anemia.
   b. A distinctive feature of iron metabolism is the absence of a specific mechanism for excretion from the body.
   c. Intestinal iron absorption is largely a passive process.
   d. Iron is normally excreted from the body as a component of bilirubin, a heme decomposition product.
   e. The majority of iron in plasma is normally bound to erythropoietin.

5. **The principal function of iron in the body involves:**
   a. Acid/base balance.
   b. Blood pressure regulation.
   c. Lipoprotein clearance from the circulation.
   d. Oxygen transport.
   e. Insulin biosynthesis and release.

6. **Which one of the following is a macromineral?**
   a. Iron
   b. Vanadium
   c. Phosphorus
   d. Zinc
   e. Cobalt

7. **Approximately what percentage of iron in the body is normally associated with hemoglobin?**
   a. 5%
   b. 28%
   c. 50%
   d. 70%
   e. 95%

ANSWERS

7. d
6. c
5. d
4. b
3. c
2. e
1. a

Chapter 49

Zinc

Overview

- Zinc is an active component of several important enzymes (including collagenase), and it is widely distributed in animal tissues.
- Zinc affects the metabolism of the pancreas, skin, and male reproductive organs.
- Zinc toxicity affects several different organ systems.
- Zinc is normally absorbed from the intestine at an approximate 20-30% efficiency, and competes with Cu^{++}, Fe^{++}, and Ca^{++} for absorption.
- Zinc deficiency results in growth retardation, delayed wound healing, photophobia, scaly dermatitis, and sometimes loss of taste and smell.
- Vitamin A metabolism is partially Zn^{++}-dependent.

Zinc (Zn^{++}) is a widely distributed trace element in animal tissues, concerned with the function of enzymes in several different areas of metabolism. As indicated in **Table 49-1**, this element is associated with enzymes involved in protein biosynthesis and degradation, nucleic acid biosynthesis, carbohydrate and energy metabolism, acid/base balance (CA), cellular protection against free radicals (SOD), inter-conversion of trans-retinal with retinol, heme biosynthesis, dihydrotestosterone production, and several other reactions. Approximately 50% of Zn^{++} in the body is complexed in bone, and therefore generally unavailable for metabolism. High concentrations are found in the integument, retina, testes, and prostate.

Apart from muscle/organ meats, the richest sources of dietary Zn^{++} are found in whole grains (the germ), seeds, nuts, eggs, and roots of leafy vegetables. Zinc is absorbed from the upper small intestine at an approximate **20-30% efficiency**. It moves by facilitated diffusion into mucosal cells, and from there a portion is transferred across the basolateral membrane (to blood and interstitial fluid) by an energy-dependent process. This is also the site where Ca^{++}, Cu^{++}, and Fe^{++} are absorbed, and these ions, when present in excess amounts, can compete with Zn^{++} for uptake. Iron supplements can cause Zn^{++} deficiency, and Zn^{++} supplements are known to cause Fe^{++} and Cu^{++} deficiency. Cereal diets high in Ca^{++} and phytate (which binds Zn^{++}), and marginally low in Zn^{++}, can promote Zn^{++} deficiency (**Table 49-2**). Following transfer to blood, Zn^{++} is primarily bound to albumin. In contrast to Fe^{++}, Zn^{++} is not stored by the body to any great extent, and therefore exhibits a greater turnover.

The most obvious effects of zinc occur on the metabolism, function, and maintenance of the skin, pancreas, and male reproductive organs. Since it is involved in important roles in many different cell types, deficiencies result in broad metabolic changes, including diminished growth (see **Table 49-2**). Zinc is associated with the abundantly secreted **exo-** and **endo-peptidases** of the exocrine pancreas, which are necessary for dietary protein digestion

Copyright © 2015 Elsevier Inc. All rights reserved.

Table 49-1	
Zinc	
Tissue Distribution	**Enzymes**
Bone (50%; not available)	Protein metabolism
Muscle	Proteases (collagenase)
Integument (skin, hair & nails)	Peptidases
Retina	Protein synthesis (skin)
Testes & Prostate	Nucleic acid metabolism
Liver	Aspartate transcarbamylase
Kidney	Thymidine kinase (sperm, skin)
Pancreas	RNA & DNA polymerases
Stomach (parietal cells)	Carbohydrate metabolism
Erythrocytes	Aldolase
	Pyruvate carboxylase
	Several dehydrogenases
	Others
	Carbonic anhydrase (CA)
	Superoxide dismutase (SOD)
	Alkaline phosphatase
	Retinal reductase
	Δ-ALA dehydratase (heme)
	5-α-Reductase (dihydrotestosterone)

Modified from Linder MC, 1991.

(see Chapter 7). It is also associated with stored insulin, although it does not appear to play a direct role in insulin action. The sloughing of intestinal mucosal cells and the pancreatic exocrine pathway are thought to be major routes for **zinc elimination** from the body. Smaller amounts are lost through sweat, hair, and skin, while lactation and pregnancy impose additional losses.

Zinc is necessary for **testicular development**, acting as a component of **5-α-reductase**, the enzyme that converts **testosterone** to **dihydrotestosterone (DHT)**. It also plays a role in the formation of testosterone by testicular Leydig cells, thereby indirectly affecting spermatogenesis. Circulating levels of both testosterone and DHT are reduced in zinc deficiency.

The role of zinc in skin and fibrous connective tissue metabolism involves effects on collagen synthesis and degradation (**collagenase** is a zinc-containing enzyme), as well as other proteins. Among the most important zinc-containing enzymes is **carbonic anhydrase (CA)**, abundantly present in erythrocytes, renal tubular epithelial cells, gastric parietal cells, and pancreatic and biliary ductular epithelial cells. **Superoxide dismutase (SOD)**, which requires both Zn^{++} and Cu^{++}, plays a defensive role in the cytosol of many cells by disposing of damaging superoxide anions (see Chapter 30). **Retinal reductase** (or dehydrogenase) is also a Zn^{++}-containing enzyme, that plays a role in vitamin A and thus visual pigment metabolism (see Chapter 44). Moreover, Zn^{++} is necessary for the synthesis of **retinol binding**

Table 49-2
Zinc Deficiency

Causes	Symptoms
Increased body loss	Anorexia
Starvation	Impaired taste
Burns	Growth retardation
Diabetes Mellitus	Hypogonadism (males)
Ketoacidosis	Delayed wound healing
Diuretic treatment	Nystagmus
Kidney disease	Photophobia
Proteinuria	Night blindness
Dialysis	Skin lesions
Liver disease	Nail loss
Intravascular hemolysis	Diarrhea
Chronic blood loss	Scaly dermatitis
Parasitism	Thinning & depigmented hair coat
Exfoliative dermatitis	Weight loss and depression
Excessive sweating	
Inadequate dietary intake	
Protein-calorie deficiency	
IV feeding	
High Ca^{++} and phytate diets	
Malabsorption	
Pancreatic insufficiency	
Inflammatory bowel disease	

Modified from Linder MC, 1991.

protein (**RBP**) in the liver, which is required for distribution of vitamin A via blood. These connections to vitamin A metabolism help to explain the photophobia and night blindness associated with Zn^{++} deficiency, as well as the general integrity of the skin. Additionally, the expression/production of growth hormone, which is turned-on by the **retinoic acid-hormone receptor complex**, and the expression of **triiodothyronine** (**T$_3$**), whose activity on certain proteins is complimented by vitamin A, can be affected in Zn^{++} deficiency.

Zinc may also be required for the activity of **adrenocorticotropic hormone** (**ACTH**), and for the capacity of erythrocytes, platelets, and other cell membranes to secrete **eicosanoids** (see Chapters 68 and 69). Reports indicate that Zn^{++} is fundamental to **T-cell function** in immunity, and a deficiency of this trace element can lead to thymic atrophy and decreased lymphokine production. Additionally, there may be a depression of natural killer cell and lymphocyte activities, and delayed hypersensitivity. There may also be a direct action of Zn^{++} in antibody production by B-cells. From these observations, it is evident that Zn^{++} is fundamental to growth and remarkable for its broad involvement in metabolism.

Zinc Toxicity

Although Zn^{++} is generally one of the least toxic trace elements, Fe^{++} and Cu^{++} deficiency and pancreatic, intestinal, hepatic, and renal damage are known to occur in animals receiving high doses. Additionally, hemolytic anemia has also been reported. Signs of depression, abdominal discomfort, anorexia, vomiting and diarrhea may precede the appearance of red urine and icterus. Although Zn^{++} **poisoning** has been described in several different animal species (**Table 49-3**), the dog appears to be affected most often. Animals can gain access to excessive amounts of Zn^{++} through **pennies** minted since 1983 (96% Zn^{++}), zinc nuts on collapsible transport cages (e.g., those sometimes used by the airline industry), zinc oxide ointments, zinc phosphide rodenticides, and zinc sulfate solutions used as fungicides on plants and in sheep foot-baths. Zinc is eroded from pennies or metal nuts during retention in the acid milieu of the stomach. Stomach acid also releases phosphine gas from zinc phosphide baits. Phosphine is apparently responsible for the acute effects of this rodenticide toxicity, and then Zn^{++} contributes to hepatic and renal changes present in animals that live several days.

Zinc Therapy in Copper Toxicosis

As indicated above, the **intestinal absorption** of Cu^{++} can be reduced by dietary Zn^{++}. This interaction forms the basis for Zn^{++} therapy in Cu^{++} toxicosis (see Chapter 50). Zinc induces synthesis of **metallothionein**, a protein that binds both Cu^{++} and Zn^{++} in mucosal cells of the small intestine. Thus, Cu^{++} absorption is reduced and the complexed Cu^{++} is held within mucosal cells of the gut until they are sloughed. The binding of Zn^{++} to metallothionein is apparently weaker than the binding of Cu^{++}, therefore this protein does not significantly reduce Zn^{++} absorption.

Zinc Deficiency

Zinc deficiency is not uncommon, and may occur because of **inadequate intake** or availability, **malabsorption**, or **increased rates of loss** from the organism. Growth retardation, skin lesions, and impaired sexual development in young males are recognized symptoms. Loss of taste and smell and impaired wound healing have also been described. The mechanism of Zn^{++} involvement in taste is unclear, but it may be a preneural event involving Zn^{++} attachment to **gustin**, a salivary protein. As indicated above, Zn^{++} plays a promotional role in skin and connective tissue metabolism, which has been recognized since ancient times when **calamine lotion (99% ZnO + 1% Fe_2O_3)** was first used on the skin. The connection is thought to be in the conversion of retinol to retinal, and then to

Table 49-3		
Zinc Toxicosis		
Susceptible Animals	**Causes**	**Symptoms**
Dogs	Zinc oxide ointments	Anorexia
Cats	Zinc sulfate solutions	Vomiting
Birds	Zinc metal nuts	Diarrhea
Ferrets	Zinc pennies (minted after 1983)	Hemolytic anemia
	Zinc phosphide	Renal dysfunction
		Hepatic dysfunction
		Pancreatic dysfunction

retinoic acid (retin A), which is necessary for epithelial cell differentiation (see Chapter 44), and also in the action of Zn^{++} as a cofactor for collagenase.

Increased Zn^{++} losses occur in **burn victims**, and in patients with **kidney damage**. In the latter, glomerular leakage of Zn^{++} attached to albumin is thought to be the main factor. Similarly, patients may lose substantial amounts of Zn^{++} during renal dialysis, and those on total parenteral nutrition (IV feeding) may receive less Zn^{++} than they require if the trace element has not been added to administered fluids. A variety of conditions and treatments, including inflammation, stress, cancer, glucocorticoid disturbances or treatments, starvation, or hemolysis may affect serum levels of Zn^{++}. Several of these factors apparently act by inducing hepatic **metallothionein**, which, like the same compound in mucosal cells of the gut, is thought to withdraw some of the metal from blood and hold it intracellularly. This acute phase may help subdue bacterial infections, for infusion of Zn^{++} into the blood of experimental animals brought into an inflammatory state by endotoxin has been shown to enhance virulence of the inflammatory process.

OBJECTIVES

- Explain why iron supplements can cause zinc deficiency, and compare iron to zinc absorption and storage by the body.

- Identify the tissue distribution of zinc, as well as the enzymes which depend upon this trace element as a cofactor.

- Discuss and understand the varied causes and symptoms of zinc deficiency, and know why zinc is fundamental to growth, development and optimal activity of the organism.

- Explain why zinc is used in calamine lotion.

- Understand how zinc affects the reproductive physiology of males.

- Know the relationship between zinc and metallothionein, and understand why zinc may be therapeutic in copper toxicity.

- Outline the causes and symptoms of zinc toxicosis, and identify animals that appear to be most susceptible.

- Summarize the relationships between zinc and the pancreas.

- Explain the connection between zinc deficiency and photophobia.

QUESTIONS

1. **Which one of the following is NOT a zinc-containing enzyme?**
   a. Glycogen phosphorylase
   b. Collagenase
   c. Carbonic anhydrase
   d. Retinal reductase
   e. 5-α-Reductase

2. **Which one of the following competes with zinc for intestinal absorption?**
   a. Cl$^-$
   b. K$^+$
   c. Na$^+$
   d. Alanine
   e. Cu^{++}

3. **Zinc is closely associated with which one of the following vitamins?**
   a. A
   b. B$_{12}$
   c. C
   d. D
   e. E

4. **Ingestion of which one of the following would most likely cause zinc toxicosis?**
   a. A new pencil
   b. A newly minted penny
   c. A small section of electrical cord
   d. A pair of plastic sunglasses
   e. An old sock

5. **Which one of the following proteins is known to bind zinc?**
   a. Fibrinogen
   b. Sucrase
   c. Prothrombin
   d. Metallothionein
   e. Tropomyosin

5. d
4. b
3. a
2. e
1. a
ANSWERS

Chapter 50

Copper

Overview

- Intestinal Cu^{++} absorption appears to be enhanced by dietary protein.
- Ceruloplasmin carries Cu^{++} in plasma, and also assists circulating transferrin in the reception of hepatic iron.
- Superoxide dismutase is the most abundant Cu^{++}-containing enzyme.
- Copper is a component of the mitochondrial electron transport chain.
- Copper is required for both catecholamine synthesis and degradation.
- Lysyl oxidase is a Cu^{++}-containing enzyme that aids in the crosslinking of elastin and collagen.
- Copper toxicity causes hepatic necrosis, methemoglobinemia, and hemolysis.
- Copper deficiency is associated with anemia, skin and hair depigmentation, CNS disturbances, and vascular degeneration.
- Wilson's-like disease is prevalent in Bedlington terriers.

Copper (Cu^{++}) is not as abundant as Zn^{++} or Fe^{++} in animals, however, disease states associated with Cu^{++} deficiency or Cu^{++} excess do exist. Like several other trace elements, the Cu^{++} content of the body is thought to be homeostatically regulated, with normally little storage of excess. High concentrations are found in **hair** and **nails**, but in terms of mass, the skeleton, muscle, liver, brain, and blood account for most of the Cu^{++} (although it is present in most all cells and tissues; **Table 50-1**).

Copper, like many other trace elements, is abundant in the germ of whole grains, therefore herbivores and omnivores usually show little or no evidence of Cu^{++} deficiency. Legumes, liver, and fish are also good dietary sources. Muscle meats (except duck) are generally less abundant in Cu^{++}.

Copper is normally absorbed at about **50% efficiency** from the upper small intestine by an energy-dependent process, and may compete with Zn^{++} and Fe^{++} for absorption. **Copper absorption** appears to be enhanced by dietary protein and amino acids. Evidence from studies in rats indicates that the concentration of **metallothionein** in mucosal cells determines how much Cu^{++} is free to proceed on into portal blood or stay behind attached to this small,

Table 50-1
Copper Content of Various Tissues
Bone > Muscle > Liver > Hair and Nails > Brain (substantia nigra) > Blood > Kidney > Heart > Lung > Spleen

Data modified from Linder MC, 1991.

Copyright © 2015 Elsevier Inc. All rights reserved.

high-cysteine protein. Conditions that increase metallothionein production (such as a high Zn^{++} intake), decrease overall Cu^{++} transfer into blood (see Chapter 49).

Copper is initially bound to **albumin** in portal blood, where it is carried to the liver. Within hepatocytes, Cu^{++} is incorporated into **ceruloplasmin** and other proteins/enzymes (e.g., **metallothionein**, **superoxide dismutase**, **cytochrome c oxidase**), and some undergoes canalicular excretion into bile. Only small amounts are normally excreted into urine. Ceruloplasmin, like other plasma proteins made by the liver, moves into blood and transports Cu^{++} to cells throughout the organism. Incorporation of Cu^{++} into ceruloplasmin may be necessary for Cu^{++} homeostasis in primates, since Cu^{++} accumulates in the liver when there is inadequate ceruloplasmin synthesis (see **Wilson's disease** below).

In addition to ceruloplasmin, Cu^{++} is associated with several intra- and extracellular enzymes (**Fig. 50-1**). **Cytochrome c oxidase,** the terminal component of the mitochondrial electron transport chain, transfers a pair of electrons from each of two cytochrome c molecules and a Cu^{++}-containing enzyme to O_2 (see Chapter 36). The cytochrome c

Reactions Catalyzed by Copper-Containing Enzymes

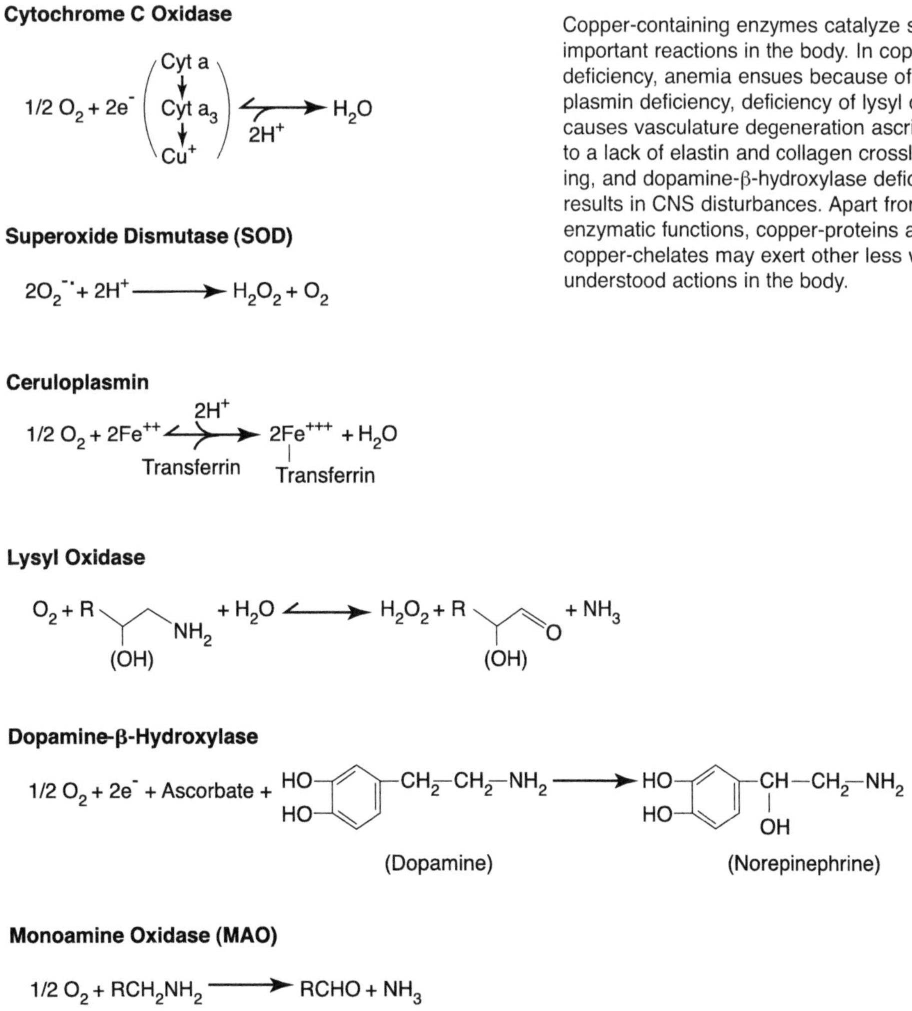

Copper-containing enzymes catalyze several important reactions in the body. In copper deficiency, anemia ensues because of ceruloplasmin deficiency, deficiency of lysyl oxidase causes vasculature degeneration ascribable to a lack of elastin and collagen crosslinking, and dopamine-β-hydroxylase deficiency results in CNS disturbances. Apart from enzymatic functions, copper-proteins and copper-chelates may exert other less well understood actions in the body.

Figure 50-1

Copyright © 2015 Elsevier Inc. All rights reserved.

oxidase complex contains heme iron as part of cytochromes a and a_3, and during electron transfer, the Cu^{++}-containing enzyme changes charge on reduction from +2 to +1.

The most abundant Cu^{++}-containing enzyme is **superoxide dismutase (SOD)**, which also contains Zn^{++} (see Chapter 49). This enzyme is concerned with the disposal of potentially damaging superoxide anions throughout the body (see Chapter 30).

Ceruloplasmin, which carries about **60% of Cu^{++} in plasma**, is also a weak, yet broad-based oxidase. It functions in copper transport, and in **antioxidant** defense as an extracellular scavenger of superoxides and other oxygen radicals. Serum Cu^{++} concentrations and ceruloplasmin are often increased in inflammatory conditions, indicating a positive role for these agents in the healing process and in connective tissue repair. Many types of cell-mediated inflammatory processes are propagated by the production of superoxide anions and other oxygen radicals. Ceruloplasmin, playing the role of a free-radical scavenger, is thought to help combat these processes.

Ceruloplasmin also plays a role in allowing the flow of **iron** from storage sites in the liver to **transferrin**, for transport to bone marrow and other sites (see Chapter 48). Specifically, **ceruloplasmin** is necessary for **oxidation** of Fe^{++} (which leaves ferritin where it is stored), **to Fe^{+++}** in order to allow attachment to plasma transferrin (see **Fig. 50-1**). In Cu^{++}-deficient rats that accumulate liver Fe^{++}, studies have shown that infusion (or liver perfusion) of ceruloplasmin causes release of liver iron to circulating transferrin. This finding helps to explain the similar symptomatology of Fe^{+++} and Cu^{++}-deficiency anemias, although energy availability for hematopoiesis, through oxidative phosphorylation and cytochrome c oxidase, may also be rate-limiting.

Another important Cu^{++}-containing enzyme is **lysyl oxidase**, which is secreted by connective tissue cells to aid in the crosslinking of elastin and collagen. This enzyme is essential for the health and maintenance of connective tissue and blood vessels, for in its absence, vascular degeneration occurs (**Table 50-2**).

Dopamine-β-hydroxylase is a Cu^{++}-containing, vitamin C-dependent enzyme needed for catecholamine production in the brain, adrenal medulla, and sympathetic postganglionic neurons (see **Fig. 50-1**). In addition to catecholamine production, one enzyme involved in the degradation of catechola-

Table 50-2	
Copper	
Deficiency Symptoms	**Toxicity Symptoms**
Anemia	Acute
Osteopenia	Hemolysis
Hypercholesterolemia	Methemoglobinemia
Neutropenia	Hepatic necrosis
Vascular degeneration	Wilson's-like disease
Skin and hair depigmentation	Chronic hepatitis and cirrhosis
CNS disturbances	Lethargy
Decreased immune function	Ascites
Hyperextension of distal phalanges	Weight loss
	Jaundice

mines, namely **monoamine oxidase (MAO)**, also contains Cu^{++}. Two isozymes of MAO have been described, **MAO-A** in neural tissue, and **MAO-B** in extraneural tissue (i.e., effector cells, the liver, stomach, kidney, and intestine). Monoamine oxidase inhibitors are drugs that would obviously prolong the action of the monoamines (namely serotonin, histamine, and the catecholamines, (dopamine, norepinephrine, and epinephrine)), including those in the CNS, and have thus been used to treat depression.

Apart from enzymatic functions, Cu^{++}-proteins and Cu^{++}-chelates are thought to possess other less well understood roles. For example, nonsteroidal antiinflammatory drugs like aspirin exert their actions as Cu^{++}-chelates, and Cu^{++} deficiency is associated with immune dysfunction (see **Table 50-2**). Other Cu^{++}-chelates are thought to possess limited anticancer activity, and Cu^{++} may play a role in the development of blood vessels and capillaries (i.e., angiogenesis).

Copper Deficiency

Although rare, **Cu^{++} deficiency** in the dog has been associated with hyperextension of the distal phalanges, and tissue Cu^{++} decreases in hair (with depigmentation), liver, kidney, and heart muscle (see **Table 50-2**). Osteopenia, lameness, and bone fragility have also been reported in dogs on Cu^{++}-deficient diets. Additional symptoms of Cu^{++}-deficiency include an anemia similar to that seen in Fe^{++}-deficiency, neutropenia, and a degeneration of the vasculature ascribable to lack of elastin and collagen crosslinking (via lysyl oxidase). In rats, CNS disturbances, probably due to alterations in catecholamine production and degradation, and hypercholesterolemia have been associated with Cu^{++}-deficiency.

Although Cu^{++}-deficiency can occur from a deficient diet or from deficient parenteral feeding, excess loss of Cu^{++} through renal dialysis can also occur. Unexplained observations have been made in experimental animals that excessive fructose (or sucrose) intake may exacerbate Cu^{++}-deficiency.

Copper Toxicity

Acute Cu^{++}-toxicity can follow ingestion of $CuSO_4$ solutions used as fungicides, algicides, and sheep foot-baths. Copper is corrosive to mucus membranes of the digestive tract, and causes hepatic necrosis, methemoglobinemia, and hemolysis (see **Table 50-2**).

Hepatic Cu^{++} accumulation can be associated with significant hepatic injury in animals, resulting in **chronic hepatitis** and **cirrhosis**. It is one of the few well-documented causes of chronic hepatitis in the dog. Certain breeds, including the West Highland white terrier (WHWT), Doberman pincher, American and English cocker spaniel, keeshond, Skye terrier, standard poodle, and Labrador retriever appear to have higher mean values for hepatic Cu^{++} than other breeds. An inherited metabolic defect in biliary Cu^{++} excretion causes chronic hepatitis in **Bedlington terriers**. A similar disorder occurs in WHWTs, however it is not identical, and is of lower magnitude. Because Cu^{++} is normally excreted in bile, hepatic Cu^{++} accumulation can also theoretically occur secondary to any cholestatic disorder.

In affected dogs, Cu^{++} is initially sequestered in hepatic lysosomes, and hepatic damage is reportedly minimal. However, with progressive accumulation of Cu^{++}, hepatic injury appears to become significant. The disease has been recognized in Bedlington terriers around the world, but primarily occurs in the United States, where the **prevalence** has been reported as **over 60%**. This disorder is similar in many respects to **Wilson's disease** in humans, and both disorders are transmitted by an autosomal recessive inheritance. Dogs, however, do not apparently show evidence of CNS or corneal Cu^{++} accumulation, and plasma Cu^{++} and ceruloplasmin concentrations are reportedly normal in affected dogs (rather than decreased, as in Wilson's disease of humans).

OBJECTIVES

- Contrast and compare the normal copper content of various body tissues.

- Identify the protein of intestinal mucosal cells that regulates copper absorption, and the primary plasma protein that binds this trace element.

- Recognize six important cellular reactions catalyzed by copper-containing enzymes.

- Explain the dual role of ceruloplasmin in copper and iron transport, and the dual role of copper in catecholamine production and degradation.

- Identify and explain the copper deficiency symptoms.

- Discuss common causes of acute copper toxicity in animals, and explain the symptoms.

- Contrast and compare the causes and symptoms of Wilson's disease in humans to hepatic copper accumulation in Bedligton terriers.

QUESTIONS

1. Which one of the following exocrine secretions of the body is the major route for Cu^{++} elimination?
   a. Saliva
   b. Pancreatic juice
   c. Bile
   d. Sweat
   e. Urine

2. Approximately 60% of Cu^{++} in circulating plasma is normally complexed with:
   a. Albumin.
   b. Fibrinogen.
   c. Transferrin.
   d. Ceruloplasmin.
   e. Lysyl oxidase.

3. What is the most abundant Cu^{++}-containing enzyme in the body?
   a. Hexokinase
   b. Superoxide dismutase
   c. Glutathione reductase
   d. Monoamine oxidase
   e. PEP carboxykinase

4. Which one of the following Cu^{++}-containing compounds helps to transfer hepatic iron to circulating transferrin?
   a. Cytochrome c oxidase
   b. Superoxide dismutase
   c. Ceruloplasmin
   d. Lysyl oxidase
   e. Dopamine-β-hydroxylase

5. Which one of the following symptoms is best associated with Cu^+-deficiency?
   a. Hypocholesterolemia
   b. Polycythemia
   c. Hepatitis
   d. Polyuria
   e. Vascular degeneration

6. Hepatic Cu^{++} accumulation resulting in hepatitis and cirrhosis is best associated with:
   a. Ponies.
   b. Female cats.
   c. Dairy cows.
   d. Goats.
   e. Bedlington terriers.

7. Select the TRUE statement below regarding copper:
   a. Copper is normally absorbed from the upper small intestine by an energy-dependent process.
   b. Amino acids are known to compete with copper for intestinal absorption.
   c. Copper is normally associated with intracellular enzymes, but not extracellular enzymes.
   d. Activation of lysyl oxidase, a Cu^{++}-containing enzyme, causes vascular degeneration.
   e. Unlike catecholamine synthesis, catecholamine degradation involves a Cu^{++}-containing enzyme.

7. a
6. e
5. e
4. c
3. b
2. d
1. c

Chapter 51

Manganese and Selenium

Overview

- Mn^{++} is associated with a wide variety of mammalian enzymes, including carboxylases, dehydrogenases, and transferases.
- Mn^{++} is best associated with mucopolysaccharide and lipopolysaccharide metabolism.
- Mn^{++} is absorbed from the small intestine at a rather low 3-4% efficiency, and is generally an abundant plant element.
- Mn^{++} is an additive in many pet food products.
- Mn^{++} is needed for normal bone metabolism.
- Se is an essential component of glutathione peroxidase, and is normally absorbed from the digestive tract at a rather high (50-90%) efficiency.
- The chemistry of Se appears to resemble that of sulfur.
- Some soils are Se-deficient, while others grow plants that when consumed can cause Se-toxicity.
- Se toxicity is associated with "blind staggers" and "alkali disease" in farm animals.
- Both vitamin E deficiency and Se deficiency lead to nutritional muscular degeneration, as well as other symptoms.

Manganese (Mn^{++})

Manganese is associated with a wide variety of enzymes in several different areas of metabolism (**Table 51-1**). In contrast to several of the other trace elements, however, a deficiency in this mineral is uncommon in animals, and does not appear to have broad effects (probably because magnesium (Mg^{++}), the much more abundant intracellular divalent cation, can substitute for Mn^{++} in many of its various enzyme-related activities). The roles of this element in **mucopolysaccharide** and **lipopoly-saccharide** formation would appear, however, to be the most vital in animals (**Table 51-2**).

Manganese is considerably less abundant than Fe^{++}, Zn^{++}, Cu^{++}, Ca^{++}, or Mg^{++} in the body, and is absorbed from the small intestine at a rather low **3-4% efficiency** (**Fig. 51-1**). Intestinal absorption is **hindered** by Ca^{++}, $PO_4^{\equiv}$, Fe^{++}, and **phytate**, yet **aided** by **histidine** and **citrate**. Blood distribution is thought to be **aided** by **transferrin**, the same plasma protein that carries iron (see Chapters 48 and 50). Blood concentrations of Mn^{++}, however, are normally about **1-2%** that of Fe^{+++}, Zn^{++}, or Cu^{++}. Tissue concentrations are highest in the **pineal**, **pituitary**, and **lactating mammary glands**, which are generally higher than those found in bone, liver, and pancreatic tissue. Manganese concentrations are generally low in lung and muscle tissue. As with Fe^{++}, Zn^{++}, and Cu^{++}, normal excretion occurs via intestinal sloughed cells, bile, and pancreatic juice, with urinary concentrations normally being low unless excessive intake occurs.

Copyright © 2015 Elsevier Inc. All rights reserved.

Table 51-1
Manganese-Dependent Processes

Glycolysis	Mitochondrial superoxide dismutase
Citric acid cycle	(protect mitochondrial membranes)
Hepatic urea synthesis	Maintain connective tissue & cartilage
Hexose monophosphate shunt	Blood clotting
	Bone mineralization & demineralization
Production of:	Lactation
Mucopolysaccharides	Fetal development
Glycoproteins	Pancreatic function
Lipopolysaccharides	Brain function
Hyaluronic acid	Ear otolith development
Heparin	
Chondroitin sulfate	
Melanin	
Dopamine	
Lipids	**ATP**

Table 51-2
Mn^{++}-Deficiency Symptoms

*Decreased mucopolysaccharide production	Pancreatic β-cell granulation
*Decreased lipopolysaccharide formation	Impaired lactation
Reduced glucose tolerance	Impaired fetal development
Defective otolith production	Hypocholesterolemia

*Primary

Rich **food sources** of Mn^{++} include nuts and whole grains, where the major portion is in the germ. Leafy vegetables are also good sources, where Mn^{++} is associated with a **thylakoid membrane binding protein** (the O_2-evolving complex) in the photosynthetic electron transport chain, and Mg^{++} is associated with the chlorophyll component of chloroplasts. Meats, fish, milk, and poultry are generally considered to be poor sources of Mn^{++}. For this reason Mn^{++} is usually added to feeds and pet food products.

Both Mn^{++} and Mg^{++} can serve as components of **ATP** or **ADP**, and as cofactors for several different types of enzymes, including carboxylases and decarboxylases, hydrolases and dehydrogenases, and transferases. Of particular significance are **pyruvate carboxylase**, **biotin carboxylase** and **acetyl-CoA carboxylase** (see Chapters 27, 42, and 56), and **isocitrate dehydrogenase** in the mitochondrial tricarboxylic acid cycle (see Chapter 34). The mitochondrial form of **superoxide dismutase**, which is thought to help protect mitochondrial membranes, is also associated with Mn^{++}, whereas the cytoplasmic form is best associated with Zn^{++} and Cu^{++}. Thus, Mn^{++}, which attains high levels in mitochondria, helps to reduce the oxidative stress associated with high mitochondrial O_2 consumption. **Arginase**,

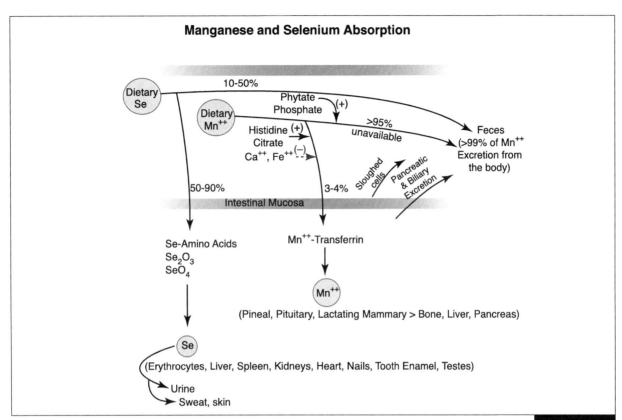

Manganese and Selenium Absorption

Figure 51-1

the terminal enzyme in urea production found in the cytoplasm of liver cells, also contains Mn^{++} (see Chapter 10). Other cytosolic Mn^{++}-containing enzymes are involved in the **hexose monophosphate shunt**, **glycolysis** (e.g., hepatic glucokinase; see Chapter 22), and **serine metabolism** (hydroxymethyl-transferase). This trace element is also associated with some of the enzymes of mucopolysaccharide, glycoprotein, and lipo-polysaccharide production, including **galactose transferase** and other membrane-bound glycosyltransferases. Manganese deficiency has negative effects on the production of **hyaluronic acid**, **chondroitin sulfate**, **heparin**, and other forms of mucopolysaccharide that are important for growth and maintenance of connective tissue, cartilage, and bone metabolism (see Chapter 19). Reports also indicate that Mn^{++} is involved with γ-carboxylation of the glutamate side chains of certain vitamin K-dependent proteins involved in blood clotting (see Chapter 47).

The connection between Mn^{++} and bone metabolism in rats indicates that both **osteoblastic** and **osteoclastic** activities are reduced in Mn^{++}-deficiency, resulting in osteoporosis and osteopenia (reduced bone cell number). Manganese also appears to be involved in melanin and dopamine production, in lipid biosynthesis (i.e., acetyl-CoA carboxylase), and in the formation of membrane phosphatidylinositol. Manganese deficiency can cause hypocholesterolemia, which may be due to a need for Mn^{++} by cholesterolgenic enzymes. It is also associated with nucleic acids, and, as indicated above, a portion is present in the mineral compartment of bone.

Selenium (Se)

Selenium is an essential component of **glutathione peroxidase**, a rather ubiquitous enzyme in mammalian tissues, that plays a role in the detoxification of peroxides and free radicals (see Chapter 30). These peroxides and free radicals can exert damaging effects on cell membranes of erythrocytes, hepatocytes, and other tissues in Se deficiency. While **vitamin E** appears to be the first line of defense against peroxidation of unsaturated fatty acids contained in cell membrane phospholipids, **catalase** and **glutathione peroxidase** constitute the second line of defense (see Chapter 46).

The chemistry of Se appears, in some regards, to resemble that of **sulfur**. Glutathione peroxidase is a 4-subunit enzyme with one Se per subunit in the form of **Se-cysteine**, required for activity. Other mammalian proteins containing Se-cysteine include 3 in the testes involved with spermatogenesis, and a **deiodinase** found in many thyroid hormone target cells that converts tetraiodothyronine (T_4) to the more active triiodothyronine (T_3) (**Table 51-3**). It is possible that Se-methionine, the main form of Se in plants, is sometimes incorporated into mammalian proteins in place of sulfur-containing methionine, and this insertion may be random and nonspecific.

From **50-90%** of dietary Se can be **absorbed** from the small intestine, with plant sources making it more readily available. Once absorbed, the highest tissue concentrations are found in erythrocytes, the liver, spleen, kidneys, heart, nails, tooth enamel, and the testes (see **Fig. 51-1**). Unlike Mn^{++}, Se is lost from the body primarily via urine in the form of di- and trimethylselenonium compounds.

Soils, and therefore their pastures, vary widely in their Se content. In general, soils derived from rocks of recent origin (e.g., the granitic and pumice sands of New Zealand), are notably Se deficient, as are soils in Finland, England, parts of Australia and China, and in the Pacific northwest and eastern seaboards of the USA. Alkaline soil encourages plants to absorb Se, yet a high sulphur content competes for absorption sites with Se in both plants and animals. The Se content of pastures is usually lowest in the spring when rainfall is heavy.

Selenium Deficiency

Both **vitamin E deficiency** and **Se deficiency** lead to nutritional muscular degeneration (i.e., **white muscle disease**; see Chapter 46). Additional symptoms of Se deficiency are included in **Table 51-4**. Within the past few years there has been increasing interest in the role of deficiencies of specific nutritional components in the etiology of myocardial cell destruction. The most important of these appears to be the amino acid **taurine** (see Chapters 3 and 62), and **carnitine**, the compound required to shuttle long-chain fatty acids across mitochondrial membranes (see Chapter 55). Vitamin E deficiency and Se deficiency are also associated with acute **myocardial necrosis** in farm animals, as well as in experimental dogs. There may also be other connections between Se deficiency and cardiovascular disease. The mechanism may involve increased aggregability of platelets and production of thromboxane A_2 (TXA_2), with less production of prostacyclin (PGI_2) from vascular endothelial cells (see Chapters 68 and 69 for more information on the functions of these eicosanoids). Heavy neonatal mortality in farm animals, chronic diarrhea in calves, infertility due to fetal resorption in ewes, and dietetic hepatosis in swine are complications that reportedly respond to dietary supplementation with this element.

Lastly, the possibility that adequate Se intake may prevent or retard tumor formation indicates a need for an optimal Se intake in both humans and animals.

Table 51-3
Selenium-Dependent Processes
Cell membrane protection
Peroxide detoxification
Conversion of T_4 to T_3
Retard tumor development
Spermatogenesis

Table 51-4	
Selenium	
Se-Deficiency Symptoms	**Se-Toxicity Symptoms**
Growth retardation	"Blind Staggers" or "Alkali
Cataract formation	Disease" in ruminant animals
Decreased spermatogenesis	Neurological damage
Placental retention	Rough hair
Myositis	Alopecia
Muscular degeneration	Hoof abnormalities
Cardiomyopathy	GI Disorders

Selenium Toxicity

Selenium poisoning can occur in areas where soils are derived from particular rock formations containing a high content of Se. It has been recorded in certain areas of North America, Ireland, Israel, Canada, and parts of Australia and South Africa. Acute Se-toxicity is known colloquially as **"blind staggers,"** because affected animals appear to be blind, wander aimlessly (often in circles), and display head pressing. The appetite may be depraved, with abdominal pain evident. The terminal stage is one of paralysis with death due to respiratory failure. Essentially the same picture can be produced by the experimental oral dosing of sheep with sodium selenite. Chronic poisoning ("alkali disease"), is reportedly manifested by dullness, emaciation, lack of vitality, stiffness, and lameness. In cattle, horses, and mules, the hair at the base of the tail and switch is lost, and in pigs there may be general alopecia. There are also hoof abnormalities, with deformity or separation and sloughing of the hooves reported in several species. Newborn animals whose dams have received diets containing an excess of Se may also be born with congenital hoof deformities.

In summary, Mn^{++} and Se are two important dietary trace elements associated with several metabolic processes. However, since Mg^{++} can substitute for Mn^{++} in many of these processes,

Mn^{++} deficiency symptoms are uncommon. In contrast, dietary Se deficiency can lead to significant muscular degeneration as well as other debilitating conditions.

Manganese is considerably less abundant than other divalent cations in the body, and is absorbed from the digestive tract at a rather low efficiency. Both Mn^{++} and Mg^{++} serve as cofactors for several different types of enzymes, particularly **carboxylases** and **decarboxylases.**

Selenium is an essential component of **glutathione peroxidase,** a rather ubiquitous enzyme that participates in the detoxification of peroxides and free radicals, it is a component of the **deiodinase** that converts thyroxine (T$_4$) to its more active form (T$_3$), it is a component of three different enzymes involved in **spermatogenesis,** and it may retard tumor development. In contrast to the low intestinal absorption efficiency of Mn^{++} (3-4%), the absorption efficiency of Se is quite high (50-90%). While Se is lost from the body primarily via urine, most Mn^{++} exits the body in feces. Selenium poisoning may occur in areas of the world where soils are rich in this trace element, and affected animals may exhibit classic symptoms of "blind staggers." This syndrome, also known as "alkali disease," occurs because alkaline soils encourage plants to absorb Se at a high rate.

OBJECTIVES

- Compare the normal intestinal absorption efficiency of Mn^{++} to that of Fe^{++}, and identify factors that could enhance or hinder Mn^{++} absorption.

- Identify the second most abundant intracellular cation, and explain its relationship to Mn^{++}.

- Explain how Mn^{++} is transported in plasma, identify tissues containing the most Mn^{++}, and indicate how Mn^{++} usually exits the organism.

- Show how Mn^{++} and ATP are structurally related, and discuss the involvement of Mn^{++} and/or Mg^{++} in cytoplasmic and mitochondrial energy metabolism.

- Summarize the Mn^{++}-dependent processes of the body, and recognize the signs and symptoms of Mn^{++} deficiency.

- Indicate how Se is involved in peroxide and free radical detoxification (see Chapters 30 & 46).

- Contrast the efficiency of intestinal Se absorption to that of the other trace elements, and identify areas of the world where Se-deficient and Se-rich soils exist.

- Compare signs and symptoms of Se deficiency to those of vitamin E, and recognize the causes and effects of "blind staggers," or "alkali disease."

QUESTIONS

1. **Which one of the following divalent cations is known to substitute for Mn^{++} in many of its various enzyme-related activities?**
   a. Ca^{++}
   b. Fe^{++}
   c. Cu^{++}
   d. Zn^{++}
   e. Mg^{++}

2. **Which one of the following is a Mn^{++}-dependent process?**
   a. Ear otolith development
   b. Peroxide detoxification
   c. Conversion of T_4 to T_3
   d. Spermatogenesis
   e. Retard tumor development

3. **Most Mn^{++} is normally excreted from the body through:**
   a. Feces.
   b. Urine.
   c. Respiration.
   d. Sweat.
   e. Hair loss.

4. **Which one of the following is a Mn^{++}-deficiency symptom?**
   a. Growth retardation
   b. Cataract formation
   c. Decreased mucopolysaccharide production
   d. Placental retention
   e. Cardiomyopathy

5. **Selenium is an essential component of:**
   a. Catalase.
   b. Vitamin E.
   c. Vitamin C.
   d. Glutathione peroxidase.
   e. Superoxide dismutase.

6. **Select the TRUE statement below regarding selenium (Se):**
   a. Acidic soil encourages plants to absorb Se.
   b. Se-deficiency causes "blind staggers" in farm animals.
   c. Spermatogenesis is a Se-dependent process.
   d. Se-toxicity is associated with cataract formation.
   e. Se-deficiency is associated with hoof abnormalities.

7. **Selenium chemistry resembles that of:**
   a. Manganese.
   b. Sulfur.
   c. Zinc.
   d. Copper.
   e. Cobalt.

8. **Most Se is normally excreted from the body through:**
   a. Feces.
   b. Urine.
   c. Sweat.
   d. Hair loss.
   e. Respiration.

Answers
8. b
7. b
6. c
5. d
4. c
3. a
2. a
1. e

Chapter 52

Iodine and Cobalt

Overview

- Dietary iodine is largely converted to iodide in the digestive tract, where it is normally absorbed at >95% efficiency.

- Iodide is required for thyroid hormone biosynthesis, and is extensively recycled within the organism.

- The thiocarbamide group of plant-derived goitrogens possesses antithyroid activity.

- ^{131}I is sometimes used to treat animals with hyperthyroidism.

- Cobalt is an important constituent of vitamin B_{12}.

- Cobalt deficiency in ruminant animals results in an inability to metabolize propionate, as well as an inability to reform tetrahydrofolate from N^5-methyl-H_4 folate.

- Several physiologic processes require inorganic cobalt.

- Cobalt and iron may share the same transport system for intestinal absorption.

- Iron deficiency enhances intestinal Co absorption.

Iodine (I)

The importance of **iodine** to mammalian thyroid metabolism has been recognized for over one-hundred years, and it is here that this trace element exerts its primary physiologic action. Within the thyroid iodine becomes a part of the **tri-** and **tetraiodothyronines**, which are the active thyroid hormones (**T_3** and **T_4**, respectively), as well as **reverse T_3** (**rT_3**), the inactive hormone (**Fig. 52-1**). The metabolism and endocrinology of the these hormones, as well as discussions of hypo- and hyperthyroidism in animals, can be reviewed in **Engelking LR: Metabolic and Endocrine Physiology, Jackson, WY: Teton NewMedia**.

Goiter, an enlargement of the thyroid gland with hypertrophy and/or hyperplasia of the follicular epithelium, occurs largely in response to iodine deficiency. Young animals born to dams on iodine-deficient diets are more likely to develop severe thyroid hyperplasia, and demonstrate signs of hypothyroidism. Iodine is a fairly benign trace element, which is thought to cause little harm at 10-20 times the recommended daily allowance. However, animal studies indicate that greater amounts can cause **toxicity** by ultimately inhibiting thyroid hormone biosynthesis, and causing goiter.

The richest natural **dietary sources** of iodine are fish, vegetables, meats, and eggs, while processed feeds and pet food products routinely include iodine as a supplement. **Dietary iodine** is normally converted to iodide (I^-) in the digestive tract, which is then actively absorbed from the small intestine at a **>95%** efficiency. Iodide circulates in plasma largely bound to plasma proteins (though some is free). This element is actively concentrated from

Copyright © 2015 Elsevier Inc. All rights reserved.

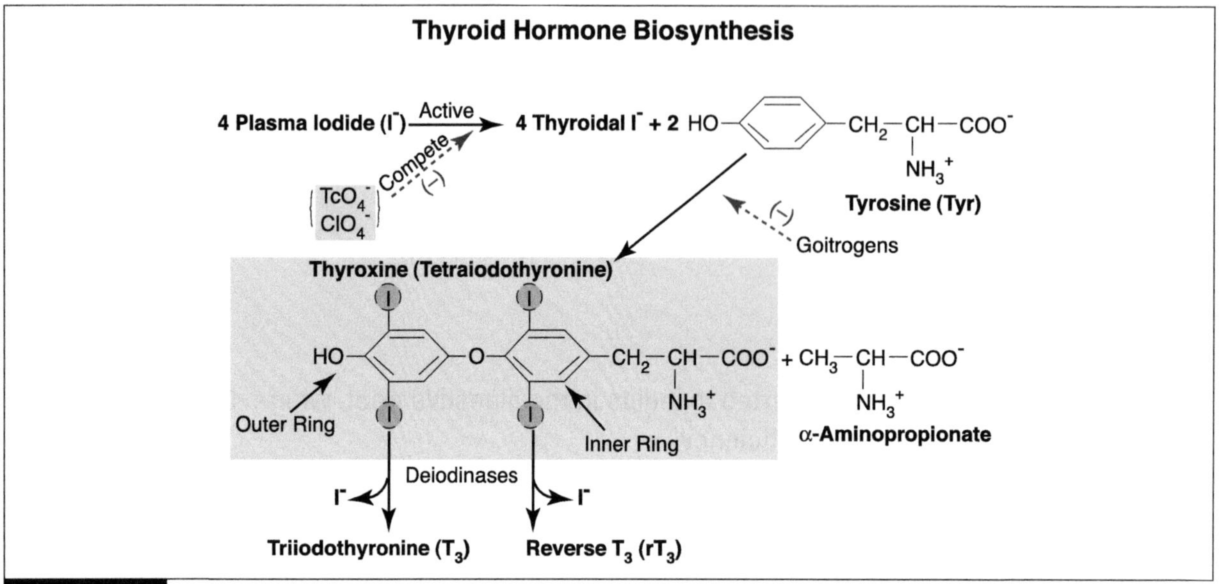

Thyroid Hormone Biosynthesis

Figure 52-1

blood not only by follicular cells of the thyroid gland (which normally contain **>90% of I⁻ in the body** and maintain a **25:1 thyroid-to-plasma I⁻ concentration ratio**), but also by cells of the salivary and lactating mammary glands, as well as the choroid plexus (which produces cerebrospinal fluid). It is extensively recycled within the organism, and excreted mainly via urine.

Goitrogens

While **iodine deficiency** has been associated with **goiter** in animals, it is not the only factor precipitating this condition. Genetic defects in thyroid hormone biosynthesis (thyroxinogenesis) occur, and **goitrogens**, which interfere with this process, are present to a variable degree in a number of plants from the family *Brassicacceae*. **Thiocyanates**, which can be produced by ruminal degradation of cyanogenic glucosides from plants such as white clover (*Trifolium*), couch grass, and linseed meal, and by degradation of glucosinolates of *Brassica* crops, are associated with hyperplastic goiter in ruminant animals. **Goitrin (5-vinyloxazolidine-2-thione; Fig. 52-2)**, derived from these glucoinolates, inhibits thyroxinogenesis (i.e., thyroid hormone biosynthesis). The **thiocarbamide group** of goitrin (see **Fig. 52-2**) is

essential for antithyroid activity, and is used in drugs prescribed for the treatment of thyrotoxicosis. **Propylthiouracil (PTU)** and **methimazole (Tapazole)** are the two thiourylene antithyroid drugs available for use in the USA, while **carbimazole** is the alternative to PTU in Europe (where methimazole is generally unavailable). Animals rapidly convert carbimazole to methimazole *in vivo*, so that only methimazole is detectable in serum and the thyroid gland following carbimazole administration.

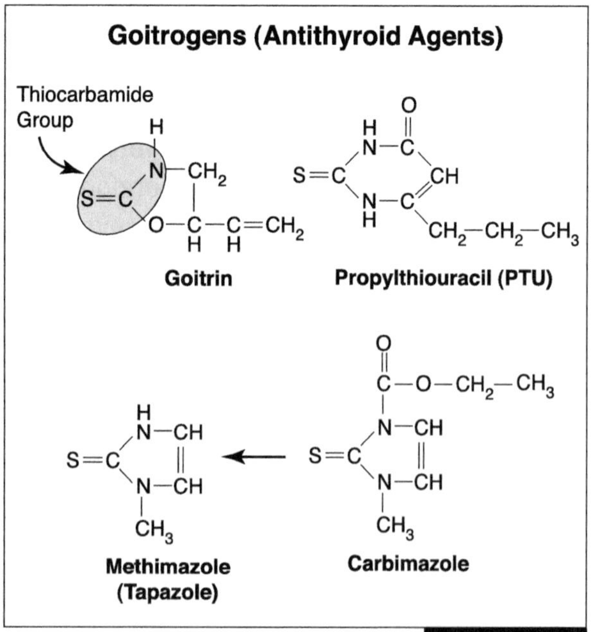

Goitrogens (Antithyroid Agents)

Figure 52-2

These compounds are actively concentrated by the thyroid gland, where they are thought to act to **inhibit thyroxinogenesis** through the following mechanisms:

1) By blocking incorporation of I^- into the tyrosyl groups in thyroglobulin (which is the thyroidal protein that normally binds thyroid hormones until they are secreted);

2) By preventing the coupling of iodotyrosyl groups (mono- and diiodotyrosines) into T_4 and T_3; and

3) Through direct interaction with the thyroglobulin molecule.

Antithyroid drugs are not thought to interfere with the thyroid gland's ability to concentrate or trap inorganic I^-, nor are they thought to block release of stored thyroid hormone into the circulation. Monovalent anions such as **perchlorate (ClO_4^-)** and **pertechnetate (TcO_4^-)** compete with I^- for thyroidal uptake, and they are also concentrated, like I^-, against a gradient (see **Fig. 52-1**). Technetium-labeled TcO_4^- is used clinically to monitor thyroid activity via thyroid scans.

Radioactive iodine provides a simple, effective, and relatively safe treatment for animals (mainly cats) with hyperthyroidism. The basic principle is that follicular cells of the thyroid do not differentiate between stable and radioactive iodine; therefore, radioiodine, like stable iodine, is concentrated by the thyroid gland following administration. The radioisotope most frequently used is ^{131}I, which has a half-life of 8 days, and emits both β and γ radiation. The β particles are locally destructive, but usually spare adjacent hypoplastic thyroid tissue, parathyroid glands, and other cervical structures. The goal of ^{131}I **therapy** is to restore euthyroidism with a single dose without producing hypothyroidism. With careful dosing, this therapy can be highly effective.

Cobalt (Co)

Cobalt is an important constituent of **vitamin B_{12}** (see Chapter 43), and as such, is required for two enzymatic reactions central to mammalian metabolism:

1) Synthesis of methionine from homocysteine to reform tetrahydrofolate (H_4 folate) from N^5-methyl-H_4 folate, and thus allow the normal flow of folate metabolism (and thymidine synthesis; see Chapters 14-16 and 43); and

2) The rearrangement of methylmalonyl-CoA to succinyl-CoA which is important for converting propionate from microbial cellulose digestion, the terminal 3 carbons of odd-chain fatty acids from mitochondrial β-oxidation, β-aminoisobutyrate from pyrimidine degradation, and several amino acids from protein degradation, to a member of the tricarboxylic acid cycle (see Chapters 37 and 42).

Because **vitamin B_{12}-deficiency** impairs the **methionine synthase** reaction, and thus DNA synthesis, cell division in bone marrow is impaired and **pernicious anemia** develops. This B-complex vitamin has also been associated with **myelin** and **carnitine** formation (**Table 52-1** and Chapters 43, 55, and 59).

Bacteria synthesize this vitamin, and **ruminant animals** appear to absorb only this form of Co. **Cobalt deficiency**, which causes anemia, anorexia, and wasting, is important in Australia, New Zealand, the UK and USA, and probably occurs in other areas of the world. The essential defect in ruminant Co deficiency is an inability to metabolize propionate. Where the deficiency is extreme, large tracts of land have been found to be unsuitable for raising ruminant animals, but apparently not horses.

Most Co absorbed has not been found to be in the form of vitamin B_{12} in non-ruminant animals; likewise, only about 1/10th to 1/12th of Co in the mammalian organism is in vitamin

Table 52-1
Cobalt-Dependent Processes

Involving Vitamin B$_{12}$

*Rearrangement of methylmalonyl-CoA to succinyl-CoA

  Propionyl-CoA entry into gluconeogenesis

Antipernicious anemia factor

*Transfer of a methyl group from N^5-methyl-H$_4$ folate to homocysteine in methionine formation

  Purine, pyrimidine, and nucleic acid biosynthesis

  Reformation of folic acid

Myelin formation

Carnitine formation

Independent of Vitamin B$_{12}$

Stimulate glycylglycine dipeptidase activity

Maintain erythropoiesis

Stimulate bradykinin release

  Lower blood pressure

Thyroxinogenic

*Primary

form, and it is conceivable that inorganic Co may be released from B$_{12}$ during metabolism to enable other actions. For example, the activity of **glycylglycine dipeptidase** appears to require Co (**Table 52-1**). Additionally, **bradykinin** is released into the circulation in response to Co salt administration, and the blood pressure is lowered. Co salts also enhance proliferation of bone marrow **erythropoietic cells**, independent of vitamin B$_{12}$. Large daily doses of **CoCl$_2$** have been used to treat anemias that are refractory to Fe^{++}, folate, and vitamin B$_{12}$.

Studies in rats indicate that inorganic Co may help to support **thyroxinogenesis**. However, high doses of Co oxides and/or sulfides have been shown to produce proliferation of otherwise normal cells, and to cause cancer in animals at the injection site, or in muscle and follicular thyroid tissue. Although rare in animals, **Co-toxicity** is reportedly increased by thiamin and protein deficiencies.

Cobalt salts are soluble in neutral and alkaline environments, and are thus more easily **absorbed** across the intestinal mucosa. Cobalt is thought to use the same intestinal transport systems as Fe^{++}/Fe^{+++} (see Chapter 48). Iron deficiency enhances Co absorption, and in rats iron appears to be the main regulator of intestinal Co uptake. Cobalt absorbed into intestinal mucosal cells is not invariably transferred to blood, and may be lost when intestinal cells are sloughed. In blood, inorganic Co is distributed attached to **albumin**. It initially deposits in liver and kidney tissue, and later in bone, spleen, pancreas, intestine, and other tissues. Liver, heart, kidney, and bone are considered to have the highest Co concentrations, which contrasts with the preferential accumulation of B$_{12}$-cobalt in the liver, where 50-90% of this vitamin is stored (see Chapter 43). The **intestinal absorption efficiency** for Co is **63-97%**, and it is mainly eliminated through urine.

Radiation delivered from a distance is called teletherapy or external beam radiotherapy. Orthovoltage x-ray and **cobalt-60** (^{60}**Co**) machines have been extensively used in veterinary oncology. ^{60}Co emits γ-rays with energy approximately six-times greater than orthovoltage x-ray units. ^{60}Co units are in some cases being replaced by more modern linear accelerators.

OBJECTIVES

- Understand why iodine is employed as a pet food supplement, and compare the intestinal absorption efficiency of this trace element to others.

- Explain how and why iodide is concentrated in the thyroid gland, and why it is also actively removed from blood by salivary and mammary glands.

- Identify dietary sources of iodine.

- Recognize normal routes of iodine elimination from the body.

- Explain how iodide is carried in blood.

- Describe the dietary and therapeutic sources of goitrogens, and identify how they are thought to interfere with thyroid function.

- Explain relationships between perchlorate, pertechnetate and iodide.

- Discuss the basis of using ^{131}I to treat hyperthyroidism.

- Identify the structural relationship that exists between cobalt and vitamin B_{12}, and recognize two key enzymatic reactions that are dependent upon this complex.

- Recognize cobalt-dependent processes that do not involved B_{12}.

- Identify areas of the world possessing cobalt-deficient soils.

- Explain how cobalt is absorbed from the intestine, transported in plasma, stored and excreted from the body.

- Compare cobalt and iodide intestinal absorption efficiency to that for zinc, iron, copper, manganese and selenium.

QUESTIONS

1. **All of the following are known to actively concentrate I⁻ from blood, EXCEPT the:**
   a. Thyroid gland.
   b. Pituitary gland.
   c. Choroid plexus.
   d. Salivary glands.
   e. Lactating mammary glands.

2. **Which one of the following is an essential component of antithyroid drugs, propylthiouracil, methimazole, and carbimazole?**
   a. Cobalt
   b. Reverse T_3
   c. Selenium
   d. Thiocarbamide group
   e. Iodotyrosyl group

3. **Antithyroid drugs act by blocking:**
   a. Tyrosine uptake into thyroid tissue.
   b. Thyroglobulin synthesis.
   c. Incorporation of I⁻ into tyrosyl groups in thyroglobulin.
   d. I⁻ uptake into follicular cells of the thyroid.
   e. Release of stored thyroid hormone into the circulation.

4. **The primary defect in ruminant Co-deficiency is the inability to:**
   a. Produce bile.
   b. Metabolize propionate.
   c. Produce urine.
   d. Metabolize acetate.
   e. Synthesize thyroxine.

5. **CoCl₂ therapy is sometimes used to treat:**
   a. Hyperthyroidism.
   b. Fatty liver syndrome.
   c. Muscular dystrophy.
   d. Vitamin E deficiency.
   e. Anemia.

6. **Co appears to use the same intestinal transport system as:**
   a. Sodium.
   b. Potassium.
   c. Iron.
   d. Glucose.
   e. Tyrosine.

6. c

5. e

4. b

3. c

2. d

1. b

ANSWERS

Section IV Examination Questions

1. **The process of moving iron from hepatic storage sites to transferrin, and then to plasma, involves:**

   a. Hephaestin.
   b. Pyridoxine (vitamin B_6).
   c. Ceruloplasmin.
   d. Phylloquinone.
   e. Elimination of phenoxy free radicals.

2. **In hydroquinone deficiency:**

   a. Scotopic vision would be compromised.
   b. Dicoumarol could be effectively used for therapy.
   c. Steatorrhea may be present.
   d. Excessive lipid peroxidation of membranes would be expected.
   e. Selenium, B_{12} and KCl could be effectively used for therapy.

3. **Which one of the following is considered to be the LEAST toxic?**

   a. Copper
   b. Iron
   c. Selenium
   d. Vitamin D
   e. Vitamin E

4. **Which of the following are INCORRECTLY paired?**

   a. Collagen : Zinc and vitamin C
   b. Thiaminase : Raw fish
   c. 5′-Deoxyadenosylcobalamin : Conversion of methylmalonyl-coA to succinyl-CoA
   d. ↓NADH : Thiamin and/or Niacin deficiency
   e. Coenzyme A•SH : Pyridoxine

5. **Retinoic acid is best associated with which of the following functions?**

   a. Assist in the conversion of retinol to retinal
   b. Essential role in scotopic vision
   c. Facilitate renal Ca^{++} reabsorption
   d. Growth and differentiation of epithelial tissues
   e. Inhibition of spermatogenesis

Copyright © 2015 Elsevier Inc. All rights reserved.

6. Metallothionein:

a. Is also known as intrinsic factor.
b. And ceruloplasmin are Cu-binding proteins.
c. Is directly involved in the transfer of a methyl group from N^5-methyl-H_4 folate to homocysteine in methionine formation.
d. Inhibits thyroxinogenesis, and thus can be effectively used to treat hyperthyroidism.
e. Production is inhibited by Zn^{++}.

7. Irradiation of milk is a commercial means of producing:

a. Vitamin E from vitamin D.
b. γ-Carboxyglutamate from glutamate.
c. Carbimazole from cyanocabalamin.
d. Hydroquinone from menadione.
e. None of the above

8. Which one of the following is the best dietary source of cobalamin in the feline diet?

a. Skin
b. Chocolate
c. Liver
d. Kidney
e. Muscle

9. Select the FALSE statement below:

a. Vitamin D excess can be successfully treated with biotin and vitamin B_{12}.
b. The rate-limiting reaction in fatty acid biosynthesis requires biotin as a cofactor.
c. Pyridoxamine phosphate plays an important role in nonessential amino acid formation.
d. Retinoic acid deficiency can retard placental and fetal development.
e. Vitamin D acts on osteoblasts to increase the synthesis and secretion of osteocalcin, which prevents bone mineralization.

10. L-Ascorbate:

a. Is a fat-soluble vitamin found in chylomicrons (CMs).
b. Is synthesized from Glc 6-phosphate in fish, but not in dogs and cats.
c. Can be regenerated from dehydroascorbate in astrocytic supporting cells of the CNS.
d. Transport across mucosal cells of the intestine is similar to that of vitamin E.
e. Is normally derived (biosynthetically) from ergosterol.

Matching (each answer used once):

   a.   **Selenium**
   b.   **Copper**
   c.   **Iodide**
   d.   **Iron**
   e.   **Manganese**

11. ___ Pertechnetate is handled by the body similarly to this element.

12. ___ Associated with enzymes involved with both the biosynthesis (DA β-hydroxylase) and degradation (MAO) of catecholamines.

13. ___ This element is a component of the deiodinase that converts thyroxine(T_4) to T_3.

14. ___ Hepatic arginase, a key urea cycle enzyme, is best associated withthis cation.

15. ___ Excessive accumulation of this element may promote large hemosiderin deposits.

----

16. Which of the following are INCORRECTLY paired?

   a. Selenium toxicity : Blind staggers or alkali disease
   b. Vitamin B_{12}-IF dimers : Ileal receptors
   c. Dietary I⁻ : High intestinal absorption efficiency ($\approx 95\%$)
   d. Pernicious anemia : Cobalt and/or folic acid deficiency
   e. Propylthiouracil : Stimulate thyroxine biosynthesis

17. Select the FALSE statement below:

   a. Reverse T3 (rT3) is an iodinated compound.
   b. More iodine is normally excreted in urine that in feces.
   c. The thiocarbamide group of goitrogenic compounds provides them with antithyroid activity.
   d. Specific Se-dependent deiodinases can either up- or down-regulate T_4 in target cells.
   e. Dietary iodine normally has about the same intestinal absorption efficiency as dietary manganese.

18. Select the TRUE statement below:

   a. The active form of vitamin D, $1,25(OH)_2D$, is known to stimulate formation and activity of renal 24-hydroxylase.

b. NMN (and thus NAD$^+$) can be formed from the aromatic amino acid, tyrosine.

c. Phosphopantetheine is an intermediate in hepatic biotin formation.

d. HMG-CoA is used in the plasma transport of pyridoxaldehyde.

e. Methylcobalamin is used to convert methylmalonyl-CoA to succinyl-CoA in the cytoplasm of liver cells.

19. Select the FALSE statement below:

a. The most common finding in animals with biotin deficiency is pernicious anemia.

b. 1,25-Dihydroxycholecalciferol helps to promote Ca^{++} absorption in the upper part of the small intestine.

c. Ascorbate helps to regenerate the active form of vitamin E in plasma membranes.

d. In some natural animal diets, the "niacin equivalents" from dietary Trp may be greater than those from dietary niacin.

e. Hypothyroidism may lead to a β-carotenemia in herbivores.

20. Pantothenic acid (vitamin B$_5$) is properly associated with all of the following, EXCEPT:

a. HMG-CoA (cholesterol and ketone body biosynthesis).

b. 4-Phosphopantetheine (prosthetic group in the acyl carrier protein (ACP) used in fatty acid biosynthesis).

c. Succinyl-CoA (heme biosynthesis).

d. UDP-glucose (uronic acid pathway).

e. Cytoplasmic long-chain fatty acid activation (acyl-CoA formation).

21. Select the FALSE statement below:

a. Selenium is associated with the regeneration of ascorbate from dehydroascorbate, and the oxidation of GSH.

b. Unlike Mn^{++}, I$^-$ is mainly excreted from the body through urine.

c. The vitamin deficiency best associated with muscle degeneration is cholecalciferol.

d. The "3 Ds," diarrhea, dermatitis and dementia, are associated with niacin deficiency.

e. Cobalt salts are known to enhance proliferation of bone marrow erythropoietic cells, independent of vitamin B$_{12}$.

22. Select the TRUE statement below:

a. 1,25(OH)$_2$D receptors in osteoblasts are down-regulated by retinoic acid.

b. α-Tocopherol tends to be concentrated in structures exposed to high partial pressures of O$_2$.

c. The most common coenzyme forms of niacin are thioesters of CoA•SH.

d. Superoxide dismutase, a cobalt-containing enzyme, is used to convert H$_2$O$_2$ to O$_2$ and H$_2$O.

e. Thiamin is the fat-soluble vitamin required in the methylation of homocysteine.

23. Manganese:

a. And selenium are normally absorbed from the digestive tract at a similar efficiency ($\approx$ 50-90%).
b. Is normally far more abundant than Mg^{++} in muscle cells.
c. Is associated with enzymes involved with protein metabolism, but not those associated with carbohydrate metabolism.
d. Deficiencies are common in the domestic small animal population.
e. Is best associated with muco- and lipopolysaccharide formation.

24. All of the following are properly associated with vitamin B_{12} deficiency, EXCEPT:

a. Gastrectomy.
b. Polycythemia.
c. The trapping of N^5-methyl-H_4 folate.
d. Dietary cobalt deficiency.
e. Subnormal intestinal mucosal cell activity.

25. Select the TRUE statement below regarding vitamin C:

a. Deficiency of this vitamin may lead to a reduction in bile formation due to bile acid deficiency.
b. It converts peroxyl to phenoxy free radicals.
c. It can regenerate vitamin D in glial cells of the brain.
d. It can be synthesized from tryptophan in a 2-step reaction sequence.
e. Deficiency of this vitamin leads to symptoms of pellagra.

26. Which vitamin deficiency and related function below would most likely be associated with steatitis?

a. Cholecalciferol and Ca^{++} homeostasis
b. Retinoic acid and epithelial cell differentiation
c. α-Tocopherol and antiperoxidation
d. Hydroquinone and γ-carboxylation
e. Riboflavin and facilitation of transketolase reactions

27. Which of the following are INCORRECTLY matched?

a. Menadione : Used to treat hypoparathyroidism
b. Wilson's-like disease : Hepatic Cu accumulation
c. Warfarin : Inhibits hepatic 2,3-epoxide reductase
d. Transferrin : Carries both Fe and Mn in plasma
e. Lysyl oxidase : Copper-containing enzyme

28. The conversion of pyruvate to acetyl-CoA involves which 3 vitamins?

   a. Biotin, ascorbate and α-tocopherol
   b. Thiamin, niacin and pantothenate
   c. Folic acid, pyridoxine and cobalamin
   d. Vitamins A, D and E
   e. Riboflavin, hydroquinone and 4-phosphopantetheine

29. Which of the following are PROPERLY matched?

   a. Catalase : 2,3-Epoxide reductase
   b. Muscle degeneration : Vitamin E deficiency
   c. Manganese : Cytoplasmic SOD
   d. Scotopic vision : Vitamin D
   e. Thyroid hormones : Contain selenium

30. Flavin adenine dinucleotide (FAD):

   a. Usually contains zinc or selenium.
   b. Is formed from vitamin C in the HMS.
   c. Is a mitochondrial succinate dehydrogenase cofactor.
   d. Is the coenzyme derivative of niacin.
   e. Is a cofactor for erythrocytic transketolase reactions.

31. Which one of the following is an iron-containing enzyme?

   a. Retinal reductase
   b. Carbonic anhydrase
   c. Lysyl oxidase
   d. Ribonucleoside diphosphate reductase
   e. Superoxide dismutase

Matching (each answer used once):

   a. Carboxylation   **d. Hydroxylation**
   b. Transmethylation   **e. Decarboxylation**
   c. Oxidation/Reduction

32. ___ Ascorbate

33. ___ Biotin

34. ___ Riboflavin

35. ___ Thiamin

36. ___ Cobalamin

---

37. CoCl$_2$ therapy is sometimes used to treat:

a. Vitamin E deficiency due to liver disease.
b. Anemias that are refractory to Fe^{++}, folate or vitamin B$_{12}$ administration.
c. hyperthyroidism due to excessive pituitary TSH secretion.
d. Nutritional muscular degeneration that does not respond to menadione therapy.
e. Fatty liver syndrome caused from cholecalciferol toxicity.

38. Which of the following are INCORRECTLY paired?

a. Riboflavin coenzyme : NADP$^+$
b. Low serum folate : Proximal bowel dysfunction
c. Glucocorticoids : Interfere with 1,25(OH)$_2$D intestinal action
d. Monoamine oxidase : Cu-containing enzyme
e. White muscle disease : α-Tocopherol deficiency

39. Select the FALSE statement below regarding zinc:

a. Zinc-deficiency symptoms are similar to those for cholecalciferol-deficiency.
b. Pancreatic exocrine secretions normally contain zinc.
c. Iron supplements can cause zinc-deficiency.
d. Cereal diets high in Ca^{++} and phytate can promote zinc-deficiency.
e. In contrast to iron, zinc is not stored by the body to any great extent.

40. Which one of the following enzymes is vitamin K-dependent?

a. Galactose transferase
b. Retinal reductase
c. Glutamate carboxylase
d. Superoxide dismutase
e. Propionyl-CoA carboxylase

41. Pyridoxine (vitamin B$_6$) is involved in which enzymatic reaction?

a. γ-Carboxylation of Gla proteins
b. Methylation of norepinephrine in the formation of epinephrine

    c. Oxidative decarboxylation of pyruvate to acetyl-CoA

    d. Deamidation of nicotinamide to niacin

    e. Generation of glucose 1-phosphate units from glycogen

42. All of the following are properly matched, EXCEPT:

    a. Pellagra : Niacin.

    b. Scurvy : Ascorbate.

    c. Rickets : Cobalamin.

    d. Ruminant polioencephalomalacia : Thiamin.

    e. Intestinal bacterial over-growth : High serum folate.

43. Which of the following statements regarding iron is TRUE:

    a. It is normally absorbed into intestinal mucosal cells in the ferrous (Fe^{++}) state.

    b. it is an essential cofactor for retinal reductase.

    c. it competes with Na^+ for intestinal absorption.

    d. It is an essential component of lysyl hydroxylase.

    e. it is a major component of both heme and bilirubin.

44. This vitamin is a reducing agent that helps to convert dopamine to norepinephrine in sympathetic neurons:

    a. Vitamin A

    b. Vitamin B_{12}

    c. Vitamin C

    d. Vitamin D

    e. Vitamin E

Answers

1. c	12. b	23. e	34. c
2. c	13. a	24. b	35. e
3. e	14. e	25. a	36. b
4. e	15. d	26. c	37. b
5. d	16. e	27. a	38. a
6. b	17. e	28. b	39. a
7. e	18. a	29. b	40. c
8. c	19. a	30. c	41. e
9. a	20. d	31. d	42. c
10. c	21. c	32. d	43. a
11. c	22. b	33. a	44. c

Addendum to Section IV

The solubility of a vitamin influences its mode of action, storage and toxicity. Generally speaking, and with the exception of vitamin B_{12}, water-soluble vitamins are not stored in the body, they enter the body freely, are present in intra- and extracellular fluids, and generally exit via urine, relatively unchanged. Most function as coenzymes in specific reactions of protein/amino acid, nucleic acid, carbohydrate and lipid metabolism. Fat-soluble vitamins possess more individualized actions (with the exception of vitamin E, a broad-spectrum lipid antioxidant). They are readily stored, but with the exception of vitamin E, are not absorbed or excreted as readily as the water-soluble vitamins. With the exception of vitamin K, they are not coenzymes in the sense of the B vitamins, and two of them (vitamins A and D) are sometimes considered hormones.

Although knowledge of the importance of trace elements to animal health and disease is increasing, clearly the roles of iron, iodine and selenium are best established. Because trace element deficiencies and toxicities are being increasingly associated with chronic health problems, veterinarians must continue to monitor animal diets and environmental exposures in order to make proper associations between specific trace elements and pathophysiologic symptoms.

Introduction to Section V

Section V now deals with **lipid metabolism**. The importance of **acetyl-CoA** as an intermediate in the generation of **ATP** has been described in previous sections. As well as arising from carbohydrates (via glycolysis and the pyruvate dehydrogenase reaction), we will see that acetyl-CoA can also be derived from lipids. A classification frequently used for **lipids** is into those that are ultimately **oxidized** for ATP generation (e.g., fatty acids and ketone bodies), and those which serve a largely **structural** role (e.g., phospholipids and other compound lipids). However, this distinction is somewhat misleading since fatty acids are also components of structural lipids. This functional distinction is also underscored by the location of these lipids within the cell and within the animal. For example, structural lipids are largely found as components of cell membranes, whereas oxidizable lipids are located in plasma and in intracellular lipid stores, especially adipose tissue.

This section begins with an overview of lipid metabolism (Chapter 53), followed by a chapter dealing with the biochemical and physiologic properties of saturated and unsaturated fatty acids (Chapter 54). Processes of fatty acid oxidation and biosynthesis are next detailed (Chapters 54 & 55), with processes of triglyceride and phospholipid biosynthesis and degradation following (Chapters 57 & 58). Sphingolipid biochemistry (Chapter 59) and lipid digestion are next discussed (Chapter 60), then cholesterol and bile acid biosynthesis and metabolism become the focus of Chapters 61 & 62. Problems inherent in the transport of hydrophobic lipid-fuels in the aqueous medium of blood become the focus of Chapters 63-66 (i.e., lipoproteins). The importance of eicosanoids to biomedical science are discussed in Chapters 68 & 69, control of lipolysis, ketone body formation and utilization are presented in Chapters 70 & 71, and some clinical situations that are a consequence of abnormal fat metabolism are discussed in Chapters 67 (hyperlipidemias) and 72 (steatosis).

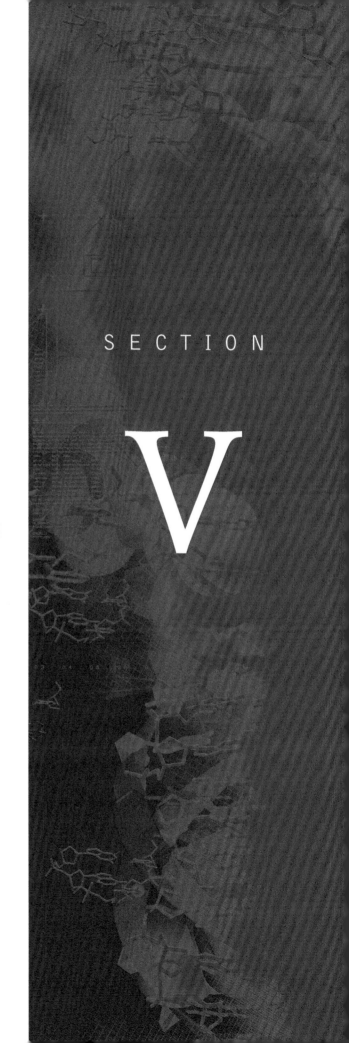

Lipid Metabolism

Chapter 53

Overview of Lipid Metabolism

Overview

- The caloric value of triglyceride is over twice that of glycogen or protein (per unit mass).
- Triglyceride provides more metabolic water upon oxidation than glycogen.
- When body fat content is high, total body water content is low.
- Lipids act as both thermal and electrical insulators.
- Some lipids help to regulate biologic reactions.
- Acetyl-CoA is the source of carbon atoms in all mammalian lipids, and it can be derived from carbohydrate, lipid, or amino acid catabolism.
- Major structural lipids include cholesterol and the charged lipids, such as glycerophospholipids and sphingolipids.
- Saturated and unsaturated fatty acids are simple lipids, phospholipids are complex lipids, and bile acids are derived lipids.

Lipids (from the Greek *lipos*, meaning "fat") are a heterogenous group of compounds which have the common property of being relatively insoluble in water, yet soluble in nonpolar solvents such as ether, chloroform, and benzene. Those of metabolic significance in mammalian organisms include the **fat-soluble vitamins, triglycerides (TGs), phospholipids (PLs), glycolipids, lipoproteins (LPs), sphingolipids, steroids, saturated** and **unsaturated fatty acids (SFAs** and **UFAs), eicosanoids,** and **ketone bodies (KB⁻s)**. In addition, herbivorous animals ingesting cellulose convert this carbohydrate polymer (via microbial fermentation) to **volatile fatty acids (VFAs**; primarily **acetate, butyrate,** and **propionate)**, then absorb these short-chain lipids for energy purposes (see Chapters 19 and 54).

Simple lipids are the SFAs and UFAs, as well as esters of these FAs with various alcohols, such as **glycerol** (e.g., mono-, di- and triglyc- erides; **Table 53-1**). Because they are uncharged, glycerides (acylglycerols) are sometimes referred to as neutral fat. **Waxes,** which are esters of FAs with higher molecular weight monohydric alcohols, are also simple lipids, and fat in the liquid state is an **oil. Phospholipids** are **complex lipids,** and contain a phosphoric acid residue in addition to esterified FAs and an alcohol. Phospholipids also contain nitrogen-containing bases or other substituents. In many PLs (e.g., the glycerophospholipids), the alcohol is **glycerol** (see Chapters 57 and 58), but in others (e.g., the sphingophospholipids like sphingomyelin), it is **sphingosine**. Nonphosphate-containing sphingolipids include the **glycolipids** (i.e., the cerebrosides, sulfatides, globosides, and gangliosides (see Chapter 59)).

Glycerophospholipids are amphipathic molecules (i.e., they possess both hydrophilic and hydrophobic properties), that fulfill several important roles in the organism. For example,

Copyright © 2015 Elsevier Inc. All rights reserved.

Table 53-1
Lipid Classification
Simple Lipids
Saturated and unsaturated fatty acids
Neutral fats (triglycerides)
Waxes
Complex Lipids
Phospholipids
Sphingolipids
Glycolipids
Lipoproteins
Sulfolipids
Aminolipids
Derived Lipids
Eicosanoids
Steroids
Ketone bodies
Fat-soluble vitamins

they act as major constituents of membranes and the outer layer of LPs, as surfactant in the lung, as precursors of second messengers, and as important constituents of nervous tissue (see Chapter 57). **Glycolipids** are compounds of fatty acids with carbohydrate, containing nitrogen, but no phosphoric acid (see Chapter 20). They are also important complex lipids, and serve as constituents of nervous tissue (particularly in the brain), and the outer leaflet of the cell membrane (where their carbohydrates project onto the cell surface). Other complex lipids include **lipoproteins** (i.e., chylomicrons, high-, low-, intermediate-, and very low-density lipoproteins), sulfolipids, and aminolipids. **Derived lipids** are lipids that cannot be neatly classified into either group above, and are generally derived from other lipids by hydrolysis. These include **steroids** (e.g., bile acids and the steroid hormones), **eicosanoids** (prostaglandins, thromboxanes, and leukotrienes), **ketone bodies** (acetoacetic acid, β-hydroxybutyric acid, and acetone), and the **fat-soluble vitamins** (A, D, E, and K).

For many years **triglycerides** present in fat cells were considered to be inactive storehouses of calorigenic material, called upon only in times of energy shortage. However, investigators showed by experiments in which labelled FAs were fed to mice in caloric equilibrium, that in only four days a considerable proportion of depot fat (i.e., TG) had been formed from dietary lipid. Since the total mass of TG in depot fat remained relatively constant, a corresponding amount of TG must have been mobilized during this period. These investigations demonstrated the dynamic state of body fat, a concept that forms the basis of our present understanding of lipid metabolism today.

A variable amount of dietary carbohydrate and protein is usually converted to TG before it is utilized for the purpose of providing energy. As a result, FAs derived from TG may be the major source of energy for many tissues; indeed, there is evidence that in certain organs, fatty acids may be used as fuel in preference to carbohydrate and protein (see Chapters 74 and 75).

The major site of **TG accumulation** in mammals is in the cytoplasm of adipose cells (fat cells or adipocytes). Droplets of TG coalesce to form a large globule, which may occupy most of the cell volume. Adipose cells are specialized for the assembly, synthesis, and storage of TG, and for its mobilization into fuel molecules (i.e., **free fatty acids (FFAs)**), that are transported to other tissues bound to albumin in blood.

As the principal form in which energy is stored in the body, **triglyceride** has advantages over carbohydrate and protein. Its caloric value per unit mass is over twice as great (i.e., **9 kCal/gm** for TG compared to about **4 kCal/gm** for carbohydrate and protein). The basis for this large difference in caloric yield is that FAs are far more anhydrous than proteins and carbohydrates, and they are also more highly reduced. They are nonpolar, whereas proteins and carbohydrates are much more polar, which makes them more highly hydrated. In fact, a gram of dry glycogen binds about two grams of water. Consequently, a gram of nearly anhydrous fat stores more energy than a gram of hydrated glycogen, which is probably the reason why TG, rather than glycogen, was evolutionarily selected as the major energy reserve. Fatty acids also provide more **metabolic water** upon oxidation than other biologic fuels, which is advantageous to hibernating animals and those occupying dry environments.

The relationship between **body fat, total body water (TBW)**, and the **lean body mass (LBM)** (i.e., TBW plus solids), is held within narrow limits in most normal, adult animals (**Fig. 53-1**). Total body water may vary with age and sex, due principally to the amount of body fat an animal possesses, which is on average about **18%** of the body weight. Since fatty

tissue contains less water per unit weight, obese animals have relatively less water than lean animals. Total body water per unit body weight is usually less in females after puberty than in males. The ratio of TBW to body weight varies inversely with the amount of fatty tissue present. When the fat content is high, TBW (i.e., **intracellular water (ICW)** plus **extracellular water (ECW)**) is low. Conversely, when fat content is low, fluid content is high.

Lipids are important dietary constituents not only because of their high **energy** value, but also because of **fat-soluble vitamins** and **essential FAs** contained in the fat of natural foods (**Table 53-2**). Lipids serve to **protect** cells from infection as well as excess loss or gain of water, and adipose tissue is frequently found around vital organs (e.g., kidneys), where it serves to cushion them against physical damage. Lipids also act as **thermal insulators** in subcutaneous tissue and around certain

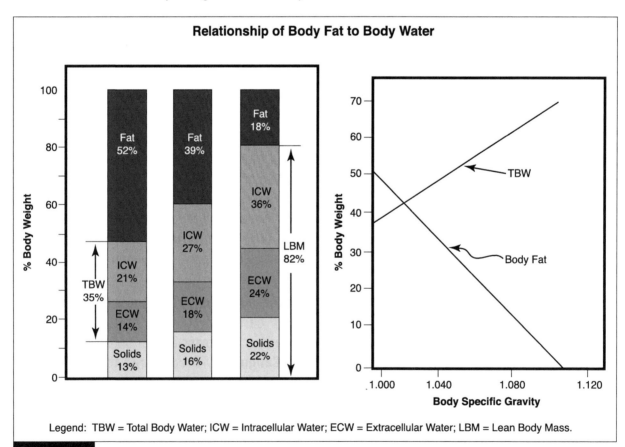

Relationship of Body Fat to Body Water

Legend: TBW = Total Body Water; ICW = Intracellular Water; ECW = Extracellular Water; LBM = Lean Body Mass.

Figure 53-1

Table 53-2
Primary Lipid Functions

Energy sources

Structural components of membranes

Protection against physical trauma

Thermal insulators

Metabolic regulators

Digestive aids

Electrical insulators

and they serve as a means of transporting lipids in blood. **Phospholipid** (lecithin) and lipid metabolites, such as **bile acids**, are important components of bile, and help to facilitate digestion and absorption of dietary lipid. A number of lipid metabolites also play important roles in **regulating biologic reactions**. Among these are the **steroid hormones** and derivatives of essential fatty acids known as **prosta-glandins**. In addition, some fat-soluble vitamins are metabolized in the body, and can then function to help regulate metabolism (e.g., **1,25-DHC**; see Chapter 45).

An overview of major metabolic pathways in lipid metabolism is shown in **Fig. 53-2**. Many metabolic products within cells are processed to a common product, **acetyl-CoA**, which is then either completely oxidized in the tricarboxylic acid (TCA) cycle, or is used in the biosynthesis of lipids. For example, the source of long-chain FAs within cells is either dietary lipid, that

organs, and nonpolar lipids act as **electrical insulators**, allowing rapid propagation of depolarization waves along myelinated nerves. The fat content of nerve tissue is particularly high. Lipids are important components of all biological membranes. Major **structural lipids** are cholesterol and the charged lipids, such as glycerophospholipids and sphingolipids.

Combinations of fat and protein (i.e., **lipoproteins**) are also important cellular constituents,

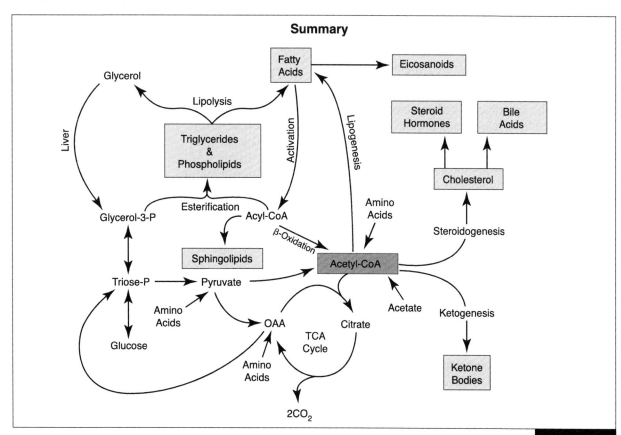

Figure 53-2

arriving from adipocytes following lipolysis of TG, or that derived from *de novo* biosynthesis from acetyl-CoA. Unsaturated FAs (e.g., arachidonic acid), derived from the 2-position of membrane bound PLs, are used in the biosynthesis of eicosanoids (see Chapters 68 and 69). Acetyl-CoA, in turn, can be derived from carbohydrate, lipid, or amino acid catabolism. Fatty acids may also be oxidized to acetyl-CoA (β-oxidation) within cells, or, following activation to acyl-CoA, be esterified to acylglycerols (e.g., TGs and PLs), or sphingolipids (see Chapter 59). Acetyl-CoA is also the source of all carbon atoms in cholesterol, steroid hormones, and bile acid molecules (see Chapter 61). In the liver, acetyl-CoA can also be used to form KB⁻s, which are alternative water-soluble fuels that become important sources of energy under certain metabolic conditions (e.g., starvation; see Chapters 73-76).

OBJECTIVES

- Give examples of derived, complex and simple lipids, and discuss their physiologic differences.

- Explain what is meant by "the dynamic state of body fat."

- Contrast the caloric yields of equimolar amounts of triglyceride, glycogen and protein oxidation, and explain the differences.

- Outline the physiologic advantage to hibernating animals of oxidizing fat over carbohydrate.

- Recognize the direct and inverse relationships between lean body mass, total body water and body fat content.

- Know and understand the seven primary functions of lipids.

- Identify the singular compound that appears to be common to all lipids, and discuss the dynamic nature of this relationship.

- Identify the intersection where carbohydrate, amino acid and lipid metabolism converge.

- Explain why animals occupying dry environments gain an advantage by oxidizing fat.

QUESTIONS

1. **Which one of the following is a simple, neutral lipid?**
   a. Phospholipid
   b. Triglyceride
   c. Glycolipid
   d. Sphingolipid
   e. Acetoacetate (a ketone body)

2. **Which one of the following is TRUE of glycogen (compared to triglyceride)?**
   a. It is more anhydrous.
   b. It is more reduced.
   c. It has a higher caloric value (per unit mass).
   d. It provides more metabolic water upon oxidation.
   e. It is more polar.

3. **Which one of the following is the compound that can serve as the source of all carbon atoms in mammalian lipids?**
   a. Glucose
   b. Glycerol
   c. Arachidonic acid
   d. Acetyl-CoA
   e. Succinyl-CoA

4. **The principal form in which energy is stored in the body is as:**
   a. Triglyceride.
   b. Cholesterol.
   c. Glycogen.
   d. Protein.
   e. Phospholipid.

5. **Select the TRUE statement below:**
   a. Since fatty tissue contains more water per unit weight than muscle tissue, obese animals possess relatively more water than lean animals.
   b. Eicosanoids are lipids derived from essential fatty acids.
   c. The caloric value of glycogen is over twice that of triglyceride (per unit mass).
   d. Glycerophospholipids possess a sphingosine backbone.
   e. Although lipids possess caloric value, they do not perform a role in regulating biologic reactions.

ANSWERS

1. b
2. e
3. d
4. a
5. b

Chapter 54

Saturated and Unsaturated Fatty Acids

Overview

- Over 100 different types of fatty acids have been isolated from various lipids of animal, plant, and microbial origin.
- Carbon atoms of the fatty acid hydrocarbon chain may be numbered from either end.
- Rumen microbes have a tendency to form saturated and trans-unsaturated fatty acids from dietary cis-unsaturated fats.
- Mammals cannot synthesize linoleic acid, therefore it must be provided in the diet.
- Mammals can synthesize γ-linolenic and arachidonic acid from dietary linoleic acid.
- Essential fatty acids are necessary precursors in the biosynthesis of eicosanoids.
- Omega-3 (ω-3) unsaturated fatty acids are components of fish oils.
- The fat of ruminant animals is high in trans-unsaturated fatty acids.

Fatty acids (FAs) are carboxylic acids obtained mainly from the hydrolysis of triglycerides (TGs), which are uncharged glycerol esters (i.e., neutral fats). Dietary TGs are usually the major form of lipid delivered to the small intestine, and represent the primary form in which lipid is stored in mammalian organisms. In addition to their biochemical role as a storage form of energy, FAs serve as building blocks of glyco- and phospholipids, molecules which play important structural and physiologic roles in the organism. A FA in the unesterified form is referred to as a **free fatty acid (FFA)**, a **non-esterified fatty acid (NEFA)**, or an **unesterified fatty acid**, and is a transport form bound to albumin in plasma.

Over 100 different types of FAs have been isolated from various lipids of animal, plant, and microbial origin. Fatty acids occurring in animals are usually unbranched, straight-chain derivatives, with an even number of carbon atoms (because they are synthe-

sized from acetyl-CoA, which is a 2-carbon fragment). Few branched-chain FAs have been isolated from plant and animal sources. The hydrocarbon chain may be **unsaturated** (containing one or more double bonds), or **saturated** (containing no double bonds), with a terminal carboxyl group (**Fig. 54-1**). Carbon atoms may be numbered from either end of the chain. When numbered from the **carboxyl end** (carbon No. 1), the adjacent carbon (No. 2) is the α-**carbon**, and carbons No. 3 and 4 are the β- and γ-**carbons**, respectively. The terminal methyl carbon is referred to as the **omega (ω-)** or **n-carbon**.

Various conventions are used for indicating the number and position of double bonds in **unsaturated fatty acids (UFAs)**. For example, when counting from the carboxyl end, $\Delta^{9,12}$**18:2** or **18:2;9,12** indicate that the UFA has 18 carbon atoms with two double bonds, located to the **left** of carbon atoms 9 and 12 (linoleic acid; see **Fig. 54-1**). When counting from the methyl end,

Copyright © 2015 Elsevier Inc. All rights reserved.

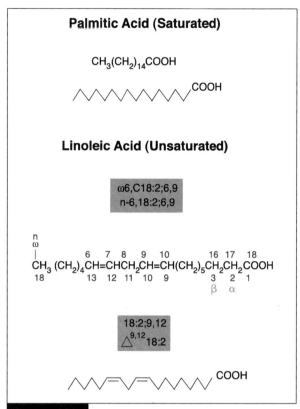

Figure 54-1

ω6,C18:2;6,9 or **n-6,18:2;6,9** indicate the double bonds to be located to the **right** of carbon atoms 6 and 9. Fatty acids produced and stored by mammalian cells generally contain between 14 to 24 carbon atoms (even numbers), with 16- and 18-carbon atom compounds (C_{16} and C_{18}) being the most prevalent. Fatty acids that are C_{10} or less are rarely present in animal lipids. Bacteria contain fewer and simpler types of FAs, namely C_{12} and C_{18} saturated fatty acids (SFAs), but also C_{16} and C_{18} monounsaturated FAs are found. Fatty acids with two or more double bonds (i.e., polyunsaturated fatty acids (PUFAs)), usually contain an even number of carbon atoms. Those with an odd number occur in only trace amounts in terrestrial animals, but in higher amounts in marine organisms.

Saturated fatty acids, which include the common short-chain volatile fatty acids (VFAs) derived from microbial carbohydrate fermentation, may be viewed as based upon **acetic acid** as the first member of the series (**Table 54-1**). Lipogenesis in higher animals involves formation of **palmitate** (C_{16}), from which other saturated and unsaturated fatty acids can be formed (see Chapter 56).

Unsaturated fatty acids generally predominate over the saturated ones, particularly in higher plants and in animals occupying cold environments (**Table 54-2**). Adding a double bond to a linear, SFA promotes twisting of the molecule. Thus, UFAs occupy more space and therefore are less tightly packed in biologic membranes. Their loose structure imparts a lower melting point to membranes (compared with more tightly packed, linear SFAs). Thus, as animals prepare for winter, desaturation of structural membrane FAs (i.e., addition of double bonds) is associated with lowering the melting point, and increasing membrane fluidity.

Variation in the location of the double bonds in UFA chains produces isomers. **Geometric isomerism** depends on the orientation of atoms or groups around the axis of double bonds. If the groups being considered are on the same side of the bond, the compound is called **cis**; if on opposite sides, **trans** (**Fig. 54-2**). Oleate, for example, is bent at the cis double bond, giving it somewhat of an L-configuration, whereas elaidate remains fairly straight at its trans double bond. Most naturally occurring unsaturated long-chain fatty acids (LCFAs) are of the cis-configuration. For example, **arachidonate** has 4 cis double bonds, and is U-shaped. Since it is found in the 2-position of many membrane-bound glycerophospholipids, its configuration influences molecular packaging and thus membrane fluidity.

Some **trans-UFAs** are found in ruminant fat, where they arise from the action of rumen microbes. The presence of large amounts of these FAs in partially hydrogenated vegetable oils (e.g., margarine) has raised the question of their safety as food additives (note: they "taste"

Table 54-1
Naturally Occurring Saturated Fatty Acids (a partial listing)

NAME	CARBON ATOMS	IMPORTANCE
Short-Chain		
Acetate	2	Major end-product of carbohydrate fermentation by symbiotic microbes
Propionate	3	Gluconeogenic end-product of carbohydrate fermentation by symbiotic microbes
Butyrate	4	End-product of carbohydrate fermentation by symbiotic microbes
Valerate	5	Minor end-product of carbohydrate fermentation by symbiotic microbes
Medium-Chain		
Caproate	6	Minor end-product of carbohydrate fermentation by symbiotic microbes
Caprylate (Octanoate)	8	Found in fats of plant origin
Laurate	12	Found in sperm
Long-Chain		
Myristate	14	Found in fats of plant origin
Palmitate	16	Common in plants and animals
Sterate	18	Common in plants and animals
Arachidate	20	Common in peanut oil
Behenate	22	Common in seeds

good). Their long-term effects in humans, where up to 15% of tissue FAs have been found to be in the trans-configuration of various individuals at autopsy, are being evaluated. They appear to affect membranes more like SFAs (since they are more linear). Dietary trans-UFAs (like SFAs) promote cholesterol absorption from the intestinal tract. They tend to raise blood LDL and lower HDL levels, and thus are associated with hypercholesterolemia, atherosclerosis, and coronary artery disease (see Chapters 65-67). Since they apparently lack essential FA activity, it is possible that they may antagonize the metabolism of the essential FAs.

Essential Fatty Acids

Over sixty years ago, investigators noticed that rats fed a purified nonlipid diet to which vitamins A and D were added exhibited a reduced growth rate and reproductive deficiency. Later work showed that the deficiency syndrome was cured by the addition of **linoleic**, **γ-linolenic**, and **arachidonic acids** to the diet. Further diagnostic features of the syndrome include scaly skin, lessened resistance to stress, impaired lipid transport, necrosis of the tail, and lesions in the urinary system. However, the condition is not fatal. Fatty acids required in the diet of mammals are termed

Table 54-2
Naturally Occurring Unsaturated Fatty Acids (a partial listing)

Name	Carbon Atoms and Double Bond Positions	Series	Location
Palmitoleate	16:1;9	ω7	Found in most fats
Oleate	18:1;9 (cis)	ω9	Common in natural fats
Elaidate	18:1;9 (trans)	ω9	Ruminant fats
Vaccenate	18:1;11	ω7	Formed by bacteria
Linoleate	18:2;9,12	ω6	Plants and animals
γ-Linolenate	18:3;6,9,12	ω6	Plants and animals (Derived from linoleate)
α-Linolenate	18:3;9,12,15	ω3	Fish oils
Eicosatrienoate (Dihomo γ-Linolenate)	20:3;8,11,14	ω6	Derived from dietary linoleate and γ-linolenate
Arachidonate (Eicosatetraenoate)	20:4;5,8,11,14	ω6	Animal phospholipids (Derived from Eicosatrienoate)
Timnodonate (Eicosapentaenoate)	20:5;5,8,11,14,17	ω3	Fish oils
Eruate	22:1;13	ω9	Mustard seed oil
Clupanodonate (Docosapentaenoate)	22:5;7,10,13,16,19	ω3	Fish oils, brain phospholipid
Cervonate (Docosahexaenoate)	22:6;4,7,10,13,16,19	ω3	Fish oils, brain phospholipid
Nervonate (Tetracosenoate)	24:1;15	ω9	Cerebrosides

essential FAs, with the most abundant being **linoleate**, which normally makes up from 10 to 20% of the total FA content of their TGs and glycerophospholipids. Linoleate cannot be synthesized by mammals, and must be obtained from meat or plant sources, in which it is quite abundant. Although linoleate is a precursor in mammals for the biosynthesis of **γ-linolenate**, **eicosatrienoate**, **arachidonate** and **eicosapentaenoate** (an ω3 UFA which can also be synthesized from dietary α-linolenate; see Chapter 68), these conversions may not occur in quantitatively significant amounts, thus rendering all of these UFAs essential.

Dietary essential FA deficiency can normally be prevented by an essential FA intake of only 1 to 2% of total caloric requirement. This quantity is normally obtained in the diet in all but the most extraordinary circumstances. Essential FAs are necessary precursors in the biosynthesis of a rather ubiquitous group of derivatives known as **eicosanoids** (prosta-glandins, thromboxanes, and leukotrienes), which are hormone-like compounds that in trace amounts have profound influences on a number of important physiologic activities (see Chapters 68 and 69).

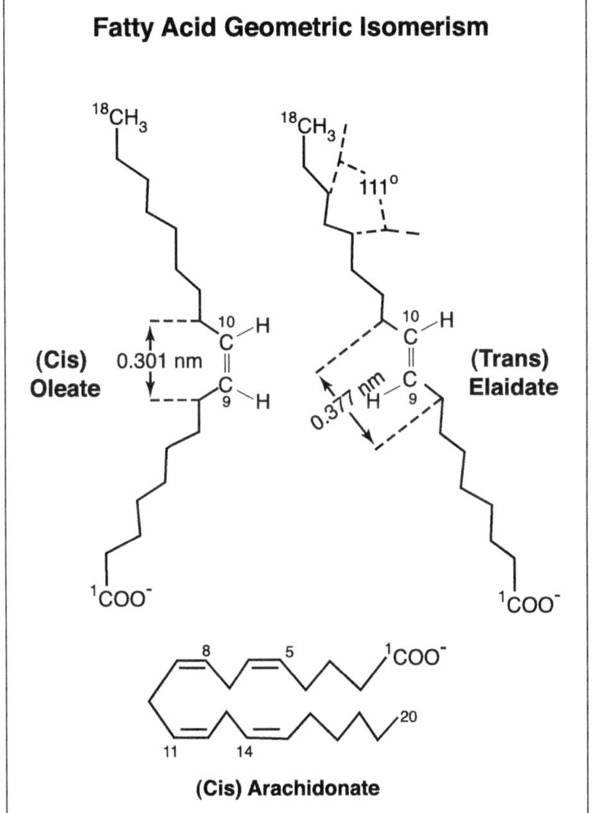

Fatty Acid Geometric Isomerism

(Cis) Oleate

(Trans) Elaidate

(Cis) Arachidonate

Figure 54-2

Animals can add **double bonds** to FA chains (i.e., **desaturation**) at positions **4**, **5**, **6**, **8**, and **9**; however, most mammals lack the ability to desaturate FA chains beyond C_9. **Desaturases** are found associated with liver microsomes, where O_2 and either NADPH or NADH are necessary for reaction. The enzymes appear to be those of a typical monooxygenase system involving cytochrome b_5 (hydroxylase). For example, if a double bond is introduced in the Δ^9 position of sterate, oleate is formed, or in palmitate, palmitoleate would be formed. The **cat** may be the only mammal lacking the enzyme that desaturates FAs at C_6.

It is a common finding in the husbandry of animals that the degree of saturation of the fat laid-down in depots can be altered by dietary means. If, for example, an animal is fed a diet containing a large quantity of vegetable oil (i.e., a high proportion of UFAs), the animal may lay down a soft type of depot fat. The converse situation may be found, however, in ruminant animals, where a characteristic hard, saturated fat is laid down as a result of the action of rumen microbes (which tend to saturate many UFAs of the diet, or convert them to the trans-configuration). Additionally, rumen microbes carry-out extensive lipolysis of dietary TG.

In summary, FAs occurring in animals are usually unbranched, straight-chain derivatives with an even number of carbon atoms, and a significant proportion (>50%) of them found in mammalian TGs are unsaturated. Although **desaturation** (that is conversion of -CH_2CH_2- to -CH=CH-) can be carried out by enzymes present in the liver, certain PUFAs which are required by the organism cannot be formed in this way, and therefore must be provided in the diet. Unsaturated fatty acids are necessary precursors in the biosynthesis of eicosanoids.

Saturated and unsaturated FAs are esterified to glycerol in TGs, the major lipid storage form, and they also are found in phospholipids and glycolipids, compounds which play important structural and physiologic roles in the organism. Fatty acids in the unesterified form are referred to as FFAs, NEFAs, or unesterified fatty acids, and are transported in blood bound to albumin.

Saturated FAs, which include the short-chain VFAs derived from microbial carbohydrate fermentation, have acetic acid (or acetate) as the first member of the series. Fatty acid biosynthesis in mammals involves formation of palmitate (C_{16}), from which other saturated and unsaturated fatty acids can be produced (see Chapter 56).

Several systems are used to name the UFAs. One system names them according to the number of carbon atoms present with the addition of suffixes, **-enoic**, **-dienoic**, **-trienoic**, etc. indicating one, two, three, or more double bonds. Double bond positions are indicated by the symbol Δ, with superscript numbers specifying the carbon atoms (numbered from the carboxyl end) involved in the double bonds. In a second, more abbreviated nomenclature, three sets of numbers are used to specify, respectively, the number of carbon atoms, the number of double bonds and their positions. For example, **linoleic acid** is described as **18:2;9,12** (see **Fig. 54-1**). A third system counts from the methyl end, with **n-6,18:2;6,9** or **ω6,C18:2;6,9** indicating the double bonds of linoleic acid to be located to the right of carbon atoms 6 and 9.

OBJECTIVES

• Explain differences in naming UFAs from the carboxyl end as opposed to the methyl end.

• Provide reasons for the prevalence of UFAs in cell membranes of animals preparing for winter.

• Name the common short-chain, medium-chain and long-chain FAs found in plants and animals, and explain why most are even-numbered.

• Explain differences between cis- and trans-UFAs, discuss their formation, and recognize why dietary trans-UFAs promote intestinal cholesterol absorption..

• Identify the essential FAs, and show how they can be (partially) interconverted.

• Name the common UFAs, identify the positions of their double bonds, and note dietary sources.

• Show how FA desaturation and elongation from palmitate can result in the biosynthesis of many UFAs found in the CNS.

QUESTIONS

1. **$CH_3(CH2)_7CH=CH(CH_2)_7COO^-$:**
   a. Arachidonic acid
   b. ω6,C18:2;6,9
   c. 18:1;10 (Palmitate)
   d. $\Delta^9 18:1$
   e. n-3,18:1;11

2. **Which one of the following is an omega-3 (ω-3) unsaturated fatty acid?**
   a. Palmitoleate (16:1;9)
   b. γ-Linolenate (18:3;6,9,12)
   c. α Linolenate (18:3;9,12,15)
   d. Eicosatrienoate (20:3;8,11,14)
   e. Arachidonate (20:4;5,8,11,14)

3. **Which one of the following is an essential fatty acid?**
   a. Linoleate
   b. Oleate
   c. Eruate
   d. Nervonate
   e. Butyrate

4. **Select the TRUE statement below:**
   a. Most naturally occurring unsaturated long-chain fatty acids are of the trans-configuration.
   b. Polyunsaturated fatty acids are commonly found in bacteria.
   c. Short-chain volatile fatty acids are normally found in triglycerides of ruminant animals.
   d. Unsaturated fatty acids generally predominate over the saturated ones in higher plants.
   e. Unsaturated fatty acids impart a higher melting point to cell membranes of animals.

5. **Select the FALSE statement below:**
   a. Mammals are capable of synthesizing linoleate from arachidonate.
   b. The liver is capable of converting a saturated fatty acid to an unsaturated form.
   c. Propionate is a short-chain, saturated fatty acid.
   d. Lipogenesis in mammals typically results in the formation of palmitate, from which other fatty acids can be formed.
   e. Arachidonate is primarily found in animal phospholipids.

6. **Select the TRUE statement below:**
   a. Cats cannot desaturate fatty acids.
   b. Animals generally possess more SFAs than UFAs.
   c. n-6, 18:2;6,9 is a SFA.
   d. Trans-UFAs promote intestinal cholesterol absorption.

6. d
5. a
4. d
3. a
2. c
1. d

ANSWERS

Chapter 55

Fatty Acid Oxidation

Overview

- Fatty acids are oxidized in mitochondria, and synthesized in the cytoplasm.
- Long-chain fatty acids must form an active intermediate, fatty acyl-CoA, before being oxidized inside mitochondria.
- Carnitine shuttles long-chain fatty acids across mitochondrial membranes.
- Fatty acids are both synthesized from and oxidized to a common compound, acetyl-CoA.
- Important enzymes in mitochondrial membranes that help prepare LCFAs for mitochondrial β-oxidation include acyl-CoA synthetase, carnitine palmitoyltransferase I (CPT-1), carnitine-acylcarnitine translocase (CAT), and carnitine palmitoyltransferase II (CPT-2).
- Peroxisomal β-oxidation is reserved for very long-chain fatty acids (C_{20} or longer), that are poorly handled by mitochondria.
- Oxidation is uncoupled from phosphorylation in peroxisomes.
- Cytoplasmic malonyl-CoA inhibits the rate-limiting enzyme in fatty acid oxidation.

Fatty acids (FAs) are both synthesized from and oxidized to a common compound, **acetyl-CoA**. Although the starting material for one process is identical to the product of the other, FA oxidation (i.e., β-oxidation) is not the simple reversal of FA biosynthesis. **β-Oxidation** is a **mitochondrial** process, whereas **FA biosynthesis** occurs in the **cytoplasm**. Additionally, physiologic conditions that promote FA biosynthesis largely inhibit β-oxidation, and *vice versa*, thus preventing futile cycling.

Fatty acids are an important source of energy for muscle, kidney, and liver tissue. The term "**free fatty acid**" (FFA) usually refers to long-chain fatty acids (LCFAs) that are in the unesterified state. They are also sometimes referred to as **unesterified fatty acids**, or **non-esterified fatty acids** (**NEFAs**). The LCFAs in plasma are bound to serum albumin, and inside cells they are attached to a FA binding protein (or Z protein). Consequently, they are seldom free, and are usually in lipoprotein form. As previously discussed, short- and medium-chain fatty acids (SCFAs and MCFAs) are more water-soluble, and therefore are less likely to be combined with protein.

As with glucose, FAs that enter responsive cells must first be converted in a reaction with ATP, in this case to an **active intermediate** (**fatty acyl-CoA**) that interacts with enzymes responsible for further metabolism (**Fig. 55-1**). This is the only step in the catabolism of FAs that requires energy from ATP, and it is **physiologically irreversible** (like the conversion of phosphoenolpyruvate to pyruvate, see Chapter 26). The enzyme responsible for this activation is **acyl-CoA synthetase**, which is

Copyright © 2015 Elsevier Inc. All rights reserved.

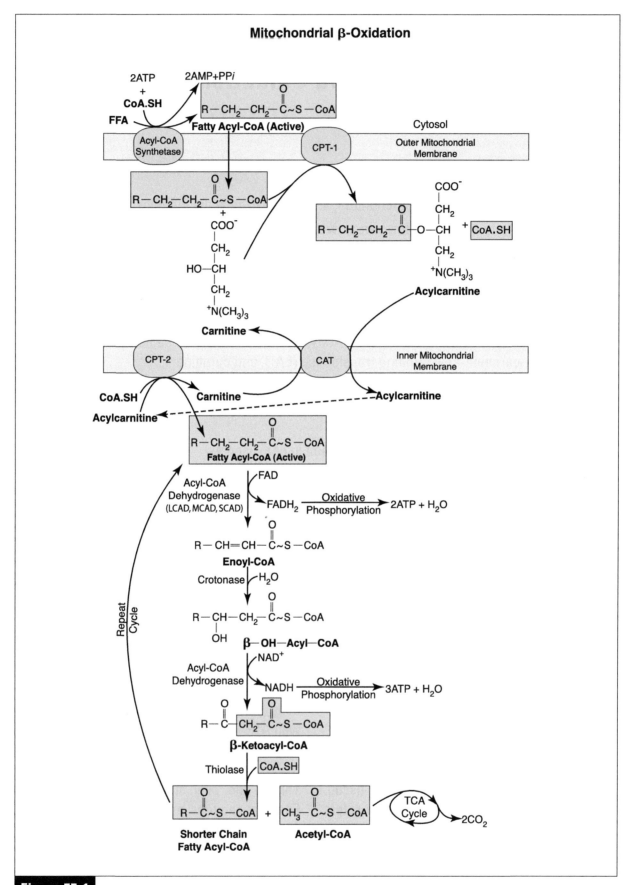

Mitochondrial β-Oxidation

Figure 55-1

found in the endoplasmic reticulum, inside mitochondria, and on outer mitochondrial membranes. Several acyl-CoA synthetases have been described, each specific for FAs of different chain lengths, and they are all vitamin B_5-dependent (see Chapter 41).

Carnitine

Carnitine (β-OH-γ-trimethylammonium butyrate), a compound synthesized from methionine and lysine in the liver and kidney, is widely distributed, particularly in mitochondrial membranes of muscle tissue. It is also found in high concentrations in milk, particularly during the early stages of lactation. The immediate precursor to carnitine is γ-**butyrobetaine**, which can be obtained from muscle, liver or kidney (**Fig. 55-2**).

Activation of MCFAs, and their oxidation within mitochondria, is thought to occur independent of carnitine (**Fig. 55-3**), but long chain fatty acyl-CoA units will not penetrate the inner mitochondrial membrane and become oxidized unless they form acylcarnitines. **Carnitine palmitoyltransferase I** (**CPT-1**), an enzyme present in the outer mitochondrial membrane, converts long-chain fatty acyl-CoA units to **acylcarnitine**. This compound next gains access to the inner mitochondrial matrix and β-oxidation system of enzymes through the action of **carnitine-acylcarnitine translocase** (**CAT**), an enzyme which acts as an inner mitochondrial membrane carnitine exchange

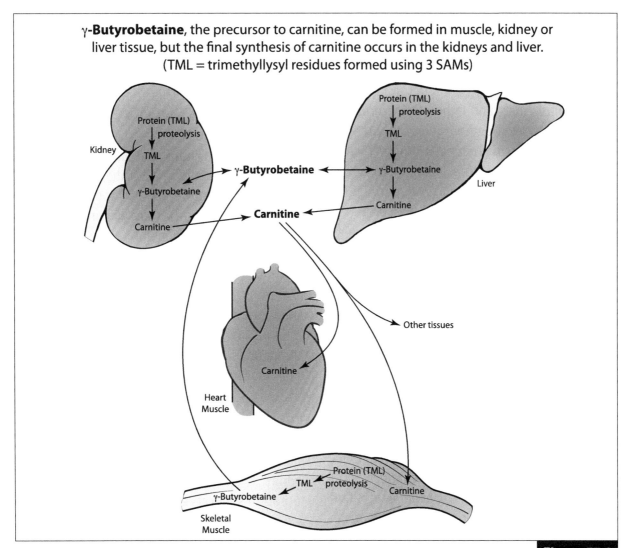

γ-**Butyrobetaine**, the precursor to carnitine, can be formed in muscle, kidney or liver tissue, but the final synthesis of carnitine occurs in the kidneys and liver. (TML = trimethyllysyl residues formed using 3 SAMs)

Figure 55-2

transporter. While acylcarnitine is transported in, one molecule of carnitine is transported out. Acylcarnitine then reacts with CoA, catalyzed by **carnitine palmitoyltransferase II (CPT-2)**, located on the inside of the inner mitochondrial membrane. Fatty acyl-CoA is reformed in the mitochondrial matrix, and carnitine is released. Another enzyme, **carnitine acetyltransferase**, is thought to facilitate transport of acetyl groups through certain mitochondrial membranes, and together with fructose and lactate, may be important for support of sperm motility.

Relationships between **carnitine** and **cardiovascular** health have been examined, with some studies indicating that administration of this compound may actually reduce the risk of death following at heart attack, for ischemic events (e.g., myocardial infarction) can deplete tissue carnitine levels. Other studies, however, indicate that carnitine contained in red meat, fish, poultry, milk and other dairy products allows intestinal proatherogenic microbes to convert it to **trimethylamine-N-oxide (TMAO)**, which some investigators believe can accelerate atherosclerotic processes, perhaps by reducing **reverse cholesterol transport** (see Chapter 66).

Mitochondrial β-oxidation

The mitochondrial β-oxidation of LCFAs is a cyclic process, involving four key enzymes that sequentially remove acetyl-CoA from the carboxyl end of the FA molecule. Each cycle of β-oxidation produces one molecule each of **acetyl-CoA**, the reduced form of flavin adenine dinucleotide (**FADH$_2$**, a vitamin B$_2$-dependent coenzyme; see Chapter 40), and **NADH** (see **Fig. 55-1**). Entry of FADH$_2$ and NADH into oxidative phosphorylation leads to the synthesis of **5 ATP** for each of the first **seven acetyl-CoA molecules** formed through complete β-oxidation of palmitate (**7 x 5 = 35 ATP**). A total of **8 acetyl-CoA** units are formed, and each gives

rise to **12 mol of ATP** following oxidation in the TCA cycle, thus producing an additional **96 mol of ATP (8 x 12 = 96 ATP)**. Two must be subtracted for the initial FA activation reaction, yielding a net gain of **129 mol of ATP per mole of palmitate**.

There are three **isozymes** of **fatty acyl-CoA dehydrogenase**, the enzyme that catalyzes the first step in β-oxidation: **long-chain (LCAD)**, **medium-chain (MCAD)**, and **short-chain acyl-CoA dehydrogenase (SCAD)**, that work on **C$_{14}$-C$_{18}$**, **C$_6$-C$_{12}$**, and **C$_4$ FAs**, respectively. The complete oxidation of LCFAs requires all three isozymes, with MCAD and SCAD becoming preferred isozymes as the FA becomes progressively shorter. Genetic deficiencies in each have been reported.

Fatty acids with an **odd number** of carbon atoms are oxidized by the pathway of β-oxidation until the final 3-carbon **propionyl-CoA** residue remains. This compound is then converted to succinyl-CoA, a constituent of the tricarboxylic acid (TCA) cycle (see Chapter 37). **Unsaturated fatty acids** are similarly oxidized by β-oxidation until the unsaturation point (double bond) is encountered. Most double bonds in naturally occurring FAs are **cis**, in contrast to the trans-configuration formed in the normal FA oxidation cycle (see **Fig. 55-1**). Therefore, in order for oxidation of UFAs to proceed beyond the steps where the cis configurations are encountered, they must first be **isomerized** (cis to trans), and then **epimerized** (D to L) into compounds which satisfy structural requirements for further mitochondrial oxidation.

Peroxisomal β-oxidation

A secondary form of β-oxidation occurs in microbodies (**peroxisomes**) of the liver and kidney. It differs from its mitochondrial counterpart in several respects:

- **It is quantitatively less important.**
- **Entry of fatty acyl-CoA does not require the carnitine shuttle, for peroxisomes lack**

carnitine palmitoyltransferase I (CPT-I).

- **Oxidation is catalyzed by different enzymes, such as oxidases that require a high oxygen tension, and produce H_2O_2 as a byproduct.**
- **Catalase is a prevalent enzyme in peroxisomes.**

Peroxisomes possess a modified form of β-oxidation that ultimately leads to the formation of acetyl-CoA, octanoate, and H_2O_2, with the latter being broken down by catalase (see Chapter 30). The system begins with an initial activation by a **very long-chain acyl-CoA synthetase** that facilitates the oxidation (but not the phosphorylation) of very LCFAs (e.g., C_{20}-C_{22}) that are poorly handled by mitochondria. Peroxisomal enzymes are induced by high-fat diets, and in some species by hypolipidemic drugs such as clofibrate (see Chapter 67). These enzymes do not attack MCFAs or SCFAs, and the β-oxidation sequence ends at octanoyl-CoA. Octanoate (a MCFA) and acetyl groups subsequently leave peroxisomes to be further oxidized in mitochondria.

Since oxidation is uncoupled from phosphorylation in peroxisomes (like in brown fat tissue), these organelles function in thermoregulation as well as in disposal of potentially harmful lipid peroxides from very LCFAs (see Chapters 30 and 46). Another function of peroxisomes is to shorten the side chain of cholesterol in bile acid formation (see Chapter 62). **Zellweger syndrome**, which is characterized by early death from liver and kidney abnormalities, is reportedly associated with a defect in the importation of enzymes into peroxisomes.

In summary, the enzyme-catalyzed reactions constituting metabolism include some in which large molecules are broken down into smaller ones, and others in which small molecules are used from the biosynthesis of larger, more complex molecules. The former can be generally grouped under the category of **catabolism,** and the latter under the category of **anabolism.** Catabolism involves two major

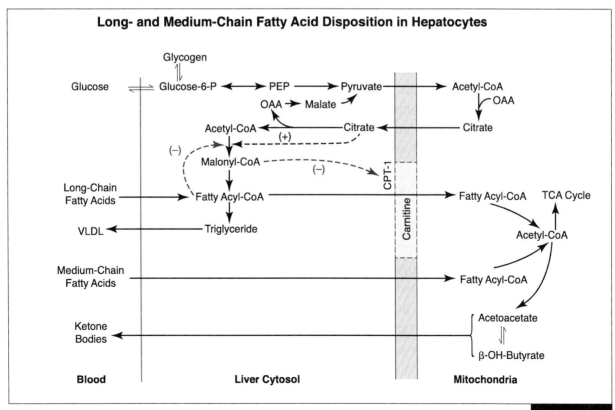

Long- and Medium-Chain Fatty Acid Disposition in Hepatocytes

Figure 55-3

phases. First, there is digestion in the intestinal tract, which serves to break large organic molecules present in the diet into smaller ones which can be absorbed, distributed throughout the organism, and enter cells in various tissue beds. The digestion of protein, carbohydrate, and lipid, as well as the absorption of products of these processes are described in Chapters 7, 38, and 60, respectively.

Absorbed constituents may themselves be further degraded in individual cells of the body. When this occurs, some of the chemical energy present in these compounds is transferred to other molecules, notably ATP, which can then be used directly to provide energy for processes that require it. In addition, the same products of digestion are used to synthesize new molecules in anabolic processes. These are required for a variety of purposes:

1) **to achieve growth and repair of the organism,**

2) **to replace molecules which have broken down, and**

3) **to replenish storage fuel which can subsequently undergo catabolism.**

The above metabolic generalizations cannot be properly elaborated without considering the chemical nature of the molecules involved, in this case FAs. Long-chain fatty acids are made available from stored TG when glucose is unable to provide sufficient energy for the needs of many active tissues of the body. This can be a frequent occurrence; LCFAs become the major respiratory fuel during starvation or prolonged exercise (see Section VI). β-Oxidation of FAs is largely a mitochondrial process. Fatty acids must first be converted to an active intermediate (**fatty acyl-CoA**) before undergoing β-oxidation. Long-chain fatty acids are activated in the cytoplasm or on the outer mitochondrial membrane surface, whereas MCFAs may be activated within the mitochondrial matrix (**Fig. 55-3**). **Carnitine** transports long-chain fatty acyl-CoA subunits across mitochondrial membranes, while MCFAs require no special transport system. Their ultimate fate inside hepatocytes is **β-oxidation** to **acetyl-CoA,** with the concomitant reduction of **NAD⁺** to **NADH,** and **FAD** to **FADH₂ (see Fig. 55-1**). Once acetyl-CoA is formed, it may then enter the TCA cycle, or, depending upon the degree of FA influx, enter ketone body (acetoacetate and β-OH-butyrate) biosynthesis.

The rate-limiting step in FA oxidation is catalyzed by **CPT-1,** which is allosterically inhibited by **malonyl-CoA.** Thus, accumulation of malonyl-CoA simultaneously provides substrate for cytoplasmic FA biosynthesis, and blocks entry of FAs into mitochondria where β-oxidation occurs. The rate-limiting step in FA biosynthesis is catalyzed by **acetyl-CoA carboxylase,** the enzyme that synthesizes malonyl-CoA. Acetyl-CoA carboxylase is allosterically activated by **citrate,** which carries acetyl-CoA from mitochondria into the cytosol. **Palmitoyl-CoA,** the normal end product of FA biosynthesis, inhibits acetyl-CoA carboxylase (see Chapter 56).

As a final note, the liver cell may re-esterify some incoming LCFAs into TG, and export it into the circulation as a part of very low density lipoprotein (VLDL; see Chapter 65). In contrast, MCFAs are not normally utilized for TG biosynthesis within the cytosol of hepatocytes.

OBJECTIVES

- Identify solubility differences between SCFAs, MCFAs and LCFAs, and provide the biochemical basis for the inability of animals to convert FAs into glucose.

- Discuss the physiological significance of the Acyl-CoA Synthetase reaction, and identify locations where this reaction occurs.

- Explain why milk, during the early stages of lactation, is high in carnitine.

- Summarize and understand how the rates of

hepatic triglyceride formation, FA biosynthesis and β-oxidation are controlled.

- Recognize why the β-oxidation of MCFAs is independent of carnitine, and why MCFAs are not found in circulating lipoproteins.

- Identify two outer and two inner membrane enzymes that participate in the movement of LCFAs from the cytoplasm to the mitochondrial matrix.

- Itemize net ATP production from the complete oxidation of one mole of palmitate.

- Predict what would most likely happen to the propionyl-CoA produced in bovine hepatocytes following β-oxidation of an odd numbered fatty acid.

- Compare the process of UFA oxidation to that of SFA oxidation.

- Compare and contrast peroxisomal β-oxidation to mitochondrial β-oxidation, and discuss the role played by peroxisomes in thermoregulation.

- Know why cytoplasmic malonyl-CoA allosterically inhibits CPT-1.

QUESTIONS

1. The only step in fatty acid catabolism that requires ATP is:
a. Fatty acyl-CoA —> Enoyl-CoA.
b. FFA —> Fatty acyl-CoA.
c. Enoyl-CoA —> β-Hydroxyacyl-CoA.
d. β-Hydroxyacyl-CoA —> β-Ketoacyl-CoA.
e. β-Ketoacyl-CoA —> Acetyl-CoA + Fatty acyl-CoA.

2. Which one of the following enzymes is found on the outer mitochondrial membrane?
a. Carnitine palmitoyltransferase I
b. Carnitine palmitoyltransferase II
c. Carnitine acetyltransferase
d. Acyl-CoA dehydrogenase
e. Thiolase

3. How many moles of ATP are generated per mole of palmitate completely oxidized through mitochondrial β-oxidation and the TCA cycle?
a. 19

b. 39
c. 89
d. 109
e. 129

4. Peroxisomal β-oxidation:
a. Is identical to mitochondrial β-oxidation.
b. Is quantitatively more important than mitochondrial β-oxidation.
c. Requires the carnitine shuttle.
d. Will not completely oxidize fatty acids.
e. Has preference for short-chain fatty acids (C_2-C_4).

5. Carnitine palmitoyltransferase I is allosterically inhibited by:
a. Citrate.
b. Oxaloacetate.
c. Acetyl-CoA.
d. Malonyl-CoA.
e. Palmitoyl-CoA.

6. Carnitine is synthesized from:
a. Malonyl-CoA.
b. Lysine and methionine.
c. Acetyl-CoA.
d. Proline and alanine.
e. Urea.

7. Fatty acids with an odd number of carbon atoms are β-oxidized to:
a. Succinyl-CoA.
b. CO_2 and H_2O.
c. Propionyl-CoA.
d. CoA.SH.
e. Acetyl-CoA.

8. Malonyl-CoA:
a. Blocks entry of FAs into mitochondria.
b. Is formed in the cytoplasm via carboxylation of acetyl-CoA.
c. Is a substrate for cytoplasmic palmitate biosynthesis.
d. Coordinates FA biosynthesis with β-oxidation.
e. All of the above

ANSWERS

8. e
7. c
6. b
5. d
4. d
3. e
2. a
1. b

Chapter 56

Fatty Acid Biosynthesis

Overview

- Fatty acids are normally synthesized from acetyl-CoA, a process that requires ATP, biotin, Mg^{++}, and Mn^{++}.

- Acetyl-CoA carboxylase, the rate-limiting enzyme in fatty acid biosynthesis, is inhibited by glucagon and epinephrine, and stimulated by insulin.

- Intermediates in fatty acid biosynthesis are attached to acyl carrier protein (ACP).

- Malonyl-CoA serves as an activated donor of acetyl groups in fatty acid biosynthesis.

- Propionate (C_3) may be used in place of acetate (C_2) as a priming molecule for fatty acid biosynthesis in adipocytes and in the lactating mammary gland.

- Fatty acid elongation beyond palmitate takes place in mitochondria, or on the smooth endoplasmic reticulum (SER).

- Animals are capable of synthesizing all of the FAs they need except for the essential FAs, which must be supplied through the diet.

- The hexose monophosphate shunt, malic enzyme, and cytoplasmic isocitrate dehydrogenase assist in providing NADPH for palmitate biosynthesis.

- Fatty acid biosynthesis occurs in adipocytes of ruminant animals, and in the liver of non-ruminant animals.

Fatty acids (FAs) are synthesized from **acetyl-CoA**, the same substance to which they are degraded in β-oxidation (see Chapter 55). There are, however, major differences between these catabolic and anabolic pathways. **Fatty acid β-oxidation**, for example, is a **mitochondrial** process, whereas the **biosynthesis of saturated fatty acids (SFAs)** up to C_{16} (**palmitate**), takes place in the **cytoplasm**. A mitochondrial system of enzymes catalyzes further extension of the C_{16} chain length, and the smooth endoplasmic reticulum (SER) contributes enzymes responsible for the conversion of SFAs to unsaturated fatty acids (UFAs), and also for further chain length extension. This FA biosynthetic system has been found in several different organs and tissues, including liver, kidney, brain, lung, mammary, and adipose tissue.

The synthesis of **palmitate** from **acetyl-CoA** can be described by the following equation:

$$\left[\begin{array}{c} \text{8 Acetyl-CoA} + \text{7 ATP} + \text{14 NADPH} \longrightarrow \\ \text{Palmitate} + \text{14 NADP}^+ + \text{8 CoA} \\ + \text{6 H}_2\text{O} + \text{7 ADP} + \text{7 Pi} \end{array} \right]$$

Acetyl-CoA, derived primarily from **glucose**, **amino acids**, and in herbivorous animals, **acetate**, must first be converted into **malonyl-CoA** by the enzyme, **acetyl-CoA carboxylase** (Enz; **Fig. 56-1**), a **committed step** which is **physiologically irreversible**. This rate-limiting

Copyright © 2015 Elsevier Inc. All rights reserved.

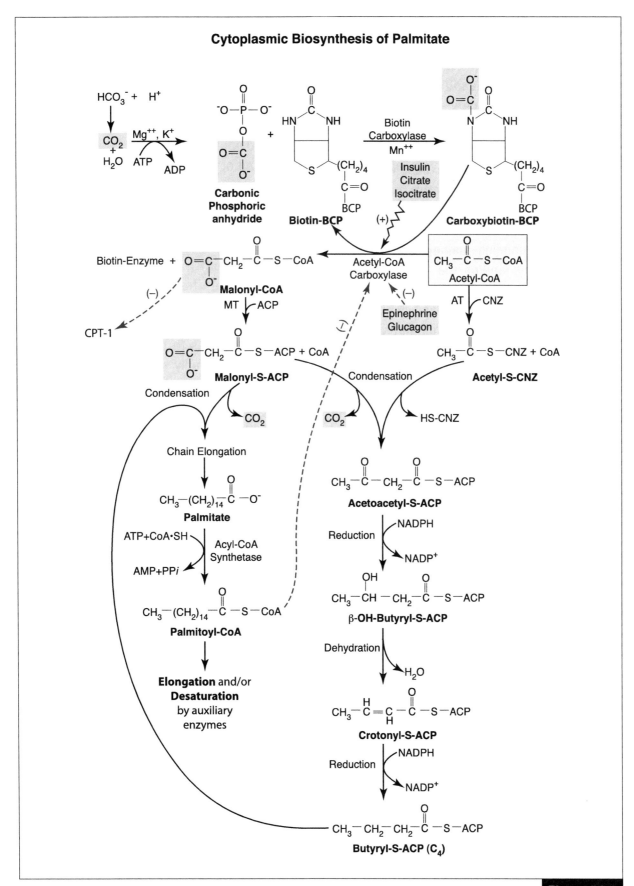

Cytoplasmic Biosynthesis of Palmitate

Figure 56-1

enzyme in fatty acid biosynthesis is allosterically activated by **citrate** and **isocitrate** (V_{max} increased 15-fold), and inhibited by **palmitoyl-CoA negative feedback**. Additionally, **insulin** stimulates acetyl-CoA carboxylase activity, whereas **glucagon** and **epinephrine**, by increasing intracellular levels of cAMP, inhibit it.

Formation of malonyl-CoA initially involves covalent binding of CO_2 to **biotin**, using energy from ATP hydrolysis (see Chapter 42). Then, using the same enzyme, CO_2 is transferred to acetyl-CoA, producing **malonyl-CoA**. Changes in diet, activity, and/or the insulin:glucagon ratio affect the amount of FA biosynthesis by affecting acetyl-CoA carboxylase synthesis and activity. For example, a diet rich in protein and carbohydrate, but low in fat, increases enzyme biosynthesis, whereas exercise, starvation, or a diet high in fat has the opposite effect. Known functions of **malonyl-CoA** include serving as an activated donor of acetyl groups in FA biosynthesis, helping to coordinate FA biosynthesis with FA β-oxidation (i.e., malonyl-CoA inhibits CPT-1), and suppressing feeding behavior (see Chapters 55 & 70).

The remaining reactions in FA biosynthesis take place on a multifunctional protein known as **fatty acid synthetase (FAS)**. Seven different enzymatic activities are contained in this single polypeptide, which has a molecular weight of about 250,000 daltons. Additionally, intermediates are attached to **acyl carrier protein (ACP)** during elongation. The FAS/ACP complex catalyzes the sequential addition of C_2 units to the carboxyl end of the growing fatty acyl chain. Each new C_2 addition requires malonyl-CoA, and is accompanied by the release of CO_2.

The initial condensation reaction involves an **acetyl group** and a **malonyl group**. However, before this reaction can occur, the acetyl group and the malonyl group must be transferred from their respective CoA derivatives to a **condensing enzyme (CNZ)** and **ACP**, respectively. These priming reactions are catalyzed by **acetyl transacetylase (AT)** and **malonyl transacetylase (MT; Fig. 56-1)**. Each subsequent elongation step consists of four core reactions (a **condensation**, **reduction**, **dehydration**, and then another **reduction**), which are repeated until the FA is completely synthesized. At the end of the first cycle, the C_4 **butyric acid** is transferred from **ACP** to the condensing enzyme, and the **ACP** is primed with another **malonyl group** in preparation for the next cycle of elongation. When the FA has reached the C_{16} chain length (i.e., palmitate), it can either be released as free palmitate or transferred to CoA, yielding **palmitoyl-CoA**. Palmitoyl-CoA can then be elongated and/or desaturated by auxiliary enzymes to provide a variety of FAs. Animals can usually synthesize all of the FAs they need except **linoleic** and γ-**linolenic acid**, which are essential components of the diet (see Chapter 54).

Fatty Acid Elongation Beyond Palmitate

Two elongation systems exist in animals, a **mitochondrial** and a **microsomal (SER)** system. Both elongate palmitate to a variety of SFAs by sequential addition of C_2 units to the carboxyl end. The mitochondrial system uses acetyl-CoA as a C_2 donor, and either NADH or NADPH as reducing agents (see Chapter 28); whereas the microsomal system uses malonyl-CoA as a C_2 donor, and NADPH as the sole reducing agent. Double bonds are introduced by a family of **microsomal desaturases** having slight differences in their active sites. These reactions are wide spread in the body, and permit formation of a variety of LCFAs that can be used for several purposes, to include nervous tissue formation, eicosanoid biosynthesis, and for altering membrane structure.

A variety of PUFAs can be formed using desaturation and elongation reactions that introduce double bonds at positions 4, 5 or 6, and elongate on the proximal (COOH) end of the FA chain.

Conversion of **α-linolenic acid** to **cervonic acid** in the **central nervous system (CNS)** is a specific example of such a reaction sequence (**Fig. 56-2**). Although cervonic acid can also be obtained from fish oils, it is needed for brain and retina development, and is supplied from maternal blood to the fetus via the placenta, and to the neonate via milk. The FA desaturation/chain elongation system has been found to be diminished with starvation, upon epinephrine or glucagon administration, and in insulin-dependent diabetes mellitus (DM).

NADPH Generation and FattyAcid Biosynthesis

NADPH, serving as a reducing equivalent donor, is an essential component of lipid biosynthesis. Oxidative reactions of the **hexose monophosphate shunt** (**HMS**; see Chapter 28) are a major source of hydrogen, and thus **NADPH**, and it is significant that tissues specializing in lipogenesis (e.g., liver, adipose tissue, and the lactating mammary gland) also possess an active HMS. Moreover, both metabolic pathways (FA biosynthesis and

---

Conversion of α-linolenic acid to docosahexaenoic acid (cervonic acid) in the brain is an example of a reaction complex using sequential desaturation and elongation reactions. Since these reactions can occur in any order, they hold potential for giving rise to many different UFAs.

$CH_3 - (CH_2 - CH = CH)_3 - CH_2 - CH_2 - CH_2 - CH_2 - CH_2 - CH_2 - CH_2 - COOH$ **Proximal (COOH) end of the FA chain**
α-Linolenic Acid 18:3;9,12,15

⎯⎯⎯▶ "Δ⁶-desaturase"

$CH_3 - (CH_2 - CH = CH)_3 - CH_2 - CH = CH - CH_2 - CH_2 - CH_2 - CH_2 - COOH$

⎯⎯⎯▶ **elongation (using mitochondrial acetyl-CoA)**

$CH_3 - (CH_2 - CH = CH)_3 - CH_2 - CH = CH - CH_2 - CH_2 - CH_2 - CH_2 - CH_2 - CH_2 - COOH$

⎯⎯⎯▶ "Δ⁵-desaturase"

$CH_3 - (CH_2 - CH = CH)_3 - CH_2 - CH = CH - CH_2 - CH = CH - CH_2 - CH_2 - CH_2 - COOH$
Timnodonate 20:5;5,8,11,14,17

⎯⎯⎯▶ **elongation (again using mitochondrial acetyl-CoA)**

$CH_3 - (CH_2 - CH = CH)_3 - CH_2 - CH = CH - CH_2 - CH = CH - CH_2 - CH_2 - CH_2 - CH_2 - CH_2 - COOH$
Clupanodonate 22:5;7,10,13,16,19

⎯⎯⎯▶ "Δ⁴-desaturase"

$CH_3 - (CH_2 - CH = CH)_3 - CH_2 - CH = CH - CH_2 - CH = CH - CH_2 - CH = CH - CH_2 - CH_2 - COOH$
Cervonic Acid (docosahexaenoic acid) 22:6;4,7,10,13,16,19

Most animal possess desaturases that add double bonds to positions 4,5,6,8 and 9; however, primates can only add double bonds at position 4,5 & 6, and cats lack the D6-desaturase.

Figure 56-2

the HMS) are extramitochondrial, thus there are no membrane barriers for the transfer of NADPH/NADP$^+$ from one pathway to the other.

Extramitochondrial **isocitrate dehydrogenase** is a substantial, if not main source of **NADPH** for FA biosynthesis in adipocytes of ruminant animals (**Fig. 56-3**). Since **acetate** (derived from cellulose fermentation in the rumen) is the primary source of extramitochondrial acetyl-CoA, there is no necessity for it to enter mitochondria and form citrate prior to incorporation into LCFAs. Ruminant animals, therefore, are less dependent on malic enzyme and citrate cleavage enzyme (see below),

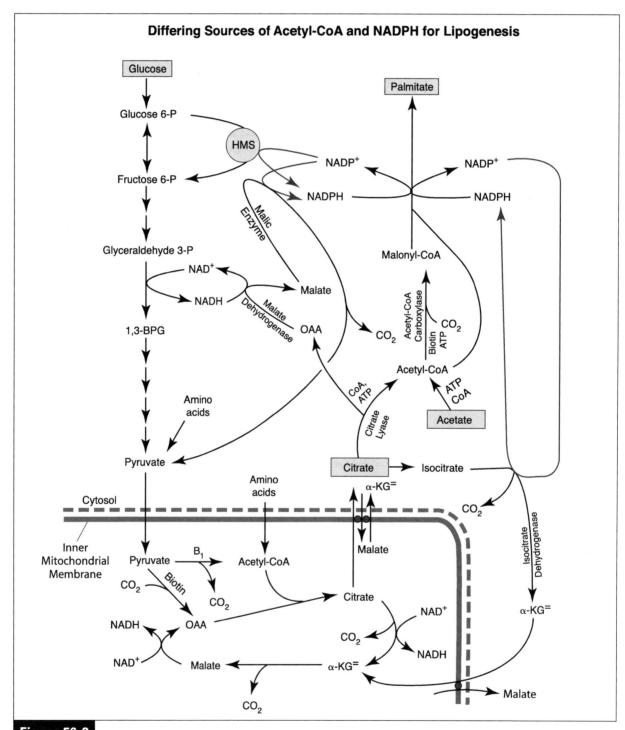

Differing Sources of Acetyl-CoA and NADPH for Lipogenesis

Figure 56-3

since they possess:

1) Increased levels of readily available acetate, and

2) They are less dependent upon glucose and amino acid catabolism as a source of mitochondrial acetyl-CoA.

Although **propionate** is primarily a gluconeogenic VFA in ruminant animals (see Chapter 37), it is also occasionally used by adipocytes and the lactating mammary gland as a priming molecule for fatty acid biosynthesis in place of acetate. When this occurs, fatty acids with an **uneven number of carbon atoms** (e.g., C_{15} and C_{17}) are produced, which explains their increased presence in ruminant fat and milk. It should also be noted that the mammary gland produces fatty acids shorter than C_{15}.

In **non-ruminant** animals, mitochondrial acetyl-CoA is primarily derived from **pyruvate** and **ketogenic amino acids** (see Chapter 8). Therefore, acetyl-CoA must be removed from mitochondria to the cytosol where FA biosynthesis occurs. A metabolic problem exists in this regard since **acetyl-CoA** is largely **impermeable** to mitochondrial membranes. Therefore, it must first be converted to **citrate** by coupling with mitochondrial oxaloacetate (OAA). Citrate is then free to diffuse across mitochondrial membranes into the cytosol, where it is next cleaved to acetyl-CoA and OAA by **citrate cleavage enzyme** (i.e., **citrate lyase**), a reaction requiring ATP. The **acetyl-CoA**, now in the cytoplasm, is free to enter into FA biosynthesis, but the **OAA** must be returned to mitochondria. Since the inner mitochondrial membrane is **impermeable to OAA**, another cytoplasmic conversion is required. This time an important intermediate is formed, malate, that is capable of:

1) Traversing mitochondrial membranes, or

2) Generating needed NADPH for lipid biosynthesis.

First, **OAA** is reduced to **malate** by **NADH** and **malate dehydrogenase**. Hydrogen is next transferred to **NADP⁺**, thus forming **NADPH**, in a decarboxylation reaction catalyzed by **malic enzyme**. The **pyruvate** formed from this reaction is free to diffuse into mitochondria, where it can next form **OAA** (or **acetyl-CoA**). **Malate** can also diffuse directly into mitochondria in exchange for **citrate** or **α-ketoglutarate** (**α-KG⁼**). Indeed, the **citrate** (**tricarboxylate**) **transporter** in the mitochondrial membrane which is involved in this exchange **requires** malate to exchange with citrate or α-KG⁼. Once inside mitochondria, malate can form OAA without being converted to pyruvate.

As indicated in **Fig. 56-3**, **glucose** and **amino acids** can be used for lipogenesis by way of **citrate**, particularly in the well-fed state of non-ruminant mammals. Lipogenesis is higher when sucrose is fed instead of glucose because fructose bypasses the **phosphofructokinase** (**PFK**) control point in the Embden-Meyerhoff Pathway, consequently flooding the hepatic lipogenic pathway (see Chapter 25). Fat in the diet (>10%) causes a reduction in hepatic lipogenesis from carbohydrate and amino acid sources. NADPH is required for lipid biosynthesis in all animals, and non-ruminant mammals generate NADPH primarily from the HMS and cytoplasmic malic enzyme, while ruminant animals generate much of it from cytoplasmic isocitrate dehydrogenase. In birds (who require hepatic-derived lipids for egg formation), and in non-ruminant mammals, the liver is a primary site for FA biosynthesis. In ruminant animals this process occurs largely from **acetate**, primarily in **adipocytes**. Although adipose tissue is involved with FA biosynthesis in all animals, it is primarily a triglyceride storage depot.

OBJECTIVES

• Discuss the allosteric control of acetyl-CoA carboxylase activity, the rate-limiting enzyme in cytoplasmic FA biosynthesis.

- Identify the main functions of mylonyl-CoA in the body, and describe its formation.

- Understand the functions of CNZ, ACP, AT and MT in FA biosynthesis, and identify differing sites of lipogenesis in ruminants, non-ruminant mammals and birds.

- Explain how endogenous FA elongation and desaturation occurs once palmitoyl-CoA is formed.

- Contrast cytoplasmic NADPH generation in ruminant adipocytes to NADPH generation in canine hepatocytes (for FA biosynthesis).

- Show how mitochondrial citrate participates in cytoplasmic FA biosynthesis and glycolysis (see Chapter 25).

- Explain how cytoplasmic conversion of OAA to malate, then malate to pyruvate moves reducing equivalents from glycolysis to hepatic FA biosynthesis.

- Describe how the Krebs Cycle can be partially cytoplasmic in ruminant adipocytes.

QUESTIONS

1. **The rate-limiting enzyme in fatty acid biosynthesis that is inhibited by glucagon and epinephrine is:**
   a. Biotin carboxylase.
   b. Also inhibited by insulin.
   c. Acetyl-CoA carboxylase.
   d. Activated by palmitoyl-CoA.
   e. Activated during periods of starvation and exercise.

2. **All of the following extramitochondrial enzymes are known to participate in reactions that provide reducing equivalents in the form of NADPH for lipogenesis, EXCEPT:**
   a. Malic enzyme.
   b. Glucose 6-phosphate dehydrogenase.
   c. 6-Phosphogluconate dehydrogenase.
   d. Isocitrate dehydrogenase.
   e. Citrate lyase.

3. **Which of the following are generally permeable to mitochondrial membranes?**
   a. OAA and acetyl-CoA

   b. NADPH and NADH
   c. Citrate and malate
   d. Pyruvate and acetyl-CoA
   e. Amino acids and NADP$^+$

4. **The primary site for fatty acid biosynthesis in ruminant animals is:**
   a. Liver.
   b. Muscle.
   c. Kidney.
   d. Adipocytes.
   e. Lung.

5. **The core reactions in fatty acid biosynthesis (i.e., condensation, reduction, dehydration, and then another reduction), take place on a multifunctional, enzymatically active protein known as:**
   a. Acyl carrier protein.
   b. Condensing enzyme.
   c. Biotin.
   d. Acetyl-CoA carboxylase.
   e. Fatty acid synthetase.

6. **Conversion of saturated to unsaturated fatty acids occurs primarily in:**
   a. Mitochondria.
   b. The cytoplasm.
   c. The smooth endoplasmic reticulum.
   d. Lysosomes.
   e. The nucleus.

7. **All of the following can be used as building blocks for fatty acid biosynthesis in the livers of dogs, EXCEPT:**
   a. Carnitine.
   b. Glucose 6-phosphate.
   c. Alanine.
   d. Citrate.
   e. Acetyl-CoA.

8. **Carbohydrates and amino acids can be used for hepatic lipogenesis by way of _____ in dogs.**
   a. Acetoacetate
   b. Citrate
   c. Malate
   d. α-KG$^=$
   e. Biotin

ANSWERS

8. b
7. a
6. c
5. e
4. d
3. c
2. e
1. c

Chapter 57

Triglycerides and Glycerophospholipids

Overview

- The cytoplasmic biosynthesis of triglyceride and glycerophospholipid begins with glycerol 3-phosphate formation.

- Acyl transferases normally add fatty acid residues to glycerol 3-phosphate in the ER.

- Phosphatidic acid serves as an intermediate in both triglyceride and glycerophospholipid biosynthesis.

- Phospholipids are divided into two classes, depending on whether the alcohol backbone is glycerol (glycerophospholipids), or sphingosine (sphingophospholipids).

- In glycerophospholipids, the phosphate esterified to the C_3 of glycerol is further linked to a hydroxy group of inositol, or the hydroxy group of one of 3 nitrogen bases (i.e., choline, serine, or ethanolamine).

- While phosphatidylcholine can be formed by several different routes, phosphatidylinositol formation requires CDP-diacylglycerol.

- Phosphatidylserine accounts for about 15% of brain phospholipid.

- Lecithin can be produced through methylation of cephalin in hepatocytes.

- Cardiolipin is a phospholipid found in the inner mitochondrial membrane.

- Platelet activating factor (PAF) is a plasmalogen.

- Dipalmitoyl phosphatidylcholine is pulmonary surfactant, and reduces alveolar surface tension.

Dietary lipids (Fig. 57-1) are mainly **triglycerides (TGs)**, which are hydrolyzed to monoglycerides and fatty acids in the intestinal lumen, then re-esterified in mucosal cells of the small intestine (see Chapter 60). Here they are packaged with protein (APO-B$_{48}$), then secreted into interstitial fluid as **chylomicrons (CMs)**, the largest of the plasma lipoproteins (see Chapters 63 & 64). Since they are too large to enter the hepatic portal circulation, they move into the lymphatic system, and thence into systemic blood. Although CMs contain other lipid-soluble nutrients, they are usually about **87% TG**. Unlike glucose and amino acids, the TG of CMs is not taken up directly by the liver, for it is first metabolized by tissues having **lipoprotein lipase (LPL)** on their capillary endothelium (namely adipose and aerobic muscle fibers). This enzyme, which is activated by **insulin**, hydrolyzes TG in circulating lipoproteins, releasing free fatty acids (FFAs) that are either incorporated into tissue lipids, or oxidized as fuel. Chylomicron remnants, which remain (see Chapter 64), are normally cleared by the liver.

Adipose tissue TG is the main fuel reserve of the body. When it is hydrolyzed (lipolysis), glycerol and FFAs are released into the circulation. Glycerol is a substrate for both hepatic

Copyright © 2015 Elsevier Inc. All rights reserved.

and renal gluconeogenesis. The FFAs are transported bound to serum albumin; they are taken up by most tissues (but not the brain or erythrocytes), and either esterified to TG for storage, or oxidized as fuel. Newly synthesized TGs in the liver, and those from CM remnants, are secreted into the circulation in **very low density lipoprotein (VLDL)**. The TG of VLDL undergoes a fate similar to that of CMs (see Chapter 65). Partial oxidation of FFAs in the liver leads to **ketone body (KB⁻)** production (ketogenesis, Chapter 71). Ketone bodies are exported to extrahepatic tissues where they either act as a fuel in prolonged starvation, or as substrates for lipogenesis.

Triglycerides

Triglycerides (triacylglycerols), as indicated above, are the major form of storage lipid in animals, and therefore normally constitute the bulk of dietary lipid. They are often referred to as neutral fats, and contain three esters of the alcohol glycerol with fatty acids (**FAs**; R_1, R_2, and R_3; **Fig. 57-2**). The proportion of TG molecules containing the same FA residue in all three ester positions is normally small, for most TGs are mixed acylglycerols. In addition to TGs, both monoglycerides and diglycerides can be found in tissues of the body, but they are rarely found in plasma. Diglycerides (e.g., 1,2-diacylglycerol) serve as important intermediates in TG biosynthesis, and monoglycerides

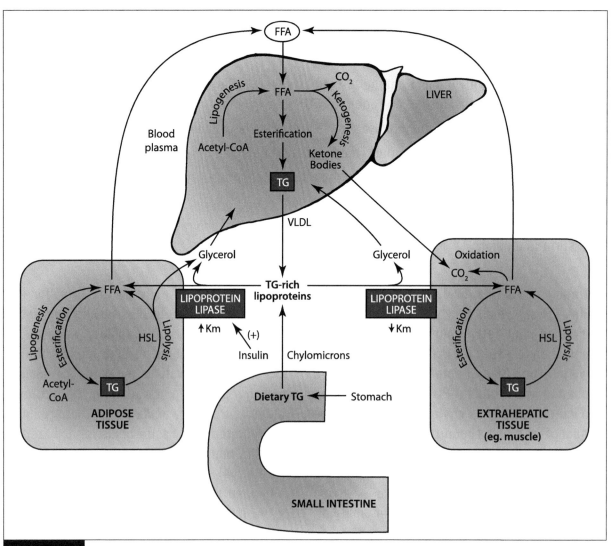

Figure 57-1

play a similar role in mucosal cells of the small intestine (see Chapter 64).

The biosynthesis of TG and glycerophospholipids in most cells begins with **glycerol 3-phosphate** (**Fig. 57-2**). This compound may be formed via **anaerobic glycolysis** (i.e., the Embden Meyerhoff Pathway (EMP)) or via **glyceroneogenesis** of pyruvate through the reduction of **dihydroxyacetone phosphate**, or it may be formed by the action of **glycerol kinase**,

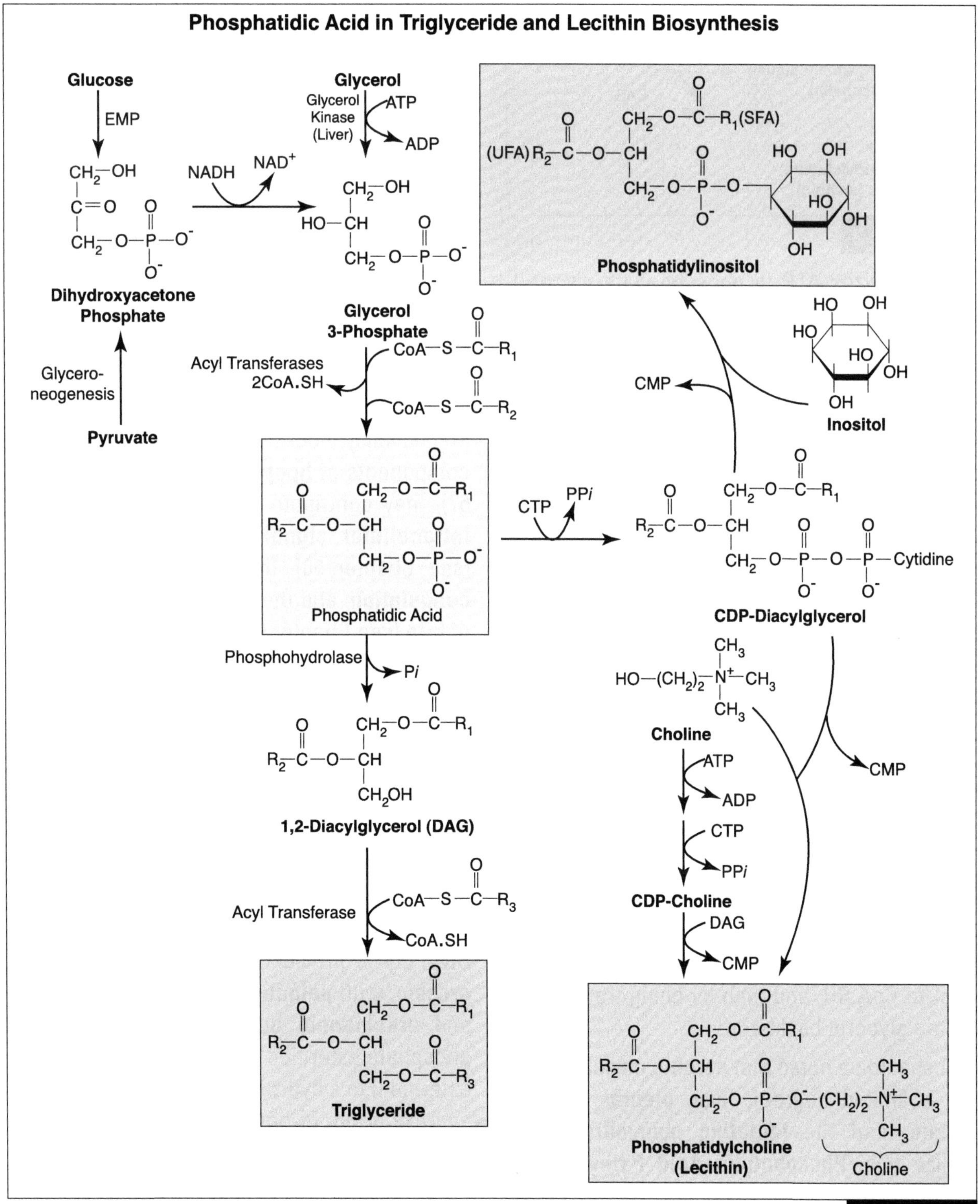

Phosphatidic Acid in Triglyceride and Lecithin Biosynthesis

Figure 57-2

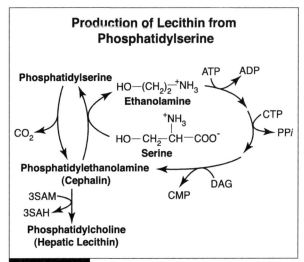

Production of Lecithin from Phosphatidylserine

Figure 57-3

which utilizes ATP to phosphorylate glycerol. Glycerol kinase is virtually absent in **adipose tissue,** but is an active enzyme in **liver and kidney tissue.** Therefore, since glycerol cannot serve as a precursor for TG biosynthesis in adipose tissue, glycerol 3-phosphate must be generated from stored glycogen, from incoming plasma glucose, or from pyruvate (see Chapter 70).

Acyl transferases add fatty acids to the 1 and 2-positions of glycerol 3-phosphate in the endoplasmic reticulum, thus producing **phosphatidic acid (phosphatidate).** These transferases require the activated form of these derivatives (i.e., fatty acyl-S-CoA). Next, the phosphate (Pi) at position 3 is removed by a **phosphohydrolase,** yielding 1,2-diacylglycerol (DAG). Finally, a third acylation results in the formation of **triglyceride.**

As indicated above, all FA transfers in this biosynthetic scheme are made from fatty acyl-S-CoA derivatives. **Free fatty acids,** therefore, can enter TG synthesis only by being attached first to CoA.SH, and then by being transferred to the glycerol backbone.

It should be noted that **aerobic muscle fibers** (to a limited degree), **liver, pleural adipose tissue,** and the **lactating mammary gland** utilize this "**Phosphatidic Acid Pathway**" for TG biosynthesis. Mucosal cells of the small intestine, however, are capable of assem-

bling dietary long-chain fatty acids into TG via acylation of absorbed 2-monoglycerides (without utilizing glycerol 3-phosphate and phosphatidic acid (see Chapter 64)). However, between meals and in the starvation state, intestinal mucosal cells can utilize the phosphatidic acid pathway for TG synthesis (as a "**scavenger pathway**").

Glycerophospholipids

Phospholipids (PLs) are substances in which both FAs and phosphate-containing compounds are esterified to alcohols. They are thus compound lipids, divided into two classes depending on whether the alcohol backbone is **glycerol** (glycerophospholipids), or **sphingosine** (sphingophospholipids; see Chapter 59). Phospholipids exhibit important structural and functional roles within the organism. They are building blocks of **membranes,** important components of **lipoproteins** (see Chapters 63-67), they contribute substituents to important **intracellular signal transduction pathways** (see Chapter 58), they play a part in **blood coagulation,** and they are functional elements of **bile** (see Chapter 62), and **lung surfactant.**

Glycerophospholipids far outnumber sphingophospholipids, and they constitute a diverse family with varying structures. They contain FAs esterified to both the C_1 and C_2 carbon positions of the glycerol backbone, and a phosphate ester is found at C_3 (**Fig. 57-2**). The FA esterified to C_1 (R_1) is usually a **saturated fatty acid (SFA),** and that esterified to C_2 (R_2) is usually an **unsaturated fatty acid (UFA).** Diversity is provided by variation in the FAs present, with **palmitic, stearic, oleic, linoleic,** and **arachidonic** being most common. The phosphate esterified to C_3 is further linked (i.e., esterified) to a hydroxy group of **inositol,** or the hydroxy group of one of three nitrogen bases (i.e., choline, serine, or ethanolamine).

As indicated above and in **Fig. 57-2, phosphatidic acid** becomes an intermediate in both

TG and PL biosynthesis. Esterification of one of the hydroxyl-containing compounds mentioned above to the phosphate group on phosphatidic acid can occur by activation of either the compound to be added, or the phosphatidic acid residue itself with **cytidine triphosphate (CTP)**. If activation occurs with phosphatidic acid, **cytidine diphosphate (CDP)-diacylglycerol** is formed. Condensation with **choline** or **inositol** with loss of **cytidine monophosphate (CMP)** yields **phosphatidylcholine (lecithin)** and **phosphatidylinositol**, respectively. Although this is the only pathway for phosphatidylinositol formation, **choline** can be phosphorylated by **ATP** and then activated by reaction with CTP as a secondary means of producing phosphatidylcholine. The resulting **CDP-choline** that forms through these sequential reactions can react with **DAG**, thus forming lecithin. Although **phosphatidylinositol** normally accounts for **less than 5%** of total lipid in plasma membranes, it plays an important role in intracellular signal transduction (see Chapter 58).

In a third mechanism, **phosphatidylcholine** can be produced from **phosphatidylserine**, a compound which accounts for about **15% of brain phospholipid (Fig. 57-3)**. Decarboxylation of phosphatidylserine results in phosphatidylethanolamine formation, which in turn can be converted to phosphatidylcholine by three successive methylation reactions, each requiring **S-adenosylmethionine (SAM)** as an active methyl group donor. **SAM** also acts as a common denominator for other important methylation reactions, including:

- **Each methyl group in the (CH₃)₃ of choline.**
- **Each methyl group in the (CH₃)₃ of carnitine.**
- **The methyl in epinephrine.**
- **The methyls in catecholamine degradation products.**
- **The methyl for modifying purine and pyrimidine bases in DNA and RNA.**
- **The methyl for the 5' cap of eukaryotic messenger RNA.**

Methionine is typically formed from homocysteine, a vitamin B_{12}-dependent reaction (see Chapters 16 and 43). To serve as an activated methyl donor, this sulfated amino acid first undergoes a reaction with ATP to convert its methyl sulfide function (H_3C-S-R) to a more reactive methyl sulfonium function. The specific sulfonium product is **SAM** (**Fig. 16-2**), produced by the transfer of an **adenosyl** group from **ATP** to the **S** of **methionine**. With its active methyl group SAM can now serve as a universal methyl donor for several different transmethylase enzymes, including those for the conversion of cephalin (phosphatidylethanolamine), to lecithin (phosphatidylcholine). Phosphatidylethanolamine and phosphatidylserine are readily interconverted in an exchange reaction that occurs between **serine** and **ethanolamine** polar head groups. Phospholipids containing ethanolamine (**cephalins**) are originally produced by phosphorylation of ethanolamine with **ATP** and activation with **CTP**, followed by reaction with **DAG** (a pathway similar to that described for choline above). Most lecithin found in bile is derived via hepatic phosphatidylethanolamine methylation.

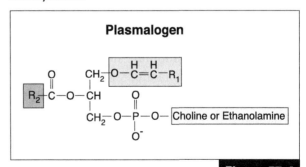

Figure 57-4

Plasmalogens are PLs that contain a long-chain fatty alkyl group attached to the **C₁** of glycerol by an **ether** linkage (**Fig. 57-4**). The alkyl groups are usually unsaturated, containing a cis-double bond. The most common plasmalogens contain **ethanolamine** or **choline** as nitrogen bases, as found, respectively, in **myelin** and **cardiac muscle mitochondrial**

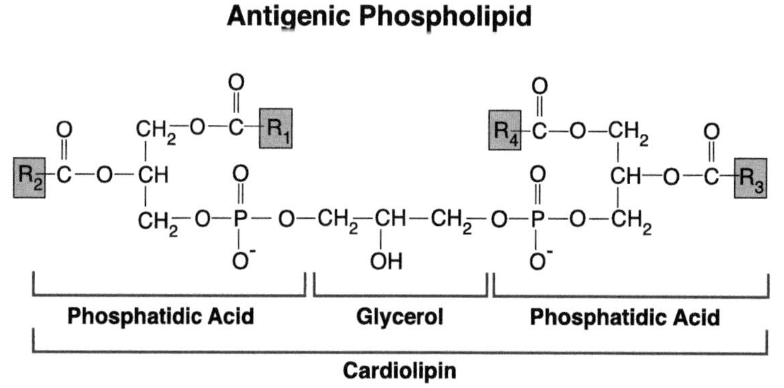

Antigenic Phospholipid

Phosphatidic Acid   Glycerol   Phosphatidic Acid

Cardiolipin

Figure 57-5

membranes. **Platelet-activating factor (PAF)**, produced by polymorphonuclear leukocytes, platelets, and vascular endothelial cells, is a plasmalogen, and a potent mediator of inflammatory responses. It also promotes platelet aggregation and vascular smooth muscle relaxation (see Chapter 68).

Cardiolipin, formed by two molecules of phosphatidic acid connected to the 1 and 3 positions of glycerol, is found in bacterial membranes as well as inner mitochondrial membranes of mammals, and is the only PL known to be antigenic (**Fig. 57-5**).

Choline is important in nervous transmission not only as a part of membrane-bound PLs, but also as a part of **acetylcholine (ACh)**, a key neurotransmitter. It also functions as a store of labile methyl groups. **Dipalmitoylphosphatidylcholine**, another important PL, is produced in large quantities by type II epithelial cells of lung alveoli where, together with sphingomyelins and protein, it forms pulmonary **surfactant**, an effective surface-active agent. Surfactant is secreted from these cells onto the exterior surface of alveoli where it reduces alveolar surface tension. High surface tension, otherwise present, tends to oppose the opening of alveoli as the lungs expand (in much the same way as two glass slides being held together by water). When soap (i.e., surfactant) is added, surface tension is reduced and the slides tend to separate more easily. Surfactant production by the newborn

is stimulated by **catecholamines** (epinephrine and norepinephrine), and by **glucocorticoids** (cortisol), the stress-related hormones.

Regardless of structure, all **PLs** are markedly **polar**, possessing both a hydrophobic tail (the two hydrocarbon chains which come to lie close together), and a hydrophilic head (containing the phosphate group and the second alcohol which rotates away from the tail). They are thus well-suited for their roles in membranes, and as surfactants.

It should be noted that **phosphatidylcholine (lecithin)** is a **zwitterion**: it possesses both a **negative charge (phosphate)**, and a **positive charge (choline)**, which provide for greater interaction with water (see **Fig. 57-2**). Since lecithin possesses both lipid and water solubility, it can stabilize LPs and emulsions in aqueous surroundings; for instance, stabilizing fat emulsion droplets in the gut by attaching lecithin molecules to the outside of those droplets in preparation of the fats being absorbed into mucosal cells (see Chapter 62). Phosphatidylcholine is also the major lipid found in membranes of the endoplasmic reticulum (ER).

In summary, TGs and glycerophospholipids are formed primarily through acylation of glycerol 3-phosphate, removal of the phosphate, and acylation of the final hydroxyl group. All FAs are transferred from CoA. Phospholipids contain 1,2-diacylglycerol phosphate attached with an ester bond to choline, ethanolamine, serine, or inositol, and are formed from CDP carriers of the glycerol- or hydroxyl-containing portions of the molecule. Lecithin can also be produced by methylation of cephalin, particularly in hypatocytes. Cardiolipin is a glycerophospholipid consisting of two phosphatidic acids, and plasmalogens contain choline or ethanolamine, and an acyl glycerol phosphate with a fatty alcohol in position 1. Cardiolipin is the only phospholipid known to be antigenic, and PAF, a potent mediator of inflammatory repsonses, is a plasmalogen.

OBJECTIVES

- Identify potential sources of the Glycerol 3-Phosphate used by hepatocytes and adipocytes in the biosynthesis of TGs and PLs (see Chapter 70).

- Recognize how Phosphatidic Acid is used in the biosynthesis of both TGs and PLs.

- Describe differences between the "Phosphatidic Acid Pathway" of hepatocytes and adipocytes for synthesizing TG, and the "Monoglyceride Acylation Pathway" of intestinal mucosal cells (see Chapter 64).

- Understand basic biosynthetic and functional differences between glycerophospholipids and sphingophospholipids (see Chapter 59).

- Know how structural and functional diversity is established among and between the various PLs.

- Outline the role of nucleoside triphosphate in glycerophospholipid biosynthesis.

- Show the involvement of cephalin, SAM and ethanolamine in the formation of lecithin from phosphatidylserine.

- Explain how SAM is formed, and outline the methylation reactions in which it is involved (see Chapter 16).

- Compare and contrast plasmalogen formation and function to that of cardiolipin.

- Understand the role of dipalmitoylphosphatidylcholine in respiratory physiology.

QUESTIONS

1. **The "Phosphatidic Acid Pathway":**
   a. Is a primary route for triglyceride synthesis in adipocytes and hepatocytes.
   b. Is used for phospholipid, but not triglyceride synthesis.
   c. Is used for triglyceride, but not phospholipid synthesis.
   d. Requires glycerol kinase activity as the sole source of glycerol 3-phosphate.
   e. Is used for the formation of phosphatidylinositol, but not phosphatidylcholine.

2. **Select the FALSE statement below:**
   a. Glycerophospholipids normally outnumber sphingophospholipids in mammalian organisms.
   b. The fatty acid esterified to C_1 in membrane-bound glycerophospholipids is usually saturated.
   c. Lecithin can be produced from cephalin in hepatocytes.
   d. Cardiolipin is found in bacterial membranes.
   e. Plasmalogens consist of two molecules of phosphatidic acid connected through their phosphate groups by glycerol.

3. **Dipalmitoyl phosphatidylcholine is a part of:**
   a. Cardiolipin.
   b. A plasmalogen.
   c. Pulmonary surfactant.
   d. A neurotransmitter.
   e. A sphingolipid.

4. **Phospholipids that contain a long-chain fatty alkyl group attached to C_1 of glycerol (by an ether linkage) are called:**
   a. Lecithin.
   b. Sphingolipids.
   c. Plasmalogens.
   d. Triglyceride.
   e. Monoglycerides.

5. **Which one of the following compounds is used to activate phosphatidic acid for subsequent glycerophospholipid synthesis?**
   a. Choline
   b. Cytidine triphosphate (CTP)
   c. Inositol
   d. Adenosine triphosphate (ATP)
   e. 1,2-Diacylglycerol (DAG)

ANSWERS

5. b
4. c
3. c
2. e
1. a

Chapter 58

Phospholipid Degradation

Overview

- Phospholipases act as digestive enzymes, and they also generate highly active intracellular signal molecules.
- Membrane-bound PLs are continually being formed and degraded by living cells.
- Hydrolysis of membrane-bound PL by PLC yields DAG and IP_3.
- G-stimulatory (G_s) and G-inhibitory (G_i) proteins, which function through GTP/GDP or ATP/ADP, couple certain hormone receptors to PLC.
- Part of the antiinflammatory action of the glucocorticoids is accounted for by their ability to inhibit PLA_2.
- Some bacterial toxins act through stimulating the activity of PLC.
- Some tumor promoters are potent activators of PKC.
- Cytoplasmic Ca^{++} can be derived from extracellular fluid, or from mitochondria and the endoplasmic reticulum from within cells.
- IP_3 is a water-soluble inducer of Ca^{++} release from the endoplasmic reticulum.
- Ca^{++} binds intracellularly to calmodulin (CAM), and this complex activates specific and multifunctional CAM kinases.

Membrane-bound phospholipids (PLs) undergo **turnover**; that is, they are continually being formed and degraded. Degradation involves enzyme-catalyzed hydrolysis of the various ester bonds found in these compounds, and release from the membrane of some hydrolyzed constituents. Enzymes controlling this activity are called **phospholipases (PLases)**, and they are present in many different types of cells. Phospholipases show catalytic specificity for ester bonds within a glycerophospholipid substrate (**Fig. 58-1**), and they exhibit **two** general functions:

1) They act as **digestive enzymes** (e.g., pancreatic, bacterial, and snake venom secretions), and

2) They can generate highly active **signal molecules** (or their immediate precursors).

For example, **phospholipase A_2 (PLA_2)** may release arachidonate, an eicosanoid precursor, and **phospholipase C (PLC)** unleashes two important intracellular messengers in the phosphphoinositide cascade.

Although PLs are actively degraded in plasma membranes, each portion of the molecule turns over at a different rate (e.g., the turnover time of the phosphate group is different from that of the 1-acyl group). This is due to the presence of enzymes that allow partial degradation followed by resynthesis. For example, **PLA_2** catalyzes hydrolysis of the ester bond in position 2 of membrane-bound glycerophos-

Copyright © 2015 Elsevier Inc. All rights reserved.

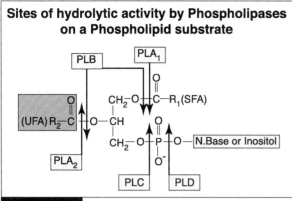

Sites of hydrolytic activity by Phospholipases on a Phospholipid substrate

Figure 58-1

pholipids to form a **polyunsaturated (free) fatty acid (UFA)**, and **lysophospholipid**, which in turn may be reacylated via acyl-CoA in the presence of an acyltransferase (**Fig. 58-2**). Long-chain **saturated fatty acids (SFAs)** are found predominantly in the 1 position of glycerophospholipids, whereas the polyunsaturated fatty acids (i.e., precursors of the **eicosanoids**), are incorporated more into the 2 position. The phospholipase present in **pancreatic exocrine secretions** is PLA_2, thus giving rise to a free fatty acid plus lysolecithin in the intestinal lumen. Absorbed products (i.e., a long-chain free fatty acid plus lysolecithin) are normally reassembled inside mucosal cells to complete PLs via a molecular mechanism similar to that of the "**Monoglyceride Reacylation Pathway**" (see Chapter 64). Also, the presence of PLA_2 in **snake venom** hydrolyzes PLs to lysophospholipids, compounds that are good detergents, and result in lysis of erythrocytes in snakebite victims. Lastly, the antiinflammatory activity of **glucocorticoids** is partially explained by their ability to inhibit plasma membrane-bound PLA_2 (see Chapter 69).

Lysolecithin may also be formed by an alternate route involving **lecithin:cholesterol acyltransferase (LCAT)**. This enzyme, found in circulating plasma and possibly in hepatocytes, catalyzes transfer of a FA residue from the 2-position of lecithin to cholesterol, thus forming a **cholesterol- (or cholesteryl-) ester (CE)**. The enzymatic action of circulating LCAT is considered to be responsible for much of the CE present in circulating lipoproteins (see Chapter 66).

Incorporation of FAs into lecithin can occur through complete synthesis of the PL, by transacylation between CE and lysolecithin, or by direct acylation of lysolecithin by acyl-CoA. Thus, a continuous exchange of FAs is possible, particularly with regard to introducing essential UFAs into PL molecules.

Lysophosphatidylcholine (e.g., **lysolecithin**) in cell membranes is sometimes additionally attacked by **lysophospholipase A_1** (also called **phospholipase A_i; PLA_1**), removing the remaining 1-acyl group and forming the corresponding glycerophosphocholine, which in turn may be split by a **hydrolase**, liberating glycerol 3-phosphate plus choline (**Fig. 58-2**). **Phospholipase C (PLC)** attacks the ester bond in position 3, liberating **1,2-diacylglycerol (DAG)** plus a phosphoryl base, and this enzyme is a major bacterial toxin. **Phospholipase D (PLD)**, an enzyme found mainly in plants, hydrolyzes the nitrogenous base or inositol from PLs, and **phospholipase B (PLB)**, which is not well described, appears to possess both PLA_1 and PLA_2 activity (**Fig. 58-1**).

Ca⁺⁺ Signaling

Calcium is derived at the cellular level from both external and internal sources (**Fig. 58-3**). It can enter from outside the cell by passing through Ca^{++}-specific voltage- or ligand-gated channels that span the plasma membrane, or it can be released from internal stores in mitochondria and the endoplasmic reticulum (or sarcoplasmic reticulum). When a Ca^{++} channel opens, a concentrated plume of Ca^{++} forms around its mouth, then via diffusion dissipates rapidly after the channel closes. Such localized signals, which can originate from channels in the plasma membrane or on internal organelles, represent the **elementary events** that occur in Ca^{++} signaling, and are closely associated with membrane excitability, mitochon-

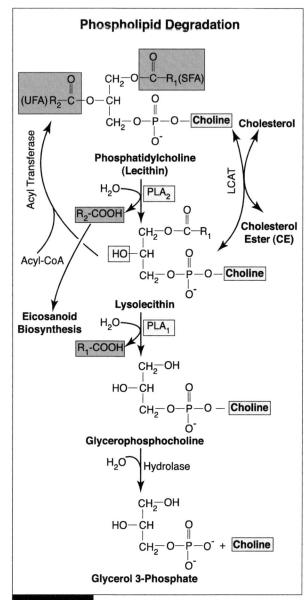

Figure 58-2

drial metabolism, vesicle secretion, smooth muscle relaxation and mitosis. These elementary signals have two basic functions: they can activate highly localized cellular processes in the immediate vicinity of the channels, primarily through enzyme phosphorylation, or, by recruiting channels throughout the cell, they can activate processes at a more global level. In smooth muscle, for example, Ca^{++} increases that arise locally near the plasma membrane activate K$^+$ channels, thus causing K$^+$ to diffuse out of the cell, causing it to hyperpolarize and relax. Yet when elementary Ca^{++} release events deeper in the cell are coordinated to create

a **global Ca^{++} signal**, the muscle contracts. This is an example of how spatial organization enables Ca^{++} to activate opposing cellular responses in the same cell. For sites of elementary Ca^{++} release to produce intracellular global events, individual channels must communicate with one other to set up Ca^{++} waves (e.g., fertilization, muscle contraction, liver metabolism, gene transcription and cell proliferation). If cells are connected, such intracellular waves can spread into neighboring cells and become intercellular waves that cause responses within tissues (e.g., wound healing, ciliary beating, glial cell function, insulin secretion, bile flow and endothelial NO synthesis (blood vessels)).

Elementary Ca^{++} signaling begins when certain hormones or neurotransmitters interact with their plasma membrane receptors (e.g., catecholamines interacting with α_1-adrenergic receptors, or acetylcholine (ACh) interacting with muscarinic receptors). This sets up a chain of events through the Ca^{++} messenger system.

Phospholipids and the Ca^{++} Messenger System

Many ligands (e.g., hormones and neurotransmitters) bind to cell-surface receptors, as stated above, and some cause an elevation in the **cytosolic calcium (Ca^{++})** concentration, even when Ca^{++} is absent from the surrounding extracellular medium (**Fig. 58-4**). During the initial stages of stimulation, Ca^{++} is released into the cytosol from the **endoplasmic reticulum (ER)** or other intracellular vesicles (e.g., mitochondria), not from the extracellular medium. How this occurs became clear in the early 1980s, when it was demonstrated that a rise in cytosolic Ca^{++} is preceded by the hydrolysis of membrane-bound **phosphatidylinositol 4,5-bisphosphate**, one of several inositol PLs found in the cytoplasmic leaflet of the plasma membrane. Hydrolysis of the PL

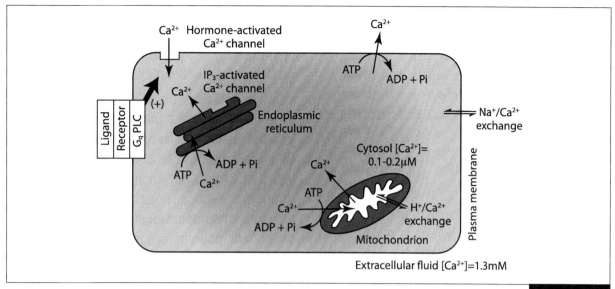

Figure 58-3

by plasma membrane-bound **PLC** yields two important products: **DAG**, which remains in the membrane, and the water-soluble **inositol 1,4,5-triphosphate (IP₃)**, which is released into the cytosol.

A **G-stimulatory (G$_s$) protein** (sometimes referred to as G$_q$ protein), is known to couple various hormone receptors to PLC, and treatment of cells with pertussis toxin, which inactivates **G$_s$**, abolishes activation of PLC. Four subtypes of mammalian PLC have been identified (α, β, γ, and δ), and all but the γ-isoform, which appears to be coupled to **tyrosine kinase**, are coupled to **G$_s$** protein. In contrast to **G$_s$** (or **G$_q$**) protein, **G-inhibitory (G$_i$) protein** is known to inhibit membrane-associated enzyme activities. The **G proteins** function through either **GTP/GDP**, or **ATP/ADP**.

Once liberated from membrane-bound phosphatidylinositol 4,5-bisphosphate, **IP₃** diffuses to the ER surface, where it binds to a specific receptor (**IP₃-R**). This receptor binding, in turn, induces opening of specific **Ca⁺⁺** channel proteins, which allows Ca⁺⁺ to exit the ER lumen into the cytosol where it binds to a **Ca⁺⁺-dependent regulatory protein** known as **calmodulin (CAM)**. Although cells contain other phosphorylated inositols, only **IP₃** has been found to cause **Ca⁺⁺** release by binding to

the ER receptor protein. Within approximately one second following its formation, most **IP₃** is hydrolyzed to **inositol 1,4-bisphosphate**, an inactive intermediate, which usually terminates release of **Ca⁺⁺** unless more **IP₃** is formed via **PLC** activation. Inositol 1,4-bisphosphate is now available as a substrate for the resynthesis of membrane-bound phosphatidylinositol 4,5-bisphosphate.

Calmodulin has four Ca⁺⁺ binding sites, and full occupancy of these leads to a conformational change which allows the **Ca⁺⁺-CAM** complex to activate both specific and multifunctional **CAM kinases**. These kinases, in turn, govern the phosphorylation of various intracellular proteins associated with enzymatic activity and ion movement. **Ca⁺⁺-CAM** regulated processes include smooth muscle contraction and glycogenolysis (see Chapter 23).

The other intracellular second messenger formed in this process, **DAG**, stays within the plasma membrane. Diacylglycerol and IP₃ work in tandem to increase the activity of **protein kinase C (PKC)**, which before activation remains free in the plasma membrane. **Inositol 1,4,5-triphosphate** increases the cytoplasmic **Ca⁺⁺** concentration, and **Ca⁺⁺** helps to facilitate both **PLC** and **PKC** activity. **Diacylglycerol** binds directly to a specific domain on the **PKC** molecule to increase its activity, and also anchor this enzyme to the membrane. Once activated by **DAG** and **Ca⁺⁺**, **PKC** phosphorylates serine and threonine residues in a number of target enzymes that have been shown to play important roles in several aspects of growth and metabolism.

Tumor Promoters and PKC

Through investigations of the actions of **PKC**, investigators discovered that some **tumor promoters** are potent activators of this enzyme. Tumor promoters are lipid-soluble chemicals, isolated from several sources (mainly plants), that play a part in transforming normal cells

into malignant ones capable of uncontrolled growth. Carcinogenesis generally requires both an **initiator** and a **promoter**, and many carcinogens are apparently capable of acting as both.

Since **PKC** acts as a "**receptor**" for a class of tumor promoters (not initiators) known as **phorbol esters**, activation of this enzyme may be a key event in cell proliferation. The active agent of croton oil, for example, is a mixture of phorbol esters. These oils are derived from a genus of euphorbiaceous shrubs, some of which are popular ornamentals. C capitatus, for example, is known to be poisonous to livestock. The most active phorbol ester known

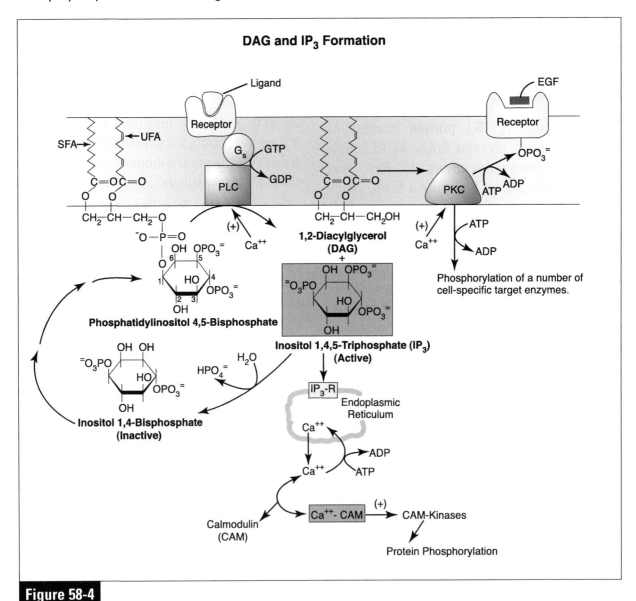

Figure 58-4

is **12-O-tetradecanoylphorbol-13-acetate (TPA)**, which has numerous reported effects.

Natural substrates of **PKC** include cell-surface receptors for several **growth factors** (i.e., somatomedins), such as **epidermal growth factor (EGF)**. Phosphorylation of the **EGF-receptor** by **PKC** decreases its affinity for **EGF**, which normally modulates the growth-stimulating activity of cells. Overproduction of **PKC** in normal fibroblasts causes cells to grow unattached to an extracellular matrix, as do many tumor cells. (Note: Normal cells grow when attached to an extracellular matrix.) Clearly, the appropriate degradation of membrane-bound **phosphatidylinositol 4,5-bisphosphate** to **DAG** and **IP$_3$**, and the subsequent ability of **DAG** to stimulate **PKC**, is of fundamental importance in controlling cell growth.

OBJECTIVES

- Describe how dietary lecithin is digested and absorbed (see Chapter 60).

- Discuss the roles of PLA$_1$, PLB, PLA$_2$, PLC and PLD in membrane-bound PL degradation.

- Explain how lysolecithin and cholesterol esters are formed, and how lysolecithin is normally reacylated.

- Describe roles played by PLs in the intracellular Ca^{++} messenger system.

- Outline the phosphoinositide cascade.

- Describe effects of IP$_3$ on Ca^{++} sequestration, and the effects of DAG on PKC. Also identify mechanistic and functional relationships between IP$_3$ and DAG.

- Identify three separate sources of Ca^{++} for the cytoplasm of a smooth muscle cell.

- Discuss the biochemical and physiologic effects of phorbol esters.

- Identify the role played by CAM in the Ca^{++} messenger system.

- Draw the structure of Phosphatidylinositol 4,5-Bisphosphate.

- Show how G$_q$ is involved in the regulation of PLC activity.

QUESTIONS

1. **The antiinflammatory activity of glucocorticoids is partially explained by their ability to inhibit:**
   a. PLA$_1$.
   b. PLA$_2$.
   c. PLC.
   d. PLD.
   e. Glycerophosphocholine hydrolase.

2. **Lecithin:cholesterol acyltransferase (LCAT) is a circulating plasma enzyme that catalyzes formation of:**
   a. Cholesterol ester and lysolecithin.
   b. Lysolecithin and 1,2-diacylglycerol (DAG).
   c. Inositol 1,4,5-triphosphate (IP$_3$) and DAG.
   d. Cholesterol and phosphatidylcholine.
   e. Glycerol 3-phosphate and choline.

3. **Which one of the following enzymes is a major bacterial toxin?**
   a. PLA$_1$
   b. PLA$_2$
   c. PLC
   d. PLD
   e. PLB

4. **Which one of the following is a water-soluble inducer of Ca^{++} release from the endoplasmic reticulum?**
   a. PLA$_2$
   b. PKC
   c. DAG
   d. IP$_3$
   e. PLC

5. **The phospholipase present in pancreatic exocrine secretions is:**
   a. PLA$_1$.
   b. PLA$_2$.
   c. PLB.
   d. PLC.
   e. PLD.

6. **The cellular EGF receptor is phosphorylated through the action of PLC.**
   a. True
   b. False

6. b

5. b

4. d

3. c

2. a

1. b

ANSWERS

Sphingolipids

Overview

- Sphingolipids are found in membranes of both plant and animal cells.

- Sphingosine, which can be formed from palmitoyl-CoA and serine, is used by cells to form ceramides.

- Ceramides are the basic structural units of all sphingolipids, and are formed through the union of a very long-chain fatty acids with sphingosine.

- Sphingomyelins are the only phospholipids that do not contain a glycerol backbone.

- Cerebrosides are formed through linkage of either glucose or galactose to the C_1 of ceramide.

- Globosides and gangliosides are synthesized from glucocerebrosides, and sulfatides are synthesized from galactocerebrosides.

- Inherited deficiencies in sphingolipid-specific lysosomal hydrolases lead to sphingolipidoses (lipid storage diseases).

- Sphingolipid activator proteins (saposins) are required for maximal hydrolysis of sphingolipids by their respective lysosomal hydrolases.

Sphingolipids are complex lipids found as membrane components of both plant and animal cells. They are present in cell membranes of nerve and other tissues, however, only trace amounts are found in depot fat. **Sphingolipids** are generally classified as either sphingomyelins (which are true **phospholipids (PLs)**), or **glycolipids** (i.e., ceramide-containing **gluco-** and **galactocerebrosides**, **sulfatides**, **globosides**, and **gangliosides**, each of which possesses one or more monosaccharide residue linked to the sphingosine backbone rather than phosphate (**Fig. 59-1** and Chapter 20)).

Ceramide is the basic structural unit of all sphingolipids, and **sphingosine**, a precursor to ceramide, is a long-chain amino alcohol formed from **palmitoyl-CoA** and **serine** (**Fig. 59-**

1). Sphingosine is one of 30 or more different long-chain amino alcohols found in nature. It is the major base of mammalian sphingolipids, in higher plants and yeast phytosphingosine is the major base, and in marine invertebrates doubly unsaturated bases such as 4,8-sphingadiene are found.

Ceramide (N-acylsphingosine) is formed in the endoplasmic reticulum through the amide linkage of a **long-chain fatty acid** moiety to the amino group of **sphingosine**. It has two **hydroxyl** groups, one attached to C_3 that remains unsubstituted, and the other attached to C_1 that is substituted in the formation of **sphingomyelins** and the **glycolipids** (i.e., glycosphingolipids). Characteristically, very long-chain (i.e., C_{24}) fatty acids occur in ceramides, particularly those found in glycosphingolipids of the brain

Copyright © 2015 Elsevier Inc. All rights reserved.

Sphingomyelin and Glycolipid Formation

Sphingosine and ceramide (the basic structural unit of all sphingolipids), are produced from palmitoyl-CoA and serine. Sphingomyelins, which are found in erythrocytic membranes and the myelin sheath of nerves, are formed by the addition of phosphocholine to the C_1 of ceramide, and they are the only PLs not attached to the glycerol backbone.

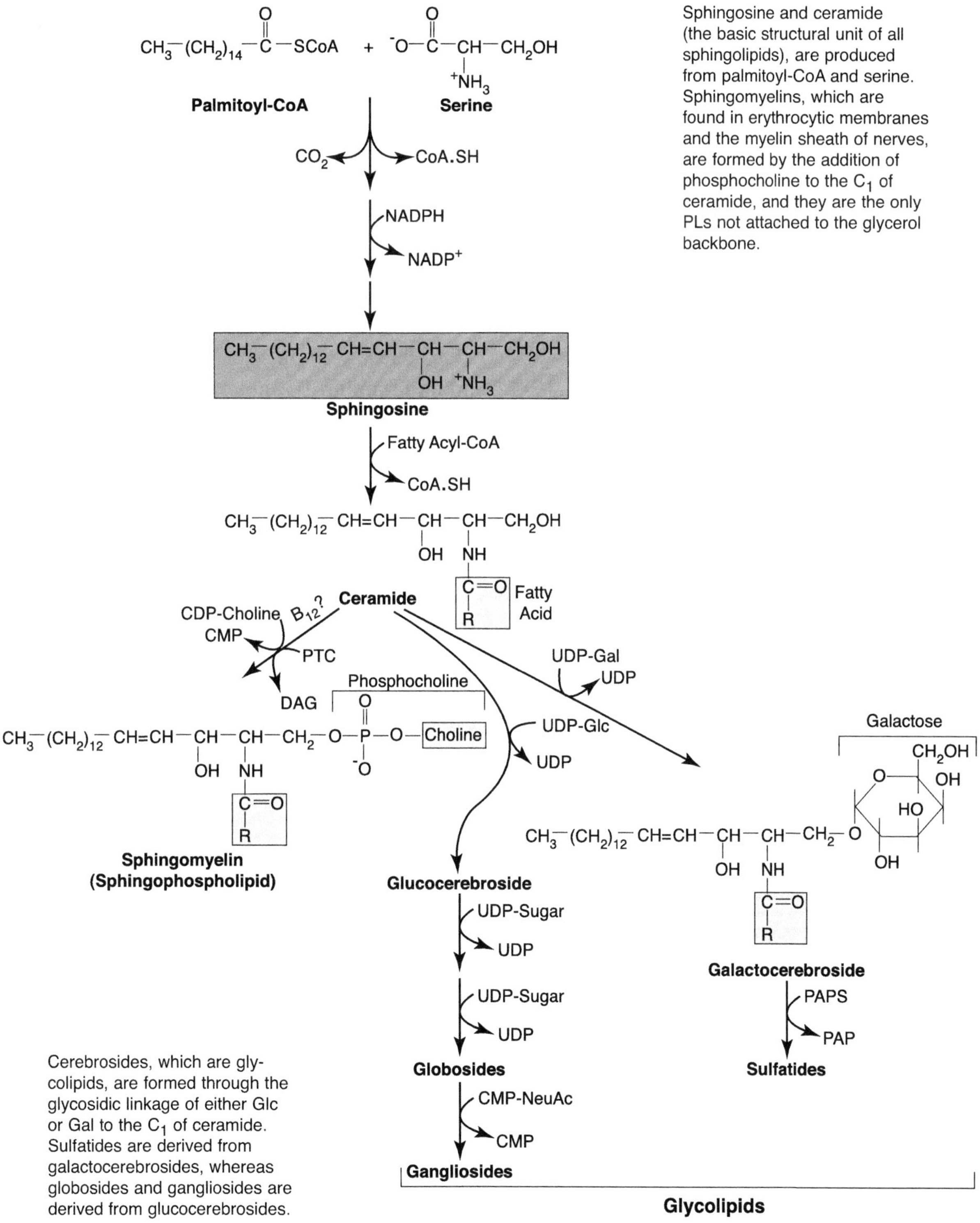

Cerebrosides, which are glycolipids, are formed through the glycosidic linkage of either Glc or Gal to the C_1 of ceramide. Sulfatides are derived from galactocerebrosides, whereas globosides and gangliosides are derived from glucocerebrosides.

Figure 59-1

(e.g., lignocerate, and nervonate; see Chapter 54). There is also evidence that ceramide may act as a lipid mediator (second messenger), activating a protein kinase that opposes some actions of diacylglycerol (DAG; see Chapter 58).

Sphingomyelins are formed by the addition of **phosphocholine** to the C_1 of ceramide, which can occur following transfer from either **CDP-choline**, or **phosphatidylcholine (PTC; Fig. 59-1)**. This process occurs mainly in the Golgi apparatus, and to a lesser extent in plasma membranes. Sphingomyelins are found in high concentrations in erythrocytic membranes, and in the myelin sheath that surrounds nerves of the central nervous system (CNS). **They are the only PLs not attached to the glycerol backbone** (see Chapter 57).

Cerebrosides are formed through the glycosidic linkage of either **glucose (Glc)** or **galactose (Gal)** to the C_1 of ceramide (rather than phosphocholine; see Chapter 20 and **Fig. 59-1**). **Galactocerebrosides** are found in high concentration in nerve tissue, whereas **glucocerebrosides** are found primarily in extraneural tissues (where they serve as intermediates in the synthesis of more complex glycolipids). Cerebroside biosynthesis requires **uridine diphosphate (UDP) derivatives** of **Gal** and **Glc**.

Sulfatides, found primarily in nerve tissue, are galactocerebrosides in which the monosaccharide contains a sulfate ($SO_4^=$) ester (**Fig. 59-2**). Sulfate addition requires **3'-phosphoadenosine-5'-phosphosulfate (PAPS)** as the activated $SO_4^=$ donor (**Fig. 59-1**). PAPS is also involved in the biosynthesis of other sulfolipids, such as steroid sulfates. Sulfated

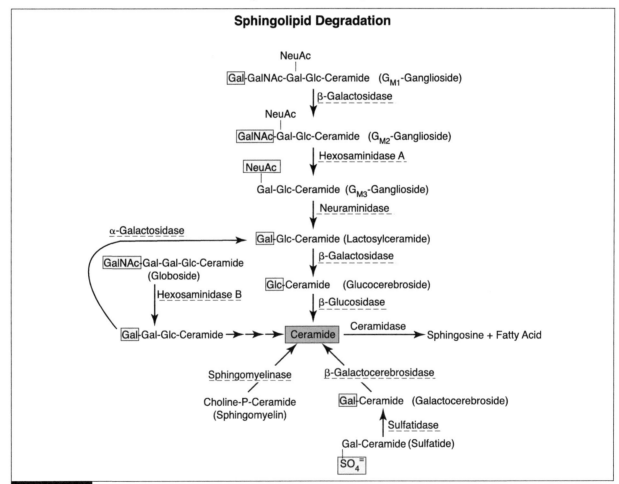

Sphingolipid Degradation

Figure 59-2

galactocerebrosides account for about 25% of cerebrosides in the brain.

Monosaccharides can be added to glucocerebrosides, resulting in a family of more complex glycosphingolipids, with each addition being catalyzed by a specific glycosyltransferase. **Globosides** contain **Glc**, **Gal**, and **N-acetylgalactosamine** (**GalNAc**), they are important components of erythrocytic membranes, and they contain determinants of several blood group systems.

Gangliosides are synthesized from **glucocerebrosides** by sequential addition of monosaccharides from their activated donors (see **Fig. 59-1**). They contain a variety of oligosaccharides consisting of Glc, Gal, GalNAc, and a sialic acid, which is usually **N-acetylneuraminic acid** (**NeuAc** or **NANA**; see Chapter 20). Gangliosides function, for example, as receptors for bacterial toxins, which subsequently activate adenyl cyclase. Different members of each ganglioside class are identified by subscript number (e.g., G_{M1}, G_{M2}, G_{M3}, or G_{M4}), which indicate the specific sequence of oligosaccharide groups attached to ceramide (**Fig 59-2**). G_{M1} is one of the most common gangliosides.

Gangliosides are abundant in the gray matter of the brain, where they constitute about **6% of total lipids**. Because they are especially abundant in nerve endings, it has been proposed that they function in the transmission of nerve impulses across synapses. They are also present at receptor sites for acetylcholine and other neurotransmitters.

In addition to contributing to the structural rigidity of membranes, **glycolipids** are implicated in several cellular functions mediated at the level of the cell surface. They:

- **Provide antigenic chemical markers for cells.**
- **Act as chemical markers identifying various stages of cell differentiation.**
- **Regulate normal growth patterns of cells.**
- **Allow cells to react with other bioactive substances such as bacterial toxins (bind G_{M1}), glycoprotein hormones, interferons, and viruses.**

Table 59-1
Sphingolipidoses

Disease	Enzyme Defect	Intermediate Accumulated	Most Affected Organs
Farber's	Ceramidase	Ceramide	Brain
Fabry's	α-Galactosidase	Ceramide trihexoside	Skin, kidney, extremities
G_{M1}-Gangliosidosis	β-Galactosidase	G_{M1}-Ganglioside	Brain, visceral organs
Gaucher's	β-Glucosidase	Glucocerebroside	Liver, spleen, long bones, pelvis
Krabbe's (Globoid leukodystrophy)	β-Galactocerebrosidase	Galactocerebroside	Brain (demyelination & death)
Metachromatic leukodystrophy	Sulfatidase	Sulfatide	Brain (white matter), kidney
Niemann-Pick	Sphingomyelinase	Sphingomyelin	Liver, spleen, brain
Sandhoff (Variant A) (G_{M2}-Gangliosidosis)	Hexosaminidase (A and B)	G_{M2}-Ganglioside	Brain, visceral organs
Tay Sachs (Variant B) (G_{M2}-Gangliosidosis)	Hexosaminidase A	G_{M2}-Ganglioside	Brain, visceral organs

The above listing covers processes that are physiologically significant, and also overlap with several functions associated with **glycoproteins**. This correlation emphasizes the importance of the **oligosaccharide chains** associated with the cell surface (see Chapter 20).

Sphingolipid Degradation

During normal cell turnover, **sphingolipids are degraded** by a series of specific **lysosomal hydrolases** which require **activator proteins** (**Fig. 59-2**). More than 70 lysosomal hydrolases participate in the turnover of macromolecules, including not only sphingolipids, but also proteins, complex carbohydrates, and nucleic acids. **Inherited deficiencies** in several of the sphingolipid-specific lysosomal hydrolases, or their activator proteins, have been reported in both animals and humans, leading to a group of diseases commonly known as **sphingolipidoses**, or **lipid storage diseases (Table 59-1)**. Failure to remove a substituent interferes with subsequent steps in the degradative pathway, and in each disease a characteristic intermediate accumulates, frequently in the nervous system.

The extent to which the activity of the affected enzyme is decreased is similar in most tissues of affected animals. For example, sphingomyelinosis (or sphingomyelin lipidosis) is associated with visceral and neuronal accumulation of a variety of sphingomyelins. This condition is commonly known as **Niemann-Pick disease**, and is seen in cats, dogs, mice, and humans. Signs in affected animals reportedly include absent conscious proprioception, severely depressed to absent spinal reflexes, muscle tremors (especially in the pelvic limbs), plantigrade stance, and moderate hepato-splenomegaly. Pain perception and cranial nerve function are reportedly normal, yet motor nerve conduction velocities are said to be depressed. Peripheral nerves apparently show widespread myelin degeneration associated with many vacuolated macrophages. There is widespread infiltration of virtually every body system. Affected animals reportedly demonstrate severe reductions in CNS and visceral lysosomal sphingomyelinase activity. The prognosis in these animals is poor, and currently there is no effective treatment available.

Gaucher's disease has been reported in dogs, sheep, and pigs, **metachromatic leukodystrophy** in geese, G_{M1}-**gangliosidosis** in dogs, cats, cattle, and sheep, and G_{M2}-**gangliosidoses** (**Sandhoff** and **Tay Sachs disease**) in deer, cats, dogs, pigs, and the black bear. It is likely that hereditary lipid storage diseases are present in other species as well.

Sphingolipid activator proteins (e.g., saposin A, B, C, and D) are small glycoproteins which are required for maximal hydrolysis of sphingolipids by their respective lysosomal hydrolases. For example, sphingolipid activator protein 1 (saposin B) activates hydrolysis of sulfatides, G_{M1} ganglioside, and globosides. It binds lipid substrates, thereby solubilizing them for hydrolysis. Sphingolipid activator protein 2 (**saposin C**) activates the hydrolysis of gluco- and galactocerebrosides. It interacts with the corresponding enzyme to increase V_{max} and decrease K_m. Sphingolipid activator protein 3 (**saposin A**) activates hexosaminidase A, which degrades G_{M2}-ganglioside. Deficiencies in the sphingolipid activator proteins can also lead to sphingolipidoses, even though the corresponding enzymes they activate are normal.

In summary, sphingolipids are complex lipids found as membrane components of both plant and animal cells. They are generally classified as either sphingomyelins (sphingophospholipids), or the nonphosphate-containing glycosphingolipids (or glycolipids). Both groups contain ceramide as their basic structural component. Cerebrosides (which are glycolipids including the glyco-and galactocerebrosides, globosides, gangliosides, and sulfatides) are formed through the glycosidic linkage of either Glc or Gal to the C_1 of ceramide. During normal cell turnover, sphingolipids are degraded by a series of specific lysosomal hydrolases, which when deficient lead to a group of hereditary diseases commonly known as sphingolipidoses, or lipid storage diseases.

OBJECTIVES

- Differentiate structural and functional differences between sphingomyelins and glycolipids.

- Outline the biochemical pathway from Palmitoyl-CoA, Ser and Fatty Acyl-CoA to N-Acylsphingosine.

- Identify primary locations for sphingomyelins and cerebrosides in the body.

- Show how globosides, gangliosides and sulfatides can be formed from gluco- and galactocerebrosides.

- Explain how complex glycosphingolipids are formed and degraded.

- Define GM_1 - GM_4, and know where they are found.

- Summarize the general cellular functions of glycolipids.

- Recognize the different lysosomal hydrolases that participate in sphingolipid degradation.

- Identify the different lipid storage diseases of animals, and discuss their causes.

- Discuss the role of sphingolipid activator proteins in glycolipid turnover.

- Draw the structure of the largest common component in all sphingolipids.

QUESTIONS

1. The basic structural unit of mammalian sphingolipids is:
   a. Glycerol.
   b. Ceramide.
   c. Choline.
   d. Cerebroside.
   e. Globoside.

2. Which of the following sphingolipids are phospholipids?
   a. Gangliosides
   b. Sulfatides
   c. Galactocerebrosides
   d. Sphingomyelins
   e. Globosides

3. Sphingolipid activator proteins are:
   a. Glycoproteins required for maximal hydrolysis of sphingolipids by their respective lysosomal hydrolases.
   b. Enzymes, such as β-galactosidase, required to hydrolyze various sphingolipids in lysosomes.
   c. Used in the biosynthesis of sphingolipids from ceramide.
   d. Usually overactive in various sphingolipidoses, such as Gaucher's disease.
   e. Allow cells to react with bioactive substances, such as bacterial toxins.

4. Which one of the following is formed from a galactocerebroside?
   a. Sphingomyelin
   b. Globoside
   c. Ganglioside
   d. Sulfatide
   e. Ceramide

5. Which one of the following is a ganglioside, abundant in the gray matter of the brain?
   a. G_{M1}
   b. Sphingosine
   c. Palmitoyl-CoA
   d. Ceramide
   e. Galactoceramide

6. Which one of the following diseases is associated with sulfatide accumulation in the brain?
   a. Niemann-Pick
   b. Sandhoff
   c. Tay Sachs
   d. Farber's
   e. Metachromatic leukodystrophy

6. e
5. a
4. d
3. a
2. d
1. b

ANSWERS

Chapter 60

Lipid Digestion

Overview

- Lipid digestion begins in the mouth and stomach, but most occurs in the small intestine.

- Bile salts are required for the emulsification of most dietary lipids in the duodenum.

- Pancreatic lipase, unlike the proteases, is secreted in an active form.

- Pancreatic phospholipase is a PLA_2.

- Pancreatic cholesterol esterase hydrolyzes fatty acids from dietary cholesterol esters in the intestinal lumen.

- Lipid absorption requires "mixed micelles".

- Pancreatic lipase, unlike lipoprotein lipase, releases fatty acids from carbons 1 and 3 of triglyceride, but leaves a fatty acid in the 2 position.

- The "monoglyceride acylation pathway" is the major route for triglyceride formation within jejunal mucosal cells.

- Short- and medium-chain fatty acids are not resynthesized into triglyceride in mucosal cells of the small intestine.

- Steatorrhea may occur secondary to either liver, pancreatic, or small bowel disease.

Triglycerides (TGs) are usually the major constituents of dietary fat, with lesser amounts of **cholesterol (CH)**, **cholesterol esters (CEs)**, **phospholipids (PLs)**, and **fat-soluble vitamins** present. Digestion of TG begins in the mouth and stomach, but most occurs in the small intestine. To gain entry into blood, fat must first be emulsified, and then hydrolyzed by pancreatic enzymes. Long-chain fatty acids (LCFAs), 2-monoglycerides, lysolecithin, and CH produced by lipolysis, along with the fat-soluble vitamins (A, D, E, and K), must also be solubilized in **bile salt solution** in order to penetrate jejunal enterocyte membranes. Once inside these cells, **LCFAs** and **2-monoglyceride** are resynthesized into **TG**, and, along with **lecithin**, **CH**, and **CE**, incorporated into **chylomicrons (CMs)** for transport into lymph. **Short-chain fatty acids (SCFAs)** and **medium-chain fatty acids (MCFAs)**, each containing 12 or fewer carbon atoms (see Chapter 54), are not incorporated into **CMs**, and therefore are absorbed directly into the hepatic portal circulation.

Dietary fat requires **emulsification** before it can be efficiently degraded by pancreatic enzymes in the lumen of the small intestine. Lipolysis of ingested fat is regulated by two hormones, **cholecystokinin (CCK)**, and **secretin**, both of which are secreted into blood by endocrine cells of the duodenum following the appearance of fat in the lumen. In response to **CCK**, the exocrine pancreas releases several digestive enzymes into the duodenum, and the gallbladder contracts releasing bile as

Copyright © 2015 Elsevier Inc. All rights reserved.

well. The major function of gallbladder bile is to provide **bile salts** and **lecithin**, which are important emulsifying agents. In response to **secretin** (i.e., "nature's antacid"), biliary and pancreatic ductular cells release a solution enriched in **NaHCO₃**, which helps to neutralize the acidic contents of chyme passing into the small bowel from the stomach.

Emulsification of Dietary Fat

Fat globules entering the small intestine are generally too large to be effectively hydrolyzed by pancreatic enzymes. They are finely emulsified by the detergent action of **biliary bile salts**, which are generally considered to be poor emulsifying agents without the added assistance of biliary lecithin, lysolecithin and monoglycerides (formed in the duodenum during lipid digestion). This emulsification process provides a larger surface area upon which the pancreatic digestive enzymes can operate.

At low concentrations (**< 2 mM**), conjugated bile salts begin to form molecular solutions, and as the concentration rises they form micelles or spherical aggregates (approximately 33 bile salts per micelle, 3-10 nm in diameter). Their polar hydroxy, peptide bond, and carbonyl or sulfate portions face outward, and their non-polar steroid nuclear portions form a hydrophobic core (see Chapter 62). Each lipid-incorporated micelle, referred to as a "**mixed micelle**", contains, on average, approximately **44 LCFAs, 20 monoglycerides, 2 lysolecithins, 2 CHs**, and usually 1 **fat-soluble vitamin** in addition to the **33 bile salt molecules** on the surface. Mixed micelles, containing the products of lipid digestion, act as carriers of digestive products to the brush borders of mucosal cells for absorption.

Enzymatic Hydrolysis of Dietary Lipids

Enzymes that are of significance to lipid digestion in the small intestine are listed in **Table 60-1**, and in the **Appendix**. Small amounts of TG are hydrolyzed to fatty acids (FAs) and diglyceride (DG) by **lingual lipase**, an enzyme secreted by salivary glands and swallowed into the stomach. Lingual lipase and **gastric lipase** are both acid-stable, with pH optima around 4.0. Hydrolysis of fat in the mouth and stomach is normally slow since it has not been emulsified. These acid-stable lipases account for about 10% of TG hydrolysis in adult animals, and, along with **milk lipase**, as much as 40-50% in infants. Homogenized milk contains TG that is highly emulsified, but the homogenization process usually inactivates milk lipase.

Pancreatic lipase, unlike the proteases (see Chapter 7), is secreted in an active form rather than as a proenzyme. It has a pH optimum of approximately 8.0, and is denatured by a pH below 3.0. A small protein, **colipase**, is also released from the pancreas along with lipase, and is required for the binding of lipase to small dispersed lipid particles without being washed

Table 60-1

Enzymes Involved in the Digestion of Dietary Fat

Enzyme	Source	Substrate	Products
Milk lipase	Mammary glands	Triglyceride	Diglyceride + fatty acid
Lingual lipase	Salivary glands	Triglyceride	Diglyceride + fatty acid
Gastric lipase	Stomach/Abomasum	Triglyceride	Diglyceride + fatty acid
Pancreatic lipase	Pancreas	Triglyceride & diglyceride	2-Monoglyceride + fatty acid
Cholesterol esterase	Pancreas	Cholesterol ester	Cholesterol + fatty acid
Phospholipase A₂	Pancreas	Phospholipid	Lysophospholipid + fatty acid

off by bile salts. Colipase also changes the pH optimum for lipase from 8.0 to 6.0. Pancreatic lipase releases FAs from carbons 1 and 3 of the triglyceride backbone, forming **2-monoglyc-**

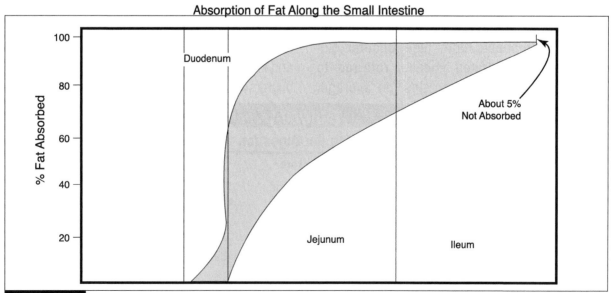

Triglyceride → Pancreatic Lipase + H₂O → 2-Monoglyceride + FFA + FFA

eride plus **two free fatty acids (FFAs)**. Since the FA in position 2 usually remains intact, only small amounts of free glycerol are formed during intestinal lipid digestion.

Phospholipase A₂ releases the FA from carbon 2 of dietary PLs, generating **lysophospholipids** and one **FFA** (see **Figs. 58-1 & 58-2**), and **CH esterase** releases the FA from CE, forming **CH** and one **FFA** (see **Fig. 66-2**).

Lipid Absorption in the Small Intestine

The end-products of lipid digestion are delivered to the luminal surface of jejunal mucosal cells by mixed micelles. Upon arrival, the micelles dissociate, and their lipid-incorporated products diffuse through the brush-border, across the apical membrane and into the cell, leaving the bile salts behind. The

bile salts return to the lumen to incorporate more highly lipophilic dietary material, and transport it again to the jejunal surface. Short-and MCFAs, because of their relatively high water-solubility, are not dependent upon mixed micelles for mucosal uptake.

Most fat absorption occurs in the jejunum **(Fig. 60-1)**, and in dogs begins to increase about **2 hours** following a meal. Another 2-3 hours are usually required before the serum TG concentration returns to basal levels. Normal stools contain about 5% fat, but most fecal fat is probably generated from sloughed cells and microbes rather than from the diet. Processes of fat absorption are not fully developed at birth, consequently infants normally fail to absorb about 10-15% of ingested fat.

Mucosal Resynthesis of Dietary Lipids

The major pathway to **TG resynthesis** within mucosal cells is the "**monoglyceride acylation (or reacylation) pathway**". Absorbed LCFAs are

Absorption of Fat Along the Small Intestine

Figure 60-1

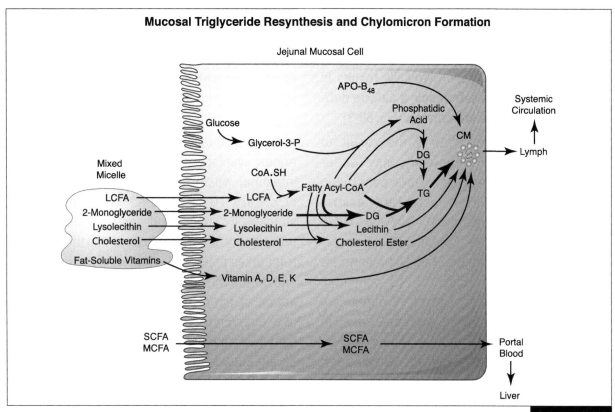

Mucosal Triglyceride Resynthesis and Chylomicron Formation

Jejunal Mucosal Cell

Figure 60-2

first converted to fatty acyl-CoA derivatives, reactions catalyzed by a family of **fatty acyl-CoA synthetases** that recognize fatty acids of different chain lengths. 2-Monoglycerides are then joined with two fatty acyl-CoA derivatives, thus forming TG + 2 CoA (**Fig. 60-2**). A second, minor "**scavenger**" pathway of TG resynthesis in mucosal cells involves the "**phosphatidic acid pathway**" (see Chapter 57), which uses glucose as a source of glycerol 3-phosphate. Long-chain fatty acyl-CoA is also used by mucosal cells for the resynthesis of **lecithin**, and, through the action of **acyl-CoA:cholesterol acyltransferase (ACAT)**, CEs. All of these resynthesized products, plus **free CH** and any absorbed **fat-soluble vitamins** present, are next packaged into the core of **CMs**, which are released by exocytosis into interstitial fluid, and then move into the lymphatic system.

An inability of mucosal cells to synthesize **apoprotein B₄₈,** which is needed to form the CM lipoprotein, leads to accumulation of TG in jejunal mucosal cells, and to low circulating levels of CMs (see Chapter 64). Abnormalities in the synthesis (or resynthesis) of TG via the monoglyceride acylation pathway has not been reported in animals.

Abnormalities in Lipid Digestion and Absorption

Although several conditions can lead to impaired lipid absorption and **steatorrhea** (excess fat in the feces), the most common causes of steatorrhea are related to **bile salt deficiency, pancreatic enzyme deficiency, defective CM synthesis**, or **lymphatic obstruction**.

Bile salts are synthesized in the liver, and stored in the gallbladder (in animals that have a gallbladder; see Chapter 62). A bile salt deficiency can result from liver disease that impairs synthesis, an obstruction of the bile duct that impairs delivery, an ileal defect that impairs their active return to the enterohepatic circulation (EHC), or an overgrowth of intestinal bacteria which convert them to bile acids, thus diminishing their efficacy as emulsifying agents. Lipid malabsorption typically results in a **fat-soluble vitamin deficiency, a calorie deficiency, diarrhea**, and **steatorrhea (Fig. 60-3)**.

Lipid malabsorption that is the result of **pancreatic enzyme deficiency** may occur secondary to **pancreatitis**, where pancreatic ducts may be partially obstructed, or to **gastric hyperacidity**, where pancreatic enzymes fail to reach their pH optima. Supplemental pancreatic enzyme preparations can sometimes be taken orally to help alleviate this condition. Additionally, therapeutic use of TG containing MCFAs (instead of LCFAs) is sometimes used in lipid malabsorption syndromes. Since MCFAs are not reesterified to TG in mucosal cells, and they are not incorporated into CMs, they move directly into the portal circulation where they can provide added calories to needy patients.

In summary, normally 95-98% of the dietary lipid consumed by animals is TG, with the remainder consisting of PLs, CH, CEs, and fat-soluble vitamins. Upon oxidation, lipids yield more energy per gram than do carbohydrates or proteins. Lipids are also thought to possess a greater **satiety value** than carbohydrates or proteins, since they tend to remain in the stomach longer, and are digested more slowly.

Dietary lipids must be broken down into simpler molecules (e.g., LCFAs, 2-monoglycerides, lysolecithin, CH) to facilitate efficient intestinal absorption. Only a small fraction of dietary TG is digested in the mouth and stomach by lingual and gastric lipase. Hence, the majority of TG hydrolysis within the digestive tract takes place in the lumen of the duodenum and jejunum via the hydrolytic digestive action of pancreatic lipase. To facilitate this, two obstacles must be overcome. **First,** the lumen of the upper small bowel consists of an aqueous medium, and TGs are insoluble in water. **Second,** digestive enzymes are proteins, and are dissolved in the aqueous environment of the intestinal lumen. Thus, TGs must be solubilized in the aqueous phase before digestion can occur. This requires a coordinated and efficient series of mechanical and chemical interactive steps that are, for convenience,

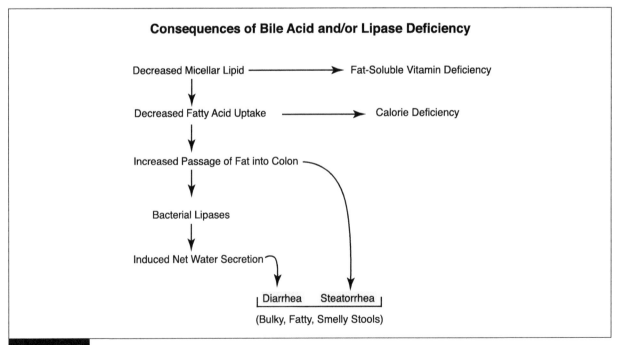

Consequences of Bile Acid and/or Lipase Deficiency

Decreased Micellar Lipid ⟶ Fat-Soluble Vitamin Deficiency

Decreased Fatty Acid Uptake ⟶ Calorie Deficiency

Increased Passage of Fat into Colon

Bacterial Lipases

Induced Net Water Secretion

Diarrhea   Steatorrhea

(Bulky, Fatty, Smelly Stools)

Figure 60-3

divided into **intraluminal** (i.e., mixing patterns of the bowel, emulsification by biliary bile salts and lecithin, and enzymatic hydrolysis), and **intracellular** (i.e., passive absorption, mucosal lipid resynthesis, and CM formation) events. Impaired lipid absorption leads to **steatorrhea,** which can be caused by bile salt deficiency, pancreatic enzyme deficiency, defective CM formation, or lymphatic obstruction.

OBJECTIVES

- Describe how dietary fat is solubilized, digested and absorbed by the gastrointestinal tract.

- Identify the enzymes involved in digestion of dietary fat, as well as their sources, substrates and hydrolytic products. Contrast and compare to lipoprotein lipase (LPL), and hormone sensitive lipase (HSL; see Chapter 70).

- Explain how emulsification of dietary fat occurs, and why it is necessary.

- Contrast the formation, content, location and function of a "mixed micelle" to that of a "lipoprotein."

- Compare the primary and secondary means of synthesizing TGs in mucosal cells of the jejunum to the pathway normally found in hepatocytes and adipocytes (see Chapter 57).

- Recognize how APO-B$_{48}$ is utilized by jejunal mucosal cells, and identify the normal contents of a CM.

- Outline the most common causes and results of steatorrhea.

- Explain why the MCFA content of portal blood would be expected to be higher than the CM content following a lipid meal.

QUESTIONS

1. **Mixed micelles in the intestinal lumen generally contain all of the following, EXCEPT:**
   a. Cholesterol.
   b. Fat-soluble vitamins.
   c. Medium-chain fatty acids.
   d. Monoglycerides.
   e. Lysolecithin.

2. **Which one of the following is NOT a source of lipase?**
   a. Saliva
   b. Stomach/abomasum
   c. Breast milk
   d. Bile
   e. Pancreatic juice

3. **Colipase:**
   a. Is also known as gastric lipase.
   b. Is the inactive proenzyme of pancreatic lipase.
   c. Normally reduces the pH optimum of pancreatic lipase.
   d. Is a phospholipase secreted by the pancreas.
   e. Cleaves cholesterol esters into cholesterol and free fatty acid.

4. **Most lipids are absorbed from the:**
   a. Stomach.
   b. Duodenum.
   c. Jejunum.
   d. Ileum.
   e. Large intestine.

5. **The major pathway to triglyceride resynthesis with mucosal cells of the small intestine is:**
   a. The phosphatidic acid pathway.
   b. The Embden Meyerhoff pathway.
   c. The hexose monophosphate shunt.
   d. β-Oxidation.
   e. The monoglyceride acylation pathway.

6. **All of the following are correctly associated with steatorrhea, EXCEPT:**
   a. Bile duct obstruction.
   b. Pancreatic dysfunction.
   c. Gastric hyperacidity.
   d. Defective chylomicron synthesis.
   e. Hepatic phospholipase deficiency.

7. **Digestive action of pancreatic lipase on a TG molecule in the duodenum produces:**
   a. 2-Monoglyceride and 2 FFAs.
   b. Glycerol and 3 FFAs.
   c. 1,2-Diglyceride and a FFA.
   d. 1,3-Diglyceride and a FFA.
   e. Non of the above

7. a
6. e
5. e
4. c
3. c
2. d
1. c

ANSWERS

Chapter 61

Cholesterol

Overview

- Although cholesterol is largely a product of animal metabolism, it is found in small amounts in plants.

- Cholesterol is distributed into 3 body pools, and not all CH molecules are equally accessible to metabolic interactions or to exchange.

- Bile is the major route for cholesterol elimination from the body.

- Acetyl-CoA is the source of all carbon atoms in cholesterol.

- HMG-CoA reductase is the rate-limiting enzyme in cholesterol biosynthesis.

- Glucagon and cortisol, hormones typically elevated during starvation, inhibit hepatic CH biosynthesis.

- Insulin and thyroid hormones tend to enhance the hepatic clearance of CH from the circulation, as well as hepatic synthesis of CH.

- The cholesterol biosynthetic scheme in plants seems to end at squalene.

- The cyclopentanoperhydrophenanthrene nucleus of CH is common to all animal steroids.

- The liver synthesizes more cholesterol than any other organ.

- Cholesterol, coprostanol, and cholestanol are generally the primary sterols found in feces.

Cholesterol (CH; Gr. "**Chole**" - bile, "**stereos**" - solid) is essential for the survival of all animals, where it becomes a structural component of membranes, and the parent molecule that gives rise to other important steroids (**Fig. 61-1**). The greater part of CH arises from *de novo* biosynthesis, particularly in herbivores. Indeed, virtually all nucleated eukaryotic cells have the capacity to synthesize CH, but the quantitatively most important are **hepatocytes**, and, to a lesser extent, **intestinal mucosal cells**. Cholesterol is largely a product of animal metabolism (not plant or microbial), and occurs therefore in foods of animal origin such as liver, meat, milk, and egg yolk (a particularly rich source). In carnivores and omnivores, about one-third of the body CH pool is generally provided through the diet.

More than 250 steroids (phytosterols) have been described in plants, but of these **β-sitosterol** (otherwise known as **β-prostate**), which differs from CH by an added ethyl grouping at position 24, is the most common (**Fig. 61-2**). Plants contain small amounts of CH, both free and esterified, where it is a component of plant membranes. While CH averages about **0.05 gm/kg total lipid** in plants, it is much higher in animals. Although the total CH pool varies minimally in normal adult animals, tissue contents can vary from about **0.5 gm/kg muscle tissue**, to about **15 gm/kg brain tissue**. The average CH content is about **1.4 gm/kg tissue**

Copyright © 2015 Elsevier Inc. All rights reserved.

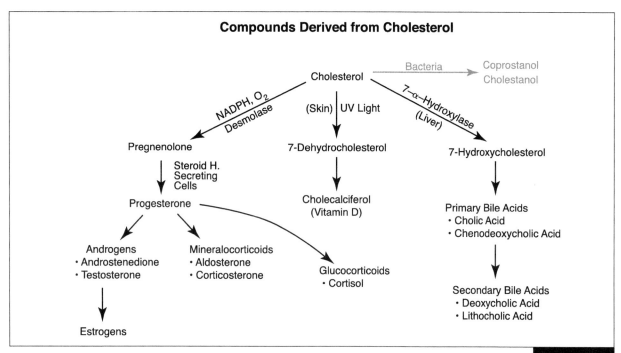

Compounds Derived from Cholesterol

Figure 61-1

for the body as a whole. Thus, a **35 kg mammal** contains about **50 gm of CH**.

In general, CH is distributed into **three** body pools, and not all are equally accessible to metabolic interactions, or to exchange with CH carried in blood:

1) **Rapidly miscible pool** -- Cholesterol absorbed from the diet, and CH in blood, the small intestine, lung, liver, and spleen.

2) **Slowly exchangeable pool** -- A portion of CH in adipocytes, skeletal muscle, and skin.

3) **Slowly miscible, or nonexchangeable pool**

--Cholesterol present in brain and bone, and some in skin, skeletal muscle, and adipocytes.

Dietary CH usually takes several days to equilibrate with that in plasma, and several weeks to equilibrate with that in tissues. Hepatic turnover is rapid compared with the half-life of total body CH, which is several weeks. Cholesterol in plasma and that in liver usually equilibrate in a matter of hours.

Wide variations in the **plasma CH concentrations** of animals have been reported (e.g.,

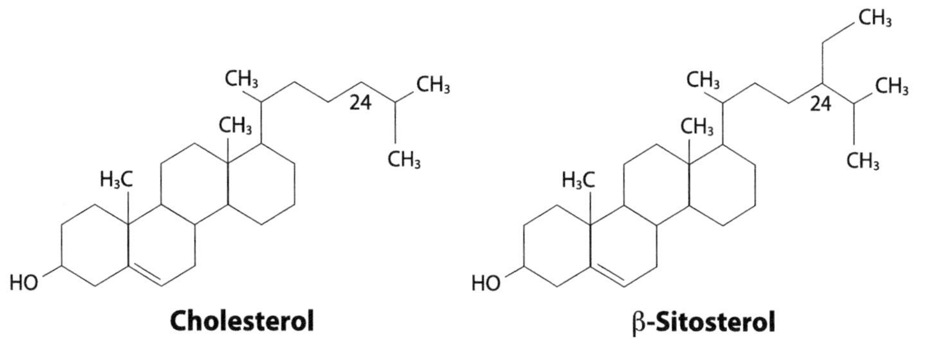

Cholesterol

β-Sitosterol

Figure 61-2

horses 75-150 mg/dl, **cows** 80-120, **goats** 80-130, **sheep** 52-76, **pigs** 36-54, **dogs** 125-250, and **cats** 95-130). The greater part of plasma CH (80-90%) is usually esterified with unsaturated fatty acids (Chapter 60). It is transported in plasma lipoproteins, with the highest proportion found in **low-density lipoprotein (LDL)**, which is formed from **very low-density lipoprotein (VLDL)** and **intermediate-density lipoprotein (IDL)** in plasma (see Chapter 65). However, under conditions where VLDL are quantitatively predominant (e.g., ponies), increased proportions of CH will continue to reside in

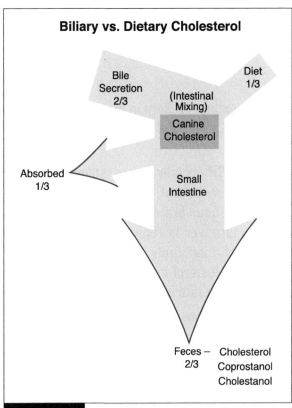

Biliary vs. Dietary Cholesterol

Bile Secretion 2/3

Diet 1/3

(Intestinal Mixing)

Canine Cholesterol

Absorbed 1/3

Small Intestine

Feces — Cholesterol
2/3   Coprostanol
Cholestanol

Figure 61-3

this fraction. Ultimately, LDL are removed from plasma, mainly by the liver (see Chapters 65, 66, and 67).

There are several pathways for removal of CH from the body; through the sloughing of oily secretions and cells from the skin, through desquamation of cells from the stomach, small intestine, and colon, and through the movement

of CH into pancreatic, gastric, intestinal, and biliary canalicular secretions. Cholesterol can also be metabolized to other steroids, such as **bile acids** and **steroid hormones**, which in turn are excreted from the body either through bile (the bile acids), or urine (the glucuronide conjugates of steroid hormones). Of these various routes, hepatic conversion of **CH to bile acids**, and secretion of both **bile acids** and **CH directly into bile** are of greatest quantitative importance (see Chapter 62).

Dietary CH in dogs is normally less than one-half of total body CH synthesis per day, and the amount used in steroid hormone biosynthesis is only 10% of that used in bile acid biosynthesis. Although bile acid synthesis and excretion by the liver can potentially be a significant way in which to remove CH from the body, this route of elimination is normally negated through extensive BA reabsorption in the ileum. Biliary CH excretion is, however, a significant way in which the body rids itself of CH, for **approximately 60% of CH normally present in the small intestine at any one time is that derived from bile (Fig. 61-2)**. The amount of CH absorbed from the intestine (both biliary and dietary CH), is dependent upon several factors, including the amount of dietary saturated fat which tends to increase intestinal CH absorption.

Cholesterol Biosynthesis

Acetyl-CoA is the source of all carbon atoms in CH, with the overall reaction sequence described as follows:

$$\left[\begin{array}{c} 18 \text{ Acetyl-CoA} + 18 \text{ ATP} + 16 \text{ NADPH} \\ + 4 \text{ O}_2 \longrightarrow \text{Cholesterol} + 9 \text{ CO}_2 + \\ 16 \text{ NADP}^+ + 18 \text{ ADP} + 18 \text{ Pi} \end{array} \right]$$

Cholesterol biosynthesis is generally subdivided into four stages:

Stage 1. Formation of **mevalonate**,

Cholesterol Biosynthesis

$$CH_3-\overset{\overset{\displaystyle O}{\|}}{C}\sim S-CoA$$
2 Acetyl-CoA

| Thiolase | ⤵ CoA.SH

$$\overset{\overset{\displaystyle CH_3}{\|}}{C}=CH_2-\overset{\overset{\displaystyle O}{\|}}{C}\sim S-CoA$$
Acetoacetyl-CoA

$$CH_3-\overset{\overset{\displaystyle O}{\|}}{C}\sim S-CoA$$
Acetyl-CoA

H_2O

| HMG CoA Synthase | ⤵ CoA.SH

$$^-OOC-CH_2-\overset{\overset{\displaystyle CH_3}{|}}{\underset{\underset{\displaystyle OH}{|}}{C}}-CH_2-\overset{\overset{\displaystyle O}{\|}}{C}\sim S-CoA$$
3-Hydroxy-3-methylglutaryl-CoA (HMG-CoA)

2NADPH

| HMG CoA Reductase |

2NADP$^+$ + CoA.SH

$$^-OOC-CH_2-\overset{\overset{\displaystyle CH_3}{|}}{\underset{\underset{\displaystyle OH}{|}}{C}}-CH_2-CH_2-OH$$
Mevalonate

$(+)$ $(-)$

Insulin
Thyroxin

Glucagon
Cortisol
Cholesterol
"Statins"

3ATP   3ADP CO_2 H_2O

$$CH_3-\overset{\overset{\displaystyle CH_3}{|}}{C}=CH-CH_2$$
Isoprenoid Unit

↓ x6

Squalene (C$_{30}$)

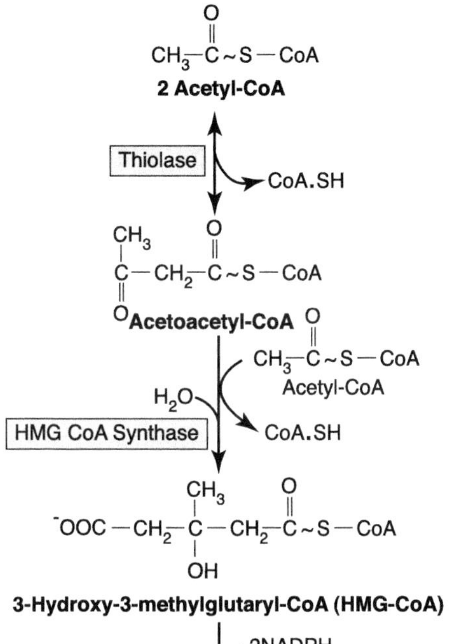

The four stages of cholesterol biosynthesis are shown (formation of mevalonate, conversion to isoprenoids, condensation to squalene, and conversion to cholesterol). Stage one possesses the rate-limiting reaction, conversion of HMG-CoA to mevalonate. This reaction is catalyzed by HMG-CoA reductase, an important enzyme known to be regulated by a number of endogenous and exogenous factors. It should be recognized that insulin and thyroxine not only increase hepatic cholesterol biosynthesis, but they also enhance liver clearance of cholesterol from the circulation by stimulating hepatic LDL receptor synthesis and activity (see Chapters 66 & 67). In diabetes mellitus (DM) and hypothyroidism, plasma cholesterol levels rise.

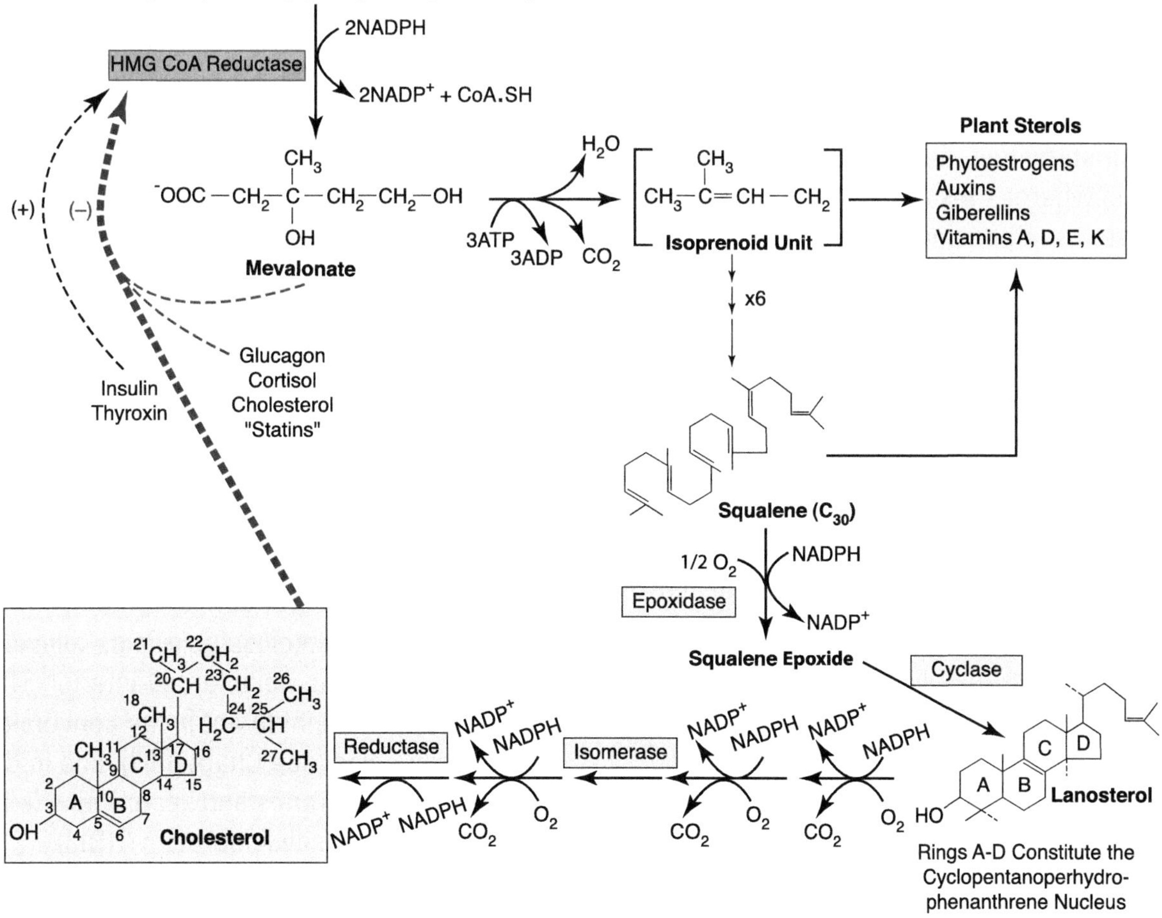

Plant Sterols

| Phytoestrogens |
| Auxins |
| Giberellins |
| Vitamins A, D, E, K |

$1/2 O_2$ NADPH

| Epoxidase |

NADP$^+$

Squalene Epoxide | Cyclase |

| Reductase | NADP$^+$ NADPH | Isomerase | NADP$^+$ NADPH   NADP$^+$ NADPH

NADP$^+$ NADPH CO_2 O_2 CO_2 O_2 CO_2 O_2 HO

Cholesterol

$^{21}CH_3$ $^{22}CH_2$
^{20}CH $^{23}CH_2$ $^{26}CH_3$
$^{18}CH_3$ $^{24}H_2C$ ^{25}CH
^{12} $^{27}CH_3$
^{11}CH ^{13}C ^{17} ^{16}
^{1} $^{19}CH_3$ C D
^{2} A ^{10}B ^{8} ^{14} ^{15}
^{3} ^{4} ^{5} ^{6} ^{7}
OH

Lanosterol

Rings A-D Constitute the Cyclopentanoperhydrophenanthrene Nucleus

Figure 61-4

Stage 2. Conversion of mevalonate to active **isoprenoids**,

Stage 3. Condensation of isoprenoids to **squalene**, and

Stage 4. Conversion of squalene to **CH** (**Fig. 61-4**).

In **stage 1**, HMG-CoA, a C_6 intermediate, is formed by sequential condensation of 3 acetyl-CoA molecules. This cytoplasmic process follows the same reaction sequence described for the biosynthesis of ketone bodies in mitochondria (see Chapter 71). Next, **HMG-CoA** is converted to **mevalonate** via reduction by **2 NADPH**, catalyzed by the microsomal enzyme, **HMG-CoA reductase**. This **rate-limiting step** in cholesterol biosynthesis is **inhibited** by **mevalonate**, the immediate product, by **CH**, the main product of the pathway, by certain fungal metabolites (i.e., the cholesterol-lowering "statin" drugs), and by **starvation**. **Glucagon** and **cortisol**, hormones typically elevated during starvation, depress hepatic CH biosynthesis, whereas **insulin** and **thyroid hormones** elevate it. Opposing effects of insulin and glucagon in the liver are mediated by the dephosphorylation and phosphorylation of HMG-CoA reductase, respectively, whereas the effects of thyroxin and cortisol are mediated by the induction and repression of HMG-CoA reductase biosynthesis. Although CH feeding decreases hepatic CH biosynthesis, increased dietary carbohydrate or triglyceride intake augments it primarily by increasing availability of acetyl-CoA and NADPH. The pathways that supply these substrates from carbohydrate in the liver, particularly glycolysis (see Chapters 23-27), the pyruvate dehydrogenase reaction (see Chapter 27), and the NADPH-generating steps of the hexose monophosphate shunt (see Chapter 28), are activated by **insulin**.

In **stage 2**, formation of activated **isoprenoid** units from mevalonate requires 3 ATP, and results in the loss of H_2O and CO_2. These isoprenoid units now become building blocks not only of the basic steroid skeleton, but also of other derivatives, such as **dolichol** and **ubiquinone** (**coenzyme Q**; see Chapter 36).

In **stage 3**, six isoprenoid units condense in a series of 3 reactions to form the intermediate, **squalene**. NADPH is required in the final condensation reaction. **Plant sterols** (phytosterols like the **phytoestrogens**, the **auxins** and **giberellins** (steroid-like plant hormones), and the **fat-soluble vitamins**), are **polyprenoids** largely derived from mevalonate and/or squalene. Plant sterols differ from animal sterols in the nature of their side-chains, and generally cannot be utilized by animal cells. **Less than 5%** of dietary plant sterols are normally absorbed by the mammalian digestive tract (compared to **30%** of available CH). The liver normally clears plant sterols efficiently from portal blood, and excretes them into bile.

In **stage 4**, squalene is first converted to an oxide, and then to lanosterol, which possesses the C_{17} **cyclopentanoperhydrophenanthrene nucleus** common to all animal steroids. Formation of CH from lanosterol takes place in the endoplasmic reticulum, and involves changes in the steroid nucleus and side chain. Five steps are thought to be involved, in which three methyl groups are released from the ring as CO_2, and a double bond in the side chain is reduced. The exact order in which these steps occur is not known with certainty.

Hepatic intermediates from squalene to CH are attached to a **sterol carrier protein**, which binds sterols and other insoluble lipids, allowing them to react in the soluble cell interior. This protein is also thought to act in the conversion of CH to bile acids (see Chapter 62), and in the formation of membranes and lipoproteins (e.g., VLDL and high-density lipoprotein (HDL)).

Following movement of CH into bile, and then into the intestine, **coprostanol** is formed via bacterial reduction of the double bond of CH between C_5 and C_6. **Cholestanol**, a precursor to coprostanol, is formed in the liver as well as in the intestinal tract, and these two stanols, together with **CH**, normally become primary fecal sterols.

Abnormalities in the Plasma Cholesterol Concentration

Animals that exhibit hypo- or hypercholesterolemia most likely have a dietary or endocrine abnormality, a drug-induced disorder, kidney disease, or an intestinal, pancreatic or hepatobiliary abnormality. **Hypocholesterolemia** may occur with hyperthyroidism, by the administration of drugs that reduce endogenous CH biosynthesis or its absorption from the intestine, or by intestinal diseases which adversely affect bile acid and CH absorption. When intestinal CH absorption is compromised, increased amounts of CH are used by the liver for bile acid biosynthesis, thus lowering the plasma CH concentration (see Chapters 62, 66, and 67). **Hypercholesterolemia** in animals is usually recognized as a secondary hyperlipidemia, most often associated with decreased hepatic removal of LDL from the circulation, as seen in hypothyroidism, hypersomatotropism, diabetes mellitus, Cushing's disease or syndrome, and various forms of hepatobiliary disease. Additionally, ultra-high fat diets, drug-induced disorders (e.g., megestrol acetate administration to cats), pancreatitis, and kidney disease are also associated with hypercholesterolemia in animals (see Chapters 63 and 67).

In summary, CH is a structural component of cell membranes, adding to membrane rigidity, but varies in concentration from 0-40% of total membrane lipid. Although virtually all nucleated eukaryotic cells can synthesize CH, hepatocytes do so to the largest degree. **HMG-CoA reductase** is the rate-limiting enzyme in CH biosynthesis, whose activity is controlled via several endogenous and exogenous agents. Cholesterol can be metabolized to other steroids, such as bile acids and steroid hormones, which maintain considerable physiologic activity. Plasma CH concentrations vary among animal species, and hypercholesterolemia is usually recognized as a secondary hyperlipidemia (see Chapter 67). It is normally secreted into bile, not urine, and usually exits the body through feces and the sloughing of oily secretions and cells from the skin.

OBJECTIVES

- Identify the primary sources of CH for animals, show how tissue contents vary, and how plasma concentrations differ between animal species.

- Explain how CH is absorbed by the intestinal tract, transported in plasma, and excreted from the body.

- Distinguish each of the four stages of CH biosynthesis.

- Identify the rate-limiting reaction in CH biosynthesis, and discuss its endogenous and exogenous control.

- Recognize common causes of hypo- and hypercholesterolemia, and explain why hypercholesterolemia is usually viewed as a secondary hyperlipidemia (see Chapter 67).

- Explain the relationship of CH to bile acid metabolism (see Chapter 62).

- Compare plant sterol to CH absorption from the gut, and clearance from the body.

- Provide an accounting of the relative amounts of CH normally excreted into bile, converted to bile acids or used in the biosynthesis of steroid hormones.

- Understand what is meant by "good" and "bad" CH (see Chapter 66).

- Recognize the fecal forms of cholesterol.

- Understand structural differences between cholesterol and β-sitosterol.

- Explain why and how starvation reduces hepatic cholesterol biosynthesis, yet increased dietary carbohydrate and/or triglyceride intake increases it.

- Understand why and how dietary saturated fat intake, or trans-unsaturated fat intake, increases the plasma CH concentration.

- Recognize the difference between CH and CE, and discuss how essential UFAs are conserved.

QUESTIONS

1. **Cholesterol present in which one of the following organs becomes part of the rapidly-miscible cholesterol pool?**
   a. Skeletal muscle
   b. Skin
   c. Adipocytes
   d. Brain
   e. Liver

2. **The greater part of cholesterol in plasma is:**
   a. Associated with VLDL.
   b. Esterified with unsaturated fatty acids.
   c. Bound to albumin.
   d. Found as free cholesterol in circulating lipoproteins.
   e. Bound to steroid hormones.

3. **The primary excretory route for cholesterol is:**
   a. Bile.
   b. Urine.
   c. Expired air.
   d. Sweat.
   e. Hair loss.

4. **HMG-CoA reductase, the rate-limiting enzyme in cholesterol biosynthesis, is inhibited by:**
   a. Insulin.
   b. Thyroid hormones.
   c. Squalene.
   d. Cholesterol.
   e. Triglyceride ingestion.

5. **Select the TRUE statement regarding plants:**
   a. Plants, unlike animals, do not contain cholesterol.
   b. Less than 5% of most dietary plant sterols (phytosterols) are normally absorbed by the mammalian digestive tract.
   c. Since herbivores cannot synthesize cholesterol, they derive all of their needs for this steroid directly from plants.
   d. β-Sitosterol, a plant steroid, is what animals use to synthesize cholesterol.
   e. Plants and animals contain about the same amount of cholesterol per gram.

6. **Which one of the following is formed from cholesterol in the intestine via bacterial reduction?**
   a. Coprostanol
   b. Cholecalciferol
   c. Cortisol
   d. Lanosterol
   e. Mevalonate

7. **Hypocholesterolemia is best associated with:**
   a. Diabetes mellitus.
   b. Cushing's disease.
   c. Hyperthyroidism.
   d. Kidney disease.
   e. Biliary obstruction.

8. **The primary phytosterol of plants is:**
   a. Cholesterol ester (CE).
   b. Squalene.
   c. Lanosterol.
   d. Cholestanol.
   e. β-Sitosterol.

9. **Hypercholesterolemia in mammals:**
   a. Is not recognized, and therefore not important.
   b. Is usually a secondary hyperlipidemia.
   c. Is generally a result of hyperthyroidism.
   d. Causes lipemia.
   e. Rarely occurs in patients with diabetes mellitus.

10. **All carbon atoms in CH can be derived from:**
    a. Palmitate.
    b. NADPH.
    c. Acetyl-CoA.
    d. Deoxycholate.
    e. HMG-CoA reductase.

11. **Which cell type is associated with the bulk of CH biosynthesis?**
    a. Neuron
    b. Adipocyte
    c. Testicular Leydig cell
    d. Hepatocyte
    e. Skeletal myocyte

11. d
10. c
9. b
8. e
7. c
6. a
5. b
4. d
3. a
2. b
1. e

Chapter 62

Bile Acids

Overview

- Bile acids are hydroxylated steroids, synthesized in the liver from cholesterol.
- Peroxisomal enzymes assist in the hepatic biosynthesis of BAs.
- Bile acids are normally conjugated in the liver to the amino acids, Gly and Tau, or sulfate.
- The active transport of BAs across canalicular membranes of hepatocytes is a primary driving force for bile flow.
- Duodenal emulsification of fat requires BA micelles and biliary lecithin.
- The distal ileum actively reabsorbs BAs, against their concentration gradient.
- As BAs return to the liver through the EHC, they inhibit cholesterol biosynthesis, as well as further BA biosynthesis.
- Secondary BAs, deoxycholate and lithocholate, are formed through bacterial 7α-dehydroxylation of the primary BAs, cholate and chenodeoxycholate, respectively.
- Glucose, insulin and glucagon also appear to help regulate BA biosynthesis.
- BAs are ligands for the nuclear farnesoid X receptor (FXR), as well as a G protein-coupled receptor (TGR5).
- BAs appear to inhibit hepatic PEP carboxykinase activity, while stimulating glycogen synthase activity.
- BA therapies may be used in the treatment of various forms of liver disease, DM and obesity.

Bile acids (BAs) are synthesized in the liver (only) from **cholesterol (CH)**, and are conjugated with the amino acids **glycine** and **taurine**, and in some animals, sulfate (**SO$_4^=$**). Conjugation adds polar constituents to these fat-soluble compounds, thus rendering them more water-soluble and less likely to precipitate in a watery medium (e.g., bile). The active transport of these osmotically active steroidal compounds across canalicular membranes of hepatocytes provides a primary driving force for bile flow, and following their secretion into bile they are normally concentrated and stored in the gallbladder (in animals possessing a gallbladder). When the gallbladder contracts following movement of lipid-rich chyme into the duodenum, BAs and other constituents of bile are delivered to the small bowel, where they function to emulsify dietary lipid, promote its digestion and absorption, and also assist in the absorption of CH and the fat-soluble vitamins. Other important lipids secreted into bile are **lecithin** (phosphatidylcholine), free **CH** (i.e. **biliary cholesterol (BC)**) and **bile pigments** (e.g., **bilirubin-glucuronides**, **-glucosides**, and **-xylosides**; see Chapter 33).

Copyright © 2015 Elsevier Inc. All rights reserved.

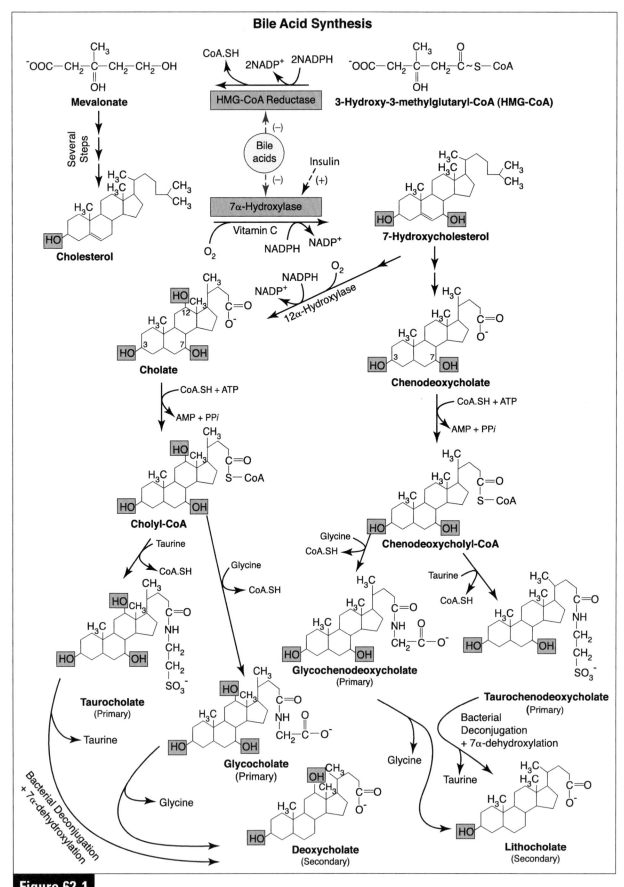

Figure 62-1

Sodium and K⁺ salts of the BAs are referred to as **bile salts (BSs)**, and form in the alkaline milieu of bile. Although the two terms, BAs and BSs, are generally used interchangeably, BSs are generally considered to be better emulsifying agents than BAs (see Chapter 60).

Hepatic BA Biosynthesis

The primary types of reactions involved in BA biosynthesis from CH are **hydroxylation, side-chain cleavage**, and **conjugation (Fig. 62-1)**. Hydroxyl groups are added to carbons 7 and 12 of **CH** by specific hydroxylases found in the smooth endoplasmic reticulum of liver cells. The **7α-hydroxylase** requires the presence of **vitamin C** (see Chapter 39), and both hydroxylases require **molecular O_2, NADPH**, and **cytochrome P-450**. The cytochrome P-450 enzymes are members of a large family of enzymes known as **mixed function oxygenases**.

Following **7α-hydroxylation** of CH, a three-carbon fragment is cleaved from the side chain by peroxisomal enzymes, and the terminal carbon is oxidized to a carboxylic acid. Conjugated BAs are formed through the addition of either **glycine** or **taurine** (and sometimes **sulfate** (not shown)) to the carboxy terminals. Prior to conjugation, the side-chain carboxyl group is activated by reaction with CoA, forming a high-energy thioester bond. The energy in the thioester bonds is used to form the amide linkage between the BA and either amino acid.

Of these two amino acids, **glycine** can be easily derived from protein turnover, however **taurine** is a nonprotein amino acid that must either be derived directly from the diet, or synthesized from cysteine (**Fig. 62-2**). Although fish, reptiles, birds and most mammals are capable of meeting biologic demands through *de novo* taurine biosynthesis, **cats** have difficulty synthesizing enough of this amino acid to meet body needs, and therefore require a continual dietary supply (see Chapter 3). The primary biosynthetic pathway for taurine begins with the oxidation of cysteine to cysteine sulfinate by **cysteine dioxygenase**. This enzyme requires **Fe^{++}, NAD^+**, and **O_2**, and basically inserts two atoms of oxygen into the cysteine molecule. Decarboxylation next yields **hypotaurine**, which

Taurine Biosynthesis

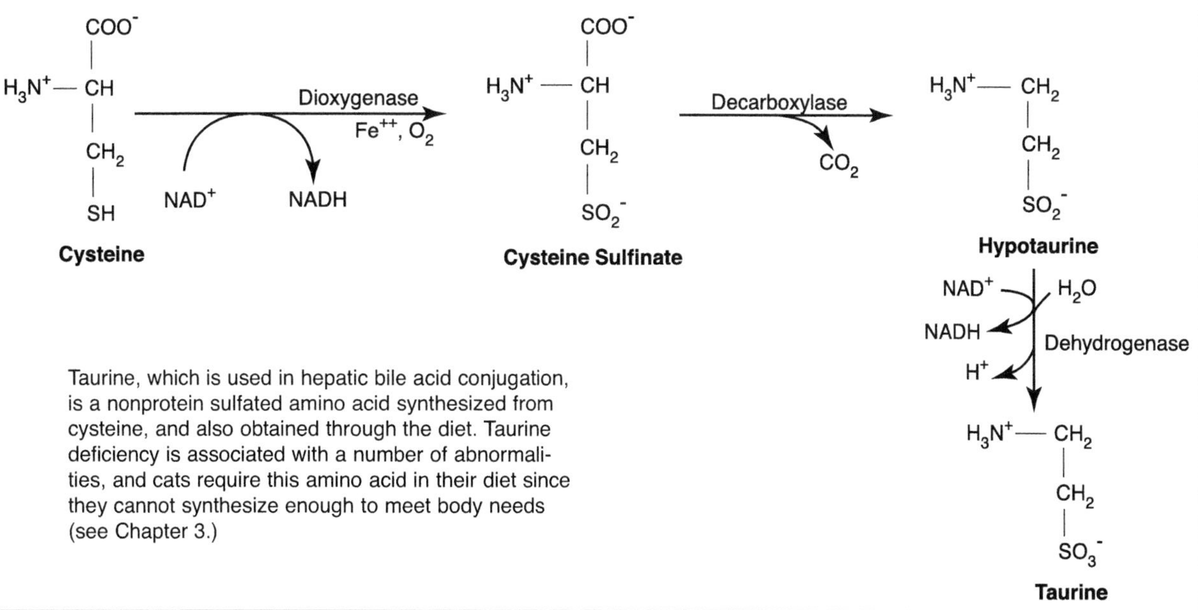

Taurine, which is used in hepatic bile acid conjugation, is a nonprotein sulfated amino acid synthesized from cysteine, and also obtained through the diet. Taurine deficiency is associated with a number of abnormalities, and cats require this amino acid in their diet since they cannot synthesize enough to meet body needs (see Chapter 3.)

Figure 62-2

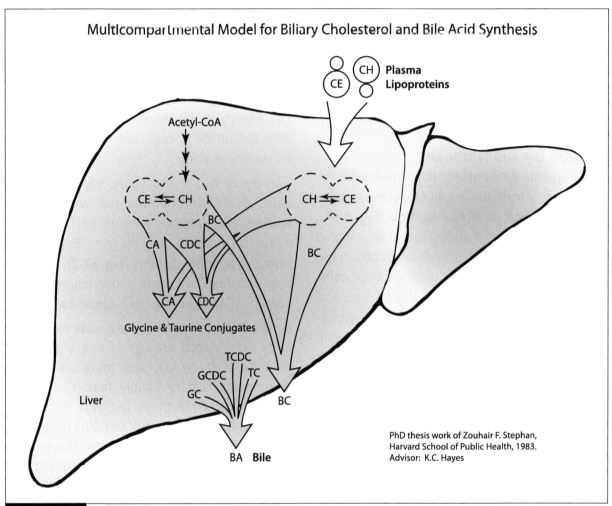

Multicompartmental Model for Biliary Cholesterol and Bile Acid Synthesis

Plasma Lipoproteins

Acetyl-CoA

CE ⇌ CH

CH ⇌ CE

BC

CA CDC

BC

CA CDC

Glycine & Taurine Conjugates

TCDC

GCDC TC

GC

BC

Liver

BA **Bile**

PhD thesis work of Zouhair F. Stephan, Harvard School of Public Health, 1983. Advisor: K.C. Hayes

Figure 62-3

can then be **oxidized** to **taurine**.

Hepatic **CH** and **BA** biosyntheses are regulated by the enzymes **HMG-CoA reductase** and **7-α-hydroxylase**, respectively. The activities of these two enzymes change in parallel, making it difficult to ascertain whether control of BA synthesis takes place primarily at the HMG-CoA reductase step, or at the 7α-hydroxylase reaction. Bile acids do not seem to regulate these enzyme activities by a direct allosteric mechanism. However, studies indicate that these two enzymes can be controlled by covalent phosphorylation-dephosphorylation. In contrast to HMG-CoA reductase (see Chapter 61), it is the **phosphorylated** form that results in increased activity of **7α-hydroxylase**.

Hepatic BA synthesis from CH results in the formation of two primary BAs, **cholate (CA)**, a 3,7,12-trihydroxy BA, and **chenodeoxycholate (CDC)**, a 3,7-dihydroxy BA. The ratio of these two varies with species, with CA predominating in primates, and CDC predominating in several domestic animals studied. Conjugation of **CA** and **CDC** with **glycine (G)** or **taurine (T)** produces **glycocholate (GC)**, **taurocholate (TC)**, **glycochenodeoxycholate (GCDC)**, and **taurochenodeoxycholate (TCDC)**, respectively. Studies in monkeys have shown that the greater proportion of CH entering the liver from plasma goes directly into bile, whereas much of that synthesized in the liver goes into BA formation (mainly CDC; **Fig. 62-3**). Conversely, plasma CH entering the liver from circulating lipoproteins appears to favor CA biosynthesis over CDC. Thus, increased dietary CH favors biliary cholesterol (BC), GC and TC excretion, which apparently gives primates, carnivores

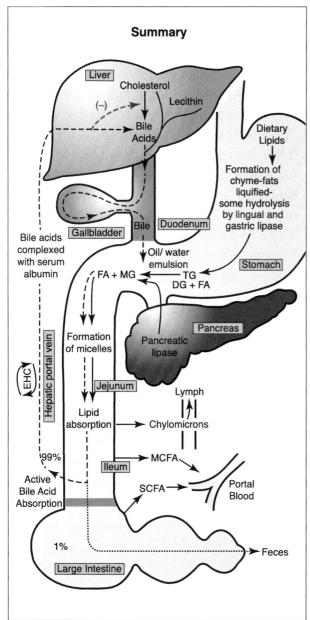

Summary

Liver
Cholesterol
Lecithin
(–)
Bile Acids
Dietary Lipids
Formation of chyme-fats liquified-some hydrolysis by lingual and gastric lipase
Bile acids complexed with serum albumin
Gallbladder   Bile   Duodenum
Oil/ water emulsion
Stomach
FA + MG ← TG
DG + FA
Hepatic portal vein
EHC
Formation of micelles
Pancreas
Pancreatic lipase
Jejunum
Lymph
Lipid absorption → Chylomicrons
99%
Active Bile Acid Absorption
Ileum
→ MCFA
SCFA →
Portal Blood
1%
⋯→ Feces
Large Intestine

Figure 62-4

and omnivores less protection against CH gallstone formation since GC and TC are less effective than GCDC and TCDC at solubilizing biliary CH.

Bile Acid Actions in Bile, and in Luminal Contents of the Intestine

As indicated above, bile acids are the major osmotic force producing water movement across canalicular membranes of hepatocytes. Thus, under normal circumstances, **bile flow** is directly proportional to the amount of **BA**

excreted. Additionally, BAs may also modulate fluid absorption from the terminal ileum. **Milk lipase** and **pancreatic lipase** are somewhat dependent upon the primary BAs (particularly the taurine conjugates) as cofactors, and BAs have also been reported to protect a number of other luminal lipolytic enzymes (e.g., **lingual lipase**, **gastric lipase**, and **phospholipase A₂**) from proteolytic digestion, and prevent surface denaturation (see Chapter 60).

Following biliary excretion, **BAs**, **lecithin**, and **biliary CH** are concentrated in the gallbladder (of animals possessing gallbladders). The relative proportions of these lipid solutes in hepatic bile are normally about **67%**, **22%**, and **4%**, respectively. When chyme from the stomach enters the duodenum, the hormone **cholecystokinin** (**CCK**) is released into blood, and it stimulates release of digestive enzymes from the pancreas, and also stimulates contraction of the gallbladder. Gallbladder contraction delivers its contents to the duodenum, where the lipids contained therein function to emulsify dietary fat (**Fig. 62-4**). Lecithin and the BAs, particularly, have considerable ability to lower surface tension. This enables them to function as emulsifying agents, and to dissolve **long-chain fatty acids** (**LCFAs**) and water-insoluble soaps. When fat digestion is impaired due to their absence, other foodstuffs are also poorly digested, and since fat covers food particles and prevents enzymes from attacking them, they pass into the large intestine where the activity of intestinal bacteria can cause considerable putrefaction and production of gas, diarrhea, and steatorrhea (see Chapter 60).

Bile acids and lecithin form **mixed micelles** in the intestine, which are spherical aggregates of about 3-10 nM diameter. The polar hydroxy groups of the BAs, as well as their peptide bonds and carboxyl groups face outward, and their non-polar steroid nuclear portions, as well as the FA chains of the PLs, form a hydrophobic core (see **Fig. 62-5**).

Each **mixed micelle** normally incorporates **LCFAs, monoglycerides, lysolecithin, CH,** and **fat-soluble vitamins**. The BA concentration at which micelles begin to form (i.e., the **critical micelle concentration (CMC)**), is about **2 mM**. Gallbladder concentration via NaCl and H_2O reabsorption usually causes the BA concentration therein to approximate **300 mM**, which indicates that biliary obstruction, or significant liver disease resulting in greatly reduced BA output, would need to be present before micelles were excluded from duodenal content.

Lipids that are relatively soluble in water need not be carried in mixed micelles. For example, **short-chain (SCFAs)** and **medium-chain fatty acids (MCFAs)** are sufficiently soluble in molecular form so that diffusion through the unstirred layer of the intestine is rapid enough to allow their absorption without micelle incorporation. Additionally, **SCFAs** and **MCFAs** absorbed from the intestine are not packaged into chylomicrons in the cytoplasm of mucosal cells, and therefore are allowed to pass directly into portal blood (see Chapter 60).

Intestinal Bile Acid Reabsorption and Enterohepatic Cycling

Once conjugated **BAs** appear in luminal contents of the **distal ileum**, approximately **99%** are **reabsorbed** by an energy-dependent **active transport** mechanism. Once they enter the hepatic portal circulation they are returned to the liver, where they are efficiently (**70-90%**) removed from blood. For this reason low systemic concentrations (**2-4 μmol/L**) are normally observed compared with rather high hepatic portal concentrations (**60-80 μmol/L**), unless liver disease is present. As BAs return to the liver through this **enterohepatic circulation (EHC)**, they feedback negatively on **HMG-CoA reductase**, as well as on **7α-hydroxylase** (see **Figs. 62-1** and **62-4**). Therefore, they not only regulate their own biosynthesis from CH, but they also partially regulate *de novo* hepatic CH biosynthesis.

In animals with gallbladders, BAs have been found to traverse the entire EHC (i.e., **liver —> bile —> intestine —> portal blood —> liver**) approximately **12 times/day**. Since approximately **1%** of the BA pool passes into the large intestine and is lost with each cycle, only about **12%** of the pool needs to be replaced each day. In contrast, animals without gallbladders (e.g., horses, or dogs having undergone cholecystectomy), maintain a much smaller BA pool, yet they cycle it more often (approximately **33 times/day**).

The small amount of the BA pool that normally enters the large intestine can be acted upon by bacteria therein. Microbes are known to **deconjugate** the primary BAs, and also to **dehydroxylate** them in the 7-position. 7α-Dehydroxylation of CA forms the **secondary BA** known as **deoxycholate**, and 7α-dehydroxylation of CDC forms **lithocholate**, also a secondary BA (see **Fig. 62-1**). Although most of the secondary BAs are excreted in feces, some are reabsorbed into portal blood and are mixed with other BAs in the EHC.

Regulation of Hepatic Bile Acid Biosynthesis

Mechanisms of BA negative feedback on its own hepatic biosynthesis (namely on **7 α-hydroxylase** activity) have been studied for

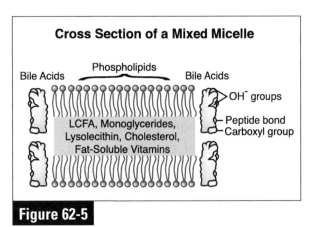

Cross Section of a Mixed Micelle

Phospholipids

Bile Acids   Bile Acids

OH^- groups

LCFA, Monoglycerides, Lysolecithin, Cholesterol, Fat-Soluble Vitamins

Peptide bond
Carboxyl group

Figure 62-5

over 50 years. Recent studies indicate that nutrients and various hormones also play roles in BA biosynthesis regulation during starvation and refeeding. At physiologic concentrations, glucose and insulin appear to induce, while glucagon represses 7 α-hydroxylase gene expression. In the hyperglycemic, diabetic, insulin-resistant state, however, these roles may be reversed. The BA pool size of diabetic patients is usually elevated, and decreased upon insulin treatment. This increased pool is not the cause of diabetes, but a consequence of BA metabolism dysregulation and altered homeostasis. Indeed, in recent years plasma BA levels have become not only biomarkers for the diagnosis of liver disease, but also diabetes and obesity.

Bile Acid Signaling

Bile acids appear to be endogenous ligands for the **nuclear farnesoid X receptor (FXR)**, and stimulation of this receptor inhibits hepatic 7 α-hydroxylase gene expression. The FXR appears to play a critical role in the EHC of BAs not only by regulating BA biosynthesis, but also by facilitating canalicular BA secretion, ileal BA reabsorption, and hepatic BA uptake. This receptor has been shown to play a role as well in pancreatic insulin release at high plasma glucose concentrations. These revelations, and others, provide some intriguing mechanistic insights not previously suspected. An additional **BA-activated G protein-coupled receptor (TGR5)** appears to be involved with several inflammatory diseases including fatty liver infiltration, atherosclerosis, diabetes mellitus (DM) and the inflammatory bowel diseases (IBDs). Indeed, **BA-activated FXR and TGR5 signaling** have been reported to suppress inflammation in macrophages, the intestine and hepatocytes, and to increase energy expenditure in brown adipose tissue.

Integration of Bile Acid Signaling, Hepatic Carbohydrate and Lipid Metabolism

Recent studies indicate an integrated regulation of BA signaling through **FXR** and **TGR5 pathways** in **hepatic triglyceride (TG)** and **glucose formation**. Additionally, the FXR appears to induce expression of **Apo-C**, which is an activator of **lipoprotein lipase (LPL)** involved in hydrolysis of TGs carried by TG-rich lipoproteins (**VLDL** and **CMs**), and **Apo-A** (which is an activator of **lecithin-cholesterol acyl-transferase (LCAT**; see Chapters 63-66). Bile acids also appear to inhibit hepatic **PEP carboxykinase** activity (see Chapter 37), and stimulate **glycogen synthase** (see Chapter 23). Thus, **BAs** are apparently playing important metabolic roles by **inhibiting hepatic gluconeogenesis**, and **stimulating glycolysis** and **glycogenesis** to **improve glucose tolerance and insulin sensitivity**.

Bile Acids as Therapeutic Agents

Therapeutic potentials of **BA derivatives** for treating hepatobiliary diseases (such as **nonalcoholic fatty liver disease (NAFLD)** and **metabolic syndrome**) are now well-recognized. NAFLD is, unfortunately, a prevalent human chronic liver disease affecting about 30% of adults in developed countries. Metabolic syndrome is a collection of five well-known pathophysiologic conditions including **hypertension, hyperglycemia, hypertriglyceridemia, insulin resistance** and **obesity**. It also contributes to chronic heart disease, atherosclerosis, adult-onset DM (type II), and NAFLD. Dyslipemia produces insulin resistance and inflammation, with this pathogenesis leading to NAFLD. Therefore, NAFLD becomes a spectrum of chronic liver abnormalities from simple steatosis to nonalcoholic steatohepatitis (NASH), which can cause cirrhosis. About 10-20% of NASH patients reportedly progress to cirrhosis and hepatocellular carcinoma, which are end stages of liver disease.

Feeding a **BA sequestrant** like **cholestyramine** (see Chapter 67) will reduce the BA pool, which is enlarged in diabetes. It also increases hepatic BA biosynthesis, which in turn inhibits gluconeogenesis and lipogenesis (while favoring glycolysis, FA oxidation and glycogenesis). Indeed, BA sequestrants have glucose-lowering potential and reportedly improve glycemic control in some diabetics. Additionally, increasing BA synthesis stimulates *de novo* cholesterol biosynthesis, but increases biliary cholesterol secretion without increasing intestinal cholesterol absorption. Thus it helps to maintain whole body cholesterol homeostasis.

A **synthetic BA derivative (obeticholic acid (OCA, 6-ethyl-CDCA** or **INT-747))**, is a potent and selective **FXR agonist** with anticholestatic properties. In recent animal and human studies, OCA was found to increase insulin sensitivity, inhibit gluconeogenesis, inhibit hepatic lipogenesis, and exhibit anti-inflammatory and anti-fibrotic actions. It appears to ameliorate high fat diet-induced obesity and insulin resistance in mice, and insulin resistance and fatty liver infiltration in rats. Another BA derivative (**6α-ethyl-23(S)-methyl-CA (EMCA** or **INT-777))** appears to be a selective and potent **TGR5 agonist**. It improved glucose tolerance in animal studies, reduced atherosclerotic lesions by reducing macrophage inflammation and lipid loading, and appeared to reduce hepatic and intestinal inflammation.

Basic research into BA metabolism and signaling over the past 50 years has significantly improved our knowledge and understanding of liver physiology and pathology. Indeed, at this point in time the future of BA-based therapies appears to retain considerable potential for assisting in the treatment of diabetes and various forms of liver disease.

OBJECTIVES

- Identify the primary types of reactions involved in hepatic BA biosynthesis, as well as the substrates and cofactors involved.

- Describe the enterohepatic cycling of BAs, and identify the two hepatic rate-limiting reactions they regulate.

- Know the taurine biosynthetic pathway, as well as the species-specific dietary requirement of this compound.

- Note how the structure of CH differs from that of CA, CDC, deoxycholate and lithocholate.

- Explain how primary BAs become secondary BAs, and understand why this transition occurs.

- Understand the physiologic roles of BAs in bile and micelle formation, and in dietary lipid emulsification and transport to the intestinal brush border.

- Summarize the movement of BAs through the EHC, and predict how ileal disease, cholecystectomy or liver dysfunction might affect this cycling, as well as the systemic BA concentration.

- Identify the types of lipophilic compounds that are normally incorporated into a mixed intestinal micelle.

- Recognize what is meant by the bile acid-dependent and bile acid independent fractions of bile flow.

- Understand the origins of biliary cholesterol (BC; hepatic biosynthesis vs. that derived from lipoproteins).

- Explain how the bulk of biliary CA is derived (as opposed to CDC).

- Understand what a rise in the peripheral plasma bile acid concentration signifies (from a clinical perspective).

- Summarize the control of hepatic 7 α-hydroxylase activity.

- Recognize and understand the importance of the BA signaling pathways.

- Discuss the roles played by Bas in hepatic carbohydrate and lipid metabolism.

- Know what obeticholic acid is, and why it may become an important therapeutic agent.

QUESTIONS

1. Which one of the following is NOT a physiological action of bile acids?
a. Driving force for bile formation
b. Inhibit hepatic cholesterol and bile acid biosynthesis
c. Emulsify dietary fat in the duodenum
d. Stimulate gallbladder contraction
e. Cofactors for pancreatic lipase

2. Approximately how many times per day does the bile acid pool traverse the enterohepatic circulation of a horse?
a. Once
b. Four times
c. Twelve times
d. Thirty-three times
e. Seventy times

3. Which one of the following would be found in a mixed micelle, but not in a chylomicron?
a. Taurocholate
b. Long-chain fatty acid
c. Phospholipid
d. Cholesterol
e. Fat-soluble vitamin

4. The normal hepatic extraction efficiency for bile acids is about:
a. 10%.
b. 25%.
c. 40%.
d. 55%.
e. 75%.

5. Lithocholate is:
a. A primary bile acid.
b. Formed from chenodeoxycholate in the large bowel.
c. A pancreatic enzyme.
d. Used by the liver to synthesize lecithin.
e. A polypeptide.

6. The rate-limiting enzyme in hepatic bile acid synthesis is:
a. Hexokinase.
b. 7-α-Hydroxylase.
c. 12-α-Hydroxylase.
d. Carboxypeptidase.
e. Deoxycholate.

7. Removal of which part of the intestine would have the greatest influence upon bile acid reabsorption, and thus, potentially, lipid digestion?
a. Stomach
b. Jejunum
c. Terminal ileum
d. Colon
e. Rectum

8. Which one of the following amino acids can be used in the biosynthesis of taurine?
a. Phe
b. Ala
c. Ile
d. Cys
e. Pro

9. Increased dietary CH favors:
a. Biliary CH excretion.
b. Hepatic CDC formation over CA formation.
c. Luminal digestion by cholesterol esterase.
d. Enhanced pancreatic insulin release.
e. Intestinal BA absorption.

10. Hepatic 7α-Hydroxylase activity is influenced by:
a. CA.
b. Insulin.
c. CDCA.
d. Glucagon.
e. All of the above

11. In addition to being emulsifying agents for dietary fat, BAs appear to be involved in the regulation of:
a. Hepatic PEP carboxykinase activity.
b. Insulin sensitivity.
c. Hepatic lipogenesis.
d. Bile flow.
e. All of the above

ANSWERS

1. d
2. d
3. a
4. e
5. b
6. b
7. c
8. d
9. a
10. e
11. e

Chapter 63

Lipoprotein Complexes

Overview

- Free fatty acids (FFAs), triglycerides (TGs), phospholipids (PLs), cholesterol (CH), and cholesterol esters (CEs) are normally transported in blood in close association with protein.

- CMs and VLDL transport the greatest proportion of TG in blood, while LDL and HDL transport the greatest proportion of CH.

- Apo-B is the main integral apoprotein of CMs, VLDL and LDL, while Apo-C and Apo-E are peripheral apoproteins of HDL.

- HDL is also known as α_1-lipoprotein.

- The clearance of lipoproteins from blood is facilitated by insulin-sensitive LPL, which is found on the capillary endothelium servicing adipocytes and other tissues.

- Lipoproteins do not normally cross the blood-brain-barrier, or the placental barrier.

- Lipoprotein complexes are not normally found in bile, or urine.

- FFAs are also known as non-esterified fatty acids (NEFAs).

- The transport of FFAs in blood is facilitated by albumin.

- FFAs are considered to be the most metabolically active lipids in blood.

- Abnormalities in lipoprotein metabolism generally occur at the site of their production, or the site of their destruction.

Although **triglycerides (TGs)** are apparently stored in cells without close association with protein, they are transported between tissues in lipoprotein complexes. This is necessary because TGs are insoluble in water, and their combination with protein confers stability in the aqueous environment of plasma. Additionally, varying amounts of **phospholipids (PLs**; mostly lecithin and sphingomyelin), and **cholesterol (CH**; both free and **esterified (CE)**; see Chapter 66), are also transported in lipoprotein complexes (**Fig. 63-1**).

Although there are four main classes of circulating lipoproteins in animals, only two are quantitatively important in TG transport, **chylomicrons (CMs)**, and **very low-density lipoproteins (VLDLs)**. The other two, **low-density (LDL)** and **high-density lipoprotein (HDL)**, play an important role in the transport of CH and CEs in blood. These four types of lipoproteins are usually separated and classified according to either their **density**, which is a function of their lipid/protein ratio, or their **electrophoretic mobility**, which is more dependent upon the composition of their apoproteins (**Fig. 63-2**).

Chylomicrons are derived from the intestinal absorption of fat (see Chapter 60); **VLDL** (also called α_2 or **Pre-β** lipoproteins), are derived

Copyright © 2015 Elsevier Inc. All rights reserved.

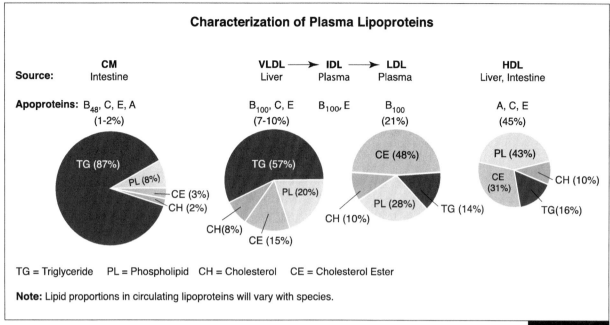

Characterization of Plasma Lipoproteins

	CM	VLDL ⟶ IDL ⟶ LDL	HDL

Source: CM — Intestine   VLDL — Liver   IDL — Plasma   LDL — Plasma   HDL — Liver, Intestine

Apoproteins: B₄₈, C, E, A (1-2%)   B₁₀₀, C, E (7-10%)   B₁₀₀, E   B₁₀₀ (21%)   A, C, E (45%)

CM: TG (87%), PL (8%), CE (3%), CH (2%)

VLDL: TG (57%), PL (20%), CH (8%), CE (15%)

LDL: CE (48%), PL (28%), CH (10%), TG (14%)

HDL: PL (43%), CE (31%), CH (10%), TG (16%)

TG = Triglyceride   PL = Phospholipid   CH = Cholesterol   CE = Cholesterol Ester

Note: Lipid proportions in circulating lipoproteins will vary with species.

Figure 63-1

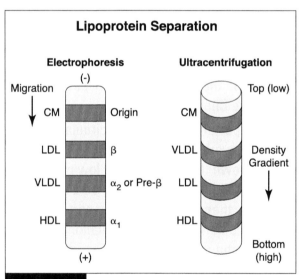

Lipoprotein Separation

Electrophoresis

Migration ↓

(-)

CM — Origin
LDL — β
VLDL — α₂ or Pre-β
HDL — α₁

(+)

Ultracentrifugation

Top (low)

CM
VLDL
LDL
HDL

Density Gradient ↓

Bottom (high)

Figure 63-2

largely from the liver for export of TG into blood; **LDL** (or β-lipoproteins) represent a final stage in the catabolism of VLDL, and are major carriers of CH and CEs; and **HDL** (or α₁-lipoproteins), which are derived largely from the liver, are involved with both CM and VLDL metabolism, and function in CH exchange and esterification. A fifth group, not generally classified with other lipoproteins, are circulating **long-chain fatty acids (LCFAs)** bound to serum **albumin**.

Abnormalities in lipoprotein metabolism generally occur at the site of their production, or at the site of their destruction, thus causing various hypo- or hyperlipoproteinemias. One common form of **hyperlipoproteinemia** in animals is caused by the relative **insufficiency of insulin** in **diabetes mellitus (DM)**. This leads to excessive mobilization of LCFAs from adipocytes, and to a **hypertriglyceridemia** due to decreased removal of circulating **CMs** and **VLDL**. **Hypercholesterolemia** in animals is most often associated with decreased hepatic removal of circulating **LDL**, as seen in **hypothyroidism**, **DM**, and various forms of **liver disease** (see Chapter 67).

Apoproteins

Plasma lipoproteins are characterized by their content of one or more proteins known as **apoproteins**, or **apolipoproteins**. **Peripheral** apoproteins, which lie on the surface of circulating lipoproteins, can be exchanged between lipoprotein complexes, whereas **integral** apoproteins cannot (**Fig. 63-3**).

The main **integral apoprotein** of CMs, VLDL, and LDL is **Apo-B**. Apo-B of CMs (**Apo-B₄₈**) is smaller than that of VLDL and LDL (**Apo-B₁₀₀**),

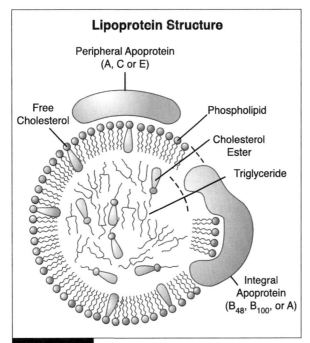

Lipoprotein Structure

Peripheral Apoprotein
(A, C or E)

Free Cholesterol

Phospholipid

Cholesterol Ester

Triglyceride

Integral Apoprotein
(B_{48}, B_{100}, or A)

Figure 63-3

and has a different amino acid composition. Apo-B_{48} is synthesized in the intestine, while Apo-B_{100} is synthesized in the liver. Apo-B_{48} was so named because its molecular weight is 48% of the liver's B_{100} apoprotein, which has one of the longest single polypeptide chains known (4536 amino acid residues). Carbohydrates account for about 5% of Apo-B, and include mannose, galactose, galactosamine, fucose, glucose, glucosamine, xylose, and sialic acid. Thus, some lipoproteins are also glycoproteins (see Chapter 20).

The two peripheral apoproteins of **HDL** are **Apo-C** and **Apo-E**, which are also synthesized in the liver. These apoproteins appear to be freely exchangeable between HDL and CMs, and between HDL and VLDL. Apo-E is arginine-rich, and binds to specific receptors on the surface of liver cells, whereas **Apo-C**, which has three subtypes, is a cofactor for **lipoprotein lipase (LPL)** found on the endothelial surface of capillaries servicing adipose and other tissues. This enzyme is activated by **insulin**, and hydrolyzes lipoprotein **TG** to **FAs** and **glycerol** (see Chapter 64). **Apo-A**, which has three or four subtypes, is an integral apoprotein of **HDL** and a peripheral

apoprotein of CMs. It activates **lecithin-cholesterol acyltransferase (LCAT)**, an enzyme that participates in the esterification of CH, as well as its transfer from **HDL** to **LDL** (see Chapter 66).

FFA-Albumin Complexes

In order to distinguish unesterified FAs from those esterified in TG, the former have been variously termed **non-esterified fatty acids (NEFAs)**, or **free fatty acids (FFAs)**. Arguments can be forwarded to defend either of these terms, but in recent years the term FFA seems to have gained wide acceptance, and will be used here when it is necessary to make a distinction. Except where otherwise qualified, the term **FFA** is taken to mean **LCFA**, that is, one having **14 or more carbon atoms**, and to include both FAs in solution (minor amounts), and those bound to serum albumin (see Chapter 54). The term **fatty acyl-CoA**, or simply **acyl-CoA**, is used to describe the coenzyme A derivative of those acids (see Chapters 41 and 55).

The importance of FFAs as circulating lipid-fuel was overlooked until the mid-1950s, when it was shown that despite their low concentration in blood (**0.3-2.0 mM**), they had a **short half-life (less than 2 min)**. Their importance was further indicated by the observation that their concentration increased during starvation, and decreased after feeding, which indicated that they were not involved in lipid transport from the gut, but in mobilization of lipid stores when required. Of the total amount of **LCFAs** present in blood, on average **45%** are associated with **TG** (largely in CMs and VLDL), **35%** are in **PLs**, **15%** in **CEs**, and less than **5%** are present as **FFAs**. During starvation, however, the flux of FFAs through blood can be considerable, and can account for a high proportion of total caloric requirement. Indeed, FFAs are now known to be metabolically the most active of the plasma lipids.

Free fatty acids arise from hydrolysis of **TG** within **adipocytes**, and are not released into blood from any other tissues. They are removed from blood and oxidized primarily by the **liver**, **kidneys**, **heart**, **aerobic skeletal muscle fibers**, and **brown adipose tissue**. In uncontrolled diabetes mellitus, their blood concentration will rise, and in meal eaters their concentration normally falls after feeding, and rises again prior to the next meal. In rather continual feeders, like ruminant animals, the blood FFA concentration normally remains within narrow limits, and at low levels.

The transport of FFAs in blood is facilitated by **albumin**, a small soluble protein produced by the liver, which normally accounts for about **60% of plasma protein**. At least **10 FFA** molecules may be bound by each **albumin** molecule, although this capacity is seldom reached under physiologic conditions. Three of these 10 sites are high-affinity sites, and when they are occupied (at a total FFA concentration of about **2 mM**), the concentration of FFA not bound to albumin increases. The formation of FFA complexes with protein facilitates their transport in blood, but restricts their movement across the **blood-brain-barrier** as well as the **placental barrier**. These complexes also prevent FFAs from forming potentially destructive micelles. Micellar solutions of FAs can act as detergents, disrupting protein conformation and disorganizing membranes, thus potentially causing tissue damage.

Following dissociation of the **FA-albumin** complex at the plasma membrane of receptive cells, FAs bind to a **membrane FA-binding protein** that apparently acts as a transmembrane cotransporter with **Na$^+$**. Entry into cells will only occur, however, if the FFA concentration in the extracellular fluid is greater than that in cells. It is likely that the low intracellular concentration necessary for inward diffusion is maintained by the presence of a **cytosolic FA-binding protein** (or **Z-protein**), which has high affinity for **LCFAs** (much like albumin does in blood). Such a protein is known to occur in absorptive cells of the intestine, as well as in hepatocytes. This cytosolic binding protein would be expected to play a role in muscle cells similar to that of myoglobin, which increases the rate of O_2 diffusion by reducing the concentration of free O_2 within these cells. Indeed, there is some evidence that in cardiac muscle cells, myoglobin may also function as a FA-binding protein.

In summary, excess calories are ingested in the anabolic phase of the feeding cycle, followed by a period of negative caloric balance. It is during this latter period that animals draw upon their carbohydrate and fat stores for energy purposes. Lipoprotein complexes help to mediate the lipid part of this cycle by transporting lipids from the intestines as CMs, from the liver as VLDL, to most tissues for oxidation as FFA-albumin complexes, and to adipose tissue for storage (e.g., TG contained in CMs and VLDL). **Lipoprotein complexes (i.e., CMs, HDLs, LDLs,** or **VLDLs**), consist of a core of hydrophobic compounds surrounded by a shell of more polar lipids and proteins. They are generally classified according to their density, or their electrophoretic mobility. The various hyperlipoproteinemias observed in animals are thought to arise from either an imbalance in lipoprotein formation, or an imbalance in destruction. One of the more common lipoprotein abnormalities occurs in animals with **diabetes mellitus,** where insulin deficiency causes excessive FFA mobilization from adipose tissue, and decreased clearance of CMs, VLDL and LDL from blood, thus leading to hypertriglyceridemia and hypercholesterolemia. **Hypothyroidism**, **hypersomatotropism** and **Cushing's-like syndrome** can also cause hyperlipidemia, as can various other abnormalities (see Chapter 67).

OBJECTIVES

- Identify the five primary lipoprotein complexes normally found in plasma, and describe the formation of each.

- Know the relative proportion of apoprotein, PL, CH, CE and TG normally found in CMs, VLDL, LDL and HDL, and recognize the apoproteins normally associated with each lipoprotein.

- Indicate how lipoprotein separation via ultracentrifugation differs from that via electrophoresis.

- Explain why FFAs have difficulty crossing the blood-brain-barrier and the placental barrier (see Chapter 90).

- Distinguish the various apoproteins in terms of origin, lipoprotein complex and function.

- Show how LCFAs are carried in blood, and know what compounds they are best associated with.

- Understand why the plasma FFA concentration of dogs and cats falls following a meal, but rises between meals.

- Explain how and where FFAs are normally oxidized (see Chapter 55).

- Give the two most common general causes of hyperlipoproteinemia.

QUESTIONS

1. Apoprotein B$_{48}$ is synthesized in:
   a. Mucosal cells of the small intestine.
   b. Adipocytes.
   c. Hepatocytes.
   d. Proximal renal tubular epithelial cells.
   e. Skeletal muscle tissue.

2. Which of the following lipoprotein complexes are most quantitatively important in the transport of triglyceride (TG) in blood?
   a. LDL and VLDL
   b. HDL and CM
   c. TG-Albumin and HDL
   d. VLDL and CM
   e. HDL and LDL

3. The β-lipoprotein fraction of plasma determined through electrophoresis is also known as:
   a. FFA-Albumin.
   b. CM.
   c. VLDL.
   d. LDL.
   e. HDL.

4. The most metabolically active plasma lipids are:
   a. Steroid hormones.
   b. Triglycerides.
   c. Lipoproteins.
   d. Phospholipids.
   e. Free fatty acids.

5. Which one of the following does not oxidize free fatty acids?
   a. Hepatocyte
   b. Erythrocyte
   c. Cardiac muscle cell
   d. Skeletal muscle cell
   e. Renal tubular epithelial cell

6. Of the total amount of long-chain fatty acids in blood, most are usually associated with:
   a. FFA-Albumin complexes.
   b. Phospholipids.
   c. Cholesterol esters.
   d. Sphingolipids.
   e. Triglyceride.

7. Of the total amount of LCFAs present in blood, 45% are usually associated with:
   a. PLs.
   b. CEs.
   c. TGs.
   d. FFAs.
   e. LDL.

8. The primary way in which cholesterol is carried in blood is:
   a. In VLDL.
   b. As a phospholipid.
   c. In esterified form.
   d. In chylomicrons.
   e. As free cholesterol.

ANSWERS

8. c
7. c
6. e
5. b
4. e
3. d
2. d
1. a

Chapter 64

Chylomicrons

Overview

- Chylomicrons (CMs) arise solely from the intestine, and contain TG primarily of dietary origin.

- CMs also contain PLs, CH, CEs, and usually one fat soluble vitamin.

- High amounts of CMs in blood cause serum to become lactescent.

- Nascent CMs must acquire Apo-C from circulating HDL in order to become effective substrates for LPL (i.e., clearing factor lipase).

- Once parent CMs have become depleted of most of their TG by extrahepatic tissues, they become CM remnants.

- Remnant CMs require Apo-E in order to become cleared by the liver.

- The LPL associated with cardiac muscle has a low K_m for TG, whereas that associated with adipose tissue has a high K_m.

- When CMs are administered intravenously, approximately 80% of their labeled FAs are found in extrahepatic tissues.

Chylomicrons (CMs) are found in the "**chyle**" which moves into the lymphatic system draining the intestine. It has been known since the work of the French physiologist, Claude Bernard (1813-1878), that absorbed fat enters blood via the **lymphatic system**, and that this route helps to assure that adipose tissue will be exposed to a higher concentration of CMs than would be the case if they first passed through the liver.

Chylomicrons arise solely from the intestine, and contain **triglyceride (TG)** of primarily dietary origin (**Fig. 64-1**). In addition to **TG**, CMs usually contain **phospholipids (PLs)**, **cholesterol (CH)**, **cholesterol esters (CEs)**, and at least one **fat-soluble vitamin** (see Chapters 60 and 63). The major apoprotein components of CMs are **Apo-B₄₈** and **Apo-A**. Apo-A is a peripheral apoprotein, and therefore exchangeable with other lipoproteins in plasma, whereas Apo-B₄₈ is an integral apoprotein, and therefore non-exchangeable (see Chapter 63). Apo-B₄₈ is essential for intestinal CM formation, and Apo-B₁₀₀ is used for very low-density lipoprotein (VLDL) formation in hepatocytes (see Chapter 65). Defects have been reported in which lipoproteins containing Apo-B are not properly formed (i.e., **abetalipoproteinemia**), thus resulting in lipid accumulation in hepatocytes and/or mucosal cells of the small intestine.

Labelled CMs are cleared rapidly from blood, with their circulatory half-life being a few minutes in **rodents**, yet somewhat longer in **dogs** in whom it is still under one hour. However, digestion and absorption of lipid and formation and secretion of CMs takes several hours in dogs; therefore, **lipemia** (i.e., **lactescent**

Copyright © 2015 Elsevier Inc. All rights reserved.

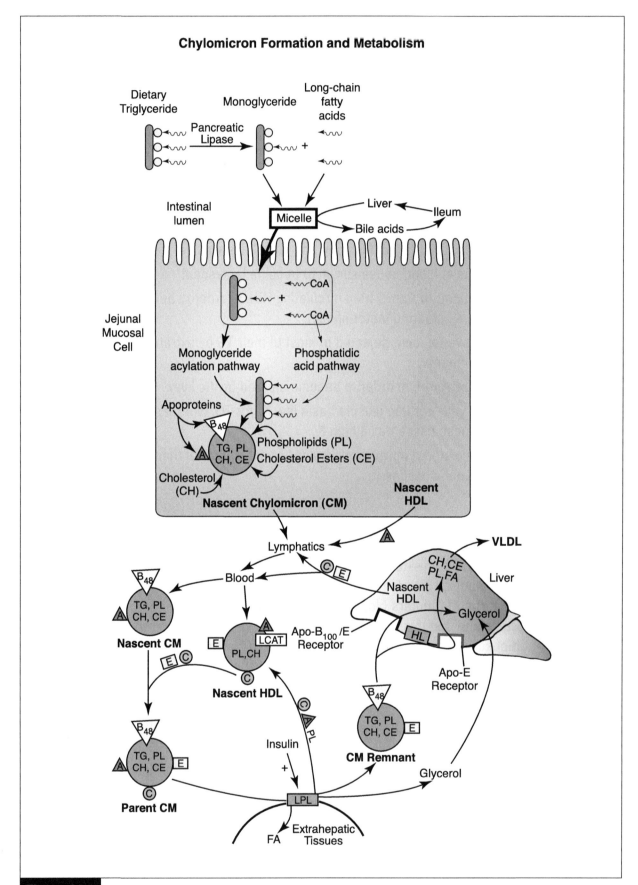

Figure 64-1

serum) generally develops slowly following a fatty meal (see Chapter 67). Chylomicrons will usually disappear from blood overnight if animals do not eat. When CMs are administered intravenously, some 80% of their labelled fatty acids (FAs) are found in extrahepatic tissues (i.e., adipocytes, heart, and aerobic skeletal muscle fibers), and approximately 20% in liver.

Smaller and denser particles having physical characteristics of VLDL are also found in chyle. However, their apoprotein composition resembles CMs rather than VLDL, indicating that they should probably be regarded as small CMs. Their formation is more constant than that of CMs derived from dietary sources, and during starvation they may transport up to 50% of lymphatic TG and CH, with their lipids being derived mainly from bile and intestinal secretions (i.e., sloughed cells).

Chylomicron formation increases with the load of TG digested and absorbed. Triglyceride, whether in the form of CMs or other lipoproteins, is not generally taken up directly by extrahepatic tissues (an exception being receptor-mediated endocytosis of some lipoproteins). It is typically hydrolyzed outside the cell to FAs and glycerol, which then become available to either enter the cytoplasm of the cell (FAs), or continue on in the circulation (glycerol). This hydrolysis is carried out by **lipoprotein lipase** (**LPL**), also known as **clearing-factor lipase**. In extrahepatic tissues, including adipose tissue, skeletal muscle, heart, lung, and lactating mammary tissue, this enzyme is found on **outer surface endothelial cells lining capillaries**, anchored by proteoglycan chains of **heparan sulfate** (see Chapter 19). Although normal blood plasma does not contain significant quantities of LPL, following an injection of **heparin**, some LPL is released from its heparan sulfate binding sites into blood where it remains active. In sinusoids of the liver, some **hepatic lipase** (**HL**) is also released following

heparin injection, but this enzyme has properties different from those of extrahepatic LPL, and appears to react only with CM remnants, VLDL remnants, and high-density lipoprotein (HDL; see Chapter 66).

Interestingly, LPL is synthesized within extrahepatic tissues, and then transported to endothelial surfaces of capillaries servicing those tissues. In adipose tissue, **insulin** enhances LPL synthesis and its translocation to the luminal surface of the capillary endothelium. **Insulin** is also required to **activate LPL** on capillary endothelial tissue servicing adipocytes, and thus serves an important role in the clearance of TG and other lipids from the circulation.

Chylomicrons obtained directly from the lymphatic duct (i.e., **nascent CMs**) are generally considered to be poor substrates for LPL. In order to become effective substrates, these nascent particles must acquire **Apo-C** (namely Apo-C_{II}) from circulating **HDL**. **Apo-E**, also obtained from circulating HDL, is added as a second peripheral apoprotein, thus establishing this lipoprotein as a **parent CM**. Apo-E is needed for eventual uptake of CM remnant particles by the liver.

Amphipathic **phospholipids** contained in the surface layer of CMs (see **Fig. 63-3**), as well as the newly acquired peripheral **Apo-C**, are cofactors for **LPL**. Apo-C contains a specific PL binding site through which it is attached to CMs. The activity of insulin-stimulated extrahepatic LPL tends to deplete CMs of almost 90% of their TG, so that their lipid/protein ratio falls. Progressive TG hydrolysis takes place while CMs are attached to endothelial LPL, forming first a diglyceride, then a monoglyceride, and finally three fatty acids (FAs) plus glycerol. Although some FAs may remain in the circulation bound to albumin, most diffuse immediately into tissues. This stepwise process results in the production of a **remnant CM** (or remnant particle), which is finally metabolized by the liver.

Remnant CMs are smaller than parent CMs, and become relatively enriched in PLs, CH, and CEs through loss of TG, and exchange with other lipoproteins. Only about 5% of the CE in CMs is lost through LPL action. The ratio of TG to CE in parent CMs is almost 30:1, whereas in CM remnants it is about 1:1. During reaction of parent CMs with LPL, Apo-C, Apo-A, and surface PL are transferred to circulating HDL, but not Apo-E and Apo-B_{48}. Additionally, as HDLs, CMs and VLDLs encounter each other in extrahepatic capillary beds, the combined action of **lecithin:cholesterol acyltransferase** (**LCAT**) on the surface of HDLs, **LPL** and **cholesterol ester transfer protein** (**CETP**) results in the transfer of **CEs** to CM remnants and VLDL remnants, thus enriching their CE content even more (see Chapter 66).

Remnant CMs are ultimately removed by the liver through **receptor-mediated endocytosis**. Vesicles formed by endocytosis fuse with lysosomes, where the remnants are degraded by acid hydrolases to a mixture of amino acids, FAs, and free CH. Uptake of CM remnants is mediated by a **receptor** specific for **Apo-E**, by another receptor which is also specific for low-density lipoprotein (LDL; the **Apo-B_{100}/E receptor**), and by **HL**. Hepatic lipase appears to act by hydrolyzing remaining TG and PL in remnant CMs. Some lipids that are delivered to the liver via CM remnants are repackaged into **VLDLs**, and returned to plasma (see Chapter 65).

As indicated in the previous chapter, only two lipoprotein complexes are quantitatively important in **TG** transport, **CMs** and **VLDLs**. Although CMs are prevalent following a lipid-rich meal, and, as indicated above, some of the lipid in CM remnants is repackaged by the liver into VLDLs and exocytosed into blood, circulating VLDLs are better associated with starvation, where FFAs are being released from adipocytes in large amounts. **Lipoprotein lipase** activity in capillaries servicing **adipose tissue** is high in the fed state for the clearance of TG contained in CMs and VLDLs, however, because of its **high K_m** and dependence upon **insulin** for its activity, the activity of this enzyme is low during starvation. On-the-other hand, **LPL** on capillary endothelial cells servicing **cardiac muscle** and **aerobic (Type I) skeletal muscle** tissue has a **low K_m** for TG, some ten-times lower than that in capillary beds servicing adipocytes. As the concentration of plasma TG in CMs decreases in transition from the fed to the starved condition, the heart enzyme remains saturated with substrate while saturation of LPL in adipose tissue declines. Also, as the concentration of plasma TG in VLDL increases during starvation, the cardiac enzyme remains saturated, thus assuring plentiful aerobic substrates (i.e., LCFAs) for cardiac muscle mitochondria. Thus, in transition from the fed to starved condition, the ability to clear TG from the circulation is redirected from adipose tissue toward the heart, and other aerobic muscle fibers (see Chapter 80).

A similar redirection occurs during lactation, where adipose tissue metabolism is directed more toward TG catabolism, and **mammary gland** metabolism is directed more towards TG anabolism (from either circulating FFAs, or circulating TG in CMs or VLDLs). The **LPL** activity of the lactating mammary gland is usually high in either the fed or starved condition, thus helping to assure a rather continual uptake of LCFAs from lipoprotein TG for milk fat synthesis.

In summary, lipoprotein particles known as CMs provide a means of transporting TGs and other large, insoluble dietary lipids away from the digestive tract via lymph, and then into blood for distribution within the organism. The circulating half-life of CMs in plasma is usually short (minutes-hours), with larger particles generally cleared more rapidly than the smaller ones. Approximately 20% of the lipid present in CMs is eventually found in the liver, with 80% disappearing into extrahepatic sites. In order to become effective substrates for hepatic clearance, however, parent CMs must lose their Apo-C and Apo-A in extrahepatic vascular beds, but retain Apo-E and Apo-B_{48}. Insulin helps to clear the TG found in CMs by activating LPL on the capillary endothelium.

OBJECTIVES

- Explain how and where CM formation occurs, show their route into plasma, and know the normal content of a circulating nascent CM.

- Recognize the differences between lipemia, hyperlipidemia and abetalipoproteinemia (see Chapter 67).

- Explain why animals with DM have difficulty clearing CMs, VLDL and LDL from the circulation.

- Understand the origin of LPL, its primary location, and its relationship to insulin, feeding and starvation. Also realize why the LPL servicing adipocytes has a higher K_m than that servicing aerobic muscle fibers.

- Discuss the functions of the various peripheral and integral apoproteins associated with CMs.

- Describe the different substrate specificities of LPL and HL.

- Show how nascent CMs become parent CMs, and then remnant CMs, and know how remnant CMs are removed from blood.

- Explain the roles LCAT, LPL, CETP, HDL, CM remnants and VLDL remnants in capillary CE transfer.

- Outline how LCFA contained in a CM can end up in circulating VLDL.

QUESTIONS

1. **The major apoprotein components of nascent chylomicrons are:**
   a. Apo-B_{48} and Apo-B_{100}.
   b. Apo-C and Apo-E.
   c. Apo-A and Apo-C.
   d. Apo-B_{100} and Apo-E.
   e. Apo-A and Apo-B_{48}.

2. **Which one of the following hormones is known to enhance lipoprotein lipase (LPL) synthesis and activity, as well as its translocation to the luminal surface of the capillary endothelium?**
   a. Epinephrine
   b. Insulin
   c. Thyroxine
   d. Glucagon
   e. Estrogen

3. **Which apoprotein is needed for the clearance of remnant chylomicrons by the liver?**
   a. Apo-A
   b. Apo-B_{100}
   c. Apo-C
   d. Apo-D
   e. Apo-E

4. **What is the primary difference between parent chylomicrons, and chylomicron remnants?**
   a. Chylomicrons remnants have been depleted of most of their triglyceride.
   b. Chylomicron remnants lack cholesterol.
   c. Parent chylomicrons are poor substrates for lipoprotein lipase.
   d. Parent chylomicrons are smaller than chylomicron remnants.
   e. Chylomicron remnants are made by the liver, whereas parent chylomicrons originate from the intestine.

5. **Select the TRUE statement below:**
   a. The lipoprotein lipase associated with cardiac tissue has a lower K_m than that associated with adipose tissue.
   b. The lipoprotein lipase associated with adipose tissue has a lower K_m than that associated with lactating mammary tissue.
   c. Chylomicrons are known to be produced by both the liver and intestine.

5. a
4. a
3. e
2. b
1. e

Chapter 65

VLDL, IDL, and LDL

Overview

- Mucosal cells of the small intestine synthesize and secrete TG-rich CMs, whereas hepatocytes synthesize and secrete TG-rich VLDLs.

- Excess VLDL (and CMs) in the circulation gives plasma a turbid appearance.

- After the TG in VLDL has been largely removed by the action of extrahepatic LPL, the VLDLs become IDLs.

- IDLs can be converted to LDLs in hepatic sinusoids.

- LDLs deliver CH to the liver, as well as to extrahepatic tissues.

- IDLs are usually cleared from the circulation more rapidly than LDLs.

- LDLs are removed from the circulation by receptor-mediated endocytosis, and by scavenger receptors.

- LDLs remain in the circulation longer than VLDLs, and their excess may lead to atherosclerotic lesions.

- Investigations on WHHL rabbits have helped to explain a puzzling feature of FH.

Very Low-Density Lipoprotein (VLDL)

Although chylomicrons (CMs) generally contain the majority of circulating **triglycerides (TGs)**, the second major source of plasma TG is that synthesized endogenously (i.e., that not arising directly from dietary fat). Although several different tissue types are capable of esterifying fatty acids (FAs) with glycerol to form TG, only intestinal mucosal cells and hepatocytes contribute TG-containing lipoproteins to the circulation in physiologically significant amounts (see Chapter 63). Mucosal cells secrete TG-rich **chylomicrons (CMs)**, and hepatocytes secrete TG-rich **VLDLs**. Lesser amounts of TG are also contained in **high-density lipoproteins (HDLs)** secreted by these tissues (see Chapter 66).

Very low-density lipoproteins transport TG, cholesterol (CH), cholesterol esters (CEs), and phospholipids (PLs) **away from the liver** into blood. These compounds are derived from *de novo* biosynthesis, and from degradation of **CM remnants**. Fatty acids can be synthesized *de novo* in hepatocytes from acetyl-CoA derived from lactate, amino acids, or glucose, or they can be taken-up as such from outside the cell where they arise either from the hydrolysis of TGs in remnant CMs, or as free fatty acids (FFAs) that moved into blood following TG hydrolysis inside adipocytes (i.e., lipolysis; see Chapters 64 and 70). The **phosphatidic acid pathway** of TG synthesis is used by hepatocytes rather than the monoglyceride acylation pathway of small intestinal mucosal cells (see Chapter 64). In reassembling TGs, the liver is able to alter their FA composition. Thus, TG hydrolyzed from liver-derived VLDL can provide adipocytes with a different FA compo-

Copyright © 2015 Elsevier Inc. All rights reserved.

sition than that contained in CMs, which was derived largely from the diet.

Very low-density lipoproteins are somewhat **smaller** than CMs, and they possess a **lower lipid/protein ratio**, thus giving them a slightly higher density (see Chapter 63). Like **CMs**, high concentrations of **VLDL** give plasma a **turbid, milky** appearance, however unlike CMs, VLDL will not separate on standing to form a creamy layer (see Chapter 67). Both CMs and VLDLs contain apolipoprotein B (Apo-B) as their major lipoprotein component; however, **CMs** contain **Apo-B$_{48}$,** while **VLDLs** contain **Apo-B$_{100}$.** Only one molecule of Apo-B is present in each of these lipoprotein particles. Like nascent CMs (see Chapter 64), **nascent VLDLs** must acquire

Apo-C from circulating **HDL** before becoming effective substrates for extrahepatic **lipoprotein lipase (LPL)**, and **Apo-E** before becoming effective substrates for clearance by the liver (**Fig. 65-1**).

The fate of TG in VLDL appears to be similar to that in CMs, that is, extracellular hydrolysis by capillaries possessing **LPL**, with subsequent **LCFA** diffusion into tissues. In the fed state, uptake will be predominantly into adipose tissue, and in the starved state, uptake will be predominantly into aerobic muscle fibers (e.g., cardiac myocytes; see Chapters 64 and 80). The circulating **half-life of VLDLs (2-4 hrs)** has been found to be somewhat longer than that of CMs.

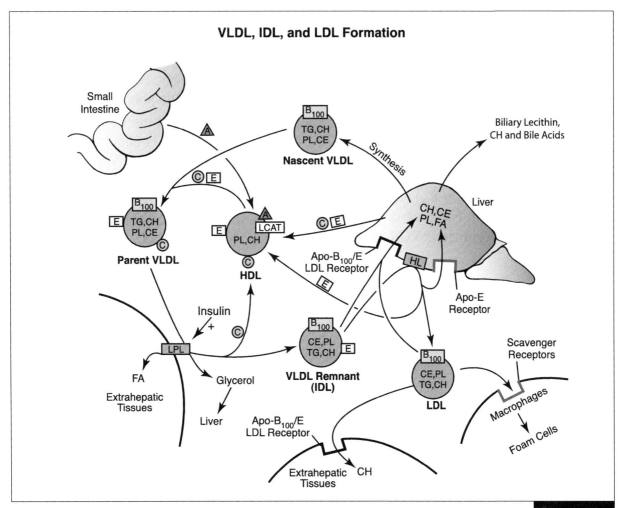

VLDL, IDL, and LDL Formation

Figure 65-1

Intermediate-Density (IDL), and Low-Density Lipoprotein (LDL)

After the **TG** in **VLDL** has been largely hydrolyzed by the action of extrahepatic **LPL**, **VLDLs** become **IDLs**, also known as **VLDL remnants**. Intermediate-density lipoproteins, like LDLs, contain about equal proportions of CH and TG in dogs and cats.

Two possible fates now await **IDL**, either clearance by the liver or conversion to **LDL**. Although the mechanism through which **IDL** conversion to **LDL** occurs remains largely unknown, it appears to require **hepatic lipase (HL)**, an extracellular enzyme located on sinusoidal surfaces of the liver. Since **LDLs** contain only **Apo-B$_{100}$**, transformation of **IDLs** to **LDLs** obviously involves the return of **Apo-E** to circulating **HDLs** (**Fig. 65-1**). In most animal species, the majority of **IDL** is removed from blood by the liver; however, a much larger proportion ($\approx$ **60-70%**) appears to be converted to **LDL** in primates.

Hepatic LDL receptors apparently recognize the **Apo-B$_{100}$** present in **IDL**, however, **Apo-E** also plays a role since this apoprotein binds more tightly to **LDL** receptors than **Apo-B$_{100}$**, thus giving **IDL** a more rapid uptake than **LDL** (which contains only **Apo-B$_{100}$**). As with CM remnants, hepatic **Apo-E receptors** may also participate in the clearance of **VLDL remnants** from blood.

Low-density lipoproteins are the **end-products of VLDL** metabolism, and their primary function is to **deliver CH to the liver, as well as to extrahepatic tissues**. All cells require CH for the synthesis of plasma membranes, and some require additional CH for the synthesis of specialized products such as steroid hormones, vitamin D (see Chapter 45), and in the case of liver, bile acids (see Chapter 62). In addition to endocytosis of LDL, cellular CH can be obtained by *de novo* biosynthesis from acetate. Most **LDL** is removed from blood by **LDL (ApoB$_{100}$/E) receptors**. Although

several different tissues are involved, the major site of removal is the liver, where approximately **75%** of receptor-mediated clearance occurs. Another **25%** of **LDL** removal from blood is thought to occur via low-affinity receptors known as **scavenger receptors**, which are found on **macrophages**. Although these receptors have a low affinity for parent LDL, they apparently have a high affinity for chemically modified forms, particularly oxidized LDL. In macrophages, unlike other cells, the synthesis of scavenger receptors is not regulated by the accumulation of intracellular CH. Therefore, oxidized LDL uptake occurs in an uncontrolled fashion. As intracellular CH accumulates, macrophages are converted to **foam cells**. Accumulation of these cells in blood vessel walls results in the formation of fatty streaks, which are not necessarily harmful, and can sometimes be reversed in early stages of their formation. However, foam cell streaks may continue to form over a long period of time, particularly in humans, where they become a part of **atherosclerotic plaques**.

Studies indicate that the normal **half-life** of **LDL** in the circulation is about **2 days**, which is much longer than that of VLDL. Two major factors regulate plasma concentrations of IDL and LDL, and thus plasma CH, as well as their half-life. One is the rate of **input of VLDL** (and thus the rate of IDL and LDL formation), and the other is the rate of **IDL** and **LDL clearance**, largely by the liver (**Fig. 65-2**). Genetic variation, species, and diet all appear to play a part in creating these differences.

Primates appear to have higher concentrations of LDL in blood than do other animal species studied. Most mammals exhibit **LDL** plasma concentrations ranging from **25-50 mg/dl (mg%)**, whereas the baseline level in adult **primates** is **75-100 mg/dl**. Synthesis rates of VLDL alone cannot explain these differences, for most vertebrate species maintain similar VLDL production rates unless they are consuming a high fat diet (e.g., dogs and cats),

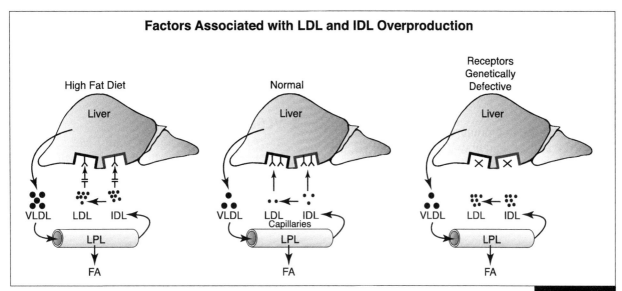

Factors Associated with LDL and IDL Overproduction

Figure 65-2

or are starving (e.g., ponies; see Chapter 67).

Dietary saturated fatty acids (SFAs), trans-UFAs and CH are known to increase circulating LDL levels, probably by decreasing the synthesis of hepatic LDL receptors. Exactly how SFAs and trans-UFAs suppress receptor synthesis and/or activity is unknown; however, it has been proposed that they may reduce fluidity of hepatocyte plasma membranes (see Chapter 54), thus reducing availability of surface receptors that assist in their removal from blood.

Genetic defects in the structure of the LDL receptor give rise to a condition known as **familial hypercholesterolemia (FH)** in humans. The heterozygous form of this condition occurs with a frequency of about **1 in 500**, and is also known to exist in animals. Yoshio Watanabe, a veterinarian from Kobe University in Japan, reported in 1973 the discovery of the **Watanabe heritable-hyperlipidemic (WHHL) rabbit**. Although initially the similarity between this rabbit colony and FH was not appreciated, in part because the WHHL rabbits also exhibited hypertriglyceridemia, it is now known that the hypertriglyceridemia relates in part to the fact that rabbit LDL contains more TG than human LDL.

The WHHL rabbits were helpful in explaining a previously puzzling feature of FH, namely that affected individuals not only degraded LDL more slowly, but they also overproduced LDL. The overproduction of LDL was found to be due largely to failure of IDL, as well as LDL, to be removed from blood. The WHHL rabbits, as well as other experimental animal models, have been instrumental in expanding our knowledge of lipoprotein metabolism. In addition, because of these and other animal studies, the primary **hyperlipoproteinemias** of **dogs** and **cats** are more easily identified and characterized (see Chapter 67).

In summary, plasma VLDL is largely of hepatic origin, and this lipoprotein serves largely as a means of transporting TG from liver to extrahepatic tissues. Studies using Apo-B$_{100}$-labeled VLDL demonstrated that **VLDL** is the precursor of **IDL,** and that **IDL** is the precursor of **LDL.** Thus, each LDL particle is derived from only one VLDL particle. Only one molecule of Apo-B$_{100}$ is present in each lipoprotein particle (VLDL, IDL or LDL), and it is conserved during these transformations. Once IDL is formed, it can be taken up directly by the liver via the LDL (Apo-B$_{100}$/E) receptor, or it can be converted to LDL. In many animal species, the majority of

IDL is removed directly by the liver, whereas in primates a larger proportion is converted to LDL.

OBJECTIVES

- Recognize the two primary sources of circulating TGs, and explain how VLDL acquires TG, CH, PL, CE and Apo-B$_{100}$.

- Know why the LCFA composition of TG found in VLDL may be different from that found in circulating CMs.

- Understand why high plasma concentrations of VLDL will not separate on standing to form a creamy layer.

- Explain how circulating VLDLs become effective substrates for LPL and hepatic clearance.

- Show how IDL is formed from VLDL, and LDL from IDL.

- Discuss the nature of hepatic LDL receptors and extrahepatic scavenger receptors.

- Recognize the variables that can lead to hypercholesterolemia and formation of atherosclerotic plaques.

- Recognize research contributions made by investigators using WHHL rabbits.

- Contrast and compare LDL to VLDL metabolism.

QUESTIONS

1. VLDLs transport all of the following lipids away from the liver into blood, EXCEPT:
   a. Cholesterol.
   b. Chenodeoxycholate.
   c. Phospholipids.
   d. Cholesterol esters.
   e. Triglycerides.

2. Triglyceride (TG) hydrolyzed from VLDLs may have a different fatty acid (FA) composition than that hydrolyzed from CMs, because:
   a. The liver may alter the FA composition of VLDLs.
   b. Unlike CMs, circulating VLDLs may pick-up additional FAs from tissues.
   c. VLDLs originate from mucosal cells of the small intestine, whereas CMs

originate from the liver.
   d. Different enzymes are involved in the hydrolysis of TG contained in VLDLs as opposed to those contained in CMs.
   e. CMs do not contain TG.

3. VLDLs:
   a. Are somewhat larger than CMs.
   b. Possess a higher lipid/protein ratio than CMs.
   c. Have a higher density than CMs.
   d. Like CMs, will separate on standing to form a creamy layer above plasma.
   e. Contain apoprotein B$_{48}$.

4. VLDL remnants:
   a. Are known as LDLs.
   b. Are normally converted to HDLs in sinusoids of the liver.
   c. Normally have a much slower clearance from blood than LDL.
   d. Are normally removed from hepatic sinusoidal blood via LDL receptors.
   e. Are known as HDLs.

5. Most LDL is normally removed from the circulation by:
   a. Scavenger receptors.
   b. The liver.
   c. Apo-B$_{48}$/C receptors.
   d. Macrophages.
   e. Adipocytes.

6. What are the two primary factors regulating plasma concentrations of IDL and LDL?
   a. The rate of input of IDL into the circulation, and the rate of VLDL and HDL clearance.
   b. The rate of input of LDL into the circulation, and the rate of VLDL and IDL clearance.
   c. The rate of input of HDL into the circulation, and the rate of IDL and VLDL clearance.
   d. The rate of input of CMs into the circulation, and the rate of LDL and VLDL clearance.
   e. The rate of input of VLDL into the circulation, and the rate of IDL and LDL clearance.

6. e

5. b

4. d

3. c

2. a

1. b

ANSWERS

Chapter 66

LDL Receptors and HDL

Overview

- Although most all nucleated mammalian cells contain LDL receptors, hepatocytes generally retain the highest concentration.

- High cytoplasmic free CH concentrations feedback negatively on *de novo* CH and LDL receptor biosynthesis.

- Insulin and thyroxine increase hepatic LDL receptor synthesis and activity, thus helping to clear CH from the circulation.

- HDL helps to shuttle excess CH from peripheral tissues to the liver, and it also acts as a repository for Apo-C and Apo-E in the circulation.

- HDL-containing LCAT aids in the formation of CEs.

- HDL_3 delivers CH and CE to remnant CMs and VLDL, which then carry these steroids to the liver.

- The incidence of coronary atherosclerosis in primates is inversely correlated with the plasma HDL_2 concentration.

- HDL_1 has been found in the plasma of diet-induced hypercholesterolemic animals.

Nature of the Low-Density Lipoprotein (LDL) Receptor

Most all nucleated cells contain **LDL receptors** that recognize the **Apo-B_{100}/E** of the LDL particle, with some containing as many as 20,000 to 50,000 receptors per cell. However, approximately **75%** of receptor-mediated LDL clearance normally occurs in the **liver** (see Chapter 65).

The receptor is a protein molecule that is highly susceptible to destruction by proteases. Receptors are present on the cell surface in discrete regions of the plasma membrane known as **coated pits (Fig. 66-1)**. These regions are indentations of the plasma membrane covered on the internal side by **clathrin**, a protein that polymerizes, thus forming a stabi-

lizing scaffold. Clathrin gives the coated pit a "fuzzy" appearance. Although coated pits cover only about 2% of the cell surface, they contain most all of the LDL receptors.

Internalization of LDL begins about 2 to 5 minutes after binding, and occurs in several stages. Endocytosis of the LDL-receptor complex results in an intracellular vesicle that is coated with clathrin. The clathrin coat is shed because of depolymerization, resulting in an **endosome**. The endosome now becomes acidified, and LDL dissociates from its receptors. The receptors segregate, and are recycled back to the plasma membrane. Next, endosomes fuse with **primary lysosomes**, forming secondary lysosomes where LDL particles are hydrolyzed. Apo-B_{100} is hydrolyzed

Copyright © 2015 Elsevier Inc. All rights reserved.

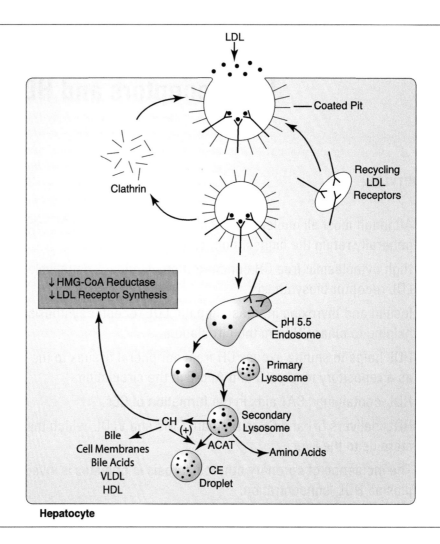

Internalization and metabolism of CH begins with LDL binding to hepatocyte membrane receptors. LDL-receptor complexes are internalized by receptor-mediated endocytosis into low pH endosomes, where receptors are then freed to recycle to the plasma membrane. Other LDL components enter lysosomes, where CH and FAs are freed from CEs, and amino acids are freed from Apo-B$_{100}$. CH is then free to 1) inhibit HMG-CoA reductase and further LDL receptor synthesis, 2) be processed in part to other CEs by ACAT, 3) enter cellular membranes, 4) enter into bile acid (BA) synthesis, 5) be incorporated into VLDL or HDL, or 6) be secreted into bile.

Figure 66-1

to amino acids, and cholesterol esters (CEs) to fatty acids (FAs) and free cholesterol (CH). Cholesterol passes out of the lysosome where it can enter cellular membranes, contribute to bile acid synthesis, be incorporated into VLDL or HDL, or be excreted into bile.

Cholesterol biosynthesis is a multi-step process controlled by the rate-limiting enzyme, **HMG-CoA reductase** (see Chapter 61). The activity of this enzyme is controlled by the level of free CH inside cells, and when most cell CH is derived from LDL, the activity of HMG-CoA reductase is suppressed. Conversely, a decrease in LDL uptake causes a rise in the activity of this enzyme. In this way, the concentration of cellular CH is optimized. However, if the level of CH in the cell rises too high, some

unesterified CH is reesterified by **acyl-CoA cholesterol acyltransferase (ACAT)**, and stored temporarily in CE droplets. The activity of ACAT is directly stimulated by increased concentrations of free cytosolic CH (**Fig. 66-2**).

LDL receptor synthesis is also under CH feedback control. When the level of free CH in the cell rises, LDL receptor synthesis is suppressed, and when cellular CH levels drop, LDL receptor synthesis increases. Cholesterol has been found to control transcription of the LDL receptor gene. Additionally, an endogenous protease known as **PCSK9 (proprotein convertase subtilisen/kexin type 9)** has been found to degrade the hepatic LDL receptor. A monoclonal antibody that inhibits PCSK9 apparently reduces circulating LDL levels, with injections every 2-4 weeks reported as

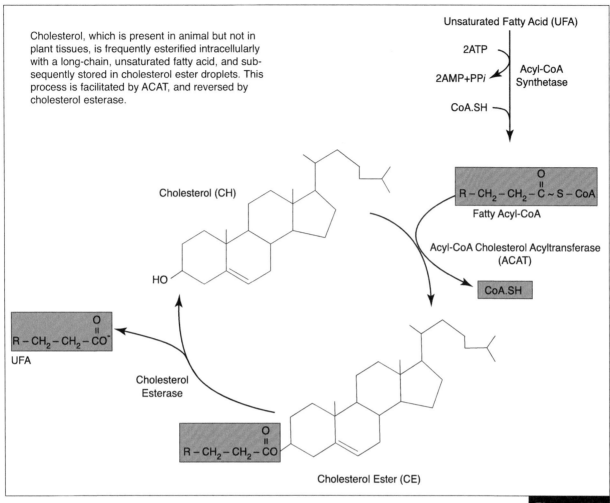

Cholesterol, which is present in animal but not in plant tissues, is frequently esterified intracellularly with a long-chain, unsaturated fatty acid, and subsequently stored in cholesterol ester droplets. This process is facilitated by ACAT, and reversed by cholesterol esterase.

Unsaturated Fatty Acid (UFA)

2ATP

2AMP+PP*i*

Acyl-CoA Synthetase

CoA.SH

Cholesterol (CH)

HO

$R-CH_2-CH_2-C \sim S-CoA$

Fatty Acyl-CoA

Acyl-CoA Cholesterol Acyltransferase (ACAT)

CoA.SH

$R-CH_2-CH_2-CO^-$

UFA

Cholesterol Esterase

$R-CH_2-CH_2-CO$

Cholesterol Ester (CE)

Figure 66-2

providing good results in the treatment of hypercholesterolemia.

Hepatic LDL receptor synthesis/activity has also been shown to be stimulated by two important endocrine factors in animals:

- **Insulin**

- **Thyroxine**

Therefore, in **diabetes mellitus** associated with insulin deficiency and in **hypothyroidism**, hepatic LDL removal from the circulation is reduced, thus resulting in **hypercholesterolemia** (and potentially, atherosclerosis). Additionally, high circulating levels of **cortisol** reduce pituitary release of **thyroid stimulating hormone (TSH)**, thus causing a **secondary hypothyroidism**. Therefore, glucocorticoid excess (e.g., **Cushing's-like syndrome** and **disease**

or **excessive glucocorticoid therapy**) is associated with reduced removal of LDL (and thus CH) from the circulation (see Chapter 67, and Engelking LR, 2012).

High-Density Lipoprotein (HDL)

It is apparent that in the steady state, CH leaves as well as enters cells. Cholesterol leaving cells is absorbed into HDL, which are lipoproteins synthesized primarily in the liver, with smaller amounts derived from the small intestine (**Fig. 66-3**). Excess surface components of circulating CM remnants and IDLs can also pinch-off in peripheral capillary beds, thus aiding in the formation of HDLs. The **two primary functions of HDL** are to:

1) **Participate in cholesterol exchange and esterification, and**

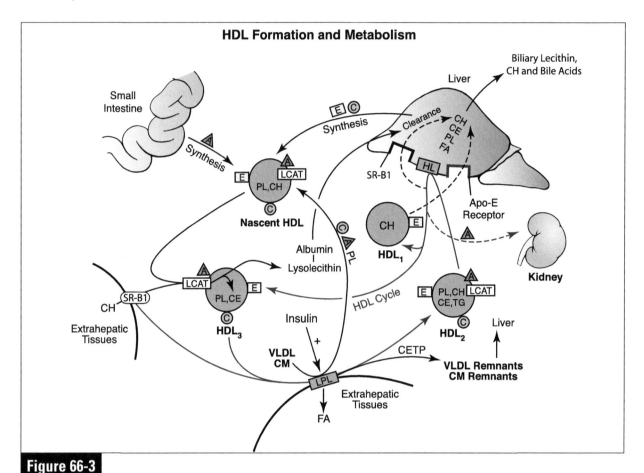

Figure 66-3

2) Act as a repository for Apo-C and Apo-E that are required in the metabolism of CMs and VLDL.

Although HDL is synthesized and secreted from both the liver and intestine, **nascent HDL** from the intestine contains **Apo-A**, but not Apo-C or Apo-E. Apo-C and Apo-E are synthesized only in the liver, and are transferred from liver HDL to intestinal HDL when the latter enter the circulation. As **nascent HDL** courses through extrahepatic tissue beds, a **class B1 scavenger receptor (SR-B1)** binds HDL via Apo-A, mediating HDL acceptance of **CH** from tissue cells. In the liver this receptor binds **HDL₂** via Apo-A, thus serving a secondary role of delivering CEs to hepatocytes (see below). Nascent HDL possesses a discoid phospholipid (PL) bilayer, containing free CH and apolipoprotein. The HDL bilayer also contains an important integral enzyme,

lecithin:cholesterol acyltransferase (**LCAT**), which is positioned immediately adjacent to its activator, **Apo-A**. As nascent HDL particles accept CH from tissues via SR-B1 in extrahepatic capillaries, LCAT catalysis transfers a fatty acid (FA) from the 2 position of lecithin (phosphatidylcholine) to the hydroxyl group of CH, thus producing **CE** and lysolecithin (see Chapters 57 and 58). As the nonpolar CEs move into the hydrophobic core of the HDL particle, **lysolecithin** is transferred onto plasma **albumin**, and nascent HDL matures into **HDL₃**. Thus, the LCAT system participates in removal of excess CH from tissues. While this process is responsible for much of plasma CE formation in primates, appreciable ACAT activity in non-primate hepatocytes usually allows for significant export of CE in nascent VLDL.

As **HDL₃, CMs**, and **VLDLs** encounter each other in extrahepatic capillary beds, the combined action of **LCAT, lipoprotein lipase (LPL)**, and **cholesterol ester transfer protein (CETP)** results in the transfer of **CEs** to **CM remnants** and **VLDL remnants (IDL)**. The space in the remnant particle that becomes filled with CEs is created by the depletion of **TG** as a result of hydrolysis by **LPL**. These remnant particles thus carry much of the CH obtained from tissues to the liver, a process sometimes referred to as **reverse CH transport**. During this capillary lipoprotein interchange, excess surface components of these remnant particles can be removed (i.e., namely **Apo-A, Apo-C**, and **PL**), thus aiding in the reformation of **nascent HDL** (see Chapter 64).

As **HDL₃** particles course through extrahepatic capillary beds, they become larger and less dense due to CH and CE acquisition, thus forming **HDL₂**. This form of HDL can next be acted upon by **hepatic lipase (HL)**. This enzyme hydrolyzes the PL and TG contained in **HDL₂**, and also allows CH and CEs to move into hepatocytes via the **class B1 scavenger receptor**. Hepatic CH and CE removal allows HDL to decrease in size and increase in density, thus reestablishing it as an **HDL₃** capable of acquiring more CH from peripheral tissues during reverse CH transport. As long as HDL remains in the circulation, this "**HDL cycle**" continues shuttling excess CH from peripheral tissues to the liver.

An **HDL₁** particle has also been found in the plasma of diet-induced hypercholesterolemic animals. It is thought to be formed from **HDL₂** following action by HL. **HDL₁** is rich in CH, and it contains only one apoprotein, **Apo-E**. This lipoprotein is apparently removed from sinusoidal blood by the **Apo-E (remnant) receptor** of hepatocytes.

Studies have shown that plasma HDL concentrations vary reciprocally with those of TG, and thus directly with LPL activity. This may be explained by the return of Apo-A, Apo-C, and surface PL to HDL during LPL-activated hydrolysis of TG contained in parent CMs, thus contributing toward nascent HDL formation. Additionally, plasma concentrations of **HDL₂** are also known to be inversely related to the incidence of **coronary atherosclerosis** in humans, possibly because they reflect the efficiency of CH scavenging from tissues. Because of this relationship, the CH contained in **HDL₂** has become known as "**good cholesterol**", as opposed to that contained in LDL (which has become known as "**bad cholesterol**"). **CETP inhibitors** tend to increase the HDL:LDL ratio, and estrogens are known to increase hepatic LDL receptor activity, while decreasing HL activity. Conversely, androgens increase HL activity, thus increasing the HDL₃/HDL₂ ratio. Thus, premenopausal women typically exhibit lower LDL/HDL₂ ratios than men, and thus demonstrate a lower incidence of coronary atherosclerosis. However, following menopause, when estrogen levels decline, the incidence of atherosclerosis in women increases.

While similar HDL subgroups occur in animals, the role of HDL subgrouping in predicting or diagnosing various hyperlipidemic states in animals has not been clearly defined. Therefore, although increased CH-containing HDL₁ particles have been found in some hypercholesterolemic animals, HDL subgroup analysis is currently considered to have little immediate clinical application in veterinary medicine.

OBJECTIVES

- Summarize how the "HDL cycle" functions in moving CH from extrahepatic tissue sites to liver.

- Explain how hepatic LDL receptor synthesis and activity are controlled, and explain relationships of these variables to various endocrinopathies.

- Identify the two primary functions of HDL.

- Discuss the SR-B1 receptor's role in CH transport.

- Summarize what occurs in the transition of nascent HDL to HDL_3, and in the transition of HDL_3 to HDL_2.

- Explain why plasma HDL concentrations vary reciprocally with those of TG, and directly with LPL activity, and why plasma HDL_2 concentrations in humans are inversely related to the incidence of coronary atherosclerosis.

- Discuss the roles of ACAT and cholesterol esterase in hepatic CE formation and reversal.

- Identify all steps involved in hepatic LDL internalization and CH metabolism/biliary excretion.

- Know the functions of LCAT and CETP, and explain what is meant by "reverse CH transport."

QUESTIONS

1. Select the FALSE statement below regarding LDL receptors:
   a. Their synthesis is reduced by elevated levels of free cytoplasmic cholesterol.
   b. They are found on all nucleated mammalian cells.
   c. They are found on cell membranes in discrete regions known as coated pits.
   d. They are typically degraded to amino acids inside lysosomes, and thus do not recycle to the plasma membrane.
   e. Their synthesis and activity are increased by insulin and thyroxine.

2. Acyl-CoA cholesterol acyltransferase (ACAT) is an enzyme:
   a. That is integral to the discoidal membrane of HDL.
   b. Whose activity is suppressed by rising levels of free cholesterol.
   c. It the rate-limiting enzyme in cholesterol biosynthesis.
   d. Needed in the biosynthesis of LDL receptors.
   e. That functions in the reesterification of cholesterol inside cells.

3. Which one of the following lipoproteins is responsible for picking-up excess cholesterol in peripheral tissue capillary beds?
   a. CM
   b. LDL
   c. IDL
   d. VLDL
   e. HDL

4. HDL is synthesized and secreted from:
   a. The kidneys.
   b. The liver and mucosal cells of the small intestine.
   c. The lungs and stomach.
   d. Circulating nascent VLDL.
   e. Circulating nascent CMs.

5. Which one of the following apoproteins activates lecithin:cholesterol acyltransferase (LCAT)?
   a. Apo-A
   b. Apo-B_{48}
   c. Apo-C
   d. Apo-B_{100}
   e. Apo-E

6. Remnant CMs and VLDL can accept cholesterol esters (CEs) from HDL_3 because:
   a. They contain Apo-C.
   b. They contain LCAT.
   c. Action by LPL has depleted their triglyceride stores.
   d. They contain LDL (i.e., Apo-B_{100}/E) receptors.
   e. Action by ACAT, which is present on capillary endothelial cells, converts free CH to CEs.

7. The process of delivering CH and CEs to the liver in CM and VLDL remnants is referred to as:
   a. The Cori cycle.
   b. The urea cycle.
   c. The enterohepatic circulation.
   d. Reverse CH transport.
   e. Remnant recycling.

8. An inverse correlation exists between:
   a. The circulating LDL/HDL_2 ratio and the incidence of coronary atherosclerosis.
   b. Circulating HDL concentrations and LPL activity.
   c. Circulating LDL concentrations and the incidence of coronary atherosclerosis.
   d. Circulating CM remnant concentrations and LPL activity.
   e. Circulating HDL_2 concentrations and the incidence of coronary atherosclerosis.

ANSWERS

8. e
7. d
6. c
5. a
4. b
3. e
2. e
1. d

Chapter 67

Hyperlipidemias

Overview

- Postprandial hypertriglyceridemia in the dog or cat may be normal.
- Serum can be hyperlipidemic without being hyperlipemic.
- Hyperlipidemia in the starved (> 12 hours) dog or cat is abnormal.
- Hyperlipidemia in the starved (> 2 days) horse is normal.
- Hyperlipidemic dogs and cats are at risk of developing significant clinical illness; therefore, specific dietary and/or drug intervention is usually warranted.
- Hypercholesterolemia is most commonly recognized as a secondary hyperlipidemia in animals.
- Hypertriglyceridemia causes lactescent serum (i.e., lipemia).
- Horses possess a greater capacity for exporting VLDL from hepatocytes than do ruminant animals.
- Several endocrine disorders are associated with hyperlipidemia in domestic animals.

Hyperlipidemias are disturbances of lipid metabolism resulting in elevated serum concentrations of nonpolar fats (i.e., namely **triglyceride (TG)** and **cholesterol esters (CEs)**). They are generally due to increased lipoprotein production, decreased clearance, or a combination of both. The defect can be primary (genetic), secondary (due to diseases or drugs), or a combination of the two.

Most investigators distinguish between a mild elevation in serum TG (up to **500 mg%**), and a great elevation (**> 500 mg%; lipemia or hyperlipemia**). Unlike chylomicrons (CMs) and very low-density lipoproteins (VLDLs), the primary CH-containing lipoproteins (i.e., low-density (LDLs) and high-density lipoproteins (HDLs)) do not refract light. Serum can thus be **hyperlipidemic** without necessarily being **lipemic** (i.e., **lactescent or cloudy**).

Hypertriglyceridemia is the most common hyperlipidemia recognized in domestic animals, and patients with **lactescent serum** may exhibit extreme elevations in **triglyceride-rich CMs** from either overproduction or retarded degradation. Chylomicrons are the largest and least dense circulating lipoproteins in animals (see **Fig. 63-2**), and they are responsible for the transport of dietary fat (mainly TG) away from the intestine (see Chapters 60 and 64). Elevations in **TG-rich VLDL**, which are somewhat smaller than CMs, can also cause **lactescent serum**, however, unlike hyperchylomicronemic serum, a cream layer will not usually separate in VLDL-rich samples left standing for 6 to 10 hours.

Most cases of mild to moderate hypertriglyceridemia in dogs and cats can be attributed to hyperchylomicronemia subsequent to a recent

Copyright © 2015 Elsevier Inc. All rights reserved.

(with 4 to 6 hours) meal (i.e., post-prandial or physiologic hyperlipidemia). Detection of an abnormal hyperlipidemia requires that a fasting lipid profile be performed, with the period of fasting being about 12-14 hours. Chylomicrons formed after the last meal should normally be cleared by this time. Since VLDL is the major carrier of TG in the starved state, modest, isolated fasting hypertriglyceridemia is due to VLDL elevation. However, severe hypertiriglyceridemia in the starved (> 12 hours) animal is usually an abnormal finding that justifies further study, with affected animals being at risk of developing additional complications (particularly acute pancreatitis in dogs (**Table 67-1**)).

Why does hypertriglyceridemia cause acute pancreatitis? Although the precise mechanism remains unclear, some investigators believe that the rather large and low density CMs and VLDLs obstruct pancreatic capillaries, leading to local ischemia and acidemia. This local damage could expose TGs to pancreatic lipase, with the FFAs produced leading to cytotoxic

Table 67-1 Hyperlipidemias in Dogs and Cats	
Causes	**Symptoms**
Postprandial	Hypertriglyceridemia
Hypertriglyceridemia	Lipemic serum
Hyperchylomicronemia	Cutaneous xanthomas (cats)
Idiopathic	Lipemia retinalis (dogs & cats)
LPL deficiency	Acute pancreatitis (dogs)
Familial (cats)	Pseudopancreatitis (dogs)
High fat diets	Hypercholesterolemia
Diabetes mellitus (insulin secretion or sensitivity)	LDL/HDL ratio
LPL activity (hypertriglyceridemia)	Atherosclerotic lesions
LDL receptor activity (hypercholesterolemia)	
Drug induced	
Megestrol acetate (progestogen)	
Diabetes mellitus (cats)	
Prolonged glucocorticoid administration	
Lipolysis	
Secondary hypothyroidism	
GH excess	
Nephrotic syndrome	
Proteinuria	
Hypothyroidism (hypercholesterolemia)	
Primary (dogs)	
Secondary	
Cushing's-like disease or syndrome	
Liver disease	

injury, resulting in further local injury that increases inflammatory mediators and free radicals, eventually manifesting as pancreatitis.

Abdominal distension and distress, a painful abdomen, diarrhea, and other signs of pancreatitis are sometimes evident in the absence of true pancreatitis. Therefore the term **pseudo-pancreatitis** has been used to describe this pattern of manifestations in hypertriglyceridemic dogs. Seizures, behavioral changes, and lipemia retinalis have also been observed in both dogs and cats. In the latter condition, retinal arterioles and veins apparently develop a pale pink color due to light scattering by large CMs. Cats with hypertriglyceridemia that exhibit this characteristic have been reported to develop **xanthomas** (i.e., deposits of lipid in the skin or other tissues (e.g., the liver, spleen, kidneys, heart, muscle and/or intestines)).

A **"gene therapy"** has been developed and applied in Europe for a rare hereditary **human LPL deficiency** resulting in hypertriglyceridemia, namely excess CMs. As with dogs, patients with this disease usually develop **pancreatitis**. The gene is packaged in an adeno-associated virus (AAV), which targets muscle tissue (where much of LPL is produced before migrating to the capillary endothelium; see Ch. 64).

Miniature Schnauzers have been shown to be the dogs most prone to hypertriglyceridemia, and an autosomal recessive disorder leading to a similar condition in certain families of **kittens** has also been reported. The insulin deficiency of **diabetes mellitus (DM)** can cause and exacerbate hypertriglyceridemia due to failure to clear TG-rich CMs and VLDLs from blood. **Insulin** is required for the activation of **lipoprotein lipase (LPL)** on the endothelium of capillaries servicing muscle and adipose tissue (see Chapters 64-66). Megestrol acetate-induced DM in cats, as well as hyperadrenocorticism and acromegaly (i.e., conditions associated with increased lipolysis and insulin resistance) are examples of secondary hypertriglyceridemias. Growth hormone and cortisol increase **hormone sensitive lipase (HSL)** activity (see Chapter 70), and lack of insulin reduces LPL activity. Long-chain fatty acids released from fat tissue are reesterified in the liver to TG, which is then discharged from hepatocytes into blood in VLDL.

Hyperlipidemia, characterized by increased serum CH or TG concentrations, has been reported in patients with proteinuria due to glomerulonephritis or amyloidosis (i.e., **nephrotic syndrome**). The cause may relate to a number of factors involving altered lipoprotein metabolism and LDL clearance from the circulation. The liver responds to hypoproteinemia by increasing its **synthesis** of plasma proteins, including VLDL, which are exocytosed and transformed into IDL and LDL in

---

Table 67-2

Factors Affecting the Serum Cholesterol (CH) Concentration

- Uptake of CH-containing lipoproteins by receptor-mediated pathways.
- Uptake of CH-containing lipoproteins by a non-receptor mediated pathway.
- CH synthesis.
- Formation and/or hydrolysis of CH esters.
- Efflux of CH from plasma membranes to HDL.
- Utilization of CH for synthesis of other steroids (e.g., bile acids and steroid hormones).
- Biliary CH excretion.
- Diet, various endocrinopathies, liver, pancreatic or kidney disease.

plasma (see Chapter 65).

Hypercholesterolemia is usually recognized as a secondary hyperlipidemia in animals, associated with an underlying disorder such as **hypothyroidism, hypersomatotropism** (growth hormone (GH) excess), **Cushing's-like disease or syndrome**, or **diabetes mellitus** (although some families of Doberman pinschers and rottweilers have been reported to exhibit a primary hypercholesterolemia). Both **insulin** and **thyroxine** are known to stimulate hepatic LDL receptor synthesis, and high circulating titers of **glucocorticoids** decrease responsiveness of the anterior pituitary to TSH-releasing hormone (i.e., TRH; see Chapter 66). Hypercholesterolemia has been found to be present in about two-thirds of hypothyroid dogs, and atherosclerotic-type lesions have occasionally been reported in diabetic animals.

Although several factors can affect the serum CH concentration (**Table 67-2**), most often hypercholesterolemia in companion animals results from **endocrinopathies, liver disease** (particularly bile duct obstruction), **ultra-high fat diets, drug-induced disorders** (in cats, megestrol acetate), **pancreatitis**, or **kidney disease.** Unlike the human population, hypercholesterolemic animals are not generally considered to be at risk clinically, and therefore specific treatments aimed at reducing serum CH alone are not usually considered.

Hypocholesterolemia, on the other hand, can be due to malnutrition, decreased liver function (e.g., porto-systemic shunts or certain hepatocellular disorders that reduce CH biosynthesis), intestinal loss (e.g., inflammatory bowel disease (IBD) or protein losing enteropathies), hyperthyroidism, or

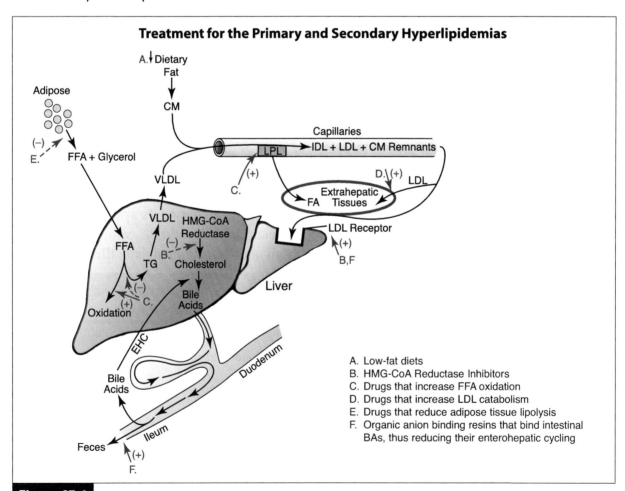

Treatment for the Primary and Secondary Hyperlipidemias

A. Low-fat diets
B. HMG-CoA Reductase Inhibitors
C. Drugs that increase FFA oxidation
D. Drugs that increase LDL catabolism
E. Drugs that reduce adipose tissue lipolysis
F. Organic anion binding resins that bind intestinal BAs, thus reducing their enterohepatic cycling

Figure 67-1

hypoadrenocorticism (Addison's-like disease). In the latter instance it is unclear why serum CH is sometimes reduced, but reduced amounts of glucocorticoid may reduce intestinal CH absorption. As discussed above, cortisol (a glucocorticoid) is also known to decrease sensitivity of the anterior pituitary to TRH, thereby reducing TSH output during times of stress (e.g., starvation). The relative hypothyroid state that follows can cause hypercholesterolemia, as T_4 (thyroxine) is one hormone that helps to maintain LDL receptor synthesis in liver cells. This feedback mechanism may be lost in hypoadrenocorticism, with resulting hypocholesterolemia.

Treatments for the Secondary Hyperlipidemias

Most secondary hyperlipidemias should resolve following adequate treatment of the underlying disease. Feeding animals a **low-fat diet** (**A.** in **Fig. 67-1**) will also help to relieve symptoms, and **medium-chain TG oils** may be considered as alternative lipid fuels since they do not contribute to CM formation (see Chapter 64). **Fish oils**, which are rich in **omega-3 polyunsaturated fatty acids** (see Chapters 68 and 69), have been shown to reduce serum TGs in humans, however they may not be effective in certain domestic animals. **Fungal metabolites** that inhibit HMG-CoA reductase (**B.** in **Fig. 67-1**) are effective in reducing cholesterol synthesis in humans, however their use in hypercholesterolemic animals may not be justified (**Fig. 67-1**). These "**statins**" also increase hepatic LDL receptor synthesis. **Cholestyramine** (**F.** in **Fig. 67-1**), an organic anion binding resin, is also sometimes administered orally in humans to bind bile acids in the intestine, thus preventing their enterohepatic cycling (EHC). This forces the liver to convert more of its CH to bile acids, thus increasing CH input into hepatocytes from the circulation. Like the HMG-CoA reductase inhibitors, cholestyramine increases hepatic

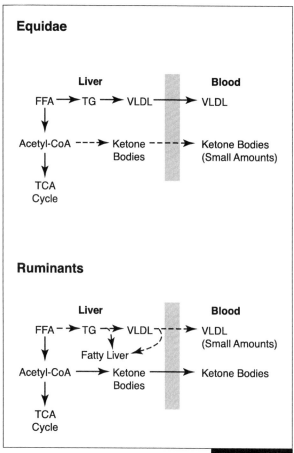

Figure 67-2

LDL receptor synthesis. **Clofibrate** and **gemfibrozil** (**C.** in **Fig. 67-1**) are examples of medications that exert their hypolipidemic effects (mainly on TG) by diverting the hepatic inflow of FFAs from pathways of esterification into those of oxidation, thus decreasing secretion of TG-containing VLDL by the liver. In addition, they are thought to increase hydrolysis of CM- and VLDL-containing TGs by LPL. **Probucol** (**D.** in **Fig. 67-1**) is an example of a drug that increases LDL catabolism via receptor-independent pathways, and **niacin** (the generic name for nicotinic acid (see Chapter 41; **E.** in **Fig. 67-1**) reduces FFA flux by inhibiting adipose tissue lipolysis, thereby reducing hepatic VLDL formation.

An elevation in the serum lipids of **horses** and **ponies** is also known to lead to cloudy plasma. Equidae are thought to possess a greater hepatic exporting capacity for lipoprotein than **ruminant animals**, as well as a greater hepatic clearance capacity for free fatty acids (FFAs; **Fig. 67-2**). The pathway for ketone body synthesis (see Chapter 71) is not as well-developed in horses as it is in ruminant animals, therefore, ketosis is infrequently observed in horses. On the other hand, ruminant animals appear to have difficulty forming and exporting VLDL from hepatocytes, therefore leaving them more prone to fatty liver infiltration (see Chapter 72).

Starved (2 to 3 days), pregnant, or lactating **ponies** have been found to be particularly susceptible to hyperlipidemia characterized by an increase in circulating FFAs as well as the TG-rich VLDL fraction of serum. Few differences have been noted in their LDL and HDL fractions. It appears that the equine liver is capable of closely matching the hepatic uptake of FFAs to re-esterification with glycerol, lipoprotein production (VLDL), and export into blood. Consequently, equine hypertriglyceridemia during starvation is not necessarily an abnormal finding.

In summary, disturbances of lipid metabolism sometimes lead to **hyperlipidemias,** which are elevations in the serum concentrations of nonpolar fats (usually **TG** and/or **CEs**). **Hypertriglyceridemia** is the most common hyperlipidemia recognized in domestic animals. Affected animals usually exhibit **lactescent (lipemic)** serum from extreme elevations in **CMs** or **VLDL,** and many are at risk of developing additional complications, particularly acute pancreatitis. Hypertriglyceridemia is common in starved horses, where the hepatic uptake of FFAs is carefully matched to re-esterification with glycerol, lipoprotein production (VLDL), and export into blood.

Hypercholesterolemic animals have usually developed this condition secondarily to other disorders. **Endocrinopathies** such as hypothyroidism, hypersomatotropism, diabetes mellitus, or Cushing's-like syndrome can cause

this condition, as can **liver disease**, various **drug-induced disorders, pancreatitis, kidney disease,** or consumption of **ultra-high fat diets.** Unlike the human population, however, hypercholesterolemic animals are not usually considered to be at risk clinically.

Most secondary hyperlipidemias should resolve following adequate treatment of the underlying disease. Additional therapies available are fish oils containing omega-3 polyunsaturated fatty acids, various fungal metabolites that inhibit HMG-CoA reductase, medications that exert their hypolipidemic effects by diverting the hepatic inflow of FFAs from pathways of esterification to those of β-oxidation, and others that increase LDL catabolism, or reduce FFA flux by inhibiting adipose tissue lipolysis.

OBJECTIVES

- Explain basic differences between hyperlipidemia and hyperlipemia.
- Name the most common cause of lipemia in domestic animals.
- Recognize the numerous causes, signs and symptoms of hyperlipidemia in dogs and cats.
- Know differences between primary and secondary hypertriglyceridemias, and understand how insulin withdrawal or nephrotic syndrome could lead to hyperlipidemia.
- Explain and understand variables affecting the serum CH concentration, and common causes of hypo- and hypercholesterolemia in domestic animals.
- Discuss the various rationales for treating primary and secondary hyperlipidemias.
- Understand why ponies develop hypertriglyceridemia when starved, while cows develop steatosis (see Chapter 72).
- Explain how hypertriglyceridemia leads to pancreatitis.
- Discuss the link between hypoadrenocorticism, TSH and hypocholesterolemia.

QUESTIONS

1. An elevated concentration of which of the following would cause lactescent or cloudy serum?
a. HDL
b. Cholesterol
c. Free fatty acids
d. Chylomicrons
e. LDL

2. Diabetes mellitus can cause and/or exacerbate hypertriglyceridemia because:
a. Insulin is required for the activation of hormone sensitive lipase, which helps to clear TGs from the circulation.
b. Insulin stimulates pancreatic lipase release.
c. Diabetic animals secrete excess quantities of triglyceride-rich HDL into the circulation from adipocytes.
d. Triglyceride-rich CMs and VLDLs are not properly cleared from blood when insulin levels are low.
e. Insulin-sensitive mucosal cells of the small intestine overproduce triglyceride-rich CMs in diabetic animals.

3. All of the following are known to contribute to hyperlipidemia in dogs and cats, EXCEPT:
a. Prolonged glucocorticoid administration.
b. Hyperparathyroidism.
c. Liver disease.
d. Diabetes mellitus.
e. Nephrotic syndrome.

4. Hypercholesterolemia:
a. Is most often recognized as a secondary hyperlipidemia in animals.
b. Is associated with hyperthyroidism in cats.
c. Can be effectively treated with glucocorticoids.
d. Is the most common form of hyperlipidemia in animals.
e. Leads to lactescent serum.

5. Which one of the following is known to reduce the flux of free fatty acids by inhibiting adipose tissue lipolysis, thereby reducing hepatic VLDL formation?
a. Probucol
b. Niacin
c. Clofibrate
d. Omega-3-polyunsaturated fatty acids
e. Gemfibrozil

6. Cortisol:
a. Excess usually leads to hypocholesterolemia.
b. Is used by the liver to synthesize cholesterol.
c. Reduces sensitivity of the adenohypophysis to TRH.
d. Increases LDL catabolism via receptor-independent pathways.
e. And cholestyramine can be used effectively to treat animals with hypertriglyceridemia.

7. Starvation will, within 24 hours, lead to lipemia in:
a. Dogs.
b. Cattle.
c. Cats.
d. Ponies.
e. Pigs.

8. Nephrotic syndrome and proteinuria may lead to hyperlipidemia because:
a. Of increased hepatic VLDL biosynthesis.
b. Pancreatitis usually develops.
c. Hormone sensitive lipase is wasted in the urine.
d. Growth hormone and TSH secretion are suppressed.
e. None of the above

9. Hypocholesterolemia is associated with all of the following, EXCEPT:
a. Inflammatory bowel disease (IBD).
b. Hyperthyroidism.
c. Addison's-like disease.
d. Porto-systemic shunts.
e. Diabetes mellitus.

9. e
8. a
7. d
6. c
5. b
4. a
3. b
2. d
1. d

Chapter 68

Eicosanoids I

Overview

- Eicosanoids are synthesized from dietary essential fatty acids.
- Naturally occurring eicosanoids are rapidly degraded in the body, thus permitting only local, cell-specific actions.
- Most eicosanoid metabolites are excreted in urine.
- NSAIDs inhibit the cyclooxygenases (COX-1 and COX-2), but not the lipoxygenases.
- Glucocorticoids reduce PLA_2 activity, thus reducing the availability of arachidonate for eicosanoid formation, and they also reduce COX-2 activity.
- TXs are produced in platelets, and upon release cause platelet aggregation and constriction of vascular smooth muscle.
- Prostacyclin (PGI_2), produced by vascular endothelial cells, normally blunts the action of thromboxanes.
- Thrombin, PGI_2, and PAF promote vascular smooth muscle relaxation.
- Omega-3 polyunsaturated fatty acids give rise to series 3 prostanoids, and the series 5 leukotrienes.
- Thrombin, collagen, ADP, 5-HT, and PAF promote platelet aggregation.

Eicosanoids, physiologically and pharmacologically active compounds known as **prostaglandins (PGs)**, **thromboxanes (TXs)**, and **leukotrienes (LTs)**, can be synthesized from dietary essential fatty acids (FAs; e.g., **linoleate, arachidonate**, and **α-linolenate**; see Chapter 54). Their biosynthesis has been demonstrated in all major organ systems of the body. The term "eicosanoid" is used in referring to these compounds since they all derive from the same **eicosaenoic (eicosa** (20-carbon); **enoic** (containing double bonds) **acid** precursors. Although arachidonate can be formed from linoleate or α-linolenate in most mammals, it is a dietary essential FA in the cat family. It is present in most all mammalian membranes, accounting for about **5-15%** of **FAs** in **PLs**.

There are **three basic groups** of eicosanoids, each comprising PG, TX, and LT, that give rise to **five series** of active compounds (**Fig. 68-1**). Subscripts for the eicosanoids denote the total number of double bonds in each molecule, as well as the series to which they belong. The basic structure common to all **PGs** and **TXs**, referred to as **prostanoic acid**, consists of a cyclopentane ring with 2 aliphatic side chains (**Fig. 68-2**). The series 1 **prostanoids** (PGs and TXs) and series 3 **LTs** are derived from **dietary linoleate** through the formation of **eicosatrienoate**. The series 2 **prostanoids** and series 4 **LTs** are derived directly from **arachidonate**. Series 3 **prostanoids** and series 5 **LTs** are derived directly from **eicosapentaenoate** (an **omega-3 (ω-3** or **n-3) polyunsaturated**

Copyright © 2015 Elsevier Inc. All rights reserved.

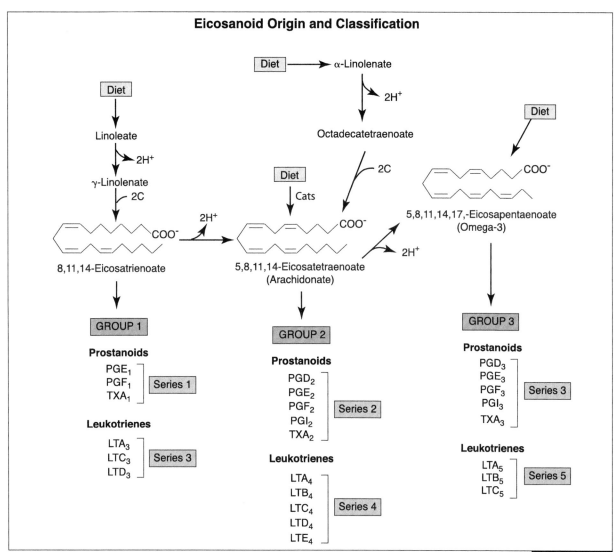

Eicosanoid Origin and Classification

Figure 68-1

fatty acid; **UFA** or **PUFA**), which is particularly prevalent in **fish oils** (see Chapter 54).

Arachidonate is normally derived from the 2-position of phospholipids in the plasma membrane as a result of **phospholipase A₂** (**PLA₂**) activity (see Chapter 58). Although various stimuli are known to activate this enzyme (e.g., thrombin, immunoglobulins, physical trauma, microbial products, angiotensin II, and epinephrine), anti-inflammatory **glucocorticoids** reduce it (**Fig. 69-1**). Glucocorticoids act by stimulating the synthesis of a protein known as **lipocortin**, which in turn inhibits PLA₂ activity, thus reducing arachidonate release required to activate subsequent enzymatic pathways in eicosanoid biosynthesis.

Pathways of arachidonate metabolism are divergent, with the synthesis of **PG₂** and **TX₂** series prostanoids competing with the synthesis of **LT₄** for the arachidonate substrate. These two pathways are known as the **cyclooxygenase** and **lipoxygenase** pathways, respectively (**Fig. 69-1**). Subsequent eicosanoids generated are products of a series of membrane-bound and cytosolic enzyme-catalyzed reactions, with appropriate stimuli for their activities varying with cell type. **Glutathione** (**GSH**) participates in eicosanoid biosynthesis by serving as a reductant in the cyclooxygenase reaction, and as a substrate in **peptidyl-leukotriene** (**LTC₄**) biosynthesis (see Chapters 12 and 69).

Cyclooxygenase converts **arachidonate**

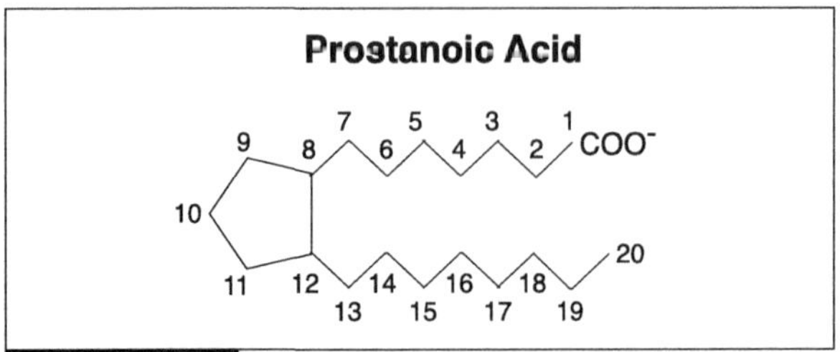

Prostanoic Acid

Figure 68-2

to an unstable prostaglandin intermediate (**PGG$_2$**), which is then peroxidized to **PGH$_2$**. This compound is next metabolized through appropriate enzymatic activity to more stable prostaglandins (**PGE$_2$, PGD$_2$, PGF$_{2\alpha}$,** and **PGI$_2$** (also called **prostacyclin**)), or to the **thromboxanes** (**TXA$_2$** and **TXB$_2$**). **Prostaglandins** were so named because they were first discovered in seminal fluid, and were thought to originate from the **prostate gland**. Unfortunately, the later discovery that they are not produced and secreted by the prostate but by the seminal vesicles (and many other tissues of the body), came too late to influence their name. **Thromboxanes** were so named because they were first discovered in **aggregating platelets** (**thrombocytes**), and contain an **oxane ring**.

Another group of enzymes collectively called **lipoxygenases**, convert arachidonate to a series of **hydroperoxyeicosatetraenoic acids (HPETEs), hydroxyeicosatetraenoic acids (HETEs)**, and **LTs**. These substances lack the cyclic structure of the prostanoids, and contain a hydroperoxy or hydroxyl substitution on the eicosatetraenoic (arachidonic) acid backbone (**Fig. 69-1**). Leukotrienes were so named because they were discovered in **white blood cells (leukocytes)**, and possess at least three alternating double bonds.

Nonsteroidal anti-inflammatory drugs (**NSAIDs**; e.g., aspirin, indomethacin, ibuprofen), block both PG and TX biosynthesis by **inhibiting cyclooxygenase** activity, but do not affect the lipoxygenases. There are two isoforms of cyclooxygenase (dubbed **COX-1** and **COX-2**). While COX-1 is constitutively expressed in most cells, COX-2 is less prevalent, but its

activity may be induced by certain serum factors, cytokines, growth factors, and/or endotoxins (an effect that is inhibited by both steroidal anti-inflammatory agents (e.g., the glucocorticoids), or the NSAIDs; **Fig. 69-1**). Blocking COX-1 and/or COX-2 activity alone (without blocking PLA$_2$), could lead to increased arachidonate utilization through the lipoxygenase pathways, which could further lead to increased production of the smooth muscle-stimulating LTs (see Chapter 69). False substrates like **5,8,11,14-eicosatetraynoic acid** are known to inhibit both the cyclooxygenases and lipoxygenases.

In the stomach, COX-1 activation is best associated with suppression of HCl release, whereas COX-2 activation is associated with reparative processes of the GI mucosa (see Chapter 69).

Eicosanoid Degradation and Activity

While naturally occurring eicosanoids are rapidly degraded in the body, synthetic analogs, which are metabolized more slowly, are now being produced by the pharmaceutical industry. Although many tissues contain enzymes necessary for eicosanoid degradation, **pulmonary, hepatic**, and **renal vascular beds** are particularly active. The **circulating half-life** of most endogenous eicosanoids is **less than 1 minute**, and their metabolites about 8 minutes. Roughly **90%** of eicosanoid metabolites are normally excreted in **urine**, with about **10%** appearing in **feces**. Presence of the enzyme **15-hydroxyprostaglandin dehydrogenase** (**15-PGDH**) in most mammalian tissues is thought to be responsible for **PG degradation**, and indomethacin, a COX inhibitor, up-regulates this enzyme.

Rapid eicosanoid inactivation within the organ in which they are synthesized, or within the systemic circulation, permits **local, cell-specific actions**. Thus, the physiologic actions of eicosanoids become valid only when responding tissues or cells under consideration are specifically identified. Thus, in most cases, they should not be looked upon as circulating hormones.

Thromboxanes

Thromboxanes are synthesized primarily in platelets (i.e., thrombocytes), and upon release cause **platelet aggregation** and **constriction of vascular smooth muscle** (**Fig. 68-3**). Conversely, **PGI$_2$**, produced by vascular endothelial cells, is a potent inhibitor of platelet aggregation, and causes vascular smooth muscle to relax (**Table 69-1**). Other stimulators of platelet aggregation include **thrombin**, **adenosine diphosphate** (**ADP**), **serotonin** (**5-HT**), **collagen**, and **platelet activating factor** (**PAF**; a phospholipid produced by leukocytes, platelets, and vascular endothelial cells). Although 5-HT is also a vasoconstrictor, PAF is a potent vasodilator, it contracts gastrointestinal, uterine and pulmonary smooth muscle, and it is a potent ulcerogen.

Under normal circumstances the production of PGI$_2$ is sufficient to override the effects of TXA$_2$, so that there is no significant vasoconstriction and platelet aggregation (which are the initial stages of hemostasis). However, **damage to the arterial wall** normally reduces the rate of PGI$_2$ release, so that the effects of TXA$_2$ predominate. The presence of **atherosclerotic plaques** has been shown to reduce the rate of PGI$_2$ production and release, thereby increasing the risk of thrombus formation in an otherwise undamaged artery (with possible fatal consequences).

The low incidence of heart disease, diminished platelet aggregation, and prolonged clotting times in Greenland Eskimos have been attributed to their high intake of fish oils containing **20:5 omega-3 fatty acids** (which give rise to the group 3 prostanoids and LTs; **Fig. 68-1**). The series 3 prostanoids inhibit release of arachidonate from membrane-bound phospholipid, and therefore formation of group 2 prostanoids

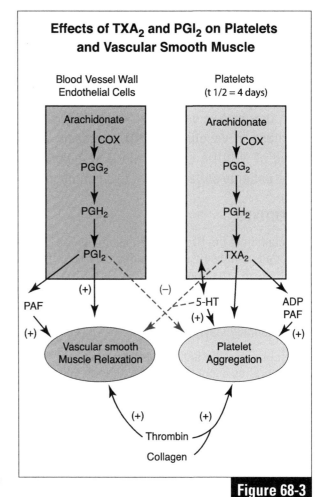

Effects of TXA$_2$ and PGI$_2$ on Platelets and Vascular Smooth Muscle

Figure 68-3

and LTs. **PGI$_3$** is as potent an antiaggregator of platelets as **PGI$_2$**, but **TXA$_3$** is a weaker aggregator than **TXA$_2$**; thus, the balance of activity is shifted towards nonaggregation.

The cyclooxygenase of platelets is highly sensitive to **NSAIDs**, whereas the same enzyme in vascular endothelial cells regenerates within a few hours following drug administration. Thus, the overall balance between TXA$_2$ and PGI$_2$ can be shifted in favor of the latter with low doses of NSAIDs, which oppose platelet aggregation.

TXA$_2$ is perhaps the most potent of the eicosanoids as a bronchoconstrictor (**Table 69-1**). It is a potent constrictor of umbilical vessels, and it also constricts coronary and renal vascular beds when released locally from ischemic tissues. Although TXB$_2$ is not a bronchoconstrictor, it has been shown in animals to have modest constrictor activity in pulmonary vascular beds. The hypertensive condition of **pre-eclampsia** in the final stage of pregnancy may in part be the result of placental vascular cell dysfunction, resulting in an **increase** in vasoconstricting **TXA₂** and a **decrease** in vasodilating **PGI₂** synthesis. Low doses of NSAIDs can therefore be helpful to some patients suffering from this condition.

OBJECTIVES

- Explain how the Group 1-3 Eicosanoids are formed, and discuss differences between the Series 1-3 Prostanoids, and the Series 3-5 Leukotrienes.

- Show where essential dietary UFAs enter eicosanoid biosynthesis.

- Discuss how the activity of PLA₂ in plasma membranes is controlled (see Chapter 58).

- Summarize differences between the lipoxygenase and cyclooxygenase pathways of eicosanoid metabolism, and show how GSH is involved.

- Explain how the PGs, TXs and LTs received their names.

- Understand why NSAIDs block PG and TX biosynthesis, and potentially lead to increased LT biosynthesis.

- Summarize the manner in which eicosanoids are degraded and excreted by the body, and explain why they are considered autocrine agents rather than hormones.

- Identify the differing sites of PGI₂ and TXA₂ synthesis in the vasculature, and discuss their differing activities.

- Recognize the multiple factors involved in vasomotion and platelet aggregation, and explain why atherosclerotic plaques may lead to thrombus formation.

- Explain why 20:5 omega-3 FAs and NSAIDs may diminish platelet aggregation, and why the latter may be useful in pre-eclampsic patients.

QUESTIONS

1. The basic structure common to all prostaglandins and thromboxanes is:

a. Prostanoic acid.
b. Arachidonic acid.
c. Leukotriene.
d. Hydroperoxyeicosatetraenoic acid.
e. Hydroxyeicosatetraenoic acid.

2. Glucocorticoids inhibit PLA₂ activity by increasing the biosynthesis of which protein?

a. Albumin
b. IGA
c. Insulin
d. Lipocortin
e. Lipoprotein lipase

3. NSAIDs may increase leukotriene formation through which mechanism?

a. Decrease lipoxygenase activity
b. Increase cyclooxygenase activity
c. Increase arachidonate utilization through cyclooxygenase pathways
d. Increase arachidonate utilization through lipoxygenase pathways
e. Increase arachidonate utilization through prostacyclin synthase pathway

4. Eicosanoids are not generally considered to be circulating physiologic humoral agents (i.e., hormones), because:

a. Receptors for these compounds are found only in cells that produce them.
b. They are not found in the systemic circulation.
c. They are rapidly degraded in the body, mainly by the vasculature and by tissues that produce them.
d. They are unstable.
e. They are tightly protein-bound in blood.

5. TXA₂ is best associated with:

a. Bronchodilation.
b. Vascular smooth muscle relaxation.
c. Platelet aggregation.
d. Dietary 5,8,11,14,17-eicosapentaenoate. (an omega-3 unsaturated fatty acid).
e. Dietary α-linolenate.

6. Subscripts for the eicosanoids denote:

a. The total number of double bonds in each molecule.
b. The group to which they belong.
c. The total number of carbon atoms in each molecule.
d. The total number of amino acids in each molecule.
e. Which omega-3 fatty acid they were derived from.

ANSWERS

6. a
5. c
4. c
3. d
2. d
1. a

Chapter 69

Eicosanoids II

Overview

- 5-HETE produces slow contractions of isolated bronchiolar smooth muscle.
- 15-HPETE is used to synthesize lipoxin A and lipoxin B.
- LTC_4 and LTD_4 are potent constrictors of bronchiolar smooth muscle.
- $PGF_{2\alpha}$ contracts smooth muscle within the reproductive tract, and causes luteolysis.
- Some PGs relax respiratory smooth muscle, while others cause it to contract.
- PGs are generally considered to be vasodilators.
- PGs participate in the induction of fever, and sensitize pain receptors to the algesic effects of other stimuli.
- Omega-3 fatty acid supplementation is associated with both positive and negative physiologic effects.

Hydroperoxyeicosatetraenoic Acids (HPETEs) and Hydroxyeicosatetraenoic Acids (HETEs)

The **12-, 15-,** and **5-HPTEs** are largely synthesized in the lung from arachidonate, and are rapidly reduced to their respective **HETE derivatives (Fig. 69-1** and Chapter 68). Although respiratory 12-HETE appears to be inactive, 5-HETE has been reported to produce slow contractions of isolated bronchiolar smooth muscle (**Table 69-1**), with a potency comparable to that of histamine. Neither compound appears to affect vascular smooth muscle, and little has been reported about their catabolism. Likewise, the effects of these substances on gastrointestinal smooth muscle, as well as smooth muscle of the reproductive tract, are not well documented. 15-HPETE is used to synthesize **lipoxin A** and **B**, which reportedly inhibit the cytotoxic effects of natural killer cells. Lipoxin A is also known to dilate the microvasculature.

Leukotrienes (LTs)

The **slow-reacting substance of anaphylaxis (SRS-A)** appears to be a mixture of LTs (which is 100-1000 times more potent than **histamine** or $PGF_{2\alpha}$ as a constrictor of bronchiolar smooth muscle). Animal studies show that NSAIDs can indirectly block the effects of LTs on airway smooth muscle, indicating that PGs and/or TXs are also involved. Indeed, **constriction of the guinea pig airway** by **LTB₄** appears to be augmented by induced local release of **TXA₂**. The **LTs** also **increase vascular permeability**, and appear to aid in the inflammatory response by enhancing phagocytosis and other bactericidal actions of neutrophils.

Before their structural elucidation, SRS-A activity, now known to be caused primarily by **LTC₄** and **LTD₄**, was determined on smooth muscle preparations from **guinea pig ileum**, which shows a slow onset but prolonged contractile response. However, this effect appears to be species specific since rat ileum

Copyright © 2015 Elsevier Inc. All rights reserved.

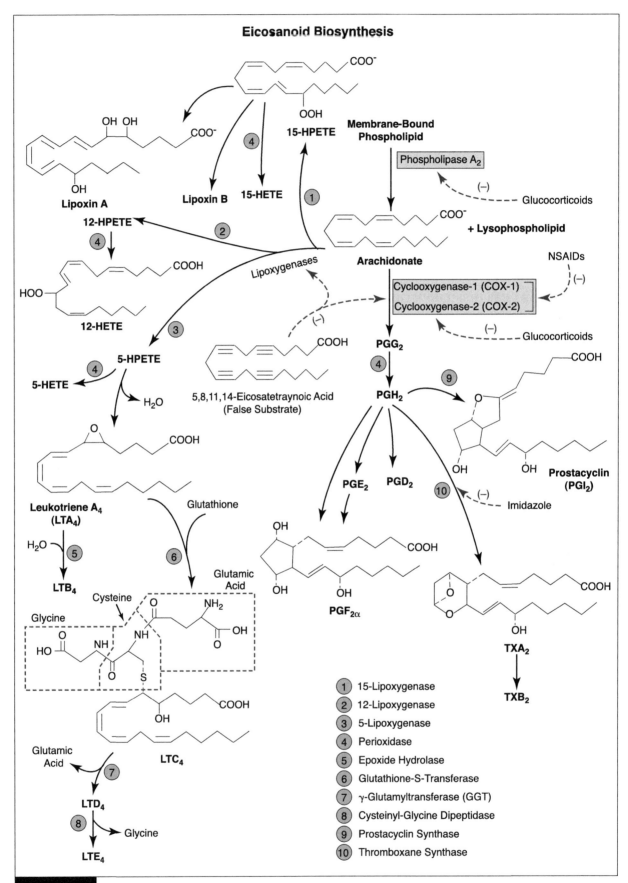

Eicosanoid Biosynthesis

① 15-Lipoxygenase
② 12-Lipoxygenase
③ 5-Lipoxygenase
④ Perioxidase
⑤ Epoxide Hydrolase
⑥ Glutathione-S-Transferase
⑦ γ-Glutamyltransferase (GGT)
⑧ Cysteinyl-Glycine Dipeptidase
⑨ Prostacyclin Synthase
⑩ Thromboxane Synthase

Figure 69-1

fails to respond. In contrast, the **rat colon** and **stomach fundus** are reported to be highly responsive to **LTC$_4$** and **LTD$_4$**.

In animal studies, **cutaneous arterioles** have been found to constrict initially and then relax in response to LTC$_4$ and LTD$_4$, and only LTC$_4$ causes constriction of mucosal venules. Therefore, regional specificity appears to exist for the effects of **LTs** of vascular smooth muscle.

Prostaglandins (PGs)

Reproductive tract: Animal studies indicate that **PGE$_2$** contracts the proximal quarter of the uterine tubes, but relaxes distal segments. In contrast, **PGF$_{2\alpha}$** exerts a **contractile effect on all segments**. Additionally, the ability of **PGF$_{2\alpha}$** to induce **luteolysis** in cows, ewes, mares, and other species is sometimes used to control the onset of estrus.

PGF$_{2\alpha}$ is also involved in the process of **parturition**. Estrogens and progesterone act through receptor-mediated processes to stimulate and suppress, respectively, the synthesis of mRNA essential for the production of endometrial cyclooxygenase. At the time of parturition, a local decrease in the effectiveness of progesterone associated with continued production of estrogen serves to **increase** the **estrogen/progesterone ratio**, thus increasing **PGF$_{2\alpha}$**

concentrations in uterine fluid. In addition, **PGF$_{2\alpha}$** production is increased by **oxytocin**, which also acts via receptor binding to the endometrium. Oxytocin and PGF$_{2\alpha}$ then act directly on uterine smooth muscle to stimulate contractions. The role of PGs during parturition is supported by the observation that **NSAIDs prolong gestation**, thus resulting in protracted labor. In the fetus, however, PGs act to keep the **ductus arteriosus** open. Therefore the clinical usefulness of NSAID administration during pregnancy is restricted by their propensity to close the ductus arteriosus, leading to fetal pulmonary hypertension.

Respiration: There is evidence that PGs participate in regulating **bronchopulmonary tone**. **PGF$_{2\alpha}$** and **PGD$_2$** are potent **constrictors** of tracheal and bronchial smooth muscle, whereas **PGE$_2$** and **PGI$_2$** **relax** these structures (**Table 69-1**).

Vascular Smooth Muscle: Although PG activity is diverse among different vascular beds, PGs are generally considered to be **vasodilators**. In the dog, PGs of the E and A groups dilate **coronary vessels**, whereas PGF$_{2\alpha}$ has no significant effect. PGF$_{2\alpha}$ has, however, been found to cause venoconstriction in some vascular beds. It is thought that local synthesis of PGs by both vascular endothelial and smooth

Table 69-1

Effects of Eicosanoids on Smooth Muscle

	Vascular	Respiratory	Gastrointestinal	Reproductive
PGE$_2$	R	R/C	R/C	C
PGF$_{2\alpha}$	R/C	C	C	C
PGI$_2$	R	R	C	R
TXA$_2$	C	C	–	–
TXB$_2$	C	O	–	–
5-HETE	O	C	–	–
LTB$_4$	O	C	*C	–
LTC$_4$	C	C	*C	–
LTD$_4$	C	C	–	–

R = Relaxation; C = Constriction; O = No Effect; - = Unknown; * = Species Specific

muscle cells contributes to the maintenance of normal vascular tone.

One effect of locally produced **prostacyclin** (**PGI$_2$**) on the kidney is to increase **renin** release from **juxtaglomerular** (**JG**) **cells** of afferent arterioles. This results in an increase in circulating **angiotensin II**, which in turn stimulates **aldosterone** release from the adrenal cortex. Aldosterone increases renal **Na$^+$ retention** (and **K$^+$ excretion**), favoring an increase in blood volume and pressure. On the other hand, PGs have been shown to reduce the effect of **antidiuretic hormone** (**ADH**) on collecting ducts of the kidney, thus causing diuresis and a lowering of the blood volume and pressure.

Gastrointestinal Tract: Animal studies indicate that PGs **increase gastrointestinal smooth muscle activity**, largely by affecting longitudinal smooth muscle. However, intravenous infusion of PGE$_2$ decreases lower esophageal sphincter (LES) and pyloric sphincter tone, and may reduce circular smooth muscle contraction in other areas of the digestive tract as well. PGF$_{2\alpha}$ appears to contract both **longitudinal** and **circular** smooth muscle, however, in the gallbladder this PG contracts longitudinal smooth muscle but has no effect on circular smooth muscle. PGI$_2$, in contrast to PGE$_2$, contracts circular smooth muscle. Intravenous administration of PGE$_2$ or PGF$_{2\alpha}$ leads to colicky cramping, indicating that the overall effect is to increase gastrointestinal smooth muscle tone.

Prostaglandin injections also **decrease gastric HCl** and **pepsin secretion**, and are thought to be involved in the natural regulation of the stomach's secretory response to food, gastrin, and histamine. Nonsteroidal anti-inflammatory drugs (NSAIDs) inhibit secretion of both **mucus** and **HCO$_3^-$** by epithelial cells in the necks of gastric glands. Mucus forms a gel on the luminal surface of the mucosa, thus protecting it from mechanical damage, and the alkaline fluid it entraps protects the mucosa against damage by HCl. Mucus and HCO$_3^-$ are a significant part of the **gastric mucosal barrier** that prevents damage to the mucosa by gastric contents. NSAIDs, by inhibiting PG production, thus render an animal susceptible to **gastric ulceration.**

Pain: Prostaglandins are thought to **sensitize pain receptors** to the algesic effects of other stimuli. This hyperalgesic action persists for some time after PG concentrations in blood have been significantly reduced.

Fever: The temperature regulatory center in the hypothalamus is affected by PGs. Raising the concentration of PGs in this center **induces fever**, and pyrogens raise the concentration of PGs in cerebrospinal fluid. For this reason, NSAIDs are sometimes used to reduce the body temperature.

Omega-3 Polyunsaturated Fatty Acids: Dietary supplementation with omega-3 polyunsaturated fatty acids is sometimes recommended for dogs and cats. Although few long-term studies have been reported, potential benefits have been reported as follows:

- **Alleviate pain associated with hip dysplasia**
- **Help control pruritus**
- **Suppress inflammation**
- **Improve hypertriglyceridemia**
- **Decrease thrombi formation**
- **Retard tumorigenesis**

Although few studies have focused on toxicity, tolerability, and safety issues, potential adverse effects include the following:

- **Increased intake of omega-3 fatty acids without adequate antioxidant protection could result in increased free radical formation with consequent lipid-peroxidation.**
- **Risk of bleeding**
- **Vomiting and diarrhea**
- **Immunoincompetence**
- **Reduce insulin release**

Further studies are needed before unconditional recommendations can be made regarding long-term supplementation, or alterations in the ratio of omega-3:omega-6 (i.e., n-3:n-6) fatty acids provided in the diets of dogs and cats.

OBJECTIVES

- Recognize where the HPETEs and HETEs are produced in eicosanoid biosynthesis, and discuss their physiologic roles.

- Identify the eicosanoids associated with the SRS-A, and discuss regional specificity of these compounds.

- Compare and contrast the effects of PGs, TXs and LTs on smooth muscle contractility of the vasculature, respiratory, GI and reproductive tracts.

- Explain how estrogen, progesterone and oxytocin influence endometrial $PGF_{2\alpha}$ production, and thus the onset of parturition.

- Discuss the relationship between PGI_2, renal renin production and arterial blood pressure.

- Understand the association of NSAIDs with gastric ulceration, and PGs with pain and fever.

- Recognize the potential benefits and risks of dietary omega-3 UFA supplementation.

- Explain why NSAID therapy to the pregnant animal may lead to fetal pulmonary hypertension.

- Trace the eicosanoid biosynthetic pathway, and show how LTE_2, TXA_2, $PGF_{2\alpha}$ and PGI_2 are produced.

QUESTIONS

1. **The slow-reacting substance of anaphylaxis (SRS-A) is a mixture of:**
   a. Steroids.
   b. Thromboxanes.
   c. Prostaglandins.
   d. Leukotrienes.
   e. Peroxidases.

2. **$PGF_{2\alpha}$ has been associated with all of the following, EXCEPT:**
   a. Luteolysis.
   b. Closing the ductus arteriosus.
   c. Parturition.
   d. Constrict bronchiolar smooth muscle.
   e. Constrict longitudinal and circular smooth muscle of the digestive tract.

3. **In general, prostaglandins are considered to be:**
   a. Vasodilators.
   b. Omega-3 fatty acids.
   c. Inhibitors of renin release in the kidney.
   d. Stimulators of gastric HCl secretion.
   e. Agents that break the gastric-mucosal barrier.

4. **Which one of the following does NOT appear to have vasoactive properties?**
   a. PGE_2
   b. TXA_2
   c. 5-HETE
   d. LTC_4
   e. LTD_4

5. **Which one of the following sensitizes pain receptors to the algesic effects of other stimuli?**
   a. PGs
   b. 12-HPETE and 12-HETE
   c. LTs
   d. TXs
   e. Omega-3 fatty acids

6. **Which compound below inhibits the cytotoxic effects of natural killer cells?**
   a. TXB_2
   b. Lipoxin B
   c. PGI_2
   d. 5-HETE
   e. LTE_4

7. **Which one of the following is used in the conversion of LTA_4 to LTC_4?**
   a. Arachidonate
   b. Eicosapentaenoate (an Omega-3 fatty acid)
   c. Ergosterol
   d. Creatine
   e. Glutathione

7. e
6. b
5. a
4. c
3. a
2. b
1. d

ANSWERS

Chapter 70

Lipolysis

Overview

- Epinephrine, thyroxine, and cortisol are potent lipolytic hormones in most domestic animals.
- Interleukin-6 may be an important lipolytic agent released from exercising muscle tissue.
- Insulin and insulin-like growth factor-1 possess antilipolytic activity.
- Adipocytes contain several gluconeogenic enzymes.
- Glyceroneogenesis modulates FFA release from adipocytes during starvation.
- Adipocyte lipolysis and glyceroneogenesis are largely cAMP-dependent processes.
- Glucocorticoids promote lipolysis through a cAMP-independent pathway.
- Leptin is a protein hormone produced by adipocytes that promotes satiety.
- Brown adipose tissue oxidizes FFAs, and is specialized for heat production.

Triglyceride (TG) stores in adipocytes are continually undergoing **lipolysis** and **reesterification**, with these two processes utilizing separate pathways, reactants, and enzymes (**Fig. 70-1**). The extent to which **free fatty acids (FFAs)** are mobilized from adipose tissue into blood is dependent upon the resultant of these two processes, which in turn is governed by several nutritional and hormonal influences.

Triglyceride is formed from **acyl-CoA** and **glycerol 3-phosphate** (glycerol 3-P) in adipose tissue (see Chapter 57). Because adipocyte glycerol kinase activity is low, glycerol released through lipolysis cannot be reutilized to any great extent in the reesterification of acyl-CoA inside adipocytes (**Note:** the activity of this enzyme is high in liver, kidney, and the lactating mammary gland). Therefore, adipocytes are dependent upon the input of **glucose** from plasma for the provision of glycerol 3-P in TG

formation, or they can generate it through glyceroneogenesis (see below). Glucose entry into adipocytes is facilitated by the **GLUT-4** transporter, which is translocated from the Golgi complex to the plasma membrane in the presence of **insulin** (see Chapter 22).

In addition to glycerol 3-P formation in adipocytes, glucose can be stored as **glycogen** (see Chapter 23), oxidized to CO_2 and H_2O through the TCA cycle (see Chapter 34), or oxidized through the **hexose monophosphate shunt (HMS)** to form the **NADPH** needed for **long-chain fatty acid (LCFA)** and thus **acyl-CoA** formation (see Chapter 56 and **Fig. 70-1**). When glucose utilization is high, a larger proportion of this carbohydrate is converted to **acetyl-CoA** and then to **LCFAs**, or oxidized in the **HMS** and the **TCA** cycle. However, as total glucose utilization decreases, the greater proportion of glucose is directed to the formation of **glycerol**

Copyright © 2015 Elsevier Inc. All rights reserved.

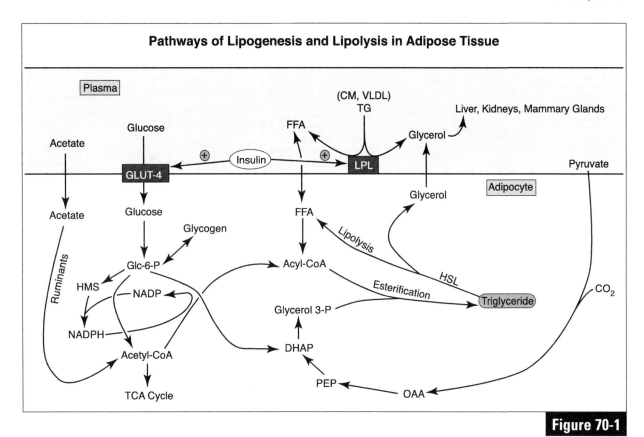

Pathways of Lipogenesis and Lipolysis in Adipose Tissue

Figure 70-1

3-P and **TG**, which helps to minimize **FFA** efflux.

In addition to facilitating translocation of the **GLUT-4** transporter to the plasma membrane of adipocytes, **insulin** also activates **lipoprotein lipase (LPL)** on the capillary endothelium servicing adipocytes (see Chapters 64 and 65). This results in the hydrolysis of TG contained in circulating **chylomicrons (CMs)** and **very low-density lipoproteins (VLDLs)**, thus providing FAs for esterification with glycerol 3-P inside adipocytes. In **ruminant animals**, hydrocarbons for lipogenesis are provided largely through **acetate**, which is generated from rumen microbial cellulose digestion (**Fig. 70-1**). In **birds**, estrogen-stimulated lipogenesis is largely confined to the liver, where it is important in producing lipids for egg formation (see Chapter 56).

Endocrine Control of Lipolysis

The rate of FFA release from adipocytes is affected by several hormones (**Fig. 70-2**).

However, the most potent appear to be the **catecholamines** (epinephrine and norepinephrine), **thyroid hormones** (T_3 & T_4), and the **glucocorticoids** (e.g., cortisol). Others circulating mediators reported to augment the effects of these primary stimuli include growth hormone (GH), glucagon, thyroid stimulating hormone (TSH), α- and β-melanocyte-stimulating hormone (MSH), adrenocorticotropic hormone (ACTH), antidiuretic hormone (ADH), and interleukin-6 (IL-6).

Catecholamines act rapidly in promoting lipolysis by stimulating membrane-bound **adenyl cyclase**, thus increasing intracellular levels of cyclic-AMP (**cAMP**). This mechanism is similar to that described for **glycogenolysis** (see Chapter 23). Through increasing the number of β_1-**adrenergic receptors** on adipocytes, **thyroid hormones** become synergistic with the **catecholamines** in promoting lipolysis (**Fig. 70-3**).

By stimulating a **cAMP-dependent protein**

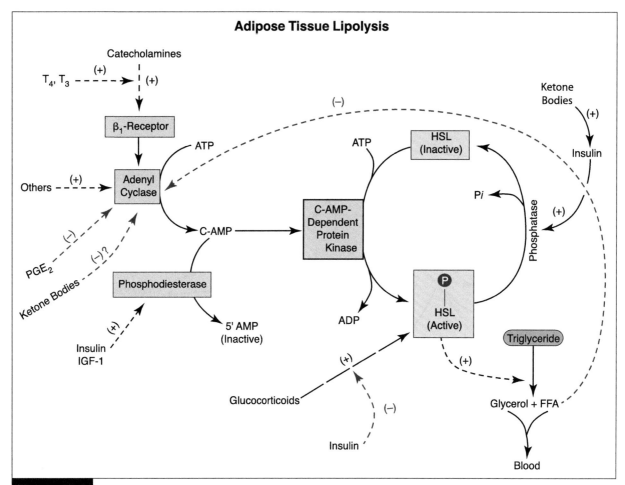

Adipose Tissue Lipolysis

Figure 70-2

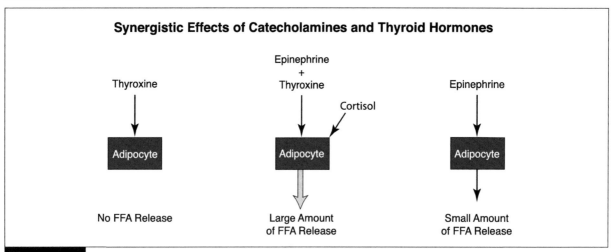

Synergistic Effects of Catecholamines and Thyroid Hormones

Figure 70-3

kinase (i.e., **protein Kinase A**), increasing cytoplasmic levels of cAMP are responsible for converting inactive **hormone-sensitive lipase (HSL;** also called **adipolytic triglyceride lipase)**, into its active, phosphorylated form, thus promoting lipolysis. Intracellular cAMP levels are reduced through the activity of a **cAMP-dependent phosphodiesterase**, which is known to be activated by **insulin** and **insulin-like growth factor-1 (IGF-1)**, and inhibited by caffeine. Other **antilipolytic** agents include **prostaglandin E₂ (PGE₂)** and the **ketone bodies,** which are thought to reduce adenyl cyclase activity and thus cAMP generation. β-Hydroxybutyrate (the most reduced ketone body) also mildly stimulates pancreatic insulin release, and increases insulin sensitivity on adipocytes (see Chapters 71 & 75).

Glucocorticoids promote lipolysis through a **cAMP-independent pathway**. They promote **gene transcription**, and following receptor stimulation in adipocytes promote an increase in HSL biosynthesis. **Insulin** exerts part of its antilipolytic action by opposing the action of **glucocorticoids on HSL** biosynthesis, and by promoting the **dephosphorylation of HSL**, thus returning it to its **inactive** form (**Fig. 70-2**).

Not all mammals respond similarly to the lipolytic and antilipolytic actions of the above hormones. For example, there is only a modest lipolytic response to circulating catecholamines in the **rabbit**, **guinea pig**, **pig**, and **chicken**. In **birds**, glucagon appears to have a pronounced lipolytic effect, while insulin exerts little antilipolytic action. Most glucagon receptors (>90%) in **mammals** are located in the liver, not on adipocytes.

Studies indicate that exercise results in a marked increase in circulating **IL-6**, and that most is derived from the contracting limb. In addition to promoting hepatic glycogenolysis, this cytokine may have potent lipolytic properties.

Glyceroneogenesis

It has been shown that **pyruvate** can be converted to **glycerol 3-P** in both adipocytes and hepatocytes due to the presence of **phosphoenolpyruvate carboxykinase (PEP carboxykinase)** and certain other gluconeogenic enzymes (**Fig. 70-1**). This pathway of glyceroneogenesis is thought to **modulate FFA release** during periods of **starvation** (Beale EG, et al, 2002; see Chapter 75). Exercise and starvation are associated with lower circulating insulin levels and an increased cAMP concentration

in adipocytes, which leads to increased PEP carboxykinase activity. As **dihydroxyacetone phosphate (DHAP)** is converted to **glycerol 3-P**, **up to 60% of FFAs released from TG may be re-esterified with glycerol 3-P** (while virtually 100% of the glycerol released from TG enters blood). Thus, FFA release from adipocytes may be modulated in the presence of high intracellular cAMP levels through glyceroneogenesis.

Satiety

Leptin (Gr. thin), a protein of 167 amino acid residues, is produced by **adipocytes** and acts on the hypothalamus to suppress feeding behavior. Receptors are found in arcuate nuclei that lie slightly outside the BBB. Leptin is normally increased in the plasma of obese animals in direct proportion to the amount of body fat, and since it may be an important compound promoting satiety, it is possible that leptin receptors may be defective, they may be down-regulated, or there may be a defect in the 2nd messenger systems activated by these receptors in obese animals.

The decrease in plasma leptin observed during food deprivation has also been associated with delayed puberty in rodents, which is thought to be an adaptive response to calorie deficiency.

Hypothalamic malonyl-CoA also serves an important role in modulating energy balance. When glucose levels rise, more malonyl-CoA is produced (see Chapter 56), and food intake is reduced. Since ATP stimulates acetyl-CoA carboxylase, and since less ATP is ostensibly produced from fructose than from glucose oxidation, this feedback mechanism is apparently less effective in promoting satiety following fructose ingestion.

Polypeptides from the digestive tract (e.g., **ghrelin** and **obestatin**) have also been implicated as messengers controlling feeding behavior.

Lipolysis in Brown Adipose Tissue

Brown adipose tissue is metabolically active at times when heat generation is necessary (i.e., in nonshivering thermogenesis). Thus, this tissue is active in newborn animals, in those arising from hibernation, and in those exposed to the cold. Brown adipose tissue is characterized by a well-developed **blood supply** and a high content of **mitochondria** and **cytochromes**, but low activity of ATP synthetase. Metabolic emphasis is placed on oxidation of both **glucose** and **fatty acids**.

Catecholamines (epinephrine and norepinephrine) and **thyroid hormones (T_3 and T_4)** are important in increasing lipolysis in both white and brown adipose tissue. Experiments indicate that under the influence of these hormones, oxidation and phosphorylation are uncoupled in mitochondria of brown adipose tissue, thus, oxidation produces much heat, and little free energy is trapped as ATP (see Chapter 36). It appears that the proton gradient, which is normally present across the inner mitochondrial membrane of coupled mitochondria, becomes dissipated in brown adipose tissue by a unique thermogenic uncoupling protein known as **thermogenin**.

In summary, the TG stores of fat tissue are continually undergoing lipolysis (hydrolysis), and reesterification (**Fig. 70-1**). These two processes, for the most part, utilize separate reactants and enzymes, thus allowing several nutritional, metabolic, and endocrine factors that regulate the metabolism of this tissue type to act upon one process or the other. The resultant of these two processes will ultimately determine the size of the FFA pool, which can exert profound effects upon the metabolism of other tissues, most notably the kidneys, liver and muscle tissue (see Chapter 75). Therefore, factors operating in adipose tissue to regulate the outflow of FFAs may exert influences far beyond the tissue itself.

The initial event in the mobilization of FFAs from adipocytes is the hydrolysis of TGs by **HSL,** an event referred to as **lipolysis.** The HSL of fat tissue is activated by **epinephrine** and **cortisol,** and the **thyroid hormones** (T_4 and T_3) are synergistic with epinephrine since they stimulate the synthesis of β_1-**adrenergic receptors** on the plasma membranes of adipocytes. Epinephrine acts through cAMP, which in turn activates an intracellular cAMP-dependent protein kinase A. This kinase aids in the phosphorylation of HSL, which is required for it's activation. **Insulin,** in turn, reduces the rate of FFA release from adipocytes through 3 separate mechanisms: **1)** it stimulates the activity of a cAMP-dependent phosphodiesterase (which reduces the intracellular cAMP concentration), **2)** it activates a specific phosphatase which dephosphorylates HSL, thus returning it to its inactive form, and **3)** it antagonizes the actions of glucocorticoids on adipocytes. The **glucocorticoids** (e.g., cortisol) promote lipolysis through a cAMP-independent pathway by promoting gene transcription. Following intracellular receptor stimulation, HSL biosynthesis is increased.

The pathway of **glyceroneogenesis** (i.e., conversion of pyruvate to glycerol 3-P) is also thought to help govern the degree of FFA release from adipocytes during periods of starvation, since it provides the compound needed for FA reesterification. Since this process is also cAMP-dependent, FFA release from adipocytes may be modulated in the presence of lipolytic hormones.

Because the **glycerol kinase** activity of adipocytes is low, glycerol released through lipolysis cannot be reutilized to any great extent in the FA reesterification process. Therefore, adipocytes are dependent on either glyceroneogenesis from **pyruvate** (as indicated above), or on the input of glucose from plasma for the provision of glycerol 3-P. **Glucose** entry into adipocytes is facilitated by the **GLUT-4** transporter, which is insulin-sensitive. Glucose can also be stored as glycogen in adipocytes, partially oxidized through the EMP and HMS, or completely oxidized through the TCA cycle and oxidative phosphorylation.

OBJECTIVES

- Since glycerol released through lipolysis cannot be rephosphorylated inside adipocytes, explain how glycerol 3-P is formed for TG formation.

- Identify the glucose transporter in cell membranes of adipocytes, as well as the intracellular pathways available to glucose.

- Know how insulin, epinephrine, thyroxine and glucocorticoids affect adipose cell lipolysis and/or TG deposition.

- Identify and discuss the endocrine control of LPL, adenyl cyclase, HSL, cyclic-AMP-dependent phosphodiesterase and protein kinase activity.

- Explain how and why much of the FFA released during adipocyte lipolysis is used in the reesterification of glycerol 3-P during starvation.

- Discuss proposed roles of leptin, ghrelin and obestatin in feeding behavior.

- Indicate how the metabolism of brown adipose tissue assists in thermoregulation.

- Identify and discuss the factors operating in or on adipose tissue that help to regulate the plasma FFA concentration.

QUESTIONS

1. **Select the FALSE statement below:**
   a. The processes of lipolysis and triglyceride reesterification utilize separate pathways, reactants, and enzymes.
   b. Because adipocyte glycerol kinase activity is low, glycerol release through lipolysis cannot be utilized to any great extent in the reesterification of acyl-CoA in adipocytes.
   c. Insulin promotes glucose uptake into adipocytes.
   d. Adipocytes normally lack glycogen stores.
   e. Insulin activates LPL on the capillary endothelium servicing adipocytes.

2. **All of the following hormones are known to promote lipolysis, EXCEPT:**
   a. Insulin-like growth factor-1.
   b. Growth hormone.
   c. Epinephrine.
   d. Cortisol.
   e. Thyroxine.

3. **Activation of which one of the following enzymes is best associated with triglyceride hydrolysis in adipose tissue?**
   a. Adenyl cyclase
   b. Phosphatase
   c. Hexokinase
   d. cAMP-dependent phosphodiesterase
   e. Monoamine oxidase

4. **Which one of the following humoral agents, produced by adipocytes, promotes satiety in a negative feedback fashion?**
   a. IGF-1
   b. IL-6
   c. ADH
   d. Leptin
   e. cAMP

5. **Which one of the following hormones stimulates β_1-adrenergic receptor synthesis on adipocytes?**
   a. ACTH
   b. T_4
   c. Glucagon
   d. Cortisol
   e. Epinephrine

6. **Which one of the following is a circulating lipolytic factor produced by exercising muscle fibers?**
   a. IGF-1
   b. Insulin
   c. T_3
   d. cAMP
   e. IL-6

7. **Pyruvate can be converted to glycerol 3-P in hepatocytes, but not in adipocytes:**
   a. True
   b. False

7. b
6. e
5. b
4. d
3. a
2. a
1. b

ANSWERS

Chapter 71

Ketone Body Formation and Utilization

Overview

- Ketone bodies (acetoacetate and β-OH-butyrate) are small, water-soluble circulating lipids produced by the adult liver, that cross the blood-brain-barrier, as well as the placental barrier.

- Acetone, the third ketone body, is metabolically inert, but causes the fruity smell on the breath and in the urine of severely ketotic patients.

- Ketoacidosis is associated with displacement of Cl^- and HCO_3^- from the circulation.

- The flux-generating step in ketogenesis appears to be lipolysis of TG in adipocytes.

- Ketone bodies stimulate modest insulin release, and they also feedback negatively on hormone sensitive lipase (HSL) activity in adipocytes.

- Ketone bodies are preferred fuel in the brain during starvation.

- AcAc and β-OH-butyrate can be used in the biosynthesis of glycerophospholipids, sphingolipids, and sterols.

- β-OH-Butyrate may not be detected in urine through some rapid field tests.

The compounds **acetoacetate (AcAc; CH_3CO CH_2COO^-)** and **β-OH-butyrate ($CH_3CH(OH)CH_2$ COO^-)** are freely soluble, circulating lipids. They are known as ketone bodies (KB⁻s), and arise from the partial β-oxidation of fatty acids (FAs) in liver mitochondria. They can be removed from blood and used by most aerobic tissues (e.g., muscle, brain, kidney, mammary gland, small intestine, and fetal liver), but they cannot be oxidized by the adult liver from which they are produced (**Fig. 71-1**). These small particles become important lipid fuel during starvation (see Chapter 75), and important substrates for both fetal and neonatal complex lipid biosynthesis.

It should be noted that in physiologic solution acetoacetic acid and β-OH-butyric acid act as strong acids (**pK ≈ 3.5**), donating their protons in promotion of a **ketoacidosis**. The remaining ketone body anions (AcAc and β-OH-butyrate) tend to displace **Cl^-** and **HCO_3^-** from extracellular fluid, thus increasing the **anion gap** (see Chapters 86-88).

The **KB⁻s** are misnamed, for they are not "bodies", and β-OH-butyrate is not a ketone. There are also several ketones in blood that are not referred to as KB⁻s (e.g., pyruvate and fructose). The appellation ketone was given to these compounds over 100 years ago by German physicians who found that the urine of diabetic patients gave a positive reaction with reagents used to detect ketones. In those circumstances, **AcAc** and **acetone (CH_3COCH_3; propanone)** contributed to the reaction, but not **β-OH-butyrate**. **Acetone**, which is metabolically inert, is formed by the **spontaneous decarboxylation of AcAc**, and is only detectable when the concentration of the latter is

Copyright © 2015 Elsevier Inc. All rights reserved.

abnormally high (**Fig. 71-2**). Acetone is usually excreted through the lungs and kidneys, where it accounts for the characteristic **sweet** or **fruity smell** on the breath and in the urine of severely ketotic patients.

The flux-generating step in **KB⁻ formation** is usually **lipolysis of triglyceride (TG)** in adipocytes, so that the physiologic pathway spans more than one tissue, and includes **adipocytes**, **blood**, and **liver**. Free fatty acids (FFAs) in liver tissue are first activated through **fatty acyl-CoA**

formation, then transported into mitochondria via the carnitine shuttle (see Chapter 55). Following β-oxidation, conversion of **excess acetyl-CoA** to **KB⁻s** requires 4 more stages; **1) condensation** of 2 molecules of acetyl-CoA to form AcAc-CoA, **2) acylation** of AcAc-CoA to HMG-CoA, **3) deacylation** of HMG-CoA to AcAc, and **4) reduction** of some of the resulting AcAc to β-OH-butyrate (**Fig. 71-2**). This pathway of KB⁻ formation was named after Lynen, the Nobel laureate who discovered it.

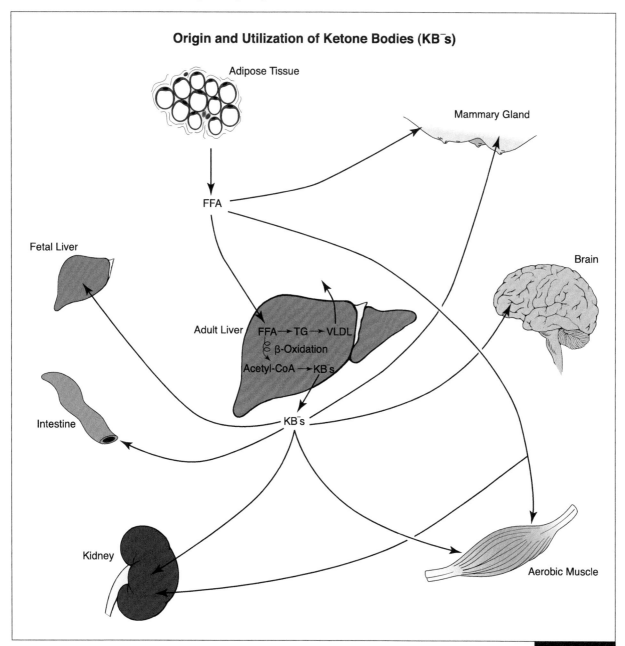

Origin and Utilization of Ketone Bodies (KB⁻s)

Figure 71-1

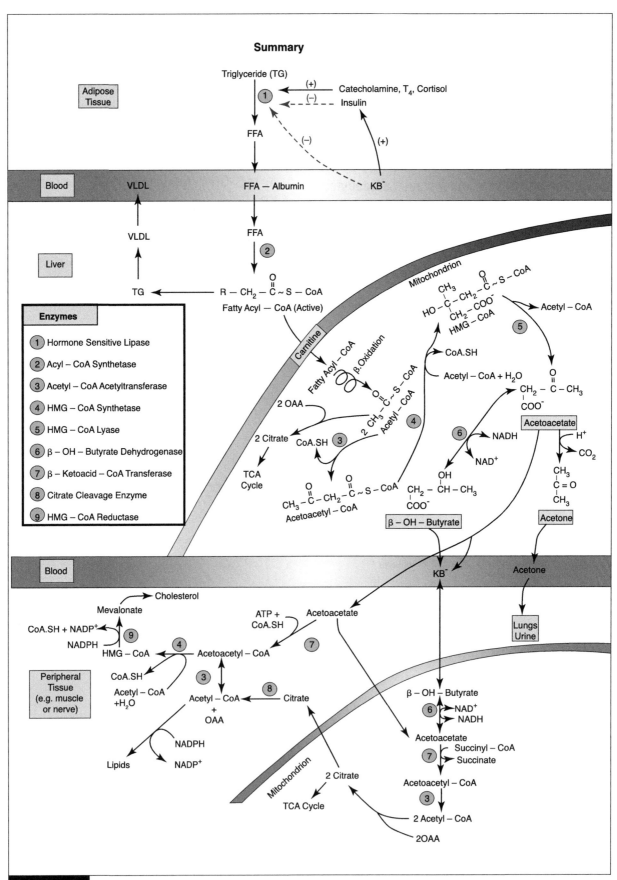

Figure 71-2

Why two, rather than one, metabolizable KB⁻s are produced is not entirely clear. Comparative studies indicate that **AcAc** is the more primitive KB⁻, as it occurs in the absence of β-OH-butyrate in **invertebrates** and some **fish**. However, β-OH-butyrate is the better fuel since it is more reduced. Conversion of AcAc to β-OH-butyrate requires **NADH** (see **Fig. 71-2**). Although the mitochondrial **NAD⁺/NADH ratio** is normally maintained at about **10:1**, this ratio continues to decline during extensive fat oxidation, thus favoring β-OH-butyrate formation. A frequently used rapid field test for ketonuria (which uses ®**Clinistix** or similar material) detects only **AcAc** and **acetone** (the ketones), which can sometimes result in serious **underestimates** of the extent of ketonuria since it does not detect β-**OH-butyrate**.

In answer to the above questions, producing only one or the other KB⁻ could render KB⁻ synthesis independent of the hepatic redox state. However, since a steady state mixture of the two KB⁻s can be produced in mammals, changes in the redox state of liver mitochondria (i.e., the **NAD⁺/NADH** ratio) simply alters the **AcAc:β-OH-butyrate ratio**.

Since their discovery in the urine of diabetic patients in the latter part of the 19th century, KB⁻s have had a checkered history, being variously regarded as indicators of disturbed metabolism, and as important fuels. Their association with diabetes mellitus branded them initially as undesirable metabolic products. However, the demonstration in the 1930s that various tissues could oxidize KB⁻s led to the suggestion that they could be important lipid fuel. Thus, in a review in 1943, MacKay wrote:

> "**Ketone bodies may no longer be looked upon as noxious substances, which are necessarily deleterious to the organism.**"

In 1956, however, evidence was presented that long-chain FAs provided important lipid-fuel in starvation, so that KB⁻s were again overlooked, and primarily associated in the minds of biochemists, physiologists, and clinicians with pathologic conditions. Indeed, in a review article in 1958, Fredrickson and Gordon did not mention KB⁻s. Even in 1968, doubts were expressed in a review of FA catabolism by Greville and Tubbs who stated:

> "**Clearly it is not obvious in what way ketogenesis in starvation is a good thing for the whole animal; should the liver be regarded as providing manna for extrahepatic tissues, or does it simply leave them to eat up its garbage?**"

Research since that review was written has provided strong support for the view that KB⁻s have an important role to play in the transport of lipid-fuel in blood. Furthermore, recent developments indicate that KB⁻s play an important role in integrating fuel mobilization and utilization in the whole animal (see Chapter 75).

Why Should one Lipid Fuel be Converted to Another in the Liver?

In order to answer this question it is necessary to look at the limitations of FFA as fuel. In plasma, FFAs must be bound to albumin in order to render them soluble. This has an important drawback; the FFA-albumin complex is restricted from crossing the **blood-brain-barrier (BBB)**, and also the **placental-barrier**. Conversely, KB⁻s are freely soluble in plasma, they do not require albumin for their transport, and they diffuse across these barriers easily. Therefore, unlike FFAs, KB⁻s can provide important fuel for the CNS and for the fetus during starvation, and also serve as building blocks for essential lipids (see **Fig. 71-2**).

There are a number of **muscles** in the body which carry out vital physiologic actions (e.g., heart, diaphragm, smooth muscle in the digestive tract, and myometrial smooth muscle). If hypoglycemia coincided with a reduction in blood supply to these tissues (e.g., during hypotension or hypoperfusion), insufficient fuel (i.e., FFAs and glucose) would be provided, resulting in fatigue of these muscles, hence precipitating a potentially life-threatening situation. In such incidences, a high concentration of readily diffusible KB⁻s could help to replace these fuels. Such vital muscles do indeed possess high activities of the enzymes for KB⁻ utilization.

Additionally, measurements of arterio-venous differences across the BBB of obese patients undergoing therapeutic starvation have shown that KB⁻s progressively displace glucose as a fuel for the brain. During the intermediate stages of starvation, the brain derives most (≈ **60%**) of its energy from KB⁻ oxidation, with the remainder provided by the oxidation of glucose (see Chapter 75).

Ketone bodies also **feedback negatively** on fat tissue during starvation, thus restraining lipolysis (see **Fig. 71-2**). By stimulating **modest insulin release**, KB⁻s are assisted in this regard, since insulin also inhibits lipolysis (see Chapter 70).

Ketone Body Utilization

Ketone bodies can be utilized by a number of aerobic tissues for both **oxidative** and **anabolic** purposes. For example, it appears that during the early postnatal period, AcAc and β-OH-butyrate are **preferred** over glucose as substrates for the biosynthesis of glycerophospholipids, sphingolipids, and sterols in accordance with requirements for **brain growth** and **myelination**. In the **lung**, AcAc has been found to serve better than glucose as a precursor for the synthesis of **lung phospholipids** in **neonates**. These synthesized lipids, particularly dipalmitoylphosphatidyl choline (DPPC), are incorporated into pulmonary surfactant. Finally, it has been shown that **KB⁻s inhibit *de novo* biosynthesis of pyrimidines** in fetal rat brain slices (Williamson DH, 1985). Thus, during maternal

starvation, KB⁻s may maximize chances for fetal survival by restraining cell replication.

The primary pathway for KB⁻ utilization in extrahepatic tissues may involve an initial oxidation of **β-OH-butyrate** to **AcAc,** utilizing **NAD⁺** as cofactor, and then activation of **acetoacetate** to **acetoacetyl-CoA** (process 7 in **Fig. 71-2**), a reaction requiring succinyl-CoA and the enzyme, **β-ketoacid-CoA transferase** (also known as **succinyl-CoA-acetoacetate CoA transferase**). The **acetoacetyl-CoA** is then split into **two acetyl-CoAs** by **acetyl-CoA acetyltransferase** (also known as thiolase), and then coupled with oxaloacetate (OAA) to form citrate in mitochondria, or used in lipid biosynthesis in the cytoplasm. Ketone bodies are generally metabolized in extrahepatic tissues in proportion to their blood concentration. When the concentration is elevated, the metabolism of ketone bodies increases until, at a concentration of approximately 12 mmol/L, they saturate metabolic machinery.

Most evidence indicates that ketonemia is caused by an increased hepatic production of KB⁻s rather than decreased extrahepatic oxidation. Results of experiments on pancreatectomized animals support this hypothesis, since insulin removal takes away one of the important modulators of lipolysis, and thus KB⁻ production (see Chapter 70). Although KB⁻s are freely filtered by the kidneys, they are extensively reabsorbed in proximal portions of the nephron. Therefore, even in severe ketonemia, the loss of KB⁻s in urine is only a few percent of total KB⁻ production and utilization. The preferred method of assessing the severity of ketosis is therefore measurement of the blood KB⁻ concentration, and not that of urine.

In summary, the control of ketogenesis is exerted initially in adipose tissue, for ketonemia will not occur unless there is an increase in the level of circulating FFAs that arise from lipolysis of TG in adipose tissue. Free fatty acids are the

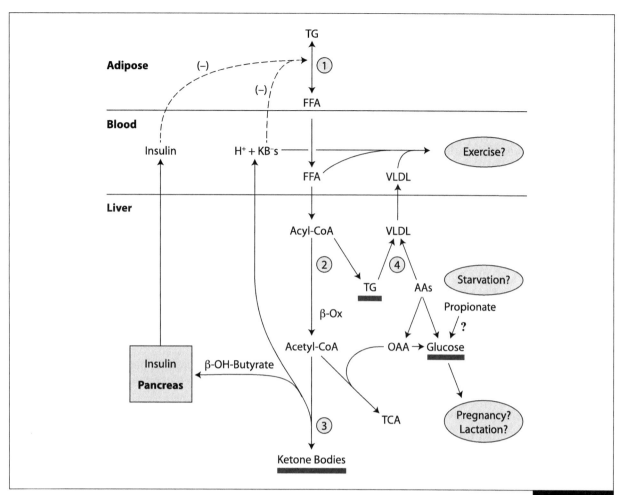

Figure 71-3

precursors to KB⁻s in adult hepatocytes, and the liver generally extracts about 30% of the FFAs contained in the perfusate of it's sinusoids. Therefore, factors regulating lipolysis are of paramount importance in controlling ketogenesis. The catecholamines, thyroid hormones and glucocorticoids enhance lipolysis, while KB⁻ accumulation and insulin suppress it (**Fig. 71-3**).

As FFAs enter the circulation from adipocytes, their removal from plasma will be enhanced by aerobic muscle fibers when the animal exercises, thus reducing their entry into hepatocytes. Since the K_m of LPL servicing aerobic muscle fibers is low (see Chapter 64), TG clearance from VLDL and CMs will also be enhanced by these muscle fibers. Thus, exercise removes pressure on the liver to

oxidize fatty acids, and produce KB⁻s.

After uptake by hepatocytes, FFAs are first activated by acyl-CoA synthetase, then the acetyl-CoA formed via β-oxidation is oxidized in the TCA cycle, or it can enter the pathway of ketogenesis to form KB⁻s. Acyl-CoA may also be esterified to TGs and PLs in the cytoplasm, which then should be transferred out of the liver in VLDL. Regulation of FFA entry into β-oxidation is by CPT-1 (see Chapter 55), whose activity is low in the fed state, leading to depression of FA oxidation, and high in starvation, allowing FA oxidation to increase. Malonyl-CoA, the initial intermediate in FA biosynthesis, is a potent inhibitor of CPT-1. Lack of malonyl-CoA inhibition during starvation, and a low insulin:-glucagon ratio usually enhance entry of FAs into hepatic β-oxidation. However, in some animal species (e.g., equine), the pathway for ketogenesis is not as well developed as it is in others, therefore ketonemia is infrequently observed (see Chapter 67). However, these animals retain considerable ability to export VLDL into blood during starvation, with a resultant hypertriglyceridemia. Ruminant animals, however, have difficulty forming and exporting VLDL from hepatocytes, therefore leaving them more prone to fatty liver infiltration (excess TG retention), and ketonemia.

During starvation, herbivores will have little or no propionate entering the hepatic portal circulation in support of gluconeogenesis, therefore they will be forced to use glycerol and glucogenic amino acids to maintain their plasma glucose levels. If these animals are pregnant or lactating, there will be fewer amino acids available to liver cells for Apo B_{100} biosynthesis, therefore VLDL production will decrease, and TG will remain in liver cells (steatosis). Additionally, as oxaloacetic acid (OAA) is siphoned-off into gluconeogenesis, less will be available to couple with acetyl-CoA in the condensing reaction of the TCA cycle, thus favoring ketogenesis. As more and more FAs undergo β-oxidation, the hepatic NADH:NAD$^+$ ratio will rise, and more and more acetoacetate will be converted to β-OH-butyrate. This ketone body stimulates modest pancreatic insulin release, increases sensitivity of adipose tissue to insulin, and further reduces the rate of adipose tissue lipolysis (see Chapter 75). Thus, this negative feedback mechanism will help to reduce hepatic ketogenesis.

Ketone bodies are not bound to albumin in the circulation, therefore they freely cross the blood-brain barrier and the placental barrier. Although KB$^-$ production causes a metabolic acidosis, these small, water-soluble lipids can be used by a number of aerobic tissues for both oxidative and anabolic purposes.

OBJECTIVES

- Explain how and where KB$^-$s are produced and oxidized, and identify how the anion gap (AG) is altered in ketoacidosis (see Chapters 86-88).

- Discuss why the physiologic pathway of KB$^-$ production involves adipocytes, blood and liver.

- Identify the four stages of hepatic KB$^-$ biosynthesis.

- Describe metabolic differences between AcAc, β-OH-butyrate and acetone, and show how the hepatic mitochondrial NAD$^+$/NADH ratio affects the AcAc:β-OH-butyrate ratio.

- Indicate how KB$^-$s serve to integrate fuel mobilization and utilization in the whole animal during starvation (see Chapter 75).

- Compare FFAs to KB$^-$s as sources of fuel for the adult brain, and as potential substrates for complex lipid formation in the fetal brain.

- Explain why the brain appears to oxidize KB$^-$s more readily than glucose during starvation (see Chapter 75).

- Identify the primary causes of ketonemia, and discuss why measurement of the serum KB$^-$ concentration is preferred over that of urine in assessing the severity of ketosis.

QUESTIONS

1. **Which one of the following enzymes is used in the formation of AcAc in mitochondria of adult liver cells?**
   a. HMG-CoA lyase
   b. Citrate cleavage enzyme
   c. β-Ketoacid transferase
   d. HMG-CoA reductase
   e. Hormone sensitive lipase

2. **The flux generating step in KB$^-$ formation is:**
   a. Conversion of HMG-CoA to mevalonate.
   b. Lipolysis of triglyceride in adipocytes.
   c. The spontaneous decarboxylation of AcAc to acetone.
   d. Fatty acid activation in hepatocytes.
   e. Acetone excretion in urine.

3. **Ketone bodies are known to feedback negatively upon:**
   a. Insulin release from the pancreas.
   b. HMG-CoA reductase activity in hepatocytes.
   c. Hormone sensitive lipase activity in adipocytes.
   d. Surfactant biosynthesis in type II alveolar epithelial cells of the lungs.
   e. Sphingolipid biosynthesis in the fetal brain.

4. **The frequently used rapid field test for ketonuria (using ®Clinistix or similar material) detects:**
   a. AcAc and β-OH-butyrate.
   b. Acetone only.
   c. β-OH-butyrate and acetone.
   d. AcAc, β-OH-butyrate, and acetone.
   e. Acetone and AcAc.

5. **Select the FALSE statement below:**
   a. β-OH-butyrate is the more primitive ketone body.
   b. During the early postnatal period, AcAc and β-OH-butyrate are preferred over glucose as substrate for lipid biosynthesis in accordance with requirements for brain growth and myelination.
   c. Ketone bodies have been shown to inhibit de novo biosynthesis of pyrimidines in fetal rat brain slices.
   d. In the brain, ketone bodies may be preferred fuel over glucose during starvation.
   e. Ketone bodies can be oxidized by the fetal liver.

6. **The hepatic mitochondrial NADH:NAD⁺ ratio is usually in direct proportion to which KB⁻ ratio?**
   a. β-OH-Butyrate:Acetone
   b. AcAc:β-OH-Butyrate
   c. Acetone:AcAc
   d. β-OH-Butyrate:AcAc
   e. Acetone:β-OH-Butyrate

7. **Which of the following suppress lipolysis?**
   a. β-OH-butyrate and insulin
   b. Exercise
   c. Lactation
   d. Catecholamines and thyroid hormone
   e. All of the above

8. **When FFAs are activated by acyl-CoA synthetase in hepatocytes:**
   a. They may be esterified with glycerol.
   b. They may be β-oxidized in mitochondria.
   c. They may ultimately leave hepatocytes (esterified to glycerol in VLDL).
   d. They may enter into ketone body formation.
   e. All of the above

9. **Since ketone bodies are not protein-bound in the circulation, they cannot:**
   a. Be oxidized by peripheral nerve fibers.
   b. Be filtered by the kidneys.
   c. Be excluded from crossing the placental barrier.
   d. Cross the BBB.
   e. Serve as building blocks for complex lipids.

10. **What percentage of FFAs contained in the sinusoidal liver perfusate are normally extracted (i.e., taken-up) by the liver?**
    a. 2%
    b. 30%
    c. 50%
    d. 80%
    e. 100%

11. **Which of the following tissues are known to oxidize ketone bodies?**
    a. Cardiac myocytes
    b. Uterine myometrial smooth muscle
    c. Diaphragm
    d. Smooth muscle cells of the digestive tract
    e. All of the above

12. **The more primitive ketone body is:**
    a. Acetone.
    b. β-OH-butyrate.
    c. Acetoacetate.
    d. Acetyl-CoA.
    e. HMG-CoA.

12. c
11. e
10. b
9. c
8. e
7. a
6. d
5. a
4. e
3. c
2. b
1. a

ANSWERS

<u>Chapter 72</u>

Fatty Liver Syndrome (Steatosis)

Overview

- Fatty liver syndrome (i.e., steatosis) is due to excess accumulation of TG in liver cells.
- As a general rule, most hepatotoxins that cause steatosis do so mainly through interference with lipid egress from hepatocytes.
- High fat diets can cause steatosis.
- Some agents cause steatosis through inhibition of VLDL formation.
- Hepatotoxins that interfere with hepatic Apo-B$_{100}$ formation can cause steatosis.
- Studies have shown that VLDLs are virtually absent from the blood of choline-deficient rats.
- Orotic acid, an intermediate in pyrimidine biosynthesis, can cause steatosis.
- Cats with taurine deficiency can develop cholestasis, which leads to steatosis.
- Hyperbilirubinemia, hypoalbuminemia, hyperammonemia, and abnormal coagulation parameters are also associated with steatosis.

Fatty liver syndrome, also referred to as **hepatic lipidosis** or **steatosis**, is due to excessive accumulation of triglyceride (TG) in the liver, which is usually stored in vacuoles. This pathophysiologic condition most often results from **nutritional**, **metabolic**, **hormonal**, **toxic**, or **hypoxic** disturbances.

Comparisons between hepatectomized and intact animals demonstrated many years ago that the liver is an important source of plasma lipoproteins derived from endogenous sources (see Chapters 65-67). Triglycerides incorporated into hepatic **very low-density lipoprotein (VLDL)** are derived from **two** sources:

1) Synthesis within the liver from acetyl-CoA derived from carbohydrate or amino acid sources (see Chapters 56 and 57), and

2) Uptake of free fatty acids (FFAs) from the circulation.

The first source is predominant in the well-fed state, when FA synthesis is high and circulating levels of FFAs are low. Since TG does not normally accumulate in the liver under this condition, it must be inferred that it is transported away from the liver as rapidly as it is formed. On the other hand, during starvation, during the feeding of high-fat, low carbohydrate diets, or in diabetes mellitus (DM), circulating levels of FFAs are raised, and more are extracted from plasma into liver cells. Under these conditions, FFAs become a primary source of TG in liver cells and subsequently circulating VLDL because lipogenesis from acetyl-CoA is limited. Factors that enhance both TG synthesis and secretion of VLDL by the liver include:

1) Feeding diets high in fat (or carbohydrate), and

Copyright © 2015 Elsevier Inc. All rights reserved.

2) Conditions in which high concentrations of circulating FFAs prevail (e.g., starvation or DM; see Chapters 70 and 71).

Although a modest degree of accumulated fat in the liver may be normal (up to **35%** in some animals), when it becomes excessive it is regarded as a pathophysiologic state. Chronic conditions cause fibrotic changes to occur that can progress to cirrhosis and impaired liver function. In general, fatty livers fall into three general categories.

The **first type** of fatty liver is associated with elevated concentrations of plasma FFAs as a result of mobilization of TG from adipocytes (**Fig. 72-1**). The production of hepatic VLDL in this case may not keep pace with FFA influx, thus allowing TG to accumulate. The quantity of TG present in hepatocytes is significantly elevated during starvation, and the ability to secrete VLDL may also be impaired. In light of the heightened requirement for amino acids to enter the gluconeogenic pathway during starvation, hepatic apoprotein production may become reduced with this condition, thus limiting VLDL formation. In uncontrolled DM, pregnancy toxemia of ewes, or ketosis of cattle, fatty infiltration may be sufficiently severe to cause visible pallor and enlargement of the liver (hepatomegaly). Additionally, these conditions usually precipitate a ketoacidosis, since the ability of hepatic mitochondria to oxidize all of the acetyl-CoA generated from fatty acid β-oxidation is exceeded (see Chapters 37 and 71).

The **second type** of fatty liver develops when animals are fed a high fat diet (or hens are fed a high carbohydrate diet; see Chapter 56). The output of VLDL from liver cells simply cannot keep pace with the amount of fatty acyl-CoA generated.

The **third type** of fatty liver is usually due to a metabolic block in production of VLDL, or the inability to move VLDL across a structurally or functionally damaged plasma membrane. Theoretically, the metabolic block may be due to one (or more) of four causes:

1) Impaired fatty acid β-oxidation (e.g., a carnitine deficiency or damaged hepatic peroxisomes secondary to oxidative stress (see Chapter 55)).

2) Defective synthesis, or abnormal glycosylation of the integral apoprotein needed to form VLDL (Apo B_{100}).

3) Defective assembly of VLDL (i.e., a failure in the union of Apo B_{100} with phospholipid, cholesterol, and TG).

4) Impaired egress of the assembled VLDL across a damaged plasma membrane (i.e., failure in the secretory mechanism itself), or a combination of the above (see **Fig. 72-1**).

Defective Apo B_{100} synthesis appears to be responsible for the steatosis produced by a number of **hepatotoxins**. This may result from destruction of the cellular site of protein synthesis (i.e., the rough endoplasmic reticulum (RER) and its ribosomes), or result from introduction of a selective biochemical lesion in any one of the step of protein synthesis. Destruction of the RER appears to be one means by which **carbon tetrachloride (CCl₄)** interferes with the biosynthesis of protein. **Puromycin** inhibits protein synthesis by attaching itself to ribosomes, supplanting activated tRNA that would normally be attached. Accordingly, this leads to an abbreviated peptide chain and inhibition of protein formation. **Tetracycline** and its congeners are antibiotics which inhibit synthesis of protein, in bacterial and other cells, apparently by binding to tRNA. **Ethionine** inhibits protein formation by depleting the cell of available ATP, and perhaps by interfering with other steps of the biosynthetic pathway. **Galactosamine** inhibits protein synthesis by the somewhat analogous mechanism of depleting the cell of UTP. These and other possible mechanisms by which defects in the

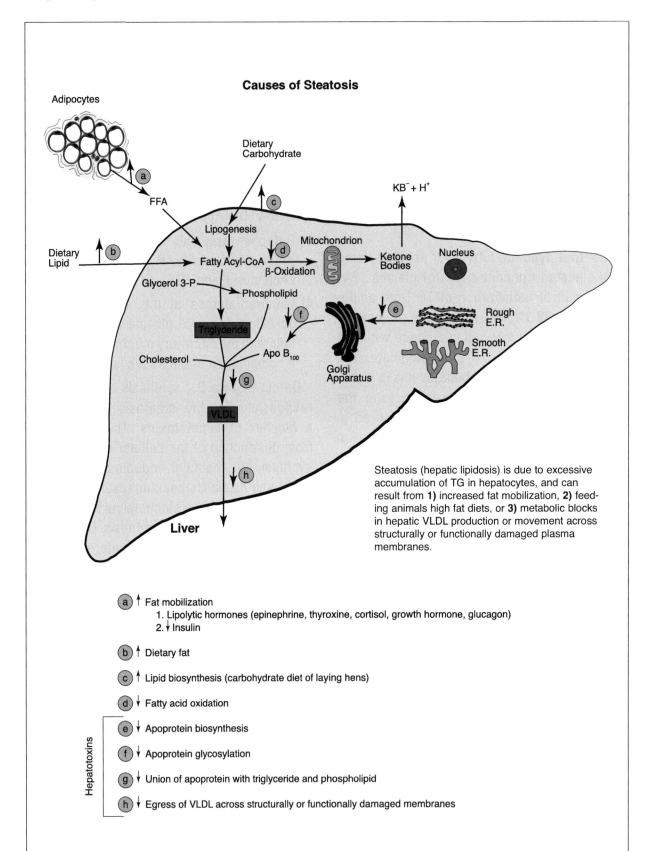

Causes of Steatosis

Steatosis (hepatic lipidosis) is due to excessive accumulation of TG in hepatocytes, and can result from **1)** increased fat mobilization, **2)** feeding animals high fat diets, or **3)** metabolic blocks in hepatic VLDL production or movement across structurally or functionally damaged plasma membranes.

(a) ↑ Fat mobilization
    1. Lipolytic hormones (epinephrine, thyroxine, cortisol, growth hormone, glucagon)
    2. ↓ Insulin

(b) ↑ Dietary fat

(c) ↑ Lipid biosynthesis (carbohydrate diet of laying hens)

(d) ↓ Fatty acid oxidation

(e) ↓ Apoprotein biosynthesis

(f) ↓ Apoprotein glycosylation

(g) ↓ Union of apoprotein with triglyceride and phospholipid

(h) ↓ Egress of VLDL across structurally or functionally damaged membranes

Hepatotoxins

Figure 72-1

cellular machinery for the synthesis of hepatic protein occur, can lead to inhibited formation of Apo B$_{100}$, and thus steatosis.

Protein synthesis inhibitors which produce little or no increase in hepatic fat include **cycloheximide** and **actinomycin D**. Some investigators believe that these agents simultaneously inhibit TG synthesis, thereby nullifying the effect of inhibited hepatic protein synthesis.

An agent that does not inhibit hepatic protein synthesis, yet leads to fatty liver infiltration by interfering with egress of lipid, is **orotic acid (OA)**, an intermediate in pyrimidine biosynthesis (see Chapter 14). This agent causes steatosis by facilitating formation of a defective, nonglycosylated apoprotein component of VLDL, that apparently cannot function in the transport of lipids. Increased OA formation may occur in animals with **arginine (Arg)** deficiency. This amino acid is essential for normal **urea cycle** function, and impaired action of the urea cycle is associated with production of excess **carbamoyl phosphate**, an OA precursor (see Chapters 10 and 14). Arginine is an essential amino acid in cats, and it is sometimes elevated in the blood of animals with hepatic lipidosis. Additionally, **taurine (Tau)** deficiency in cats may contribute to intrahepatic cholestasis since Tau is needed for bile acid conjugation (see Chapter 62). Cholestasis, in turn, can lead to steatosis.

Defective assembly of the **VLDL** can also contribute to accumulation of fat in the liver. The interference with exit of lipid begins too soon after a dose of **CCl$_4$** to be attributed solely to impairment of protein synthesis (which this agent also produces). This has led investigators to suggest that impaired assembly of the complex may be responsible for the critical accumulation of fat.

Damage to the plasma membrane also contributes to accumulation of fat in hepatocytes. The effect of toxic agents on the plasma membrane may lead to prompt interference with movement of VLDL out of the cell, and their effect on synthesis of VLDL may be responsible for the later stage of lipid accumulation.

Another type of fatty liver that has been studied extensively is due to a **deficiency of choline** (i.e., a lipotropic factor). Since choline is synthesized using labile methyl groups donated by **S-adenosylmethionine (SAM)** in the process of transmethylation (see Chapter 57), the deficiency is basically due to a shortage of the type of methyl group donated by SAM. Studies have shown that VLDL are virtually absent from the blood of choline-deficient rats, indicating that the defect lies mainly in hepatic VLDL formation.

Any one of the pathophysiologic lesions above, or a combination thereof, could lead to inhibited transport of fat from the liver, and thus **steatosis**. This can occur even though lipid is available to the liver in normal amounts. As a general rule, most agents that lead to steatosis do so mainly through interference with **egress** of lipid from hepatocytes.

Cytosolic fatty vacuolation (steatosis) usually exhibits either a macro- or microvesicular pattern, and may accompany toxic hepatitis. Microvesicular steatosis may further reflect severe metabolic disruptions associated with abnormal mitochondrial function, and when diffuse may accompany inflammation and hepatocellular necrosis.

Other Symptoms of Steatosis

Additional findings in animals with steatosis include **anemia**, **hyperbilirubinemia** (and **bilirubinuria**; see Chapter 33), increased serum concentrations of enzymes that leak from liver cells (e.g. **alkaline phosphatase (SAP or ALP)**, **aspartate aminotransferase (AST)**, and **alanine aminotransferase (ALT**; see Chapter 9)), **hypoalbuminemia**, **hyperammonemia**, and **abnormal coagulation parameters**. Fasting and postprandial **serum bile acid** concentrations may also be elevated (see Chapter 62).

OBJECTIVES

- Identify the three basic types of fatty liver infiltration, and discuss the origin of each.

- Explain relationships between VLDL formation and egress from liver cells, and steatosis.

- Understand why carnivores and ruminant animals are more prone to steatosis than equine (see Chapter 67).

- Indicate how hepatotoxins are thought to promote steatosis.

- Recognize relationships between glucogenic amino acid availability to liver cells, APO-B$_{100}$ formation, starvation and steatosis.

- Explain how dietary Arg or Tau deficiency might lead to fatty liver syndrome (see Chapters 10, 14 & 62).

- Know why DM predisposes animals to fatty liver disease.

- Understand why fatty liver infiltration promotes various signs and symptoms of liver disease.

QUESTIONS

1. Fatty liver syndrome is usually due to the accumulation of which one of the following in hepatocytes?
  a. Free fatty acids
  b. Fatty acyl-CoA
  c. Triglyceride
  d. High-density lipoprotein
  e. Phospholipid

2. As a general rule, most agents that lead to steatosis do so mainly through:
  a. The cAMP messenger system.
  b. The Ca^{++} messenger system.
  c. Inhibition of triglyceride biosynthesis.
  d. Inhibition of VLDL biosynthesis.
  e. Interference with egress of lipid from hepatocytes.

3. The intermediate in pyrimidine biosynthesis that causes steatosis is:
  a. Deoxyuridine diphosphate.
  b. Orotic acid.
  c. Glutamine.
  d. Carbamoyl phosphate.
  e. Aspartate.

4. Choline deficiency causes steatosis though decreased hepatic:
  a. VLDL formation.
  b. Triglyceride synthesis.
  c. Apo B$_{100}$ biosynthesis.
  d. Ketone body formation.
  e. Fatty acid oxidation.

5. A carnitine deficiency could lead to steatosis by impairing hepatic:
  a. Triglyceride formation.
  b. Fatty acyl-CoA formation.
  c. Apo B$_{100}$ biosynthesis.
  d. β-Oxidation of fatty acids.
  e. VLDL egress.

6. All of the following may be symptoms of hepatic lipidosis, EXCEPT:
  a. Hyperbilirubinemia.
  b. Elevated serum AST and ALT.
  c. Steatorrhea.
  d. Hypoalbuminemia.
  e. Anemia.

7. Select the FALSE statement below:
  a. A modest degree of fat accumulation in the liver is normal.
  b. A high carbohydrate diet can cause hepatic lipidosis in laying hens.
  c. Cats with taurine deficiency can develop hepatic lipidosis.
  d. Reduced hepatic protein synthesis can cause steatosis.
  e. Hepatic lipidosis is usually associated with a reduction in hepatic TG formation.

8. Hepatotoxins are known to promote steatosis by decreasing any or all of the following, EXCEPT:
  a. Lipid biosynthesis.
  b. VLDL egress.
  c. The union of apoprotein with TG and PL.
  d. Apoprotein glycosylation.
  e. Apoprotein biosynthesis.

ANSWERS

8. a
7. e
6. c
5. d
4. a
3. b
2. e
1. c

Addendum to Section V

The property of being **hydrophobic** is a defining characteristic of lipids, since they do not associate well with water. Most lipids contain, or are derived from, **fatty acids**. They possess several important functions, including being **triglycerides**, major storage and dietary fuels of the body. Other lipids, including **phospholipids**, **glycolipids** and **cholesterol**, are crucial constituents of biological membranes, for the unique surface active properties of these molecules allow them to form the membrane backbone, separating and defining aqueous compartments within cells. Surface active lipids have other important functions, including maintenance of alveolar integrity in the lungs and solubilization of nonpolar substances in body fluids. Finally, some lipids are important signaling molecules. For example, the **steroid hormones** and **eicosanoids** (i.e., the prostaglandins, thromboxanes and leukotrienes), serve important roles in intercellular communication, and the **bile acids** (which are also steroids) serve important roles in bile formation and intestinal lipid emulsification, which aids digestion.

In meal-eating animals, excess calories are ingested in the anabolic phase of the feeding cycle, followed by a period of negative caloric balance when the organism draws upon its carbohydrate, lipid and protein stores. **Lipoproteins** mediate this cycle by transporting lipids from the intestines as **CMs** during the anabolic phase, and from the liver as **VLDL** during the catabolic phase. **Lipoprotein lipase** located on the capillary epithelium of many tissue sites can hydrolyze TG contained in these circulating lipoproteins, thus aiding fatty acid oxidation during the catabolic phase, and helping to facilitate TG reassembly and storage in adipose tissue during the anabolic phase. **Triglyceride** is mobilized from adipose tissue as glycerol and FFAs during starvation, with the latter attaching to plasma albumin. Abnormalities of lipoprotein metabolism lead to various **hypo-** or **hyperlipoproteinemias**, with the most common being **diabetes mellitus**, where insulin deficiency causes excessive FFA mobilization and underutilization of CMs and VLDL, leading to **hypertriglyceridemia**. Excessive accumulation of TG in the liver leads to **steatosis**, and several **endocrinopathies** (e.g., hypothyroidism, hypersomatotropism, Cushing's-like syndrome or diabetes mellitus) lead to **hypercholesterolemia**.

Introduction to Section VI

Basic principles of physiological chemistry will next be emphasized through examination of metabolic fluxes occurring during sequential phases of **starvation** and **exercise**. The only freely soluble circulating lipids, the **ketone bodies**, will become important elements of this approach to intermediary metabolism, helping to modulate fat breakdown (**lipolysis**), while at the same time providing fuel to important aerobic tissues of the body. **Glycogen**, **FFA** and **amino acid** catabolism will also prove to be essential in meeting tissue demands. To satisfy the metabolic needs of tissues during starvation and exercise, pathways previously described must be precisely adjusted through actions of the nervous, cardiovascular and endocrine systems. To that end, biochemical homeostasis can be maintained, active organs and tissues can be supplied with essential nutrients, and waste products of metabolism can be properly conveyed to the lungs, liver, kidneys and skin for excretion.

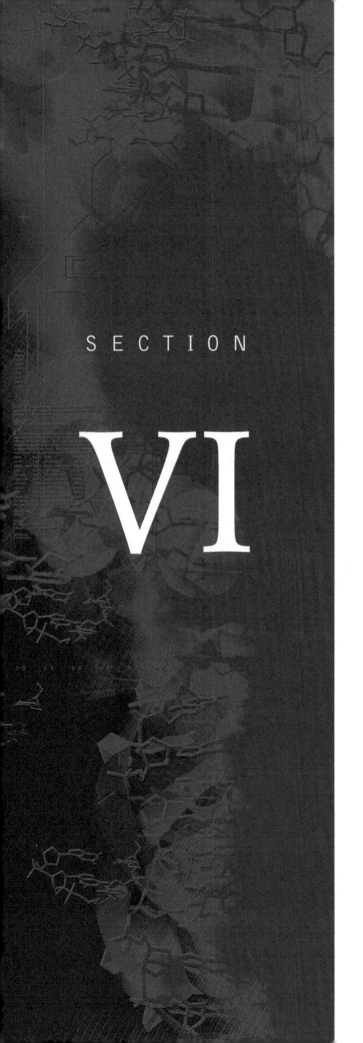

Starvation & Exercise

Chapter 73

Starvation
(Transition into the Postabsorptive Phase)

Overview

- Insulin and glucagon are usually released in an inverse fashion.
- With low carbohydrate, high protein diets, both insulin and glucagon secretion may increase.
- The metabolic phase following a meal varies as a function of the fuel ingested.
- Exquisitely sensitive physiologic mechanisms respond to small changes in the blood glucose concentration.
- The immediate postabsorptive phase of starvation is associated with decreasing hepatic glycogen reserves.

When an animal consumes food, the **first** priority is to provide for immediate metabolic requirements, displacing endogenous fuels. The **second** priority is to expand modest glycogen reserves in liver, fat, and muscle tissue (**Figure 73-1**), and also to replace the amount of protein broken-down in various tissues since the last meal. The **third** priority is to convert the excess, be it carbohydrate, protein, or fat, into triglyceride (TG), and to store this energy largely in adipose tissue and liver.

In transition from the fed to starved condition, priorities are reversed. The body undergoes a series of endocrine changes that selectively draw upon its extensive energy reserves, yet spare breakdown of vitally needed protein which includes enzymes in vital structures like cardiac and nerve tissue.

In animals that absorb high amounts of dietary **glucose** from the small intestine, this compound is incorporated into **glycogen**, and also oxidized in the Embden Meyerhoff pathway (**Figure 73-1** and Chapters 21-17).

Pyruvate is subsequently converted to acetyl-CoA, which is used for energy needs in the liver, and for fatty acid (FA) synthesis (see Chapter 56). Subsequently, FAs are incorporated into TG and then exported to adipose tissue in **very low-density lipoprotein (VLDL)**. During this period of glucose excess, the brain continues to oxidize glucose as fuel, as it did before the meal was consumed. Because of elevated insulin levels due to glucose excess (**Figure 73-2**), muscle preferentially utilizes glucose to replenish its glycogen reserves, as well as for fuel. This preferential glucose metabolism in muscle is also a function of low circulating levels of free fatty acids (FFAs), a result of insulin's effect on adipose tissue (see Chapters 70 and 71). **Insulin** also stimulates glucose uptake and its conversion to glycogen and glycerol 3-phosphate in adipose tissue, with the latter becoming the TG backbone that accepts fatty acids from blood. Insulin inhibits **hormone sensitive lipase (HSL)** activity in adipocytes, thus reducing levels of circulating FFAs (see Chapter 70).

Copyright © 2015 Elsevier Inc. All rights reserved.

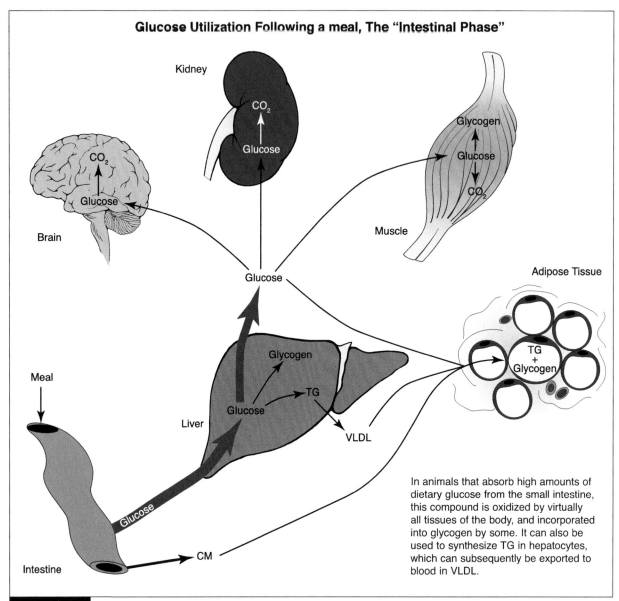

Glucose Utilization Following a meal, The "Intestinal Phase"

In animals that absorb high amounts of dietary glucose from the small intestine, this compound is oxidized by virtually all tissues of the body, and incorporated into glycogen by some. It can also be used to synthesize TG in hepatocytes, which can subsequently be exported to blood in VLDL.

Figure 73-1

Ingested fats from the meal enter the blood stream as **chylomicrons (CMs)** via lymphatics (see Chapter 64), while simple sugars and amino acids are absorbed into the hepatic portal circulation. **Glucogenic amino acids**, for the most part, as well as certain essential and aromatic amino acids like Phe and Trp, are removed and metabolized by the liver (see Chapter 3). The three branched chain amino acids, Leu, Ile, and Val, are removed mainly by extrahepatic tissues (see Chapter 8). Because of the availability of insulin and the increase in circulating amino acids, peripheral protein, particularly in muscle, is replenished. The CMs serve to transport absorbed fat directly to adipose tissue, where FAs are released and then incorporated into new TG.

The Insulin:Glucagon Ratio

Should the meal be **high in protein but low in carbohydrate**, such that the rate of carbohydrate entry into blood from the intestine is less than that required by the brain and other obligatory glucose users, insulin is still released by the pancreas at a greater rate than basal **(Figure 73-3)**. This metabolic situation is normal

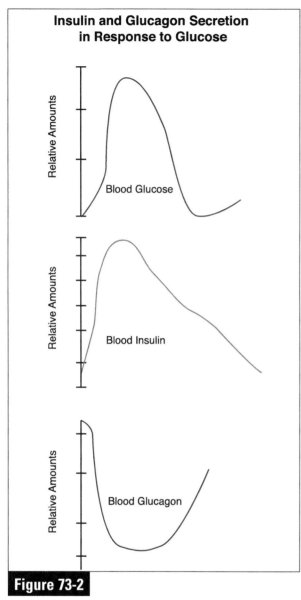

Insulin and Glucagon Secretion in Response to Glucose

Figure 73-2

in carnivores, and also in ruminant animals absorbing minimal amounts of carbohydrate. In this dietary situation, **insulin release** is promoted by certain amino acids, through the paracrine actions of **glucagon**, and by certain **GI hormones** (e.g., gastric inhibitory polypeptide, glucagon-like peptide-1, gastrin and cholecystokinin; Engelking LR, 2012). Secretion of both insulin and glucagon is coordinated with secretion of exocrine pancreatic enzymes, with secretion of both being stimulated by entry of nutrients into the GI tract.

This greater rate of insulin release allows initiation of peripheral protein synthesis, especially in muscles, stimulation of lipoprotein lipase (LPL) activity on capillary endothelial cells servicing adipose tissue (see Chapters 64 and 65), and also stimulation of TG formation in adipose tissue. However, blood glucose levels must be maintained in the presence of this increased insulin, and thus the liver must be poised toward glucose production in spite of the insulin presence. This is the physiologic role of **glucagon**, whose secretion is promoted not only by low plasma glucose levels, but also by glucogenic amino acids, particularly Arg. Therefore, **extrahepatic tissues** receive the "**fed**" signal to take up circulating fuels, yet the **liver** remains in the **starvation mode** (i.e.,

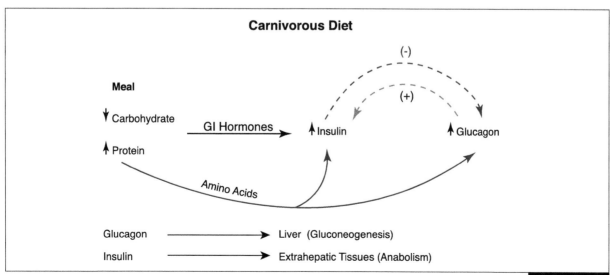

Figure 73-3

gluconeogenic mode) in order to maintain the blood glucose concentration. Thus, whether the liver is gluconeogenic or glycogenic-glycolytic is a function of the **insulin:glucagon ratio**. In the absence of amino acid, if sufficient carbohydrate is absorbed from the GI tract to displace the need for hepatic glucose production, then the rise in blood glucose increases pancreatic insulin release, and suppresses glucagon release. As a result, hepatic glucose production is suppressed. In addition, this slight increase in the blood glucose concentration markedly synergizes insulin-secreting pancreatic cells to produce even more insulin as a response to increased amino acids, so that the insulin:glucagon ratio increases even more.

In summary to this point, the intestinal phase following a meal varies as a function of the fuel ingested, and if deficient in carbohydrate, the liver is hormonally signaled to produce glucose as if no meal had been ingested. Should this type of diet be continued for long periods of time, the liver becomes more and more poised toward gluconeogenesis due to increased activities of the rate-limiting enzymes involved (see Chapter 37), and the liver is thus metabolically similar to that in total starvation.

Glucose Availability

Total body free glucose amounts to only about 0.25 gm/kg body weight, or about one hour's worth of fuel for basal energy needs. If an animal permits a physiologic excursion from the starvation level of about 80 to 60 mg% in plasma, that would be 15 minutes fuel for the whole body, or if it was limited to brain's needs, about 45 minutes. Thus, there must be exquisitely sensitive mechanisms to respond to small changes in the blood glucose concentration, which can result in increased or decreased rates of glucose production by the liver.

Although **liver** cell membranes have a specific glucose transporter (**GLUT-2**; see Chapter 22), it is **insulin-independent**. However, the glucose phosphorylating enzyme in liver cells (**glucokinase**) has a **low affinity (high K_m)** for glucose, but one within the physiologic range following a carbohydrate-rich meal, and this enzyme is **insulin-sensitive**. This is different from the mechanism in **muscle** and **adipose tissue**, where the **GLUT-4** transporter is **insulin-dependent**, and the phosphorylating enzyme (**hexokinase**) has a **high affinity (low K_m)** for glucose. Thus, hyperglycemia results in increased glucose phosphorylation in liver and thereby in glucose uptake, this being a function of the glucose concentration as well as the presence of insulin. In other tissues, permeability, as controlled by insulin in muscle and adipose tissue, is rate-limiting, as is **glucose 6-phosphate negative feedback** upon the activity of **hexokinase** (see Chapter 22).

The Initial Postabsorptive Phase of Starvation

As glucose absorption decreases at the end of the intestinal phase, insulin levels fall, and the liver gradually stops removing glucose (see Chapter 22). A few hours following a meal (and perhaps longer if it was a very large meal), the liver begins to return its stored glycogen as free glucose back into blood to provide fuel needs, mainly for the kidneys and the central nervous system (CNS; **Figure 73-4**). This time period is generally referred to as the **postabsorptive** state. The signals are twofold; **1**) lower insulin levels, and **2**) lower levels of portal blood glucose. **Glucagon** release, which is no longer suppressed, usually increases during this brief time period in order to turn the liver into a glycogenolytic/gluconeogenic organ.

During this phase of decreasing glucose and insulin levels, peripheral tissues such as muscle and adipose tissue progressively diminish glucose utilization, so that soon

Postabosorptive Phase

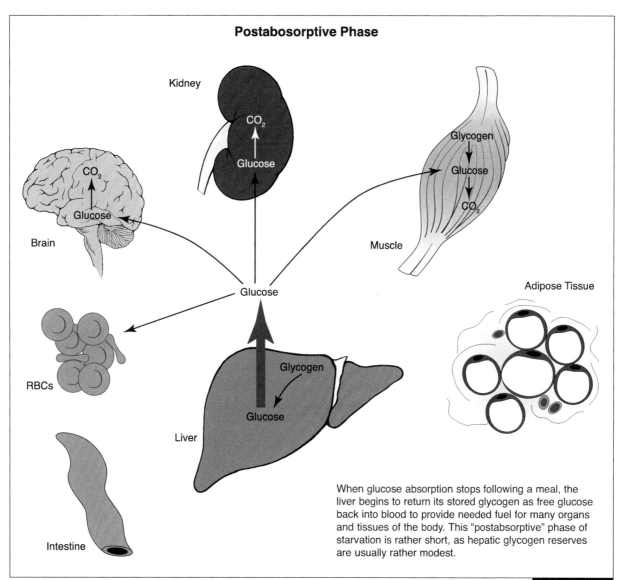

When glucose absorption stops following a meal, the liver begins to return its stored glycogen as free glucose back into blood to provide needed fuel for many organs and tissues of the body. This "postabsorptive" phase of starvation is rather short, as hepatic glycogen reserves are usually rather modest.

Figure 73-4

muscle fuel needs will become met by FFA oxidation (see Chapter 74). Circulating FFA levels increase as insulin levels fall, primarily due to increased cAMP-mediated TG lipolysis in adipose tissue. Whereas in liver insulin is usually pitted against glucagon in controlling the cAMP concentration (see Chapter 23), in adipose tissue it is generally insulin versus epinephrine (see Chapter 70).

In summary, during the initial postabsorptive phase of starvation skeletal muscle uses primarily glucose to satisfy its energy demands. As carbohydrate stores become depleted, FFAs are mobilized from adipocytes, and their rate of oxidation increases. This occurs despite the fact that the blood glucose concentration falls only marginally, and remains higher than the FFA concentration. Since both fuels are available in the bloodstream at the same time, the question arises as to how muscle, either at rest or during exercise, utilizes FFAs in preference to glucose? The answer to this question lies in the substance of several important physiologic control mechanisms, which will be outlined in greater detail in Chapters 74 and 75.

OBJECTIVES

- Identify the manner and directions in which insulin directs metabolism during periods of dietary bounty, and dietary distress.

- Explain why the high protein diet of carnivores promotes both insulin and glucagon release into blood, and discuss effects of this dual pancreatic release on liver, muscle and fat metabolism.

- Discuss the role of the insulin:glucagon ratio in directing intermediary metabolism.

- Explain how and why hyperglycemia normally results in increased hepatic glucose phosphorylation.

- Recognize how the transition from hepatic glycogenesis to hepatic glycogenolysis is controlled during the postabsorptive phase of starvation (see Chapter 23).

- Understand why the insulin:glucagon ratio is important in controlling the cAMP concentration of liver cells, while the insulin:epinephrine ratio is similarly important to adipocytes (see Chapters 23 & 70).

- Explain why the postabsorptive phase of starvation is usually rather short.

- Understand the relationship between the K_m of hepatic glucokinase (180 mg% glucose), and the plasma concentration at which glucose begins to appear in urine (180 mg%).

QUESTIONS

1. A meal that is high in protein, yet low in carbohydrate will result in:
   a. The inhibition of both insulin and glucagon release.
   b. Simulation of insulin yet inhibition of glucagon release.
   c. Stimulation of glucagon yet inhibition of insulin release.
   d. Stimulation of both glucagon and insulin release.
   e. Hypoglycemia.

2. The glucose transporter associated with the plasma membrane of liver cells, is:
   a. GLUT-4.
   b. Insulin-independent.
   c. Involved with changing the cell membrane permeability to glucose.
   d. Glucagon-dependent.
   e. Nonexistent.

3. A high insulin:glucagon ratio:
   a. In blood would be expected soon after a meal that is low in carbohydrate, but high in protein and fat.
   b. Is generally associated with the early, postabsorptive phase of starvation.
   c. Is associated with the initiation of peripheral protein synthesis, particularly in muscles.
   d. Favors hepatic gluconeogenesis.
   e. Helps to assure that extrahepatic tissues remain in the starvation mode.

4. Select the TRUE statement below:
   a. In liver tissue, the cAMP concentration is normally elevated by insulin.
   b. In adipose tissue, insulin is normally pitted against cortisol in controlling the cAMP concentration.
   c. Hexokinase has a low affinity (high K_m) for glucose.
   d. Total body free glucose amounts (on average) to about one day's worth of fuel for basal energy needs.
   e. The liver is continually either removing or adding glucose to the circulation.

5. Select the FALSE statement below:
   a. A protein-rich diet normally results in the inhibition of heptic gluconeogenesis.
   b. Glucose is a preferred fuel in erythrocytes.
   c. Glucose can be used by the liver as a substrate for TG formation.
   d. A significant rise in the insulin:glucagon ratio will inhibit heptic gluconeogenesis.
   e. The liver is the primary target organ for glucagon.

6. Insulin inhibits activity of:
   a.   GK.
   b.   HSL.
   c.   HK.
   d.   LPL.
   e.   All of the above

ANSWERS

6. b
5. a
4. e
3. c
2. b
1. d

Chapter 74

Starvation (The Early Phase)

Overview

- Hepatic glycogenolysis and gluconeogenesis, and adipose tissue lipolysis provide most of the fuel for the early stages of starvation.

- The rate-control of hepatic gluconeogenesis during starvation is governed largely by substrate availability, as well as by the insulin:glucagon ratio.

- Muscle tissue reduces its dependency upon glucose during starvation as FFAs and KB⁻s become available.

- Circulating VLDLs tend to rise during the early-intermediate stages of starvation.

- As OAA is used for gluconeogenesis in liver tissue, less is available to couple with acetyl-CoA for citrate formation. Excess acetyl-CoA may then be forced into KB⁻ formation.

- The progressive increase in serum KB⁻ levels during starvation is due largely to reduced oxidation of these substrates by muscle tissue.

- The brain appears to function well when using predominantly KB⁻s as fuel.

- Significant reductions in pituitary gonadotropin output may occur in starved animals.

The pattern of fuel production during **early starvation** (2-3 days) is characterized by **hepatic glycogenolysis** and **gluconeogenesis**, and **adipose tissue lipolysis (Fig. 74-1)**. Nervous tissues continue to be an important sink for glucose, yet mitochondrial-rich tissues like kidney and muscle, that can utilize protein-bound free fatty acids (FFAs) from the circulation, begin oxidizing them, along with glucose for energy purposes. Knowledge of the contribution of muscle glycogen to fuel utilization as a direct result of starvation without exercise is limited.

The glycogen content of muscle after a meal may be as high as **1%**. With exercise it is rapidly diminished, and, if starvation ensues, it is not easily replenished. At a glycogen content of about 1%, the 28 kg of muscle in a normal

65 kg mammal would provide only about 286 gm of glycogen (or 1143 kCals of energy). This is a nominal supply, thus requiring muscle to import alternative fuels for energy purposes during starvation, or alternately to breakdown its extensive protein and modest triglyceride (TG) reserves.

An overall accounting of energy as either glycogen or TG is necessary at this point in order to provide perspective into fuel economy and mobilization during starvation (**Fig. 74-2**). A normal active mammal utilizes energy at the rate of about 21-26 kCal/day/Kg body weight (BW). With increased physical activity and/or cold exposure, this could increase to 70-90 kCal/day/Kg BW. Thus, the approximate 100,286 kCals available in the neutral fat of adipocytes can potentially provide several weeks of

Copyright © 2015 Elsevier Inc. All rights reserved.

Early Starvation

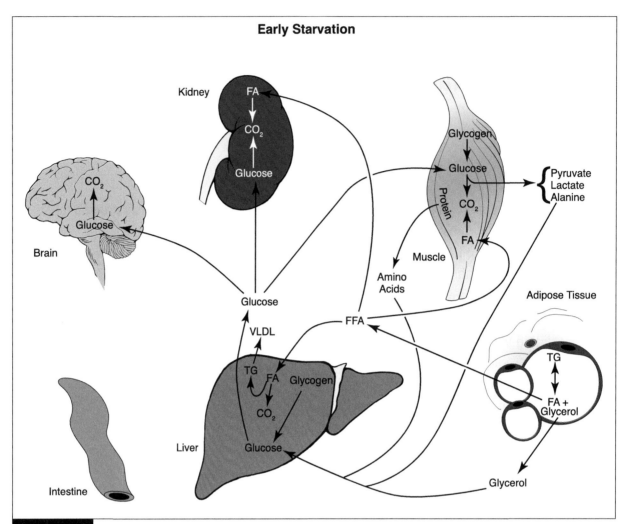

Figure 74-1

survival-fuel, depending upon physical activity. In obese animals, survival would be much longer.

Although **Fig. 74-1** indicates that **hepatic glucose production** during early starvation is primarily due to **glycogenolysis**, **gluconeogenesis** also contributes. From about 4-24 hours following the postabsorptive phase, gluconeogenesis continues to increase, reaching a peak level at about 2 days (**Fig. 73-4**). The time-scale for these transitional periods would, however, be species-dependent.

The Gluconeogenic Phase of Starvation

Although there are few data concerning liver glycogen levels during starvation, extrapo-

lating backward from the approximate 4-5% glycogen content in liver derived from a carbohydrate-rich meal, the amount may be as much as 10% or more in animals like dogs who store significant amounts of hepatic glycogen. In ruminant animals and carnivores, hepatic glycogen levels are negligible (see Chapter 23). Thus, liver glycogen in some animals could be expected to maintain the blood glucose concentration for about 12-16 hours, and studies in the postabsorptive state show glycogen providing 75% of splanchnic glucose output; gluconeogenesis provides the remainder (see **Fig. 74-3**).

As **gluconeogenesis** is initiated during the early phase of starvation, a number of metabolic adjustments occur in liver. These

Figure 74-2				
Approximate Energy Stores in a 65 Kg Mammal				
Tissue	**Glycogen**		**Triglyceride**	
	gm	kCal	gm	kCal
Adipose	23	93	11,143	100,286
Liver	70	278	9	83
Striated Muscle	286	1,143	232	2,136
Total	379	1,514	11,384	102,505

changes result from two processes; the first involves a **lower insulin:glucagon ratio** due to a significant decrease in pancreatic insulin output, thereby increasing levels of hepatic cAMP (see Chapter 23). In the second process, incoming free fatty acids (FFAs) undergo fat oxidation, thus producing increased levels of fat-derived products such as acetyl-CoA and fatty acyl-CoA (see Chapter 55). As enzyme and co-factor activities change over the ensuing 12-24 hours, the rate of liver glucose output begins to be controlled by the level of substrate coming to it, and the rate-control is thus transferred from liver to the release of precursors

from peripheral tissues. Notable among these are muscle-derived amino acids, whose mobilization must be diminished if the animal is to survive.

Over the ensuing 2-3 days of starvation, **muscle** tissue becomes progressively more efficient by **decreasing** its **glucose utilization**, both by blocking glucose uptake, and as a further check, by reducing glycolysis of glucose to pyruvate (see Chapter 75). Mitochondrial conversion of pyruvate to acetyl-CoA in muscle is also reduced because of the inhibition of **pyruvate dehydrogenase** by ever increasing mitochondrial levels of

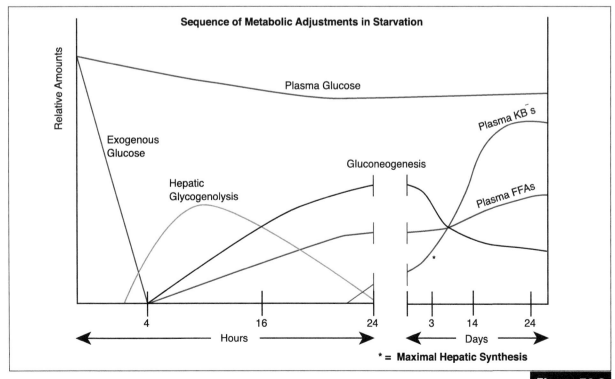

Figure 74-3

NADH, ATP, and **acetyl-CoA** derived from the **β-oxidation of FAs** (see Chapters 27 and 55). Therefore, what little pyruvate is formed is either immediately exported to the circulation, transaminated to alanine and then sent back to the liver for gluconeogenesis (see Chapter 9), or, depending upon the amount of aerobic muscle contraction, it may be converted to oxaloacetic acid (OAA) in order to couple with acetyl-CoA (from β-oxidation) in the formation of citrate. Thus, under non-exercising conditions, muscle progressively reduces its carbohydrate utilization in starvation, sparing this compound for use by the nervous system and other obligatory glucose users.

During starvation FFA removal by the liver results in some reesterification to TG with consequent **very low-density lipoprotein (VLDL)** formation. Thus, circulating VLDL, particularly in horses, tends to rise during early starvation (see Chapter 67). Energy needs of the liver are largely met through β-oxidation of the remaining FAs, but as starvation progresses, and hepatic OAA is utilized more for gluconeogenesis (see Chapter 37), less is available for tricarboxylic acid (TCA) cycle activity. Furthermore, what OAA is available is reduced to malate and the liver can be flooded with FAs and reducing equivalents resulting from their oxidation. Thus, as FAs are dehydrogenated and split into 2 carbon units of acetyl-CoA, the diminished acceptor, OAA, decreases **acetyl-CoA** entry into the TCA cycle, and the liver's alternative is to export the acetyl-CoA, two at a time, as **acetoacetate, β-hydroxybutyrate**, and **acetone** (the **ketone bodies (KB⁻s)**; see Chapter 71). Hepatic KB⁻ production thus appears to be the result of three phenomena during starvation:

1) The insulin:glucagon ratio being low, thus increasing activity of the enzymatic machinery for KB⁻ production;

2) An increase in the delivery of FFAs from adipose tissue to the liver; and

3) Use of OAA for gluconeogenesis.

To summarize the **gluconeogenic phase** of early starvation, muscle proteolysis and adipose lipolysis provide needed substrates as liver enzymes become poised toward glucose synthesis, and incoming FAs are reesterified to TG and exported as VLDL, or they become hepatic fuel. But eventually FAs are only partly oxidized to acetate units, and then exported to the periphery two at a time as acetoacetate, β-hydroxybutyrate, and acetone (see Chapter 75). Adipose tissue hydrolyzes its TG to FFAs and glycerol (lipolysis), the former serving as fuel for muscle, kidney, and liver, and the latter, like amino acids, lactate, and pyruvate from muscle, serving as gluconeogenic substrates for the liver.

By approximately the third day of starvation, KB⁻ production by the liver is usually maximal, but blood levels continue to increase progressively until the end of the second week when a plateau is achieved. It has been shown that this progressive increase is mainly a function of decreased KB⁻ catabolism by muscle as starvation progresses (see Chapter 75). Thus, by the third or fourth day of starvation, KB⁻ concentrations in blood may be 1-2 mM, but by the second week they may reach 6-10 mM (species dependent). Serum HCO_3^- concentrations are reduced accordingly, and there is a mild but compensated **metabolic acidosis** which can lead to increased urinary excretion of Ca^{++} and $H_2PO_4^-$ beyond that lost from catabolism of lean tissue (see Chapter 87). Thus, modest amounts of bone mineral may be dissolved.

A beneficial result of KB⁻ production during starvation is the concentration gradient that is established for the facilitated diffusion of these water-soluble fat products across the placental- and blood-brain-barriers (see Chapter 71). Again, this production of KB⁻s by the liver is a result of a **low insulin:glucagon ratio**, and an increase in FA release from adipose tissue, another result of low insulin.

The **brain** appears to function well when using predominantly **KB⁻s as fuel**. Intellectual function in humans appears to remain intact, but emotional alterations have been noted, such as depression or lability. Animals have been observed to reduce spontaneous activity, obviously as a means of sparing calories, but physical activity, when required, need not be compromised. The hypothalamic area of the brain appears to be significantly altered in the ketosis of prolonged starvation. Appetite is diminished, which may be part of the success of ketogenic diets used to treat obesity. More dramatically, the desire for fluid intake is diminished, and libido is reduced. Significant reductions in pituitary gonadotropin (LH and FSH) output have been reported, and females may become anovulatory. In males, there may be decreased testicular function.

OBJECTIVES

- Compare the glycogen and TG contents of the liver to those of adipose and muscle tissue.

- Explain why muscle conversion of pyruvate to acetyl-CoA is reduced during the early phase of starvation (see Chapter 27).

- Know why hepatic gluconeogenesis is increased during the postabsorptive phase, but then decreased during the early-intermediate phases of starvation.

- Explain why equine plasma becomes lactescent during the early phase of starvation (see Chapter 67).

- Recognize how and why the insulin:glucagon ratio and the hepatic availability of OAA influence KB⁻ production during starvation.

- Identify the primary gluconeogenic substrates during the early phase of starvation.

- Explain how and why the ketoacidosis of starvation results in increased urinary Ca⁺⁺ and $H_2PO_4^-$ excretion, an increased plasma AG, and a decreased plasma HCO_3^- concentration (see Chapters 85-87).

- Understand why fluid intake is usually reduced during the early to intermediate stages of starvation (see Chapter 55).

QUESTIONS

1. **The pattern of fuel production during early starvation is characterized by:**
   a. Hepatic glycogenolysis and adipose tissue lipolysis.
   b. Ketone body production and hepatic glycolysis.
   c. Adipose tissue lipolysis and hepatic chylomicron formation.
   d. Gluconeogenesis in the liver and in muscle tissue.
   e. FFA oxidation in brain tissue.

2. **During starvation, hepatic glycogen could be expected to maintain the plasma glucose concentration for about:**
   a. One hour.
   b. Fifteen hours.
   c. Two days.
   d. Twenty days.
   e. Two months.

3. **As muscle tissue begins oxidizing fatty acids for fuel during starvation:**
   a. The mitochondrial conversion of acetyl-CoA to acetoacetic acid increases.
   b. VLDL production by muscle cells increases.
   c. Mitochondrial conversion of pyruvate to acetyl-CoA increases.
   d. Carbohydrate utilization decreases.
   e. Mitochondrial NADH and ATP production decreases.

4. **The rise in the serum ketone body concentration as starvation progresses is primarily due to:**
   a. Increased hepatic production.
   b. Decreased muscle utilization.
   c. Decreased brain utilization.
   d. Increased lipolysis.
   e. Decreased hepatic gluconeogenesis.

5. **Select the FALSE statement below:**
   a. The brain appears to function well when using predominantly ketone bodies as fuel.
   b. Physical activity is impeded when using ketone bodies as fuel.
   c. Ketogenic diets may diminish the appetite.
   d. Ketosis is associated with a decrease in pituitary gonadotropin output.
   e. Hepatic gluconeogenesis usually declines somewhat following a few days of starvation.

ANSWERS
1. a
2. b
3. d
4. b
5. b

Chapter 75

Starvation (The Intermediate Phase)

Overview

- The intermediate phase of starvation is best characterized by the carbohydrate- and nitrogen-sparing effect of fat.
- During starvation KB⁻s help to restrain glucose utilization by a number of tissues, and they also help to restrain muscle proteolysis, and adipose tissue lipolysis.
- KB⁻s can provide up to 70% of the brain's energy supply during starvation.
- As AcAc is removed by muscle tissue during starvation, it is sometimes returned to the circulation in a more reduced state as β-OH-butyrate.
- During starvation β-OH-butyrate stimulates modest pancreatic insulin release, and it increases sensitivity of adipose tissue to insulin.
- Insulin helps to restrain proteolysis in muscle tissue, and lipolysis in adipose tissue during starvation.
- The basal metabolic rate (BMR) normally decreases during starvation due to a decrease in the active forms of thyroid hormones.

The basic chemistry of carbohydrate and lipid metabolism has been described in previous chapters. With starvation progressing into the **intermediate phase**, it now becomes possible to discuss the integration of metabolic pathways between several different tissues; specifically the integration of carbohydrate and lipid metabolism in **brain, muscle, adipose tissue,** and **liver**. A more complete description of metabolic integration must necessarily include consideration of protein and amino acid metabolism, and extend the discussion to other tissues as well, including the **kidneys.**

The intermediate phase of starvation is best characterized by the **carbohydrate- and nitrogen-sparing effect of fat.** Explanations for these effects lie in unique interrelationships between **glucose, free fatty acid (FFA), ketone body (KB⁻),** and **muscle protein metabolism (Fig. 75-1).**

As mentioned in the previous chapter, early in starvation KB⁻s serve as muscle fuel, but with more prolonged starvation they are less well utilized. As **acetoacetate (AcAc)** is removed by muscle, it is sometimes returned back into blood as **β-OH-butyrate,** signifying a more reduced state of muscle mitochondria secondary to FFA utilization (see Chapter 71). Thus, FFAs appear to take preference over KB⁻s as muscle fuel during the intermediate stages of starvation (**Fig. 75-2**). This effect spares KB⁻s for utilization by nervous tissue, a superb overall survival process.

The reduced state of **muscle** mitochondria due to FFA oxidation also results in a lower rate of amino acid release from proteolysis. A direct inhibitory effect of KB⁻s on muscle proteolysis has also been demonstrated. Thus, **fat,** in the form of both FFAs and KB⁻s, **spares oxidation of amino acids in muscle tissue,** which appears

Copyright © 2015 Elsevier Inc. All rights reserved.

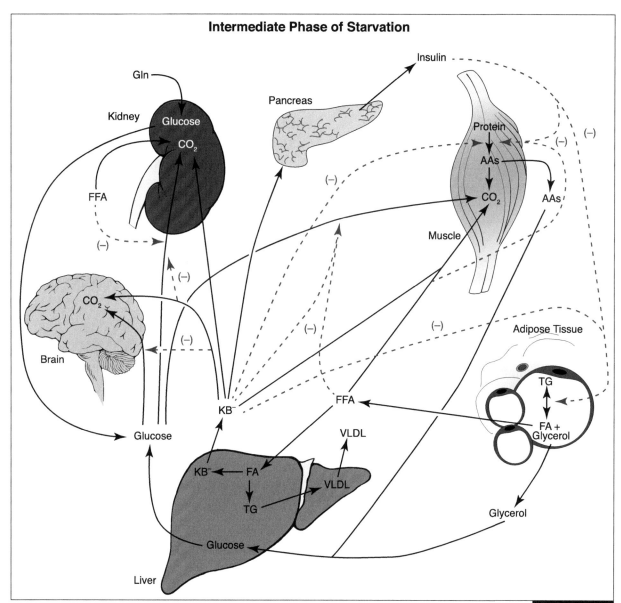

Intermediate Phase of Starvation

Figure 75-1

to play a central role in maintaining muscle protein reserves during starvation. The **blood urea nitrogen (BUN)** concentration is typically **lower** during the intermediate than during the early gluconeogenic phase of starvation, and total urinary nitrogen excretion decreases (**Fig. 75-3**). Feeding only a small amount of energy as protein can sometimes replace the nitrogen depletion of otherwise total starvation, a concept emphasized by investigators who propose that endogenous fat, if allowed to be mobilized by not giving carbohydrate, may be efficient in sparing nitrogen in patients receiving parenteral alimentation.

Ketone bodies and **FFAs** also help to **restrain the uptake and utilization of glucose by muscle, the renal cortex, lactating mammary gland, small intestine,** and **nerve tissue.** Through this reduction in glucose utilization, gluconeogenic organs like the liver and kidneys are under less pressure to convert gluconeogenic (and proteogenic) amino acids to glucose. Under normal circumstances the kidneys provide less than 10% of glucose production. However, in prolonged starvation, as overall gluconeogenesis decreases (see Chapter 74), the component provided by liver decreases dramatically while that of the kidneys increases (**Figs. 75-4 &**

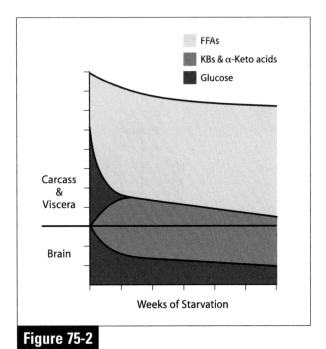

Figure 75-2

75-5). The explanation for this shift involves ketoacidosis, an increase in hepatic nitrogen production as **glutamine (Gln)** rather than **urea** (see Chapter 10), and renal utilization of Gln for gluconeogenesis and urinary **NH$_4^+$** disposal (see Chapter 11).

The combination of increased blood KB$^-$ levels and the derepression of enzymes required for their utilization as fuel (**β-hydroxybutyrate dehydrogenase, β-ketoacid-CoA transferase**, and **acetyl-CoA acetyltransferase**; see Chapter 71) enables the brain in starvation to use KB$^-$s instead of (and in addition to) glucose. Studies indicate that KB$^-$s (particularly **β-OH-butyrate**) can provide up to **70% of the energy requirements of the brain** during prolonged starvation, with a substantial reduction in simultaneous glucose utilization **(Fig. 75-6)**.

Because of this reduced dependency upon glucose catabolism, the threshold concentration at which hypoglycemic shock will ensue is also reduced. However, a diabetic, hyperglycemic and ketotic animal may not experience this reduced threshold concentration since the brain continues to oxidize glucose over KB$^-$s in hyperglycemia. Thus, the enzymes required for KB$^-$ oxidation are less active. If hypoglycemia is induced through insulin injection in diabetic, ketotic, hyperglycemic animals, the brain cannot convert to KB$^-$ utilization rapidly enough to prevent hypoglycemic shock from occurring at slightly higher plasma glucose levels.

During starvation it is important that the rate of FFA mobilization from adipose tissue be precisely related to the ability of aerobic tissues to remove and metabolize these substrates (particularly muscle and liver tissue). Undoubtedly, changes in the concentrations of several hormones play an important role in the regulation of FFA mobilization, but it is unlikely that they alone can provide the precision necessary to meet rapid changes in energy demand. The rate of **glyceroneogenesis** from pyruvate in adipocytes is thought to play an important role in the re-esterification of FFAs generated in lipolysis, thus modulating their release when cAMP levels are high. Up to 60% of the FFAs generated during lipolysis may enter into re-esterification via this pathway (see Chapter 70).

Investigators have determined that KB$^-$s, particularly **β-OH-butyrate**, also play an

Figure 75-5				
Early vs. Intermediate - Late Starvation				
Starvation Phase	**Glucose Production**			**Urinary Nitrogen Excretion**	**Fuel for Brain**
	Rate	**Liver**	**Kidney**		
Early (Postabsorptive)	High	> 90%	< 10%	Urea > NH$_4^+$	Glucose > KB$^-$s
Intermediate - Late	Lower	55%	45%	NH$_4^+$ > Urea	KB$^-$s > Glucose

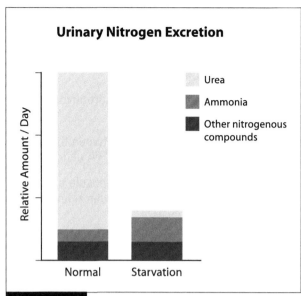

Figure 75-3

important role in preventing excessive lipolysis during starvation. As **NADH:NAD⁺** ratios in liver and muscle mitochondria begin to increase because of excessive FFA β-oxidation, increased quantities of β-OH-butyrate are generated from acetoacetate. **β-OH-Butyrate** has been shown to exert three important metabolic effects which tend to modulate blood levels of FFAs:

- **Reduce the rate of adipose tissue lipolysis.**
- **Stimulate modest pancreatic insulin release.**
- **Increase sensitivity of adipose tissue to insulin.**

Although all three effects could simultaneously reduce the rate of lipolysis, arguments have been forwarded to suggest that the increase in insulin sensitivity may be quantitatively the most important.

As shown over 50 years ago, in addition to facilitating glucose uptake into muscle and adipose tissue, and inhibiting lipolysis, **insulin** also initiates uptake of certain amino acids into muscle, and facilitates their incorporation into protein. A more recently described effect of insulin is to decrease muscle proteolysis. Thus insulin, like KB⁻s, appears to help control the rate of net muscle proteolysis during starvation.

A decrease in the **basal metabolic rate** (**BMR**) also occurs with starvation. Part of this reduction is explained by the progressively decreasing body mass. However, selective use of fat as fuel, everything else being equal, would be expected to increase O_2 consumption, since fat, unlike glycogen, is a strictly "aerobic" fuel. However, **total O_2 consumption** normally **decreases** about **10-15%** during the intermediate phase of starvation. The explanation for this decrease is found in circulating thyroid hormone levels, most notably **triiodothyronine** (**T_3**, the most active form), and **reverse T_3** (**rT_3**, the inactive form). Circulating **T_3** levels progressively decrease during starvation, while **rT_3** levels increase. When starved animals are appropriately refed with carbohydrate, these triiodothyronine levels reverse, and the active

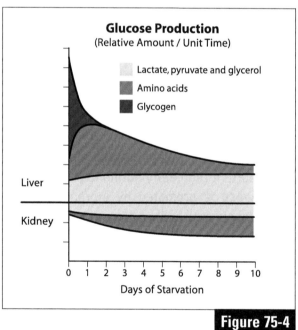

Figure 75-4

form returns within the euthyroid range.

An additional parameter that may change during starvation is the plasma unconjugated bilirubin (UCB) concentration. The liver exhibits a reduced capacity to remove UCB from plasma during starvation, particularly in equine species, thus promoting an **unconjugated hyperbilirubinemia** (see Chapter 33).

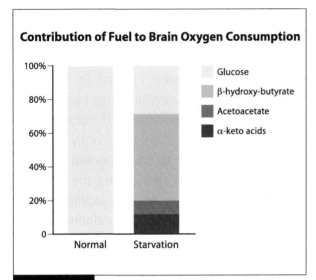

Contribution of Fuel to Brain Oxygen Consumption

Legend:
- Glucose
- β-hydroxy-butyrate
- Acetoacetate
- α-keto acids

X-axis: Normal, Starvation

Figure 75-6

In summary, starvation entails a progressive selection of fat as body fuel. KB⁻s become elevated during the early-intermediate phases of starvation, and tissues preferentially use these fuels for energy purposes, thus reducing their dependency upon glucose. Additionally, KB⁻s also inhibit muscle proteolysis. Nevertheless, starvation is usually associated with a negative nitrogen balance that can be partially nullified by amino acid or protein supplementation. **Insulin** appears to be an important regulatory hormone in starvation. Its release is promoted by increased circulating levels of **β-hydroxybutyrate**, and it helps to spare both fat and protein from excessive breakdown. Studies indicate that decreased circulating levels of active T_3 (and increased levels of the inactive rT_3) may play a role in sparing otherwise obligatory caloric depletion through decreasing the BMR. Once the supply of FFAs from depot fat has been consumed, the basal energy requirement of the organism must be met from **body protein**. It is at this time that the muscle mass, the largest single source of protein, can be heavily depleted. This third and final phase of starvation will have a most unfortunate, inevitable end unless nutritional intervention occurs.

OBJECTIVES

- Discuss the integration of carbohydrate, protein and lipid metabolism during the intermediate phase of starvation.

- Explain what is meant by the "carbohydrate- and nitrogen-sparing effect of fat."

- Understand the relationship between the AcAc:β-OH-butyrate ratio and FFA utilization.

- Identify the effects of KB⁻s on muscle proteolysis and glucose utilization, insulin release and adipose tissue lipolysis.

- Know how and why the BUN concentration changes throughout starvation.

- Explain why the proportion of glucose produced by the kidneys increases during the intermediate stage of starvation (see Chapter 11).

- Understand why the hypothalamic hypoglycemic threshold of a non-diabetic animal is reduced during the intermediate stage of starvation.

- Describe factors involved in the rate of FFA mobilization from adipocytes during starvation (see Chapter 70).

- Discuss the control of insulin release and the effects of this hormone on different tissues throughout starvation.

- Show how and why the BMR changes throughout starvation.

- Explain how and why urinary nitrogen excretion changes throughout starvation (see Chapter 11).

- Explain what an α-keto acid is (see Chapter 9).

- Understand how and why lipid fuels feedback negatively on muscle protein catabolism during starvation.

- Explain how the "rate" of lipolysis is controlled during starvation (see Chapters 70 & 71).

- Recognize why the rate of renal gluconeogenesis increases during the intermediate stages of starvation (see Chapter 11).

- Understand how and why the β-OH-butyrate concentration influences fat mobilization and muscle protein catabolism.

- Identify the tissues that experience reduced uptake and utilization of glucose when the plasma KB⁻ and FFA concentrations are elevated.

QUESTIONS

1. During the intermediate phase of starvation, FFAs:
a. Become preferred fuel over KB⁻s in muscle tissue.
b. Restrain muscle proteolysis.
c. Restrain the uptake of utilization of glucose by muscle tissue.
d. Restrain the uptake of utilization of glucose by the renal cortex.
e. All of the above

2. The threshold concentration at which hypoglycemic shock will occur is reduced during the intermediate phase of starvation, because:
a. Starvation reduces pancreatic insulin output.
b. The brain is deriving most of its metabolic fuel from circulating FFAs.
c. The brain has an increased dependency upon glucose.
d. The enzymes required for KB⁻ oxidation in the brain are inactive.
e. Glucose utilization by the brain is normally reduced during this phase.

3. During starvation the reduced state of muscle mitochondria secondary to FFA utilization is exemplified by:
a. Conversion of acetoacetate to β-OH-butyrate.
b. Increased aerobic oxidation of glucose.
c. Increased ketone body oxidation.
d. Reduced oxygen utilization.
e. The breakdown of muscle protein.

4. Approximately what percentage of the brain's energy requirement can be met through ketone body oxidation during the intermediate stages of starvation?
a. 2%
b. 20%
c. 45%
d. 70%
e. 95%

5. β-OH-Butyrate has been shown to exert all of the following metabolic effects, EXCEPT:
a. Stimulate hepatic and renal gluconeogenesis.
b. Reduce the rate of adipose tissue lipolysis.
c. Stimulate modest pancreatic insulin release, and increase sensitivity of adipose tissue to insulin.
d. Reduce glucose utilization by muscle, nerve, and renal tissue.
e. Reduce muscle proteolysis.

6. Select the FALSE statement below regarding the intermediate stages of starvation:
a. Reverse T₃ levels in plasma typically increase.
b. Derepression of enzymes required for ketone body oxidation in the brain occurs.
c. Oxygen consumption normally decreases (during resting stages).
d. Amino acids are not used by the liver for gluconeogenic purposes.
e. T₃ levels in plasma typically decrease.

7. During the intermediate stages of starvation, ketone bodies and free fatty acids help to restrain the uptake and utilization of glucose by all of the following, EXCEPT:
a. Erythrocytes.
b. Renal cortex.
c. Small intestine.
d. Nerve tissue.
e. Lactating mammary gland.

8. Compared to the early postabsorptive phase, the intermediate-late phase of starvation shows:
a. Greater glucose utilization by the brain.
b. Greater urinary urea excretion.
c. A lower rate of hepatic glucose production.
d. A lower rate of hepatic KB⁻ production.
e. A higher BMR.

ANSWERS

8. c
7. a
6. d
5. a
4. d
3. a
2. e
1. e

Chapter 76

Starvation (The Late Phase)

Overview

- There is a definite sequence in which body proteins are lost during starvation in order to preserve the blood glucose concentration.
- The late phase of starvation is characterized by increased muscle loss, an increased BUN, and an increase in hepatic and renal gluconeogenesis.
- Muscle tissues prefer to oxidize the BCAAs during the late phases of starvation.
- The brain oxidizes both glucose and branched-chain amino acids during the late phase of starvation.
- Intestinal mucosal cells prefer to oxidize Gln during periods of repair, and during starvation.
- Catabolic processes (e.g., proteolysis, lipolysis, and glycogenolysis) are generally exacerbated in cachexic animals.
- Death from starvation can result from azotemia, pneumonia, or hypovolemic shock.
- The hibernating black bear appears to remain in near perfect water and nitrogen balance, with only a modest fall in body temperature.

The length of the **intermediate phase** of starvation will be dependent upon the amount of physical exercise the animal experiences, the general health of the animal, the species of animal, and also the amount of fatty tissue available. As outlined in the previous chapter, this period of starvation is characterized by rather low rates of carbohydrate and protein metabolism, and a higher rate of fat metabolism. Once fat reserves become depleted, circulating **free fatty acid (FFA)** and **ketone body (KB⁻)** concentrations **decline**. Finally, the most severe phase of starvation commences. The basal energy requirement of the organism must now be met from **body protein**. Mobilized **amino acids** are used for energy purposes by either being **oxidized directly**, or by being converted to **glucose** in either the liver or kidney. It is at this time that the muscle mass, the largest single source of protein, is heavily catabolized. This characterizes the **late phase** of starvation, which ends with either refeeding, or death.

Sequence of Body Protein Depletion

There is a definite sequence in which body proteins are lost during starvation in order to preserve the blood glucose concentration. Those lost first are the **digestive enzymes** secreted by the stomach, pancreas, and small intestine (see **Appendix**); they are no longer needed, nor are other enzymes and proteins that are required to synthesize digestive enzymes. Also lost early are various enzymes of the **liver** that normally process incoming nutrients from the intestine, and convert them into plasma

Copyright © 2015 Elsevier Inc. All rights reserved.

proteins, lipids, and lipoproteins. Then begins the drain on **muscle proteins**, not only those of contractile fibers, but also the glycolytic enzymes. When muscle proteins begin to be catabolized, starved animals become physically inactive, another **beneficial** physiologic adaptation. Thus, the starved organism makes a series of calculated choices in the utilization of body protein, largely for the benefit of the nervous system, which is either partially (intermediate phase) or largely dependent (late phase) upon glucose.

Muscle: At least six amino acids can be oxidized by muscle, namely **alanine (Ala)**, **aspartate (Asp)**, **glutamate (Glu)**, and the three branched-chain amino acids (**BCAAs**), **leucine (Leu)**, **isoleucine (Ile)**, and **valine (Val)**. Studies have shown that during the late phase of starvation, about **60%** of amino acid release from muscle consists of **Ala** and **Gln** (**Fig. 76-1**), which was inexplicable due to their content in contractile proteins (which is about 10%). Further studies indicated that as the BCAAs are oxidized by muscle tissue, their amine groups are transaminated onto pyruvate and α-ketoglutarate (α-KG$^=$), thus forming Ala and Gln, respectively. **Alanine** becomes a primary gluconeogenic substrate for the **liver**, and **Gln** serves a similar purpose in the **kidney**.

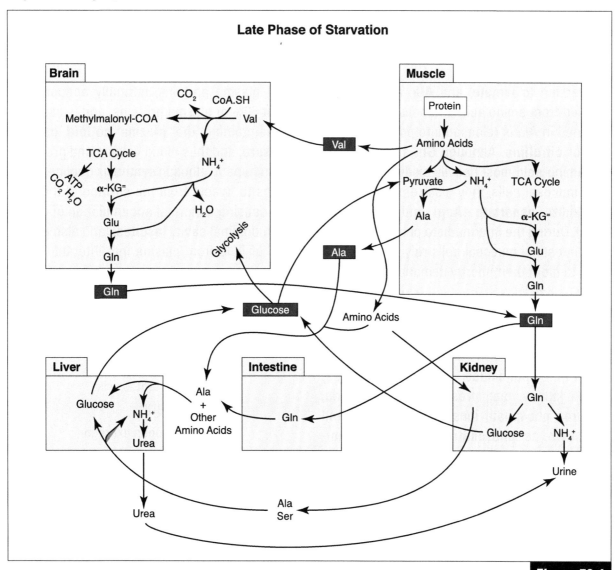

Late Phase of Starvation

Figure 76-1

Brain: During the late phase of starvation, the brain must revert to using **glucose** as a primary energy substrate; however, it can also oxidize amino acids for energy purposes. The primary amino acid oxidized appears to be **Val**, which is mobilized from muscle tissue. This process requires transamination of the amine group from Val onto α-KG$^=$, thus forming Glu, or transamination onto Glu, thus forming Gln (see Chapter 9). Indeed, **Gln formation** appears to be the primary mechanism by which the brain rids itself of **ammonia (NH$_3$)**.

Intestine: Although the intestine is less active during starvation than it is during feeding, intestinal mucosal cells still retain a fairly high metabolic rate and thus turnover rate, which requires energy. Metabolism during the late phase of starvation appears to consist largely of amino acid oxidation to CO_2 and H_2O, or conversion to **lactate**, and **Ala**. Any nitrogen derived from amino acid catabolism that is not retained in Ala is released into portal blood as **NH$_3$** or **citrulline**. Although **Gln** appears to be the amino acid most heavily oxidized by intestinal mucosal cells, they also retain the ability to oxidize **aspartate (Asp)** and **asparagine (Asn)**. During the intermediate phase of starvation, intestinal mucosal cells rely heavily upon **KB$^-$s** to meet their energy demands.

Kidney: The ammonium ion (**NH$_4$$^+$**) appearing in urine during starvation is derived largely from the breakdown of **Gln** in proximal renal tubular cells (see Chapter 11). There are two separate glutaminase enzymes (isozymes) in the kidney; one, present in mitochondria, requires phosphate ions for full activity and is known as **phosphate-dependent glutaminase**; the other is active in the absence of phosphate, and is localized on the luminal side of proximal tubular epithelial cells. The mitochondrial enzyme is considered to play a major role in NH$_4$$^+$ production from Gln. The α-**KG$^=$** generated from deamination (and deamidation) of Gln either enters renal **gluconeogenesis**, or

it can be converted through the TCA cycle and oxidative phosphorylation to **CO$_2$** and **H$_2$O**.

Liver: The liver is quantitatively the most important sink for amino acids during the late phase of starvation. It must maintain **glucose output** from glucogenic amino acids in order to meet the needs of glucose-dependent tissues (e.g., erythrocytes and the CNS), and it must perform this task while also attempting to maintain its output of **plasma proteins**. Therefore, amino acids in the liver compete for these two primary processes (i.e., **gluconeogenesis** and **protein synthesis**). As the amino acid pool declines and glucogenic amino acids are siphoned off into glucose production, the **blood urea nitrogen (BUN)** concentration increases (as does urinary urea excretion), and the plasma protein pool begins to decline (**Fig. 76-2**). **Albumin**, one of several liver-derived plasma proteins, normally accounts for about 60% of plasma proteins, and is essential in maintaining the **plasma colloid osmotic pressure**, and thus blood volume and pressure. When hepatic albumin synthesis declines, fluid begins to "weep" from the surface of the liver, thus causing abnormal accumulation of fluid in the abdominal cavity (**ascites**), and also **edema** (loss of fluid from plasma into interstitial fluid

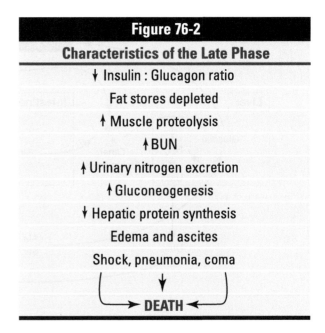

Figure 76-2

Characteristics of the Late Phase

↓ Insulin : Glucagon ratio

Fat stores depleted

↑ Muscle proteolysis

↑ BUN

↑ Urinary nitrogen excretion

↑ Gluconeogenesis

↓ Hepatic protein synthesis

Edema and ascites

Shock, pneumonia, coma

→ DEATH ←

spaces of the body). This, in turn, can cause a severe **blood volume depletion**.

Starvation and Death

There are several reasons why starvation eventually leads to death. One common cause is **pneumonia**. Because of loss of myofibrillar protein from the diaphragm and intercostal muscles, there may be inadequate removal of fluid from the bronchioles and lungs so that the lung is open to infection. Furthermore, the immunological response to an infection may be reduced. Another cause of death can be **shock** from blood volume depletion, which may be secondary to hypoproteinemia, edema, and ascites. A third cause may be from **coma** caused by blood volume depletion and azotemia (**Fig. 76-2**).

Starvation vs. Cachexia

Cachexia, unlike physiologic starvation, is associated with ill-health rather than simple food deprivation, and can lead to severe malnutrition, depression, and emaciation in a shorter period of time. Biochemical catabolic processes (e.g., glycogenolysis, lipolysis, and proteolysis) in cachexic animals are generally exacerbated compared to the more controlled processes described above. Pathophysiologic disorders associated with cachexia include heart disease, cancer, pituitary and/or thyroid disorders, parasitic infection, liver and kidney disease.

The Survivors

With their increased metabolic rate per unit mass, **small animals** adapt to starvation by dropping their body temperatures. This strategy has been observed in **rodents**, **bats**, **marsupials**, and **insectivores**, and even nocturnally in the **hummingbird**. Some larger animals, however, can meet the starvation challenge without much or any drop in metabolic rate. Three superbly studied examples are the **bear**, the **elephant seal**, and the **emperor penguin**.

The black bear dens throughout the winter months with only a minimal fall in body temperature, and neither eats, drinks, urinates, nor defecates. In other words, this animal appears to be in near perfect water balance with near 100% nitrogen economy. Additionally, only modest degrees of ketosis are observed in the hibernating black bear. It is believed that the **glycerol** generated from **lipolysis** provides an adequate supply of hydrocarbons for hepatic gluconeogenesis in this animal, which has a rather low brain/carcass ratio (see Chapter 37). A similar phenomenon is found in the elephant seal pup who fasts for 2-3 months after weaning, and the nesting male emperor penguin who waddles many miles to the nesting site, breeds with the female, waits two months for her to produce an egg, and then incubates the egg for another two months while he (and the chick) wait for the mother to return with a meal. It is no wonder this animal is called an "emperor."

In summary, for convenience starvation has been subdivided, somewhat arbitrarily, into 4 phases; the **postabsorptive stage,** then the **early, intermediate,** and **late phases.** The postabsorptive stage begins a few hours after the last meal, when contents of the small intestine have been absorbed. During this period blood glucose and insulin concentrations will initially rise, then fall resulting in a reduction in glucose uptake by the liver and peripheral tissues. Hepatic glycogen synthase activity will decline, and phosphorylase activity will increase; thus, hepatic glycogen will be broken down for a short period of time as the animal now moves into the early phase of starvation. During this period FFAs will be released from adipocytes, and the rate of glucose oxidation will be reduced in some tissues. During the intermediate phase of starvation, complex adjustments occur in the fuel supply. The rate of gluconeogenesis decreases as the KB^- concentration rises. The length of this phase will be somewhat dependent upon the basal metabolic rate and the size of the fat stores, for their ultimate depletion heralds the late phase of starvation. This phase is characterized by a higher rate of protein catabolism in order to meet the carbohydrate needs of tissues who no longer have FFAs and KB⁻s to oxidize. For every gram of glucose synthesized from amino acids during this phase, approximately 1.75 gm of protein must be degraded. The body is unable to withstand a loss of more than half of its muscle protein before serious consequences ensue. Therefore, prolonged starvation ends with either refeeding, or death.

OBJECTIVES

• Recognize the biochemical and physiologic variables that determine the length of starvation.

• Explain why digestive enzyme and hepatic protein formation decrease before muscle proteins are lost during the late phase of starvation.

• Know the relationship between ascites and hepatic protein synthesis.

• Understand why the bulk of amino acid release from muscle is in the form of Ala and Gln during the late phase of starvation.

• Know why Gln is the favored renal gluconeogenic amino acid.

• Describe how astrocytes of the brain buffer excess ammonia.

• Explain why Gln is frequently added to parenteral fluids following intestinal surgery.

• Identify the quantitatively most important sink for amino acids during the late phase of starvation.

• Provide physiologic reasoning for developing hypotension during the late phase of starvation.

• Understand differences between physiologic starvation and cachexia.

• Summarize primary features of the postabsorptive stage following the last meal, then the early, intermediate and late phases of starvation.

QUESTIONS

1. **The most severe, late phase of starvation is generally characterized by:**
   a. The nitrogen-sparing effect of fat.
   b. Decreased carbohydrate utilization by muscle and nerve tissue.
   c. Ketoacidosis.
   d. Muscle protein wasting.
   e. An increased insulin:glucagon ratio.

2. **In order to preserve the blood glucose concentration, proteins are sacrificed during starvation. Those lost first are the:**
   a. Immunoglobulins.
   b. Digestive enzymes secreted by the GI tract.
   c. Muscle proteins.
   d. Glycoproteins in plasma membranes.
   e. Mitochondrial enzymes associated with the TCA cycle.

3. **Which one of the following amino acids would most likely be oxidized by skeletal muscle tissue during the late phase of starvation?**
   a. Ile
   b. Ala
   c. Phe
   d. Gln
   e. Trp

4. **Which one of the following amino acids would most likely be oxidized by the brain during the late phase of starvation?**
   a. Gln
   b. Ala
   c. Val
   d. Glu
   e. Tyr

5. **The amino acid most heavily oxidized by intestinal mucosal cells during starvation is:**
   a. Ala.
   b. Phe.
   c. His.
   d. Met.
   e. Gln.

ANSWERS

5. e

4. c

3. a

2. b

1. d

Chapter 77

Exercise (Circulatory Adjustments and Creatine)

Overview

- Cardiac ejection efficiency increases remarkably during exercise.
- CPK isozymes (i.e., CPK_m, CPK_c, and CPK_g) participate in "shuttling" high energy phosphates toward myosin filaments during exercise.
- Creatinine is a natural break-down product of $C{\sim}PO_3$ in muscle tissue.
- Muscle typically contains two to three times more $C{\sim}PO_3$ than ATP.
- Adenosine, a break-down product of AMP during muscle contraction, is a vasodilator.
- ADP, AMP, Pi, and NH_4^+ activate phosphofructokinase (PFK) in exercising muscle.
- Arg, Gly, and Met are used by the kidney and liver in the biosynthesis of creatine.
- Uric acid is a purine degradation product formed in muscle tissue during exercise.
- Creatinine concentrations increase in plasma when the glomerular filtration rate is compromised.
- Creatinuria (i.e., increased amounts of creatine in urine) occurs with excessive muscle breakdown.

Since **exercise** involves not only the intricate neuromuscular coordination of body movement, but also many complex adjustments of metabolism, respiration, and circulation, practically the entire organism becomes involved in the physiologic adaptations to work (i.e., physical labor). The actual coordination of movement depends upon the nervous system, while excitation-contraction coupling processes in muscle require several complex physical changes fueled by both anaerobic and aerobic reactions. Tissues must be supplied with O_2, while CO_2 and other waste products of metabolism are removed in amounts proportional to the energy requirements for work. To meet those demands, the respiratory and circulatory systems must be precisely adjusted through local autoregulatory mechanisms, as well as by the actions of the nervous and endocrine systems. These systems may be set into action by a number of chemical, thermal, and mechanical stimuli associated with neuromuscular activity itself. Thus, an exercising animal must supply muscles with metabolic material from lungs, liver, adipose tissue and the intestine by way of the circulatory system, and maintain biochemical homeostasis by transporting metabolites of work to the lungs, liver, kidneys and skin for excretion.

Under favorable conditions the mammalian organism performs work (i.e., exercise) with an overall mechanical efficiency of about **20-30%**. Although animals may be considered efficient machines compared to the steam engine which has an approximate **10%** work efficiency, they are apparently about equal to the gasoline engine.

Copyright © 2015 Elsevier Inc. All rights reserved.

Circulatory Adjustments to Exercise

A large proportion of energy released by working muscles takes the form of heat, which can be dissipated through the skin and respiratory tract. The conduction of heat from exercising muscles to skin requires an increase in cutaneous blood flow, which places an added burden on the circulation, particularly if an animal is exercising in a hot environment. **Fig. 77-1** shows the character and magnitude of the circulatory adjustments that occur from rest to strenuous exercise. The **increased cardiac output (CO)** supplies an enormously elevated skeletal muscle blood flow, which provides for blood gas exchange and conducts away excess heat and waste products of metabolism. **Cutaneous blood flow** increases to favor body cooling, however, with maximal effort cutaneous vasoconstriction can overcome thermoregulatory vasodilator responses, and the core body temperature can rise. In dogs instrumented with regional flowmeters, brief bouts of exhausting exercise are found to cause a reduction in **gastrointestinal (GI) blood flow**, but little decrease in **renal blood flow**. With prolonged strenuous exercise, **visceral blood flow** (namely GI) eventually returns to restore adequate organ function. Perfusion of **brain** tissue does not change much following the onset of exercise.

Cardiac Adjustments to Exercise

Animals are capable of achieving remarkable adjustments in circulation by increasing **CO** to several times its resting level (**Fig. 77-2**

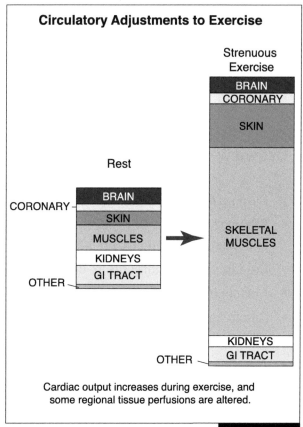

Circulatory Adjustments to Exercise

Cardiac output increases during exercise, and some regional tissue perfusions are altered.

Figure 77-1

and Chapter 81). In a 70 kg mammal at rest, **end-diastolic volume** may be **120-130 ml**, and following ventricular contraction almost 50% of this volume remains in the ventricles (i.e., **end-systolic volume**). However, during strenuous exercise **cardiac contractility** increases significantly (as does **heart rate**), and only about **10%** of the **end-diastolic volume** remains in the ventricles following systole. Thus, **cardiac ejection efficiency** increases markedly during exercise.

Figure 77-2		
Cardiac Adjustments to Exercise		
70 Kg Mammal		
Ventricular	**Rest**	**Strenuous Exercise**
End-diastolic volume	120-130ml	220-250 (2x)
Stroke volume	70	190-220 (3x)
End-systolic volume	50-60	10-30

Cardiac output (ml/min) = Heart rate (beats/min) x Stroke volume (ml/beat)

Creatine Phosphate (C~PO₃)

When an animal exercises, it normally expends energy at a rate proportional to the rate of movement. If it were to begin running at maximal speed, it could sustain that speed for only a few seconds, during which time it would be utilizing readily available **ATP** in muscle, and also new ATP derived from **anaerobic glycolysis** (i.e., muscle glycogenolysis; **Fig. 77-3**). As running speed begins to quickly recede, high energy phosphate from **C~PO₃** reserves in muscle are thought to sustain exercise in transition to the largely **aerobic phase** (see Chapter 78). Muscle typically contains two to three times more C~PO₃ than ATP. As the aerobic phase is achieved, exercise can be continued for a longer period of time at a relatively constant, yet slower pace, supplied by high energy phosphates generated from **oxidative phosphorylation** (see Chapter 36).

C~PO₃ is formed from **ATP** and **creatine** (**Fig. 77-4**) during exercise, and also at times when the muscle is relaxed and demands for ATP are reduced. The enzyme catalyzing phosphorylation of creatine in the **mitochondrial** inter-membrane space is **creatine phosphokinase** (**CPKₘ**; also called **creatine kinase (CKₘ)**), and a second **cytoplasmic CPK** (**CPKc** or **CKc**) nearer the myosin filaments catalyzes the reverse reaction. During aerobic exercise these two isozymes are thought to help **shuttle** high energy phosphates generated in mitochondria to myosin filaments for contraction. Thus, **C~PO₃** can function as a back-up source of high energy phosphates, and also as a shuttler of high energy phosphates during exercise. An additional **CPK** (**CPKg** or **CKg**) is thought to participate in the transfer of high energy phosphates from ATP to C~PO₃ during **anaerobic glycolysis** (see Chapter 80).

CPK is found in high concentration in brain tissue and in aerobic muscle fibers, and following **myocardial infarction** this enzyme leaks into the circulation (along with lactate dehydrogenase, **LDH**), and therefore specific isozymes of both are used as plasma markers for diagnosis of this condition (see Chapter 6).

As ATP is hydrolyzed by **myosin ATPase** in muscle contraction, **myokinase (adenylyl kinase)** sometimes catalyzes formation of one molecule of **ATP** and one of **AMP** from two molecules of **ADP** (**Fig. 77-4**). The AMP produced can be deaminated by **AMP deaminase**, forming **IMP** and **NH_4^+** (see Chapter 17). Thus, muscle is a source of NH_4^+, to be disposed of by the hepatic urea cycle (see Chapter 10). **Adenosine monophosphate** can also be acted upon by **5'-nucleotidase**, hydrolyzing the phosphate and producing **adenosine** (see Chapter 17). Adenosine, in turn, is a substrate for **adenosine deaminase**, producing **inosine** and **NH_4^+**. **Adenosine**, as a physiologically significant molecule itself, acts as a **vasodilator**, increasing blood flow and supply of nutrients to muscle. This autoregulatory form of vasomotion is thought to be particularly important to the **coronary vasculature**. Additionally, **ADP, AMP, Pi**, and **NH_4^+** formed during the various reactions described above activate **phosphofructokinase (PFK)**, thus increasing the rate of glycolysis in rapidly exercising muscle, such as during a short sprint (see Chapter 25).

The remaining steps in the degradative pathway of adenosine and IMP are those

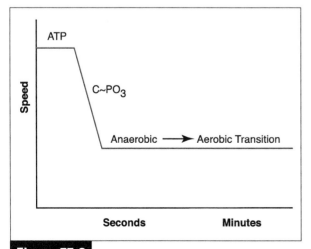

Figure 77-3

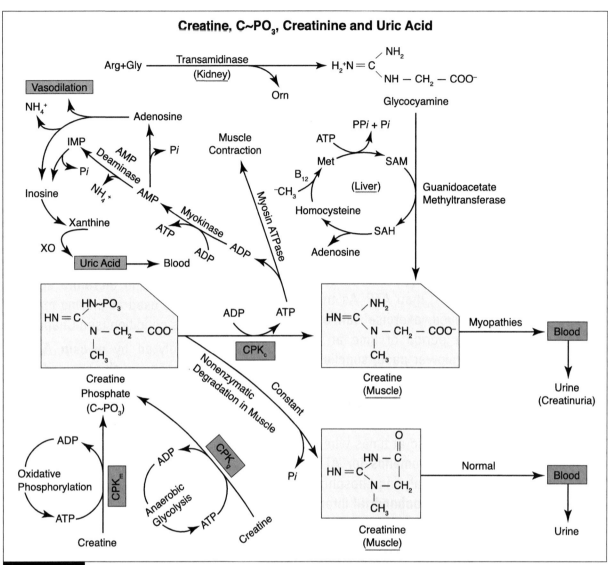

Figure 77-4

associated with **purine degradation**, with the ultimate end-product being **uric acid** (see Chapter 17). During exercise, uric acid production by muscle tissue increases, as does the production of certain reactive oxygen species. It has been hypothesized that uric acid may be beneficial in helping to scavenge some of these potentially harmful free radicals.

Creatinine and Creatine

Creatinine (creatine anhydride) is formed in muscle from **C~PO₃** by irreversible, nonenzymatic **degradation** and loss of **inorganic phosphate (Pi)**. The 24-hour urinary excretion

of creatinine is normally fairly constant, and proportionate to muscle mass. Creatinine is filtered by the kidneys, and when this process is compromised, creatinine concentrations in plasma increase.

Arginine (Arg), **glycine (Gly)**, and **methionine (Met)** all participate in **creatine** biosynthesis. Transfer of a guanidino group from Arg to Gly occurs in the **kidney**, thus forming **glycocyamine** (guanidoacetate). This compound passes into blood and is removed by the liver where creatine biosynthesis is completed by methylation of glycocyamine, using **SAM (S-adenosylmethionine)** as the methyl donor (see Chapter 57). This reaction is catalyzed by **guanidoacetate methyltransferase**. Methionine (and thus SAM) is normally reconstituted through methylation of homocysteine in the **vitamin B$_{12}$** requiring **methionine synthase** reaction involving **N^5-methyl-H$_4$ folate** (see Chapter 16).

Following **creatine** formation, the liver normally returns this compound to the circulation, where it can next be removed by muscle (and other tissues). During this process small amounts of creatine are normally filtered by the kidney, and thus appear in urine. When creatine is taken orally to boost athletic performance, or when heavy meat diets are consumed, greater amounts of creatine may appear in urine. **Creatinuria** (i.e., high amounts of **creatine** (not creatinine) in **urine**) is also sometimes seen in adolescence, in the dam during and immediately after pregnancy, and occasionally in nonpregnant females. Creatinuria is pathophysiologically associated with extensive muscle breakdown (e.g., the late stage of starvation, glucocorticoid excess, thyrotoxicosis, poorly controlled diabetes mellitus, and the various "**myopathies**").

In summary, the hemodynamic changes accompanying exercise are summarized in **Figs. 77-1** and **77-2**. The response begins immediately with the onset of exercise, although a new steady state may not be achieved for several minutes. Principal changes can be attributed to **1)** sympathetic adrenergic discharge to the heart and blood vessels, and **2)** vasodilation in vascular beds servicing exercising skeletal muscles produced largely by local, autoregulatory metabolic stimulants. Tachycardia and an increase in cardiac contractility are largely sympathetic effects, and at maximal work rates CO is controlled largely by sympathetic stimuli. Alterations in regional resistance following the onset of exercise are graded, so that blood flow to the kidneys, brain, mesentery, and coronary vasculature change little. Despite vasoconstriction in nonmuscular beds, total peripheral resistance falls, and there is some evidence that this fall may precede the increase in CO.

A major metabolic problem in the control of **glycolysis** in exercising skeletal muscles during a sprint is that it must quickly increase from a low resting rate, to a much higher rate needed to satisfy the energy demands of exercise. This increase is at least **1000-fold** (about 0.05 μmol/min/gm at rest to about 60 μmol/min/gm during a sprint). Rate control appears to be at the level of **phosphofructokinase (PFK)**, and the controlling agents appear to be **C~PO$_3$, epinephrine, H$^+$, NH$_4^+$, AMP** and **ADP** (see Chapter 25 and **Fig. 77-4**). The intracellular regulator undergoing the largest change in concentration is **C~PO$_3$**, whose level falls about 4-fold in the first few seconds of a sprint. **Creatine phosphate** is formed from **ATP** and **creatine,** with the enzymes catalyzing formation being **CPK$_m$** in mitochondria, and **CPK$_g$** in the cytoplasm. The enzyme that removes high energy phosphate from **C~PO$_3$** near myosin filaments is **CPK$_c$. Creatinine** is formed in muscle from C~PO$_3$ by irreversible, nonenzymatic degradation and loss of Pi, and its 24-hour urinary excretion is normally within narrow limits and proportional to muscle mass. However, when the glomerular filtration rate decreases, plasma creatinine levels will rise. **Creatine,** which is used in C~PO$_3$ formation, is synthesized by the kidneys and liver from the amino acids Arg, Gly, and Met (a methyl donor). Following final synthesis in the liver, it is normally returned to the circulation where it can be extracted by various tissues of the body. **Creatinuria** (i.e., high amounts of creatine in urine) is associated with several pathophysiologic conditions, with the most notable being myopathies.

Oxygen consumption, the respiratory quotient (RQ), muscle fiber types, substrate utilization and endocrine adjustments during exercise will be examined in Chapters 79 and 80, whereas muscle fatigue, athletic animals and the benefits of conditioning will be examined in Chapter 81.

OBJECTIVES

- Distinguish changes in renal, CNS, GI, cutaneous, coronary and skeletal muscle blood flow following the onset of exercise (e.g., running).

- Understand the cardiac adjustments to exercise (i.e., changes in stroke volume, ventricular end-diastolic and end-systolic volumes).

- Explain differences in reactions catalyzed by the creatine phosphokinases of exercising muscle fibers (i.e., CPK_m, CPK_c and CPK_g).

- Know how and why NH_4^+, Pi, AMP and adenosine are produced by exercising muscle fibers, and explain their effects upon vasomotion and/or PFK activity (see Chapter 25).

- Recount how uric acid is produced by exercising muscle fibers (see Chapter 17).

- Differentiate between creatinuria and an elevation in the creatinine content of urine, and identify potential causes of creatinuria.

- Describe the steps involved in creatine formation, and show how vitamin B_{12} and SAM are involved.

- Recognize dietary sources of creatine, and predict whether they would be potentially beneficial or harmful.

- Discuss the manner in which $C\sim PO_3$ helps to control energy flow in exercising muscle fibers.

- Recognize the most common cause of an elevated plasma creatinine concentration.

QUESTIONS

1. **Which one of the following enzymes is thought to remove the high energy phosphate from $C\sim PO_3$, thus forming ATP which becomes immediately available to myosin ATPase.**
   a. CPK_c
   b. Myokinase
   c. CPK_m
   d. AMP deaminase
   e. Transamidinase

2. **Select the FALSE statement below:**
   a. Blood flow to the brain normally changes little during exercise.
   b. End-diastolic volume normally increases following the onset of exercise.
   c. End-systolic volume normally increases following the onset of exercise.
   d. Blood flow to the GI tract normally decreases following the onset of exercise.
   e. Muscle typically contains two to three times more $C\sim PO_3$ than ATP.

3. **When 5'nucleotidase acts upon AMP in strongly contracting cardiac muscle tissue, the following coronary vasodilator is produced:**
   a. IMP
   b. Xanthine
   c. Uric acid
   d. Adenosine
   e. NH_4^+

4. **All of the following are known activators of phosphofructokinase (PFK) in muscle tissue, EXCEPT:**
   a. $C\sim PO_3$.
   b. NH_4^+.
   c. Pi.
   d. AMP.
   e. ADP.

5. **Which one of the following is a natural break-down product of $C\sim PO_3$ in muscle tissue?**
   a. Uric acid
   b. Creatine
   c. Glycocyamine
   d. Creatinine
   e. CPK_m

6. **Select the TRUE statement below:**
   a. Both the liver and kidney participate in the biosynthesis of creatine.
   b. Creatinuria is a condition whereby increased amounts of creatinine appear in urine.
   c. Uric acid is a product of purine degradation in the liver, but not in muscle tissue.
   d. The three amino acids used to synthesize creatine are Orn, Phe, and Arg.

ANSWERS

6. a
5. d
4. a
3. d
2. c
1. a

Chapter 78

Exercise ($\dot{V}O_{2(max)}$ and RQ)

Overview

- $\dot{V}O_{2(max)}$ is a reliable indicator of physical conditioning.
- O_2 utilization by working muscles is a physiologic variable that can limit the exercise performance of normal animals.
- The excess O_2 consumed during recovery from exercise is in payment of the "oxygen dept" incurred during exercise.
- Lactate tolerance can be improved with physical conditioning.
- The RQ is the ratio of CO_2 produced to O_2 utilized, and it is a useful indicator of the types of fuels being utilized during exercise.
- In normal fed animals, the RQ of the brain during exercise is usually close to unity.
- ^{31}P NMR has been used to measure concentrations of phosphorylated metabolic intermediates in tissues during exercise.

Oxygen Consumption

Oxygen consumption ($\dot{V}O_2$) by the organism does not rise instantly when an animal begins to exercise, nor does it necessarily return quickly to the resting state when it stops (**Fig. 78-1**). However, $\dot{V}O_2$ does rise soon after the onset of an exhausting run, with $\dot{V}O_{2(max)}$ (i.e., the maximal rate of oxygen consumption) achieved when energy requirements exceed the maximal capacity for aerobic metabolism. $\dot{V}O_{2(max)}$ is considered to be a reliable physiologic indicator of an animal's physical conditioning. For most animals, $\dot{V}O_{2(max)}$ is remarkably constant from day to day, although it may be decreased by long periods of inactivity, and increased again following an extended period of proper conditioning (**Fig. 78-2**).

Three important variables that can limit $\dot{V}O_{2(max)}$ are listed in **Table 78-1**. Oxygen transport in blood depends on cardiac output and the O_2 carrying capacity of blood (i.e., the hemoglobin

(Hb) concentration). The blood gas diffusion capacity of the lungs is a function of the amount of lung tissue available for gas exchange, and other ventilation/perfusion considerations. In the absence of hematopoietic, cardiovascular, or pulmonary abnormalities, these first two variables are not usually considered rate-limiting. However, the third factor, O_2 utilization by working muscles, is a physiologic variable that can alter exercise performance of normal,

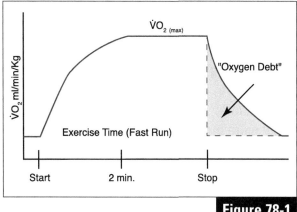

Figure 78-1

Copyright © 2015 Elsevier Inc. All rights reserved.

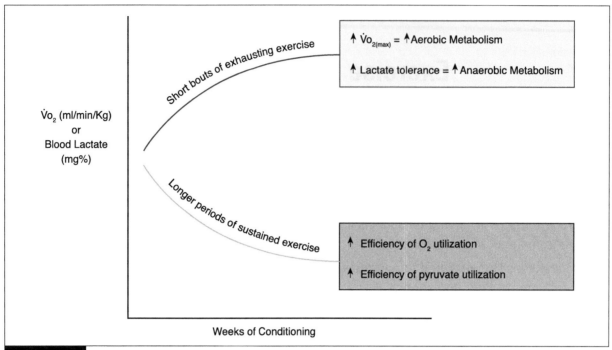

Figure 78-2

Table 78-1
Physiologic Variables that limit $\dot{V}O_{2(max)}$

1. Rate of O_2 transport by the circulation
2. O_2 diffusing capacity of the lungs
3. O_2 utilization by muscle tissue

healthy animals (see Chapter 81).

As indicated above, after a period of exercise an animal's O_2 consumption does not immediately return to the pre-work level. It declines as a logarithmic function of time, and thus during recovery the animal normally consumes O_2 at a faster rate than during rest. This excess O_2 consumed during recovery is in payment of the "**oxygen debt**" incurred during exercise. Extra oxidation produces energy required for restoring high-energy systems in muscles which have released energy anaerobically during work, and therefore the phenomenon might more properly be called an "**energy debt**". The total O_2 debt paid in recovery is considered to have four components (**Table 78-2**), with primary debts paid being items 2 and 4.

The most abundant metabolite of anaerobic muscle metabolism is **lactate**, which accumulates in working muscles and diffuses into blood. There is generally a reasonable correlation between the O_2 debt and the lactate concentration of blood shortly after strenuous exercise. An animal may sprint 100 yards in 10 seconds or less, and perhaps use 30 Kcal of energy in the process; but he/she may only consume 0.5 L of O_2 during the run. Since the consumption of 1 L of O_2 leads to the production of enough ATP to support about 5 Kcal of work, this anaerobic sprinter accumulated an O_2 debt of 5.5 L, that must be paid during recovery.

From a practical standpoint, the anaerobic mechanisms of energy release in muscle are important in maximal work where they permit an animal to expend energy far in excess of the

Table 78-2

Components of the Oxygen Debt that must be Repaid

1. A small net diminution in O_2 stored in venous blood and muscle myoglobin (which is a minor part of the metbolic energy debt), is paid off quickly in recovery.

2. The rapid resynthesis of high-energy phosphate bonds in ATP and $C \sim PO_3$.

3. The slow removal and oxidation of lactate formed from pyruvate in the anaerobic glycolysis of working muscles.

4. The conversion of lactate to glucose in liver, with subsequent removal of glucose from blood by muscle and reconversion to glycogen.

capacity for carrying on oxidative metabolism. Even in moderate work, anaerobic reactions in muscle are important in that they allow an animal to release required energy instantly without the delay involved in mitochondrial O_2 utilization. However, the amount of energy an animal may borrow from these anaerobic mechanisms is limited by its tolerance for **acidosis** (i.e., its lactate tolerance; see Chapter 81). **Lactate tolerance** (or more accurately, **H⁺ tolerance**), is a variable that can be increased through conditioning (**Fig. 78-2**). Additionally, physical conditioning has been shown to increase muscle perfusion, and also the total number of mitochondria in muscle fibers. Therefore, during sustained, moderate aerobic exercise, blood lactate (and H⁺) concentrations can be reduced in well-conditioned athletes, the efficiency of pyruvate utilization can be increased, and thus the efficiency of O_2 utilization can be improved.

The Respiratory Quotient (RQ)

Both carbohydrate and fat are normally oxidized by working muscles during exercise, whereas protein is usually spared. The **RQ**, which is the rate of **CO_2 production ($\dot{V}CO_2$)** divided by the rate of **O_2 utilization ($\dot{V}O_2$; Fig. 78-3)**, may be determined by collection and analysis of inspired and expired air at rest, or exercising at a steady rate. If an animal is metabolizing **carbohydrate only** (e.g., during the anaerobic phase), the **RQ** should be close to 1.0. However, if the exercising animal is catabolizing both **carbohydrate and fat** (e.g. during the aerobic phase), then the **RQ** should be somewhere between **0.7** and **1.0**.

The RQ for the complete combustion of various substrates is readily determined from the chemical equation for their oxidation. That for **protein** oxidation, however, is not easily

$$\text{Respiratory Quotient (RQ)} = \frac{\dot{V}CO_2}{\dot{V}O_2}$$

Carbohydrate

$$C_6H_{12}O_6 + 6\,O_2 \longrightarrow 6\,CO_2 + 6\,H_2O$$
(Glucose)

$$\boxed{RQ} = \frac{6\,\dot{V}CO_2}{6\,\dot{V}O_2} = \boxed{1.0}$$

Fat

$$C_{16}H_{32}O_2 + 23\,O_2 \longrightarrow 16\,CO_2 + 16\,H_2O$$
(Palmitate)

$$\boxed{RQ} = \frac{16\,\dot{V}CO_2}{23\,\dot{V}O_2} = \boxed{0.7}$$

Protein

$$\text{Average}\quad \boxed{RQ} = \frac{77.5\,\dot{V}CO_2}{96.7\,\dot{V}O_2} = \boxed{0.8}$$

Figure 78-3

determined because proteins are incompletely catabolized *in vivo*, and the oxidation of various amino acids yields different RQ values. However, an empirically derived **average RQ** of about **0.8** is generally applied to protein oxidation. If an animal has consumed a high-protein meal before exercise, amino acid oxidation may be contributing to the overall energy pool. A better indicator of the degree to which amino acids are being oxidized for energy purposes would be urinary nitrogen output rather than RQ.

It should also be recognized that **nonmetabolic CO_2 production** occurs during exercise through bicarbonate buffering (see Chapters 85 & 87). Thus, as anaerobic exercise conditions prevail and metabolic acidosis ensues, this nonmetabolic CO_2 production may contribute significantly to an elevation in the RQ.

In laboratory animals, use has been made of the RQ in **perfused organs** by looking at arterio-venous blood gas differences across these structures. This technique involves the surgical placement of catheters into the two major blood vessels that supply and drain the organ being investigated. This type of study is useful in that it allows simulation of specific physiologic conditions, but suffers from the disadvantage that the contribution from differing types of cells within the organ being investigated cannot be assessed. Important discoveries, like the use of ketone bodies by the brain during starvation (see Chapter 75), and the release of alanine and glutamine from skeletal muscle during starvation (see Chapter 76), could only have been made through such studies. Unless an animal is ketotic, the **RQ** of the **brain** is usually close to unity, even during exercise, indicating that it is primarily catabolizing glucose. Similar studies, which have looked at the **stomach**, for example, indicate that when it is readily secreting **HCl** into the gastric lumen, utilizing CO_2 from blood for this process, the RQ is quite low.

Alternative Techniques for Determining Fuel Utilization During Exercise

Another technique which examines muscle fibers directly during short- or long-duration exercise involves removal of small samples of tissue via **biopsy**. Surface pain receptors in the skin overlying the muscle being investigated can be anaesthetized, and a small incision made. The subject then begins to exercise (on a treadmill, for example), and at an appropriate time a biopsy needle is inserted through the incision. This needle, a few millimeters wide, is hollow with a rounded tip. Just above the tip is a small window through which a piece of muscle intrudes. This muscle piece is rapidly severed by a plunger, which is pushed down the needle. The entire needle is then removed and, if metabolite assays are to be performed, plunged into liquid nitrogen. In this way, a few milligrams of muscle can be sampled with minimal (15-30 sec) interruption of exercise. Several samples can be taken from the same subject, with, it is claimed by humans, no more after-effect than a mildly aching muscle the next day.

The ideal research tool, however, would be a means for quantitating flux through metabolic pathways by a **non-invasive technique**, and without altering function of the tissues being investigated. Flux through glycolysis under anaerobic conditions has been measured in such a way by **nuclear magnetic resonance (NMR)** imaging. This technique depends upon the measurement of specific radio-frequency emissions produced in a strong magnetic field from molecules containing nuclei with magnetic properties (e.g., ^{31}P, the most prevalent isotope of phosphorus). ^{31}P **NMR** has been used to measure the concentrations of phosphorylated metabolic intermediates in living tissues both *in vitro* and *in vivo*. These intermediates include **ATP, creatine phosphate (C~PO_3)**, and **inorganic phosphate (Pi)**. Furthermore, there is a shift in the position of the signal peak emitted from **Pi** as $HPO_4^=$ is converted to $H_2PO_4^-$, or *vice versa*, with a change in **pH** (see Chapter 85).

Hence, it is possible to follow the increase in **proton accumulation** in muscle by ^{31}P NMR, which is usually proportional to lactic acid formation during exercise.

OBJECTIVES

- Explain why $VO_{2(max)}$ is a reliable indicator of physical conditioning, and identify three physiologic variables most likely to affect it.

- Recognize why the "oxygen debt" of exercise can also be considered an "energy debt."

- Identify four components of the oxygen debt that must be repaid immediately following exercise.

- Explain what is meant by "lactate tolerance" and "pyruvate utilization," and discuss how these factors may change with physical conditioning (i.e., repeated aerobic exercise).

- Justify relationships between muscle mitochondrial number, muscle perfusion and physical conditioning.

- Know the meaning of RQ, recognize how it is determined, and discuss what a value of 0.8 might represent in an exercising athlete.

- Show how nonmetabolic CO_2 is produced during exercise (see Chapters 85-87).

- Describe some non-invasive techniques for examining exercise muscle metabolism.

QUESTIONS

1. **Which of the following are inappropriately matched?**
   a. "Oxygen debt" : "Energy debt"
   b. O_2 utilization by muscle tissue : Limit $\dot{V}O_{2(max)}$
   c. "Oxygen debt" : Lactate accumulation
   d. ↑Lactate tolerance : ↑Tolerance for anaerobic metabolism
   e. ↑$\dot{V}O_{2(max)}$: ↓Efficiency of pyruvate utilization

2. **If both carbohydrate and fat were being oxidized by exercising muscle fibers, the RQ would be expected to be:**
   a. 0.2.
   b. 0.4.
   c. 0.6.
   d. 0.8.
   e. 1.0.

3. **In normal, fed animals, the RQ of the brain during exercise would be expected to be:**
   a. 0.2.
   b. 0.4.
   c. 0.6.
   d. 0.8.
   e. 1.0.

4. **^{31}P NMR can be used to determine which exercise variable:**
   a. Proton accumulation by exercising muscle.
   b. The plasma glucose concentration.
   c. Uptake of free fatty acids by aerobic muscle fibers.
   d. The plasma HCO_3^- concentration.
   e. Respiratory blood gas exchange.

5. **In normal, healthy animals, which one of the following would have the greatest influence on $\dot{V}O_{2(max)}$?**
   a. The O_2 diffusing capacity of the lungs
   b. Cardiac output
   c. The rate at which lactate is converted to glucose by the liver
   d. Oxygen utilization by muscle tissue
   e. Creatine phosphate (C~PO_3) reserves in muscle tissue

6. **The rate of anaerobic glycolysis is controlled by which one of the following enzymes in exercising muscle tissue?**
   a. Hexokinase
   b. Aldolase
   c. Phosphofructokinase
   d. Glucose 6-phosphatase
   e. PEP carboxykinase

7. **The best indicator of the degree of amino acid oxidation during exercise would be the:**
   a. RQ.
   b. Urinary nitrogen excretion.

7. b
6. c
5. d
4. a
3. e
2. d
1. e

ANSWERS

Chapter 79

Exercise
(Substrate Utilization and Endocrine Parameters)

Overview

- Anaerobic metabolism provides little energy for the aerobic athlete.

- Major muscle fuels during prolonged exercise are usually glucose (33%) and free fatty acids (66%) (i.e., "fats are burned in a carbohydrate flame").

- The plasma glycerol concentration is a better indicator of the extent of lipolysis occurring during aerobic exercise than is the plasma FFA concentration.

- Liver glycogen stores may be depleted during exercise in about 30 minutes.

- During exercise ketone body and lipoprotein production by the liver are normally reduced, and their clearance from the circulation is usually increased.

- Circulating levels of cortisol, epinephrine, norepinephrine, and glucagon are all normally increased during exercise, while those of insulin are reduced.

- The insulin sensitivity of muscle and adipose tissue is normally increased for a few hours following a healthy bout of aerobic exercise.

- Exercise can be an effective tool for reducing the insulin requirement of a diabetic animal.

The metabolic response to **exercise** resembles that to **starvation**, in that mobilization and generation of fuels for oxidation are dominant factors. The type and amounts of substrate vary with the intensity and duration of exercise (**Fig. 79-1**). For intense, short-term exercise (e.g., a 10 to 15 second sprint for a dog), stored **creatine phosphate (C~PO$_3$)** and **ATP** provide most of the energy (see Chapter 77). When those stores are depleted, additional intensive exercise for up to three minutes can be sustained by breakdown of **muscle glycogen** to glucose 6-phosphate, with anaerobic glycolysis yielding the necessary energy. This phase is not necessarily limited by depletion of muscle glycogen at this point, but rather by the accumulation of **protons** (H$^+$ from lactic acid) in exercising muscle and in the circulation.

For less intense but longer periods of exercise, aerobic oxidation of substrates is required to produce necessary energy. After a few minutes **glucose uptake** from plasma increases dramatically, up to **thirty-fold** in some muscle groups. To offset this drain, **hepatic glucose production** increases up to **five-fold**. Initially this is largely from glycogenolysis. However, with exercise of long duration, gluconeogenesis becomes increasingly important as liver glycogen stores become depleted. Except for increases in circulating pyruvate and lactate resulting from enhanced muscle glycolysis, the pattern of change in plasma substrates is similar to that seen in starvation, only telescoped in time.

It has been established that anaerobic metabolism provides little energy for the aerobic

Copyright © 2015 Elsevier Inc. All rights reserved.

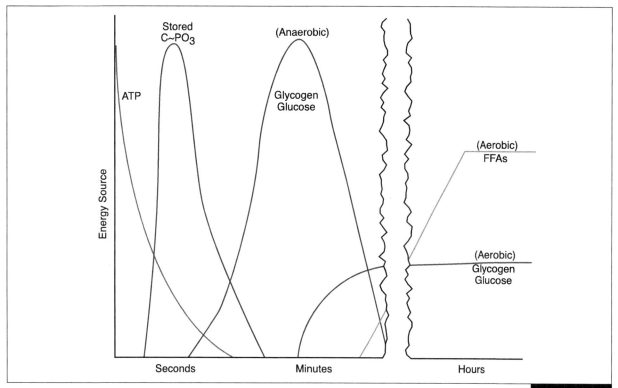

Figure 79--1

athlete. Major muscle fuels during **prolonged exercise** appear to be **glucose** plus **free fatty acids (FFAs)** obtained from the blood stream, as well as stored glycogen and **triglyceride (TG)** obtained from within muscle cells themselves. It is important to realize that these fuels are not used indiscriminately. The rate of utilization of any one fuel is controlled not only in relation to the energy demands of muscle, but also in relation to the rates of utilization of other fuels.

Any glucose taken by muscles from blood must be replaced by hepatic glycogenolysis, or by gluconeogenesis. Although differences exist among and between species, experiments demonstrate that **liver glycogen stores** are depleted during sustained exercise in about **30 minutes**. As exercise proceeds, it is apparent that **glucose uptake by muscle** continues to equate with about **30%** of oxygen uptake, with **more than 60%** being accounted for by **oxidation of fatty acids**. Glucose oxidation is required aerobically in order to provide enough oxaloacetic acid (OAA) to match the acetyl-CoA that is being generated through

β-oxidation. In other words, "**fats are burned in a carbohydrate flame**," and leaks in the TCA cycle result in insufficient OAA formation from malate to meet muscle needs. The significance of this observation was discussed in Chapter 35.

Although there is little doubt that the concentration of FFAs in blood is increased during sustained exercise, the extent of this increase is variable (**Table 79-1**). The plasma FFA concentration depends upon a variety of factors, including previous dietary history, the type, severity, and duration of exercise, as well as the fitness of the animal. Since the concentration of FFAs in blood represents a balance between the rate of mobilization by lipolysis in adipose tissue (see Chapter 70), and their β-oxidation by exercising muscle tissue, the actual plasma steady state concentration tells us little about rates of utilization. Furthermore, control of the rates of these two processes (FFA release by adipocytes and β-oxidation by muscle) may be better integrated in highly conditioned athletes, so that the increase in steady state concentration is less than in more sedentary animals. For all of these reasons, the increase in the plasma **glycerol** concentration (i.e., glycerol released from adipose tissue during lipolysis) may be a better index of the rate of lipolysis during exercise than increases in the plasma **FFA** concentration. In general, the concentration of **glucose** remains fairly stable or slightly decreases during sustained exercise, whereas the plasma glycerol and FFA concentrations are progressively elevated (**Table 79-1**). Although glycerol is not reutilized by adipose tissue following TG breakdown, up to **60%** of the **FFAs** initially released from TG can be re-esterified to newly synthesized glycerol 3-phosphate in adipocytes through **glyceroneogenesis** (see Chapter 70). Thus, depending upon the activity of this re-esterification pathway, the rate of glycerol release by adipose tissue during exercise may be similar to that of FFAs (e.g., 1 FFA molecule released per glycerol molecule), it may be less (e.g., 3 FFA molecules released per glycerol molecule), or it may be greater (e.g., glycerol is

Table 79-1
Plasma Substrate and Hormone Concentrations during Exercise

Animal	Time of Exercise (min)	Substrate			
		Glucose (mM)	Lactate (mM)	Fatty Acid (mM)	Glycerol (mM)
Dogs	0	6.1	1.6	0.6	--
	60	5.5	2.4	1.1	--
	120	5.5	1.6	1.8	--
	180	5.0	1.6	1.9	--
	240	5.0	1.9	2.1	--
Horses	0	4.4	0.7	0.29	0.14
	360	2.0	1.6	1.4	0.49
Humans	0	4.5	1.1	0.66	0.04
	40	4.6	1.3	0.78	0.19
	180	3.5	1.4	1.57	0.39
	240	3.1	1.8	1.83	0.48

Animal	Time of Exercise (min)	Hormone				
		Cortisol (nmol/L)	Epi (ng/cm^3)	NE (ng/cm^3)	Glucagon (pg/cm^3)	Insulin (μU/cm^3)
Horses	0	186	--	--	17	36
	360	695	--	--	170	6
Humans	0	--	0.1	0.4	80	14
	40	--	0.2	--	80	12
	180	--	0.7	1.5	200	7
	240	--	--	--	400	6

Legend: Epi = epinephrine; NE = norepinephrine. The workload of dogs was progressively increased over a 4 hr period; horses were exercised on a hilly course at an average speed of 12.5 km/hr; and human subjects were exercised on a bicycle at a workload of about 30% maximum. Table modified from Newsholme EA, Leech AR, 1983. Data from Ahlborg G, Felig P, Hagenfeldt L, et al, 1974; Felig P, Wahren J, 1975; Galbo H, Holst JJ, Christensen NJ, 1975; Galbo H, Holst JJ, Christensen NJ, 1979; Hall Gm, Lucke JN, Masheter K, et al, 1981; and paul P, Holmes WL, 1975.

released but all FFA molecules are re-esterified through glyceroneogenesis).

Although the K_m of **lipoprotein lipase** on the capillary endothelium servicing muscle is lower than that servicing adipocytes, TG contained in circulating lipoproteins is not generally believed to provide a significant amount of fuel for exercising muscle. Additionally, since muscle is oxidizing circulating FFAs for energy purposes, uptake of FFAs by the liver is reduced during sustained aerobic exercise, and therefore **VLDL formation is reduced**. Since the liver is importing fewer FFAs during exercise, **ketone body (KB⁻) formation** is also **reduced**. Therefore, KB⁻s and circulating lipoproteins are usually insignificant contributors to energy production in muscle during sustained exercise. However, if an animal were ketotic or hypertriglyceridemic before exercise commenced, then circulating KB⁻s would be readily available fuel for exercising muscle, and TG contained in circulating lipoproteins could be slowly hydrolyzed. Thus, exercise can be an effective means of reducing the plasma TG and KB⁻ concentrations of hypertriglyceridemic, ketoacidotic, **diabetic** patients.

The flux-generating step in hepatic glycogenolysis is **glycogen phosphorylase** (see Chapter 23). This enzyme is activated by increased blood levels of **epinephrine (Epi)**, **norepinephrine (NE)**, and **glucagon**, and decreased by increased blood levels of **insulin**. As seen in **Table 79-1**, circulating levels of Epi, NE, and glucagon all increase during exercise, while those of insulin normally decrease. Catecholamines (Epi and NE) directly reduce pancreatic insulin release, while stimulating glucagon release. This hormonally-mediated mechanism may be supplemented by a direct effect of glucose, the concentration of which falls, if only slightly, during exercise. The activity of **hormone sensitive lipase (HSL)** in adipose tissue, the flux-generating step in lipolysis (see Chapter 70), is increased by high circulating levels of Epi, NE, and cortisol, and low levels

of insulin (**Table 79-1**). Increased circulating FFAs will lead to an increase in the activities of **acyl-CoA synthetase**, the **carnitine palmitoyltransferases**, and the enzymes of **β-oxidation** in muscle tissue through internal regulation (see Chapter 55). It is unlikely that the activities of these enzymes are directly affected by high circulating titers of the catecholamines (Epi and NE), cortisol, or glucagon.

A frequently asked question is, "**will exercise reduce the insulin requirement of a diabetic animal, and is exercise an effective way of reducing his/her hyperglycemia?**" Not only are **insulin** levels **reduced** during exercise, but the entry of glucose into skeletal muscle is **increased** (in the absence of insulin). This effect is due to **three** primary factors:

1) **There is less glucose 6-phosphate negative feedback on hexokinase in exercising muscle since this substrate is utilized as fast as it is formed (see Chapter 22).**

2) **A by-product of muscle contraction is heat, which increases membrane fluidity of muscle cells.**

3) **Exercise causes an insulin-independent increased insertion of GLUT-4 transporters into the plasma membranes of skeletal muscle fibers.**

This last effect may persist for several hours into the post-exercise period, and exhibit itself as an **increase** in **sensitivity to insulin**, thus helping muscle to replenish its energy reserves. Indeed, exercise has been reported to precipitate hypoglycemia in diabetics not only because of the increase in post-exercise insulin sensitivity, but also because absorption of injected insulin is more rapid during exercise. For this reason the insulin dosage of diabetic patients should be reduced if they exercise.

Another question frequently posed regarding exercise is, "**why do some individuals (and animals) find it difficult to run even one mile?**" The answer is simple--lack of fitness means poor aerobic capacity (i.e., loss of mitochondria and hence the enzymes of the TCA cycle and β-oxidation), so that energy required for exercise is provided mainly by the anaerobic conversion of glycogen to lactic acid. Since this is an inefficient means of producing ATP, it has to occur at a high rate and readily causes fatigue either through proton accumulation or, if the exercise is more gradual, depletion of muscle glycogen. Even mild aerobic training, however, can dramatically improve aerobic capacity, thus increasing the distance that can be run by "non-athletes."

OBJECTIVES

- Explain what is meant by the saying that "fats are burned in a carbohydrate flame" during aerobic exercise (see Chapters 27 & 35).

- Identify the different sources of endogenous energy for the exercising athlete.

- Know why the plasma glycerol concentration is a better indicator of the rate of lipolysis during exercise than the plasma FFA concentration.

- Understand why exercise is recommended for the diabetic hyperglycemic animal, and why it would be imprudent to inject insulin immediately preceding exercise.

- Show how plasma concentrations of glucose, lactate, FFAs and glycerol change during sustained exercise.

- Recognize the normal alteration in the insulin:-glucagon ratio during sustained exercise, and explain why this change occurs.

- Indicate how insulin sensitivity changes following exercise, and provide physiologic reasoning for this change.

QUESTIONS

1. **During sustained, aerobic exercise, the pattern of change in plasma substrates is similar to that seen in:**
   a. The 2-hour, postprandial state.
   b. Cachexia.
   c. Uncontrolled diabetes mellitus.
   d. Physiologic starvation.
   e. Pregnancy.

2. **The plasma concentration of which hormone below normally decreases during exercise?**
   a. Insulin
   b. Epinephrine
   c. ACTH
   d. Glucagon
   e. Cortisol

3. **During sustained aerobic exercise:**
   a. VLDL production by the liver usually increases.
   b. The plasma glycerol concentration usually decreases.
   c. Ketone body production by the liver usually increases.
   d. The plasma glucose concentration usually increases.
   e. The activity of HSL in adipocytes usually increases.

4. **Select the FALSE statement below:**
   a. Exercise causes an insulin-independent increased insertion of GLUT-4 transporters into the plasma membranes of skeletal muscle cells.
   b. Circulating levels of cortisol usually increase during exercise.
   c. The plasma FFA concentration provides a better index of the rate of lipolysis during exercise than does the plasma glycerol concentration.
   d. Exercise would tend to reduce circulating KB^- levels.
   e. Exercise would tend to reduce the plasma glucose concentration of a hyperglycemic, diabetic animal.

5. **Select the TRUE statement below:**
   a. Unlike muscle glycogen reserves, those in liver are seldom depleted through sustained, aerobic exercise.
   b. After 60 minutes of moderate, sustained aerobic exercise, an animal will be aerobically oxidizing both glucose and FFAs in muscle fibers.
   c. Insulin stimulates the activity of glycogen phosphorylase in skeletal muscle fibers.
   d. The catecholamines (epinephrine and norepinephrine) stimulate pancreatic insulin release.

1. d
2. a
3. e
4. c
5. b

Chapter 80

Exercise
(Muscle Fiber Types and Characteristics)

Overview

- Type I skeletal muscle fibers are aerobic, and generally exhibit long contractile periods.

- White type IIB skeletal muscle fibers are used for escape purposes.

- The total number of type I and type IIB muscle fibers is thought to be fixed at an early stage of life.

- The quarterhorse has proportionally more type IIB anaerobic muscle fibers than the thoroughbred.

- The K_m of LPL contained in the coronary vasculature is low.

- Lactate, produced by skeletal muscle fibers during exercise, is used by cardiac muscle for energy purposes.

- The LDH of skeletal muscle fibers normally functions as a pyruvate reductase, whereas that in cardiac muscle functions in lactate oxidation.

- Cardiac muscle cells generally contain more mitochondria (and therefore fewer myosin filaments) than skeletal muscle fibers.

- Cardiac muscle tissue is normally 99% aerobic, and does not function well under anaerobic conditions.

- Citrate inhibits PFK in muscle cells.

In order to provide a biochemical explanation for why some animals are better athletes than others, and why exhaustion occurs with exercise, it is necessary to consider muscle fiber type characteristics and ratios, and fuels normally utilized by those different fiber types during exercise.

Skeletal Muscle Fiber Types

Although skeletal muscle fibers differ somewhat in both their anatomic and physiologic properties, they have been roughly categorized into three basic **types (I, IIA,** and **IIB**; **Table 80-1**). **Type I fibers** are known as **slow-twitch fibers** since the time course of their maximal twitch is longer than that found in **type II fibers**, which are known as **fast-twitch fibers**. Important metabolic differences also exist between fiber types. **Type I** fibers have a **high aerobic capacity** (i.e., high activities of enzymes of the TCA cycle, of β-oxidation, and of oxidative phosphorylation), a high content of triglyceride (TG), cytochrome-containing mitochondria and myoglobin (thus giving this muscle type a **red appearance**), yet low glycolytic capacity. **Type I** fibers are also known to be **fatigue resistant**, a property that depends upon their capacity for maintaining a low NADH level, and thus a low level of lactic acid formation. **Type IIA fibers are intermediate** between **type I** and **type IIB** fibers. For example, they possess both oxidative and glycolytic

Copyright © 2015 Elsevier Inc. All rights reserved.

Table 80-1			
Muscle Fiber Types and their General Characteristics			
	Type I	Type IIA (Intermediate)	Type IIB
Contractile Speed	Slow-twitch (oxidative)	Fast-twitch (oxidative/glycolytic)	Fast-twitch (glycolytic)
Color	Red	--	White
Metabolism	Oxidative (aerobic)	--	Glycolytic (anaerobic)
Glycogen	Less	--	More
Glycogen phosphorylase	Less	--	More
Hexokinase	More	--	Less
Triglyceride	More	--	Less
Protein turnover	More	--	Less
Myoglobin	More	--	Less
Mitochondria	Many	--	Few
Blood flow	More	--	Less
Resistance to fatigue	More	--	Less
Sarcotubular system	Less	--	More
Ca^{++} uptake & release	Slow	--	Fast
Myosin ATPase activity	Less	--	More
Contraction time	Long	--	Short

capacity, and have an intermediate content of TG. **Type IIB** fibers possess low oxidative yet **high glycolytic capacity**, and possess fewer mitochondria than type I fibers. The space that would otherwise be filled with mitochondria in type IIB fibers contains comparatively more contractile filaments, thus giving this fiber type greater force-generating capacity.

Examples of "**white**" **muscles** containing only **type IIB** fibers are **lobster** abdominal muscle, **fish** white abdominal muscles, **game bird** pectoral muscles (including the domestic fowl), and the psoas muscles of the **rabbit**. All these muscles can contract rapidly and vigorously, but for short periods of time; not surprisingly, they power **escape reactions** (e.g., the tail flick of the lobster, and the flight of the pheasant). They possess a poor blood supply, so that the provision of plasma glucose and other blood-borne nutrients as fuel is submaximal. They

rely instead on **intracellular glycogen** degradation, and therefore possess high activities of all glycolytic enzymes, except hexokinase (HK). Approximately 95% of the muscle mass in most fish is composed of white type IIB fibers. The red muscle is located just under the skin along the lateral line, and is used for "cruise" swimming. If the fisherman were to deal with a fish possessing a greater ratio of type I fibers (e.g., bluefish), he/she would most certainly wait longer to "land the catch."

The type of metabolism used to restore and utilize **creatine phosphate (C~PO$_3$)** in muscle contraction depends upon the fiber type. For example, fast-twitch, glycolytic fibers rely largely on **anaerobic glycolysis**, and thus utilize both **glycolytic CPK (CPK$_g$)** and **cytoplasmic CPK (CPK$_c$)** to deliver high energy phosphates to myosin ATPase (see Chapter 77). Although anaerobic glycolysis is generally an inefficient use of glucose (only **3.5%** of the total possible energy in glucose can be generated through anaerobic glycolysis alone), it has the advantage of producing energy rapidly, and in the complete absence of O_2. In contrast, slow-twitch, oxidative muscle fibers (like cardiac muscle), rely on aerobic metabolism of **fatty acids**, **glucose** and **lactate (Fig. 80-1)**.

High energy phosphates can be transferred from readily available creatine phosphate ($C \sim PO_3$) to myosin ATPase in the cytoplasm about 10 times more rapidly than they can be transferred through oxidative phosphorylation. Additionally, it requires longer to obtain ATP from aerobic metabolism, which ceases completely when O_2 becomes unavailable. However, muscle cells capable of oxidative aerobic metabolism can sustain activity for long periods of time (such as in the heart which has only brief periods of rest), if the

activity is not too rapid or too severe. Rapid or severe contractions, however, sometimes deplete energy supplies more rapidly than they can be replenished, and then muscles become exhausted and have to stop. Therefore, myosin in aerobic muscle fibers is adapted so that it contracts slowly, and its use of ATP is less likely to exceed the cell's capacity to restore its supplies.

Although the actual number of **type I** and **type IIB** muscle fibers in any one muscle group is thought to be fixed at an early stage of life, and therefore the ratio is also fixed, the size of individual fibers can be increased with training (i.e., **physiologic hypertrophy**). **Table 80-2** compares muscle fiber type ratios in **dogs**, **horses**, and various **human subjects** studied. The muscles of **greyhounds**, for example, which lie in animals that have been bred for several hundred years for performance over a short distance, contain

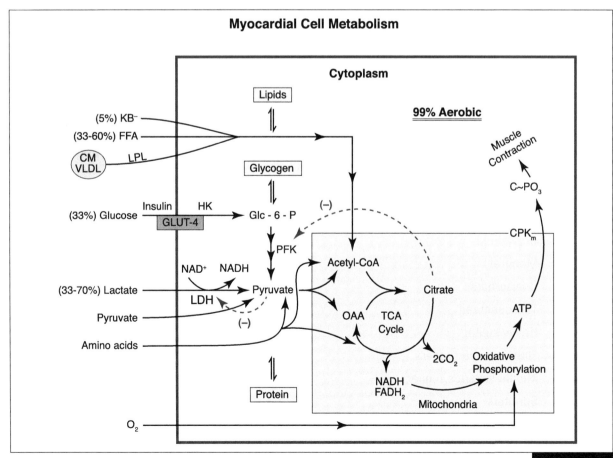

Myocardial Cell Metabolism

Figure 80-1

a high percentage of **type IIB** fibers. In contrast, those in mongrels are substantially lower. The **American Racing Quarterhorse** is a breed developed to race about 500 yards, and has **93% type IIB** skeletal muscle fibers, whereas **heavy hunters**, which are known for their stamina rather than speed, have a relatively high percentage of **type I** fibers.

Type IIB skeletal muscle fibers are innervated by **large motor neurons** which transmit impulses rapidly, and action potential transmission along the muscle membrane is also rapid. Additionally, innervation of type IIB fibers is such that the contraction and relaxation of many fibers is readily achieved, and interaction between actin and myosin filaments is short. These characteristics provide the speed necessary for rapid escape, and high power output is achieved by the possession of a large number of contractile filaments per cell, a large number of sarcomeres per fiber, and a large number of fibers per muscle group.

Muscle hypertrophy occurs readily in this fiber type through weight-bearing exercise, and anaerobic glycolytic capacity is increased.

Muscles That Do Not Accumulate an O₂ Debt

Fig. 80-1 is a schematic of a **myocardial cell**, illustrating major pathways of metabolism. Most substrates are taken up more or less in proportion to their arterial concentration. Major reactions of glycolysis take place in the cytoplasm, however, heart muscle is normally **99% aerobic**, and therefore requires considerable mitochondrial support. Mitochondria contain enzymes of the tricarboxylic acid (TCA) cycle and oxidative phosphorylation, and are responsible for aerobic **pyruvate, amino acid, free fatty acid (FFA)**, and **ketone body (KB⁻)** oxidation. The contribution of TG contained in **circulating chylomicrons (CMs)** and **very low-density lipoprotein (VLDL)** to energy generation in myocardial cells is generally considered to be low; however, the **K$_m$** of **lipoprotein lipase (LPL)** contained in the coronary vasculature is fairly low, thus adding **FFAs** from circulating lipoproteins as a source of energy. **Glucose** is transported across the cell membrane under the control of insulin in the resting state; however, during exercise **insulin** levels are low. Nonetheless,

Table 80-2
Skeletal Muscle Fiber Composition

		Percentage	
		Type I	Type IIB
Dog	Greyhound	3	97
	Mongrel	31	69
Horse	Quarterhorse	7	93
	Thoroughbred	12	88
	Heavy hunter	31	69
Human	Elite distance runners	79	21
	Middle distance runners	62	38
	Sprinters	24	76
	Untrained	53	47

Table modified from Newsholme EA, Leech AR, 1983. Data for dogs from Guy PS & Snow DH, 1977; data for horses from Snow DH & Guy PS, 1980; and data for humans from Costill DL, Daniels J, et al, 1976, and Fink WJ, Costill DL, et al, 1977.

GLUT-4 transporters are inserted into muscle membranes during exercise, thus helping to facilitate glucose entry (see Chapter 79). Phosphorylation of glucose by HK is generally maintained during exercise because the glucose 6-phosphate (Glc-6-P) produced is quickly utilized in glycolysis, thus reducing its negative feedback effects (see Chapter 22).

The metabolism of **FFAs** and **KB⁻s** by heart muscle increases the intracellular concentration of **citrate** if enough **oxaloacetic acid (OAA)** is available. Since citrate is permeable to mitochondrial membranes, it diffuses into the cytoplasm where it reduces further glycolysis by **inhibiting phosphofructokinase (PFK)** activity, the rate-limiting enzyme in glycolysis (see Chapter 25). **Lactate** is normally removed from blood and converted to pyruvate in myocardial cells, thus reducing dependency on glucose oxidation. Studies conducted on trained cyclists indicate that up to **70%** of myocardial energy consumption can be accounted for through lactate oxidation. The **lactate dehydrogenase (LDH)** that catalyzes conversion of lactate to pyruvate in myocardial cells (the H isozyme) is slightly different from the LDH isozyme in skeletal muscle (which is usually working in the reverse direction; see Chapters 5 and 6). The LDH of skeletal muscle (consisting of four **M** subunits) functions as a **pyruvate reductase** in order to re-oxidize the NADH produced from anaerobic glycolysis (see Chapter 27). The heart isozyme (but not that in skeletal muscle), is inhibited by high concentrations of pyruvate. Both LDH and **creatine phosphokinase (CPK)** are released into blood following a heart attack (see Chapter 6), and thus are used for diagnostic purposes.

Cardiac muscle hypoxia is not well tolerated. Hypoxia inhibits many mitochondrial functions, including the TCA cycle and oxidative phosphorylation, with resultant decreases in ATP and C~PO_3 production. Hypoxia also stimulates glycogen phosphorylase, and thus glycogenolysis. Under hypoxic conditions, energy production by the heart becomes severely compromised.

Muscle Atrophy during Immobilization

Muscle atrophy can be a significant problem for animals who are confined to small spaces for sustained periods of time, or are immobilized through sickness. Muscle wastage can delay rehabilitation, and have a negative impact in food animals. Although all three muscle types atrophy during immobilization, a greater proportion of **type I** fibers are generally affected.

In summary, physiological chemists discovered many years ago that mammalian skeletal muscles differed in color, and in the microscopic appearance of their fibers. It was noted that fibers of small diameter were darker and more granular than larger fibers in the same muscle group. Small fibers predominated in red muscles, and large fibers in white muscles. Accordingly, they were later classified by various techniques as "red", oxidative, slow-twitch fibers (type I), and "white", glycolytic, fast-twitch fibers (type IIB). Oxidative/glycolytic type IIA fibers were found to be intermediate between the classical red and white fibers. Although all three muscle types are lost through immobilization, type I oxidative fibers are generally lost at a higher rate.

The same muscle may vary in its histochemical composition in different animal species, depending on animal size, speed of movement, and metabolic rate. For example, **small animals** with a high basal metabolic rate and fast breathing rate have **diaphragm's** composed largely of small, red, type I skeletal muscle fibers rich in mitochondria. The diaphragm's of large mammals, such as the **cow,** with slow, phasic respiration, consist chiefly of large, white, type IIB fibers with a low mitochondrial content. In contrast, mammals of intermediate size, such as the cat, have various mixtures of the three fiber types that are most appropriate for their particular patterns of respiratory activity.

The **heart** is composed primarily of type I aerobic muscle fibers that are specialized in the oxidation of FFAs, KB⁻s, glucose, lactate, and amino acids. Although the **LDH** of skeletal muscle functions as a pyruvate reductase in order to reoxidize the NADH produced from anaerobic glycolysis, the **LDH** of cardiac muscle catalyzes conversion of lactate to pyruvate, and it is inhibited by high concentrations of pyruvate.

OBJECTIVES

- Explain why type IIB skeletal muscle fibers generally possess greater force-generating capacity than type I fibers.

- Compare the biochemical and physiologic characteristics of type I skeletal muscle fibers to type IIB fibers, and explain the differences.

- Provide biochemical rationale for the importation of lactate by cardiac muscle during exercise, and its exportation by type IIB skeletal muscle fibers (see Chapters 5 & 6).

- Understand how the PFK activity of cardiac muscle is regulated during exercise.

- Recognize the skeletal muscle fiber type compositions of canine and equine athletes.

- Explain why the force-generating capacity of heart muscle is generally less than that of skeletal muscle.

- Discuss differences between pathologic and physiologic hypertrophy.

QUESTIONS

1. Which one of the following characteristics applies more to type I skeletal muscle fibers than to type IIB fibers?
   a. Glycolytic/anaerobic
   b. Low hexokinase activity
   c. Well perfused
   d. Low resistance to fatigue
   e. Fast-twitch

2. Which one of the following animals would be expected to have the highest proportion of type I skeletal muscle fibers?
   a. Fish
   b. Quarterhorse
   c. Greyhound
   d. Thoroughbred
   e. Mongrel

3. Select the FALSE statement below:
   a. The myosin ATPase of type I skeletal muscle fibers is comparatively more active than that found in type IIB fibers.
   b. The ratio of type I to type IIB muscle fibers in an animal is fixed at an early stage of development.
   c. Heart muscle uses lactate for energy purposes.
   d. Type IIB skeletal muscle fibers are innervated by comparatively large motor neurons, which transmit impulses rapidly.
   e. Heart muscle is normally 99% aerobic.

4. Select the TRUE statement below:
   a. Characteristics of type IIA skeletal muscle fibers are generally "intermediate" between those of type I and type IIB fibers.
   b. Game bird pectoral muscles are highly aerobic.
   c. Type IIB skeletal muscle fibers contain low levels of hexokinase because they seldom degrade glycogen.
   d. It generally takes longer to obtain ATP from anaerobic metabolism than it does from aerobic metabolism.
   e. Ketone bodies cannot be oxidized by myocardial cells.

5. The lactate dehydrogenase of type IIB skeletal muscle fibers normally functions as a:
   a. Glucose transporter (GLUT-4).
   b. Pyruvate reductase.
   c. Kinase.
   d. Intracellular second messenger.
   e. TCA cycle enzyme.

6. Type I skeletal muscle fibers usually have a more extensive sarcotubular system than type IIB fibers.
   a. True
   b. False

6. b
5. b
4. a
3. a
2. e
1. c

ANSWERS

Chapter 81

Exercise (Athletic Animals)

Overview

- Aerobic training increases $\dot{V}O_{2(max)}$, muscle perfusion, mitochondrial number, oxidative capacity, and insulin sensitivity following exercise.
- Anaerobic training can increase glycogen storage capacity, and hypertrophy of type IIB muscle fibers can increase force generation.
- Splenic contraction helps to increase the O_2 carrying capacity of blood in exercising horses and dogs.
- Exercising horses and dogs maintain a high arterio-venous O_2 difference in exercising muscle beds, as well as a high cardiac output.
- Soda loading may not help the sprinter, or the aerobic athlete.
- Muscle fatigue is caused by proton accumulation, glycogen depletion, and/or dehydration.
- Proton accumulation during exercise lowers the V_{max} of PFK, reduces Ca^{++} release from the SR, reduces myosin ATPase activity, and negatively affects the conformation of contractile proteins.

Muscle Fatigue

Fatigue of muscles during exercise is a phenomenon we have all experienced. A primary cause appears to be proton accumulation in both intra- and extracellular fluid, not lactate. This fact has been demonstrated by infusing sodium lactate into blood, and observing that fatigue does not follow. An **increase of protons** (i.e., **H^+; ↓intracellular pH**) can affect the function of muscle in a number of ways:

- **Lower the V_{max} of phosphofructokinase (PFK).**
- **Reduce Ca^{++} release from the sarcoplasmic reticulum (SR).**
- **Reduce activity of myosin ATPase.**
- **Negatively affect conformation of contractile proteins.**

Another contributor to fatigue is **muscle glycogen depletion**. When muscle glycogen stores are depleted, only fatty acid oxidation can provide energy, which will only satisfy about 50% of maximal power output (see Chapter 79). The final factor contributing to muscle fatigue is intra- and extracellular **dehydration**, which severely compromises many cellular metabolic processes.

Soda loading consists of ingesting **$NaHCO_3$** in an attempt to buffer production of protons during exercise. Since HCO_3^- remains largely in extracellular spaces, it is unlikely to have any effect in **short sprints** since most protons produced in that effort remain in muscle during the sprint. However, it may be of some benefit in **mid-distance exercise**, since protons will be moving out of muscle and into blood. It would appear to be of little benefit, however,

Copyright © 2015 Elsevier Inc. All rights reserved.

in **aerobic exercise**, where the production of lactic acid is minimized.

Athletic Animals

Domestic animals selectively bred and trained for athletic prowess (e.g., **racehorses** and racing **greyhounds**) exhibit the combined effects of heredity and physical conditioning. Undoubtedly, natural selection accomplishes the same thing, perhaps even more rigorously, for feral species such as the **cheetah** and **pronghorn antelope** can attain amazing overland speeds in excess of 60 mph.

The **greyhound**, whose history dates back some 5000 years, is perhaps the fastest canine sprinter. In contrast, **sled dogs** (huskies) retain considerable endurance capacity, and are therefore **type I** aerobic athletes. The **thoroughbred**, which has been selectively bred for only about 300 years, is also a good aerobic athlete, whereas the **American Quarter Horse** is a good 400 meter sprinter (see Chapter 80).

Remarkable alterations in **physiologic variables** with exercise in horses and dogs are presented in **Table 81-1**. Most striking among the variables is the enormous **increase in maximal O_2 consumption** that can reach 40-times the resting value in exercising horses. In contrast, O_2 consumption only increases about 10-fold in unconditioned mongrel dogs, and up to 20-fold in humans with maximal exercise (values not shown). Although the 14-fold increase in cardiac output of the horse cannot entirely account for this, an increased hemoglobin (Hb) concentration of blood and an increased arteriovenous O_2 difference in muscle are thought to contribute. The 7- to 8-fold increase in heart rate with exercise in horses, and the 10-fold increase in greyhounds is remarkable since in humans only about a 3-fold increase can occur. Average heart weight to body weight ratios are about 9 and 12 gm/kg for horses and greyhounds, respectively, but only about 4 gm/kg in humans.

The **spleen** acts as an erythrocyte reserve. Release of stored erythrocytes into the systemic circulation is under the influence of the **sympathetic nervous system (SNS)**, which innervates the spleen, and circulating catecholamines also help to promote contraction. Factors increasing

Table 81-1						
Comparative Cardiovascular Characteristics						
	Horse		**Greyhound**		**Unconditioned Mongrel**	
	Rest	**Exercise**	**Rest**	**Exercise**	**Rest**	**Exercise**
Heart wt/body wt (gm/kg)	9	--	12	--	9	--
Heart rate (beats/min)	30	210-250	29-48	290-420	61-117	220-325
Stroke vol (ml)	825	1475	55	--	27	--
Cardiac output (L/min)	25	310-369	1.6-2.6	--	1.6-3.2	--
Mean arterial blood pressure (mmHg)	105	170	118	--	98	--
Hemoglobin (gm/dl)	12-14	17-24	19-20	23-24	14	--
Hematocrit (%)	40-50	60-70	50-55	60-65	45	--
O_2 consumption (ml/min/kg)	3-4	120-160	--	180	8	85

Legend: Data from Alonso FRA, 1972; Carew TE, Covell JW, 1978; Courtice FC, 1943; Cox RH, Peterson LH, Detweiler KD, 1976; Detweiler KD, Cox RH, Alonso FR, et al, 1974; Donald DW, Ferguson DA, 1966; Erickson HH, 1993; Evans DL, Rose RJ, 1988; Hopper MK, 1989; Hopper MK, Pieschl RL, Pelletier NG, et al, 1983; Meixner R, Hornicke H, Ehrlein H, 1981; Rose RJ, 1985; Snow DH, Harris RC, Stuttard E, 1988; Staaden RV, 1980; and von Engelhardt W, 1977.

SNS activity, such as asphyxia, hemorrhage, excitement, and exercise, will thus promote **splenic contraction**, and thus increase the hematocrit (Hct). As seen in **Table 81-1**, the Hct of both horses and dogs increases substantially following the onset of exercise, which is partially a result of splenic contraction. However, there is an additional reduction in plasma volume during exercise attributed to a small fluid shift from intravascular to extravascular spaces, and sweating also causes dehydration. Although there is a net gain in the O_2 carrying capacity of blood, these combined influences have a tendency to increase blood viscosity, which can put undue stress on the heart, and if it continues, also reduce capillary perfusion.

Total Hb and blood volume levels increase in these athletes with training, heart weight to body weight ratios continue to increase, and the heart rate and blood pressure decrease for a given work load.

In summary to this point, the horse can increase O_2 consumption during exercise far more than other animals studied because of:

1) **The great increase in heart rate,**

2) **Mobilization of erythrocytes from the spleen that can substantially increase the Hb concentration, and**

3) **A marked increase in the arterio-venous O_2 difference of perfused tissues.**

Table 81-2
Physiologic Benefits of Training

Increase aerobic capacity (↑ $\dot{V}O_{2(max)}$)

O_2 Uptake by skeletal muscle can be increased up to 20-fold.

O_2 Uptake by cardiac muscle can be increased up to 4-fold.

↑ Muscle myoglobin (store more O_2)

↑ Cardiac output

Muscle hypertrophy

Sarcomeres will increase in both series and in parallel, muscle fibers will branch, but they will not increase in number.

Cardiac hypertrophy, and a decrease in the resting heart rate of some animals.

Increase in blood vessel size

↑ Capillary beds in exercising muscle

More O_2 delivery to aerobic tissues

Increase in mitochondrial number and size

More enzymes of the TCA cycle and oxidative phosphorylation

Greater aerobic potential

Less formation of lactate and more efficient utilization of pyruvate

Increased KB^- and FFA oxidation

Increase in glycogen storage capacity (up to 2-fold)

Increased insulin sensitivity after exercise

Increase in glycolytic enzymes

Increase in lactate tolerance

↑ in H^+ tolerance

The cardiovascular characteristics distinguishing greyhounds from mongrels also show distinct features (**Table 81-1**). Clearly, these specialized hemodynamic, cardiac and hemic features of greyhounds, like those in horses, favor maximal O_2 delivery to muscles during exercise. The relatively large heart is congenital, but may be further developed (physiologic hypertrophy) through physical conditioning. The large stroke volume, compared to mongrels, permits greater cardiac output at slower heart rates. The Hb concentration and Hct are generally reported to be higher at rest than in mongrels, and both increase substantially following the onset of exercise. Those two factors, together with the high cardiac output, are thought to account for the much higher maximal O_2 uptake of greyhounds vs. mongrels. A **higher carotid sinus operating point** in greyhounds allows for a higher mean arterial blood pressure to be maintained during exercise. It is possible that the higher greyhound blood pressure has evolved through selective breeding for racing performance as an extra safety factor to ensure adequate perfusion of all organs when the drop in peripheral resistance with exercise is extreme.

Benefits of Conditioning

There are a number of mechanisms by which training can improve both anaerobic and aerobic exercise performance (see **Table 81-2**). However, the effects of training will differ according to whether an untrained animal is being trained, or whether a trained athlete is being brought up to peak performance. Considerably more is known about the former situation since experimental animals can help to provide such insights, but some speculative suggestions will be made regarding the athlete.

Perhaps the most important benefit for the unconditioned animal is the **increase in $\dot{V}O_{2(max)}$** that can be attained through aerobic training. This variable reflects **improved cardiac output** and **improved removal of O_2** by exercising muscles. There is also an important effect on the distribution of blood after training. Somewhat surprisingly, less of the cardiac output is directed towards muscles so that more goes to the liver and other organs. This improved hepatic blood flow following exercise may be particularly important in permitting the liver to assist in the overall metabolic reparative process. Additionally, **capillary beds** in exercising muscle tissues may be **increased** by as much as **60%** with endurance training, thus providing more area for gas, nutrient, and waste exchange.

Increased mitochondrial number and **size** provides increased capacity for metabolizing substrates through aerobic oxidative pathways. As would be expected, these changes are found in **type I** muscle fibers, so that the aerobic capacity and, in particular, the capacity to oxidize fatty acids is increased. **Insulin sensitivity** following exercise is also increased with endurance training. For anaerobic athletes, **glycogen storage capacity is increased**. There is also an increase in the amount of **glycolytic enzymes**, and **type IIB** muscle **hypertrophy** increases force generation.

In well-conditioned athletes, continued training maintains the above attributes, and also further increases **capacities of key enzymes** in both anaerobic and aerobic metabolism. **Lactate tolerance** (i.e., H^+ tolerance) is **increased**, but perhaps more importantly, well-conditioned athletes **form less lactate**, and therefore **utilize pyruvate more efficiently**.

In conclusion, discussions of exercise metabolism featured in the previous five chapters are important not only because they provide biochemical/physiologic reference to the contents of chapters throughout this text, but because they have considerable implications in both human and veterinary medicine. For example, there are a number of conditions characterized by fatigue upon mild exercise (e.g., ageing, post-surgical recovery, or peripheral vascular disease), which can markedly reduce the quality of life. In addition, the enormous increase in the number of people and animals throughout the world participathing in sporting and other endurance activities raises the question of the potential health benefits these activities possess, which can only be properly considered with a knowledge of metabolic changes that occur during exercise.

OBJECTIVES

- Identify six primary factors that could contribute to muscle fatigue during exercise.

- Recognize why $NaHCO_3$ loading is unlikely to help the short-sprint anaerobic or aerobic endurance athletes.

- Explain why the Hct of horses and dogs usually increases following the onset of exercise.

- Identify and discuss physiologic alterations in canine and equine athletes that lead to increased O_2 consumption during exercise.

- Explain why heart rate and blood pressure are usually low (for a given work load) in the well-trained athlete.

- Know why the arterio-venous O_2 difference of highly perfused exercising muscle tissues is high in well-trained athletes.

- Provide an explanation for the high carotid sinus operating point of greyhounds during exercise.

- Discuss the physiologic benefits of aerobic training in terms of $VO_{2(max)}$, muscle perfusion and hypertrophy, cardiac output, mitochondrial number and size, glycogen storage capacity, lactate tolerance and pyruvate utilization.

QUESTIONS

1. All of the following contribute to muscle fatigue, EXCEPT:
a. A reduced PFK V_{max}.
b. Lactate.
c. Dehydration.
d. Muscle glycogen depletion.
e. Proton accumulation.

2. The exercising horse appears to exhibit a:
a. Remarkable increase in heart rate.
b. Significant decrease in mean arterial blood pressure.
c. Slight decrease in stroke volume.
d. Two-fold increase in the hematocrit.
e. Decreased arterio-venous O_2 difference in exercising muscle beds.

3. Which one of the following organs adds erythrocytes to the circulation during exercise?
a. Brain
b. Kidney
c. Lung
d. Spleen
e. Duodenum

4. Soda loading may help the athlete in which way?
a. Add HCO_3^- to intracellular fluids.
b. Buffer extracellular protons in short sprints.
c. Buffer intracellular protons in aerobic exercise.
d. Buffer extracellular protons in mid-distance runs.
e. Enhance Ca^{++} absorption from the digestive tract.

5. Select the TRUE statement below regarding training:
a. Less of the cardiac output is directed towards muscles after aerobic training so that more goes to vital organs.
b. Training has no effect on lactic acid tolerance.
c. Physiologic training has no effect on heart size.
d. Blood vessel size will not change with training.
e. Trained athletes utilize pyruvate less efficiently.

6. Which one of the following factors allows greyhounds to maintain a high mean arterial blood pressure during exercise?
a. Increased heart rate
b. Low peripheral resistance
c. Increased cardiac output
d. A higher carotid sinus operating point
e. A high Pco_2

6. d
5. a
4. d
3. d
2. a
1. b

ANSWERS

Sections V and VI Examination Questions

1. **In order to become effective substrates for lipoprotein lipase (LPL), nascent CMs and VLDL must:**

   a. First become remnant particles.
   b. Acquire Apo-A from HDL in blood.
   c. Pick up cholesterol esters from HDL in blood.
   d. Acquire Apo-B_{100} and Apo-B_{48}, respectively, from LDL in blood.
   e. None of the above

2. **Microbial deconjugation and 7-α-dehydroxylation of taurochenodeoxycholate in the large bowel leads to formation of:**

   a. Cholestanol.
   b. Mevalonate.
   c. Lithocholate.
   d. Progesterone.
   e. Glycocholate.

3. **Select the FALSE statement below:**

   a. Hypertriglyceridemic animals are also hyperlipidemic and usually lipemic.
   b. Most LDL is removed from blood by high-affinity hepatic scavenger receptors.
   c. The liver responds to hypoproteinemia by increasing its synthesis of plasma proteins, including the APO-B_{100} of VLDL.
   d. Dietary medium-chain TG oils do not contribute to TG and CM formation in jejunal mucosal cells.
   e. The Watanabe Heritable-Hyperlipidemic (WHHL) rabbit is hypercholesterolemic.

4. **Select the TRUE statement below regarding eicosanoids:**

   a. TXB_2 is typically conjugated with taurine in the cytoplasm of hepatocytes, and then secreted into bile.
   b. Non-steroidal anti-inflammatory drugs (NSAIDs) are thought to act by inhibiting the activity of COX-2, thereby reducing leukotriene formation.
   c. Prostaglandins are known to disrupt the gastric mucosal barrier.
   d. Leukotrienes, formed through the lipoxygenase pathway, promote contraction of bronchiolar smooth muscle.
   e. Thromboxanes are known to relax vascular smooth muscle.

Copyright © 2015 Elsevier Inc. All rights reserved.

5. **The compound to the right is:**

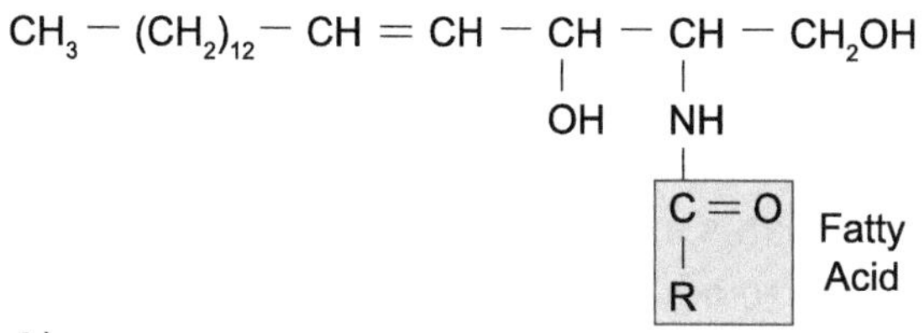

$$CH_3 - (CH_2)_{12} - CH = CH - CH - CH - CH_2OH$$

with OH and NH substituents, the NH connected to $C = O$ (R) labeled Fatty Acid.

   a. A sphingophospholipid.
   b. Pulmonary surfactant.
   c. β-Hydroxybutyrate.
   d. The substrate for ceramidase.
   e. 3-Hydroxy-3-methylglutaryl-CoA (HMG-CoA).

6. **Select the FALSE statement below regarding exercise:**

   a. Aerobic training is associated with cardiac muscle hypertrophy.
   b. The plasma glucose concentration is partially maintained during exercise through muscle glycogen breakdown (i.e., Muscle Glycogen → Glc-1-P → Glc-6-P → Glucose → Blood).
   c. During anaerobic exercise, lactate⁻ (from skeletal muscle) may be oxidized by cardiac muscle tissue.
   d. Ketone body formation is normally low during exercise.
   e. Fat contains less oxygen than carbohydrate, and requires more oxygen for oxidation. Therefore, the RQ of an athlete burning primarily fat is lower than that of one burning primarily carbohydrate.

7. **The following reaction:**

                              Enz.

CH₃-CO~S-CoA + Carboxybiotin --------> HOOC-CH₂-CO~S-CoA + Biotin

   a. Demonstrates how LCFA gets activated on the outer mitochondrial membrane surface before entering mitochondria.
   b. Shows the conversion of acetyl-CoA to acetoacetyl-CoA.
   c. Is the committed step in cytoplasmic fatty acid biosynthesis.
   d. Is catalyzed by malic enzyme.
   e. Occurs only in mucosal cells of the small intestine.

8. **VLDLs:**

   a. Are somewhat larger than CMs.
   b. Possess a higher lipid:protein ratio than CMs.
   c. Contain apoprotein B_{48}.
   d. Have a lower density than CMs.
   e. Unlike CMs, will not separate in a blood sample left standing to form a creamy layer on top of plasma.

9. **Acyl-CoA cholesterol acyltransferase (ACAT):**

   a. Functions in the intracellular esterification of cholesterol.
   b. Is the rate-limiting enzyme in cholesterol biosynthesis.
   c. Circulates in plasma and is generally known as cholesterol ester transfer protein.
   d. Helps to add fatty acids to glycerol-3P in the Phosphatidic Acid Pathway.
   e. Is an integral enzyme present in the circulating HDL bilayer.

10. **The compound to the right is:**
    (Note: The boxed structure is a LCFA.)

    a. Lysolecithin.
    b. Acylcarnitine.
    c. Taurocholate.
    d. Creatinine.
    e. Monoglyceride.

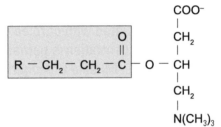

11. **The degradation of purine nucleotides in exercising muscle tissue gives rise to several compounds that stimulate PFK activity. Which one of the following is an EXCEPTION?**

    a. Adenosine
    b. ADP
    c. Pi
    d. AMP
    e. NH_4^+

12. **The pre-mortal rise in urinary nitrogen excretion that occurs during the late stages of starvation is largely due to an increase in:**

    a. Liver glycogen depletion.
    b. Adipocyte lipolysis.
    c. The blood urea nitrogen (BUN) concentration.
    d. The creatinine concentration of plasma.
    e. Urinary amino acid excretion.

13. **The structure below is:**

    $$CH_3COCH_2COO^-$$

    a. Acetoacetate.
    b. Palmitate.
    c. Lecithin.
    d. PGI$_2$.
    e. Ceramide.

14. Select the FALSE statement below:

a. Muscle normally contains 2-3 times more $C\sim PO_3$ than ATP.
b. Ketone bodies are synthesized from HMG-CoA.
c. $VO_{2(max)}$ may be decreased through long periods of physical inactivity.
d. Blood glucagon and cortisol levels normally decline during sustained, aerobic exercise.
e. It requires longer to generate ATP from aerobic metabolism in muscle tissue than from anaerobic glycolysis.

15. Malonyl-CoA:

a. Enters gluconeogenesis at the level of succinyl-CoA, and activates CPT-1 in hepatocytes.
b. Is found in adipocytes of carnivores, but not in those of ruminant animals.
c. Is an allosteric activator of CPK_m, the rate-limiting enzyme in fatty acid β-oxidation, and acts in the hypothalamus to stimulate hunger.
d. Formation initially involves covalent binding of acetyl-CoA to vitamin B_{12}.
e. Serves as an activated donor of acetyl groups in fatty acid biosynthesis.

16. A carnitine deficiency could lead to steatosis by impairing hepatic:

a. Triglyceride formation.
b. Cytoplasmic fatty acyl-CoA formation.
c. VLDL egress.
d. β-Oxidation of fatty acids.
e. All of the above

17. Glucose entering adipocytes can potentially take all of the following   intracellular metabolic routes, EXCEPT:

a. Complete oxidation through the Embden-Meyerhoff Pathway (EMP) and TCA Cycle.
b. Conversion to fatty acyl-CoA.
c. Ketone body formation.
d. Conversion to glycerol-3-phosphate, the backbone for TG formation.
e. Glycogen formation.

18. On average, 60% of the cholesterol normally present in the lumen of the canine small intestine is that derived from:

a. Desquamated mucosal cells of the GI tract.
b. Bile.
c. Endogenous VLDL.
d. The diet.
e. The pancreas.

19. Unesterified long-chain fatty acids (LCFAs):

a. Circulating in peripheral blood either entered there from mucosal cells of the small intestine, or the liver.
b. Are hydrophilic.
c. Generally remain unesterified in adipocytes.
d. Are rarely, if ever, found circulating in plasma.
e. Are generally referred to as free fatty acids (FFAs).

20. Creatinuria:

a. Refers to the presence of increased amounts of creatinine in urine.
b. Occurs only when the kidneys can no longer adequately filter plasma.
c. Is best associated with hypercholesterolemia, and high circulating titers of LDL.
d. May represent an abnormal condition in which increased amounts of creatine appear in urine due to muscle breakdown (i.e., myopathy).
e. A and C above

21. Compared to the resting state, exercise (i.e., running) normally decreases:

a. Stroke volume (of the heart).
b. End-systolic volume.
c. Skeletal muscle perfusion.
d. End-diastolic volume.
e. Respiratory rate.

22. A gluco- or galactosylceramide is:

a. By definition, sphingomyelin.
b. A phospholipid formed when sphingosine reacts with diacylglycerol.
c. A cerebroside.
d. A phospholipid formed when ceramide reacts with phosphatidylcholine.
e. A normal membrane component derived from cholesterol.

23. Select the FALSE statement below:

a. Less than 5% of dietary plant sterols are normally absorbed by the mammalian digestive tract.
b. An elevation in the cytoplasmic cholesterol concentration of hepatocytes reduces hepatic LDL receptor synthesis.
c. Fatty acids must first undergo a reaction catalyzed by Acyl-CoA Synthetase before being esterified to Glycerol 3-Phosphate in phospholipid biosynthesis.
d. Fatty acids can be elongated beyond palmitate in brain mitochondria using acetyl-CoA as a sequential two-carbon donor.

e. Fatty acyl-CoA is used for mitochondrial β-oxidation, but not for cytoplasmic esterification during triglyceride formation.

24. **CH$_3$(CH$_2$)$_3$CH=CH(CH$_2$)$_3$CH=CH(CH$_2$)$_6$COOH** can be described as:

a. 9:2;3,6
b. Δ^{3,6}18:2
c. An omega-5 UFA
d. Palmitate
e. n^{18}:5;8,13,15,17,18

25. **Select the FALSE statement below:**

a. Phosphatidylethanolamine is found in cell membranes.
b. Renal tubular epithelial cells can oxidize ketone bodies.
c. Both ACAT and LCAT participate in the formation of cholesterol esters.
d. If the hepatocyte is actively engaged in FA biosynthesis, CPT-1 activity in the outer mitochondrial membrane is usually reduced.
e. When ketone bodies accumulate in the circulation, they have a tendency to reduce the anion gap.

26. **The compound to the right is:**
   (Note: R$_2$ = Fatty acid hydrocarbon chain)

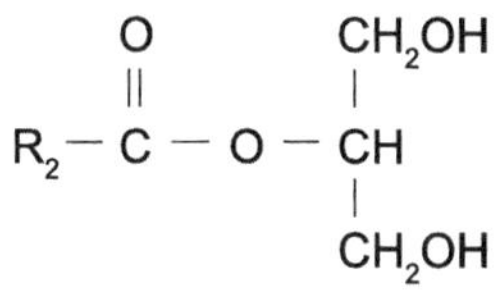

$$\begin{array}{ccc} & O & CH_2OH \\ & \| & | \\ R_2 - & C - O - & CH \\ & & | \\ & & CH_2OH \end{array}$$

a. Normally found in circulating VLDL.
b. A product of hormone sensitive lipase action, normally found circulating in 2-Monoglyceride plasma.
c. Diacylglycerol.
d. A product of pancreatic lipase digestive action, normally found in the lumen of the small intestine.
e. A ceramide.

27. **The cellular concentration of which one of the following is best associated with the rate at which pyruvic acid is converted to lactic acid in an exercising skeletal muscle cell?**

a. Glucose
b. NADH
c. Glycogen
d. Oxaloacetic acid
e. Carnitine

28. Which of the following are INCORRECTLY matched?

  a. Na^+/K^+-ATPase : Symport
  b. Phosphatidylethanolamine → Phosphatidylcholine : 3 SAMs
  c. Ascites : Hypovolemia
  d. Lysophosphatidylcholine → Glycerophosphocholine + SAF : PLA_1
  e. Leukotrienes : Slow reacting substances of anaphylaxis

29. Which one of the following compounds is used to activate phosphatidic acid for subsequent glycerophospholipid biosynthesis?

  a. Inositol
  b. cAMP
  c. Cytidine triphosphate
  d. 1,2-Diacylglycerol
  e. Choline

30. Which one of the following would correlate best with the degree of protein catabolism during exercise?

  a. The respiratory quotient (RQ)
  b. Urinary nitrogen excretion
  c. The plasma NH_4^+ concentration
  d. The plasma glycine concentration
  e. $VO_{2(max)}$

31. Which of the following are INCORRECTLY matched?

  a. Long-chain fatty acyl-CoA : Reduces acetyl-CoA carboxylase activity
  b. Adenosine : Vasodilator
  c. MCFAs : Hepatic β-oxidation
  d. γ-Butyrobetaine : Carnitine
  e. HDL : 57% Triglyceride

32. Hepatic cytoplasmic malate dehydrogenase- and malic enzyme-catalyzed reactions involved in fatty acid biosynthesis:

  a. Are vitamin B_1 and B_{12}-dependent reactions, respectively.
  b. Ultimately transfer reducing equivalents from NADH to $NADP^+$.
  c. Help to convert acetyl-CoA to acetate.
  d. Effectively transfer hydrocarbons from acetyl-CoA to malonyl-CoA.
  e. Use pyruvate and citrate, respectively, as substrates.

33. Select the TRUE statement below regarding cell membranes:

a. Practically all integral membrane proteins are derived from circulating lipoproteins.
b. Membrane components are routinely internalized, digested and replaced by the cell.
c. Carbohydrates occur in plasma membranes in association with protein, but not lipid.
d. Cholesterol is found in the cytoplasm of cells, but not in membranes.
e. Although plasma membranes contain lipids, they do not possess enzymes.

34. Select the FALSE statement below:

a. Linoleate is an unsaturated fatty acid.
b. Gangliosides are synthesized from glucocerebrosides by sequential monosaccharide addition.
c. Circulating eicosanoids, like circulating plasma proteins, are slowly degraded in the body.
d. Insulin stimulates hepatic HMG-CoA reductase activity, and thus hepatic cholesterol biosynthesis.
e. Glycolipids provide antigenic chemical markers for cells.

35. The number of low-density lipoprotein (LDL) receptors on hepatocytes has been found to DECREASE in response to high plasma LDL concentrations, thus preventing them from internalizing more of the LDL's cholesterol than is needed by these cells:

a. This animal clearly has Diabetes Mellitus.
b. An example of receptor down-regulation.
c. This animal has hypothyroidism.
d. An example of the effects of an HMG-CoA reductase inhibitor.
e. This animal has liver disease.

36. Which one of the following contains peptide bonds?

a. Lecithin
b. Propionate
c. Hormone sensitive lipase
d. Leukotriene D_4
e. Progesterone

37. Which of the following are INCORRECTLY matched?

a. Cyclooxygenase : Uterine endometrial cells
b. Glycosphingolipids : Plasma membranes
c. 15-HPETE : Prostanoid
d. Steatosis : Inhibition of hepatic VLDL formation
e. Leukotrienes : Slow-reacting substance of anaphylaxis (SRS-A)

38. **Slow-twitch, oxidative skeletal muscle fibers (in contrast to type IIB fast-twitch fibers), normally possess:**

    a. A far more extensive sarcotubular system.
    b. More CPK_g and glycogen phosphorylase activity.
    c. Greater myosin ATPase activity.
    d. A greater resistance to fatigue.
    e. Larger cell diameters.

39. **The rate-limiting enzyme in hepatic fatty acid biosynthesis that is inhibited by glucagon and epinephrine, is:**

    a. Acetyl-CoA carboxylase.
    b. Also inhibited by insulin.
    c. HMG-CoA reductase.
    d. COX-2.
    e. Protein kinase C (PKC).

40. **The most metabolically active circulating lipids are:**

    a. CMs.
    b. LDLs.
    c. NEFAs.
    d. HDLs.
    e. VLDLs.

41. **The compound to the right is:**

    a. An intracellular second messenger involved with the stimulation of lipolysis.
    b. A known activator of protein kinase C (PKC).
    c. An eicosanoid which causes platelet aggregation and vasoconstriction.
    d. A sphingolipid found in circulating LDL.
    e. The substrate for CPK_c found in skeletal muscle fibers.

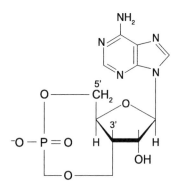

42. **Select the FALSE statement below regarding FA biosynthesis:**

    a. Carbohydrates and amino acids can be used for hepatic FA biosynthesis by way of citrate in dogs.
    b. NADPH is required for FA biosynthesis in all animals.
    c. Intracellular cAMP is a positive allosteric activator of the rate-limiting enzyme in cytoplasmic FA biosynthesis.

d. Intermediates in FA biosynthesis are attached to a protein possessing an important prosthetic group formed from pantothenic acid (vitamin B_5).

e. Propionate is occasionally used by the lactating mammary gland of ruminant animals as a priming molecule for FA biosynthesis.

43. Select the FALSE statement below regarding ketone bodies (KB's):

a. KB formation is similar to that of cholesterol, since both are cytoplasmic processes using HMG-CoA as an intermediate.

b. Although KB's are freely filtered by the kidneys, they are extensively reabsorbed in proximal portions of nephrons.

c. KB's are preferred fuel over glucose in the CNS during the intermediate stage of starvation.

d. Acetone is formed through the spontaneous decarboxylation of AcAc, and may be detected in urine when the plasma concentration of AcAc is high.

e. KB's, like volatile fatty acids (VFAs), are freely soluble in plasma, and therefore are not protein-bound.

44. Bile:

a. Acid synthesis is regulated by LCAT.

b. Is an important vehicle for cholesterol & conjugated bilirubin excretion.

c. Normally contains high concentrations of cholesterol-rich LDL.

d. Salts are Na^+ and/or K^+ salts of bilirubin.

e. Acids are normally excluded from the enterohepatic circulation (EHC).

45. Select the FALSE statement below:

a. The rate of hepatic gluconeogenesis usually declines during the intermediate stage of starvation.

b. Approximately 60% of amino acid release from muscle consists of alanine and glutamine during the late stages of starvation.

c. The kidney develops an increased dependence upon glutamine as a gluconeogenic substrate during the intermediate stages of starvation.

d. There is usually a pre-mortal increase in gluconeogenesis during the late stages of starvation (following body fat depletion).

e. As β-OH-butyrate is taken up by muscle during the fat-mobilizing, intermediate stages of starvation, it may be returned to blood as acetoacetate (thus signifying a high intracellular $NADH/NAD^+$ ratio).

46. A feline patient with an elevated plasma triglyceride, yet normal cholesterol concentration, would be expected to exhibit:

a. An apoprotein B_{48} or B_{100} deficiency.
b. Abnormally elevated plasma LDL levels.
c. Diarrhea.
d. Decreased plasma HDL and LDL levels.
e. Increased plasma Pre-β and/or CM lipoprotein levels.

47. Select the FALSE statement below regarding cell membrane transport:

a. Ion transport across cell membranes helps to maintain both intra- and extracellular pH.
b. Plasma membranes contain water-filled channels through which water and hydrated inorganic ions can diffuse.
c. Polar compounds are generally more permeable to cell membranes than non-polar compounds.
d. Symport proteins in the cell membrane require the binding of two substances, and both are transported across the membrane together.
e. Facilitated diffusion of molecules across cell membranes does not require metabolic energy.

48. Select the FALSE statement below:

a. Horses are thought to possess a greater hepatic exporting capacity for VLDL than ruminant animals.
b. Sphingolipidoses in animals are generally classified as lipid storage diseases.
c. CPK_g participates in the dephosphorylation of $C{\sim}PO_3$ in skeletal muscle cells.
d. Dogs and cats possess the 4 main classes of lipoprotein found in primates (i.e., VLDL, LDL, HDL & CMs).
e. TXA_2 is best associated with platelet aggregation.

49. Select the FALSE statement below regarding mitochondrial FA β-oxidation:

a. Medium-chain FAs normally undergo β-oxidation in the cytoplasm of liver cells.
b. FAs with an odd number of C atoms are oxidized until a 3-C propionyl-CoA residue remains.
c. Five high-energy phosphate bonds are generated (via oxidative phosphorylation of 1 NADH and 1 $FADH_2$) for each acetyl-CoA produced in β-oxidation.
d. β-Oxidation of both saturated and unsaturated FAs occurs in mitochondria of skeletal muscle tissue.
e. LCFAs use the carnitine transporter system in mitochondrial membranes before becoming substrates for β-oxidation.

50. Select the FALSE statement below:

a. The liver is the primary source of plasma lipoproteins derived from endogenous sources.
b. Low density lipoproteins (LDLs) are primarily composed of apoprotein-B_{100}, cholesterol esters and phospholipids. Smaller amounts of free cholesterol and triglyceride are also present.
c. The fatty acid (18:3;9,12,14) is unsaturated, and has 3 double bonds, one each located to the left of carbon atoms 9, 12 and 14 (note: the carboxyl group is on the right side of this structure).
d. After palmitoyl-CoA is formed in lipogenesis, it can be further elongated and/or desaturated by a variety of enzymes.
e. Intestinal mucosal cells, like hepatocytes of the adult animal, do not normally oxidize ketone bodies.

51. Select the TRUE statement below:

a. Phosphatidylinositol is usually the most plentiful phospholipid in plasma membranes.
b. Cardiolipin is a plasmalogen.
c. Pyrimidines are involved in nucleic acid biosynthesis, but not phospholipid biosynthesis.
d. COX-2 activation is associated with reparative processes of the digestive tract.
e. All phospholipids possess a glycerol backbone.

52. The compound below:

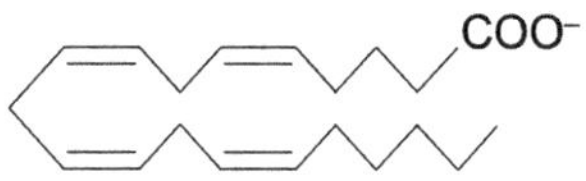

a. Is synthesized in hepatocytes from cholesterol.
b. Forms PGG_2 when acted upon by COX-1.
c. Is sphingosine, the precursor to ceramide.
d. Is an omega-5 polyunsaturated fatty acid.
e. Is a ketone body that alters the anion gap as it accumulates in plasma.

53. The structure below is:
(Note: R_1 & R_2 = FA chains)

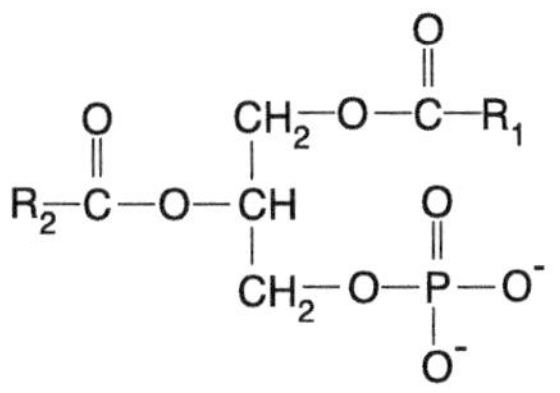

a. Prostanoic acid.
b. An intermediate in both hepatic TG & PL biosynthesis.
c. Diacylglycerol.
d. Normally found in circulating CMs.
e. Hormone sensitive lipase (HSL).

54. Which one of the following DOES NOT normally traverse the inner mitochondrial membrane?

a. Acetoacetate
b. Propionate
c. Acetyl-CoA
d. ATP
e. Citrate

55. Hepatic conversion of cholesterol to bile acid is normally reduced by:

a. Return of taurocholate (and other bile acids) through the EHC.
b. Vitamin C.
c. HMG-CoA lyase.
d. Cholesterol entering hepatocytes from circulating LDL.
e. Plamalogen.

56. All of the following enzymes are directly linked to NADPH generation, EXCEPT:

a. Citrate cleavage enzyme (citrate lyase).
b. Cytoplasmic isocitrate dehydrogenase.
c. Glucose 6-phosphate dehydrogenase (Glc-6-PD).
d. Malic enzyme.

57. Which one of the following lipases, if any, is activated (indirectly) by epinephrine?

a. Hormone sensitive lipase
b. Hepatic lipase
c. Pancreatic lipase
d. Lipoprotein lipase
e. None of the above

58. Select the TRUE statement below regarding exercise physiology:

a. Ketone body production by the liver would be expected to increase during vigorous exercise.
b. There is generally a decrease in insulin sensitivity several hours into the post-exercise period.
c. Type IIB fast-twitch muscle cells rely heavily on intracellular glycogenolysis for ATP generation.
d. C~PO_3 is a known activator of PFK in muscle tissue.
e. Circulating catecholamines (norepinephrine and epinephrine) stimulate pancreatic insulin release during exercise.

59. **During the early phase of starvation (1-3 days), you would expect the Respiratory Quotient (RQ = VCO$_2$/VO$_2$) to be _____ than/from that seen during the intermediate phase:**

   a. Lower
   b. Higher
   c. Unchanged

60. **All of the following factors are thought to significantly contribute to death of the organism in the final phase of starvation, EXCEPT:**

   a. Loss of myofibrillar protein from the diaphragm and intercostal muscles.
   b. Azotemia.
   c. Hypovolemic shock.
   d. Ketoacidosis.
   e. Decreased immunological response to an infection.

61. **Select the TRUE statement(s) below:**

   a. Starvation is associated with lower circulating insulin levels and an increased cAMP concentration in adipocytes, which leads to increased PEP carboxykinase activity.
   b. Under normal conditions, PGI$_2$ production by vascular endothelial cells is sufficient to override the effects of TXA$_2$ production by platelets, so that there is no significant vasoconstriction and platelet aggregation.
   c. If insufficient hepatic OAA is available to couple with acetyl-CoA in the condensing reaction of the citric acid cycle, ketogenesis will become favored following FA β-oxidation.
   d. The role of PGs during parturition is supported by the observation that NSAISs prolong gestation, thus resulting in protracted labor.
   e. All of the above

62. **Examine the 3 structures below:**

   1. CH_3-$(CH_2$-CH=$CH)_3$-$(CH_2)_7$-$COOH$
   2. CH_3-$(CH_2)_{14}$-$COOH$
   3. CH_3-$(CH_2$-CH=$CH)_3$-CH_2-CH=CH-CH_2-CH=CH-CH_2-CH=CH-$(CH_2)_2$-$COOH$

   a. Compound **#1** is an n-9 polyunsaturated fatty acid.
   b. Compound **#1** can be produced endogenously from compound **#3** through successive desaturation reactions.
   c. Compound **#2** is a saturated fatty acid present in pulmonary surfactant.
   d. Compound **#3** is an ω-3 fatty acid containing 4 double bonds.
   e. All of the above

63. **Which (one) of the following cannot penetrate luminal (apical) membranes of small intestinal mucosal cells?**

   a. Lysolecithin
   b. Long-chain fatty acids
   c. Monoglycerides
   d. Triglycerides
   e. All of the above

64. **Hypercholesterolemia in veterinary medicine is most often recognized:**

   a. During starvation.
   b. In association with hyperthyroidism.
   c. Following oral medium-chain triglyceride therapy.
   d. Immediately following exercise.
   e. As a secondary hyperlipidemia.

65. **Select the FALSE statement below:**

   a. Some mammals can synthesize arachidonate from linoleate.
   b. The liver is capable of converting a saturated fatty acid to an unsaturated form.
   c. All eicosanoids contain double bonds and 20 carbon atoms.
   d. Fatty acid biosynthesis (lipogenesis) in mammals typically results in the formation of palmitate, from which other fatty acids can be formed.
   e. Although hydrocarbons from amino acids can be converted to fatty acids in the canine liver, those from glucose cannot.

66. **Activation of which one of the following enzymes (if any) would most likely liberate arachidonic acid from plasma membrane-bound phospholipid?**

   a. Phospholipase C
   b. Hormone sensitive lipase
   c. HMG-CoA lyase
   d. Lipoprotein lipase
   e. None of the above

67. **Glucose entry into adipocytes is facilitated by the GLUT-4 transporter:**

   a. Fact
   b. Myth

68. Which of the following is/are TRUE regarding ketone body (KB⁻) formation and/or utilization?

a. KB⁻s are favored substrates for oxidation in brain tissue of the diabetic, hyperglycemic animal.
b. As starvation progresses, KB⁻ oxidation by muscle usually decreases, while that by brain tissue increases.
c. Insulin is known to enhance hepatic KB⁻ production.
d. KB⁻s can be used as gluconeogenic substrates, particularly during the intermediate stage of starvation.
e. All of the above

69. Select the FALSE statement below regarding the Ca⁺⁺ messenger system:

a. The turnover of plasma membrane phosphatidylinositides is a major transducing event in the Ca⁺⁺ messenger system, giving rise to 2 second-messengers -- inositol 1,4,5-triphosphate (IP_3), and diacylglycerol (DAG).
b. DAG initiates the flow of information in the protein kinase C-branch of the Ca⁺⁺ messenger system.
c. The Ca⁺⁺ and cAMP messenger systems bring about opposing actions in all cell types.
d. Although cells contain other phosphorylated inositols, IP_3 is the primary phosphorylated inositol causing Ca⁺⁺ release from the endoplasmic (or sarcoplasmic) reticulum.
e. Following PLC activation, DAG normally remains within the plasma membrane, while IP_3 is released into the cytoplasm.

70. If an otherwise normal dog that had been starved for 48 hours was suddenly fed a small meal largely devoid of carbohydrate and fat, but rich in protein:

a. Pancreatic insulin secretion would be suppressed, yet glucagon secretion would be stimulated.
b. The liver would be forced to mobilize its glycogen reserves since gluconeogenesis would now be inhibited by rising insulin levels.
c. The animal would become comatose due to a large influx of NH_3 from the digestive tract.
d. The liver would remain in the gluconeogenic mode.
e. Glucagon and insulin secretion would no longer occur, namely because the pancreas self-digested itself during the starvation period.

71. Carbohydrate is required for the complete oxidation of FFA in exercising muscle, because:

a. LCFA β-oxidation requires carnitine, which is produced from glucose.
b. Glycolysis generates oxaloacetate (from the carboxylation of pyruvate), which is needed to couple with acetyl-CoA (from FFA oxidation) in the formation of citrate.
c. Oxaloacetate is being diverted into gluconeogenesis in exercising muscle cells (via the dicarboxylic acid (DCA) shuttle).

d. ATP cannot be formed anaerobically in muscle.

e. None of the above

72. Three crucial steps that determine the magnitude of ketogenesis are:

a. Insulin availability: The rate of gluconeogenesis in muscle tissue: The plasma KB^- concentration.

b. The sensitivity of adipocytes to the negative feedback effects of β-OH-butyrate: The sensitivity of muscle and adipose tissue to insulin: Glycogen availability in liver mitochondria.

c. The responsiveness of the pancreas to ketone bodies: The sensitivity of adipocytes to circulating FFAs: The plasma lactate⁻ concentration.

d. The rate of lipolysis in adipose tissue: Triglyceride reformation in liver tissue: OAA availability in hepatic mitochondria.

e. The glycogen content of liver cells: LPL activity in adipocytes: The plasma KB^- concentration.

73. Which one of the following is NOT a function of insulin?

a. Enhance hepatic LDL receptor synthesis

b. Enhance hepatic PEP carboxykinase activity

c. Enhance hepatic and adipocyte acetyl-CoA carboxylase activity

d. Enhance adipocyte cAMP-dependent phosphodiesterase activity

e. Enhance "clearing factor lipase" activity on capillary endothelial cells

74. Increased oxygen carrying capacity in the exercising horse is largely a result of:

a. Shunting of blood away from the kidneys.

b. An increase in erythropoiesis (RBC production in bone marrow).

c. Enhanced lactate⁻ tolerance.

d. Increased eicosanoid biosynthesis.

e. Splenic contraction.

75. Cofactor requirements for fatty acid biosynthesis include all of the following, EXCEPT:

a. Vitamin C.

b. HCO_3^- (a source of CO_2).

c. NADPH.

d. ATP.

e. Mg^{++} & Mn^{++}.

76. In the normal, healthy animal, which one of the following would be expected to have the greatest influence on VO$_{2(max)}$?

a. O_2 utilization by muscle tissue
b. The O_2 diffusing capacity of the lungs
c. Cardiac output
d. The rate at which lactate- is converted to glucose by the liver
e. C~PO$_3$ reserves in muscle tissue

77. The structure below is:
(Note: R$_1$ = FA chain)

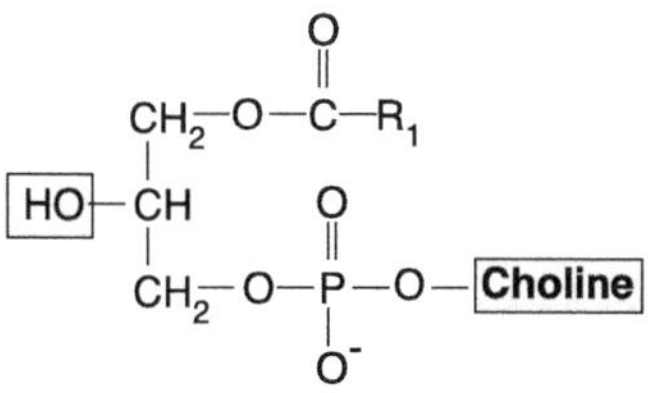

a. Diacylglycerol (DAG).
b. Found in intestinal mixed micelles, but not in blood.
c. Lysolecithin.
d. Dipalmitoylphosphatidyl choline (DPPC).
e. A lipoprotein normally found in bile.

Answers

1. e	17. c	33. b	49. a	65. e
2. c	18. b	34. c	50. e	66. e
3. b	19. e	35. b	51. d	67. a
4. d	20. d	36. c	52. b	68. b
5. d	21. b	37. c	53. b	69. c
6. b	22. c	38. d	54. c	70. d
7. c	23. e	39. a	55. a	71. b
8. e	24. c	40. c	56. a	72. d
9. a	25. e	41. a	57. a	73. b
10. b	26. d	42. c	58. c	74. e
11. a	27. b	43. a	59. b	75. a
12. c	28. a	44. b	60. d	76. a
13. a	29. c	45. e	61. e	77. c
14. d	30. b	46. e	62. c	
15. e	31. e	47. c	63. d	
16. d	32. b	48. c	64. e	

Addendum to Section VI

The roles of glucose, FFAs, ketone bodies and amino acids in processes of starvation and exercise have been emphasized in **Section VI**. It should be understood, however, that catabolic processes of starvation and exercise will be accelerated in diseased, **cachexic** animals, thus potentially leading to debilitation and emaciation in a rather short period of time.

We will now move on to **Section VII**, where biochemical and physiological aspects of acid-base balance are considered.

Introduction to Section VII

Although understanding **body fluid, electrolyte** and **acid-base disorders** may be one of the more difficult tasks facing students and clinicians, it is also one of the most essential. One problem has been that textbooks concerning these subjects are either too difficult for students to understand, too oversimplified, or more often, woefully incomplete. The following **13 chapters** attempt to make this material straightforward, relevant, well-integrated, complete, practical and as effective and understandable as possible. The underlying philosophy brought to this section is that it is possible to present the fundamental and essential concepts of fluid, electrolyte and particularly acid-base balance in a "relatively" painless fashion. Once that goal has been accomplished, then the pathophysiology underlying major clinical disorders can be understood and appreciated, and not merely memorized. Another goal is that of integrating the more traditional approach to acid-base balance with the newer approach (i.e., the **Peter Stewart** approach). To this author these two approaches seem merely to reinforce each other; a concept, unfortunately, which has not been to this date universally appreciated.

It has been assumed that students will have had a sound introduction to **cardiovascular, respiratory** and **renal physiology** before coming to this section of the text. **Why?** Because acid-base physiology involves several organ systems (i.e., the kidneys, lungs, gut, liver and blood all play essential roles). Therefore, this section of the text would be better integrated into respective **organ systems physiology courses**.

Section VII begins with a chapter on H^+ balance, followed by a chapter that identifies and describes the various strong and weak electrolytes. Next, the protein, bicarbonate, phosphate and ammonia buffer systems are developed and characterized, and then the plasma and urinary anion gaps are introduced and discussed. The traditional approach to understanding metabolic acidosis follows, with particular emphasis given to the **Davenport** diagram. Metabolic acidosis and K^+ balance in diabetes mellitus are next considered, followed by individual chapters on metabolic alkalosis, respiratory acidosis, and respiratory alkalosis. The strong ion difference (SID) is next emphasized, with several problems appropriately analyzed to give readers an appreciation for this interesting and applicable approach to acid-base balance. The book concludes with a chapter examining why certain solutions are alkalizing while others are acidifying, and another characterizing the various forms of dehydration (hypertonic, isotonic and hypotonic), and the physiologic consequences of overhydration (i.e., water intoxication) and extracellular fluid volume expansion. It is hoped that readers will particularly find this section of the text relevant to their educational goals, for relationships between biochemistry, physiology and clinical medicine are clearly evident in the substance of acid-base balance.

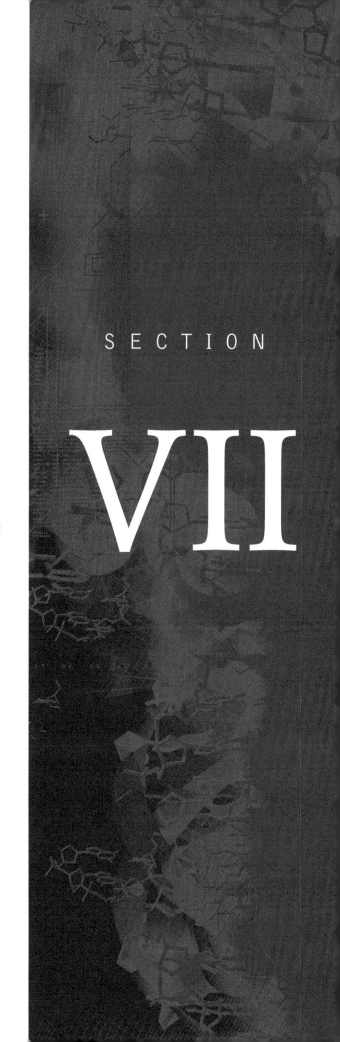

Acid-Base Balance

Chapter 82

The Hydrogen Ion Concentration

Overview

- The pH notation is an effective means of expressing the [H$^+$].
- A one unit increase in pH (e.g., from 5 to 6) represents a 10-fold decrease in the [H$^+$].
- The normal plasma pH range is 7.38 - 7.44.
- Control of H$^+$ balance differs from that of other major electrolytes.
- Volatile acid (CO_2) is normally eliminated by the lungs, while non-volatile acid (H$^+$) is excreted in urine.
- Non-volatile acids are generated through incomplete oxidation of major organic nutrients in the body.
- Dietary fruits can be a major source of alkali.
- Non-volatile acids are normally added to the body through the diet, through metabolism, and through loss of base via the feces.
- Metabolic acidosis is generally associated with diarrhea, while metabolic alkalosis results from vomiting.

The overall concept of acid-base balance is seemingly easy to understand, yet paradoxically difficult to comprehend. Simply stated, acid-base balance (i.e., defense of the extracellular hydrogen ion concentration **[H$^+$]**) is largely accomplished through coordinated activities of the **liver**, **lungs**, and **kidneys**. Over time, net H$^+$ production by the organism must be matched by excretion.

The **pH** notation is a convenient and effective means of expressing the **[H$^+$]** of plasma, since this concentration is normally low relative to those of other cations. For example, the normal plasma **[Na$^+$]** is about **145 mmol/L** (or mEq/L), yet the **[H$^+$]** is only **40 nmol/L**. The **normal plasma pH**, which is the **negative logarithm of the plasma [H$^+$]** (i.e., -log 4 x 10^{-8} mol/L), is therefore **7.4**. A decrease in **pH** of **1 unit** (e.g., from 7.4 to 6.4), represents a **10-fold** increase in the **[H$^+$]**. However, more than the concentration of any other ion in the body, the plasma [H$^+$] is usually held within narrow limits (i.e., 36 - 42 nmol/L; pH 7.44 - 7.38). However, plasma H$^+$ concentrations compatible with life cover a broader range (i.e., 20 - 100 nmol/L; pH 7.7 - 7.0; **Table 82-1**). Extremes of pH in the body can be seen in exocrine HCl secretions of the stomach, and in ductular secretions of the pancreas.

Although we tend to focus our attention on plasma pH, it is the **intracellular pH** (range **6.0-7.4**, ave. **7.0**) that is critical for cell viability, normal enzyme function, and other metabolic processes. Although cells have mechanisms by which they carefully defend their internal pH in response to changes in the pH of extracellular fluid (ECF), extreme alterations in extracellular pH can disrupt integrity of the intracellular environment, seriously interfere with metabolism, and sometimes cause cell death.

Copyright © 2015 Elsevier Inc. All rights reserved.

Table 82-1
The pH and [H⁺] of Body Fluids

Fluid	pH	[H⁺] mol/L
Plasma		
Extreme acidemia	7.0	1×10^{-7}
Normal	7.4	4×10^{-8}
Extreme alkalemia	7.7	2×10^{-8}
Pancreatic juice	8.0	1×10^{-8}
Maximal urine acidity	4.5	3×10^{-5}
Gastric juice	2.0	1×10^{-2}

Hydrogen Ion Balance

As indicated above, H⁺ balance, and thus acid-base balance, is achieved when H^+ input from the diet and endogenous metabolism and output through all routes of elimination are balanced, and the [H⁺] of extracellular fluid (ECF) remains within physiologic limits compatible with life. Control of H⁺ balance differs from that of other major electrolytes since much of the H⁺ present in the body is generated from metabolism (**Table 82-2**). Under basal conditions and while ingesting a normal diet, a **30 kg dog** will produce about **30 mEq/day** of non-volatile acid, yet almost **7,000 mEq/day** of CO_2. This volatile substance (**CO_2**), the dehydrated form of **carbonic acid (H_2CO_3)**, is eliminated by the **lungs**, while non-volatile acid (**H⁺**) is largely eliminated in **urine**. Since the magnitude of normal volatile acid excretion is far greater than that of non-volatile acid excretion, and because alveolar ventilation can be modified rapidly without substantially changing the rate of CO_2 production, respiratory compensations for metabolic acid-base disturbances can occur more rapidly, and be more effective than renal compensations for respiratory acid-base disturbances (see Chapters 87-91).

Table 82-2
Control of the [H⁺] differs from that of other Major Electrolytes

H⁺ is generated from metabolism.

H⁺ produced by the organism exists in two general forms:

    Volatile acid ($\approx$ 7,000 mEq/day)

    Non-volatile acid ($\approx$ 30 mEq/day)

H⁺ balance is achieved through the coordinated action of several different buffer systems:

Chemical	
Hemoglobin (Hb^-)	
Bicarbonate (HCO_3^-)	
Phosphate ($HPO_4^=$)	
Ammonia (NH_3)	
Protein ($Prot^-$)	
Pulmonary	Bone
Excretion of volatile acid (i.e., CO_2)	Surface
Hepatic	Na_2CO_3
$HCO_3^- + NH_4^+ \longrightarrow$ Urea	K_2CO_3
Glutamine formation	$CaCO_3$
Renal	Intracellular
Non-volatile acid (H⁺) excretion	Hydroxyapatite ($Ca_{10}(PO_4)_6(OH)_2$)
Bicarbonate (HCO_3^-) excretion	Brushite ($CaHPO_4$)

The continuous production of volatile and non-volatile acids threatens stability of the internal environment. The rate of acid production depends on the composition of the **diet**, and the rate of **metabolism**. Neither factor is under primary control of any H^+ stabilizing system. Thus, **pH balance** is achieved mainly by adjusting the **respiratory disposal of volatile acid**, CO_2, and the **renal disposal of H^+ and bicarbonate (HCO_3^-)**, a major buffer. Non-volatile H^+ generation is initially complexed with **chemical buffers** (e.g., hemoglobin, bicarbonate, phosphate, ammonia, and protein; see Chapters 83-85), but eventually excreted by the kidneys. Additionally, the **liver** becomes involved in acid-base balance through its ability to generate **urea** and **glutamine** (see Chapters 10 and 11).

Non-volatile Acid Production

Although the composition of the dietary and

Table 82-3
Examples of how Acid is Generated in the Body
Volatile Acid (CO_2) Generation (Mitochondrial)
Carbohydrate oxidation
$\quad CO_2 + H_2O$
β-Oxidation of fatty acids
$\quad CO_2 + H_2O$
Non-Volatile Acid (H^+) Generation
Carbohydrate oxidation
$\quad$ Lactic acid
β-Oxidation of fatty acids
$\quad$ Ketone bodies
Amino acid oxidation
$\quad HCO_3^- + NH_4^+ \longrightarrow$ Urea
$\quad H_2SO_4$
$\quad H_3PO_4$
$\quad HCl$
Nucleic acid oxidation
$\quad H_3PO_4$

metabolic acid load may vary among animal species, and between physiologic states, certain characteristics of non-volatile acid generation are shared by most animals.

In general, **complete oxidation** of carbohydrate and fat normally yields only CO_2 and H_2O (Table 82-3). However, **organic acids** are formed from the **incomplete oxidation** of not only dietary carbohydrates and fats, but also proteins and nucleic acids. Even during starvation, metabolism of endogenous protein results in organic acid production.

For example, hydrolysis of **phosphate esters** in some proteins as well as **nucleic acids** yields **phosphoric acid (H_3PO_4)**, and **hydrochloric acid (HCl)** is generated when **chloride salts** of the **basic hydrophilic amino acids** (i.e., **histidine**, **lysine** and **arginine**; see Chapter 2) are metabolized to neutral products. These "**strong**" acids, which significantly dissociate in solution, enter the circulation and present a major H^+ load to buffers in the ECF. Additionally, when **sulfur-containing amino acids** are metabolized, **sulfuric acid (H_2SO_4)**, another strong acid, is produced:

$$\left[\begin{array}{c} \textbf{Met or Cys} \longrightarrow \textbf{Glucose (or Fatty Acid)} \\ + \textbf{Urea} + \textbf{H}^+ + \textbf{SO}_4^= \end{array} \right]$$

Animals grazing on **pastures** that contain high **sulfate** and **phosphate residues**, or those given excessive amounts of **grain concentrates** (i.e., rumen lactic acid formation), may also experience increased acid loads. The normally modest endogenous acid production of animals can be increased under certain pathologic states as well. For example, **ketone body** formation (i.e., acetoacetic and β-hydroxybutyric acid) increases in hypoinsulinemic states (i.e., diabetes mellitus), or starvation (see Chapters 71, 74, and 75). Toxin and drug ingestion can also result in accelerated organic acid formation, such as **formic acid (HCOOH)** from **methanol (CH_3OH;** found in windshield washer fluid), **glycolic acid ($CH_2OHCOOH$)**

and **oxalic acid (HOOCCOOH)** from **ethylene glycol (CH₂OHCH₂OH**; found in antifreeze), or **salicylic acid** from **aspirin (acetylsalicylic acid)**. Additionally, ingestion of **acidifying salts** such as **ammonium chloride (NH₄Cl)** and **calcium chloride (CaCl₂)** are equivalent to adding HCl to the body (see Chapter 93).

Conversely, dietary fruits can be a major source of **alkali**. They contain **Na⁺** and **K⁺ salts** of "**weak**" **organic acids**, whose dissociated **anions** become H⁺ acceptors before becoming metabolized. **NaHCO₃** and other alkalinizing salts are sometimes administered to animals (see Chapter 93), but a more common cause of **alkalosis** is loss of acid from the body as a result of **vomiting** gastric juice rich in **HCl**. This, of course, is equivalent to adding alkali to the body.

Partial buffering of acid comes about when generated protons combine with certain anionic metabolites that are further metabolized to neutral end products. For example, **anionic amino acids (glutamate⁻** and **aspartate⁻)**, or dietary **organic anions** (e.g., **acetate⁻**, **lactate⁻**, or **citrate⁼**) typically buffer considerable quantities of H⁺ daily:

$$\left[\ \begin{array}{c} \textbf{2 NaLactate (Oral administration)} \longrightarrow \\ \textbf{2 Na}^+ + \textbf{2 Lactate}^- \textbf{(CH}_3\textbf{CHOHCOO}^-\textbf{)} \end{array}\ \right]$$

$$\left[\ \begin{array}{c} \textbf{2 CH}_3\textbf{CHOHCOO}^- + \textbf{2 H}^+ \longrightarrow \textbf{Glucose} \\ \textbf{(C}_6\textbf{H}_{12}\textbf{O}_6\textbf{; through gluconeogenesis)} \end{array}\ \right]$$

However, if the normal hepatic conversion of lactic acid to glucose is interrupted, metabolic acidosis rapidly ensues.

Non-volatile Acid Input and Loss from the Body

There are three roughly equivalent sources for the normal daily addition of H⁺ to the body, but only one major route of elimination (**Fig. 82-1**):

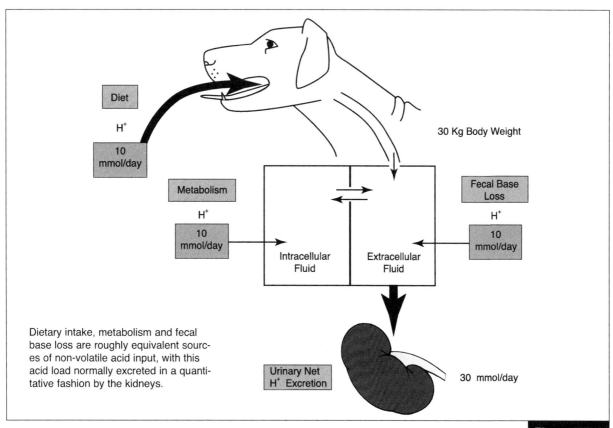

Dietary intake, metabolism and fecal base loss are roughly equivalent sources of non-volatile acid input, with this acid load normally excreted in a quantitative fashion by the kidneys.

Diet
H⁺
10 mmol/day

Metabolism
H⁺
10 mmol/day

30 Kg Body Weight

Fecal Base Loss
H⁺
10 mmol/day

Intracellular Fluid   Extracellular Fluid

Urinary Net H⁺ Excretion   30 mmol/day

Figure 82-1

Input
- **Diet**
- **Metabolism**
- **Fecal base loss**

Output
- **Urine**

The sum of the H^+ input into an average dog is about **1.0 mmol/day/Kg body weight**, and this acid load is usually quantitatively excreted by the kidneys. About **10 mmol/day of HCO_3^-** and **base equivalents** (i.e., organic anions) are normally lost in the feces, and, for every molecule of base lost via this route, one H^+ is normally retained in the ECF. Gastrointestinal base loss, therefore, adds to the systemic acid load, which can be markedly increased in certain diarrheal states. Thus, **metabolic acidosis** is generally associated with **diarrhea**, while **metabolic alkalosis** results from **vomiting** (see above).

OBJECTIVES

- Distinguish between pH and the [H^+], and recognize how much of an increase in the [H^+] it takes to decrease pH by 1 unit.

- Compare volatile to non-volatile acid generation, and recognize how each is excreted. Explain why respiratory compensations for metabolic acid-base disturbances are generally more effective than renal compensations for respiratory disturbances.

- Indicate how the "acid or alkali load" may change with the type of food ingested.

- Identify three primary sources of non-volatile acid input, and recognize how this input is normally balanced.

- Explain why metabolic acidosis is associated with diarrhea, while metabolic alkalosis is associated with vomiting (see Chapters 87 & 89).

- Name the major chemical, pulmonary, hepatic and renal buffer systems of the body.

- Explain how H^+ balance differs from that of other major electrolytes in the body.

- Understand why NH_4Cl is considered to be an acidifying agent (see cahpter 93).

- Explain what is meant by "fecal base loss," and indicate whether this is normal, or abnormal.

QUESTIONS

1. **The normal hydrogen ion concentration of plasma is about:**
   a. 140 mmol/L.
   b. 40 mmol/L.
   c. 140 μmol/L.
   d. 40 μmol/L.
   e. 40 nmol/L.

2. **A 30 Kg dog normally produces about _____ mEq/day of non-volatile acid, and about _____ mEq/day of volatile acid:**
   a. 3; 7
   b. 30; 70
   c. 3; 700
   d. 30; 7000
   e. 3; 70,000

3. **Ingestion of which one of the following would tend to elevate the plasma pH?**
   a. Sodium lactate
   b. Methanol
   c. Ammonium chloride
   d. Aspirin
   e. Calcium chloride

4. **Hydrochloric acid can be generated in the body from:**
   a. Metabolism of the chloride salt of lysine to a neutral product.
   b. Metabolism of methionine in gluconeogenesis.
   c. The complete oxidation of glucose through the Krebs cycle.
   d. The complete oxidation of ketone bodies for energy purposes.
   e. Ethylene glycol ingestion.

5. **Metabolic alkalosis is most often associated with:**
   a. Methanol ingestion.
   b. Vomiting.
   c. The oxidation of nucleic acids.
   d. Diarrhea.
   e. Ketone body formation.

1. e
2. d
3. a
4. a
5. b

ANSWERS

Chapter 83

Strong and Weak Electrolytes

Overview

- The pK's of the major buffer systems cover a wide pH range.
- The "isohydric principle" indicates that all buffer systems of the body are normally in equilibrium with each other.
- Bicarbonate is normally the major extracellular buffer of an acid load.
- An acid is a substance, charged or uncharged, that liberates protons (H^+) in solution.
- Buffer action in physiologic systems is exhibited by ions of weak acids or bases.
- Proteins are important plasma and intracellular buffers.
- In chronic acidosis, buffering by bone results in demineralization.
- The pK' for a weak acid or base is the pH at which the protonated and unprotonated species are present at equal concentrations.

An **acid** is a substance, charged or uncharged, that liberates **protons** (H^+) in solution, and a base binds protons, thus removing them from solution. In this terminology, **ammonia** (NH_3), the **acetate anion** (CH_3COO^-), and the **lactate anion** ($CH_3CHOHCOO^-$) are bases, whereas the **ammonium ion** (NH_4^+), **acetic acid** (CH_3COOH), and lactic acid ($CH_3CHOHCOOH$) are acids.

Strong electrolytes are essentially completely ionized in aqueous solution, including almost all neutral salts (**NaCl, Na$_2$SO$_4$, KBr**, etc.), strong acids (**HCl, H$_3$PO$_4$, HNO$_3$, H$_2$SO$_4$**, etc.), and strong alkalies (**NaOH, KOH**, etc.). In contrast, weak electrolytes are only partially ionized in solution, and yield a mixture of the undissociated compound (**HA**), and its ions (H^+ and the anion, **A$^-$**). Many acids (**HA**) are weak electrolytes, and only partially dissociate at physiologic pH:

$$HA \longleftrightarrow H^+ + A^-$$

It follows that an estimate of the proton concentration [H^+] of a solution of weak acid (like carbonic acid, H_2CO_3) will not give a measure of the total concentration of the acid since there is a reservoir of substrate in the form of undissociated acid (**HA**; H_2CO_3). However, if the free H^+ is titrated (i.e., the [H^+] is determined by addition of an indicator whose color change has been standardized in solutions of known [H^+]), additional H_2CO_3 will dissociate to H^+ and bicarbonate (HCO_3^-), and titration will eventually give an estimate of total acid since both the free H^+ and that derivable from the undissociated molecule will be titrated. Hence, we distinguish between **titratable acidity**, and **actual acidity** in dealing with weak acids. Since strong acids are, for the most part, completely dissociated in solution, the titratable and actual acidity are the same.

The equation above can also be regarded as that of a general **buffer system equation**. Buffered solutions resist changes in the [H^+]

Copyright © 2015 Elsevier Inc. All rights reserved.

that would otherwise result from addition of more acid or base to the system. In general, buffer action in physiologic systems is exhibited by ions of weak acids or bases. Weak electrolytes that exhibit buffer action retain important practical properties, and possess significant value for living systems since most cells can survive only within fairly narrow pH limits (see Chapter 82). Since strong acids and bases are almost completely dissociated in solution, they possess no reservoir of undissociated acid or base, and, therefore, cannot function as buffers.

Each body fluid compartment is defined spatially by one or more differentially permeable membranes. Each compartment contains characteristic kinds and concentrations of solutes, some of which are buffers (i.e., weak electrolytes) at physiologic pH. Although the concentrations of solutes in the cytoplasm of each cell type may be slightly different, most cells, with the exception of erythrocytes, are similar enough that they are considered together for purposes of acid-base balance. Thus there are, from this point of view, approximately six major compartments of the body that possess slightly different buffering characteristics (**Fig. 83-1**), and all are in equilibrium with each other (**Fig. 83-2**).

When **H$^+$** is added to the extracellular fluid (ECF) compartment from exogenous or endogenous sources, there is rapid distribution, with **ECF buffers** binding these protons, thus maintaining pH within narrow limits (**Fig. 83-3**). Intracellular distribution of these protons, however, is slower, and can take several minutes-hours to complete.

The principle ECF buffer is **HCO$_3^-$**, however, **phosphates** and **protein** (Prot$^-$) provide additional ECF buffering. The combined action of these three buffers accounts for approximately **50%** of the buffering of a nonvolatile acid load, with the remainder occurring intracellularly. Intracellular buffering involves titration of **H$^+$**

Primary Buffers in the Body

Extracellular

Plasma

$$H_2CO_3 \rightleftharpoons H^+ + HCO_3^-$$
$$HProt \rightleftharpoons H^+ + Prot^-$$
$$H_2PO_4^- \rightleftharpoons H^+ + HPO_4^=$$

Interstitial Fluid

$$H_2CO_3 \rightleftharpoons H^+ + HCO_3^-$$
$$HProt \rightleftharpoons H^+ + Prot^-$$
$$H_2PO_4^- \rightleftharpoons H^+ + HPO_4^=$$

Renal Tubular Filtrate

$$H_2CO_3 \rightleftharpoons H^+ + HCO_3^-$$
$$NH_4^+ \rightleftharpoons H^+ + NH_3$$
$$H_2PO_4^- \rightleftharpoons H^+ + HPO_4^=$$

Cerebrospinal Fluid

$$H_2CO_3 \rightleftharpoons H^+ + HCO_3^-$$

Intracellular

Tissues

$$HProt \rightleftharpoons H^+ + Prot^-$$
$$H_2CO_3 \rightleftharpoons H^+ + HCO_3^-$$
$$H_2PO_4^- \rightleftharpoons H^+ + HPO_4^=$$

Erythrocytes

$$HHb \rightleftharpoons H^+ + Hb^-$$
$$HProt \rightleftharpoons H^+ + Prot^- (other)$$
$$H_2CO_3 \rightleftharpoons H^+ + HCO_3^-$$
$$H_2PO_4^- \rightleftharpoons H^+ + HPO_4^=$$

Extracellular + Intracellular

Bone

$$\left[\begin{array}{c} Ca_{10}(PO_4)_6(OH)_2 + CaHPO_4 \\ + Na_2CO_3 + K_2CO_3 + CaCO_3 \end{array} \right]$$

$$2H_2O + 6H_2PO_4^- + 10Ca^{++} \rightleftharpoons 14H^+ + Ca_{10}(PO_4)_6(OH)_2$$
$$Ca^{++} + H_2PO_4^- \rightleftharpoons H^+ + CaHPO_4$$
$$2Na^+ + CO_2 + H_2O \rightleftharpoons 2H^+ + Na_2CO_3$$
$$2K^+ + CO_2 + H_2O \rightleftharpoons 2H^+ + K_2CO_3$$
$$Ca^{++} + CO_2 + H_2O \rightleftharpoons 2H^+ + CaCO_3$$
$$Mg^{++} + CO_2 + H_2O \rightleftharpoons 2H^{++} + MgCO_3$$

Figure 83-1

---

by either **HCO$_3^-$, phosphates**, or the **histidine** groups of **proteins** (see Chapter 84). The renal tubular filtrate and **cerebrospinal fluid (CSF)** normally have extremely low protein concentrations, therefore, protein is not considered to be a significant buffer in these fluid spaces. The protein concentration of interstitial fluid is only about 1/4th that of plasma, however, the total **interstitial fluid** space is about 4-times greater than that of **plasma**, therefore the total amount of protein buffering in interstitial fluid spaces is about equal to that in plasma. The protein concentration inside cells is about 5-times greater than that in plasma, therefore, protein becomes a major intracellular buffer.

Bone represents a complex source of both ICF and ECF buffer, and contains salts that are alkaline relative to the pH of plasma. However, in chronic acidosis, buffering by bone results in **demineralization**. Thus, Ca^{++}, H$_2$PO$_4^-$, Na$^+$, Mg$^+$,

The Isohydric Principle

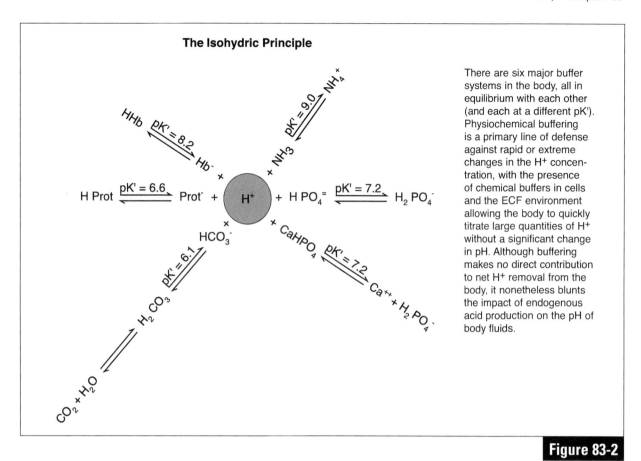

There are six major buffer systems in the body, all in equilibrium with each other (and each at a different pK'). Physiochemical buffering is a primary line of defense against rapid or extreme changes in the H⁺ concentration, with the presence of chemical buffers in cells and the ECF environment allowing the body to quickly titrate large quantities of H⁺ without a significant change in pH. Although buffering makes no direct contribution to net H⁺ removal from the body, it nonetheless blunts the impact of endogenous acid production on the pH of body fluids.

Figure 83-2

and K⁺ are all released from bone as major **buffers** (i.e., $PO_4^=$, $HPO_4^=$, OH^-, and $CO_3^=$) bind H⁺ in exchange for these minerals (see Chapter 87).

The Henderson-Hasselbalch Equation

The general equation for a buffer system is shown above (**HA <—> H⁺ + A⁻**), with **A⁻** representing any anion, and **HA** representing the undissociated acid. If an acid stronger than HA is added to a solution containing this system, the buffer equation shifts to the left. Hydrogen ions are essentially captured, thus forming more undissociated HA, such that the increase in the **H⁺** concentration is minimized. Conversely, if a base like **NaOH** is added to this system, **H⁺** and **OH⁻** react to form **H₂O**, but more HA dissociates, thus limiting the decrease in the **H⁺** concentration. If the conjugate base (e.g., **NaA**) is added, the equilibrium initially shifts to the left, with a decrease in the **H⁺** concentration. However, if the conjugate base is **NaHCO₃**, the equation will quickly shift back to the right as ventilation is

Time Course of Extracellular Fluid (ECF) and Intracellular Fluid (ICF) Buffering of an Exogenous Acid Load

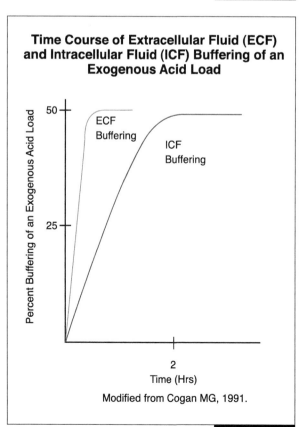

Modified from Cogan MG, 1991.

Figure 83-3

decreased, thus tending to offset the reduction in the H^+ concentration, and further elevating the HCO_3^- concentration.

By the law of mass action, the buffer system equation can also be written as follows:

$$K' = [H^+] [A^-] / [HA]$$

Where, K' = the "apparent" dissociation or ionization constant, which depends on the temperature and composition of a solution. In infinitely dilute solutions in which interionic forces are negligible, the equilibrium constant (K) is used in place of K'. However, in body fluids it is more appropriate to use the apparent dissociation constant.

In order to derive the Henderson-Hasselbalch equation, we first cross-multiply, and then divide both sides of this equation by $[A^-]$:

$$[H^+] [A^-] = K'[HA]$$
$$[H^+] = K'([HA]/[A^-])$$

By taking the logarithm of both sides:

$$\log [H^+] = \log \{K'([HA]/[A^-])\}$$
$$= \log K' + \log ([HA]/[A^-])$$

Multiplying through by -1:

$$-\log [H^+] = -\log K' - \log ([HA/[A^-])$$

Substituting **pH** for **$-\log [H^+]$**, and **pK'** for **$-\log K'$**:

$$pH = pK' - \log ([HA]/[A^-])$$

Removing the minus sign by inverting the last term gives us the **Henderson-Hasselbalch equation**:

$$pH = pK' + \log ([A^-]/[HA])$$

According to this relationship, the **pK'** is the **pH** at which the protonated (**HA**) and unprotonated (**A^-**) species of a weak acid are present at **equal concentrations**. Therefore, at half neutralization, **$[A^-]$ = [HA]**, and **pH = pK'**. Thus, a weak acid (or base) is most effective as a buffer at a pH near its pK' value. Starting from a pH one unit below to a pH one unit above the pK', ~82% of a weak acid in solution will dissociate, and therefore an amount of base equivalent to about 82% of the original acid can be neutralized with a change in pH of 2 units. Thus, the maximum buffering range for a

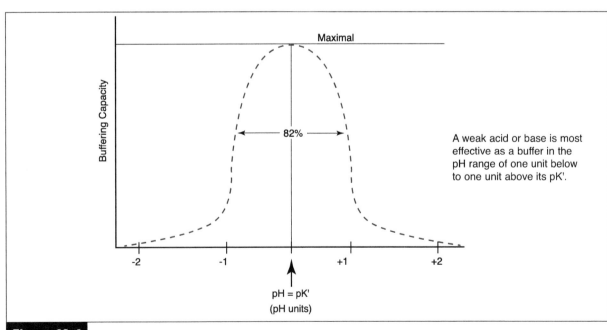

A weak acid or base is most effective as a buffer in the pH range of one unit below to one unit above its pK'.

Figure 83-4

conjugate pair is considered to be between 1 pH unit above and below its pK' (**Fig. 83-4**).

The pK's of the major buffer systems cover a wide pH range, from 6.1 for the bicarbonate buffer system, to 9.0 for the ammonia buffer system (see **Fig. 83-2**). A weak acid such as lactic acid with a pK' = 3.86, is an effective buffer in the range of pH 3-5, however it has little buffering capacity at a pH of 7.0. The **$HPO_4^=$/$H_2PO_4^-$** pair with a pK' = 7.2, and the **Prot⁻/HProt** pair with a pK' = 6.6, are, however, effective buffers at this pH. Thus, at the pH of the cell's cytosol (about 7.0), the **lactate⁻/lactic acid** pair is not an effective buffer, however, the $HPO_4^=$/$H_2PO_4^-$ and Prot⁻/HProt pairs are.

OBJECTIVES

- Distinguish an acid from a base, and titratable acidity from actual acidity.

- Describe the isohydric principle, and explain how buffers work.

- Identify primary buffers in plasma, interstitial fluid, CSF, tissues, erythrocytes, bone and the renal tubular filtrate.

- Outline the time-course of ECF and ICF buffering of an acid load, and discuss how this buffering action occurs.

- Know why strong acids and bases cannot function as physiologic buffers.

- Given the general buffer equation, K' = [H⁺][A⁻] / [HA], derive the Henderson-Hasselbalch equation and define pK'.

- Recognize why the maximum buffering range for a conjugate pair is one pH unit above and below its pK'.

- Explain why phosphates and proteins are major intracellular buffers.

- Know why the [H⁺] of ECF (≈ 40 nMol) is so much lower than the [HCO_3^-] (≈ 24 mMol; see Chapter 85).

QUESTIONS

1. Urinary titratable and actual acidity would

most likely be the same for which one of the following acids?
- a. H_2CO_3
- b. $H_2PO_4^-$
- c. CH_3COOH
- d. HNO_3
- e. NH_4^+

2. The principle buffer in the extracellular fluid compartment is:
- a. Protein.
- b. Phosphate.
- c. Hemoglobin.
- d. Bicarbonate.
- e. Ammonia.

3. Which one of the following is an important buffer system in bone?
- a. HProt <—> H⁺ + Prot⁻
- b. CH_3COOH <—> H⁺ + CH_3COO^-
- c. H_2CO_3 <—> H⁺ + HCO_3^-
- d. NH_4^+ <—> H⁺ + NH_3
- e. Ca^{++} + $H_2PO_4^-$ <—> H⁺ + $CaHPO_4$

4. The Henderson-Hasselbalch equation:
- a. Represents the pK' at which the protonated (HA) species of a weak acid is present at a concentration 10-times greater than the unprotonated (A⁻) species.
- b. pH = [HA]/[H⁺][A⁻]
- c. pK' = [H⁺][A⁻]/[HA]
- d. pH = pK' + log([HA]/[A⁻])
- e. pH = pK' + log([A⁻]/[HA])

5. Select the TRUE statement below:
- a. A weak acid (or base) is most effective as a buffer at a pH near its pK'.
- b. Starting from a pH one unit below to one unit above the pK', only 10% of a weak acid in solution would be expected to dissociate.
- c. The pK' of the ammonia buffer system is 6.1.
- d. Proteins are important buffers in cerebrospinal fluid.
- e. Intracellular buffering of an exogenous acid load normally occurs faster than extracellular buffering.

5. a
4. e
3. e
2. d
1. d

ANSWERS

Chapter 84

Protein Buffer Systems

Overview

- Proteins are plentiful buffers in the mammalian organism.
- The amino acid ionizable groups in proteins, and thus their pK's, cover a wide pH range.
- Hemoglobin is normally the most abundant protein in blood, and functions to carry O_2 to tissues, and protons to the lungs.
- Hypoproteinemia is generally associated with alkalemia, while hyperproteinemia causes acidemia.
- Deoxygenated hemoglobin (Hb⁻) is a better buffer than oxygenated hemoglobin (HbO_2^-).
- The imidazole groups of the 38 histidine residues found in Hb⁻ provide it with considerable buffering capacity.
- The venous hematocrit is normally slightly greater than the arterial hematocrit.
- HCO_3^-/Cl^- exchange normally occurs across the erythrocytic plasma membrane.
- Hb⁻ forms carbamino compounds more readily than HbO_2^-.

The control of physiologic pH is provided by certain **buffers**, which help to ensure that amino acids, nucleotides, proteins, nucleic acids, lipids, and most other important biomolecules are maintained in ionic states best suited for their structure, function, and solubility in water. As indicated in the previous chapter, **proteins** are among the most plentiful buffers in the organism, they exist in both extracellular and intracellular fluid compartments, however, their high intracellular concentrations make them far more important buffers inside cells. Since both their free carboxyl and free amino groups dissociate, protein buffering activity can occur from both ends of the polypeptide chain:

$$RCOOH <\longrightarrow H^+ + RCOO^-$$

$$RNH_3 <\longrightarrow H^+ + RNH_2$$

Protein buffering is generally represented by the overall equation below, where **HProt** represents the associated, and **Prot⁻** the dissociated anionic form of protein:

$$HProt <\longrightarrow H^+ + Prot^-$$

The amino acid ionizable groups in proteins, and thus their pK's, cover a wide pH range (**Fig. 84-1** and Chapter 2). In general, however, the combined buffer characteristics of proteins are about the same as that for oxyhemoglobin below (pK' = 6.6).

The relationship between **total plasma protein (Prot$_{tot}$)** and **H⁺** is as follows:

1) **Prot$_{tot}$ = HProt + Prot⁻**

2) **HProt <⟶ H⁺ + Prot⁻**

In plasma, **Prot$_{tot}$** does not normally exceed 20 mEq/L, but in cells about five times this amount is available. In the bicarbonate (HCO_3^-)

Copyright © 2015 Elsevier Inc. All rights reserved.

Figure 84-1
pK' Values of Amino Acid Ionizable Groups in Proteins

Amino Acid	Acid $\rightleftharpoons$ Base + H⁺		pK'
Glutamic acid (carboxyl group)	$-CH_2CH_2COOH$	$\rightleftharpoons$ $-CH_2CH_2COO^- + H^+$	3.9
Aspartic acid (carboxyl group)	$-CH_2COOH$	$\rightleftharpoons$ $-CH_2COO^- + H^+$	4.3
Histidine (imidazole group)	$-CH_2$ (imidazole, HN/NH)	$\rightleftharpoons$ $-CH_2$ (imidazole, HN/N) $+ H^+$	6.2
Cysteine (sulfydryl group)	$-CH_2SH$	$\rightleftharpoons$ $-CH_2S^- + H^+$	8.4
Tyrosine (phenolic hydroxyl group)	$-CH_2-$⟨⟩$- OH$	$\rightleftharpoons$ $-CH_2-$⟨⟩$- O^- + H^+$	10.1
Lysine (amino group)	$-(CH_2)_4NH_3$	$\rightleftharpoons$ $-(CH_2)_4NH_2 + H^+$	10.2
Arginine (guanidine group)	$-N-C$ (guanidine, H, NH₂⁺, NH₂)	$\rightleftharpoons$ $-N-C$ (guanidine, H, NH, NH₂) $+ H^+$	12.2

pK' values depend on temperature, ionic strength, and the microenvironment of the ionizable group.

buffer system, although the concentration of HCO_3^- is normally about the same (≈26 mEq/L), this compound may vary in plasma concentration by at least ten times this amount (from below 5 to over 60 mEq/L), since it takes part in an "open" system, where CO_2 may be lost or conserved through changes in ventilation (see Chapter 85). Note from the above equations that **hyperproteinemia** is associated with **acidosis**, while **hypoproteinemia** would tend to **decrease the extracellular [H⁺]** (see Chapter 89).

The Hemoglobin (Hb⁻) Buffer System

Hemoglobin is an iron-containing protein found in red blood cells (erythrocytes), and it is normally the most abundant protein in blood (see Chapters 32, 33, and 48). Normal amounts range from 14-16 gm/100 ml (% or gm%) of whole blood, representing a concentration of about 2 mMolar (mM). The primary function of this molecule is to **transport O₂** from the lungs to metabolically active, respiring tissues, and to **carry protons** from these tissues to the lungs (see **Fig. 84-2**). Thus, it is an important blood buffer.

Oxygen molecules enter erythrocytes in capillaries of the lungs, where they become bound to iron atoms (Fe⁺⁺) in hemoglobin. This oxygenated hemoglobin (HbO_2^-) is then carried in arterial blood to various tissues of the body, where O_2 is released. The deoxygenated hemoglobin (Hb⁻) then returns to the lungs in venous blood where it becomes oxygenated, and repeats another cycle.

In 1904, **Christian Bohr**, *et al*, described the regulation of **O₂** binding to **hemoglobin** by **H⁺** and **CO₂**, and in 1921 Adair GS, *et al*, speculated that some additional factor was involved in the interaction between hemoglobin and O_2. However, it was not until 1967 that **2,3-diphosphoglycerate (2,3-DPG**; or 2,3-bisphosphoglycerate, 2,3-BPG) was identified as that factor (Chanutin A, and Curnish RR, 1967; Benesch R and Benesch RE, 1967). In some animals, erythrocytic 2,3-DPG may occur at about the same concentration as hemoglobin (see Chapter 31).

When CO_2 enters the erythrocyte, most is rapidly converted to the weak acid H_2CO_3 by action of the Zn⁺⁺-dependent enzyme, **carbonic**

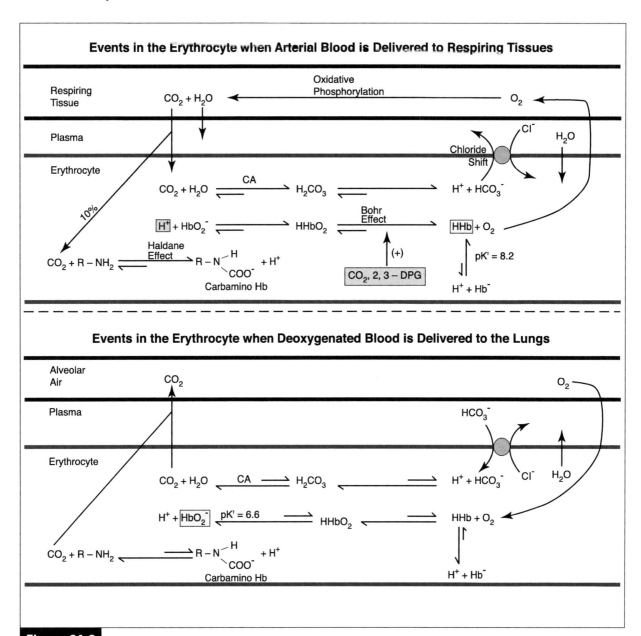

Events in the Erythrocyte when Arterial Blood is Delivered to Respiring Tissues

Events in the Erythrocyte when Deoxygenated Blood is Delivered to the Lungs

Figure 84-2

anhydrase (**CA**; see Chapter 49). Since normal erythrocytic pH is about 7.4, and the pK' of the bicarbonate buffer system is about 6.1 (see Chapter 85), 80% or more of the H_2CO_3 will be ionized to $H^+ + HCO_3^-$, and therefore most of the CO_2 will be carried in HCO_3^-. The remaining CO_2 can bind nonenzymatically to the amine (NH_2) terminal of the globin polypeptide chain (**R**) of hemoglobin, thus forming **carbaminohemoglobin** (the **Haldane effect, Fig. 84-2**). Smaller amounts of CO_2 remain dissolved in plasma, or bound to plasma proteins.

Input of CO_2 into the erythrocyte thus causes an increase in H^+ formation, which could cause a dangerous increase in blood acidity if it were not properly buffered by hemoglobin. Like every protein, hemoglobin contains ionizable groupings contributed by some of the amino acids of which it is composed. Unlike the free carboxyl and amino groups of hemoglobin, which are thought to provide little buffering capacity, the **imidazole groups of the 38 histidine residues** it contains are considered to be important buffers (see **Fig. 84-1**). On this basis, plus the fact that hemoglobin is present in large amounts in erythrocytes, hemoglobin is considered to have six times more buffering capacity than the plasma proteins, and for every H^+ that is buffered by hemoglobin, **two O_2** molecules are released.

Oxygenated hemoglobin ($H^+ + HbO_2^- \longleftrightarrow$ **$HHbO_2$**) has a pK' of 6.6, whereas that for deoxygenated hemoglobin ($H^+ + Hb^- \longleftrightarrow$ **HHb**) is 8.2. The difference between these two pK' values implies that binding of O_2 has changed a property of the hemoglobin molecule. Specifically, **$HHbO_2^-$** is a **stronger acid** than **HHb^-**, and therefore **HbO_2^-** is a **weaker buffer**. Moreover, at the pH of blood, the equilibrium concentration of each respective conjugate acid-base pair will be quite different. At **pH 7.4**, the first reaction is predominantly to the left (thus forming **$H^+ + HbO_2^-$**), and the second is to the right (thus forming **HHb**). Approximate titration curves for hemoglobin are shown in **Fig. 84-3**. The arrow from **x to y** indicates that small amounts of H^+ can be added to a solution of HbO_2 without a pH shift (through formation of **$HHb + O_2$**), and the arrow from **x to z** indicates that the pH of an **HbO_2** solution can be increased following deoxygenation.

Acid-base control by hemoglobin can now be **summarized** as follows (see **Fig. 84-2**). When **CO_2** enters the erythrocyte, the **H^+** that is generated from **H_2CO_3** will react with the predominating base form of **HbO_2^-** to form **$HHbO_2$**. The **$HHbO_2$** has a reduced affinity for **O_2** (due to **H^+**, **CO_2**, and **2,3-DPG**; the **Bohr effect**), and dissociates to yield the acid form of deoxygenated hemoglobin (**HHb**), and free **O_2**. The **O_2** diffuses from the erythrocyte and enters cells of respiring tissues. Because of its high pK' value, most of the **HHb** will not ionize at the pH of blood but, rather, will remain as **HHb**. Thus, the increased amount of **H^+** caused by the diffusion of **CO_2** into the red blood cell has been scavenged by hemoglobin.

As **CO_2** is rapidly converted to osmotically active **HCO_3^-** in erythrocytes, (while they reside in capillaries), small amounts of **H_2O** move

into these cells, thus slightly increasing their size. For this reason, the venous hematocrit is slightly greater than the arterial hematocrit. As **H^+** binds **HbO_2^-**, a negative charge on hemoglobin is lost (**$HHbO_2$**), however, it is immediately replaced by **HCO_3^-**. As **HCO_3^-** now moves down its concentration gradient into plasma, **Cl^-** enters the erythrocyte in order prevent an imbalance in the cytoplasmic ionic environment. This phenomenon, called the **chloride shift** or **Hamburger interchange** (after its discoverer, **Jakob Hamburger**, Dutch physiologist, 1859-1924), is made possible by the presence of a special **HCO_3^-/Cl^- carrier protein** in the red cell membrane that rapidly shuttles these two anions in opposite directions.

When venous blood reaches the lungs, **O_2** and **HCO_3^-** enter erythrocytes, and **Cl^-** exits. The **O_2** binds to the major hemoglobin species present, namely, **HHb**, to form **$HHbO_2$**. Now, however, the **$HHbO_2$** functions as an acid in the presence of **HCO_3^-** to yield **HbO_2^-** and **H_2CO_3**. By the action again of **carbonic anhydrase**, the **H_2CO_3** is converted to **H_2O** and **CO_2**, and the

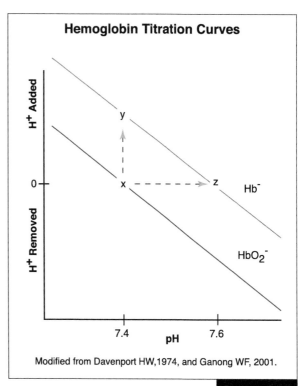

Hemoglobin Titration Curves

Modified from Davenport HW, 1974, and Ganong WF, 2001.

Figure 84-3

latter diffuses into plasma and ultimately into alveoli of the lungs. Since **Hb⁻** binds more **H⁺** than **HbO₂⁻**, and forms **carbamino compounds** more readily (the Haldane effect), binding of **O₂** to hemoglobin reduces its affinity for **CO₂**, thus making **CO₂** available for expiration.

Important events associated with erythrocytic processes by which **O₂** is delivered to and **CO₂** is eliminated from circulating blood, without any serious alteration in blood pH despite the production of **H⁺** from **H₂CO₃**, indicate the importance of hemoglobin buffering. Under normal conditions, the pH of venous blood is decreased by only a few hundredths of a pH unit, largely because of hemoglobin buffering.

OBJECTIVES

- Explain why the combined buffering characteristics of proteins give them a pK' ≈ 6.6.

- Identify the amino acid ionizable groups that provide buffering capacity, and note their pK's.

- Know why hyperproteinemia is acidifying, and hypoproteinemia is alkalinizing.

- Discuss how hemoglobin functions as a blood buffer.

- Recognize the difference between carbaminohemoglobin and carboxyhemoglobin (see Chapter 32).

- Explain the erythrocytic "Haldane" and "Bohr" effects, identify how CO₂ is carried in venous blood, and discuss how an why the "Hamburger interchange" occurs.

- Explain why the venous Hct is slightly greater than the arterial Hct.

- Recognize four key factors that reduce the HbO₂⁻ binding efficiency.

- Understand the relationship between erythrocytic CA activity and Hb⁻ buffering, and explain why Hb⁻ is a better buffer than HbO₂⁻ and HHbO₂.

QUESTIONS

1. Which one of the following is the primary
intracellular buffer in blood?
a. HCO_3^-
b. Hb^-
c. HbO_2^-
d. Carbonic anhydrase
e. $HPO_4^=$

2. Select the TRUE statement below:
a. The arterial hematocrit is usually less than the venous hematocrit.
b. Reduced affinity of hemoglobin for oxygen due to the presence of H⁺, CO₂, and 2,3-DPG is known as the Haldane effect.
c. Formation of carbaminohemoglobin in erythrocytes, and loss of CO₂ in the lungs is known as the Hamburger interchange.
d. Oxyhemoglobin is a better buffer than deoxyhemoglobin.
e. Most CO₂ is carried in blood bound to hemoglobin.

3. The Cl⁻ content of venous erythrocytes is usually greater than that of arterial erythrocytes because:
a. Venous erythrocytes contain less water than arterial erythrocytes.
b. Of erythrocytic HCO_3^-/Cl^- antiport in systemic capillaries.
c. Chloride gas is actively removed from erythrocytes by the lungs, and then expired in air.
d. Carbonic anhydrase converts CO₂ to H⁺ and Cl⁻ in venous erythrocytes.
e. Of the Bohr effect.

4. Select the FALSE statement below:
a. Because of the high pK' value of deoxygenated hemoglobin, most HHb will not ionize at the pH of blood, and therefore will remain as HHb.
b. When venous blood reaches the lungs, O₂ and HCO₃⁻ normally enter erythrocytes, and Cl⁻ exits.
c. HHbO₂ normally functions as an acid in pulmonary erythrocytes.
d. Hb⁻ normally forms carbamino compounds more readily than HbO₂⁻.
e. Under normal circumstances, the pH of venous blood is about one pH unit less than that of arterial blood.

4. e
3. b
2. a
1. b

Chapter 85

Bicarbonate, Phosphate, and Ammonia Buffer Systems

Overview

- The bicarbonate buffer system is important because the concentrations of its components are independently regulated by the lungs and kidneys.

- The total CO_2 content of blood can be estimated from the Pco_2 and the plasma bicarbonate (HCO_3^-) concentration.

- The hemoglobin and bicarbonate buffer systems are the most quantitatively important buffer systems in blood.

- The plasma ratio of $H_2PO_4^-$:$HPO_4^=$ is normally about 1:4.

- Cellular organic phosphate esters exist mainly in the dianionic form at physiologic pH.

- Inorganic phosphates are important buffers in the renal tubular filtrate.

- Most NH_4^+ destined for urinary excretion is formed in proximal renal tubular epithelial cells from the deamination of glutamine.

- The ammonia buffer system plays an important role in the rumen and large intestine.

The Bicarbonate Buffer System

Exquisite control of the **plasma hydrogen ion concentration ([H$^+$])** is accomplished largely because the **CO_2 tension (Pco_2)** and **bicarbonate concentration ([HCO_3^-])** of blood are closely regulated. Indeed, the **[H$^+$]** has been found to be correlated with the ratio of the dominant acid-base pair in blood, namely **H_2CO_3:HCO_3^-**:

$$H_2CO_3 \longleftrightarrow H^+ + HCO_3^-$$

By the law of mass action, the **[H$^+$]** is directly proportionate to **[H_2CO_3]/[HCO_3^-]**. The **H_2CO_3** is, in turn, in equilibrium with **CO_2**, catalyzed by the enzyme **carbonic anhydrase (CA)**. Therefore, the complete relationship between these acid-base species is as shown in **Fig. 85-1**.

The relationship of the bicarbonate buffer system to the pH of extracellular fluid can be given as follows:

$$pH = pK' + \log ([HCO_3^-]/[H_2CO_3])$$

Where, **K'** is the **H_2CO_3** dissociation constant. Since **H_2CO_3** is in equilibrium with dissolved **CO_2** and **H_2O**, the Henderson-Hasselbalch equation for this buffer system is most appropriately written as follows:

$$pH = pK' + \log ([HCO_3^-]/[CO_2])$$

Where, **pK'** includes the equilibrium constant for the formation of **H_2CO_3** from **CO_2** and **H_2O**. In plasma at 37°C, the relationship between the concentration of dissolved **CO_2** ([CO_2]) and the **partial pressure of CO_2 (Pco_2)**, a more conveniently measured quantity, is as follows:

$$\left[\begin{array}{c} [CO_2] \text{ (mmol/L)} = 0.03 \text{ (mmol/L/mmHg)} \\ \text{x } Pco_2 \text{ (mmHg)} \end{array} \right]$$

Where, **0.03 mmol/L/mmHg** is the **solubility coefficient (S)** for **CO_2**. The Henderson-

Copyright © 2015 Elsevier Inc. All rights reserved.

Regulation of the Bicarbonate Buffer System

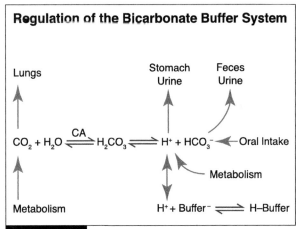

Figure 85-1

Hasselbalch equation for this buffer system can now be rewritten as follows:

$$pH = pK' + log ([HCO_3^-]/(S \times Pco_2))$$

Applying this equation to approximate normal plasma values:

$$[HCO_3^-] = 24 \text{ mmol/L (range 24-27)}$$
$$Pco_2 = 40 \text{ mmHg (range 38-42)}$$
$$[H^+] = 40 \text{ nmol/L (range 38-42)}$$
$$pH = pK' + log (24 / (0.03 \times 40))$$
$$= 6.1 + log 20$$
$$= 7.4$$

Note that the **pK'** of the **bicarbonate buffer system** is only **6.1**. Thus, this buffer system does not appear to represent an optimal body buffer. Nevertheless, the bicarbonate buffer system should be considered an important buffer system because the concentrations of its components can be **independently** regulated, CO_2 by the **lungs**, and HCO_3^- by the **kidneys**. Thus, by regulating the CO_2 and HCO_3^- concentrations, the lungs and kidneys also regulate the pH of blood (**Fig. 85-2**).

Under normal conditions, the ratio of bicarbonate to the Pco_2 of blood (times its solubility coefficient) should equal about 20:

$$[HCO_3^-]/(S \times Pco_2) = 24 / (0.03 \times 40) = 20$$

In cases of **metabolic acidosis** (i.e., a decrement in the plasma $[HCO_3^-]$), or **respiratory acidosis** (and increase in the plasma Pco_2), this ratio will be **less than 20**, and in cases of **metabolic alkalosis** (an increase in the plasma $[HCO_3^-]$), or **respiratory alkalosis** (a decrease in the plasma Pco_2), it will **exceed 20** (see Chapters 87, 89-91). Since the HCO_3^- concentration of plasma is regulated by the kidneys, while the Pco_2 is controlled by the lungs, from a physiologic perspective the above relationship can be reduced to the following:

$$pH \approx Kidneys / Lungs$$

It should be readily apparent through examination of this relationship that normal physiologic acid-base balance is a coordinate effort of both the pulmonary and renal organ

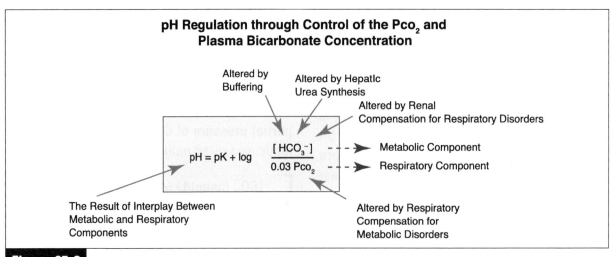

Figure 85-2

systems, and that acid-base disorders will result when one or both of these systems are overwhelmed or intrinsically disordered. However, like all overly-simplified relationships, this one also has its limitations since the plasma HCO_3^- concentration is also influenced by such factors as oral intake (e.g., $NaHCO_3$), hepatic urea synthesis, and fecal base loss (see Chapter 82).

The extracellular fluid hydrogen ion concentration (nmol/L) can also be calculated from Pco_2 and the bicarbonate concentration (mmol/L) using the following relationship:

$$[H^+] = 24\ Pco_2\ /\ HCO_3^-$$

Total CO_2 equals the sum of all forms of CO_2 in blood, including that contained in HCO_3^- and H_2CO_3, that bound to **proteins**, and that which is **dissolved** in plasma. It is generally approximated by summing the two major components, the bicarbonate concentration and the dissolved CO_2:

$$\text{Total } CO_2 \approx [HCO_3^-] + (S \times Pco_2)$$

The Phosphate Buffer System

Hydrolysis of ingested phosphoesters in the intestine, and the breakdown of phosphoproteins, nucleoproteins, and phosphatides, result in the production of **phosphoric acid** (H_3PO_4; see Chapter 82). When this strong acid, which significantly dissociates in solution, enters the circulation, it presents a major H^+ load to buffers of the body. In normal animals, the majority of phosphate is eliminated in urine.

The phosphate buffer system, which includes both **organic** and **inorganic phosphates**, is generally not considered to be an important buffer system in blood since phosphate concentrations are too low for this system to be quantitatively significant (**Fig. 85-3**). However, it plays an important role in **intracellular buffering** (particularly in muscle), and in buffering the renal **filtrate**.

The primary inorganic acid-base buffer pair for phosphate is as follows:

$$H_2PO_4^- <\!\!-\!\!-\!\!> H^+ + HPO_4^=$$

The pH range of this buffer system lies approximately 1 pH unit on either side of the **pK'** (**7.2**) (see Chapter 83). Within plasma, the ratio of $H_2PO_4^-$:$HPO_4^=$ is normally about **1:4**, with only a small fraction (**≈10%**) of inorganic phosphate bound to protein, or complexed with Ca^{++} or Mg^{++}. Thus, unlike Ca^{++}, the plasma phosphorus content is little influenced by changes in the plasma protein concentration.

If we consider the **phosphate esters** of simple sugars inside cells, represented by **glucose 6-phosphate (Fig. 85-4)**, fully protonated organic species contain two ionizable hydrogens with pK'1 = 0.94, and pK'2 = 6.11. In living cells that maintain their pH at about 7, which is far greater than 0.94 and about 1 pH unit greater than 6.11, the fully protonated form would not exist at all. About 10% would exist as the **monoanionic**, and about **90%** as the **dianionic** species.

Figure 85-3	
Quantitative Importance of the Blood Buffers	
Buffer System	**Percent Buffering in Whole Blood**
Nonbicarbonate	
Hemoglobin	35
Organic phosphates	3
Inorganic phosphates	2
Plasma proteins	7
	47
Bicarbonate	
Plasma	35
Erythrocyte	18
	53

Data from Winters Rw, and Dell R, 1965.

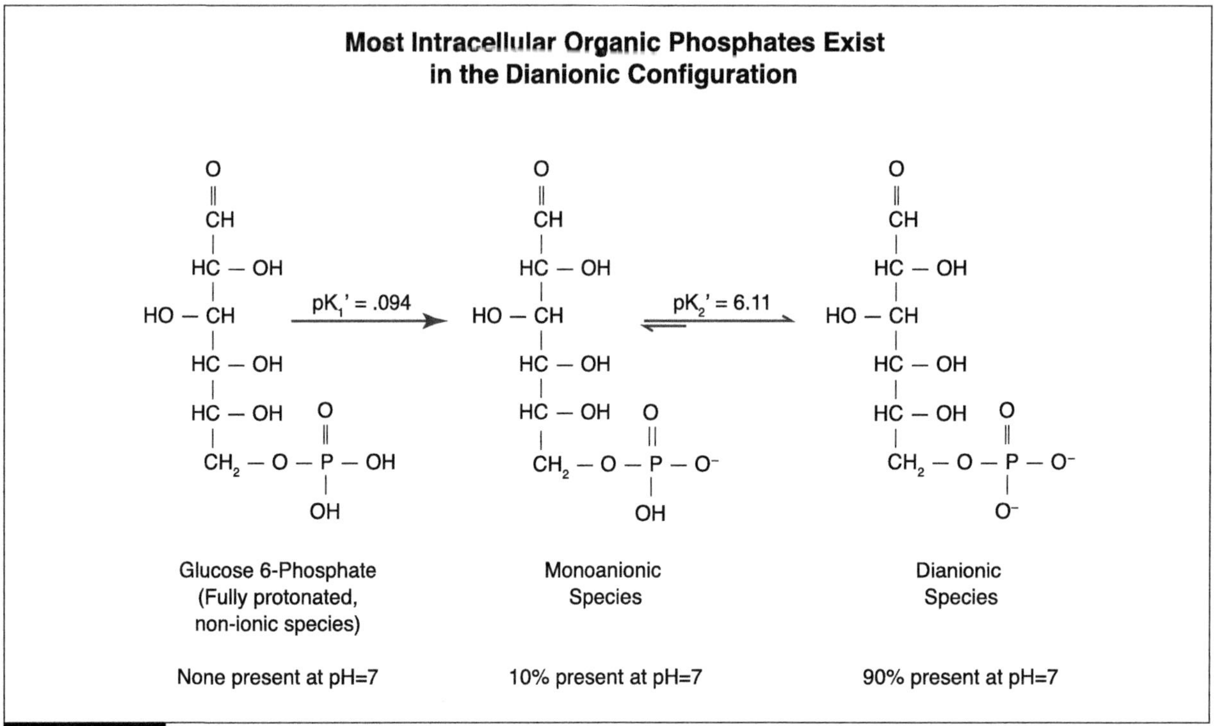

Most Intracellular Organic Phosphates Exist in the Dianionic Configuration

$$\text{Glucose 6-Phosphate} \xrightarrow{\text{pK}_1' = .094} \text{Monoanionic Species} \xrightleftharpoons{\text{pK}_2' = 6.11} \text{Dianionic Species}$$

Glucose 6-Phosphate
(Fully protonated,
non-ionic species)

None present at pH=7

Monoanionic
Species

10% present at pH=7

Dianionic
Species

90% present at pH=7

Figure 85-4

In the renal tubular filtrate, **secreted H⁺** reacts with **dibasic phosphate ($HPO_4^=$)** to form **monobasic phosphate ($H_2PO_4^-$; Fig 85-5)**. This occurs to the greatest extent in the distal tubules and collecting ducts, because it is here that the phosphate which escapes proximal reabsorption is greatly concentrated by the reabsorption of H_2O. Each H⁺ that reacts with tubular phosphate buffer contributes to urinary titratable acidity, which is measured by determining the amount of alkali that must be added to urine to return its pH to 7.4, the pH of the glomerular filtrate (see Chapter 83). However, titratable acidity obviously measures only a fraction of the acid secreted, since it does not account for the H_2CO_3 that has been converted to H_2O and CO_2. In addition, the pK' of the ammonia buffer system is 9.0, and therefore contributes very little to titratable acidity.

In alkaline urine dibasic phosphate ($HPO_4^=$) has a tendency to dissociate into $H^+ + PO_4^{-3}$. If sufficient amounts of **Mg⁺⁺** and **NH₄⁺** are present in that urine, **magnesium ammonium phosphate (struvite)** crystals can form, which leads to a

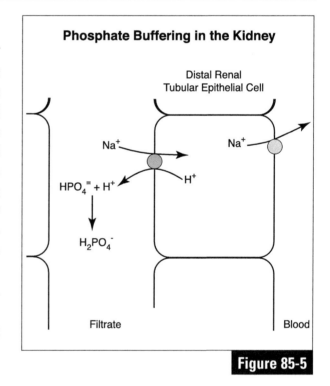

Phosphate Buffering in the Kidney

Figure 85-5

particular form of **feline lower urinary tract disease (FLUTD**; see **Case Study #6)**. A plant-based protein diet is low in sulfur-containing amino acids (which are generally acidifying), yet high in K⁺, Mg⁺⁺ and organic anions (which tend to be alkalinizing).

The Ammonia Buffer System

The acid-base pair for the ammonia buffer system is given as follows:

$$NH_4^+ \longleftrightarrow H^+ + NH_3$$

Since the pK' of this reaction is 9.0, the ratio of ammonia to ammonium ion (**NH_3:NH_4^+**) is about **1:100** at **pH 7.0**:

$$\text{pH 6} \underline{\quad\quad} 7.4 \underline{\quad\quad} 9$$
$$\quad\quad\quad NH_4^+ \quad\quad\quad NH_3 + H^+$$

The protonated NH_4^+ does not cross the blood brain barrier easily. In alkalemia more free NH_3 is created, however, that causes increased NH_3 uptake into the CNS, which can lead to coma. This may be particularly noticeable with loop diuretic overuse, where both hypokalemia and metabolic alkalosis ensue.

Most **NH_4^+** destined for urinary excretion is formed in the proximal renal tubular filtrate (see Chapter 11). Ammonia is derived largely from the deamination of glutamine in proximal renal tubular epithelial cells. As it diffuses into the proximal tubular lumen, simultaneous H^+ secretion allows protonation of NH_3 to NH_4^+.

Because of H_2O reabsorption during transit in the descending limb of the loop of Henle, the luminal $[HCO_3^-]$ rises, and the $[H^+]$ falls. This alkalinization favors dissociation of NH_4^+. The H^+ is buffered by HCO_3^- and $HPO_4^=$, while NH_3 diffuses into the medullary interstitium. It then passes into medullary collecting ducts where the low pH of the luminal fluid created by active H^+ secretion in the distal tubule allows titration of NH_3 back to NH_4^+. Thus, little NH_3 or NH_4^+ is normally detected in the distal renal tubular filtrate.

In the digestive tract NH_3 arises from the deamination of dietary proteins, and from microbial action on non-protein nitrogen, largely urea derived from saliva and the mesenteric circulation. Ammonia is incorporated into microbial protein in the rumen and large intestine, and it is absorbed into blood when lumen contents are alkaline.

OBJECTIVES

- Discuss the meaning of pH ≈ Kidneys / Lungs.

- Explain why the concentration of dissolved CO_2 in blood = 0.03 x P_{CO_2}, and why total CO_2 ≈ $[HCO_3^-]$ + dissolved CO_2.

- Know why the pK' of the bicarbonate buffer system (6.1) is considered to be somewhat "misleading."

- Explain why the plasma ratio of $[HCO_3^-]$ / (0.03 x P_{CO_2}) is far more meaningful than the absolute values of the variables.

- Identify two locations in the body where phosphate buffering is quantitatively significant.

- Describe the two most quantitatively significant buffer systems in blood, and explain why bicarbonate buffering is quantitatively more significant in plasma than in erythrocytes (see Chapter 84).

- Understand why the plasma ratio of $H_2PO_4^-$:$HPO_4^=$ ≈ 1:4 at pH = 7.4, and why most intracellular organic phosphates exist in the dianionic configuration.

- Show how the phosphate buffer system operates in the proximal renal tubular filtrate.

- Recognize why metabolic alkalosis causes increased movement of NH_3 into the CNS.

- Explain why there is usually little NH_3 or NH_4^+ detected in the distal renal tubular filtrate, and why the ammonia buffer system usually contributes little to titratable acidity.

- Identify normal sources of colonic NH_3 (see Chapter 10).

QUESTIONS

1. **An hysterical, 25-year-old student is admitted to the hospital following a physiological chemistry exam, and the following blood chemistry values are determined:**

$$[HCO_3^-] = 22.2 \text{ mmol/L}$$
$$P_{CO_2} = 30 \text{ mmHg}$$
$$P_{O_2} = 98 \text{ mmHg}$$

From these data, the hydrogen ion concentration of this patient's blood (nmol/L) would be expected to be:
   a. 17.4.

b.   22.5.

c.   28.1.

d.   32.4.

e.   36.7.

2. The following blood data were collected from an older patient:

$$[H^+] = 49 \text{ nEq/L}$$

$$P_{CO_2} = 30 \text{ mmHg}$$

$$P_{O_2} = 95 \text{ mmHg}$$

The pH would be:

a.   7.00.

b.   7.31.

c.   7.40.

d.   7.54.

e.   7.68.

3. Given the following data from an arterial blood sample:

$$pH = 7.55$$

$$P_{CO_2} = 25 \text{ mmHg}$$

$$[HCO_3^-] = 22.5 \text{ mEq/L}$$

The total CO_2 content is approximately how many mmol/L?

a.   23

b.   30

c.   41

d.   48

e.   55

4. Which one of the following expressions is CORRECT?

a.   Plasma pH $= pK' + \log ([CO_2]/[HCO_3^-])$

b. $[CO_2]$ (mmol/L) $= 0.3$ (mmol/L/mmHg) x $[HCO_3^-]$

c.   Normal plasma values for $[HCO_3^-]/(S \times P_{CO_2}) \approx 20$

d.   pH = Lungs / Kidneys

e.   Total $CO_2 = P_{CO_2} \times [HCO_3^-]$

5. Select the FALSE statement below:

a.   The phosphate buffer system, which includes both organic and inorganic phosphates, is not considered to be a quantitatively significant buffer system in blood.

b.   Most phosphate esters of organic molecules inside cells exist mainly in the dianionic form.

c.   In the renal tubular filtrate, secreted protons react with monobasic phosphate

to form dibasic phosphate.

d.   The phosphate buffer system plays an important role in buffering the renal tubular filtrate.

e.   Phosphoric acid is a strong acid generated from the intestinal digestion of phosphoesters, and the breakdown of phosphoproteins, nucleoproteins, and phosphatides.

6. The pK' of the ammonia buffer system is:

a.   5.0.

b.   6.0.

c.   7.0.

d.   8.0.

e.   9.0.

7. The filtrate in which area of the functional nephron would be expected to have the lowest concentration of ammonia (or ammonium ion)?

a.   Proximal tubule

b.   Descending limb of the loop of Henle

c.   Ascending limb of the loop of Henle

d.   Distal tubule

e.   Medullary collecting duct

8. The plasma pH is best correlated with which ratio below?

a. $NH_4^+:NH_3$

b. $H_2PO_4^-:HPO_4^=$

c. $H_2CO_3:HCO_3^-$

d. $HProt:Prot^-$

e. $HHb:Hb^-$

9. The ratio of bicarbonate to the P_{CO_2} of blood (times its solubility coefficient):

a.   Is normally about 40.

b.   Will increase in metabolic acidosis.

c.   Will decrease in metabolic alkalosis.

d.   Has little, if any, effect on the pH of blood.

e.   Will increase in respiratory alkalosis.

10. Which one of the following non-bicarbonate buffers is quantitatively most important in blood?

a.   Hemoglobin

b.   Hydroxyl anions

c.   Inorganic phosphates

d.   Plasma proteins

e.   Organic phosphates

ANSWERS

10. a

9. e

8. c

7. d

6. e

5. c

4. c

3. a

2. b

1. d

Chapter 86

Anion Gap

Overview

- The plasma anion gap is normally accounted for by proteins and other organic acids in the anionic form, phosphates, and sulfate.

- Hypoproteinemia causes a decrease in the anion gap.

- Normochloremic metabolic acidosis is usually associated with an increase in the anion gap.

- The plasma anion gap may not change in hyperchloremic metabolic acidosis.

- The urinary anion gap can be used to estimate the urine NH_4^+ concentration.

- The urinary anion gap, which is usually positive, may be negative in diarrheal disease.

- The urinary anion gap may not provide meaningful information when the urinary bicarbonate concentration in increased, or when ketone bodies appear in urine.

- Non-hypoproteinemic metabolic alkalosis is associated with an increase in the anion gap.

Presentation of the approximate normal ionic composition of blood plasma, interstitial fluid, and overall intracellular fluid can be found in **Fig. 86-1**. The **major extracellular cation** is **Na^+**, with a plasma concentration $\approx$ 152 mEq/L, and an interstitial fluid concentration $\approx$ 142 mEq/L. The **major extracellular anion is Cl^-**, with a plasma concentration $\approx$ 113 mEq/L, and an interstitial fluid concentration $\approx$ 117 mEq/L. The presence of a higher concentration of negatively charged protein (**$Prot^-$**) molecules in plasma causes an asymmetrical distribution of diffusible extracellular ions (mainly Na^+ and Cl^-) between plasma and interstitial fluid (the so-called **Gibbs-Donnan effect**). There is normally little Na^+ in intracellular fluid, and **K^+** is the **major intracellular cation**. Magnesium (**Mg^{++}**) is the second most plentiful cation inside cells. Inorganic phosphates, organic phosphate compounds, and proteins are the **major intracellular anions**. The total anion concentration equals the total cation concentration of each fluid compartment, in that there cannot be a significant deviation from **electroneutrality.**

Plasma Anion Gap (AG)

A plasma measurement that is sometimes useful in understanding acid-base disorders is the "**AG**." This term, which is somewhat of a misnomer since the total number of cations in plasma must equal the total number of anions, refers to the difference between the concentrations of **Na^+** plus **K^+**, and those of **HCO_3^-** plus **Cl^-** (**Fig. 86-1**). These electrolytes are routinely measured by most clinical laboratories, and a normal plasma AG value is about 17 mEq/L. Since the plasma **K^+** concentration usually undergoes little deviation, it is sometimes omitted from this calculation. Under these circumstances, the AG would be determined as follows:

$$\text{Plasma AG} = [Na^+] - ([HCO_3^-] + [Cl^-])$$

Copyright © 2015 Elsevier Inc. All rights reserved.

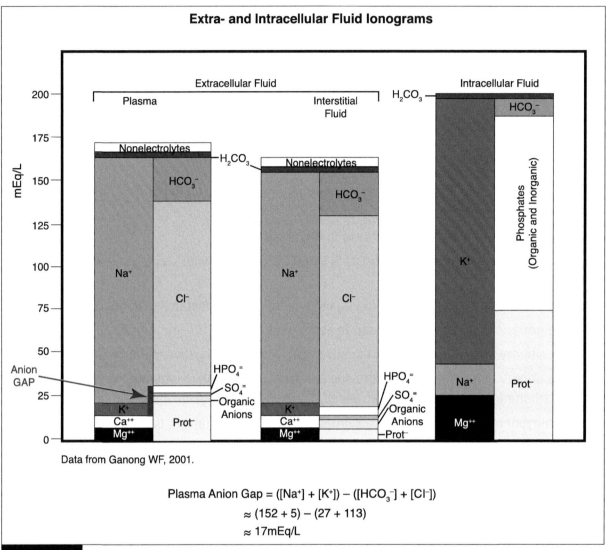

Extra- and Intracellular Fluid Ionograms

Data from Ganong WF, 2001.

$$Plasma\ Anion\ Gap = ([Na^+] + [K^+]) - ([HCO_3^-] + [Cl^-])$$
$$\approx (152 + 5) - (27 + 113)$$
$$\approx 17 mEq/L$$

Figure 86-1

A normal value would equal about **12 mEq/L**. This difference is normally accounted for by proteins and other organic acids in the anionic form, phosphates ($HPO_4^=/H_2PO_4^-$), and $SO_4^=$.

As shown in **Table 86-1**, a modest **decrease in the AG** occurs when unmeasured cations increase, as with hypercalcemia or hypermagnesemia, or when unmeasured anions decrease, as with hypoproteinemia. In **hypoproteinemic alkalosis**, both HCO_3^- and Cl^- will increase in order to fill the Prot$^-$ gap. Additionally, artifactual decreases in the AG can arise from hyperviscosity of blood (e.g., hyperglycemia or hyperlipidemia), due to an underestimation of the true plasma Na^+ concentration.

In non-hypoproteinemic alkalosis, there is an **increase in the AG**. Alkalemia increases the negative charge density on albumin by shifting the following reaction to the left:

Acidemia

$$H^+ + Albumin^- \underset{\text{Alkalemia}}{\overset{\text{Acidemia}}{\rightleftharpoons}} H\text{-Albumin}$$

Alkalemia

Since only the charged albuminate (albumin$^-$), but not the uncharged protonated form (H-albumin) contributes to the anion gap, alkalemia increases this charged component for any total amount of albumin (see Chapter 89). Conversely, acidemia has a tendency to reduce the charge equivalency of albumin, the most plentiful plasma protein.

A modest rise in the AG also occurs with a combined reduction in unmeasured cations, as occurs in hypomagnesemia or hypocalcemia, and hyperproteinemia. More importantly, the AG is elevated when **acids other than HCl or NH₄Cl** are added to plasma, so that the HCO_3^- concentration is titrated and lost as CO_2, while the Na^+ salt of the conjugate unmeasured (nonchloride) anion of the acid remains in plasma.

For example, consider the addition of 10 mEq/L of an organic acid (HA) to blood with a simplified composition as follows:

$$10\ HA + 24\ NaHCO_3 + 104\ NaCl + 12\ NaX \longrightarrow$$
$$14\ NaHCO_3 + 104\ NaCl + 10\ NaA + 12\ NaX$$

Organic anions (A⁻) are not routinely measured by the clinical lab. The AG can now be determined as follows:

$$AG = 140\ Na^+ - (14\ HCO_3^- + 104\ Cl^-)$$
$$= 22\ mEq/L$$

The increase in the AG (from 12 to 22 mEq/L) represents addition to blood of **10 mEq/L of nonchloride, nonbicarbonate anion**. The acid (HA) has simultaneously caused a fall in $[HCO_3^-]$, from 24 to 14 mEq/L, and a rise in the conjugate anion (A⁻), from 0 to 10 mEq/L. There has been no change in the $[Cl^-]$, 104 mEq/L. This type of acidosis is referred to as **increased AG metabolic acidosis** or, **normochloremic meta-bolic acidosis** (see Chapters 87 and 88). Examples are the early stages of ketoacidosis, or lactic acidosis (**Table 86-1**). Prolonged ketoacidosis, however, may cause hypochloremia.

Alternatively, consider the addition of 10 mEq/L of HCl to the same initial simplified blood solution:

$$10\ HCl + 24\ NaHCO_3 + 104\ NaCl + 12\ NaX \longrightarrow$$
$$14\ NaHCO_3 + 114\ NaCl + 12\ NaX$$

The AG becomes:

$$AG = 140\ Na^+ - (14\ HCO_3^- + 114\ Cl^-)$$
$$= 12\ mEq/L$$

In this case the conjugate anion (Cl⁻) of the acid is routinely measured, but does not change the anion gap. The rise in $[Cl^-]$, from 104 to 114 mEq/L, and reciprocal fall in $[HCO_3^-]$, from 24 to 14 mEq/L, maintains normalcy of the AG (12 mEq/L), and thus indicates that a Cl⁻-containing acid or its equivalent has been added to blood. Such a metabolic acidosis is termed **normal AG metabolic acidosis** (no change in the AG), or **hyperchloremic metabolic acidosis** (**Table 86-1**). Examples include metabolic acidoses with concomitant Cl⁻ retention, as seen with diarrhea and defects in renal acidification.

In summary to this point, a decrease in the AG is best associated with hypoproteinemic alkalosis, whereas an increase in the AG usually denotes some form of metabolic acidosis due to retention of acid other than HCl or NH₄Cl, or to a non-hypoproteinemic metabolic alkalosis, where the negative charge equivalency on albumin is increased. The plasma AG, however, can sometimes be a weak tool for understanding acid-base disturbances that combine hypoproteinemia, which lowers the AG, with any number of factors that tend to increase it. Quantitation of these types of disturbances would best be evaluated through use of the **strong ion difference** (**SID**; see Chapter 92).

Urinary Anion Gap (UAG)

The **UAG** ($Na^+ + K^+ - Cl^-$; **Fig. 86-2**) can be used to estimate the urinary ammonium ion (**NH₄⁺**) concentration. For this calculation it is assumed that the urine contains simply NaCl, KCl, and NH₄Cl. All other cations (e.g., Ca^{++} and Mg^{++}), and anions (e.g., $H_2PO_4^-/HPO_4^=$, $SO_4^=$, ketone bodies (KB⁻s), and other organic anions) are ignored. The sum of the cations ($Na^+ + K^+ + NH_4^+$) in this case is considered to be equal to the Cl⁻ concentration. The Na^+, K^+, and Cl⁻

Table 86-1
Disorders Associated with the Plasma Anion Gap (AG)

Increased AG	Decreased AG
Decreased cations (Not Na$^+$ or K$^+$)	Increased cations (Not Na$^+$ or K$^+$)
Hypocalcemia	Hypercalcemia
Hypomagnesemia	Hypermagnesemia
Increased anions (metabolic acidosis)	Decreased anions (not HCO$_3^-$ or Cl$^-$)
Hyperproteinemia	Hypoproteinemia
Hyperphosphatemia	
Ketoacidosis	Artifactual decreases in [Na$^+$]
Diabetes M.	Hyperglycemia
Starvation	Hyperlipidemia
Lactic acidosis	
Strenuous exercise	
Circulatory insufficiency	**No Change in the AG**
Cardiac failure	Hyperchloremic metabolic acidosis
Hypovolemic shock	Excess HCl, NH$_4$Cl or NaCl intake
Septic shock	HCO$_3^-$ loss
Anemia	Diarrhea
Liver disease	Pancreatic or biliary obstruction
Poisons or overdose	Renal tubular acidosis
Salicylates	CA inhibitors
Methanol	Hypoangiotensinemia
Formic acid	Hypoaldosteronism
Ethylene glycol	
Glycolic acid	
Oxalic acid	
Renal failure	
Phosphates, sulfates accumulate	
Organic anions accumulate	
Kidney fails to secrete H$^+$ & reabsorb HCO$_3^-$	
Non-hypoproteinemic metabolic alkalosis	
Increased albumin charge equivalency	

concentrations are easily measured by the clinical lab, however, the NH$_4^+$ concentration is not. By analogy with the plasma AG, **NH$_4^+$** thus represents the analytically missing component of the urine electrolytes. The term "UAG," like the plasma "AG," is another misnomer, since the major "missing" electrolyte is a cation.

Bicarbonate is normally low or absent in urine, and generally not measured, and urea is the major uncharged solute. Unlike plasma, urinary osmolarity can vary greatly depending upon such factors as the rate of NH$_4^+$ excretion, and levels of dietary salt and H$_2$O. It can be seen from **Fig. 86-2** that the additional NH$_4^+$ in the high NH_4^+ ionogram has replaced Na^+ and K^+, thus lowering the UAG from a positive to negative value. Thus, when the urinary NH_4^+ concentration is low, the UAG is positive (e.g., ≈ 30 mEq/L), and when it is high the gap is negative (e.g., ≈ -50 mEq/L). Because various minor ions in urine are not included in the UAG formula, the gap provides only a rough estimate of the urinary NH_4^+ concentration.

In the acidosis that results from **diarrheal disease**, the healthy kidney excretes larger quantities of NH_4^+, and therefore produces urine with a markedly negative UAG. However, renal NH_4^+ excretion can be impaired in **chronic renal failure**, and in **proximal renal tubular acidosis**, thus leading to a positive UAG.

In metabolic alkalosis (urinary pH > 6.5), the urinary $[HCO_3^-]$ may increase, and in ketoacidosis the urinary organic anion (A^-) concentration increases. Since HCO_3^- and ketone bodies are not normally quantitated in urine, the UAG may not provide reliable information when these anions are elevated since they both displace Cl^-.

In summary, the total number of cations are for the most part equal to the total number of anions in each body fluid compartment. In plasma, where electrolytes are routinely measured, the Na^+ concentration is generally greater than the sum of the Cl^- and HCO_3^- concentrations (**Fig. 86-1**). This difference, the **AG**, is normally accounted for by plasma proteins, other organic anions, $SO_4^=$, and $HPO_4^=/H_2PO_4^-$, and usually amounts to about 12 mEq/L (or 17 mEq/L if K^+ is included). An **increase** in the AG usually denotes some form of metabolic acidosis where acids other than HCl or NH_4Cl are added to blood (e.g., ketonemia or lactic acidemia), but it can also occur with hypocalcemia, hypomagnesemia, or non-hypoproteinemic metabolic alkalosis (**Table 86-1**). There will be **no change** in the AG with hyperchloremic metabolic acidosis. Conversely, a **decrease** in the AG occurs with hypercalcemia, hypermagnesemia, hypoproteinemic metabolic alkalosis, or artifactual decreases in the plasma $[Na^+]$ (e.g., hyperglycemia or hyperlipemia).

The **UAG** can sometimes be a useful tool, and the Na^+, K^+ and Cl^- concentrations are easily measured by clinical laboratories (**UAG = [Na⁺] + [K⁺] - [Cl⁻]**). When urinary NH_4^+ excretion increases, the UAG may become negative. However, when urinary HCO_3^- and KB^- excretion increase, the UAG may not provide reliable information since they both displace Cl^-.

OBJECTIVES

- Identify the plasma anions that account for the normal anion gap (≈ 17 mEq/L).

- Recognize the multitude of factors that can

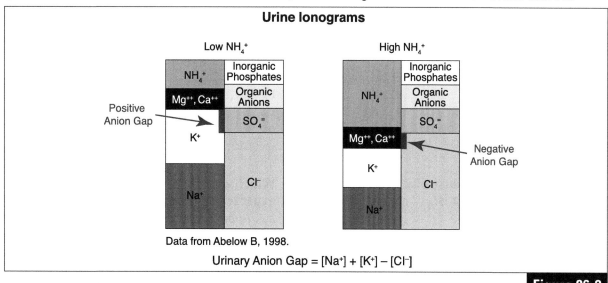

Urine Ionograms

Low NH_4^+ · High NH_4^+

Data from Abelow B, 1998.

Urinary Anion Gap = [Na⁺] + [K⁺] − [Cl⁻]

Figure 86-2

potentially increase or decrease the AG, but also realize that not all alterations in acid-base status will be reflected in the AG.

- Explain why hypoproteinemic alkalosis usually decreases the AG, while non-hypoproteinemic alkalosis increases it.

- Understand why an increase in the AG usually (but not always) denotes some form of metabolic acidosis, and explain why hyperchloremic metabolic acidosis may not change the AG.

- Discuss why hyperglycemia (alone) may cause an artifactual decrease in the AG.

- Know why hypomagnesemia and/or hypocalcemia increases the AG.

- Indicate how NH_4Cl addition to blood would alter acid-base status and the AG.

- Give an example of a normochloremic metabolic acidosis, and a normal AG metabolic acidosis.

- Explain why diarrhea may cause the UAG to become negative, and why this measurement may not be reliable if ketonuria is present.

QUESTIONS

1. **The second most plentiful cation inside cells is normally:**
   a. Ca^{++}.
   b. Mg^{++}.
   c. K^+.
   d. Na^+.
   e. Fe^{++}.

2. **The plasma anion gap is normally accounted for by:**
   a. Cl^- and HCO_3^-.
   b. Proteins and other organic acids, phosphates, and $SO_4^=$.
   c. Phosphates and Cl^-.
   d. HCO_3^-, organic acids, and proteins.
   e. Phosphates and $SO_4^=$.

3. **The most common cause of a decrease in the plasma anion gap is:**
   a. Ketoacidosis.
   b. Diarrhea.
   c. Vomiting.
   d. Hypoproteinemia.
   e. Ethylene glycol toxicity.

4. **Which one of the following causes of metabolic acidosis would not be expected to change the plasma anion gap?**
   a. Lactic acidosis
   b. NH_4Cl ingestion
   c. Ketoacidosis
   d. Salicylate toxicity
   e. Hyperproteinemia

5. **The plasma anion gap is a weak tool for estimating the degree of acidosis in the face of a:**
   a. Hypoproteinemia.
   b. Ketonuria.
   c. Methanol intoxication.
   d. Renal failure.
   e. Hyperphosphatemia.

6. **Which one of the following is normally low or absent in urine?**
   a. HCO_3^-
   b. Cl^-
   c. Na^+
   d. K^+
   e. $SO_4^=$

7. **A negative urinary anion gap would be expected with acidosis resulting from which one of the following?**
   a. Chronic renal failure
   b. Diabetes Mellitus
   c. Exercise
   d. Starvation
   e. Diarrhea

8. **The urinary anion gap can be used to estimate the urinary concentration of:**
   a. Glucose.
   b. $SO_4^=$.
   c. NH_4^+.
   d. Na^+.
   e. Ca^{++}.

9. **Which one of the following would be expected to increase the plasma AG?**
   a. Hypoproteinemia
   b. Hypocalcemia
   c. Hyperchloremia metabolic acidosis
   d. Hyperlipidemia
   e. Excess NaCl intake

ANSWERS

9. b
8. c
7. e
6. a
5. a
4. b
3. d
2. b
1. b

Chapter 87

Metabolic Acidosis

Overview

- The "processes" which cause acidemia or alkalemia are referred to as acidosis and alkalosis, respectively.

- The presence of a normal plasma pH, and an abnormal HCO_3^- or Pco_2, may represent a "mixed" acid-base disorder.

- Metabolic acidosis is the most common acid-base disorder recognized in domestic animals.

- Hydrogen ions cannot be added to the body without an accompanying anion.

- The Pco_2 may not change during the acute, uncompensated phase of metabolic acidosis.

- The bicarbonate buffer equation is shifted to the left during both the uncompensated and compensated phases of metabolic acidosis.

- Although the ventilatory response to a metabolic acidosis may help to ameliorate the effect on the plasma pH, it is insufficient to completely normalize it.

- Bone dissolution helps to mitigate a fall in pH during a metabolic acidosis.

When arterial blood **pH** is **< 7.4** (i.e., the H^+ concentration is **> 40 nEq/L**), the animal has acidemia, and when arterial blood **pH** is **> 7.4**, **alkalemia** is present. The "**processes**" causing these abnormalities are referred to as acidosis and alkalosis, respectively. **Metabolic acidosis** may arise from excess removal of HCO_3^- from the body in alkaline fluids (e.g., diarrhea or urine), from decreased acid excretion by the kidneys (e.g., kidney disease), or by excessive addition of protons to the body (e.g., NH_4Cl ingestion, or increased lactic acid or ketone body production). Conversely, **metabolic alkalosis** is associated with a rise in plasma pH, and an increase in the plasma HCO_3^- concentration. This acid-base disorder is common with vomiting (i.e., loss of H^+, Cl^-, K^+, and H_2O from the body).

Respiratory acidosis is associated with a

fall in arterial pH, resulting primarily from an increase in Pco_2 due to hypoventilation (e.g., pulmonary disease). Conversely, **respiratory alkalosis** is associated with a rise in arterial pH due to a decrease in Pco_2 through hyperventilation (e.g., excessive mechanical ventilation). During metabolic acidosis or alkalosis, the otherwise normal pulmonary system will attempt to correct the blood pH by either decreasing or increasing the Pco_2 through hyper- or hypoventilation, respectively. With respiratory acid-base disorders, the kidneys usually compensate by altering H^+ secretion and HCO_3^- reabsorption. Respiratory adjustments to metabolic acid-base disorders usually occur more rapidly, and are of greater magnitude than renal adjustments to respiratory acid-base disorders.

Given the above information, it seems that

Copyright © 2015 Elsevier Inc. All rights reserved.

the four compensated primary acid-base disturbances can be distinguished solely on the basis of the plasma **pH, Pco₂,** and **HCO₃⁻** concentration (**Table 87-1**). While this may be of some benefit, patients may also possess two or more acid-base disturbances simultaneously (e.g., an unregulated diabetic patient (ketoacidosis), who has vomited (metabolic alkalosis), and aspirated some of the vomitus into the lungs (respiratory acidosis)). These types of acid-base disorders are referred to as "**mixed**" **disturbances** (see Chapter 91). The presence of a normal plasma pH, and an abnormal HCO_3^- concentration or Pco_2, or both, may represent a mixed disturbance. Additionally, mixed disturbances are sometimes exemplified by a Pco_2 and plasma HCO_3^- concentration that move in opposite directions, or a plasma pH that moves in the opposite direction from a known primary disorder (e.g., a diabetic, ketoacidotic animal with alkalemia). The most effective way of understanding and quantitating the mixed acid-base disturbances is to make use of the **pH, Pco₂, plasma HCO₃⁻ concentration**, and **anion gap** (**AG**; see Chapter 86), and also the **Peter Stewart approach to acid-base balance** (see Chapter 92).

Metabolic acidosis is the most common acid-base disorder recognized in domestic animals. Like in respiratory alkalosis (see Chapter 91), the bicarbonate buffer equation is shifted to the left in metabolic acidosis (**Fig. 87-1**). Also, with an excess acid load or decreased urinary acid excretion, either an increased or normal plasma AG can be seen (see **Table 86-1**). What determines whether the AG will increase in metabolic acidosis? Whenever H⁺ is added to the system, HCO_3^- is consumed. The hydrogen cation cannot be added without an anion. Therefore, for each HCO_3^- consumed, a negative charge of some other type (which accompanied the H⁺) is added to body fluids. If the anion happens to be Cl⁻, no change in the AG will develop. However, if it is any other anion, the AG will be increased.

As stated above, when protons (H⁺) are initially added to the system, HCO_3^- is immediately consumed and CO_2 is generated. During this short acute, uncompensated phase, the increase in ventilation that follows a reduction in pH (i.e., carotid chemoreceptor stimulation of the respiratory center), initially keeps the Pco_2 within normal limits, even though the plasma HCO_3^- concentration falls (**arrow X —> C in Fig. 87-2**). However, since plasma pH is a function of the ratio of $[HCO_3^-]/(S \times Pco_2)$ (see Chapter 85), as the numerator decreases and the denominator stays the same in the acute phase, pH continues to decrease (**Table 87-2**). In order to bring this ratio closer to 20, and thus reestablish near-normal pH during the compensatory phase, an even stronger stimulation of the respiratory center through the carotid body reflex must occur. This stimulation is primarily a function of the low plasma pH, and as the depth and rate of respiration

Table 87-1			
Deviations in Plasma pH, Pco₂ and Bicarbonate Concentrations During the Compensatory Phases of the Four Primary Acid-Base Disturbances			
Acid-Base Disorder	**pH**	**Pco₂**	**HCO₃⁻**
Metabolic acidosis	↓	↓	↓
Metabolic alkalosis	↑	↑	↑
Respiratory acidosis	↓	↑	↑
Respiratory alkalosis	↑	↓	↓

Table 87-2			
Changes in Acid-Base Parameters During Uncompensated and Compensated Metabolic Acidosis			
Parameter	**Normal**	**Metabolic Acidosis**	
		Uncompensated	**Compensated**
HCO₃⁻(mEq/L)	24	14	11
Pco₂(mmHg)	40	40	21
HCO₃⁻/(S x Pco₂)	20	11.7	17.5
pH	7.4	7.18	7.34

Left Shift in the Bicarbonate Buffer Equation During Metabolic Acidosis

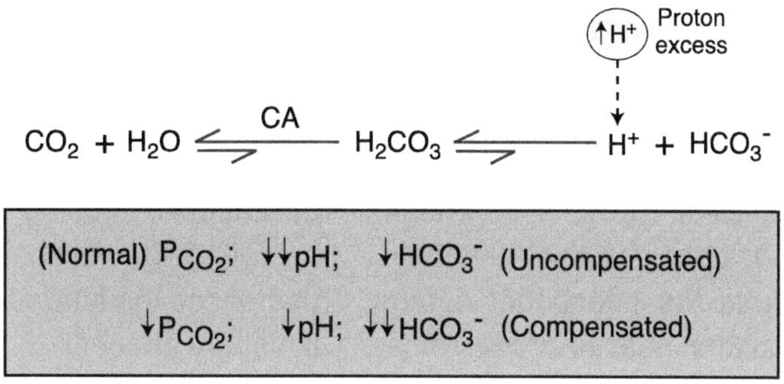

$$CO_2 + H_2O \underset{\longleftarrow}{\overset{CA}{\longrightarrow}} H_2CO_3 \underset{\longleftarrow}{\longrightarrow} H^+ + HCO_3^-$$

$(\uparrow H^+)$ Proton excess

(Normal) P_{CO_2}; $\downarrow\downarrow$pH; $\downarrow HCO_3^-$ (Uncompensated)

$\downarrow P_{CO_2}$; $\downarrow$pH; $\downarrow\downarrow HCO_3^-$ (Compensated)

Figure 87-1

Changes in the Plasma P_{CO_2}, pH and HCO_3^- Concentration During the Uncompensated and Compensated Phases of the Primary Metabolic and Respiratory Acid-Base Disturbances

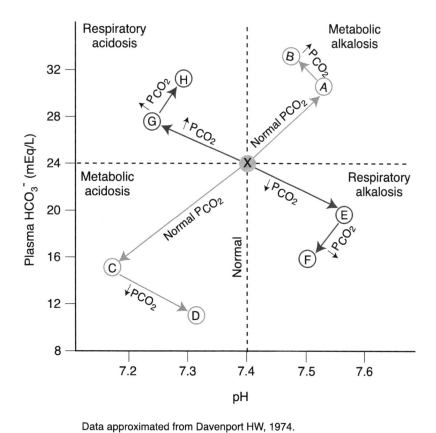

Data approximated from Davenport HW, 1974.

Figure 87-2

increases, there may be a significant decline in the Pco_2 (e.g., from 40 to 21 mmHg), and also a further (yet more modest) decline in the plasma HCO_3^- concentration (e.g., from 14 to 11 mEq/L). Even though the $[HCO_3^-]$ declines further during the compensatory phase, the $[\textbf{HCO}_3^-]/(\textbf{S} \times \textbf{Pco}_2)$ **ratio** comes closer to 20 (17.5), thus tending to normalize the plasma pH (7.34; **Table 87-2**). This effect on the plasma HCO_3^- concentration during the compensatory phase of metabolic acidosis is similar to the effect that occurs with the compensatory phases of the other three primary disturbances as well. **Whatever direction the plasma HCO_3^- concentration is headed during the uncompensated phase (arrows X —> C, X —> A, X—> E, and X —> G, Fig 87-2), it goes even further in that direction during the compensatory phase (arrows C —> D, A —> B, E —> F, and G —> H, Fig. 87-2).**

The kidneys also help to compensate during a metabolic acidosis. Acid anions (A^-) plus Na^+ are filtered to retain electrical neutrality, and renal tubular epithelial cells secrete protons into both the proximal and distal tubular filtrate in exchange for Na^+ and HCO_3^-, which are added to peritubular blood (**Fig. 87-3**). Therefore, in metabolic acidosis, the HCO_3^- concentration of urine is usually very low (unless the patient has renal tubular acidosis, RTA). Respiratory compensation for metabolic acidosis tends to inhibit the renal response in the sense that the induced drop in **Pco_2** hinders renal acid secretion. However, metabolic acidosis also decreases the filtered load of **HCO_3^-**, so its net inhibitory effect on the renal response is low.

Although extracellular and intracellular buffering mechanisms blunt the change in blood pH that would otherwise be produced from an acid load (see Chapter 83), a decrease in the plasma **HCO_3^-** concentration and a small degree of acidemia nevertheless is inevitable. However, the Henderson-Hasselbalch equation dictates that the degree of acidemia produced by a reduction in the plasma **HCO_3^-** concentration is less when there is a simultaneous decrease in the **Pco_2** (see Chapter 85).

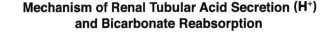

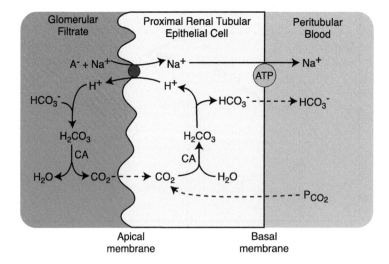

Mechanism of Renal Tubular Acid Secretion (H^+) and Bicarbonate Reabsorption

Note: Carbonic anhydrase (CA) activity is high in both the apical brush border region, and in the cytosol of these tubular cells.

Figure 87-3

Thus, respiratory compensation for metabolic acidosis blunts the change in arterial pH that would otherwise occur as a result of the reduction in the plasma HCO_3^- alone. Approximately 6-12 hours may be required for completion of this process (**Fig. 87-4**). As a general rule, the arterial Pco_2 decreases by about **1.25 mmHg** (range 0.9-1.5 mmHg) for each **mEq/L** drop in the plasma HCO_3^- concentration. It should be noted, however, that although the ventilatory response to metabolic acidosis is ameliorative, **it is insufficient to completely normalize arterial pH**. Thus, the bicarbonate buffer equation remains shifted to the left until excess acid generation ceases, and all excess protons that have been buffered by the body are properly metabolized or excreted through the lungs and kidneys.

The kidneys, under normal physiologic conditions, are continually responding to modest acid challenges in the form of dietary and metabolic acid. When the acid load increases in metabolic acidosis, there is normally an adaptive increase in renal H^+ secretion. However, the degree to which this compensation can occur is limited by several factors. First, as indicated in Chapter 82, a normal 30 Kg dog excretes only about 30 mmol of H^+ per day in urine (compared to about 7,000 mEq/day of volatile acid in the form of CO_2 through expired air). Thus, the kidneys are incapable of excreting as much acid as the lungs. Additionally, H^+ secretion into the proximal tubular filtrate is limited when the pH of the filtrate is < 6.5. However, H^+ secretion by lower capacity **α-intercalated** cells in the cortical collecting ducts may generate a steeper pH gradient (about 3 pH units less than blood), thus giving urine a low pH. The degree to which H^+ can be secreted by the proximal nephron is thus somewhat dependent upon the presence of buffers in the filtrate. Since the HCO_3^- concentration is significantly diminished in metabolic acidosis, this leaves the **$H_2PO_4^-$/ $HPO_4^=$** and **NH_3/NH_4^+** buffer systems to compensate (see Chapters 11 and 85). Lastly, the renal

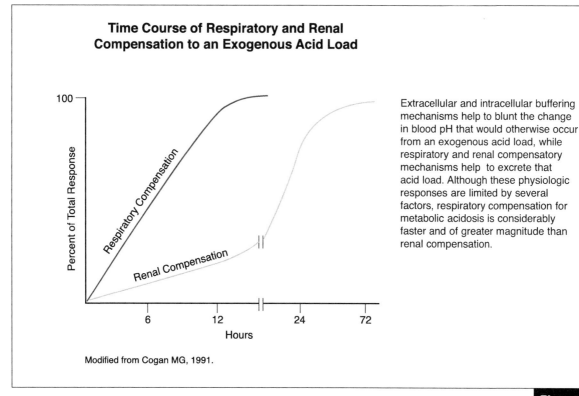

Time Course of Respiratory and Renal Compensation to an Exogenous Acid Load

Percent of Total Response

Respiratory Compensation

Renal Compensation

6   12   24   72
Hours

Extracellular and intracellular buffering mechanisms help to blunt the change in blood pH that would otherwise occur from an exogenous acid load, while respiratory and renal compensatory mechanisms help to excrete that acid load. Although these physiologic responses are limited by several factors, respiratory compensation for metabolic acidosis is considerably faster and of greater magnitude than renal compensation.

Modified from Cogan MG, 1991.

Figure 87-4

response to an acid load is much slower than the respiratory response (**Fig. 87-4**).

Effects of Chronic Acidemia on Bone

About two-thirds of the bone mass consists of inorganic minerals, especially **hydroxyapatite** ($Ca_{10}(PO_4)_6(OH)_2$), but also **brushite** ($CaHPO_4$), and the Na^+, K^+, Mg^{++}, and Ca^{++} salts of **carbonate** ($CO_3^=$; see Chapter 83). Thus, bone contains a number of proton acceptors, including $PO_4^=$, $HPO_4^=$, OH^-, and $CO_3^=$, and bone dissolution can help mitigate a fall in pH. During the acute, uncompensated phase of metabolic acidosis, $CO_3^=$ on the bone surface acts as a proton acceptor. An important mechanism appears to be the chemical exchange of free protons for Ca^{++}, Mg^{++}, Na^+, and K^+ bound to the carbonate, with a general buffering action as follows:

$$2H^+ + CO_3^= \longrightarrow CO_2 + H_2O$$

During the more chronic, compensated phase of metabolic acidosis, such as occurs with renal failure, buffering takes place through a combined exchange with bone carbonate cations, as occurs during the acute phase, and increased activity of osteoclasts, which help to mobilize additional bone mineral ($Ca_{10}(PO_4)_6(OH)_2$ and $CaHPO_4$). Both acute and chronic bone buffering lead to enhanced urinary excretion of bone Ca^{++} and $HPO_4^=/H_2PO_4^-$, which can reduce bone mass (osteopenia) and predispose animals to renal Ca^{++} stone formation. Additionally, metabolic acidosis reduces the **charge equivalency on albumin** (see Chapter 86), thus reducing the amount of Ca^{++} bound to protein in plasma (which is usually about 40% of total). This also increases the filtered load of Ca^{++}, which in turn increases urinary Ca^{++} excretion. Renal tubular acidosis (RTA), due to proximal tubular dysfunction, is a severe form of acidosis that affects Ca^{++} homeostasis.

OBJECTIVES

- Define and know the difference between the terms acidosis, acidotic, acidemic and aciduric.

- Explain why the bicarbonate buffer equation is shifted to the "left" during the initial uncompensated phase of metabolic acidosis, and then moves even further in that direction during the compensatory phase. Describe the role of the carotid body reflex in this physiologic maneuver.

- Discuss why the bicarbonate buffer equation remains shifted to the left during metabolic acidosis until all excess protons that have been buffered by the body are properly metabolized or excreted.

- Describe the proximal and distal renal compensatory processes that take place during a metabolic acidosis, and explain why urinary HCO_3^- excretion is usually low during this acid-base disorder.

- Explain why the plasma HCO_3^- concentration decreases during the uncompensated phase of a metabolic acidosis, but the Pco_2 remains unchanged.

- Recognize why respiratory compensation for metabolic acidosis is usually faster and of greater magnitude than renal compensation.

- Explain why metabolic acidosis usually increases urinary Ca^{++} excretion.

QUESTIONS

1. **Which one of the following sets of plasma parameters is best associated with metabolic acidosis?**
   a. ↑pH, ↑Pco_2, ↑HCO_3^-
   b. ↓pH, ↓Pco_2, ↓HCO_3^-
   c. ↓pH, ↓Pco_2, ↑HCO_3^-
   d. ↓pH, ↑Pco_2, ↑HCO_3^-
   e. ↑pH, ↓Pco_2, ↓HCO_3^-

2. **During the acute, uncompensated phase of metabolic acidosis:**
   a. The plasma HCO_3^- concentration usually increases.
   b. The Pco_2 of blood may not change even though the plasma HCO_3^- concentration decreases.
   c. The $[HCO_3^-]/(0.03\ Pco_2)$ ratio of blood increases.

d. The plasma HCO_3^- concentration will typically be lower than it is during the compensated phase.

e. The Pco_2 of blood will typically be lower than it is during the acute, uncompensated phase of respiratory alkalosis.

3. Select the TRUE statement below:

a. The normal kidneys are incapable of compensating for a metabolic acidosis.

b. Although the bicarbonate buffer equation is shifted to the left during the uncompensated phase of metabolic acidosis ($CO_2 + H_2O <\!\!-\!\!- H_2CO_3 <\!\!-\!\!- H^+ + HCO_3^-$), it is shifted to the right during the compensatory phase.

c. The anion gap will always be increased during a metabolic acidosis.

d. During the compensatory phase of metabolic acidosis, the Pco_2 decreases by about 12.5 mmHg for each mEq/L drop in the plasma HCO_3^- concentration.

e. Whatever direction the plasma HCO_3^- is headed during the uncompensated phases of metabolic acidosis and alkalosis, or respiratory acidosis and alkalosis, it proceeds even further in that direction during the respective compensatory phases.

4. A "mixed" acid-base disturbance would be indicated by the following:

a. A Pco_2 and plasma HCO_3^- concentration that were moving in opposite directions.

b. A decreased plasma anion gap.

c. An acidemia without a change in the plasma anion gap.

d. A diabetic, ketoacidotic animal with acidemia.

e. Acidemia with an elevated Pco_2 and plasma HCO_3^- concentration.

5. Select the FALSE statement below:

a. Although the ventilatory response to metabolic acidosis is ameliorative, it is insufficient to completely normalize the arterial pH.

b. The normal kidneys are incapable of excreting as much acid as the lungs.

c. In diabetic ketoacidosis, very little, if any, HCO_3^- should appear in urine.

d. The degree of acidemia produced by a reduction in the plasma HCO_3^- concentra-tion during metabolic acidosis will be increased by a simultaneous decrease in the Pco_2.

e. HCO_3^- reabsorption by proximal renal tubular epithelial cells, normally requires H^+ secretion in exchange for Na^+.

6. During the acute, uncompensated phase of metabolic acidosis:

a. The negative charge equivalency on albumin is increased.

b. The bicarbonate buffer equation is shifted to the right ($CO_2 + H_2O -\!\!-\!\!> H_2CO_3 -\!\!-\!\!> H^+ + HCO_3^-$).

c. The filtered load of Ca^{++} decreases in the kidneys.

d. The respiratory rate is decreased through the carotid body reflex.

e. There is a chemical exchange of H^+ for Ca^{++}, Na^+, and K^+ bound to carbonate on bone surfaces.

7. All of the following cell types typically secrete protons into the renal filtrate, EXCEPT:

a. Proximal tubular epithelial cells.

b. Epithelial cells in the ascending limb of the loop of Henle.

c. Distal tubular epithelial cells.

d. α-Intercalated cells of the cortical collecting ducts.

8. In which primary acid-base disturbance would the HCO_3^-, H^+ and Pco_2 concentrations be elevated?

a. Metabolic acidosis

b. Metabolic alkalosis

c. Respiratory acidosis

d. Respiratory alkalosis

9. During the compensatory phase of metabolic acidosis:

a. Plasma $HCO_3^-/(S \times Pco_2)$ increases.

b. The plasma HCO_3^- is typically lower than during the uncompensated phase.

c. The plasma Pco_2 decreases.

d. Renal H^+ secretion increases.

e. All of the above

9. e

8. c

7. b

6. e

5. d

4. a

3. e

2. b

1. b

ANSWERS

Chapter 88

Diabetes Mellitus
(Metabolic Acidosis and Potassium Balance)

Overview

- The first dehydration in diabetic hyperglycemia is a tissue dehydration.

- Acute hyperosmolarity redistributes K^+ and Mg^{++} out of cells.

- Metabolic acidoses which do not increase the anion gap (i.e., mineral acidoses), have a tendency to promote hyperkalemia.

- The K^+ and H^+ concentrations of blood generally increase and decrease together in metabolic acid-base disturbances.

- Insulin, epinephrine, and aldosterone promote K^+ movement into cells.

- Mg^{++} loss from cells reduces activity of the Na^+/K^+-ATPase.

- Aldosterone promotes both H^+ and K^+ secretion into the distal renal tubular filtrate.

- Ketone body anions (KB^-s) can eventually displace Cl^- from extracellular fluid.

- Metabolic acidosis may either reduce or enhance urinary K^+ excretion, depending upon the duration of the disturbance.

The primary flow of events leading to metabolic acidosis and an osmotic diuresis in **diabetes mellitus (DM)** is shown in **Fig. 88-1.** The primary event is a relative **insulin withdrawal** (or an increased requirement for insulin). With insulin lack, there is a decrease in glucose uptake by insulin-sensitive cells (particularly muscle and adipose tissue), which contributes to the developing **hyperglycemia** and mobilization of depot fat. This may sometimes result in a secondary **hypertriglyceridemia** as free fatty acids (FFAs) are synthesized into triglyceride and packaged in very low-density lipoproteins (VLDLs) by the liver (see Chapter 65). Since the liver is flooded with **FFAs**, **ketonemia** (and therefore acidemia) ensues, and as **ketone body anions (KB$^-$s)** exceed their renal threshold for reabsorption, they appear, along with accompanying **Na$^+$, K$^+$** and **Mg^{++}** cations, in urine (see Chapter 71). Added protons shift the bicarbonate buffer equation to the **left,**

thus creating a **hypobicarbonatemia**, and as the lungs compensate, a **hypocapnia ($\downarrow$Pco$_2$)** develops (see Chapter 87). The mild **ketoacidemia** alone should not alter the plasma **Cl$^-$** concentration (i.e., **normochloremic metabolic acidosis**; see Chapter 86), however, prolonged ketoacidemia can increase urinary **Cl$^-$** loss. The acute **hyperosmolarity** caused by the **hyperglycemia**, however, draws water out of cells, thus causing a tissue dehydration which dilutes the plasma **Na$^+$** and **Cl$^-$** concentrations, and **hyponatremia** and **hypochloremia** develop. Additionally, the acute hyperosmolarity causes a **shift in K$^+$ and Mg^{++} out of cells** (see below), thus mildly elevating the plasma K$^+$ and Mg^{++} concentrations. Glucose appears in urine, along with KB$^-$s and their accompanying cations (Na$^+$, Mg^{++}, and K$^+$), thus creating an **osmotic diuresis** that further depletes the ECF volume. Loss of water and electrolytes in urine leads to dehydration and hemocon-

Copyright © 2015 Elsevier Inc. All rights reserved.

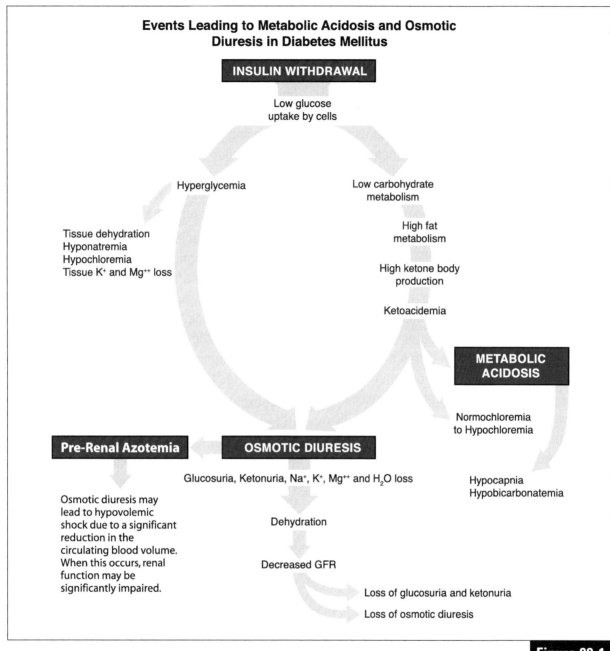

Events Leading to Metabolic Acidosis and Osmotic Diuresis in Diabetes Mellitus

INSULIN WITHDRAWAL

Low glucose uptake by cells

Hyperglycemia

Low carbohydrate metabolism

High fat metabolism

High ketone body production

Ketoacidemia

Tissue dehydration
Hyponatremia
Hypochloremia
Tissue K⁺ and Mg⁺⁺ loss

METABOLIC ACIDOSIS

Normochloremia to Hypochloremia

Pre-Renal Azotemia

OSMOTIC DIURESIS

Glucosuria, Ketonuria, Na⁺, K⁺, Mg⁺⁺ and H₂O loss

Hypocapnia
Hypobicarbonatemia

Osmotic diuresis may lead to hypovolemic shock due to a significant reduction in the circulating blood volume. When this occurs, renal function may be significantly impaired.

Dehydration

Decreased GFR

Loss of glucosuria and ketonuria

Loss of osmotic diuresis

Figure 88-1

centration, which in turn leads to peripheral circulatory failure because of the marked reduction in circulating blood volume (i.e., **hypovolemic shock**). The plasma **creatinine** and **blood urea nitrogen (BUN)** concentrations may be increased due to a combination of increased protein degradation, dehydration, and diminished renal perfusion (i.e., **pre-renal azotemia**). One characteristic feature of hypovolemic shock is hypotension, followed by a diminished renal blood flow (and glomerular filtration rate (GFR)), which can progress to the point of **anuria**. Coma may appear sometime after the appearance of peripheral circulatory failure, with death inevitable in the untreated patient.

Metabolic Acidosis and K⁺ Balance

Hyperkalemic disorders are typically associated with metabolic acidosis, and may sometimes be due to coordinately increased H⁺ production and cellular K⁺ release (e.g., states of hypercatabolism and tissue necrosis). Given the large intracellular pool of K⁺ compared to the extracellular pool (approximately 1,500 mEq versus 25 mEq in a 30 Kg dog), it is obvious that even a small shift in K⁺ distribution can exert profound influences on the plasma K⁺ concentration. For instance, the rapid movement from cells of **25 mEq of K⁺** (less than 2% of the intracellular pool), could double the plasma K⁺ content, and have serious consequences on cardiac function. Potassium is a potent stimulus for adrenal **aldosterone release**, and aldosterone promotes urinary K⁺ (and H⁺) excretion (see below). Thus, as the plasma K⁺ concentration rises, it is normally offset by an increase in urinary K⁺ excretion.

As indicated above, acute plasma **hyperosmolarity**, as occurs with hyperglycemia or strenuous exercise, redistributes both K⁺ and Mg⁺⁺ out of cells (**Fig. 88-2**). When a solute has difficulty permeating muscle cell membranes, such as glucose without insulin, the increased extracellular osmolarity causes water to flow out of these cells. Subsequent shrinkage of the cell concentrates its K⁺ and Mg⁺⁺, the two most abundant intracellular cations (see Chapter 86), and the rise in their concentrations causes them to diffuse out of the cell.

Additionally, a **decrease** in **extracellular pH** also tends to promote **K⁺** release from cells (whereas the opposite holds true with alkalemia; see Chapter 89). When protons enter cells, they bind with intracellular buffers (e.g., phosphates and anionic protein (Prot⁻)), thus momentarily decreasing electronegativity of the cell (**Fig. 88-3**). In order to reestablish normal electronegativity, either anions must be added, or cations must be lost. Since K⁺ is the most abundant intracellular cation, it may be lost from the cell.

Additionally, acidosis is a common cause of **Mg⁺⁺** loss from cells. This electrolyte is an important cofactor for many intracellular reactions, and it helps hold the Na⁺/K⁺-ATPase in position so that ATP can attach. Intracellular Mg⁺⁺ deficiencies thereby lead to a net reduction in cellular K⁺ uptake, which exacerbates the loss of K⁺ from the organism.

Hyperchloremic metabolic acidosis due to HCl or an equivalent mineral acid excess (e.g., NH₄Cl or CaCl₂), significantly increases the serum **K⁺ concentration** without altering the plasma anion gap (see Chapter 86). In fact, the serum K⁺ concentration may rise as much as **0.7 mEq/L** for each **0.1 pH unit decrease** during a mineral acidosis. Organic metabolic acidoses that increase the plasma anion gap (e.g., ketoacidosis or lactic acidosis), and respiratory acidoses have less of a direct impact on the serum K⁺ concentration.

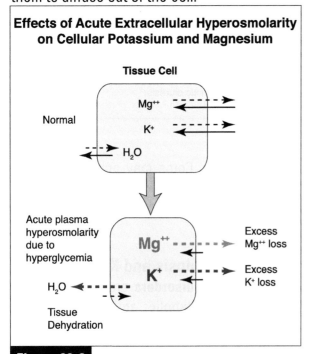

Effects of Acute Extracellular Hyperosmolarity on Cellular Potassium and Magnesium

Figure 88-2

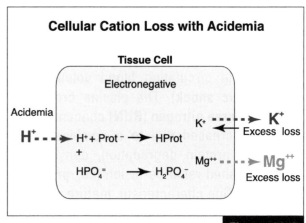

Cellular Cation Loss with Acidemia

Figure 88-3

The high extracellular H⁺ concentration during a mineral acidosis apparently **slows membrane Na⁺/H⁺ antiporter activity**, thus reducing the cytoplasmic Na⁺ concentration, which in turn depresses K⁺ uptake via the Na⁺/K⁺-ATPase (**Fig. 88-4**). Continuing K⁺ egress from the cell results in a net increase in extracellular K⁺ in response to this mineral acidosis. The anion companion, Cl⁻, has difficulty permeating the cell membrane, and thus largely remains in extracellular fluid.

In contrast, **metabolic acidosis due to an increase in organic acid (HA) production** alone (e.g., ketoacidosis or lactic acidosis), has less of a tendency to increase the serum K⁺ concentration. In these acid-base disorders, entry of H⁺ into the cell is accompanied by the organic anion (A⁻) that can easily permeate the cell membrane (bottom of **Fig. 88-4**). Intracellular release of this H⁺ coupled with ongoing acidification of the cell by metabolic processes, prevents slowing of the Na⁺/H⁺ antiporter, and hence of the Na⁺/K⁺-ATPase, and therefore induces less net cellular K⁺ loss.

At the level of the **kidney**, both H⁺ and K⁺ are effectively secreted by distal renal tubular epithelial cells, in exchange for Na⁺ (**Fig. 88-5**). As serum **aldosterone** levels rise because of **hyperkalemia**, Na⁺/K⁺-ATPase activity in basal membranes of these cells increases, thus increasing movement of Na⁺ from the tubular lumen into blood. Since Cl⁻ reabsorption lags behind Na⁺ reabsorption in this section of the nephron, the **transepithelial potential difference (PD)** increases under aldosterone stimulation (tubular lumen negative with respect to the interstitial fluid), thus attracting cations from these cells into the tubular fluid. During acidemia, protons enter these cells from blood in exchange for K⁺, thus providing relatively more H⁺ inside these cells for tubular secretion. Although protons appear to compete with K⁺ for this renal secretory mechanism, in chronic hyperkalemic acidemia the secretion of both

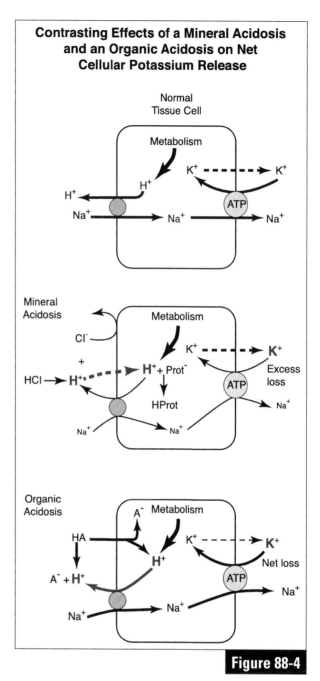

Contrasting Effects of a Mineral Acidosis and an Organic Acidosis on Net Cellular Potassium Release

Figure 88-4

H⁺ and K⁺ will usually increase in this section of the nephron, thus promoting their excretion in urine.

The K⁺ and H⁺ concentrations of plasma generally parallel each other since in acidemia, protons effectively displace K⁺ from tissue cells, and in alkalemia the opposite occurs. **Acute alkalemia** increases while acidemia decreases distal tubular K⁺ secretion. Acute acidemia effectively reduces activity of the

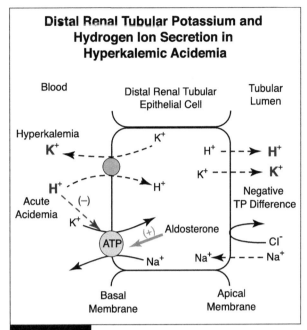

Distal Renal Tubular Potassium and Hydrogen Ion Secretion in Hyperkalemic Acidemia

Figure 88-5

aldosterone-sensitive Na⁺/K⁺-ATPase in the basal membrane of distal renal tubular epithelial cells, thereby initially reducing the cytoplasmic concentration of K⁺ in these cells, and it also promotes H⁺/K⁺ exchange across this membrane (see **Fig. 88-5**). Acute acidemia also reduces permeability of distal tubular apical membranes to K⁺, while acute alkalemia has the opposite effect. When **metabolic acidosis lasts for several days**, however, **urinary K⁺ secretion by these cells** (and excretion into urine) **is enhanced** for **three** reasons:

1) Since acidemia also reduces Na⁺/K⁺-ATPase activity in the proximal tubule, NaCl and H₂O reabsorption are reduced, hence the flow of fluid through the remainder of the nephron is enhanced. This enhanced fluid flow effectively washes-out K⁺ in the distal tubular filtrate, thus increasing the concentration gradient for K⁺ to diffuse from tubular cells into the lumen.

2) Since plasma levels of K⁺ rise with H⁺, the filtered load of K⁺ increases, and therefore more is excreted into urine.

3) Aldosterone levels rise due to an increase in the plasma K⁺ concentration and a decrease in the effective circulating volume

(due to enhanced urinary NaCl and H₂O loss). Aldosterone promotes both K⁺ and H⁺ secretion into the distal tubular filtrate by increasing the transepithelial PD.

Therefore, in **chronic metabolic acidosis**, rises in tubular fluid flow through the nephron, and rises in plasma K⁺ and aldosterone levels offset the negative effects of acute acidosis on the distal tubular cell K⁺ concentration and apical membrane permeability, and renal K⁺ secretion and urinary excretion rise. Thus, **metabolic acidosis may either reduce or enhance urinary K⁺ excretion, depending on the stage and "duration" of the disturbance**.

Endocrine Influences on K⁺ Balance

When an **oral or intravenous K⁺ load** is administered, only about 50% is excreted by the kidneys within 6-8 hours (**Fig. 88-6**). To avoid a serious hyperkalemia, K⁺ is temporarily stored in cells such as those of muscle and liver. As indicated above, the combined intracellular K⁺ reservoirs of the body are enormous (1,500 mEq, compared to an extracellular pool of 25 mEq in a 30 Kg dog). If an animal consumes 80 mEq/day of dietary K⁺, about 70 mEq should eventually be excreted in urine, about 9 mEq in the stool, and only 1 mEq from skin (**Fig. 88-7**).

As shown in **Fig. 88-8**, the entry step for K⁺ across cell membranes is via the Na⁺/K⁺-ATPase pump. Activity of this pump is controlled by several hormones, most notably

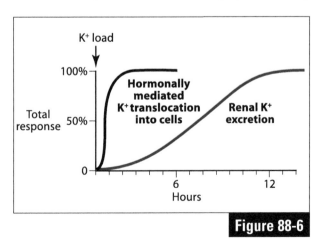

Figure 88-6

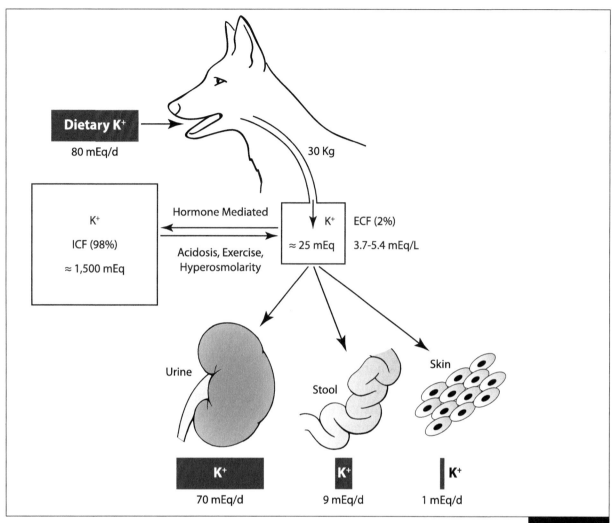

Figure 88-7

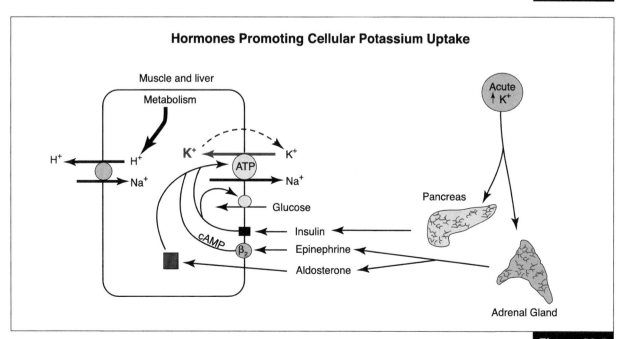

Figure 88-8

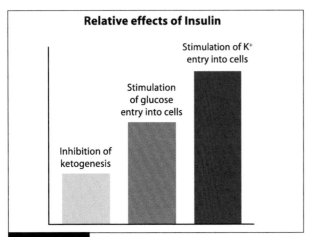

Relative effects of Insulin

Stimulation of K⁺ entry into cells

Stimulation of glucose entry into cells

Inhibition of ketogenesis

Figure 88-9

insulin, **epinephrine**, and **aldosterone**. An acute rise in the extracellular K⁺ concentration, such as may occur with **dietary intake**, **intravenous KCl administration** (see Chapter 93), **exercise**, **mineral acidosis**, or **hyperosmolarity**, stimulates secretion of these hormones, which activate Na⁺/K⁺-ATPase to allow movement of K⁺ into cells. This movement then quickly returns the extracellular K⁺ concentration to normal. As can be seen from **Fig. 88-9**, **insulin's** effect of stimulating cellular K⁺ uptake is greater than its effect on cellular glucose uptake, and on its effect of inhibiting hepatic ketogenesis.

Subnormal K⁺ uptake into cells can occur with an intracellular **Mg⁺⁺ deficiency** (i.e., reduced cofactor available for the Na⁺/K⁺-ATPase), or when there is a **deficiency** of one or more of the three hormones controlling Na⁺/K⁺-ATPase

activity: **insulin** in DM, **catecholamines** in adrenal medullary or autonomic insufficiency, or during administration of β-adrenergic blocking agents, and **aldosterone** in hyporeninism or adrenal cortical insufficiency (**Fig. 88-10**). Their **combined absence**, as may occur in **DM**, can lead to a serious elevation in the serum K⁺ concentration. During hyperglycemia, the high extracellular glucose concentration causes an extracellular hyperosmolarity which induces cell shrinkage and K⁺ extrusion from cells. The resulting hyperkalemia cannot be combated in those diabetics who coordinately lack insulin, catecholamines, and/or aldosterone. For similar reasoning, vigorous exercise in non-diabetic animals, which causes K⁺ to leave cells, may also cause a marked hyperkalemia. If said animals are also undergoing treatment with β-adrenergic blocking agents, the hyperkalemia could be severe enough to cause sudden death.

In summary, insulin withdrawal leads to hyperglycemia, tissue dehydration, hyponatremia, net K⁺ and Mg⁺⁺ loss from the body, mobilization of depot fat, hypertriglyceridemia, ketonemia, metabolic acidosis, hypocapnia, hypobicarbonatemia, glucosuria, ketonuria, osmotic diuresis, and hypotonic dehydration. Treatment should be aimed at correcting these disturbances. In untreated patients, peripheral circulatory failure can lead to coma and death.

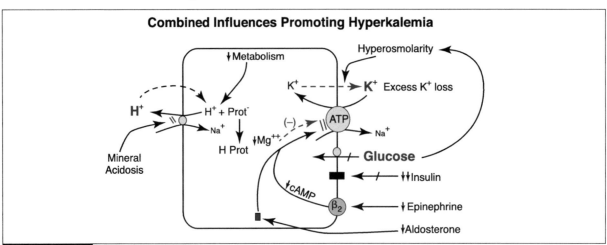

Combined Influences Promoting Hyperkalemia

Figure 88-10

OBJECTIVES

- Indicate how hyponatremia and hypochloremia develop in diabetic acidosis.

- Explain how intracellular K^+ depletion can occur in diabetic acidosis without substantially increasing the plasma K^+ concentration.

- Understand how hyperglycemia and acidemia promote intracellular K^+ and Mg^{++} depletion, and why extracellular K^+ and H^+ concentrations generally parallel each other.

- Explain why a mineral acidosis promotes more tissue K^+ loss than an organic acidosis.

- Discuss the mechanisms for increasing urinary H^+ and K^+ excretion in chronic hyperkalemic acidemia.

- Know how insulin, epinephrine and aldosterone assist in protecting the body against an oral or IV K^+ load.

- Recognize why animals being treated with β-adrenergic blocking agents may experience hyperkalemia following exercise.

- Describe the pathophysiologic events that lead to metabolic acidosis and osmotic diuresis in DM.

QUESTIONS

1. **The primary cause of hyponatremia in diabetes mellitus is:**
   a. Ketoacidemia.
   b. Hypochloremia.
   c. Hyperglycemia and tissue dehydration.
   d. Hypocapnia.
   e. Hypobicarbonatemia and glucosuria.

2. **All of the following have a tendency to favor net K^+ movement out of cells, EXCEPT:**
   a. Hyperglycemia.
   b. Mineral acidosis.
   c. Strenuous exercise.
   d. Insulinopenia.
   e. Hypermagnesemia.

3. **Which one of the following hormones promotes both H^+ and K^+ secretion by the distal nephron?**
   a. Insulin
   b. Aldosterone
   c. Glucagon
   d. Epinephrine
   e. Parathyroid hormone

4. **In hyperchloremic metabolic acidosis:**
   a. The plasma anion gap is increased.
   b. Plasma membrane Na^+/K^+-ATPase activity is increased.
   c. Glucosuria, ketonuria, and osmotic diuresis usually develop.
   d. Plasma membrane Na^+/H^+ antiporter activity is decreased.
   e. Urinary K^+ excretion is usually decreased.

5. **All of the following are generally associated with diabetic ketoacidosis, EXCEPT:**
   a. Net K^+ and Mg^{++} loss from cells.
   b. Hypocapnia.
   c. Hypertension.
   d. Increase in the plasma anion gap.
   e. Hypobicarbonatemia.

6. **During the acute phase of metabolic acidosis:**
   a. Urinary K^+ excretion dramatically increases.
   b. Protons enhance permeability of distal renal tubular apical membranes to K^+.
   c. Activity of Na^+/K^+-ATPase in basal membranes of distal renal tubular epithelial cells increases.
   d. The bicarbonate buffer equation is shifted to the right ($CO_2 + H_2O \longrightarrow H_2CO_3 \longrightarrow H^+ + HCO_3^-$), and plasma HCO_3^- levels rise.
   e. H^+/K^+ exchange across basal membranes of distal renal epithelial cells is favored.

7. **Which of the following is/are usually associated with hyperkalemia?**
   a. Intracellular Mg^{++} deficiency
   b. Hyperinsulinemia
   c. Administration of β-adrenergic agonists
   d. Metabolic alkalosis

7. a
6. e
5. c
4. d
3. b
2. e
1. c

ANSWERS

Chapter 89

Metabolic Alkalosis

Overview

- Metabolic alkalosis commonly results from excessive HCl, K^+ and H_2O loss from the stomach or through the urine.

- The plasma anion gap increases in non-hypoproteinemic metabolic alkalosis due to an increased negative charge equivalency on albumin, and the free ionized Ca^{++} content of plasma decreases.

- The $[HCO_3^-]/(S \times Pco_2)$ ratio is increased in metabolic alkalosis.

- The bicarbonate buffer equation is shifted to the right ($CO_2 + H_2O \longrightarrow H_2CO_3 \longrightarrow H^+ + HCO_3^-$) in metabolic alkalosis.

- The kidneys excrete excess HCO_3^- into urine during a metabolic alkalosis.

- Hypokalemia and kaliuresis are common complications of metabolic alkalosis.

- Patients with metabolic alkalosis are predisposed to cardiac arrhythmias.

- Post-hypercapnic metabolic alkalosis can occur in a patient with respiratory acidosis who is mechanically ventilated.

- Contraction alkalosis can occur in patients who are being treated with loop or thiazide diuretics.

- A free water deficit leads to a concentration alkalosis.

Metabolic alkalosis commonly results from **excessive HCl, K^+ and H_2O loss** from the **stomach** in the patient who is vomiting, or from the **urinary tract** in a patient receiving either **loop** or **thiazide diuretic therapy (Table 89-1)**. **Abomasal displacement** also sequesters HCl, K^+, and H_2O; thus, symptoms of this acid-base disorder resemble those of vomiting. Primary or secondary **hyperaldosteronism** will enhance renal acid loss, and **excessive alkali intake** will also precipitate a metabolic alkalosis. Usually **severe K^+ depletion** accompanies both extracellular volume and HCl depletion, therefore the association of hypokalemia with metabolic alkalosis can be multifactorial in origin.

The treatment of edema with either loop or thiazide diuretics is a common cause of metabolic alkalosis. These medications promote an extracellular fluid diuresis containing minimal amounts of HCO_3^-, and if the diuresis is uncontrolled, a condition referred to as **contraction alkalosis** (or **concentration alkalosis**) develops. Other comparable conditions that promote a free water deficit (e.g., diabetes insipidus and hypertonic dehydration) contribute to this acid-base disorder by increasing the **strong ion difference** (**SID**; see Chapter 92). As indicated in Chapter 84, the relationship between total plasma protein ($Prot_{tot}$) and H^+ is as follows:

Copyright © 2015 Elsevier Inc. All rights reserved.

Table 89-1
Common Causes of Metabolic Alkalosis

Excess HCl loss

Vomiting

   Gastric contents only

Displaced abomasum

Increased urine acidification

   K^+ wasting diuretics (loop and thiazide)

   Hyperaldosteronism

      Excessive mineralocorticoid therapy

      Hyperadrenocorticoidism (Cushing's syndrome)

      JG-cell hyperplasia (Bartter's syndrome)

   Increased impermeant anion delivery to the distal nephron

      Penicillin-derivative antibiotics

Free water deficit

Concentration alkalosis (contraction alkalosis)

   Diabetes insipidus

      Minimal urinary HCO_3^- loss

   Hypertonic dehydration

      $\uparrow SID \longrightarrow \downarrow [H^+]$

Excess alkali intake (e.g., $NaHCO_3$)

Ulcer therapy

Race horses

Treatment of metabolic acidosis

Increased sweating

Cl^- loss with anion debt filled by HCO_3^-

Severe K^+ depletion

Administration of large quantities of K^+ - free solutions

Loop and thiazide diuretic therapy

Hypoproteinemia

Post-hypercapnia

---

1)   **$Prot_{tot}$ = HProt + Prot$^-$**

2)   **HProt <——> H$^+$ + Prot$^-$**

Therefore, **hypoproteinemia** ($\downarrow Prot_{tot}$) would decrease both HProt and Prot$^-$, and since these two variables are in equilibrium with H$^+$, their reduction would result in a proton deficit.

Like in respiratory acidosis (see Chapter 90), the **bicarbonate buffer equation is shifted to the right** in metabolic alkalosis (**Fig. 89-1**). The elevated HCO_3^- displaces Cl^- from the circulation, and a relative **hypochloremia** develops. Although metabolic alkalosis is usually hypochloremic, the fall in the plasma Cl^- concentration may not be exactly ▶ the inverse

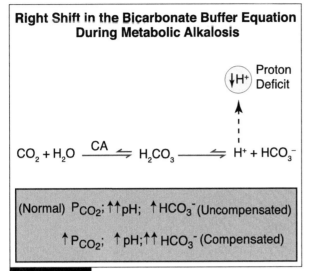

Figure 89-1

Right Shift in the Bicarbonate Buffer Equation During Metabolic Alkalosis

$CO_2 + H_2O \xrightarrow{\text{CA}} H_2CO_3 \rightleftharpoons H^+ + HCO_3^-$

$\downarrow H^+$ Proton Deficit

(Normal) P_{CO_2}; $\uparrow\uparrow$pH; $\uparrow HCO_3^-$ (Uncompensated)

$\uparrow P_{CO_2}$; $\uparrow$pH; $\uparrow\uparrow HCO_3^-$ (Compensated)

of the rise in HCO_3^- (i.e., there may be an **increase in the anion gap (AG)**; see Chapter 86). The AG predictably increases during non-hypoproteinemic metabolic alkalosis due to the greater negative charge equivalency on albumin, the major contributor to the AG.

Non-Hypoproteinemic Metabolic Alkalosis

$H^+ + \uparrow Albumin^- \longleftarrow H\text{-Albumin}$

$(\uparrow AG)$

The AG usually rises about 0.4-0.5 mEq/L for each mEq/L increment in plasma HCO_3^- during non-hypoproteinemic metabolic alkalosis. For example, if the plasma HCO_3^- concentration were 44 mEq/L (a rise of 20 mEq/L above normal), the AG ($Na^+ - (HCO_3^- + Cl^-)$) should be about 21 mEq/L (normal 12 mEq/L + (0.45 x 20)).

An AG greater than this amount would imply superimposition of a normochloremic, high AG form of metabolic acidosis (e.g., ketoacidemia).

When a **proton deficit** occurs in metabolic alkalosis, the bicarbonate buffer equation is immediately shifted to the right. During this short acute, uncompensated phase, the decrease in ventilation that follows an increase in pH (i.e., carotid chemoreceptor regulation of the respiratory center), initially keeps the Pco_2 within normal limits (**Fig. 89-1**), even though the plasma HCO_3^- concentration increases (**arrow X —> A** in **Fig. 87-2**). However, since plasma **pH** is a function of the ratio of **$[HCO_3^-]/(S \times Pco_2)$** (see Chapter 85), as the numerator increases and the denominator stays the same in the acute phase, the pH continues to rise (**Table 89-2**). In order to bring this ratio closer to 20, and thus reestablish near-normal pH during the compensatory phase, an even greater suppression of respiration through the carotid body reflex must occur. This respiratory suppression is primarily a function of the high extracellular pH, which overrides control via the increasing Pco_2 and declining Po_2. As the depth and rate of respiration decrease, there is a significant increase in the Pco_2 (e.g., from 40 to 48 mmHg), and also a further (yet more modest) increase in the plasma HCO_3^- concentration (from 30 to 32 mEq/L). Even though the plasma HCO_3^- concentration increases further during the compensatory phase, the **$[HCO_3^-]/(S \times Pco_2)$ ratio** comes closer to 20 (from 25 to 22.2),

Table 89-2

Changes in Acid-Base Parameters during Uncompensated and Compensated Metabolic Alkalosis

Parameter	Normal	Metabolic Alkalosis Uncompensated	Metabolic Alkalosis Compensated
HCO_3^- (mEq/L)	24	30	32
Pco_2 (mmHg)	40	40	48
$HCO_3^- / (S \times Pco_2)$	20	25	22.2
pH	7.4	7.5	7.44

thus tending to normalize the plasma pH (from 7.5 to 7.44). This effect on the plasma HCO_3^- concentration during the compensatory phase of metabolic alkalosis is similar to that seen during the compensatory phase of respiratory acidosis (see Chapter 90).

The **kidneys** also help to compensate during a metabolic alkalosis. When the plasma HCO_3^- concentration exceeds 28 mEq/L, HCO_3^- reabsorption decreases, and HCO_3^- begins to appear in urine. Since H^+ secretion by the nephron is also reduced (due to limited H^+ availability), this also limits renal HCO_3^- reabsorption. The rise in the Pco_2 during the compensatory phase of metabolic alkalosis tends to limit renal compensation somewhat by facilitating H^+ secretion; however, its overall effect is relatively small.

Metabolic Alkalosis and K⁺ Balance

The bicarbonate buffer equation shows that for every HCO_3^- generated, there is an equal amount of H^+. However, the normal plasma HCO_3^- concentration (24 mEq/L) is far greater than that of H^+ (40 nEq/L). The explanation for this great disparity lies in the significant

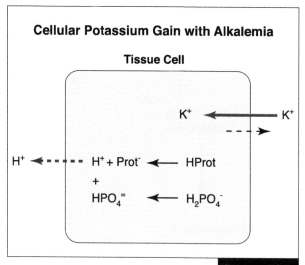

Figure 89-2

buffering actions of the body (see Chapter 83). Therefore, in cases of metabolic alkalosis there should be a tremendous reserve of H^+ available to buffer the modest extracellular proton deficit. As these protons move from cells into extracellular fluid, cellular electronegativity increases, and therefore in order to move this electronegativity back toward normal, anions must be lost or cations must be added to these cells (**Fig. 89-2**). Since protein and phosphate anions cannot easily leave cells, and Na⁺/K⁺-ATPase pump activity increases in alkalemia, **K⁺**

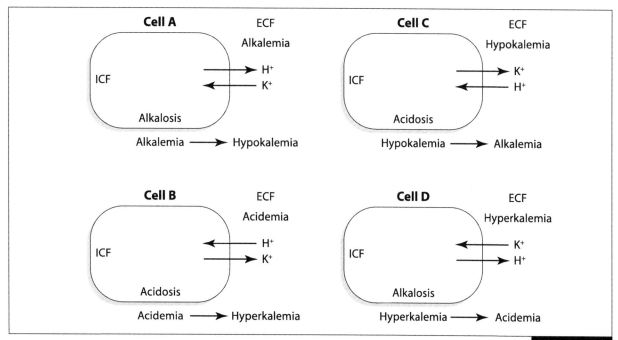

Figure 89-3

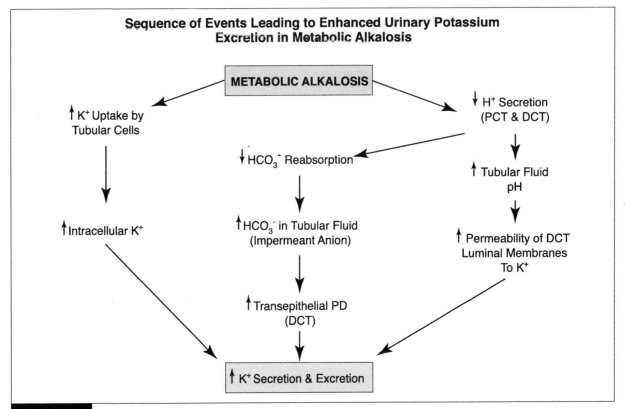

Figure 89-4

becomes the selected cation for translocation into cells. Thus, **alkalemia** tends to promote **hypokalemia (cell A** in **Fig. 89-3)**, and **acidemia** promotes **hyperkalemia (cell B** in **Fig. 89-3)**. In the reverse sense, would **hypokalemia** (of any cause) tend to promote an **alkalemia**, and **hyperkalemia** an **acidemia**? The answers to those questions are yes (**Fig. 89-3, cells C** and **D**). Additionally, since HCO_3^- reabsorption is decreased in the proximal convoluted tubule (PCT), greater amounts of HCO_3^- are delivered to distal segments of the nephron in the glomerular filtrate (**Fig. 89-4**). This alkalinization of the distal tubular filtrate increases permeability of luminal (apical) membranes to K^+, and since HCO_3^- is an impermeant anion (little H^+ available to promote reabsorption; mechanism similar to that depicted in **Fig. 87-3**), the **transepithelial potential difference (PD)** of the distal convoluted tubule (DCT) increases, thus promoting even more movement of K^+ into urine. Some impermeant penicillin anions, which are secreted by the proximal nephron,

exert a similar effect in the distal tubular filtrate.

In addition to metabolic alkalosis, **vomiting** causes dehydration (i.e., volume depletion). The natural response of the body to volume depletion is to activate the renin-angiotensin system, which in turn increases **aldosterone** release. This hyperaldosteronemia promotes even more distal tubular K^+ and H^+ loss into urine (see **Fig. 88-5**). Sometimes this effect can be so pronounced, that the urine becomes acidified (a **"paradoxic aciduria"** in the presence of an alkalemia). **Abomasal displacement** is known to produce a similar effect.

Signs and Symptoms of Metabolic Alkalosis

Adverse effects of this acid-base disorder include supraventricular and ventricular arrhythmias, decreased O_2 delivery to tissues, a decrease in the ability of hemoglobin to unload O_2, and alterations in **neuromuscular irritability (NI)**.

To a large extent, the effects of alkalemia on

neuromuscular function are similar to those of hypocalcemia. **Non-hypoproteinemic alkalemia**, which increases the negative charge equivalency on albumin, increases binding of Ca^{++} to this protein, thereby diminishing the free, ionized component of total plasma Ca^{++}.

$$H^+ + \uparrow \text{ Albumin}^- \longleftarrow \text{ H-Albumin}$$
$$\uparrow \text{ Albumin-}Ca^{++} \longleftarrow \downarrow Ca^{++}$$

As with hypocalcemia, alkalemia can provoke muscle cramping, and even tetany.

$$NI \approx (Na^+ \times K^+ \times PO_4^=) / (Ca^{++} \times H^+ \times Mg^{++})$$

Respiratory muscle paralysis may also be a manifestation of severe metabolic alkalosis. Hypokalemia, if profound, can sometimes overcome the stimulatory effects of hypocalcemia and alkalemia, thus promoting muscle weakness.

Of great importance is the predisposition to **supraventricular and ventricular arrhythmias** in metabolic alkalosis. The primary association lies less with the alkalemia, and more with the hypokalemia and decreased ionized Ca^{++} concentration. Changes in blood pressure depend on the cause of the alkalosis and directional change in extracellular volume; either hypotension or hypertension can occur.

The decrease in **ventilatory drive** induced by metabolic alkalosis results in a predictable increase in the Pco_2, and decrease in the Po_2. If a patient has preexisting pulmonary disease, this hypoventilatory response could precipitate overt respiratory failure. Additionally, alkalemia increases the affinity of O_2 for hemoglobin (see Chapter 84); however, this relationship is somewhat offset by an increase in erythrocytic 2,3-DPG synthesis (see Chapter 31). The net effect, however, can be decreased availability of O_2 to tissues. If this effect is coupled with hypovolemia and hypotension, it could seriously compromise tissue oxygenation.

The compensation for **respiratory acidosis** ($\uparrow Pco_2$) is an increase in net renal acid secretion, and an increase in renal bicarbonate generation (and return to blood; see Chapter 90). Therefore, renal compensation for respiratory acidosis elevates the plasma HCO_3^- concentration. If a patient with chronic respiratory acidosis is mechanically ventilated, and the Pco_2 is rapidly reduced to normal (or near-normal), there may be no immediate effect on the plasma HCO_3^- concentration, which remains elevated. The patient now has an **elevated $[HCO_3^-]/(S \times Pco_2)$ ratio**; and by definition, has **post-hypercapnic metabolic alkalosis**.

Volume-Resistant Metabolic Alkalosis

Metabolic alkalosis secondary to gastric fluid loss, abomasal displacement, loop and/or thiazide diuretic therapy, or diabetes insipidus (i.e., contraction or concentration alkalosis), is characteristically associated with both **hypovolemia** and **hypokalemia**. These conditions can be treated by replacing both volume and K^+ (via administration of NaCl- and KCl-containing solutions (see Chapter 93)). For this reason these common forms of metabolic alkalosis (95% of total), are sometimes referred to as being "volume-responsive," or "volume-sensitive." In contrast, there exists a group of less common "volume-resistant" entities that are not characterized by volume loss, and therefore are not usually treated with fluid therapy. The best examples are **post-hypercapnic metabolic alkalosis** and **hypoproteinemia**, discussed above, **hyperaldosteronism**, which can be caused by adrenal adenoma, adrenal hyperplasia, or by over-administration of mineralocorticoids to patients with Addison's-like disease, or **excessive alkali therapy** (e.g., ulcer therapy, or treatment of metabolic acidosis). Hyperaldosteronism leads to excessive distal tubular renal K^+ and H^+ secretion, and thus excessive loss of these molecules in urine. It also leads to excessive renal Na^+ retention, which is generally associated with hypervolemia and hypertension, not hypovolemia. Patients with Cushing's-like syndrome can also develop a volume-resistant metabolic alkalosis because excess cortisol and associated steroids (e.g., deoxycorticosterone) act as partial aldosterone agonists.

In summary, metabolic alkalosis, a disease process causing an abnormal rise in the plasma [HCO_3^-] and fall in the plasma [H^+], is generally compensated by a rise in the Pco_2 of blood (**Fig. 89-1** and **Table 89-2**). Since the kidneys excrete excess HCO_3^- into urine when there is insufficient H^+ available to support its proximal reabsorption, bicarbonaturia develops, which carries an increased amount of K^+ with it. For this and other reasons, most patients with metabolic alkalosis are also **hypokalemic.**

Like in respiratory acidosis, the bicarbonate buffer equation is shifted to the **right** in metabolic alkalosis, with the elevated HCO_3^- displacing Cl^- from the circulation. Additionally, the charge equivalency on albumin increases, which has a tendency to increase the AG. This increased charge equivalency also increases binding of Ca^{++} to this protein, thus decreasing the free, ionized component of total plasma Ca^{++}. The hypokalemia that develops in this condition along with the decrease in serum ionized Ca^{++} predisposes animals to supraventricular and ventricular arrhythmias.

The primary **causes** of **metabolic alkalosis** are generally categorized as follows:

1) Excessive loss of gastric fluid, such as occurs with vomiting or upper GI drainage.

2) Abomasal displacement.

3) Excessive thiazide or loop diuretic therapy, especially for the treatment of edema, and diabetes insipidus.

These three causes produce a metabolic alkalosis (i.e., contraction or concentration alkalosis) that can be repaired with volume (e.g., saline) plus KCl, and are therefore termed "**volume-responsive,**" or "**volume-sensitive.**"

In contrast, the fourth cause concerns a group of relatively rare entities that do not respond to volume therapy, and therefore are termed "**volume-resistant:**"

4) Patients receiving excessive alkali therapy, or those with hypoproteinemia, hyperaldosteronism or post-hypercapnic metabolic alkalosis.

This last condition can occur when patients with compensated hypercapnia experience an acute fall in Pco_2. About **95%** of metabolic alkalosis cases are said to be "volume-responsive," with only **5%** being "volume-resistant."

OBJECTIVES

- Describe common causes of metabolic alkalosis, and explain why this condition promotes hypokalemia and a symptomatic hypocalcemia.

- Draw correlations between a free water deficit and a contraction alkalosis (see Chapters 84 & 92).

- Know why the AG increases in non-hypoproteinemic metabolic alkalosis.

- Explain why the plasma [HCO_3^-] rises during the acute, uncompensated phase of metabolic alkalosis, but the Pco_2 remains unchanged.

- Indicate the mechanism for respiratory suppression during the compensatory phase of metabolic alkalosis, and know why the bicarbonate buffer equation is shifted to the right in both metabolic alkalosis and respiratory acidosis.

- Show how the kidneys compensate during metabolic alkalosis, and explain how this compensation leads to kaliuresis.

- Discuss the signs & symptoms of metabolic alkalosis, and understand differences between the volume-sensitive and volume-resistant forms.

- Explain how the hypovolemia of vomiting can lead to a "paradoxic aciduria."

- Recognize why an abnormal rise in the plasma $[HCO_3^-]$ alone is not a good way to diagnose metabolic alkalosis.

- Explain how metabolic alkalosis leads to hypochloremia.

- Explain why hyperaldosteronism is considered to be a volume-resistant form of metabolic alkalosis.

- Explain what is meant by "post-hypercapnic metabolic alkalosis."

- Understand why and how K^+ depletion occurs in both metabolic alkalosis and metabolic acidosis.

QUESTIONS

1. All of the following are common causes of metabolic alkalosis, EXCEPT:
   a. Diarrhea.
   b. Loop diuretic therapy.
   c. Hypoproteinemia.
   d. Vomiting.
   e. Excess alkali intake.

2. A non-hypoproteinemic metabolic alkalosis usually decreases the:
   a. Charge equivalency on albumin.
   b. Plasma $[HCO_3^-]/(S\ Pco_2)$ ratio.
   c. Free ionized component of total plasma Ca^{++}.
   d. Plasma anion gap (AG).
   e. Urinary HCO_3^- excretion.

3. A patient with metabolic alkalosis and severe volume depletion (e.g. following an acute bout of vomiting), might be expected to exhibit any one of the following, EXCEPT:
   a. Muscle cramping.
   b. Kaliuresis.
   c. Hyperbicarbonatemia.
   d. Paradoxic aciduria.
   e. Hyperventilation.

4. Which one of the following could precipitate a post-hypercapnic metabolic alkalosis?
   a. The decrease in ventilatory drive induced by a metabolic alkalosis.

   b. Strenuous exercise in a patient with diabetic ketoacidosis.
   c. Increased urinary HCO_3^- excretion during the compensatory phase of metabolic alkalosis.
   d. Mechanical ventilation of a patient with chronic respiratory acidosis.
   e. $NaHCO_3$ administration to a patient with metabolic acidosis.

5. Select the TRUE statement below:
   a. The bicarbonate buffer equation is shifted to the left ($CO_2 + H_2O <— H_2CO_3 <— H^+ + HCO_3^-$) during both the uncompensated and the compensated phase of a metabolic alkalosis.
   b. Adrenal insufficiency (and thus hypoaldosteronism) may be a cause of metabolic alkalosis.
   c. Both metabolic acidosis and metabolic alkalosis may be associated with a kaliuresis.
   d. Metabolic alkalosis is associated with a decrease, not an increase in the plasma anion gap.
   e. Although the plasma HCO_3^- concentration increases during the uncompensated phase of a metabolic alkalosis, it abruptly decreases during the compensatory phase.

6. Select the FALSE statement below:
   a. A free water deficit leads to a concentration alkalosis.
   b. The bicarbonate buffer equation is shifted to the right in both metabolic alkalosis and respiratory acidosis.
   c. Contraction alkalosis may follow excessive thiazide diuretic therapy.
   d. Abomasal displacement may lead to both hypokalemia and alkalemia.
   e. Hypertonic dehydration is acidifying.

6. e
5. c
4. d
3. e
2. c
1. a

ANSWERS

Chapter 90

Respiratory Acidosis

Overview

- Most causes of respiratory acidosis are due to hypoventilation, not increased CO_2 production.
- Respiratory insufficiency causes hypoxemia, which can lead to a secondary metabolic acidosis.
- The early phase of respiratory acidosis is associated with severe acidemia in acute respiratory failure.
- Renal compensation for respiratory acidosis involves enhanced bicarbonate generation and urinary proton excretion.
- The respiratory compensation for metabolic acidosis is usually faster and more complete than the renal compensation for respiratory acidosis.
- Respiratory acid-base disorders generally evoke parallel changes in the pH of CSF and blood.
- Protons have difficulty crossing the blood-brain-barrier and the blood-CSF barrier.

Respiratory ventilation refers to the movement of air into and out of the lungs, which requires a sequence of **neuromuscular events**: the respiratory center in the brain stem initiates a neural impulse that travels down the cervical spinal cord, phrenic and intercostal nerves, then across neuromuscular junctions. Respiratory muscles, especially the diaphragm, are thus stimulated to contract. These contractions cause the chest wall to expand, which lowers intrapleural pressure and causes the lungs to expand. Because this basic sequence of events is arranged in series, normal ventilation requires that every link in the chain be intact.

Respiratory acidosis can arise from a break in any one of these links. For example, it can be caused from depression of the respiratory center through drugs or metabolic disease, or from limitations in chest wall expansion due to

neuromuscular disorders or trauma (**Table 90-1**). It can also arise from pulmonary disease, cardiogenic pulmonary edema, aspiration of a foreign body or vomitus, pneumothorax and pleural space disease, or through mechanical hypoventilation. Unless there is a superimposed or secondary metabolic acidosis, the plasma **anion gap** will usually be normal in respiratory acidosis.

The arterial partial pressure of CO_2 (Pco_2) is directly proportional to the rate of CO_2 production, and inversely proportional to the rate of alveolar ventilation:

$Pco_2 \propto CO_2$ Production / Alveolar Ventilation

Based upon this relationship, one might expect that respiratory acidosis ($\uparrow Pco_2$) can be caused by either an increase in CO_2 production, or a decrease in alveolar ventilation. However, because the body normally

Copyright © 2015 Elsevier Inc. All rights reserved.

increases the depth and rate of respiration to match most increases in CO_2 production, an elevation in the numerator of this equation does not usually affect Pco_2 unless alveolar ventilation is limited (e.g., due to pulmonary disease or inadequate mechanical ventilation).

Table 90-1
Common Causes of Respiratory Acidosis

Depression of the respiratory center

  Severe CNS depression

    Metabolic disease

    Depressant drugs

      Narcotics

      Sedatives

      Tranquilizers

      Anesthetics

  Strokes

  Tumors

  Encephalitis

  Trauma

Limitations of chest wall expansion

  Neuromusclar disorders

    Myasthenia Gravis

    Rib fractures

    Hypokalemic myopathy

  Trauma and surgery

  Fixation of ribs

Pulmonary disease

  Pulmonary edema

  Chronic obstructive pulmonary disease (COPD)

  Pulmonary fibrosis

  Atelectasis

  Pneumonia

  Pulmonary thromboembolism

Cardiogenic pulmonary edema

Aspiration of foreign body or vomitus

Pneumothorax and pleural space disease

Mechanical hypoventilation

Therefore, as indicated above, most causes of **respiratory acidosis** are due to **hypoventilation**, not increased CO_2 production. Respiratory insufficiency also causes **hypoxemia** ($\downarrow Po_2$), which can lead to a **secondary metabolic acidosis**. In respiratory insufficiency, the hypoxemia that develops may sometimes be more important than the **hypercapnia** ($\uparrow Pco_2$) and subsequent respiratory acidosis.

Because normal production of CO_2 is high and chemical buffering is limited, **acute respiratory failure** is associated with **severe acidosis**, and only a modest increase in the plasma HCO_3^- concentration (**Fig. 87-2** (arrow from **X —> G**), **Fig. 90-1**, and **Table 90-2**). When respiration is compromised, the bicarbonate buffer equation is immediately shifted to the **right** ($CO_2 + H_2O —> H_2CO_3 —> H^+ + HCO_3^-$). Since plasma **pH** is a function of the **ratio** of $[HCO_3^-]/(S \times Pco_2)$ (see Chapter 85), and since the denominator increases faster than the numerator during the acute phase of respiratory acidosis, the pH abruptly falls (**Table 90-2**). Although protons are buffered both extra- and intracellularly during this phase (see Chapter 83), the major buffer system involved is the **hemoglobin** buffer system (see Chapter 84).

As with metabolic acidosis, $CO_3^=$ on the bone surface can act as a proton acceptor during acidemia, as protons exchange with

Right Shift in the Bicarbonate Buffer Equation during Respiratory Acidosis

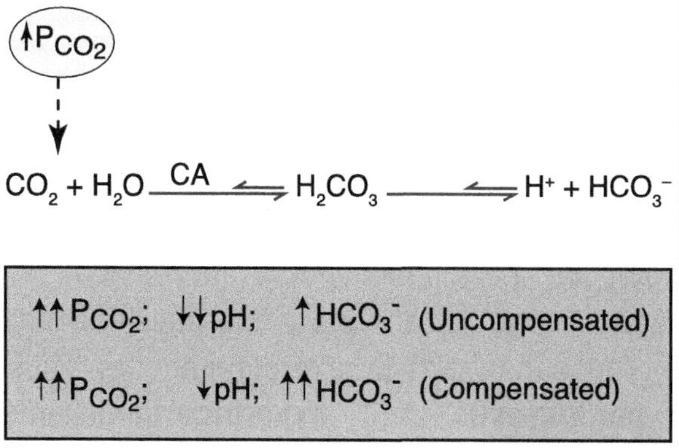

$\uparrow Pco_2$

$\downarrow$

$CO_2 + H_2O \xrightarrow{\text{CA}} H_2CO_3 \rightleftharpoons H^+ + HCO_3^-$

$\uparrow\uparrow Pco_2$; $\downarrow\downarrow pH$; $\uparrow HCO_3^-$ (Uncompensated)

$\uparrow\uparrow Pco_2$; $\downarrow pH$; $\uparrow\uparrow HCO_3^-$ (Compensated)

Figure 90-1

Table 90-2

Changes in Acid-Base Parameters during Uncompensated and Compensated Respiratory Acidosis

Parameter	Normal	Respiratory Acidosis	
		Uncompensated	Compensated
HCO_3^- (mEq/L)	24	26	30
Pco_2 (mmHg)	40	60	60
$HCO_3^- / (S \times Pco_2)$	20	14.4	16.6
pH	7.4	7.25	7.32

Ca^{++}, Na^+, and K^+ (see Chapter 87). Therefore, respiratory acidosis, like metabolic acidosis, has the potential of causing bone dissolution and enhanced urinary Ca^{++} excretion. However, most reports on bone dissolution during acid-base disorders deal with those occurring during metabolic acidosis.

In order to bring the **[HCO_3^-]/(S x Pco_2) ratio** closer to 20, and thus normalize plasma pH, the plasma HCO_3^- concentration must now increase while the Pco_2 stays the same at its new steady state value (see **Fig. 91-2**). This primarily renal **compensatory adaptation** to respiratory acidosis is a slow process. Chronic hypercapnia causes increased renal generation of new HCO_3^- at a rate of about 3 mEq/L for each 10 mmHg increment in the Pco_2, and usually takes several days to become effective (see Chapter 87). While new HCO_3^- is being generated from CO_2 in proximal renal tubular epithelial cells of the kidney, H^+ is secreted into the renal tubular filtrate at a high rate, in exchange for Na^+ (**Fig. 90-2**). During this compensatory period, the filtered load of HCO_3^- increases. However, the proximal tubular reabsorptive capacity for HCO_3^- is increased because of increased proton secretion, and little HCO_3^- appears in urine. Increased renal HCO_3^- reabsorption is, however, associated with a decrease in Cl^- reabsorption by the nephron, therefore an increase in urinary Cl^- excretion occurs. Thus, **hypochloremia** and **hyperbicarbonatemia** are common findings in patients with chronic respiratory acidosis.

In summary to this point, respiratory acidosis results from hypoventilation. Because the production of CO_2 is normally high, and chemical buffering is limited, acute (uncompensated) respiratory failure is usually associated with severe acidemia with only a nominal increase in the plasma HCO_3^- concentration (right shift in the bicarbonate buffer equation). However, chronic hypercapnia ($\uparrow Pco_2$) causes increased renal production and secretion of H^+ (in exchange for Na^+), and generation of new plasma HCO_3^- (at the rate of about 3 mEq/L of HCO_3^- for each 10 mmHg increment

Renal Compensation for Respiratory Acidosis

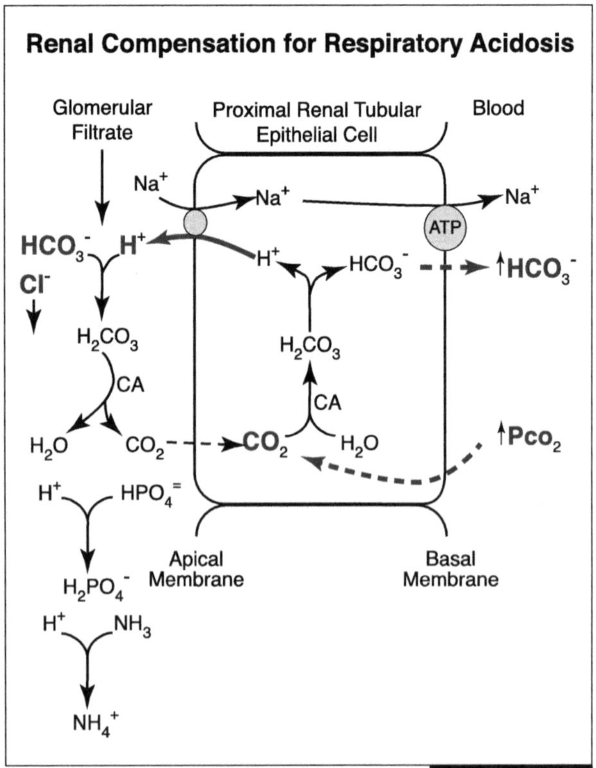

Figure 90-2

in the Pco_2). Thus, this compensatory phase of respiratory acidosis tends to normalize the $[HCO_3^-]/(S \times Pco_2)$ ratio and thus plasma pH. As HCO_3^- is slowly generated and also extensively reabsorbed by the kidneys, Cl^- is increasingly eliminated in urine. Although this mechanism tends to ameliorate the acidemia over time, it is insufficient in completely normalizing arterial pH, and the magnitude of this physiologic adjustment is less than that seen during respiratory compensation for metabolic acidosis (see Chapter 87).

Medullary Chemoreceptors

Structural and functional relationships between blood, cerebrospinal fluid (CSF), extracellular fluid (ECF; i.e., interstitial fluid), and intracellular fluid (ICF) compartments of the brain are shown in **Fig. 90-3**. A compound may enter the brain by crossing either the **blood-brain-barrier (BBB)**, or the **blood-CSF barrier**. Compounds entering via the blood-CSF barrier must cross an additional barrier, the **brain-CSF barrier**, in order to reach interstitial fluid and ICF compartments of the brain. Compounds can only enter and leave the ICF compartments (both neurons and glial cells) by the ECF compartment. Compounds are normally returned to venous blood either directly from the ECF compartment, or indirectly through CSF via arachnoid villi.

Medullary chemoreceptors are, like **peripheral carotid** and **aortic chemoreceptors**, involved with regulating pulmonary ventilation. There is a chemosensitive area located bilaterally in the brain stem (medulla) that is highly sensitive to **H^+**. It is believed that H^+ may be the most important, and perhaps only direct stimulus for these chemoreceptors (unlike peripheral chemoreceptors that are sensitive to an ↑ H^+, ↑ Pco_2, and ↓ Po_2). It should be noted, however, that in contrast to CO_2 and HCO_3^-, **H^+ has extreme difficulty crossing the BBB, and the blood-CSF barrier**. For this reason, changes

in the H^+ concentration of blood have little effect in stimulating medullary chemoreceptors. (Note: H^+ in the CNS typically escapes through venous sinus drainage of CSF into blood).

A reason for the exclusion of H^+ by the BBB and the blood-CSF barrier may be that CSF and interstitial fluid of the brain have little protein (unlike blood) to buffer protons. Importantly, during periods of **metabolic acidosis** H^+ in blood is kept away form neurons of the brain by these barriers. It should also be noted that metabolic acidosis lowers the serum HCO_3^- concentration (see Chapter 87); therefore, even though HCO_3^- can slowly cross the BBB, little would be available to buffer protons being generated in this acid-base condition. Peripheral carotid and aortic chemoreceptors, however, do sense the H^+ concentration of blood, and therefore are responsible for altering respiration during metabolic acidosis and metabolic alkalosis. The compensatory increase in respiration during metabolic acidosis, however, lowers the Pco_2, which generally evokes a mild increase in the pH of CSF ($CO_2 + H_2O \longleftarrow H_2CO_3 \longleftarrow H^+ + HCO_3^-$). Conversely, the compensatory decrease in respiration during metabolic alkalosis increases the Pco_2, thereby slightly decreasing the pH of CSF.

During periods of **respiratory acidosis** (↑ Pco_2) and **alkalosis** (↓ Pco_2), however, these situations are reversed. For example, in **respiratory acidosis** excess CO_2 readily crosses the BBB and the blood-CSF barrier, and quickly generates H^+ and HCO_3^- in CSF and interstitial fluid. The H^+ generated then attempts to stimulate an increase in both the depth and rate of inspiration. Excitation of the respiratory center by this mechanism is great during the first few hours of the acute phase, but then gradually declines over the next few days as the renal compensatory phase ensues (decreasing to about one-fifth the initial value). Since the kidneys become **"bicarbonate generators"** during renal compensation for

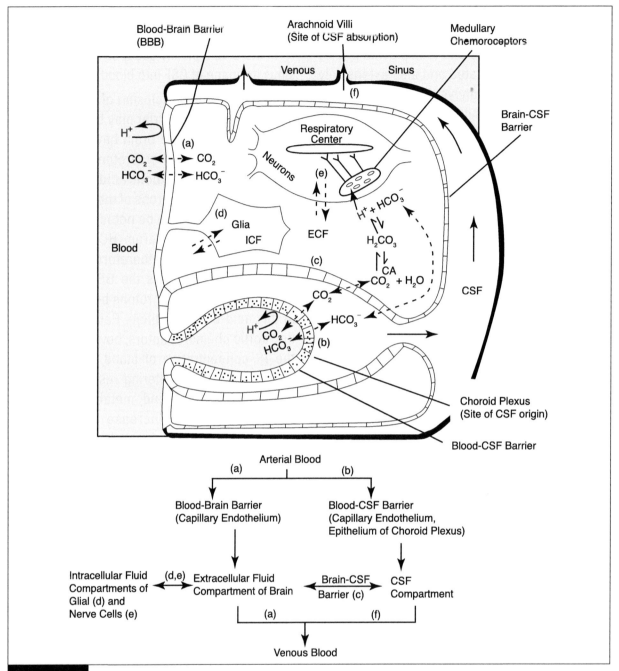

Figure 90-3

respiratory acidosis, excess HCO_3^- slowly diffuses through the BBB and the blood-CSF barrier, and combines directly with available H^+ in interstitial fluid spaces of the brain, thus increasing pH back toward normal. Therefore, an increase in P_{CO_2} has a potent acute effect on controlling respiratory drive through medullary chemoreceptors, but only a weak chronic effect after a few days of renal compensation.

In summarizing pH adjustments of peripheral blood and CSF, the important point to remember is that **metabolic acid-base disorders generally evoke opposite changes in the pH of CSF and blood, while respiratory acid-base disorders bring about parallel changes** (see Chapter 93).

OBJECTIVES

- Recognize common causes of respiratory acidosis, and explain why the AG may not change in this condition.

- Understand why a secondary metabolic acidosis frequently accompanies respiratory acidosis.

- Explain why the $[HCO_3^-]/(S \times Pco_2)$ ratio decreases in respiratory acidosis, and recognize what effect this has on the pH of body fluids (see Chapter 85).

- Show how renal HCO_3^- generation and reabsorption normally occur with this condition, and explain why urinary Cl^- excretion increases.

- Contrast the active forms of respiratory compensation during metabolic acid-base disorders to the more passive renal forms during respiratory disorders.

- Understand how and why medullary chemoreceptors are more responsive to the Pco_2 of body fluids, while peripheral carotid and aortic chemoreceptors are more responsive to the circulating $[H^+]$.

- Explain why H^+s are normally excluded from crossing the BBB.

- Explain why metabolic acid-base disorders generally evoke opposite changes in the pH of CSF and blood, while respiratory acid-base disorders bring about parallel changes.

- Know why an increase in the Pco_2 has a potent acute effect in controlling respiratory drive through medullary chemoreceptors, but a weak chronic effect after a few days of renal compensation.

- Understand why urinary Cl^- loss in respiratory acidosis can be compensatory and thus alkalinizing (see Chapter 92).

QUESTIONS

1. **Because normal production of CO_2 is high, and chemical buffering in the body is limited, acute respiratory failure will cause:**
   a. A significant decrease in the plasma bicarbonate concentration.
   b. Hypocalcemia.

   c. The bicarbonate buffer equation to shift to the left ($CO_2 + H_2O \longleftarrow H_2CO_3 \longleftarrow H^+ + HCO_3^-$).
   d. Hyperchloremia.
   e. A significant increase in the plasma H^+ concentration.

2. **Select the TRUE statement below:**
   a. Metabolic acid-base disturbances are usually compensated faster, and more completely than respiratory acid-base disturbances.
   b. The plasma anion gap is usually increased in respiratory acidosis.
   c. Renal compensation for respiratory acidosis involves excretion of increased amounts of CO_2 in urine.
   d. Most causes of respiratory acidosis are due to an increase in CO_2 production.
   e. Respiratory acidosis and metabolic acidosis seldom, if ever, coexist.

3. **Which one of the following depicts an INCORRECT relationship:**
   a. $[H^+] \propto Pco_2 / [HCO_3^-]$
   b. $Pco_2 \propto CO_2$ Production / Alveolar Ventilation
   c. $pH = pK' + \log ([A^-]/[HA])$
   d. $pH \propto [HCO_3^-] / Pco_2$
   e. In acidemia (H-Albumin $\longrightarrow$ H^+ + Albumin$^-$)

4. **During the acute phase of respiratory acidosis:**
   a. Protons are buffered by hemoglobin inside erythrocytes.
   b. The plasma HCO_3^- concentration usually rises at a faster rate than the Pco_2.
   c. There is less of a physiologic stimulus to promote respiration than there is during the compensatory phase.
   d. The serum $[HCO_3^-]/(S \times Pco_2)$ ratio is usually higher than it is during the compensatory phase.
   e. Protons readily cross the blood-brain-barrier.

Chapter 91

Respiratory Alkalosis

Overview

- Steady state blood CO_2 levels remain relatively constant in compensated respiratory acidosis and alkalosis (i.e., CO_2 in = CO_2 out).

- Uncompensated respiratory alkalosis is associated with an increased blood pH, and a modestly decreased HCO_3^- concentration.

- Renal compensation for respiratory alkalosis involves a decrease in HCO_3^- reabsorption.

- The blood pH may be within the normal range in some mixed acid-base disorders.

- A mixed acid-base disturbance is indicated when the Pco_2 and blood HCO_3^- concentration are moving in opposite directions.

- Mixed acid-base distrubances can be additive, or subtractive.

- The bicarbonate buffer equation is shifted to the left in metabolic acidosis and respiratory alkalosis.

- Respiratory alkalosis can be due to either direct or reflex hypoxemic stimulation of the respiratory center, to pulmonary disease, or to excessive mechanical ventilation.

Respiratory alkalosis is a disease process resulting in an increase in ventilation, and a fall in arterial Pco_2. The bicarbonate buffer equation is driven to the left, blood pH rises, and the serum HCO_3^- concentration falls (**Fig. 91-1**). This acid-base disorder is generally associated with fewer patho-physiologic signs than those of respiratory acidosis and the metabolic acid-base disorders.

Four basic conditions are associated with respiratory alkalosis, with each increasing the central drive for respiration (**Table 91-1**):

1) **Direct stimulation of the respiratory center**

2) **Reflex hypoxemic stimulation of the respiratory center**

3) **Pulmonary disease**

4) **Excessive mechanical ventilation**

Various CNS diseases, toxic, pharmacologic, and physical insults to the medullary respiratory center can result in **hyperventilation**. These insults include physiologic factors (e.g., elevated progesterone during pregnancy, pain,

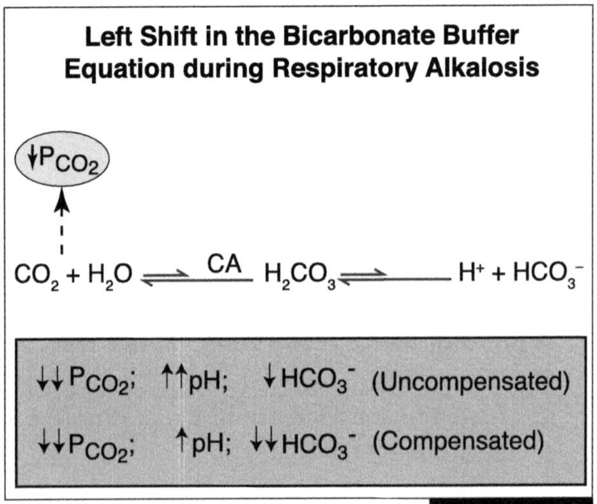

Left Shift in the Bicarbonate Buffer Equation during Respiratory Alkalosis

$\downarrow Pco_2$

$$CO_2 + H_2O \xrightleftharpoons{\ CA\ } H_2CO_3 \rightleftharpoons H^+ + HCO_3^-$$

$\downarrow\downarrow Pco_2;\ \ \uparrow\uparrow pH;\ \ \downarrow HCO_3^-$ (Uncompensated)

$\downarrow\downarrow Pco_2;\ \ \uparrow pH;\ \ \downarrow\downarrow HCO_3^-$ (Compensated)

Figure 91-1

Copyright © 2015 Elsevier Inc. All rights reserved.

Table 91-1
Common Causes of Respiratory Alkalosis

Direct stimulation of the CNS respiratory center

Psychogenic

Pain, anxiety and fear

CNS disease

Stroke

Traumatic brain injury

Neoplasia

Sepsis (gram-negative, particularly)

Inflammatory cytokines

Lesions of the brain stem

Hypermetabolic states

Fever

Thyrotoxicosis

Heatstroke

Liver failure

NH_3

Pharmacologic and hormonal stimulation

Salicylates (aspirin)

Progestins (pregnancy)

Reflex hypoxemic stimulation of the respiratory center

Anemia

Hypotension

Cyanotic heart disease

Decreased inspired O_2 tension

High altitude (Brisket disease)

CO, CN^-, and methemoglobin

Pulmonary disease

Pneumonia

Pulmonary edema

Pulmonary thromboembolism

Ventilation-perfusion inequality

Excessive mechanical ventilation

anxiety and fear), endogenous pathologic factors (e.g., un-detoxified waste products in liver disease), and exogenous agents (e.g., salicylates). Sepsis, fever, and trauma can also act to stimulate ventilation.

Peripheral arterial **hypoxemia** is sensed by chemoreceptors in the aortic arch and at the bifurcation of the carotid arteries. These receptors signal the medullary respiratory center via neural impulses to alter ventilatory drive, with hypoxemic stimulation usually increasing linearly as arterial Po_2 decreases. Thus, any fall in blood O_2 saturation tends to cause an increase in ventilation unless the Pco_2 is also low, and the arterial pH is elevated. Since carotid and aortic chemoreceptors are highly perfused tissues, conditions that cause tissue hypoxia in the absence of arterial hypoxemia (e.g., anemia, hypotension, and CO poisoning) can cause hyperventilation. Anoxia at high altitudes causes pulmonary vasoconstriction and circulatory embarrassment because O_2 requirements of tissues exceed supply. In normal animals taken to high altitudes, this is normally compensated by polycythemia and an increased cardiac output. Continued hypoxia, however, can cause severe respiratory alkalosis and myocardial weakness (i.e., Brisket disease). The development of additional factors, such as pneumonia or anemia, can exacerbate the O_2 deficit and increase susceptibility to cardiac decompensation.

Although **pulmonary disease** is often associated with respiratory acidosis (see Chapter 90), it can sometimes cause respiratory alkalosis. In mild bacterial pneumonia, for example, Po_2 may be normal or only slightly depressed. However, hyperventilation with hypocapnia and thus alkalemia may be present because of stimulation of pulmonary stretch and irritant receptors by inflammatory debris in alveoli and the airways.

Hyperventilating animals are sometimes said to be "**blowing off CO_2.**" This phrase can give the false impression that CO_2 excretion by the lungs (i.e., alveolar ventilation) is outpacing CO_2 production by tissues:

$Pco_2 \propto CO_2$ Production / Alveolar Ventilation

In reality, there is usually a steady **state for CO_2** (i.e., Pco_2; **CO_2 In = CO_2 Out**), although the steady state level is low in respiratory alkalosis, and elevated in respiratory acidosis (**Fig. 91-2**). It is only during the early, initial stages of these respiratory acid-base disorders that the Pco_2 falls or rises from its normal steady state level to a new, abnormal level.

Acute, uncompensated respiratory alkalosis is associated with an abrupt increase in the blood pH, and only a modest decrease in the HCO_3^- concentration (**Fig. 87-2 (arrow from X —> E), Fig. 91-1** and **Table 91-2**). When respiration is stimulated, the bicarbonate buffer equation is immediately shifted to the left ($CO_2 + H_2O \longleftarrow H_2CO_3 \longleftarrow H^+ + HCO_3^-$). Since plasma pH is a function of the ratio of **$[HCO_3^-]/(S \times Pco_2)$** (see Chapter 85), and since the denominator decreases faster than the numerator during the acute phase of respiratory alkalosis, the pH increases even though acute hypocapnia ($\downarrow Pco_2$) causes release of H^+ from tissue buffers in exchange for K^+ (see **Fig. 89-2** and **Table 91-2**). In order to bring the **$[HCO_3^-]/(SxPco_2)$ ratio** closer to 20 during the compensatory phase, and thus normalize plasma pH, the **plasma HCO_3^-** concentration must now **decrease** while the Pco_2 stays the same at its new steady state value. This primarily **renal compensatory adaptation** is a slow process, that involves reduced renal HCO_3^- generation, as well as reduced HCO_3^- reabsorption (**Fig. 91-3**). Even though the filtered load of HCO_3^- has decreased during the acute, uncompensated phase, since the Pco_2 has gone down, and there are also fewer protons to secrete, HCO_3^- reabsorption, which is normally about 85-90% complete in the proximal nephron, becomes compromised, and HCO_3^- appears in urine. As renal HCO_3^- reabsorption decreases, renal Cl^- reabsorption increases, thus **hyperchloremia** develops (which has an additional compensatory acidifying effect (see Chapter 92)).

Pathophysiologic signs and symptoms of respiratory alkalosis include an increase in neuromuscular irritability, which can progress to tetany (see Chapter 89). As in metabolic alkalosis, excessive amounts of K^+ can be lost in urine.

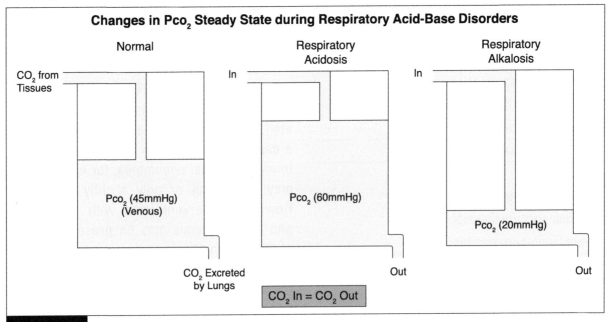

Changes in Pco_2 Steady State during Respiratory Acid-Base Disorders

Figure 91-2

Table 91-2

Changes in Acid-Base Parameters during Uncompensated and Compensated Respiratory Alkalosis

| Parameter | Normal | Respiratory Alkalosis | |
		Uncompensated	Compensated
HCO_3^- (mEq/L)	24	20	16
Pco_2 (mmHg)	40	20	20
$HCO_3^- / (S \times Pco_2)$	20	33.3	26.6
pH	7.4	7.62	7.52

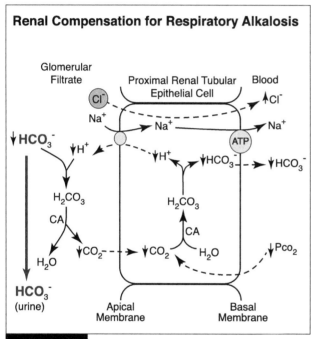

Renal Compensation for Respiratory Alkalosis

Figure 91-3

In summary to this point, it is helpful to think of the bicarbonate buffer system, perhaps the most important buffer system in the body, when thinking about primary acid-base disturbances:

$$CO_2 + H_2O \longleftrightarrow H_2CO_3 \longleftrightarrow H^+ + HCO_3^-$$

The equilibrium for this equation shows that there are three basic ways to alter the H^+ concentration:

1) **Add or remove protons directly to or from the system;**

2) **Increase or decrease the Pco_2; or**

3) **Increase or decrease the serum HCO_3^- concentration.**

The overall equilibrium for this equation is shifted to the left in metabolic acidosis and respiratory alkalosis, while it is shifted to the right in metabolic alkalosis and respiratory acidosis. The term "**respiratory**" is used when referring to any acidosis or alkalosis (i.e., acid-base processes) that results in a change in the Pco_2, while the term "**metabolic**" is usually used when referring to any acidosis or alkalosis that changes the serum HCO_3^- concentration. It should be noted, however, that the **Peter Stewart** approach to acid-base balance (see Chapter 92) does not utilize the serum HCO_3^- concentration when delineating differences within and between metabolic and respiratory acid-base disturbances.

Mixed Acid-base Disturbances

When two or more acid-base disturbances coexist, the overall condition is referred to as a mixed (or complex) acid-base disorder. Although evaluation of these disturbances can sometimes be difficult, analysis of the blood **pH, Pco_2, HCO_3^-** concentration, and plasma **anion gap (AG)** can be helpful in understanding these problems (**Table 91-3**). Two varieties of the same primary disturbance can also form a mixed disturbance, as in a mixed metabolic acidosis composed of two different pathophysiologic conditions that act independently to lower plasma HCO_3^- and pH (e.g., ketoacidosis and NH_4Cl toxicity).

Disturbance	Cause	Pco$_2$	pH	HCO$_3^-$	AG
Table 91-3					
Common Mixed Acid-Base Disturbances					
Resp. acidosis	COPD	↑	− to ↑	↑	−
Met. alkalosis	Vomiting				
Resp. acidosis	COPD	↑	↓	↓	↑
Met. acidosis	Lactic acidosis				
Resp. alkalosis	Aspirin toxicity	↓	−	↓	↑
Met. acidosis	Lactic acidosis				
Resp. alkalosis	Mechanical hyperventilation	↓	↑	↑	−
Met. alkalosis	Vomiting				
Met. acidosis	Diarrhea	−	−	−	−
Met. alkalosis	Vomiting				
Resp. alkalosis	Sepsis				
Met. acidosis	Ketoacidosis	↓	− to ↑	− to ↑	↑
Met. alkalosis	Vomiting				

COPD = Chronic Obstructive Pulmonary Disease, ↑=high, ↓=low, − = normal

When both disorders in a mixed acid-base disturbance are acidoses, or both are alkaloses, the pH changes are "**additive**." The resultant acidemia or alkalemia can be life threatening, even if the component primary disorders are not by themselves severe. Conversely, when an acidosis and an alkalosis are superimposed on each other, the net pH change is "**subtractive**" and usually modest. The final pH can be either high or low, depending on which one of the component disorders is dominant. Occasionally, effects of mixed acid-base disorders on pH may cancel, with the final pH falling within the normal range.

A mixed acid-base disturbance is usually indicated when the blood pH is normal yet the plasma HCO$_3^-$ concentration and/or Pco$_2$ are abnormal, when the Pco$_2$ and plasma HCO$_3^-$ concentration are moving in opposite directions, or when the blood pH is moving in the opposite direction from a known primary disorder (e.g., an acidemic animal who is vomiting). Occasionally, however, a simple acid-base disturbance may look like a mixed disturbance if not enough time has elapsed for physiologic compenstion (i.e., maximal renal compensation may require 2-4 days following the onset of a primary respiratory disturbance), or their may be a defect in the compensatory system which causes a true under- or over-compensation.

How does one rule-out a "masquerading" mixed disturbance? By considering the results of the Pco$_2$ in light of the history, physical examination and other lab data, a plausible "mixed explanation" for an apparently simple profile can be constructed. If an explanation cannot be found, then it becomes more likely that it is not a mixed disturbance.

Since most mixed acid/base distrubances will involve one **primary disorder** that requires treatment, focusing on the mixed nature of an acid/base problem holds potential for shifting attention away from this primary disturbance. To avoid this dilemma, ask yourself if focusing attention on the additional disturbance(s) will make a significant difference to the patient.

OBJECTIVES

- Recognize common causes and symptoms of respiratory alkalosis.

- Understand the cause of Brisket disease, and discuss involvement of the Rapoport-Luebering Shunt in this condition (see Chapter 31).

- Explain why the bicarbonate buffer equation is shifted to the left in this condition, and recognize why the plasma $[HCO_3^-]$ should not be used (alone) to evaluate the overall acid-base state.

- Discuss why the expresssion (CO_2 In = CO_2 Out) applies not only to normal animals, but also to animals with compensated respiratory acid-base disorders.

- Outline the renal compensatory mechanisms in this acid-base disorder, and explain why hyperchloremia develops. Also discuss how this change alone is considered acidifying, and thus compensatory (see Chapter 92).

- Compare alterations in K^+ balance associated with this condition to those in metabolic alkalosis (see Chapter 89).

- Explain why a mixed acid-base disturbance would be expected in a patient whose Pco_2 and $[HCO_3^-]$ were moving in opposite directions (see Table 87-1).

QUESTIONS

1. **Which one of the following hormones sensitizes the maternal respiratory center to small increases in the Pco_2 (i.e., added CO_2 from the fetus) during pregnancy?**
   a. Insulin
   b. Progesterone
   c. Cortisol
   d. Aldosterone
   e. Prolactin

2. **Which one of the following would be indicative of a mixed acid-base disturbance?**

   a. A low Pco_2 and plasma H^+ concentration
   b. A high Pco_2 and plasma HCO_3^- concentration
   c. A low Pco_2 and high plasma HCO_3^- concentration
   d. A high Pco_2 and plasma H^+ concentration
   e. A low Pco_2 and high plasma H^+ concentration

3. **Brisket disease is best associated with:**
   a. Moving ruminant animals from the lowlands to high altitudes.
   b. Respiratory acidosis.
   c. A physiologic decrease in cardiac output.
   d. Thyrotoxicosis.
   e. Excessive mechanical ventilation.

4. **Which one of the following blood values is best associated with respiratory alkalosis?**
   a. ↑ Pco_2
   b. ↓ pH
   c. ↓ HCO_3^-
   d. ↓ Albumin
   e. ↑ Po_2

5. **During the compensatory phase of respiratory alkalosis:**
   a. CO_2 excretion by the lungs is greater than total CO_2 production by the body.
   b. The bicarbonate buffer equation is shifted to the left ($CO_2 + H_2O \longleftarrow H_2CO_3 \longleftarrow H^+ + HCO_3^-$).
   c. Plasma pH $\propto$ (S x Pco_2)/ $[HCO_3^-]$
   d. The kidneys become bicarbonate generators.
   e. Plasma HCO_3^- levels slowly rise.

6. **Which one of the mixed acid-base disturbances below would be best associated with an increased Pco_2 and plasma anion gap, and a decreased plasma pH and HCO_3^- concentration?**
   a. Respiratory acidosis plus a metabolic alkalosis.
   b. A combined metabolic acidosis and alkalosis.
   c. Respiratory alkalosis plus a metabolic alkalosis.
   d. Respiratory alkalosis plus a metabolic acidosis.
   e. Respiratory acidosis plus

ANSWERS

6. e
5. b
4. c
3. a
2. c
1. b

Chapter 92

Strong Ion Difference (SID)

Overview

- The Na^+ minus the Cl^- concentration of plasma is commonly referred to as the strong ion difference (SID).

- An increase in the SID is alkalinizing, while a decrease is acidifying.

- Hyperproteinemia and hyperphosphatemia are acidifying, while hypoproteinemia is alkalinizing.

- Abnormal plasma Na^+ concentrations may indicate the presence of excess free water (i.e., hyponatremia; dilutional acidosis), or a free water deficit (i.e., hypernatremia; concentration alkalosis).

- Base excess (BE) and base deficit (-BE) are indicators of the overall nonrespiratory acid-base state, and are determined from a Siggaard-Andersen nomogram using the plasma pH and Pco_2.

- Changes in the BE can be due to free water abnormalities (i.e., Na^+ abnormalties), Cl^- abnormalities, plasma protein abnormalities, or to the presence of unidentified anions.

- Changes in the BE due to any or all of the abnormalities above can be quantitated.

Evaluation of acid-base disturbances based upon consideration of the "**Strong Ion Difference**" is a method that has gained acceptance because it:

- **Quantitates the effects of concentration alkalosis or dilutional acidosis.**

- **Reveals [Cl^-] abnormalities that otherwise would be obscured by free water abnormalities.**

- **Quantitates the effects of protein abnormalities.**

- **Quantitates the total concentration of unidentified anions.**

This method was originally developed by **Peter A. Stewart**, and is therefore sometimes referred to as the **Peter Stewart approach to**

acid-base balance. A brief summary of his approach follows.

Biological fluids contain water, weak and **strong electrolytes**. While strong electrolytes are, for the most part, fully dissociated in biological fluids, **weak electrolytes** are only partially dissociated (according to their equilibrium constants; see Chapter 83). Examples of weak electrolytes are H_2CO_3, H_2O, and protein (HProt), while strong electrolytes include NaCl, NaOH, and H_3PO_4. Water is a very weak electrolyte (Kw' = 4.4 x 10^{-14}, 37°C in plasma), and for all practical purposes dissociates little into H^+ and OH^-.

When equal amounts of two strong electrolytes (e.g., NaOH + HCl) are added to pure water, they dissociate completely as follows:

Copyright © 2015 Elsevier Inc. All rights reserved.

$$NaOH + HCl \longrightarrow Na^+ + Cl^- + H^+ + OH^-$$

Note that electroneutrality is maintained. In this system, **H⁺** and **OH⁻** are the **dependent variables**, whose concentrations cannot change unless the difference between the **Na⁺** and **Cl⁻** concentrations (i.e., the **independent variables**) changes. This difference, known as the **strong ion difference** (**SID**), is fundamental to understanding Peter Stewart's approach to acid/base balance:

$$SID = [Na^+] - [Cl^-]$$

The SID is related to the H⁺ and OH⁻ concentrations in the following way:

$$SID + [H^+] - [OH^-] = 0$$

Note from this relationship, that:

A. When the **SID increases**, the **H⁺ concentration must decrease** in order to maintain electroneutrality:

$$\uparrow [Na^+] \text{ and/or } \downarrow [Cl^-] = \uparrow SID$$

Since the Na⁺ concentration is normally greater than the Cl⁻ concentration, this relationship holds true:

$$\uparrow SID + \downarrow [H^+] - [OH^-] = 0$$

Therefore, **an increase in the SID is alkalinizing.**

B. When the **SID decreases**, the **H⁺ concentration must increase** in order to maintain electroneutrality:

$$\downarrow [Na^+] \text{ and/or } \uparrow [Cl^-] = \downarrow SID$$
$$\downarrow SID + \uparrow [H^+] - [OH^-] = 0$$

Therefore, **a decrease in the SID is acidifying.**

Plasma Proteins and Phosphates

Plasma proteins act like weak acids, with the total protein concentration ([Prot$_{tot}$]) being the sum of the ionized ([Prot⁻]) and buffered ([HProt]) forms ([Prot$_{tot}$] = [HProt] + [Prot⁻]; see Chapter 84). Thus:

Hyperproteinemia

$$\uparrow [Prot_{tot}] = \uparrow [HProt] + \uparrow [Prot^-]$$
$$\uparrow [HProt] \longleftrightarrow \uparrow [H^+] + \uparrow [Prot^-]$$

Hypoproteinemia

$$\downarrow [Prot_{tot}] = \downarrow [HProt] + \downarrow [Prot^-]$$
$$\downarrow [HProt] \longleftrightarrow \downarrow [H^+] + \downarrow [Prot^-]$$

Note from the above relationships that an increase in **[Prot$_{tot}$]** will increase both **[HProt]** and **[Prot⁻]**, thus also increasing the **H⁺ concentration**. Therefore, **hyperproteinemia is acidifying**, while **hypoproteinemia is alkalinizing** (assuming nothing else changes; see McAuliffe JJ, *et al*, 1986, and Chapter 89). The basic SID equation can now be expanded to include the ionized protein concentration [Prot⁻] as follows:

$$SID + [H^+] - [OH^-] - [Prot^-] = 0$$

Organic and inorganic phosphates ($H_2PO_4^-$/ $HPO_4^=$) normally constitute about 5% of all weak acids in plasma, while protein (**Prot⁻**) is the major weak acid present (see **Fig. 85-3**). Although hypophosphatemia usually does not cause a large enough reduction in total weak acid to cause perceptible alkalemia, **hyperphosphatemia** (like hyperproteinemia) is **acidifying**. Therefore the basic SID equation can be expanded again to include the inorganic phosphate concentration ([$H_2PO_4^-$/$HPO_4^=$], or, for convenience, [$PO_4^=$]) as follows:

$$SID + [H^+] - [OH^-] - [Prot^-] - [PO_4^=] = 0$$

Free Water Abnormalities

Free water abnormalities (i.e., over- or under-hydration of the extracellular fluid (ECF) compartment), can also be assessed in terms of SID and the acid-base state. Consider what happens to SID during a **10% free water deficit**:

Normal

$$SID = [Na^+] - [Cl^-]$$

$$140 \text{ mM} - 102 \text{ mM} = \mathbf{38 \text{ mM}}$$

10% Free Water Deficit

$$SID = [Na^+] - [Cl^-]$$

$$154 \text{ mM} - 112 \text{ mM} = \mathbf{42 \text{ mM}}$$

As **SID increases during a free water deficit** (e.g., hypertonic dehydration, all else remaining unchanged), **the [H⁺] must decrease** in order to maintain electroneutrality:

$$\uparrow SID + \downarrow [H^+] - [OH^-] = 0$$

The same reasoning can be applied to free water dissociation:

$$\downarrow H_2O \longleftrightarrow \downarrow H^+ + \downarrow OH^-$$

The result is a **concentration alkalosis**. **Free water excess decreases SID, and causes a dilutional acidosis**. The magnitude of a free water abnormality can be indirectly assessed through the Na⁺ concentration (i.e., an increase denotes a concentration alkalosis, while a decrease denotes a dilutional acidosis).

Base Excess (BE) and Base Deficit (-BE)

Base excess and base deficit are indicators of the overall non-respiratory acid-base state. These values, which are usually (**+**) in **alkalosis** and (**–**) in **acidosis**, are typically defined as **the amount of acid or base that would restore one liter of blood to normal acid-base composition at a Pco₂ of 40 mmHg**. Although this value cannot be measured directly, it can be determined from **Pco₂** and **pH** using a standard **Siggaard-Andersen Nomogram (Fig. 92-1)**. Base excess (or base deficit) is usually reported with other acid-base data by the clinical laboratory.

In summary to this point, the following definitions apply in the Peter Stewart approach to acid-base balance:

- **↓/↑ in Plasma (or blood) pH = Acidemia / Alkalemia.** (A change in pH alone will not define the acid-base subtype.)

- **↑/↓ in Pco₂ = Respiratory Acidosis / Alkalosis.** (A change in the Pco₂ may be a reflection of respiratory compensation to a metabolic disturbance, or it may result from a primary respiratory disturbance. As it increases it is acidifying, and as it decreases it is alkalinizing.)

- **↓/↑ in BE = Non-Respiratory (Metabolic) Acidosis / Alkalosis.** (This value can be obtained from Pco₂ and pH using a standard Siggaard-Andersen nomogram.)

- **↓/↑ in [Na⁺] = Indicator of Free Water Excess / Deficit.** (A free water excess represents a dilutional acidosis, while a free water deficit represents a concentration alkalosis.)

- **↑/↓ in [Cl⁻] = Hyperchloremic Acidosis / Hypochloremic Alkalosis.** (When a [Na⁺] abnormality exists, the [Cl⁻] concentration must be corrected before determining whether a Cl⁻ abnormality exists (see below).

- **↑/↓ in [Prot_tot] or Albumin [Alb] = Hyperproteinemic Acidosis / Hypoproteinemic Alkalosis.** (Since albumin normally represents about 60% of plasma protein, it is sometimes used in place of total protein to determine whether a hyper- or hypoproteinemic state exists.)

- **↑ in Unidentified Anions (UA⁻) = Lactic Acidosis, Ketoacidosis, etc.** (The presence of unidentified anions is generally synonymous with an acidification process).

Example Problem

Variable	Lab Value	Normal
pH	7.31	7.4
Pco₂	30 mmHg	40 mmHg (torr)
BE	-10 mEq/L	0 mEq/L (mM)
[Na⁺]	120 mEq/L	140 mEq/L
[Cl⁻]	91 mEq/L	102 mEq/L
[Alb]	4.5 gm%	4.5 gm% (%)

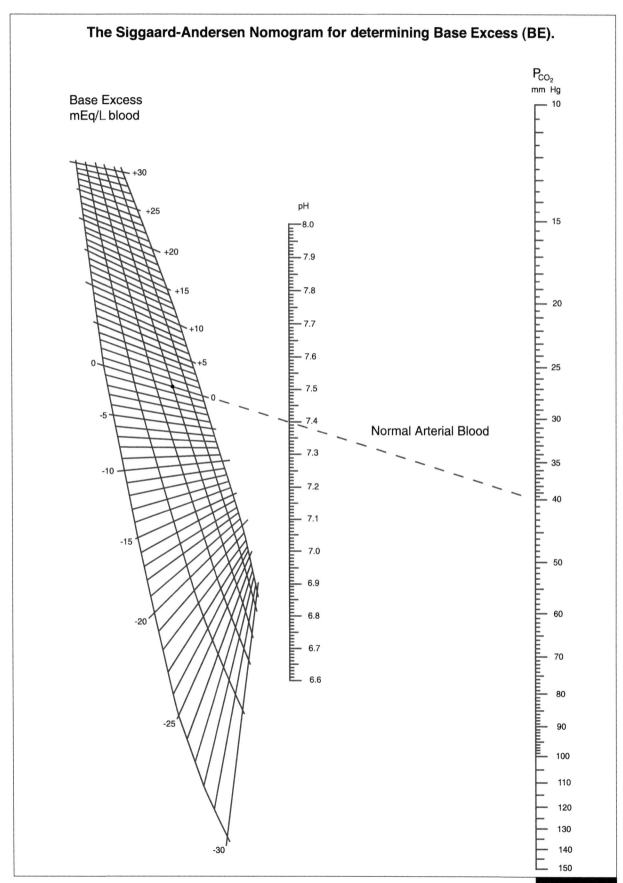

The Siggaard-Andersen Nomogram for determining Base Excess (BE).

Base Excess
mEq/L blood

pH

P_{CO_2}
mm Hg

Normal Arterial Blood

Figure 92-1

Note the following from this problem:

A. There is an **acidemia** present: the plasma pH is 7.31, which is less than the normal 7.4.

B. There is **respiratory alkalosis** in this patient: the Pco₂ is 30 mmHg, which is 10 mmHg less than the normal value of 40.

C. There is a **non-respiratory acidosis** in this patient: the BE is -10 mEq/L, which is 10 mEq/L less than the normal value of zero.

D. This may be a **mixed disturbance**; since there is acidemia present, the non-respiratory (metabolic) acidosis is most likely the primary process, with the respiratory alkalosis being the compensatory response. The non-respiratory contributions to this acid-base disturbance deserve further evaluation.

There is hyponatremia in this patient, which should be treated as an indicator of a **free water excess**; therefore a "**dilutional acidosis**" is present. **The Cl⁻ concentration must be corrected for the free water abnormality before deciding if there is also a Cl⁻ abnormality present.** The plasma **albumin** concentration is normal, and thus it is assumed that the total plasma protein concentration is also normal. Even though there is a free water abnormality present, unlike the Cl⁻ concentration, there is no need to correct the observed plasma protein concentration. Without performing a quantitative analysis, it cannot be accurately determined if an acidifying process due to unidentified strong anions is present.

Quantitative Analysis of This Problem

1) Enter the observed (reported) base excess (**BE**) in **Table 92-1** below:

$$BE\ (Obs) = -10$$

2) Enter the observed [**Na⁺**] (120 mEq/L) in the table, then calculate and enter the expected **change in BE due to the free water abnormality**:

$$BE\ (H_2O) = -6$$

3) Enter the observed [**Cl⁻**] and the observed [**Na⁺**] in the table, then calculate and enter the **change in BE due to the chloride abnormality**:

$$BE\ (Cl^-) = -4$$

Note that the following value (part of the equation) denotes the [**Cl⁻**] **corrected** for the free water abnormality:

$$[Cl^-]corr = 140\ [Cl^-] / [Na^+] = 140 \times 91/120 = 106$$

4) Enter the observed [**Alb**] in the table, then calculate and enter the expected **change in BE due to the albumin abnormality**:

$$BE\ (Alb) = 0$$

5) Sum the expected contributions of 2, 3 and 4 above, and subtract them from the observed BE; be careful of sign. Enter the result as the predicted contribution to the **change in BE due to the presence of unidentified anions**:

$$BE\ (UA^-) = 0$$

Note that the above approach uses neither the **HCO₃⁻** nor the **K⁺** concentrations (two ingredients of the plasma "**anion gap**" calculation, Chapter 86). Since changes in the base excess due to protein abnormalities can be quantitated using this method, the total concentration of unidentified anions (if present) can be more closely approximated. As stated previously, the "anion gap" is a weak tool in the presence of hypoproteinemia.

The constants used in the above equations were derived using 140 mEq/L as the normal [Na⁺], 102 mEq/L as the normal [Cl⁻], 6.5 gm/dl as the normal total plasma protein concentration, and 4.5 gm/dl as the normal albumin concentration. Ideally, normal values from the laboratory analyzing the sample should be used. The constants 3.0 for total plasma protein,

Table 92-1
Change in the Observed Base Excess (BE) Due To:

Free Water Abnormalities	$BE (H_2O) = 0.3 ([Na^+] - 140) = 0.3 (120 - 140) = -6$
Chloride Abnormalities	$BE (Cl^-) = 102 - (140 [Cl^-] / [Na^+]) = 102 - 106 = -4$
Albumin Abnormalities	$BE (Alb) = 3.7 (4.5 - [Alb]) = 3.7 \times 0 = 0$
Unidentified Anions Make up the balance	$BE (UA^-) = (-10) - ((-6) + (-4) + 0) = 0$
Total is the observed (reported) BE	$BE (Obs) = -10$

and 3.7 for albumin are based on primate *in vivo* studies, which indicate that a decrease in plasma protein of 1 gm/dl produces an increase in BE of approximately 3.0 mEq/L, and a 1 gm/dL decrease in the albumin concentration produces an increase in BE of about 3.7 mEq/L. The constant (0.3) used in the free water equation, which is equal to the SID/normal [Na$^+$], was also derived from studies in primates. This constant is 0.25 and 0.22 for dogs and cats, respectively, while the value of 0.3 is used for other animal species.

As a matter of interpretation for this example problem, it can be seen that there are combined dilutional (negative BE (H$_2$O)) and hyperchloremic acidoses (negative BE (Cl$^-$)), that account for the observed base deficit. There is no abnormality in the protein concentration, and there is no need to postulate the presence of unidentified anions since the entire observed base deficit is accounted for by the free water and chloride abnormalities ((-6) + (-4) = -10). It can also be seen from this example that the reduction in Pco$_2$ (10 mmHg) has the expected relation to the observed BE (-10) for this mixed acid/base disturbance (i.e., a **primary non-respiratory acidosis** with a **compensatory respiratory alkalosis**).

In summary, primary disturbances in acid-base balance may be classified as being either respiratory or nonrespiratory in origin.

Primary respiratory disturbances are fairly straight-forward since they affect mainly the plasma Pco$_2$ and [HCO$_3^-$] (see Chapters 90 and 91). Nonrespiratory disturbances, also called "**metabolic**" disturbances, are usually far more complex, and therefore their full categorization requires more than merely assessing changes in the plasma Pco$_2$ and [HCO$_3^-$]. Plasma is a complex solution with many interacting constituents, and **Peter Stewart** successfully employed basic physicochemical principles of aqueous solutions to write equations describing interactions among those constituents. The primary independent variables controlling the **[H$^+$]** of body fluids, and hence the **plasma pH**, were found to be the **Pco$_2$**, **SID** (**[Na$^+$] - [Cl$^-$]**), and the plasma protein concentration (**[Prot$^-$]**). When the SID increases (i.e., the [Na$^+$] increases or the corrected [Cl$^-$] decreases), the [H$^+$] will decrease; and when the [Prot$^-$] concentration increases, the [H$^+$] will increase in order to retain electroneutrality:

$$SID + [H^+] - [OH^-] - [Prot^-] = 0$$

Note that neither the **HCO$_3^-$** nor the **K$^+$** concentrations of plasma (two ingredients of the anion gap) are used in this approach to acid-base balance.

Traditional methods of evaluating the acid-base status of an animal have not previously considered SID or the plasma [Prot$^-$].

Equations developed by **Fencl** and **Leith** allow Stewart's work to be easily applied clinically for evaluating the metabolic (nonrespiratory) contribution to acid-base balance. This approach separates the net metabolic abnormality into various components, and allows one to easily detect and evaluate the nature of mixed metabolic acid-base disturbances which might otherwise remain as hidden abnormalities. This approach also provides insight into the severity of those disturbances.

OBJECTIVES

- Explain how the Peter Stewart approach to acid-base balance differs from the more traditional Henderson-Hasselbalch approach.

- Discuss what is meant by the SID (also called the fixed ion difference (FID) or in some cases the dietary cation-anion difference (DCAD)), and how it influences pH.

- Recognize the pathophysiologic implications of the following equation: $SID + [H^+] - [OH^-] - [Prot^-] - [PO_4^=] = 0$

- Explain why a free water deficit is alkalinizing, and why the plasma $[Na^+]$ is used to assess a free water abnormality.

- Show how the plasma Pco_2 and pH can be used to determine an observed base excess/deficit.

- Understand how the constants used in determining changes in the observed BE due to free water and protein abnormalities were determined.

- Recognize why hyperphosphatemia affects plasma pH more than hypophosphatemia (see Chapter 85).

- Understand why the plasma $[Cl^-]$ does not need to be corrected unless there is a free water abnormality present.

- Know why there is no reason to postulate the presence of unidentified ions when the sum of the contributions to the BE due to the free water, Cl^- and protein abnormalities equal the observed BE.

- Explain why the plasma $[HCO_3^-]$, $[K^+]$ and AG are not formerly considered in the Peter Stewart approach to acid-base balance.

- Understand how contributions to the acid-base state due to the presence of unidentified anions are quantitated.

- Understand how the constants in the free water equation and in the protein equations were derived.

- Explain how (and why) hyponatremia becomes acidifying, while hypochloremia is alkalinizing.

- Understand how and why the Peter Stewart approach to acid-base balance does not necessarily replace the more traditional approach, but merely adds to our understanding of acid-base disorders.

QUESTIONS with Explanations:

1. **Given the following information on a patient:**

Variable	Lab Value	Normal
pH	7.33	7.4
Pco_2	49 mmHg	40 mmHg
BE	0 mEq/L	0 mEq/L
$[Na^+]$	140 mEq/L	140 mEq/L
$[Cl^-]$	102 mEq/L	102 mEq/L
[Alb]	4.5 gm%	4.5 gm%

a. Unidentified anions are present.
b. Pure uncompensated metabolic acidosis accounts for this acid-base disturbance.
c. There is a net non-respiratory abnormality present.
d. Acute respiratory acidosis accounts for this acid-base disturbance.
e. The observed chloride concentration needs to be corrected since there is a free water abnormality present.

$$BE\ (H_2O) = 0$$
$$BE\ (Cl^-) = 0$$
$$BE\ (Alb) = 0$$
$$BE\ (UA^-) = 0$$
$$\overline{}$$
$$BE\ (Obs) = 0$$

Both **acidemia** (pH = 7.33) and **respiratory acidosis** (Pco_2 = 49) are present, yet there is no net non-respiratory abnormality (BE = 0). There is no abnormality of free water, so there is neither concentration alkalosis nor dilutional acidosis present. For the same reason, the observed [Cl⁻] needs no correction, and is seen to be normal. The [Alb] is normal, and there is no need to infer the presence of unidentified anions. This acid-base disturbance is primarily **respiratory acidosis**, which could be an acute condition such as hypoventilation without time for renal compensation, where hyperbicarbonatemia and hypochloremia would be expected (see Chapter 90).

2. Given the following information on a patient:

Variable	Lab Value	Normal
pH	7.31	7.4
Pco₂	30 mmHg	40 mmHg
BE	-10 mEq/L	0 mEq/L
[Na⁺]	140 mEq/L	140 mEq/L
[Cl⁻]	102 mEq/L	102 mEq/L
[Alb]	4.5 gm%	4.5 gm%

a. There is no need to postulate the presence of unidentified anions in this patient.
b. There is a dilutional acidosis present.
c. This patient has a primary respiratory acidosis.
d. The Cl⁻ concentration needs to be corrected due to the presence of a free water abnormality.
e. There is a primary non-respiratory acidosis in this patient with unidentified anions present.

$$BE\ (H_2O) = \quad 0$$
$$BE\ (Cl^-) = \quad 0$$
$$BE\ (Alb) = \quad 0$$
$$BE\ (UA^-) = -10$$
$$\overline{BE\ (Obs) = -10}$$

There is **acidemia, respiratory alkalosis** and a **net non-respiratory acidosis** present. There is no abnormality of free water, so there is neither a concentration alkalosis nor a dilutional acidosis present. For the same reason, the observed [Cl⁻] needs no correction, and is observed to be normal. The [Alb] is also normal, so these three cannot totally account for the observed base excess (-10). We must therefore postulate the presence of unidentified anions totaling 10 mEq/L.

This acid/base disturbance would be classified as a **primary non-respiratory acidosis** due to the presence of **unidentified anions**. It has apparently been present long enough (6-24 hrs) for a compensatory respiratory alkalosis to become established (**Fig. 87-2** and **Table 87-2**). What might these unidentified anions be?

If there is reason to suspect that the cardiac output is low, lactic acidosis (from anaerobic metabolism) would be likely. However, there are other possible unidentified anions as well; ketone bodies (acetoacetate or β-hydroxybutyrate that accumulate during starvation or in diabetic ketoacidosis), other organic anions generated from ingestion of toxic substances like aspirin, antifreeze or methanol, or sulfates that accumulate in chronic renal failure (see Chapters 82, 87, and 88).

3. Given the following information on a patient:

Variable	Lab Value	Normal
pH	7.40	7.40
Pco₂	40 mmHg	40 mmHg
BE	0 mEq/L	0 mEq/L
[Na⁺]	160 mEq/L	140 mEq/L
[Cl⁻]	124 mEq/L	102 mEq/L
[Alb]	4.5 gm%	4.5 gm%

a. There is a free water deficit with concentration acidosis present.
b. There is a hyperchloremic acidosis present.
c. There are unidentified anions present.
d. The observed pH is normal, so there are no acid-base abnormalities in this patient.
e. None of the above.

$$BE\ (H_2O) = +6$$
$$BE\ (Cl^-) = -6$$
$$BE\ (Alb) = \quad 0$$
$$BE\ (UA^-) = \quad 0$$
$$\overline{BE\ (Obs) = \quad 0}$$

The pH and Pco₂ are normal, and there is no net base excess. However, a **free water deficit** is present denoting a **concentration alkalosis**. The **corrected [Cl⁻]** is abnormal (i.e., 108 mEq/L, which has increased 6 mEq/L (**hyperchloremic acidosis**)), while the [Alb] is normal. These three (+6 -6 +0) account for the observed base excess (0), so there is no need to postulate the presence of unidentified anions. This appears to be a case of two offsetting metabolic conditions; a concentration alkalosis (free water deficit), offset equally by a hyperchloremic acidosis.

4. Given the following information on a patient:

Variable	Lab Value	Normal
pH	7.40	7.40
Pco_2	40 mmHg	40 mmHg
BE	0 mEq/L	0 mEq/L
[Na⁺]	140 mEq/L	140 mEq/L
[Cl⁻]	102 mEq/L	102 mEq/L
[Alb]	2.3 gm%	4.5 gm%

a. Since the pH, Pco_2 and BE are normal, there is no acid-base disturbance here.
b. This is a case of pure hypoproteinemic acidosis.
c. There is a significant acidosis from unidentified anions that is hidden.
d. This is a case of pure hypoproteinemic alkalosis.
e. None of the above.

$$BE (H_2O) = 0$$
$$BE (Cl^-) = 0$$
$$BE (Alb) = +8$$
$$BE (UA^-) = -8$$
$$\overline{\hspace{4cm}}$$
$$BE (Obs) = 0$$

The pH and Pco_2 are normal, and there is no net base excess present (BE (Obs) = 0). There is no free water abnormality, so there is neither concentration alkalosis nor dilutional acidosis present. For the same reason, the observed [Cl⁻] needs no correction, and is observed to be normal. The **[Alb]** is down by 2.2 g/dl, so there is a **hypoproteinemic** alkalosis present that would make the base excess = +8 mEq/L. To account for the observed base excess of 0 mEq/L, one must postulate the presence of

unidentified anions in the amount of -8 mEq/L.

There is a significant acidosis present from unidentified anions that is hidden (observed BE = 0) by the presence of an equally offsetting hypoproteinemic alkalosis. Since these two metabolic acid-base disturbances are equally offsetting, there is no need for respiratory compensation.

5. Given the following information on a patient:

Variable	Lab Value	Normal
pH	7.40	7.40
Pco_2	40 mmHg	40 mmHg
BE	0 mEq/L	0 mEq/L
[Na⁺]	140 mEq/L	140 mEq/L
[Cl⁻]	114 mEq/L	102 mEq/L
[Alb]	1.3 gm%	4.5 gm%

a. Unidentified anions are present as a hidden abnormality
b. The observed [Cl⁻] needs to be corrected
c. This is a marked hyperchloremic acidosis masked by a hypoproteinemic alkalosis
d. This is a marked hyperchloremic alkalosis masked by a hypoproteinemic acidosis
e. None of the above

$$BE (H_2O) = 0$$
$$BE (Cl^-) = -12$$
$$BE (Alb) = +12$$
$$BE (UA^-) = 0$$
$$\overline{\hspace{4cm}}$$
$$BE (Obs) = 0$$

A normal pH and Pco_2 are present, and there is no net base excess. There is no abnormality of free water, so there is neither a concentration alkalosis nor a dilutional acidosis present. For the same reason, the observed [Cl⁻] needs no correction, yet is seen to be 12 mEq/L above normal (a **hyperchloremic acidosis**). The [Alb] is down by 3.2 g/dl, which through quantitative analysis reveals an equally offsetting **hypoproteinemic alkalosis**. These two marked, yet offsetting metabolic abnormalities account for the entire observed base excess (0), so there is no need to postulate the presence of unidenti-fied anions.

6. **Given the following information on a patient:**

Variable	Lab Value	Normal
pH	7.40	7.40
Pco_2	40 mmHg	40 mmHg
BE	0 mEq/L	0 mEq/L
[Na^+]	160 mEq/L	140 mEq/L
[Cl^-]	123 mEq/L	102 mEq/L
[Alb]	2.3 gm%	4.5 gm%

  a. There is a concentration alkalosis present.
  b. Unidentified anions are present in the amount of -8 mEq/L.
  c. There are four hidden non-respiratory acid-base abnormalities present that offset one another.
  d. All of the above are true.
  e. There is no acid-base abnormality here since the pH and Pco_2 are normal, and there is no net base excess.

$$BE (H_2O) = +6$$
$$BE (Cl^-) = -6$$
$$BE (Alb) = +8$$
$$BE (UA^-) = -8$$
$$\overline{BE (Obs) = 0}$$

The plasma pH and Pco_2 in this patient are normal, and there is no net base excess. However, there is a **concentration alkalosis** present, as well as a **hyperchloremic acidosis** and a **hypoproteinemic alkalosis.** These sum to BE = +8, but the observed base excess is zero. One must, therefore, infer the presence of unidentified anions in the amount of -8 mEq/L. Thus, there are four hidden non-res-piratory acid-base abnormalities in this patient that tend to offset one another (i.e., the pH and Pco_2 are normal), with the net (observed) base excess being zero.

7. **Given the following information on a patient:**

$$BE (H_2O) = +0.3$$
$$BE (Cl^-) = -2$$
$$BE (Alb) = ?$$
$$BE (UA^-) = -0.1$$
$$\overline{BE (Obs) = +4.2}$$

pH = 7.37 (Normal 7.40)
Pco_2 = 53 mmHg (Normal = 40)
AG = 8.8 mEq/L

  a. The observed base excess is due solely to a free water abnormality.
  b. There is a concentration alkalosis, hyper-chloremic acidosis and hyperproteinemic acidosis present.
  c. There is a hypoproteinemic alkalosis present which causes the plasma anion gap (AG) to be subnormal.
  d. This patient has respiratory acidosis (which can account for the entire change in pH as well as the observed base excess).
  e. None of the above

Although the contribution to the base excess due to the protein abnormality (BE (Alb)) was not given in this problem, it can be calculated from the data given:

$$BE (Alb) = 4.2 - 0.3 + 2 + 0.1 = +6$$

Thus, there is a marked **hypoproteinemic alkalosis** present (BE (Alb) = +6), which causes the plasma **anion gap** to be **low** (AG = 8.8 mEq/L). This patient also has a hyperchloremic acidosis (BE (Cl^-) = -2), and a mild concen-tration alkalosis (BE (H_2O) = +0.3). There is a significant ongoing respiratory acidosis (Pco_2 = 53 mmHg), and unidentified strong anions are present which contributes to the overall net acidemia (pH = 7.37).

7. c
6. d
5. c
4. c
3. b
2. e
1. d

ANSWERS

Chapter 93

Alkalinizing and Acidifying Solutions

Overview

- Sodium lactate or NaHCO$_3$ administration increases the SID, whereas KCl and NH$_4$Cl exert the opposite effect.
- Rapid administration of NaHCO$_3$ may not allow proper equilibration of HCO$_3^-$ across the BBB.
- Alkalinizing solutions tend to reduce the serum ionized Ca^{++} concentration.
- NH$_4$Cl is a powerful acidifying agent, that can also promote hyperkalemia and NH$_3$ toxicity.
- Sodium lactate and NH$_4$Cl should not be administered to patients with liver disease.
- KCl is sometimes administered to alkalemic patients.
- Isotonic saline infusion can be used to expand the extracellular fluid volume, and it is also acidifying.
- Acetazolamide is a mild diuretic that also promotes bicarbonaturia and kaliuresis.

Various solutions, when administered either orally or intravenously, can significantly alter the volume and pH of body fluids, and in certain instances may be nutritionally beneficial. The law of electroneutrality makes it impossible to prepare and therefore administer pure solutions of Cl$^-$, HCO$_3^-$ or organic anions such as lactate$^-$, gluconate$^=$, citrate$^=$, or acetate$^-$, but these anions can be administered as "**salts**" by combining them with cations, usually Na$^+$ or K$^+$ (e.g., NaHCO$_3$, KCl, or sodium lactate). Some of these salt solutions will acidify extracellular fluids (ECFs), while others will have an alkalinizing effect. In this chapter we will examine how certain solutions exhibit these and other important actions.

Alkalinizing Solutions
Sodium Bicarbonate

Sodium bicarbonate (NaHCO$_3$) solutions are sometimes administered to patients with metabolic acidosis who have both a low plasma HCO$_3^-$ concentration, and a low plasma pH (< 7.2; see Chapters 87 and 88). Since this salt is for the most part completely dissociated in aqueous solution, **Na$^+$, HCO$_3^-$**, and **H$_2$O** are effectively added to the ECF compartment (**Fig. 93-1**). Since Na$^+$ molecules are being added without Cl$^-$, and since HCO$_3^-$ has a tendency to displace Cl$^-$ from the ECF compartment, both effects contribute to **increase the "strong ion difference" (SID)**, thus causing **alkalinization** (see Chapter 92). In addition, added HCO$_3^-$ acts as a **buffer** to accept protons, and generate **CO$_2$** and **H$_2$O**. Although the **Pco$_2$** rises by about 0.5 mmHg for each mEq/L increase in the plasma **HCO$_3^-$** concentration, assuming the lungs are normal, this excess CO$_2$ should stimulate ventilatory drive, and thus expiration. If a NaHCO$_3$ solution is administered rapidly, however, ventilatory drive may not be appropriately regulated.

Copyright © 2015 Elsevier Inc. All rights reserved.

Alkalinizing Solutions

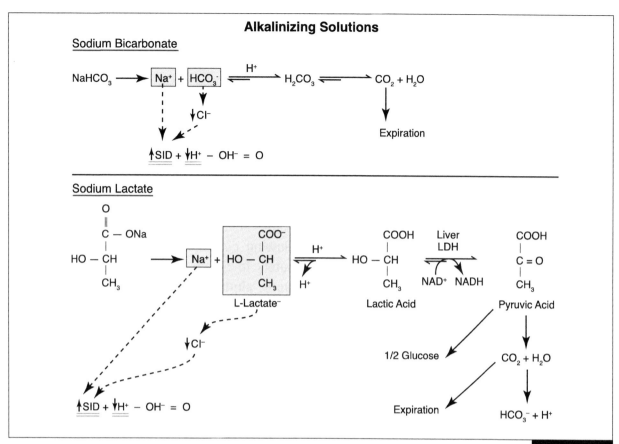

Figure 93-1

The respiratory center, which directs compensatory hyperventilation, is on the other side of the blood-brain-barrier (BBB) from plasma, and it typically slows the passage of ions (see Chapter 90). Thus, if the ECF HCO_3^- concentration is rapidly corrected through $NaHCO_3$ infusion, the brain interstitial fluid HCO_3^- concentration and pH rise only gradually, and the respiratory center may continue to "perceive" acidosis. Hyperventilation and hypocapnia may persist over time, even though the ECF HCO_3^- concentration has normalized. By definition, a low Pco_2 and normal plasma HCO_3^- concentration is respiratory alkalosis (**pH = $[HCO_3^-]/(SxPco_2)$**; see Chapters 85 and 91). If **overshoot metabolic alkalosis** occurs due to rapid or over-administration of $NaHCO_3$, the combined metabolic and respiratory (i.e., hyperventilation) alkaloses can cause a lethal alkalemia. Administering $NaHCO_3$ slowly allows for equilibration across the BBB and prevents lethal respiratory alkalosis.

Hypo- or isotonic $NaHCO_3$ solutions also have a tendency to **expand the ECF volume**. Most animals may not be harmed by this effect if they can excrete the extra Na^+ and volume load through the urine. However, if renal function is poor, or the animal is prone to pulmonary edema (e.g., poor cardiac function), expansion of the ECF volume may produce unwanted secondary effects. If a hypertonic $NaHCO_3$ solution is administered in an attempt to avoid excess fluid, hypernatremia may develop.

The alkalinizing effect of $NaHCO_3$ causes protons to dissociate from buffer sites on **plasma proteins** (e.g., albumin; see Chapter 86). This dissociation exposes additional anionic sites, which can bind Ca^{++}, thereby lowering the physiologically active free ionized Ca^{++} concentration (without affecting the total plasma Ca^{++} concentration). Thus, $NaHCO_3$ alkalinization in patients with borderline hypocalcemia can promote a **symptomatic hypocalcemia**.

The alkalinizing effect of $NaHCO_3$ also causes

K^+ to move into cells in exchange for H^+ (see **Fig. 89-2**), thus promoting a **hypokalemia**. If $NaHCO_3$ is infused rapidly, the hypokalemia which follows can cause cardiac arrhythmias. If a $NaHCO_3$ solution is infused slowly, K^+ shifts become more gradual, the plasma K^+ concentration can be monitored over time, and K^+ replacement therapy can be instituted as needed.

Sodium Lactate

Sodium lactate administration is also alkalinizing for some of the same reasons above. The lactate$^-$ anion displaces Cl$^-$ from ECF. This, combined with the addition of Na^+, **increases the SID**, which is an alkalinizing effect (Fig. 93-1). The alkalinizing action of lactate$^-$ also lies in the way in which it is metabolized. Lactate$^-$ can either be **oxidized** to CO_2 and H_2O via the TCA cycle (see Chapter 34), or it can be converted to **glucose** via hepatic gluconeogenesis (see Chapter 37).

Oxidation
$$\text{Lactate}^- + H^+ + 3\,O_2 \longrightarrow 3\,CO_2 + 3\,H_2O$$

Gluconeogenesis
$$2\,\text{Lactate}^- + 2\,H^+ \longrightarrow \text{Glucose}$$

Both of these processes utilize protons, so that they can be considered as utilizing lactic acid (lactate$^-$ + H^+ <—> lactic acid), just as anaerobic glycolysis can be considered as producing lactic acid.

The primary tissues in which oxidation occurs will be the liver, renal cortex, myocardium, and type I (aerobic) skeletal muscle fibers (see Chapters 37 and 80). However, the rate of lactate$^-$ utilization will depend upon the metabolic activity of these tissues (i.e., lactate$^-$ catabolism or anabolism will not increase only due to an increase in the blood concentration). Furthermore, it is likely that in ketoacidosis, oxidation of ketone bodies (KB$^-$s) or fatty acids will be preferred to that of lactate$^-$, since oxidation of these fuels inhibits **pyruvate dehydrogenase** (see Chapters 27, 71, and 75).

Therefore, under certain conditions associated with metabolic acidosis, the rate of lactate$^-$ oxidation could be quite low.

Hepatic gluconeogenesis, therefore, becomes the quantitatively important process for removing lactate$^-$, and hence protons from blood. **Pyruvate dehydrogenase** inhibition via enhanced mitochondrial NADH/NAD$^+$ and ATP/ADP concentration ratios allows pyruvate$^-$, formed from lactate$^-$, to be converted to oxaloacetic acid via pyruvate carboxylase, thus allowing carbon atoms to be converted ultimately to glucose via the **dicarboxylic acid cycle shuttle** (see Chapters 27 and 37). However, the rate of gluconeogenesis will depend on other factors as well, to include the blood concentrations of the gluconeogenic hormones, particularly glucagon and the glucocorticoids. In patients who are acidotic due to starvation, these hormones will already be elevated (see Chapters 73-76). Also, ruminant and carnivore livers are normally in a continual state of gluconeogenesis, and therefore are usually capable of converting lactate$^-$ to glucose (unless they are engorged with fat; see Chapter 72).

Of particular importance in the present context is the oxygen supply, which will determine the ATP/ADP concentration ratio in the liver; a decrease in the ATP/ADP ratio will inhibit gluconeogenesis, and if severe enough will increase the rate of glycolysis, thereby preventing lactate$^-$ utilization. Also, in situations where lactic acidosis prevails (e.g., tissue hypoxia, drugs, exercise, ethanol toxicity, ethylene glycol toxicity, salicylates, tumors, etc.), $NaHCO_3$ (and not sodium lactate) would be the preferred alkalinizing solution.

It is sometimes said that the alkalinizing action of sodium lactate lies in its ability to generate HCO_3^-. Although lactate$^-$ oxidation will ultimately (like the complete aerobic oxidation of glucose, fat, or amino acids) give rise to CO_2 and H_2O, the CO_2 should not be considered alkalinizing since it gives rise to both H^+ and HCO_3^-.

The sodium salt of acetate and either the sodium or calcium salt of gluconate are also used to treat metabolic acidosis. These electrolytes dissociate, as above, leaving the acetate⁻ or gluconate⁼ anions available to accept protons, thereby buffering plasma. In order for these compounds to be effective, each must be metabolized. Acetate⁻ is utilized peripherally in ruminant animals, and is not cleared by the liver. Gluconic acid is a sugar acid (and nontoxic oxidation product of glucose), that is usually readily metabolized.

Acidifying Solutions
Ammonium Chloride

Acidifying salts such as calcium chloride ($CaCl_2$), ammonium chloride (**NH_4Cl**; **Fig. 93-2**), or arginine monohydrochloride, are largely acidifying because they **lower the SID**. Also, NH_4Cl and arginine monohydrochloride **donate protons** directly to the ECF compartment, an effect which is also acidifying.

As indicated in Chapter 92, adding **Cl⁻** without accompanying **Na^+** molecules lowers the SID. Also, because the pK' of the ammonia buffer system is high (9.0 pH units; see Chapter 85), alkalemic conditions tend to favor dissociation of ammonium ion (**NH_4^+**) to ammonia (**NH_3**) and

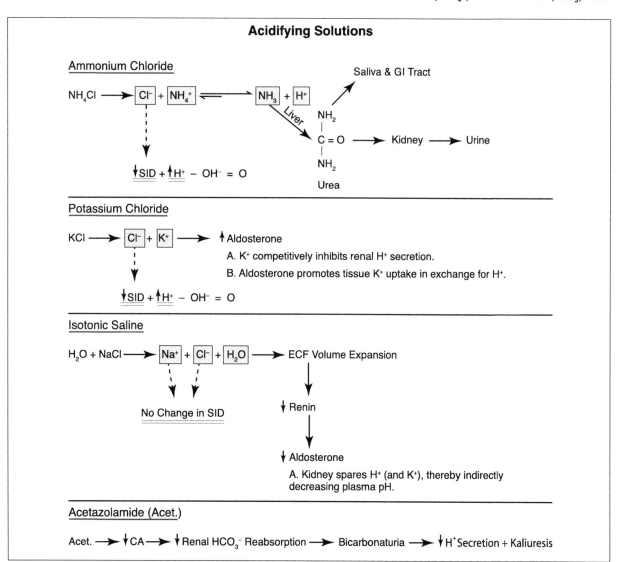

Figure 93-2

H⁺. As with the rapid infusion of alkalinizing solutions, similar precautions prevail for acidifying solutions. Ammonium chloride, if infused too quickly, can cause **NH₃ toxicity**, particularly in animals who are experiencing hepatic dysfunction. Both NH_4^+ and NH_3 are normally detoxified by the liver (see Chapters 10 and 11). Additionally, acidifying solutions cause K^+ to exit cells (**Fig. 88-3**), which can lead to a life-threatening hyperkalemia in acute situations. Therefore, acidifying solutions should be administered slowly, and attention to the plasma K^+ concentration is essential.

Potassium Chloride

Because metabolic alkalosis produces a hypokalemia (see Chapter 89), symptoms of muscle weakness and potential cardiac abnormalities can sometimes be reduced through careful and judicious potassium chloride (KCl) administration (note: high amounts of KCl can be lethal). Additionally, **KCl**, like NH_4Cl, exerts an acidifying effect by **reducing the SID** (**Fig. 93-2**). Potassium also stimulates adrenal **aldosterone** release, which promotes tissue K^+ uptake in exchange for H^+ (see Chapter 88). Since K^+ and H^+ effectively "compete" for distal renal tubular secretion under aldosterone stimulation, the slowly rising K^+ concentration will competitively inhibit H^+ secretion, thus producing an acidemic response.

Isotonic Saline

Although saline infusion will not alter the SID, it nonetheless **expands the ECF volume**, which in turn suppresses the renin-angiotensin system (**Fig. 93-2**). Since aldosterone secretion under these circumstances is reduced, distal renal tubular epithelial cells reduce their secretion of K^+ and H^+, thus helping to restore both body pools. Similarly, aldosterone antagonists such as **spironolactone**, a H^+ and K^+-sparing diuretic, would exert similar actions on the distal nephron.

Acetazolamide

Acetazolamide inhibits **carbonic anhydrase (CA)**, an enzyme required in both the proximal and distal nephron for H^+ secretion, and HCO_3^- reabsorption. Because of this combined action, it becomes acidifying (**Fig. 93-2**).

$$\left[\text{Acetazolamide} \longrightarrow \downarrow CA \longrightarrow \downarrow H^+ \text{ Secretion and } HCO_3^- \text{ Reabsorption} \longrightarrow \text{Bicarbonaturia} \right]$$

This drug also exhibits a mild diuretic action through CA inhibition. Since CA plays a key role in proximal renal tubular HCO_3^- reabsorption (see **Fig. 87-3**), acetazolamide exacerbates the bicarbonaturia of metabolic alkalosis, thereby lowering the plasma HCO_3^- concentration. However, as the HCO_3^- concentration increases in the distal tubular filtrate, this impermeant anion will further increase the transepithelial potential difference, thus promoting even more K^+ secretion and urinary excretion (i.e., kaliuresis; see **Fig. 89-3**). Therefore, KCl is usually administered prophylactically when acetazolamide is used to correct a metabolic alkalosis.

In summary, various solutions, when administered orally or intravenously, can significantly alter the volume and pH of body fluids. **Alkalinizing solutions** such as sodium bicarbonate and sodium lactate increase the SID of plasma, thus lowering the H^+ concentration. Bicarbonate and the lactate anion also act as additional buffers. A lethal overshoot metabolic alkalosis can occur with rapid or over-administration of $NaHCO_3$, therefore this alkalinizing solution should be administered cautiously. Since protons are displaced from plasma proteins upon alkalinization, these proteins tend to bind more Ca^{++}, thereby promoting a symptomatic hypocalcemia. Since alkalinization of the ECF compartment also causes K^+ to be displaced intracellularly, hypokalemia may also develop.

Acidifying salts such as calcium chloride, potassium chloride, and ammonium chloride are largely acidifying because they lower the

SID. Isotonic saline (a NaCl solution) does not alter the SID, but it expands the ECF volume, which in turn suppresses the renin-angiotensin system, thus reducing distal renal tubular H⁺ secretion. Acetazolamide is a CA inhibitor that similarly reduces renal H⁺ secretion. Since acetazolamide also causes a kaliuresis, KCl is usually administered prophylactically with this drug in the treatment of metabolic alkalosis.

OBJECTIVES

- Discuss the development of overshoot metabolic alkalosis with rapid $NaHCO_3$ administration, and explain the cause of hyperventilation (see Chapter 90).

- Explain how $NaHCO_3$ administration can promote alkalinizing changes in the SID, as well as symptomatic hypocalcemia and hypokalemia (see Chapters 89 & 92).

- Describe similarities and differences between the alkalinizing actions of $NaHCO_3$ and sodium lactate, and recognize why the later is not a recommended treatment for lactic acidosis.

- Recognize why sodium lactate's ability to generate HCO_3^- is not considered to be a (net) alkalinizing action.

- Understand why adding NH_4Cl is similar to adding HCl to the system, and recognize why NH_3 toxicity could be an unintended consequence of NH_4Cl therapy.

- Predict effects upon K⁺ homeostasis following rapid infusion of an acidifying solution.

- Discuss the adrenal effects of KCl therapy, and show how it can be acidifying.

- Explain how isotonic saline infusion can be acidifying (without altering the SID).

- Recognize why KCl is sometimes used prophylactically when acetazolamide is employed as an acidifying agent.

QUESTIONS

1. **Isotonic saline administration produces an acidifying action because it:**
   a. Increases the SID of extracellular fluid.
   b. Decreases the SID of extracellular fluid.
   c. Increases carbonic anhydrase activity in the proximal nephron.
   d. Suppresses the renin-angiotensin system.
   e. Enhances hepatic gluconeogenesis.

2. **Ammonium chloride administration would:**
   a. Be similar to adding HCl to the body.
   b. Be alkalinizing, since ammonia is a buffer.
   c. Not be expected to influence the plasma K⁺ concentration.
   d. Have a tendency to increase the plasma SID.
   e. Would have no effect on the plasma SID.

3. **Rapid $NaHCO_3$ administration to a patient with metabolic acidosis:**
   a. May decrease the blood pH even further, and cause hyperkalemia.
   b. Will quickly alkalinize interstitial fluid spaces of the brain.
   c. May promote combined metabolic and respiratory alkaloses.
   d. Would cause symptomatic hypercalcemia.
   e. Would decrease the strong ion difference (SID) of extracellular fluid.

4. **Potassium chloride administration would:**
   a. Be acidifying because it reduces the plasma SID.
   b. Suppress adrenal aldosterone release.
   c. Cause enhanced renal tubular H⁺ secretion, which would be alkalinizing.
   d. Inhibit renal tubular carbonic anhydrase, thus enhancing urinary HCO_3^- excretion.
   e. Not be dangerous if a patient possessed normal kidney function.

5. **Sodium lactate administration:**
   a. Is acidifying because lactate is being added to blood.
   b. Would increase the strong ion difference of plasma, and thus promote ECF alkalinization.
   c. Is recommended therapy for patients with lactic acidosis.
   d. Is alkalinizing because it gives rise to HCO_3^-.
   e. Is recommended therapy for patients with liver disease.

ANSWERS

1. d
2. a
3. c
4. a
5. b

Chapter 94

Dehydration/Overhydration

Overview

- Hypertonic dehydration is associated with decreases in both ECF and ICF volume, and increases in ECF and ICF tonicity.
- Hypertonic dehydration would not be expected to change the hematocrit.
- Uncontrolled diabetes insipidus would lead to a hypertonic dehydration.
- Hemorrhage is a common cause of isotonic dehydration.
- Hypotonic dehydration can lead to an increase in the ICF volume.
- Excessive administration of diuretic drugs can lead to hypotonic dehydration.
- Water intoxication in ruminant animals is not an uncommon occurrence.
- Abnormal sodium salt retention will lead to an expansion of the ECF volume.

Any net loss of water from the body can, as the common meaning of the expression implies, be termed **dehydration** (i.e., **water out > water in**). The causes, effects, and significance of dehydration, however, differ markedly between situations in which the loss is primarily a **deficit in extracellular fluid (ECF) volume** (i.e., **isotonic dehydration**), or where the loss is accompanied by a change in the osmolarity of body fluids (i.e., **hyper- or hypotonic dehydration**). If this loss results in a negative balance of electrolytes remaining in ECF spaces, then the fluid left behind is associated with hypotonic dehydration, whereas net loss of fluid in excess of electrolytes (i.e., a free water deficit) is associated with **hypertonic dehydration**.

Hypertonic Dehydration

Deficits of water alone, which are sometimes referred to as **simple dehydration** or **hydropenia**, result either from failure to replace inevitable losses of water that occur by evaporation from the lungs and skin, by excessive excretion of a hypotonic urine, or from other combinations of inadequate water intake and excessive loss. This **net loss of hypotonic fluid** diminishes both the intracellular fluid (ICF) and ECF compartments, and results in an increase in both ICF and ECF tonicity (**Fig. 94-1**).

Since simple water deficits result in increases in the osmolarity (i.e., tonicity) of body fluids, and since such increases are a normal stimulus for thirst, the persistence of hypertonic dehydration implies some break in the normal chain of events from hypertonicity through the sensation of thirst to the drinking, absorption, and retention of water. This chain may be interrupted at any of its links. Rare instances have been recorded in which a neurologic defect has led to an isolated absence of the sensation of thirst in an animal with an apparently otherwise normal sensorium. More often absence of the thirst sensation is part of the general deficit of mentation in a stuporous or comatose animal; or a patient, who presumably has the ability to feel thirst, but is physically unable to satisfy it. Perhaps

Copyright © 2015 Elsevier Inc. All rights reserved.

Figure 94-1
Changes in Hematocrit (Hct), Extracellular Fluid (ECF) and Intracellular Fluid (ICF) Volume and Tonicity in Various types of Dehydration

Dehydration	EFC Volume	EFC Tonicity	ICF Volume	ICF Tonicity	Hct
HYPERTONIC Net loss of hypotonic fluid	↓↓	↑	↓	↑	N (↑)
ISOTONIC	↓↓↓↓	N (↓)	N (↓)	N (↓)	–
HYPOTONIC Net loss of hypertonic fluid	↓↓↓↓↓	↓	↑	↓	↑↑↑
Loss of isotonic fluid replaced by an equal volume of water	↓↓↓	↓↓	↑↑	↓↓	↑↑

N=Nominal Change; – = No Change

most often the thirst sensation is present and the animal is physically capable of drinking, but an adequate supply of water is simply unavailable, or vomiting and/or diarrhea precludes its retention and absorption.

The need for water may be increased by conditions that accelerate the loss of water. Low humidity and high ambient temperatures, as well as increases in body temperature increase evaporative water loss. Excessive loss of hypotonic urine occurs when the synthesis and/or secretion of **antidiuretic hormone (ADH)** is impaired (**diabetes insipidus**), or the kidney is unable to respond to this hormone by suppressing excretion of a dilute urine (nephrogenic diabetes insipidus, NDI). The latter may occur as a congenital defect, or be induced by certain drugs (e.g., lithium salts).

Since the primary effect of water loss is an increase in osmolarity, and since membranes separating ECF from ICF are highly permeable to water, the increase in osmolarity is similar in both compartments. Consequently, loss of water is generally distributed uniformly throughout the body's fluid compartments in proportion to their initial volumes. It should be noted that because these water losses are spread more or less proportionately between cells and ECF, the erythrocyte volume is generally reduced in much the same proportion as that of plasma, so that the fraction of whole blood made up

by red blood cells (RBCs) (i.e., the **hematocrit, Hct**), remains **essentially unchanged** in even severe cases of hypertonic dehydration. It is a common error to look for a significant increase in Hct as a sign of hydropenia. An increased Hct is generally the result of a net loss of ECF (if it is not the result of an increased mass of RBCs (i.e., polycythemia)).

Isotonic Dehydration

The most common form of isotonic dehydration occurs following acute blood loss (i.e., **hemorrhage**). Compensation for this loss of vascular volume occurs due to several physiologic feedback systems, with one of the most significant being a shift of fluid from intracellular and interstitial fluid spaces into the vascular compartment (i.e., the "**capillary fluid shift**"). This occurs as a result of sympathetic-mediated arteriolar vasoconstriction, which reduces capillary hydrostatic pressure (and thus filtration pressure), and enhances reabsorption pressure. Since over 90% of available body water is located in extravascular sites, movement of fluid into the plasma compartment from these sites can help to restore blood volume and pressure, while at the same time resulting in only nominal reductions in ICF volume. One problem exists, however, in fully restoring blood volume by this mechanism without first replacing lost plasma colloids (i.e., plasma proteins); thus, infusion of saline alone following hemorrhage could result in mild edema if colloids have not been replaced.

Since hemorrhage results in an equal loss of both plasma and formed elements of blood (e.g., erythrocytes), this form of isotonic dehydration would not be expected to alter the Hct.

Some of the more dramatic instances of ECF volume depletion occur with certain forms of **diarrhea**. Diarrhea, when left untreated, may lead to shock, coma, and death. Since gastrointestinal (GI) disturbances associated with diarrhea are also often accompanied by **vomiting**, the problem is frequently compounded by not only further losses of electrolyte in the vomitus, but by the circumstance that the normal route for replacement is no longer available. Although for present purposes we are concerned mainly with depletion of volume, it is important to note that lost GI fluid may differ substantially from ECF in its specific ionic composition. Thus, loss of gastric acid may superimpose alkalemia on volume depletion (see Chapter 89), while diarrheal stools are frequently more alkaline than ECF, and thus may induce a significant degree of acidemia (see Chapter 87).

Hypotonic Dehydration

The key element in the development of hypotonic dehydration is a net negative balance of Na^+ salts. It may thus result from any disorder that leads to loss of Na^+ salts from the body when those salts are not replaced (e.g., profuse perspiration, loss of digestive secretions through vomiting or diarrhea, or hypertonic urine loss).

Exercise, particularly in hot environments, frequently produces substantial losses of salt in the form of sweat from many animals. Although the concentration of salt in sweat is normally lower than that in the ECF, especially in animals acclimatized to hot environments or in well-conditioned athletes, the concentration tends to approach isotonicity as the rate of sweating increases. Consequently, the greater the amount of sweating, the more nearly the loss approaches a depletion of ECF volume, thus producing the syndrome known as heat prostration. Since an attempt is often made to replace these losses by the drinking of pure water, dilution of body fluids with a reduction in their osmolarity is often superimposed. This dilution is partially responsible for the production of muscle cramps (known as heat cramps that are sometimes seen in less-well trained marathon runners who consume pure water instead of electrolyte replacement fluids during a marathon). Since the water ingested is distributed throughout total body water, only about one-third of it remains in the ECF compartment. Much of it moves into cells (down water's concentration gradient), thus increasing ICF volume and reducing its tonicity. This shift in body water (from extra- to intracellular spaces) will commonly result in an increase in the hematocrit (unlike hypertonic dehydration). Obviously, drinking pure water during hypotonic dehydration is an entirely inadequate means of replacing losses that have been incurred. It is these circumstances that account for the practice of ingesting salt tablets when subjects engage in heavy exercise that leads to profuse perspiration.

Normally, the only losses of Na^+ salts through the GI tract consist of minor amounts in the stools. These losses may be greatly increased by **diarrhea** or **vomiting**. Therefore, although most diarrhea and vomiting results in isotonic dehydration, some forms may lead to hypotonic dehydration.

Excessive losses of salt in **urine** are the result of impaired reabsorption of Na^+ salts in the presence of continued glomerular filtration. It is important to keep in mind the enormous volumes of fluid that are normally filtered each day (e.g., 180 L/day in a 70 kg mammal). Impaired reabsorption of Na^+ salts by renal tubular epithelial cells may be the result of abnormalities in the composition of filtered fluid.

Uncontrolled **diabetes mellitus**, for example, may lead to a marked elevation in the concentration of glucose in the glomerular filtrate, and an osmotic diuresis that carries increased amounts of salt into urine may ensue. With the supervention of **ketonemia**, filtration and excretion of ketone body anions obligates excretion of cation, much of it Na^+, promoting a reduced ECF volume as well as an acidemia (see Chapter 88). Thus, severe depletion of ECF volume is a major aspect of **diabetic ketoacidosis**. On the other hand, reduced reabsorption of salt may be the consequence of depression of the mechanisms that affect Na^+ reabsorption by renal tubules. Among the causes of such depression are loss of normal hormonal regulators of salt transport (e.g., **aldosterone** in **adrenal insufficiency**), the administration of **diuretic drugs** that have the specific property of inhibiting reabsorption of Na^+ (and thus secretion of K^+ and H^+), and intrinsic damage to renal tubules such as may be a part, particularly in the recovery phase, of acute **renal insufficiency**.

Indicators of Hypovolemia

With volume depletion, the glomeruler filtration rate (GFR) falls and renal creatinine (Cr) and urea clearances change (in the face of relatively constant rates of urea and creatinine production). The plasma concentrations of these compounds can be sensitive markers, and may be within normal limits with mild-to-moderate reductions in the GFR, but they are invariably altered when the GFR is markedly reduced. The **BUN:Cr concentration ratio** usually **increases** in hypovolemic states, because the numerator increases and the denominator decreases (i.e., as GFR declines and oliguria develops, there is a disproportionate increase in renal urea reabsorption and creatinine secretion).

During hypovolemia, the **urinary Na^+ concentration** may also be low (to unmeasurable), since more time is being allowed for renal tubular Na^+ reabsorption. **Urine specific gravity** (i.e., osmo-larity) may be elevated in some hypovolemic situations, but not all (e.g., diabetes insipidus), and urine volume may be reduced. Similarly it is important to reemphasize that disorders of Na^+ and H_2O balance are not necessarily coexistent. In particular, as explained above, the **plasma Na^+ concentration** is not an appropriate estimate of extracellular or total body Na^+ stores. The plasma Na^+ concentration is merely a reflection of total body solute divided by total body water. As a ratio, therefore, it reflects proportionate but not absolute changes in its two components. The plasma Na^+ concentration may be normal in isotonic dehydration, low in hypotonic dehydration, and elevated in hypertonic dehydration (all hypovolemic states). Thus, the determination of circulating volume depletion should not be based upon the plasma Na^+ concentration alone, nor upon the urine specific gravity alone.

Overhydration

Intake of water beyond the ability of the kidneys to excrete it leads to dilution, and a reduction in the osmolarity of body fluids. Occasionally such dilution occurs despite normal renal excretion because of massive water intake (as in psychogenic polydipsia). With rapid reduction in the osmolarity of body fluids, the syndrome of **water intoxication** may ensue with severe disturbances in CNS function and convulsions. If dilution is rapid enough, **hemolysis** may occur with subsequent loss of hemoglobin in urine. More often dilution takes place more slowly as a result of some impairment of the ability of the kidneys to excrete a dilute urine. The latter, in turn, may be the result of an intrinsic disturbance in the renal handling of salt and water, but is more often the consequence of excessive secretion of pituitary ADH in response to something other than a normal osmotic stimulus. Such "**inappropriate**" **secretion of ADH** may be due to a number of noxious stimuli, to certain drugs, or to ectopic secretion of the hormone by certain tumors. When dilution develops slowly, rather marked degrees of hyposmolarity may occur with few symptoms, although the danger of water intoxication is present. Complicating long-standing inappropriate secretion of ADH is the fact that there is a secondary negative balance of Na^+, so that there is a superimposed contraction of the ECF volume.

Water intoxication in **ruminant animals** is not an uncommon occurrence. Ruminants deprived of water will sometimes over-indulge when resupplied with fresh water, thus filling their large rumen, taking in more than they require. Since water molecules rapidly penetrate most cell membranes, the ECF compartment becomes rapidly expanded, water continues to move inside of cells, disrupting normal metabolic function and many times causing death.

Expansion of the ECF Volume

Expansion of the ECF volume as a result of Na^+ salt retention, is also a common abnormality. Such salt retention can be induced by excessive secretion of adrenalcortical hormones, as by aldosterone-secreting tumors, or more commonly by administration by clinicians. Retention of salt and expansion of the ECF volume under these conditions is ordinarily limited, as balance is usually restored by compensatory changes favoring excretion (with only modest expansion). These conditions are more prominently characterized by K^+-depletion than by overhydration. As indicated in the above discussion, excessive ADH secretion leads to a secondary negative Na^+ balance, so that a superimposed contraction of the ECF volume prevails. There is also contraction of the ECF volume with inadequate ADH secretion (i.e., diabetes insipidus), due to excessive loss of hypotonic urine, thus hydropenia (or hypertonic dehydration). Inadequate insulin secretion (i.e., uncontrolled diabetes mellitus) leads to hypotonic dehydration, with similar contraction

of the ECF volume.

Massive expansion of the ECF volume leading to accumulation of interstitial fluid (edema), and to effusions of fluid in peritoneal (ascites) and pleural cavities (pleural effusion), are characteristic of cardiac failure, nephrotic syndrome, and cirrhosis of the liver. Reduced excretion of salt by the kidneys is common among the chain of pathophysiologic stimuli leading to salt retention in these conditions.

In summary, water out is greater than water in during dehydrated states, and the ECF left behind dictates the type of dehydration (i.e., hypotonic, isotonic or hypertonic). A free water deficit creates **hypertonic dehydration**, with common causes being excessive excretion of hypotonic urine, or inadequate water intake to offset body losses. Since membranes separating ECF from ICF are freely permeable to H_2O, the increase in osmolarity becomes similar in both compartments, volume losses are spread more or less proportionately between cells and ECF, and the Hct remains unchanged. Hemorrhage and most forms of diarrhea and vomiting lead to **isotonic dehydration**, where again the Hct would not be expected to change. In hypotonic dehydration there is a net negative balance of Na^+ salts, which can result from profuse perspiration, some forms of diarrhea and vomiting, or excessive hypertonic urine loss (e.g., diabetes mellitus). When attempts are made to replace fluid/electrolyte losses with pure water, further dilution of body fluids occurs, as well as cellular hydration. This form of dehydration increases the Hct (unlike hypertonic and isotonic dehydration). The **BUN:Cr** concentration ratio can be expected to increase in **hypovolemia**, and the urinary Na^+ concentration decrease since more time is available for renal Na^+ reabsorption. **Water intoxication** and excessive pituitary ADH release lead to "free water excesses," with consequent expansion of both the ECF and ICF volumes. Excessive renal **Na^+ and H_2O retention**

leads to expansion of the ECF volume, with a common cause being excessive presence of endogenous or exogenous mineralocorticoids. Since mineralocorticoids promote renal K^+ and H^+ secretion while Na^+ and H_2O are retained, hypokalemia and alkalemia develop. ECF volume expansion also results from cardiac failure, nephrotic syndrome or cirrhosis. Renal salt and H_2O excretion are compromised, thus promoting ECF volume expansion.

OBJECTIVES

- Explain what is meant by "hypertonic" dehydration, and what a free water deficit represents.

- Recognize primary differences between the causes and effects of isotonic, hypertonic and hypotonic dehydration.

- Explain why the hematocrit may not change with hypertonic or isotonic dehydration, but usually increases with hypotonic dehydration.

- Predict changes in the plasma protein concentration with each of the three primary types of dehydration.

- Associate diarrhea, vomiting, hemorrhage, DM and diabetes insipidus with either hypotonic, hypertonic or isotonic dehydration.

- Explain how the syndrome of inappropriate ADH secretion can lead to excessive body fluid dilution.

- Recognize why ruminant animals are more prone to water intoxication than non-ruminants.

- Discuss how the BUN:Cratinine concentration ratio can be used to evalutate hypovolemic states.

- Distinguish changes in extra- and intracellular volumes and tonicities with each of the primary types of dehydration.

QUESTIONS

1. The net loss of hypotonic fluid from the body (e.g., excessive loss of hypotonic urine), results in:
   a. An increased hematocrit.
   b. Increased extra- and decreased intracellular tonicity.
   c. Decreased extra- and intracellular fluid volumes, as well as decreased extra- and intracellular tonicity.
   d. Hydropenia.
   e. Hypotonic dehydration.

2. Isotonic Dehydration (e.g., hemorrhage):
   a. Would result in an increased intracellular fluid (ICF) volume.
   b. Would cause increased sympathetic nervous system (SNS) activity, arteriolar vasoconstriction, reduced capillary hydrostatic pressure, and an enhanced reabsorption pressure.
   c. Would result in an increased extracellular fluid (ECF) volume.
   d. Would cause increased parasympathetic nervous system (PNS) activity, arteriolar vasoconstriction, increased capillary hydrostatic pressure and an enhanced reabsorption pressure.
   e. Results in an increase in capillary hydrostatic pressure following sympathetically-mediated arteriolar vasoconstriction.

3. Hypotonic Dehydration is associated with all of the following, EXCEPT:
   a. It may occur during heavy exercise if fluid losses are replaced by the drinking of pure water.
   b. It may result in an increase in intracellular fluid (ICF) volume (with a subsequent decrease in intracellular tonicity).
   c. It may result in an increase in the tonicity of the extracellular fluid (ECF) compartment.
   d. It may occur due to vomiting or diarrhea.
   e. It may occur due to loss of glucose, electrolytes and ketone body anions in the urine of diabetic patients.

4. Overhydration may:
   a. Result from undersecretion of antidiuretic hormone (ADH).
   b. Increase the osmolarity of body fluids.
   c. Stimulate the release of renin by the kidney.
   d. Result in hemolysis (with consequent appearance of hemoglobin in urine).
   e. Cause diabetes mellitus.

Section VII Examination Questions

1. **Which one of the following statements is CORRECT?**

   a. Sodium lactate can be used (effectively) to treat respiratory alkalosis.
   b. The key element in the development of hypotonic dehydration is a net negative balance of sodium salts (e.g., excessive hypertonic urine loss).
   c. A mixed acid-base disturbance is usually indicated when the plasma bicarbonate concentration and Pco_2 are abnormal, and moving in the same direction.
   d. Medullary chemoreceptors (in the CNS) appear to be more sensitive to the Pco_2 than the H^+ concentration.
   e. Patients with metabolic alkalosis would be expected to be hyperkalemic.

2. **Because non-hypoproteinemic alkalemia increases the charge equivalency on albumin, the:**

   a. Glomerular filtration rate increases.
   b. Anion gap decreases
   c. Free ionized plasma Ca^{++} concentration decreases.
   d. Ability of the proximal nephron to reabsorb glucose is enhanced.
   e. None of the above

3. **Given the following blood values:**

	Patient	Normal
$[Na^+]$	140 mEq/L	140
$[HCO_3^-]$	24	24
$[Cl^-]$	114	102
pH	7.40	7.40
[Albumin]	1.3 gm/dl	4.5

   a. The contribution to the observed base excess in this patient due to the free water abnormality is equal to that due to the protein abnormality.
   b. This is a mixed acid/base disturbance, with the hypoproteinemic alkalosis offsetting the hyperchloremic acidosis.
   c. The anion gap of this patient is well above normal.
   d. Unidentified anions are present in this patient as a hidden abnormality.
   e. All of the above

Copyright © 2015 Elsevier Inc. All rights reserved.

4. **Hypotonic dehydration is associated with:**

   a. The glucosuria and ketonuria of diabetes mellitus.
   b. A decrease in intracellular tonicity.
   c. Over-administration of loop diuretics.
   d. An increase in the hematocrit.
   e. All of the above

5. **Select the TRUE statement below:**

   a. Renal tubular acidosis (RTA) usually causes severe hypokalemia.
   b. Any patient having diuresis would also have an increased GFR.
   c. The content of the glomerular filtrate in Bowman's space normally looks more like interstitial fluid than plasma.
   d. A high protein diet normally decreases the ability of the kidneys to concentrate the urine.
   e. Metabolic acidosis is generally compensated through hypoventilation, and a subsequent rise in the P_{CO_2}.

6. **Which acid-base disorder below is most indicative of compensated respiratory alkalosis? (Note: Normal serum pH = 7.4, $[HCO_3^-]$ 24 mEq/L, and arterial P_{CO_2} = 40 mmHg)**

	$[HCO_3^-]$	P_{CO_2}	pH
a.	11	21	7.34
b.	30	60	7.32
c.	32	48	7.44
d.	16	20	7.52
e.	20	20	7.62

7. **Select the TRUE statement below:**

   a. Since H^+ and K^+ exchange across cell membranes, it stands to reason that acidemia will lead to hyperkalemia, and hyperkalemia will lead to acidemia.
   b. In general, when the plasma Cl^- concentration rises, the HCO_3^- concentration also rises.
   c. Plasma phosphates are generally more plentiful than bicarbonate in buffering an acid load.
   d. When the NH_3^+ content of urine increases, there will be a "positive" rather than "negative" UAG.
   e. Ethylene glycol toxicity would be expected to decrease the plasma AG.

8. **A previously healthy animal develops a GI illness with nausea and vomiting. Following 12 hrs of this illness, the following lab data are obtained:**

Body Wt.	35 Kg	P_{CO_2}	44 mmHg
B.P.	120/80 mmHg	Plasma $[HCO_3^-]$	32 mEq/L
Plasma pH	7.48	Urine pH	7.5

The illness continues, and 48 hrs later the following lab data are obtained:

Body Wt.	33	Pco_2	48
B.P.	80/40	Plasma $[HCO_3^-]$	36
Plasma pH	7.50	Urine pH	6.0

a. Intravenous saline infusion would help this patient because it is acidifying, and it would help to replenish the depleted ECF volume.
b. Lab data indicate a primary metabolic alkalosis with respiratory compensation.
c. Renin, angiotensin II and aldosterone levels (not reported) were likely elevated in the last blood sample drawn from this patient.
d. This patient is most likely hypokalemic.
e. All of the above

9. **Which one of the following is best associated with an increase in urinary HCO_3^- excretion?**

   a. Diarrhea
   b. Respiratory acidosis
   c. Aldosterone antagonist therapy
   d. Concentration (contraction) alkalosis
   e. Diabetic ketoacidosis

10. **Common findings in respiratory acidosis include:**

    a. Hypercapnia and hyperbicarbonaturia.
    b. A decreased AG and a left-shift in the bicarbonate buffer equation.
    c. Hyperbicarbonatemia and hypochloremia.
    d. Hyponatremia and diuresis.
    e. All of the above

11. **Compared to normal, the urinary excretion pattern in metabolic acidosis is best characterized by which one of the following?**

	NH_4^+	HCO_3^-	pH	Titratable acid
a.	↑	↓	↓	↓
b.	↑	↑	↑	↓
c.	↑	↓	↓	↑
d.	↓	↓	↓	↓
e.	↓	↑	↓	↓

12. Select the FALSE statement below:

a. Metabolic acid/base disturbances generally evoke opposite changes in the pH of CSF and blood.
b. Acetazolamide is an acidifying solution that, when administered, can lead to bicarbonaturia.
c. Although metabolic alkalosis is associated with kaliuresis, metabolic acidosis rarely (if ever) leads to this condition.
d. SID of plasma = $[Na^+]$ - $[Cl^-]$
e. Hyperproteinemia is usually acidifying.

13. During renal compensation for chronic respiratory acidosis:

a. Urinary HCO_3^- excretion increases.
b. The plasma Pco_2 remains at lower than normal levels.
c. The $HCO_3^-/(S \times Pco_2)$ ratio is lower than before renal compensation.
d. The plasma HCO_3^- concentration is higher than before renal compensation.
e. Distal renal tubular H^+ secretion is below normal.

14. Prolonged diarrhea would most likely result in:

a. An increased urinary NH_4^+ and decreased urea excretion.
b. Elevated urinary HCO_3^- excretion.
c. Increased circulating ANP levels.
d. Acidemia and hypercapnia.
e. All of the above

15. Metabolic alkalosis due to sustained vomiting (or abomasal displacement) is usually associated with all of the following, EXCEPT:

a. Bicarbonaturia.
b. Hyperchloremia.
c. An increase in the distal tubular transepithelial potential difference.
d. Kaliuresis.
e. An increase in the SID.

16. Given the following:

Plasma pH = 7.32 $\qquad$ Pco_2 = 35 mmHg

$$BE_{(H2O)} = 0$$
$$BE_{(Cl-)} = 0$$
$$BE_{(Alb-)} = \underline{0}$$
$$BE_{(Obs)} = -7$$

a. This patient clearly has metabolic acidosis.
b. Although respiratory alkalosis is present, it appears to be compensatory.
c. Unidentified anions are most likely present.
d. The bicarbonate buffer equation is shifted to the left.
e. All of the above

17. Patients with uncontrolled hyperglycemia and diabetic ketonuria may be expected to exhibit:

a. Hyperbicarbonatemia.
b. Hypertension.
c. Hypercapnia.
d. An increased total body store of K^+.
e. Hyponatremia.

18. KCl administration:

a. Would promote hypoventilation if given to an otherwise normal animal (since it is acidifying).
b. Would be expected to have no influence on the plasma SID.
c. Would promote adrenal aldosterone secretion.
d. Should not be considered for patient suffering from metabolic alkalosis.
e. None of the above

19. All of the following are generally associated with diabetic ketoacidosis, EXCEPT:

a. Hypobicarbonatemia.
b. Increase in the plasma AG.
c. Hyperchloremia.
d. Osmotic diuresis.
e. Hypocapnia.

20. A patient has hypercapnia and hyperbicarbonatemia:

a. The bicarbonate buffer equation of this patient is shifted to the left.
b. This patient has compensated metabolic acidosis.
c. This patient clearly has a primary respiratory acidosis with academia.
d. This patient could be either acidemic or alkalemic.
e. This patient clearly has compensated respiratory alkalosis with mild alkalemia.

21. **Which set of plasma values below best reflects an uncompensated metabolic alkalosis? (Normal: Pco_2 = 40, $[HCO_3^-]$ = 24)**

	PCO_2	$[HCO_3^-]$
a.	60 mmHg	30 mEq/L
b.	48	32
c.	40	30
d.	40	14
e.	21	11

22. **An animal with a low plasma Pco_2, normal pH and elevated $[HCO_3^-]$ would most likely have:**

   a. Respiratory alkalosis with appropriate physiologic compensation.
   b. A mixed acid-base disturbance.
   c. The syndrome of inappropriate ADH (SIADH) secretion.
   d. Additive acid-base disturbances (e.g., ketoacidosis, diarrhea and excessive NH_4Cl ingestion).
   e. Metabolic acidosis with appropriate respiratory and renal compensation.

23. **Given the following:**   **Plasma pH = 7.4**
                                   Pco_2 = 40 mmHg

$$BE_{(H2O)} = +3$$
$$BE_{(Cl-)} = -4$$
$$BE_{(Alb-)} = +4$$
$$BE_{(UA-)} = \underline{-3}$$
$$BE_{(Obs)} = 0$$

   This patient has:

   a. Hypochloremic acidosis.
   b. Hyperproteinemic alkalosis.
   c. A lower than normal plasma Na^+ concentration.
   d. Acidemia.
   e. None of the above.

24. **A 10% free water deficit (alone) would:**

   a. Cause the SID to increase.
   b. Be associated with hypotonic dehydration.
   c. Shift the bicarbonate buffer equation to the left.
   d. Affect the plasma Na^+, but not the plasma Cl^- concentration.
   e. All of the above

25. The following plasma data were determined (normal values in parenthesis):

$HCO_3^- = 18mEq/L$ (24) $Cl^- = 106$ mEq/L (102)
$Na^+ = 140$ (140) $Pco_2 = 25$ mmHg (40)
$K^+ = 3.55$ (4)   pH = 7.5 (7.4)

The primary acid-base disturbance present appears to be:

a. Metabolic acidosis.
b. Metabolic alkalosis.
c. Respiratory acidosis.
d. Respiratory alkalosis.

26. An hysterical, 24 year-old student is admitted to the hospital following an acid-base physiology exam, with the following blood chemistry values determined:

$[HCO_3^-] = 20$ mmol/L $Pco_2 = 20$ mmHg

From these data, the blood pH and acid-base condition of this patient would most likely be:

a. 7.32; Respiratory Acidosis from excess tobacco smoke
b. 7.42; Respiratory Alkalosis from mild hyperventilation
c. 7.52; Metabolic Alkalosis from vomiting
d. 7.62; Respiratory Alkalosis from excessive hyperventilation
e. 7.72; Metabolic Acidosis from diarrhea

27. If you were given a CO_2 value of 26 mEq/L on a serum profile, you would reasonably assume that this value represented:

a. Plasma $[HCO_3^-]$.
b. Total CO_2.
c. Pco_2.
d. Dissolved CO_2.
e. $Pco_2 + [HCO_3^-]$.

28. Which of the following is indicative of compensated metabolic alkalosis?

	$[HCO_3^-]$ mEq/L	Pco_2 mmHg	pH
a.	20	25	7.5
b.	40	46	7.56
c.	17	30	7.30
d.	34	10	7.70
e.	17	19	7.90

29. **Arrange the following abnormal blood values in the appropriate sequence of acid-base disorders: (Normal [HCO$_3$] = 24 mEq/L; Pco$_2$ = 40 mmHg)**

	[HCO$_3$]	Pco$_2$
1.	34	65
2.	10	25
3.	20	20
4.	40	45

a. 1, Respiratory Acidosis; 2, Metabolic Acidosis; 3, Respiratory Alkalosis; 4, Metabolic Alkalosis

b. 1, Metabolic Acidosis; 2, Respiratory Acidosis; 3, Metabolic Alkalosis; 4, Respiratory Alkalosis

c. 1, Respiratory Alkalosis; 2, Metabolic Alkalosis; 3, Respiratory Acidosis; 4, Metabolic Acidosis

d. 1, Metabolic Alkalosis; 2, Respiratory Alkalosis; 3, Metabolic Acidosis; 4, Respiratory Acidosis

e. None of the above

Answers

1. b	19. c
2. c	20. d
3. b	21. c
4. e	22. b
5. c	23. e
6. d	24. a
7. a	25. d
8. e	26. d
9. d	27. b
10. c	28. b
11. a	29. a
12. c	
13. d	
14. a	
15. b	
16. e	
17. e	
18. c	

Epilog

Those who teach the basic concepts of physiological chemistry in a veterinary setting owe it to their students, whose ultimate interests are in the diagnosis and treatment of disease, to emphasize those aspects of the subject matter that throw light upon common functional disorders. The biochemist can in this manner play an important role in providing students and practitioners a vantage point from which they may develop rational views of pathophysiologic processes.

I have endeavored to write a book which hopefully serves to link basic concepts in physiological chemistry with clinical medicine, and which therefore promotes continuity of physiology teaching throughout the pre-clinical and clinical years. It is hoped that when biochemical principles underlying diseased states are pointed out to veterinary students, and they are shown how a knowledge of such principles aids in the interpretation of symptoms or in directing treatment, they will take a keener interest in the subject matter. When biochemistry teachings are restricted to mere molecular aspects of the subject matter, remote from a more integrated physiologic or clinical application, students are likely to regard them as useless tasks which their professors, in their inscrutable wisdom, have condemned them to perform. Too often they develop the idea from such teachings that biochemistry is of limited utility, and come to believe that, having once passed into the clinical setting, most of what they "crammed" for examination purposes may be forgotten without detriment to their more purely clinical interests. Unfortunately, many do not realize at this point in their education that "health" of the organism depends upon many thousands of intra- and extracellular reactions proceeding at rates commensurate with survival, and that what they will need in order to practice medicine effectively is a fundamental understanding of the "substance" of those metabolic processes; something that cannot be compensated with "how to" technical skills.

The boundaries, if any, that distinguish biochemistry from much of the rest of biology have become arbitrary, indeed. Biochemistry is not only the "very stuff" and language of physiology, histology, immunology, microbiology, pharmacology and pathology, it is also the rational language for discourse in clinical medicine. However, knowledge of biochemical pathways is no more or less useful than knowledge of anatomy. Both are essential, but in themselves quite inadequate. What the student needs is a great deal of help in appreciating connections between metabolic pathways and metabolic disorders, and between form and function.

The effort to compress the ever growing body of biochemical information, and to select only those topics deemed to be of major importance to students of physiology and pathophysiology have been serious challenges to this author. Reluctantly, I found it necessary to delete much of the more popular growing body of information on molecular biology, since this appears to be the focus today of many undergraduate biochemistry courses, and since knowledge of the intriguing details of this discipline have not yet found their clinical relevance.

Relationships between biochemistry, physiology and veterinary medicine have important implications, for as long as medical treatments continue to be firmly grounded in basic science principles that evolve from well-controlled, peer reviewed research studies, the practice of medicine will have a rational basis from which to accommodate new and reasonable knowledge. This, indeed, is in apposition to the unorthodox health cults, holistic and alternative medicine practices of today that lack an intellectual basis, and are too often founded on little more than myth, smoke and mirrors, fairy tales and wishful thinking.

It is hoped that readers have discovered useful information in this text regarding biochemical phenomena -- on metabolic pathways and their control, and on the physiological significance and clinical relevance of topics discussed. Attempts were made to be clear and concise, and to convey the beauty, relevance and excitement of physiological chemistry (without getting lost in the detail).

Larry R. Engelking

Case Studies

Case Study #1: Ethylene Glycol

Oscar, a four year-old neutered male Dalmatian dog belonging to George, a local fireman, was brought to the emergency room. Physical examination revealed an animal with severe pain over both kidneys, and he appeared to be in a state of hypovolemic shock. Oscar vomited, became increasingly lethargic and ataxic while in the emergency room, and soon lapsed into a coma. George did not know what had happened to his dog, but stated that his illness seemed to appear within an hour or two after they had returned to the station from a fire. Serum analysis and urinalysis revealed the following:

Severe azotemia and metabolic acidosis, with a markedly elevated anion gap (AG). Urinary GGT and NAG were elevated, and there was a marked glucosuria. Urinary crystals were viewed under polarized light, and were thought to be calcium oxalate. High levels of urinary glycolate and bicarbonate were also evident.

Questions

1. What type of nephrotoxic agent might Oscar have ingested?
2. Are ethanol, methanol and ethylene glycol structurally similar?
3. Show the common pathway in the metabolism of these compounds.
4. How is kidney function altered by the toxic products of ethylene glycol metabolism?
5. What is renal tubular acidosis (RTA)?
6. Why was the AG gap elevated?
7. Why were urinary calcium oxalate crystals present?
8. What is the recommended treatment for ethylene glycol toxicity?

Answers

1. Several drugs and toxins can cause severe **metabolic acidosis**. Lactic acidosis is commonly associated with cardiovascular collapse, respiratory suppression, or seizures, all of which can be produced by drugs. In addition, two widely available substances in the fire station can also cause metabolic acidosis by unique mechanisms, in which lactic acidosis plays a contributory role. These substances are **methanol** and **ethylene glycol**, and they account for many ingestion-associated metabolic acidoses.

Methanol, otherwise known as wood alcohol, is widely used as an industrial solvent, and is frequently found in **windshield-washing solutions**. **Ethylene glycol**, another organic solvent, has been used for many years as an ingredient in radiator **antifreeze solutions**, and some **brake fluids**. Because it is odorless and tastes sweet, it has also been used illegally to sweeten wine. Diethylene glycol, on the other hand, is used in color film developing.

Copyright © 2015 Elsevier Inc. All rights reserved.

Dogs are attracted to ethylene glycol because of its **sweetness**. Methanol and ethylene glycol are not acids, and neither is directly toxic. However, both are metabolized to toxic acids and other products that can cause severe acidosis and other lethal effects. As little as 30 ml (2 tablespoons) of either substance can be lethal in a dog. Antifreeze that is absorbed into a concrete floor can later be resuspended by additional moisture (e.g., when the fire truck returns from a fire).

2. Chemically, both **methanol** and **ethylene glycol** have structures similar to **ethanol** (drinking alcohol):

$$CH_3 - OH$$
Methanol

$$OH - CH_2 - CH_2 - OH$$
Ethylene glycol

$$CH_3 - CH_2 - OH$$
Ethanol

3. Methanol and ethylene glycol metabolism can be understood by first studying the metabolism of **ethanol**, which occurs in two steps:

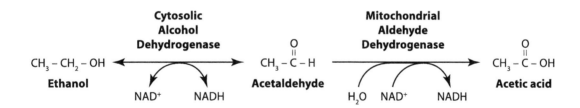

Ethanol is oxidized to **acetaldehyde** with production of **NADH** by **alcohol dehydrogenase**, an enzyme located, for the most part, in the cytosol of hepatocytes (see Chapter 24). **Acetaldehyde** next traverses the mitochondrial inner membrane for oxidation by an **aldehyde dehydrogenase** in the mitochondrial matrix. The NADH generated by this last step can be used directly in the mitochondrial **electron-transfer chain (ETC)**. However, the NADH generated by cytosolic alcohol dehydrogenase is oxidized back to NAD^+ through conversion of pyruvate to lactate, or through the malate and glycerol 3-phosphate shuttles (see Chapter 36). Thus, the capacity to oxidize alcohol is somewhat dependent on the ability of the liver to transport reducing equivalents from the cytosol into mitochondria via these shuttle mechanisms. **Acetic acid** does not usually accumulate, or cause metabolic acidosis, since it can be readily converted to acetyl-CoA, and oxidized to CO_2 and H_2O through the ETC.

The essential feature of the above process is the progression of an **alcohol** to an **aldehyde**, and then to a **carboxylic acid**. These same steps occur for **methanol** and **ethylene glycol**, with the same enzymes acting as catalysts:

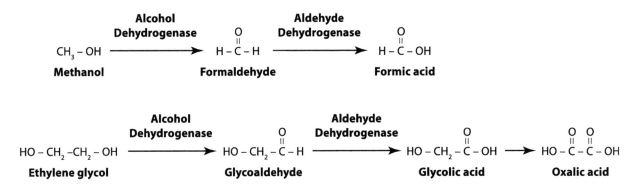

For **methanol**, the toxic end product is **formic acid**, which is not readily metabolized and, therefore, accumulates in body fluids. For **ethylene glycol**, the primary acid product is **glycolic acid**, which is further metabolized to other toxic products, including **oxalic acid**. These are strong acids that dissociate in body fluids, thus consuming HCO_3^-.

Signs of canine ethylene glycol intoxication reportedly become apparent within an hour, and are similar to those for ethanol intoxication. Coma and death can ensue soon thereafter. However, if the animal survives this initial phase, the second phase is associated with **acute renal failure**.

4. The **liver** and **kidneys** have roles to play in the biotransformation of many drugs and toxicants, usually resulting in the formation of metabolites that are less toxic than the parent compound. However, in the case of **ethylene glycol** conversion to **glycolate** and **oxalate**, they are, unfortunately, more toxic. Even with their proportionately large blood flow ($\approx 23\%$ of the cardiac output), the kidneys are susceptible to ischemia. Epithelial cells of the **pars recta** (S-3 segment of the proximal tubule), and **ascending thick limb of the loop of Henle** (LOH), are most frequently affected by hypoxemia because of **1)** their location within the outer renal medulla, and **2)** their many active transport functions and high metabolic rates. Toxicants can attach to apical or basolateral membrane sites, or to intracellular organelles. Subsequent damage to cell membranes and transport systems, along with impaired energy production and cellular respiration, can lead to excessive Ca^{++} influx, cell swelling and cell death. Disruption of active cell transport systems decreases tubular reabsorption, so that larger amounts of solute remain in the filtrate (e.g., glucose). Detection of enzymes in urine (enzymuria) that are associated with renal tubular epithelial cells (such as γ-**glutamyl transpeptidase** (**GGT**; also called γ-**glutamyl transferase**), and **N-acetyl-β-D-glucosaminidase** (**NAG**)), indicates renal tubular cell damage or necrosis, for these enzymes are too large to be normally filtered by glomeruli. Urinary **GGT** originates from the proximal tubular brush border, and **NAG** is present in proximal tubular epithelial cell lysosomes.

5. A condition in which there is impaired reabsorption of filtered HCO_3^- by the kidneys is known as **renal tubular acidosis (RTA)**. In **proximal RTA**, plasma HCO_3^- concentrations are dramatically reduced since the proximal tubules are normally the site of 90% of renal HCO_3^- reabsorption. In **distal RTA**, there is also wastage of HCO_3^- into urine, but since the distal tubules normally reabsorb only 10% of filtered HCO_3^-, this condition is generally not as dramatic as proximal RTA. In both cases, administration of $NaHCO_3$ to replenish depleted stores may be in order.

6. The **plasma AG** is about electroneutrality, which is indicative of every body fluid compartment (intra- and extracellular). That is, in each compartment the total number of cations must equal that of anions. In plasma, normally the **Cl⁻**, **HCO_3^-**, **Na⁺** and sometimes **K⁺** concentrations are taken into account when measuring the AG (see Chapter 86). When the total concentration of these two anions is compared to that for Na⁺ and K⁺, there is usually a difference of about 17 mEq/L (i.e., the AG). This gap is normally comprised of unmeasured anions, including plasma proteins, phosphates, sulfate, citrate and lactate. In a metabolic acidosis, the bicarbonate buffer equation is shifted to the **left** (Chapter 87), and therefore the plasma HCO_3^- concentration decreases. In the case of ethylene glycol toxicity, the void left by HCO_3^- is filled largely by glycolate and oxalate anions.

7. Although the excess plasma H⁺ concentration of metabolic acidosis is buffered by several different mechanisms, it has a tendency to **dissolve bone** (see Chapter 83), thus increasing urinary Ca⁺⁺ excretion. When urine becomes supersaturated with oxalate (from the ethylene glycol) and Ca⁺⁺, crystallization occurs. **Calcium oxalate uroliths**, as opposed to **Mg⁺⁺ struvite uroliths**, are difficult to dissolve medically, and therefore may require surgical removal.

8. A recommended treatment includes emesis (if the animal is alert and the gag reflex is functional), and activated carbon and/or sodium sulfate given orally to bind additional toxin. Fluid therapy for metabolic acidosis (e.g., $NaHCO_3$) can be instituted to restore blood pressure and volume, correct the acid/base disturbance, and promote toxicant excretion. Peritoneal dialysis may be a lifesaving procedure when oliguria/anuria is present. Ethanol or 4-methylpyrazole (4-MP) will competitively inhibit ethylene glycol metabolism, and therefore reduce effects of the renal phase.

Case Study #2: Phosphofructokinase (PFK)

Rusty is a 2 year-old, 50-pound, intact male English Springer Spaniel belonging to Dan, a musician. Dan purchased Rusty several months ago, with intentions of using him as a hunting dog. However, whenever Rusty becomes stressed through situations that induce hyperventilation (e.g., extensive exercise or barking while Dan plays his jazz saxophone), he appears to lose strength, become lethargic, then lose his appetite. Between stressful episodes, however, Rusty seems fine. He demonstrates no evidence of coughing, vomiting, diarrhea, altered water consumption or urinations, and his heartworm medication and vaccinations are up-to-date. Dan did mention, however, that Rusty's urine appears to be darker than normal following these stressful episodes.

Physical examination reveals pale, slightly yellow mucous membranes, an elevated body temperature and heart rate, moderate muscle wasting, and palpable (yet mild) hepatosplenomegaly. Following the exam Rusty voids a small amount of red urine, he then lies down and exhibits rapid, shallow breathing.

Laboratory tests reveal mild increases in circulating ALT and SAP, hyperkalemia, anemia (PCV = 20% (37-54)), an absolute reticulocyte count of 120,000/µl (with >60,000/µl being indicative of regeneration (i.e., an appropriate bone marrow response to anemia)), hemoglobinuria, and bilirubinuria.

Erythrocytic enzyme assays demonstrate phosphofrucokinase (PFK) activity to be 20% of normal, and pyruvate kinase activity to be normal. An *in vitro* acid-base assay shows that Rusty's erythrocytes possess increased fragility to "alkaline stress" (compared to normal).

Questions

1. What are the unique features of carbohydrate metabolism in mature erythrocytes, and why do they need ATP?

2. What was the connection between Rusty's hyperventilation, the erythrocytic Embden-Meyerhoff Pathway, and the anemia?

3. Why did Rusty have hyperkalaemia, what caused the reticulocytosis, and how does regenerative anemia differ from nonregenerative anemia?

4. What caused the yellow mucus membranes, elevations in plasma ALT and SAP, hepatosplenomegaly and sporadic pigmenturia?

5. Are there similarities in Rusty's RBC and muscle-type PFKs, thus explaining the concurrent hemolysis and muscle weakness during times of stress?

6. Would PFK-deficiency result in a "glycogen storage disease?"

Copyright © 2015 Elsevier Inc. All rights reserved.

Answers

1. Mature erythrocytes have no **mitochondria**, therefore all of their energy requirements are met through **anaerobic glycolysis**. They need **ATP** to maintain **membrane integrity**, prevent the accumulation of **methemoglobin**, and to synthesize **reduced glutathione** (**GSH**, which helps to protect them against oxidative stress). Midway through the erythrocytic **Embden-Meyerhoff Pathway** of anaerobic glucose catabolism there is an alternative route, the **Rapoport-Luebering Shunt**, where **2,3-bisphosphoglycerate (2,3-BPG)** can be generated from **1,3-BPG**. This glycolytic intermediate (**2,3-BPG**) reduces **hemoglobin-O_2 binding affinity**, thus making O_2 available to diffuse into tissue cells (see Chapters 30 and 31).

2. Stressful canine situations, such as extensive exercise, excessive barking, exposure to other dogs, or exposure to a new and unfamiliar environment (such as during a veterinary exam or boarding in a kennel), can cause dogs to **pant** (i.e., **hyperventilate**). This, in turn, can precipitate a mild **respiratory alkalosis** (see Chapter 91). This mild increase in blood pH normally increases activity of the erythrocytic **mutase** enzyme that converts **1,3-BPG** to **2,3-BPG**, which is an important adaptation to high altitude (where less O_2 is available). However, if a dog has **PFK-deficiency**, then less 2,3-DPG will be produced under alkaline conditions (because the PFK reaction is "upstream" from the Rapoport-Luebering Shunt), thus making his/her erythrocytes **"alkaline fragile."** This, in turn, makes intravascular hemolysis more likely.

3. Adult canine erythrocytes contain relatively **low K^+** and **high Na^+** concentrations (somewhat similar to plasma concentrations), because of a loss of membrane **Na^+/K^+-ATPase** activity during late erythrocyte maturation (ageing). This is thought to occur secondary to **hexokinase senility**, which is believed to be a key event in the life-span of an erythrocyte (see Chapter 31). As membrane ATPase pumps expire under normal conditions (i.e., the Na^+/K^+- and Ca^{++}-ATPases), erythrocyte cytoskeletons become rigid and, thus, red blood cells (RBCs) get stuck in narrow vessels (especially those of the spleen and liver). They then get **phagocytosed** by **reticuloendothelial (RE)** cells (also known as **fixed macrophages** or **mononuclear phagocytes**; see Chapter 33). However, **reticulocytes**, which are young RBCs, have **high K^+** and **low Na^+** concentrations. When hemolysis of these cells occurs, the plasma K^+ concentration rises (hence Rusty's **hyperkalemia**).

A metabolic block in the erythrocytic glycolytic pathway (at the level of PFK) results in **less 2,3-DPG production**, which increases the hemoglobin-O_2 binding affinity. The resulting relative peripheral tissue hypoxia stimulates renal **erythropoietin (EPO)** and thus bone marrow RBC production, which partially explains the marked **reticulocytosis**. Intravascular hemolysis also results in an appropriate bone marrow response, thus causing reticulocyte numbers to further increase. **Nonregenerative anemia**, on the other hand, may result from several chronic disease conditions, from drug or toxin exposure, or from inflammation (see Chapter 32).

4. Hyperbilirubinemia promotes **icterus**, which manifests itself as **"yellow jaundice."** Hemolysis, in turn, causes **hyperbilirubinemia** (see Chapter 33). The mild **hepatosplenomegaly** was thought to be caused by increased extravascular hemolysis (phagocytosis of RBCs), which in turn may have resulted in mild increases in circulating liver enzymes (**alanine aminotransferase, ALT**, and **serum alkaline phosphatase, SAP**). Intravascular hemolysis will increase the plasma **hemoglobin** concentration, and this protein can be filtered by renal glomeruli (thus causing the **hemoglobinuria**). Additionally, as the plasma unconjugated bilirubin concentration rises, more will be removed by the liver and conjugated. Since biliary excretion is usually the rate-limiting transport event in hepatobiliary bilirubin transport, when this step becomes saturated, **conjugated bilirubin** will reflux back into plasma. Since the more water-soluble conjugated bilirubin is only moderately bound to plasma albumin (unlike tightly bound unconjugated bilirubin), it can be filtered and excreted into urine (thus explaining the **bilirubinuria**). Together, the hemoglobinuria and bilirubinuria explain the sporadic **pigmenturia** following Rusty's hemolytic episodes.

5. Canine erythrocytic and muscle-type **PFK** appear to be similar, therefore explaining the hemolysis and loss of strength on exertion. Residual PFK activity in canine erythrocytes appears to be explained by the concurrent presence of a **hepatic-type PFK isozyme**.

6. Because glycogen deposition can be increased in skeletal muscle fibers of PFK-deficient dogs, this disorder has been referred to as a **"glycogen storage disease type VII."** Other identified canine glycogen storage diseases include a **type II** Pompe's-like disease, where glycogen storage is minimal due to a **glycogen branching enzyme deficiency**, and a **type III** Cori's-like disease where glycogen deposition is increased due to a **debranching enzyme deficiency** (see Chapter 23).

Ending

Rusty recovers without incident following five days of hospitalization, and his PCV returns to normal. He is released back to Dan with recommendations to not use Rusty for breeding purposes, and to limit his dog's stress (perhaps by considering a switch to classical piano rather than the jazz saxophone).

Case Study #3: Inflammatory Bowel Disease (IBD), Endocarditis and Cardiac Ischemia

Florence, a local school teacher, brought her cat **Zippy** in for evaluation. Zippy, it seems, has been "off" for several days, and thus has lost her zip. Florence reports that her cat is easily fatigued with exercise, and today actually collapsed during her daily run to the food bowl. She is a 9 year-old spayed female, Domestic Shorthair (DSH) with a prior diagnosis of inflammatory bowel disease (IBD). She was treated with a corticosteroid (prednisone) several weeks ago because of its antiinflammatory properties, and it appears the glucocorticoid (plus diet change) have the GI condition under control.

Zippy is quiet and a bit lethargic on physical exam. Her temperature is 103.4°F, she is tachycardic (↑ heart rate) and tachypneic (↑ breathing rate), and she has a soft cardiac murmur (with an irregular heart rhythm). Arterial pulses are variable in intensity, but mostly weak, and lung sounds are clear. Routine emergency blood tests are unremarkable, with the exception of an elevated blood glucose and modest hypertriglyceridemia. The complete blood count (CBC) shows a mild leukocytosis and monocytosis, yet a lymphopenia and erythrocytopenia (anemia).

Additional diagnostic testing documents areas of acute renal infarction (i.e., a wedge-shaped infarct on the kidney cortex is seen with ultrasound), cardiac enlargement on thoracic radiographs, and ventricular tachycardia on the electrocardiogram (ECG). An echocardiogram shows a large vegetative thrombus on the aortic valve causing both aortic stenosis (narrowing), and aortic regurgitation. There is a region of the left ventricular free wall that moves poorly, and overall cardiac contractile function is reduced.

Further plasma evaluation shows Zippy's cardiac troponin-I (cTn-I) to be markedly elevated, and her N-terminal-pro Brain Natriuretic Peptide (NT-proBNP) to be moderately elevated. Blood cultures are submitted, particularly for *Bartonella* testing.

Questions

1. Explain the hemogram (CBC), hyperglycemia and hypertriglyceridemia, and postulate a likely pathogenesis.

2. Can enzymes or other proteins that have leaked into plasma following a myocardial ischemic episode, or cardiac hypertrophy, be used as markers of cell death?

3. How is myocardial energy metabolism affected during an ischemic episode?

4. What are the intracellular forces that attempt to restore anaerobic glycolysis during cardiac ischemia, and restore perfusion?

5. How is the adenine nucleotide pool affected during an ischemic insult, and what processes are thought to be involved in subsequent cell death?

6. If reperfusion of an infarcted area of myocardium were accomplished through appropriate therapy, would there be a possibility of further "injury" to cardiac myocytes?

7. Why was blood submitted for *Bartonella* testing?

Copyright © 2015 Elsevier Inc. All rights reserved.

Answers

1. **Glucocorticoid therapy** usually promotes an increase in the number of circulating erythrocytes (↑ in the RBC mass) by stimulating their production and decreasing their destruction, thus promoting an erythrocytosis. However, in "sick" cats, an **anemia** of "chronic disease" is the usual finding. Glucocorticoids also increases the number of circulating platelets (thrombocytes), neutrophils and monocytes by either increasing their release from bone marrow, or decreasing their removal from the circulation (i.e., inhibiting diapedesis). Thus, both **neutrophilia** and **monocytosis** are expected in animals treated with glucocorticoids. Glucocorticoids also decrease the number of circulating lymphocytes (**lymphopenia**), eosinophils and basophils for 4-6 hours following a dose by distributing them away from the periphery (rather than by increasing their destruction). **Endocarditis**, for unexplained reasons, can also promote changes in the CBC, with neutrophilia, monocytosis, and thrombocytopenia being relatively common findings. How much of the neutrophilia and monocytosis can be attributed to corticosteroids vs. endocarditis in this case cannot be easily determined. However, it is possible that glucocorticoid therapy in this cat predisposed it to the development of bacterial endocarditis, because glucocorticoids can increase susceptibility to bacterial, fungal and viral infections, and allow for their dissemination. Endocarditis involves exudative and proliferative inflammatory alterations of the endocardium, usually characterized by the presence of vegetations on the endocardial surface. It commonly involves a heart valve, and sometimes affects the inner lining of cardiac chambers or the endocardium elsewhere.

Glucocorticoids also enhance hepatic glucose production by **1)** increasing the delivery of amino acids and glycerol (gluconeogenic substrates) from peripheral tissues to the liver, and by **2)** increasing the rate of gluconeogenesis through increasing the amount (and activity) of the key gluconeogenic enzymes (Chapter 37). Therefore, glucocorticoid therapy promotes a **hyperglycemia**. In dogs, glucocorticoids may also increase **hepatic glycogen deposition** in the resting state by providing, through gluconeogenesis, more available glucose 1-phohsphate (Chapter 23). However, little glycogen is stored in feline hepatocytes since the cat liver (like that of the ruminant animal) is normally in a continual state of gluconeogenesis. Through the actions of glucocorticoids on important hepatic enzymes controlling glucose homeostasis, the gluconeogenic effects of **glucagon**, **growth hormone** and **epinephrine** (the other diabetogenic hormones) are enhanced. Although not well understood, animals with **endocarditis** can exhibit either a hyper- or hypoglycemia.

Glucocorticoids also enhance **lipolysis** (Chapter 70), they promote the hepatic re-esterification of triglyceride from incoming long-chain fatty acids, and they facilitate hepatic **VLDL** formation and exocytosis (thus the **hypertriglyceridemia**; Chapters 63 & 67). When long-chain free fatty acids (FFAs) are in excess, glucocorticoids may also facilitate their hepatic conversion to **ketone bodies** (Chapter 71). By increasing hepatic **angiotensinogen** production and enhancing the vasoconstrictive actions of **angiotensin II** and **norepinephrine (NE)**, glucocorticoid excess may lead to **hypertension (HT)**. **Endocarditis** has caused **cardiac hypertrophy** in this cat, and resultant changes to the cardiovascular (CV) system (e.g., reduced cardiac output) have likely resulted in activation of the reninangiotensin-aldosterone system, as well as that of the sympathetic nervous system (SNS), causing increased circulating levels of angiotensin II, aldosterone and NE, increased peripheral resistance, and volume retention.

2. Increased presence of **cTn-I** and/or the **creatine phosphokinase (CPK)-MB** and **lactate dehydrogenase (LDH-H)** isozymes in plasma following a **myocardial infarction (MI)** can be diagnostic (Chapter 6).

The thick filament of cardiac and skeletal muscle is largely myosin, with the thin filaments consisting of actin, tropomyosin and the **troponin regulatory complex**. This complex has three subunits: **troponin-T (Tn-T)**, which attaches to "tropomyosin;" **troponin-C (Tn-C)**, which serves as a binding site for "Ca^{++}" during excitation-contraction coupling; and **troponin-I (Tn-I)**, which binds tightly to actin during relaxation, thus "inhibiting" actin-myosin interaction. The troponin regulatory complex basically holds the tropomyosin head between actin and myosin during muscle relaxation. When Ca^{++} binds to Tn-C to initiate contraction, a conformational change occurs such that the troponin-tropomysoin complex moves away from the myosin binding site on actin, thereby making actin accessible to the myosin head for binding. When Ca^{++} is removed from Tn-C, the troponin-tropomyosin regulatory complex resumes its inactivated position, thereby inhibiting myosin-actin interaction.

Both **Tn-I** and **Tn-T** are assayed in blood as **diagnostic markers** for human **MI**, because of their early release into the circulation when myocytes expire. In veterinary medicine, only the **cardiac isozyme** of Tn-I (**cTn-I**) is commonly used. In this cat, the presence of ventricular tachycardia and an area of the left ventricular myocardium that does not contract (regional myocardial hypokinesis) is likely a result of either an **MI** or **myocarditis**. In animals with bacterial endocarditis, parts of the **thrombus** can break away from the affected valve and lodge in a **coronary artery**, thus leading to an MI. Alternatively, bacteria that are shed from the vegetative lesion on the valve can travel through coronaries, thus leading to myocarditis. In either case, the resulting myocardial damage and necrosis will lead to release of **cardiac troponins**, which can be measured in blood.

Tissue damage from an ischemic myocardial area also releases **CPK** into plasma within the first few hours, with **LDH** release typically lagging behind that of CPK. Plasma levels of these enzymes today are not used as extensively as **troponin** in assessing myocardial cell damage from an **MI**.

Brain natriuretic peptide (BNP; also known as **B-type natriuretic peptide**, or assayed clinically as either **NT-proBNP** or **cardiac-BNP (c-BNP))**, was initially isolated from the porcine brain, and later found in higher concentrations in the heart. It is normally produced by cardiac atria, but in conditions leading to ventricular hypertrophy the ventricles themselves can become a major source. It is therefore used to identify cardiac enlargement, and is usually elevated due to cardiac hypertrophy or dilation in cats with congestive heart failure (CHF). However, it is apparently not as effective as **cTn-I** for detecting myocardial ischemia. NT-proBNP and c-BNP are released by the myocardium in a 1:1 ratio in response to ventricular stretch or **hypertrophy**. NT-proBNP, the inactive fragment of BNP, has a longer plasma half-life, and is rather easy to measure. C-BNP, the active molecule which creates the natriuresis and other actions of BNP, is much more difficult to measure, and a test for animals has not yet become commercially available.

3. When a **thrombus** occludes about 90% of a vessel, blood flow through that area is severely compromised, tissues otherwise served become ischemic, and capillary hemoglobin becomes rapidly oxygen-depleted. Normal myocardial metabolism is **99% aerobic** (Chapter 80), with most

ATP derived from **oxidative phosphorylation** (Chapter 36). Ischemic anoxia results in an attempted conversion to **anaerobic glycolysis** (Chapters 24-27), but it can generate (maximally) only about one-tenth the amount of ATP normally produced through mitochondrial oxidative phosphorylation. Flow of substrates into and waste products away from the myocardium is also reduced, osmotically active intracellular metabolites accumulate, and cells begin to swell. As **lactic acid** builds-up locally, the myocardial contractile force recedes (Chapter 81). The biosynthesis of **macromolecules** and **nucleotides** (Chapter 13) is also severely reduced under these abnormal metabolic conditions. **H$^+$ accumulation** inhibits **PFK** (Chapter 25), and a lack of **NAD$^+$** further inhibits glycolysis at the level of **Gl-3-P dehydrogenase** (Chapter 26). (**Note:** Since oxidative phosphorylation has ceased, **oxidized NAD$^+$** becomes deficient because it is no longer being regenerated through the **malate** and **glycerol 3-P shuttles** (Chapter 36)).

4.   As **ATP levels drop** during an ischemic insult, muscle **ADP** concentrations initially rise. **2ADPs** can now be converted by **myokinase** to **AMP** and **ATP** (for limited muscle contraction; see Chapter 77), with the **AMP** being further degraded intracellularly to **adenosine, IMP, inorganic phosphate (Pi)** and **NH$_4^+$**. Adenosine is a potent **vasodilator** of the coronary vasculature, and ADP, AMP, Pi and NH$_4^+$ activate **PFK** (Chapters 25 and 77). In this manner, an attempt is made to increase perfusion, and the rate-limiting enzyme in anaerobic glycolysis (PFK) is stimulated.

5.   **Adenosine** itself is further converted within cardiac myocytes to **inosine** and other products of **purine catabolism**, including **uric acid** (Chapter 17). All of this markedly depletes the **adenine nucleotide pool**, which is a key component of normal cell metabolism. It is known that the canine myocardial ATP level drops about 90% following 40 minutes of severe ischemia, and adenine nucleotide pool exhaustion coincides with irreversible cell damage (but it does not apparently cause it). Although changes that commit a cell to irreversible damage and death have been debated, **ATP depletion, intracellular phospholipase activation** (which produces plasma membrane damage), **protease activation**, and **intracellular Ca^{++} accumulation** all appear to be involved.

6.   **Reperfusion** of an infarcted area of myocardium is obviously optimal in resuscitating it. However, it can sometimes result in paradoxical cardiomyocyte dysfunction, a phenomenon termed **"reperfusion injury."**

The myocardium can indeed tolerate brief periods of severe and even total myocardial ischemia without resultant myocyte death. Although some cells may suffer ischemic injury, the damage is usually reversible with prompt arterial reperfusion. Indeed, such transient periods of ischemia may be encountered in angina and/or coronary vasospasm. With increasing duration and severity of ischemia, however, greater cardiomyocyte damage can and will develop, with a predisposition to a spectrum of reperfusion-associated pathologies.

Myocardial **stunning** is the best-established manifestation of reperfusion injury, defined as **"prolonged post-ischemic dysfunction of viable tissue salvaged by reperfusion."** Ischemic tissue, under this definition, reportedly goes through a period of prolonged, yet reversible, contractile dysfunction. The myocardium is basically **"stunned,"** requiring a prolonged period of time before complete functional recovery. Plasma membrane damage (e.g., to various ion pumps) may have occurred during the ischemic period, seriously altering permeability properties. This would affect the membrane potential, and it would also permit a flood of compounds (such as Ca^{++}) to enter from extracellular fluid (see below).

Superoxide and **hydroxyl radicals** (Chapter 30) are also thought to play roles in reperfusion injury, being generated by myocardial cells or circulating polymorphonuclear leukocytes. In addition to increased production, there is also a relative deficiency of the endogenous oxidant scavenging enzymes (**superoxide dismutase (SOD)**, **catalase (C)**, and **glutathione peroxidase (GP)**; Chapter 30), which further exaggerates free radical-mediated cardiac dysfunction. Free radicals damage cells by causing **lipid peroxidation of unsaturated fatty acids (UFAs)** contained in cell membrane **phospholipids** (Chapter 46), by breakage of **DNA strands**, and by oxidation of **protein SH groups**. Endothelial-dependent vasodilation can also be impaired, and responses to vasoconstrictors (such as endothelin-1 and oxygen free radicals) can be exaggerated.

The activity of myocardial **lactate dehydrogenase** (**LDH**; Chapter 80) has been found to be inhibited by about 40% after ischemia, and tends to remain depressed for up to 30 minutes after reperfusion. This places further emphasis on **anaerobic carbohydrate metabolism**, promoting **intracellular H^+ accumulation**, which activates plasma membrane **Na^+-H^+ exchange** (Chapter 88). This exchange facilitates proton extrusion while increasing Na^+ entry into cardiac myocytes. The cellular Na^+ gain next activates plasma membrane **Na^+-Ca^{++} exchange**, with Na^+ extrusion and a resultant increase in **intracellular Ca^{++}**. High levels of Ca^{++} inside these cells can wreak havoc by activating or inhibiting various enzymes in an **unregulated fashion** (Chapter 58), thus promoting further cell injury.

7. A variety of bacteria can cause **feline endocarditis**, and the diagnosis in most cases is best established via a positive blood culture. One of the bacterial organisms that seems to have a predilection for the **aortic valve** is *Bartonella*. This bacterium is often not detected by routine blood cultures, so serologic testing and/ or PCR are usually required to establish a diagnosis.

Case Study #4: Portosystemic Vascular Shunt (PSS)

Skipper, a one year-old, four pound male Yorkshire Terrier belonging to Sally, a hair dresser, was brought to your clinic late one afternoon, as the dog had become acutely ill and unarousable. Sally reported that Skipper was the "runt" of his litter, and that he has been a rather "laid-back" house pup. However, today he seemed quieter than normal (though he did eat some dry dog food earlier in the day). Shortly thereafter he became increasingly lethargic and "sleepy."

Skipper is current on all his vaccinations, and Sally does not believe he has been exposed to any of her "hair dresser toxins." He is a fussy eater, however, and has been experiencing vomiting episodes with increasing frequency. He has also been drinking more often in recent weeks, and has been having "accidents" rather than urinating on his papers. When he does attempt to urinate, she reports, it appears difficult for him.

Upon examination Skipper is slightly hypothermic, comatose, and his pupils are miotic. He also has a distended urinary bladder that cannot be expressed by palpation. Diagnostic testing reveals hyperammonemia, radiographic evidence of a small liver, and radiolucent calculi are observed in the urinary bladder. The blood urea nitrogen (BUN) concentration is below normal, the plasma bilirubin is within normal limits, plasma ALT and AST concentrations are slightly elevated, and plasma bile acid (BA) levels are severely elevated. Plasma total protein and albumin levels are low, as is total calcium. A special plasma amino acid profile reveals evidence of increased aromatic and decreased branched chain amino acids (AAAs & BCAAs).

Questions

1. What is a PSS?
2. What is hepatoencephalopathy (HE)?
3. What causes the neurological symptoms associated with HE?
4. Explain the plasma bilirubin, AST, ALT and BA findings in this patient.
5. Explain the plasma protein and Ca^{++} findings.
6. What type of urine crystalluria did this patient have, and what caused it?
7. Why did this patient exhibit signs of polyuria/polydipsia ((PU/PD)?

Answers

1. Skipper has **hepatoencephalopathy (HE)** secondary to a congenital **portosystemic vascular shunt (PSS)**. Congenital extrahepatic PSSs are usually single vessels that connect the portal venous system draining blood from the GI tract to the systemic circulation, thus bypassing the liver, and as such may arise from any portal vessel. They are found in both dogs and cats, with no apparent

Copyright © 2015 Elsevier Inc. All rights reserved.

breed disposition (although Yorkies are considered to be at increased risk). Most empty into the abdominal vena cava, but others may traverse the diaphragm prior to emptying into the thoracic vena cava or azygos vein. Many intrahepatic venous PSSs result from failure of the **ductus venosus** to close in infancy. Oxygenated blood from the placenta is normally carried from the left umbilical vein and umbilical sinus to the systemic circulation via the ductus venosus, and it is not known why the ductus fails to close in some animals. Patients with this condition, however, are frequently the **"runt of the litter."** Acquired PSS also occurs, and it is more often intrahepatic.

2.   **HE** is a neuropsychiatric disorder characterized by **augmented neural inhibition**. This syndrome is associated with acute, subacute or chronic liver failure where potentially neuroactive nitrogenous metabolites, derived from enteric bacteria and/or the wall of the intestine, accumulate in peripheral blood as a consequence of passage through portosystemic venous collateral channels, or severely compromised liver function.

3.   Although the **causes of HE** remain unsettled, **four hypotheses** have evolved to explain neurological symptoms associated with this disorder.

1) Synergistic Neurotoxins: Hepatic failure is accompanied by **HE** and ultimately coma because of the synergistic effects of accumulating toxins with coma-producing potential, as well as augmenting metabolic abnormalities. **Ammonia (NH$_3$)** is considered to play a central role in pathogenesis, as it is normally incorporated into **urea** or **glutamine (Gln)** by the liver (Chapters 8-11). Since **NH$_3$ is toxic to brain tissue**, the ability of supporting glial cells to maintain Gln formation remains integral to neuron survival, particularly during hyperammonemia. Plasma NH$_3$ becomes elevated in severe liver disease when portal hypertension causes shunting of blood away from the liver, or in congenital portosystemic vascular abnormalities. Although the **ammonium ion (NH$_4^+$)** cannot easily pass the blood-brain barrier (BBB), the highly lipophilic **NH$_3$ can**. Coincident with a rise in blood NH$_3$ concentrations during hepatic failure or PSS is usually a fall in the BUN concentration.

2) False Neurotransmitters (NTs): Since both branched-chain amino acids (**BCAAs**; Leu, Ile and Val) and aromatic amino acids (**AAAs**; Phe, Tyr and Trp) compete for uptake across the BBB, and since plasma AAAs are high relative to BCAAs in **HE**, this serves to augment entry of AAAs into the central nervous system (CNS), which in turn favors production of **false NTs** like **octopamine** and **β-phenylethanolamine** from Tyr and Phe (Chapters 2 & 8). Normal synthesis of excitatory catecholamines in the brain is also depressed. False NTs may also arise from **enteric bacterial degradation** of protein, and they cannot be cleared appropriately by the liver due to   portosystemic shunting of blood. These false (weak) NTs flood presynaptic nerve endings in the CNS, and they displace normal NTs, thus preventing nerves from responding to normal stimuli. Specially formulated amino acid mixtures that are rich in BCAAs and low in AAAs have been used to normalize the plasma amino acid profile of patients with this condition.

3) True Neurotransmitters: Currently available experimental findings are compatible with the hypothesis that the pathogenesis of HE may involve **gamma aminobutyric acid (GABA), glycine** and/or other "true" inhibitory CNS neurotransmitters, like **5-hydroxytryptamine (5-HT**; Chapter 12). The reasoning is as follows: In the presence of a PSS or liver failure, **GABA**, which is produced by enteric bacterial flora, bypasses the liver (which normally possesses up to 80% of total body **GABA-transaminase** activity, the enzyme responsible for catabolizing GABA). Enough GABA, which is the primary inhibitory NT in the CNS, can apparently traverse the BBB to bind to receptors on postsynaptic neural membranes, and thus cause encephalopathy. There also appears to be an increased sensitivity to **benzodiazepines (diazepam** and **lorazepam**), **barbiturates** and **alcohol** in chronic liver failure. This is explained by the increased number of binding sites for these ligands on the **GABA$_A$/benzodiazepine receptor/Cl$^-$** ionophore complex. It has also been postulated that other gut-derived substances with GABA potentiating properties may bypass the failing liver and contribute to the onset of HE. Substances from the gut may contribute to this condition by increasing permeability of the BBB, modulating the composition of neuronal membranes and receptors for NTs on those membranes, modulating the release of NTs from presynaptic membranes, and/or modulating neurotransmission as a consequence of their direct interaction with neurons.

Tryptophan (Trp) also increases in the brains of patients with hepatic coma. Trp is a precursor to **serotonin (5-HT)**, which is also a potent inhibitory CNS NT.

4) Endogenous Steroids: Another class of ligands for the **GABA$_A$/benzodiazepine/Cl$^-$ inophore receptor complex** seems to be **progesterone, deoxycorticosterone** (a mineralocorticoid), and maybe their metabolites. All apparently have profound positive allosteric modulatory effects on the GABA$_A$ receptor complex at nanomolar concentrations. It is possible that, through altered steroid metabolism in the liver and/or pharmacokinetic changes in these patients, there is an increase in these metabolites which, due to their hydrophobicity, easily cross the BBB to stimulate the **GABA$_A$ receptor complex**. The second possibility is that these steroids may be synthesized in excess by brain tissue. There is evidence that a peripheral-type GABA receptor complex may also be increased in HE. The precise function of this receptor is as yet undefined, however it is known that when stimulated by benzodiazepine ligands, it initiates an increase in steroid biosynthesis. Thus, either a CNS or potentially a peripheral source of these steroids, or decreased steroid conjugation by the liver could be important in the pathophysiology of this condition.

The question of **"which hypothesis above best describes symptoms associated with HE"** might be answered by assuming all four combine to create this neuropsychiatric syndrome. The most reasonable approach to ameliorating signs of HE would be to introduce therapeutic maneuvers which reduce interactions between enteric bacteria and nitrogenous substances, reduce dietary protein intake, or intervene with drugs that antagonize the GABA$_A$/benzodiazepine/Cl$^-$ inophore receptor complex.

4. When present, liver enzyme elevations (**AST** and **ALT**) are generally mild with **PSS**. However, this condition can decrease hepatic blood flow, which can cause hepatocellular atrophy. Also, **hepatotrophic factors** from portal blood, like insulin and nutrients, will also be in short supply. Fasting plasma **bile acid (BA)** and **bilirubin** levels are reportedly within the reference range in some patients, namely because there may be sufficient functional hepatic reserve capacity to extract these substances from portal blood given an adequate amount of time. However, **postprandial plasma BA concentrations** are often significantly elevated, and are therefore considered to be a good screening test for this condition (Chapters 33 & 62).

5. **Hypoproteinemia** is apparently one of the most consistent findings in dogs with **PSS,** but is reportedly much less common in cats. Hypoproteinemia is usually attributed to **hypoalbuminemia** (from impaired hepatic albumin synthesis), but it can also be secondary to an increase in blood volume associated with excessive renal Na$^+$ retention. A **hypoglobulinemia** from decreased **hepatic α-** and **β-globulin** synthesis can also occur, and when both albumin and globulin concentrations are decreased, a PSS should be differentiated from a **protein-losing gastroenteropathy**. The **hypocalcemia** in this patient is likely secondary to the low albumin level, since almost half of Ca^{++} in plasma is normally protein-bound, and potentially to the suppressive   effects of cortisol on renal Ca^{++} reabsorption.

6. Skipper had **ammonium biurate crystals** in his bladder. The persistent hyperammonemia and decreased ability to convert **uric acid** to **allantoin** in the liver (Chapter 17) favor formation of **ammonium urate uroliths**, independent of urinary pH. In some cats and dogs with PSS, the urate uroliths may be the only presenting complaint. In **Dalmatian** dogs, however, urinary ammonium biurate crystals may be a normal finding (Fig. 17-5).

7. It has been reported that about 50% of animals with PSS have dilute, **hyposthenuric urine**, and they are also **polydipsic**. The **PU/PD** has been associated with an unexplained increase in **ACTH** secretion (the "stress hormone" from the anterior pituitary), with subsequent **hypercortisolism**. Cortisol aids in the excretion of a water load by **1)** helping to maintain the **glomerular filtration rate (GFR), 2)** suppressing **antidiuretic hormone (ADH)** release from the posterior pituitary, and **3)** decreasing ADH effects on collecting ducts of the kidneys. Consequently, with cortisol excess, PU/PD is evident. Cortisol also reduces renal calcium and phosphate reabsorption.

Ending

Skipper had a coil surgically placed in his shunt, which decreased blood flow through the vessel, but should improve his liver function. Although partially successful, Skipper will require long term medical management of his liver condition. During surgery he also had a cystotomy (vesicotomy) performed to remove ammonium biurate stones from his urinary bladder. An indwelling urinary catheter was in place following surgery, and urine was flowing freely.

Case Study #5: Diabetes Mellitus (DM)

Sugar is an 8-year old spayed Doberman Pinscher with a history of recurring urinary tract infections (UTIs) that do not fully respond to antibiotic therapy.

On physical exam Sugar has bilateral cataracts, she is thin, lethargic and rather unkempt, and she has a sweet-smelling breath. There is an impression of dehydration, for she has increased skin turgor, tacky mucus membranes, and a delayed capillary refill time. She is hypotensive, and upon palpation her liver appears to be enlarged. Brutus, the owner who is a taxi cab driver, states that Sugar has lately been drinking and urinating more than usual, as well as exhibiting a few vomiting episodes. Based on his rather gruff demeanor, he apparently doesn't appreciate it when the vomiting episodes occur in the back seat of his taxi cab.

A blood sample is clearly lipemic, and the plasma profile reveals a significant hyperglycemia (465 mg/dl), hypercholesterolemia, hypertriglyceridemia, acidemia, and mild azotemia (elevated BUN and creatinine). The hematocrit (Hct.) is also elevated, as is the $HPO_4^=$, Ca^{++}, anion gap (AG), total protein, and ALT. The Pco_2 and serum Na^+, Cl^- and HCO_3^- concentrations are all lower than normal, but the K^+ and Mg^{++} concentrations are within normal limits.

The urinalysis shows an aciduria (pH 5.0), glucosuria (4+), ketonuria (3+), as well as a few white blood cells (WBCs) and bacteria (2+).

Questions

1. Why does Sugar have recurring signs of a UTI?
2. What is the cause of Sugar's cataracts?
3. Why the hyperglycemia, and should measurement of plasma fructosamine levels have been considered?
4. Why is Sugar azotemic, and should this be expected of all diabetics?
5. How is Sugar's hyperlipidemia explained?
6. Why was the liver enlarged, why are ketone bodies being produced, and why was the ALT elevated?
7. Explain the electrolyte abnormalities.
8. What acid/base disorder does Sugar exhibit, what is its etiology, and how does the body compensate?
9. What is causing the AG to increase, and is there a correlation between the amount of fixed acid produced during DKA, and the degree to which the plasma HCO_3^- concentration decreases?

Copyright © 2015 Elsevier Inc. All rights reserved.

Answers

1. A **UTI** is defined as adherence, multiplication and persistence of an infectious agent in the urogenital system. Infections often involve a bacterial organism that is present normally in the distal urogenital tract. The glucosuria of uncontrolled DM provides bacteria with an abundant amount of nutrient with which to multiply.

2. **Fructose** and **sorbitol** (a polyol also known as glucitol), are found in the lens and neurons (insulin-independent tissues), where they increase in concentration in hyperglycemic diabetic patients, and are involved in the pathogenesis of **diabetic cataract** and **neuropathies**. The sorbitol pathway from glucose (Chapter 25) is responsible for fructose formation in certain insulin insensitive tissues. Glucose undergoes reduction by NADPH to sorbitol, catalyzed by **aldose reductase (AR)**, followed by oxidation to fructose in the presence of NAD^+ and **sorbitol dehydrogenase (SDH)**. Although sorbitol and fructose can be metabolized to glycolytic intermediates, this process is slow. Additionally, sorbitol does not diffuse through cell membranes easily, and its accumulation causes osmotic damage by allowing ingress of H_2O with tissue swelling. Sorbitol also tends to reduce cellular Na^+/K^+-ATPase activity, making the osmotic damage worse by altering intracellular Na^+ and K^+ concentrations. Water ingress into the lens promotes clouding of liquid contents, probably by changing protein solubility.

3. There are numerous causes of **hyperglycemia** (e.g., Cushing's-like syndrome, insulin deficiency, glucagon excess, pheochromocytoma, hyperthyroidism, hypersomatotropism, etc.), and each should be ruled out before a definitive diagnosis is made. Given Sugar's urinary tract infection history, and the presence of **glucosuria** and **ketonuria**, **DM** is clearly a primary differential diagnosis. Glucosuria, however, does not rule out stress-induced hyperglycemia, but ketonuria strongly indicates DM. Had Sugar been on insulin therapy, blood **fructosamine** and **hemoglobin A$_{1c}$ (HbA$_{1c}$)** levels could have been measured as longer-term indicators of blood glucose control (to see if Brutus had been dosing Sugar appropriately with insulin (Chapter 21)).

4. Sugar is clearly dehydrated, and this would have a tendency to decrease the **glomerular filtration rate (GFR),** thus causing **pre-renal azotemia**. Additionally, proteins present in renal glomeruli become glycosylated in sustained hyperglycemia, which can further contribute to reductions in the GFR. However, not all patients with DM are reportedly azotemic, because the osmotic diuresis caused by hyperglycemia and ketonemia promotes medullary solute washout, leaving less time for renal urea reabsorption. Additionally, the muscle wasting of DM can lead to decreased circulating levels of **creatinine** (Chapter 77).

5. In poorly controlled insulin-dependent DM there is excessive mobilization of **triglyceride (TG)** stores from adipose tissue, which increases the plasma free fatty acid (FFA) concentration (Chapter 63). Many of these fatty acids are removed from blood by the liver, which either oxidizes them or re-esterifies them into TG, packaging these TGs into **very low density lipoprotein (VLDL),** then returning these complexes to blood (Chapter 65). Insulin deficiency leads to decreased **lipoprotein lipase (LPL)** activation, therefore the TG in circulating VLDL and chylomicrons (CMs) cannot be properly hydrolyzed so that long-chain fatty acids (LCFAs) can be removed from plasma for adipocyte TG storage (Chapter 70). Additionally, down regulation of **low density lipoprotein (LDL) receptors** in the absence of insulin results in **hypercholesterolemia** (Chapters 65-67). Post-prandial hyperlipidemia, pancreatic or liver disease, and other endocrinopathies (such as hyperadrenocorticism or hypothyroidism) may also cause abnormalities in circulating **cholesterol** and **TG** levels (Chapter 67). Hyperlipidemia may contribute to the development of pancreatitis, or it may be a consequence of pancreatitis, and in some cases multiple causes of hyperlipidemia can be present. For example, in a study of 221 canine diabetic patients, 23% had concurrent hyperadrenocorticism, 13% had acute pancreatitis, and 4% were hypothyroid.

6. In uncontrolled DM, the quantity of TG present in hepatocytes may be significantly elevated, and the ability to secrete VLDL impaired (Chapter 72). In light of the heightened requirement for glucogenic amino acids to enter the gluconeogenic pathway, hepatic **Apo B$_{100}$** production may become reduced, thus limiting **VLDL** formation. Thus, TGs get "trapped" in the liver, with the fatty infiltration being sufficient enough to cause visible pallor and enlargement (**steatosis** with **hepatomegaly**). Some hepatocytes will lose their membrane integrity, with intracellular components (e.g., ALT) moving into blood. Additionally, these conditions will also precipitate a **ketoacidosis** since the ability of hepatic mitochondria to oxidize all of the acetyl-CoA generated from FA β-oxidation through the TCA cycle is exceeded. This occurs because much of the **oxaloacetic acid (OAA)** being produced is being shunted through the **DCA shuttle** for gluconeogenic purposes, leaving little OAA to couple with acetyl-CoA for citrate formation (Chapter 37).

Alanine aminotransferase (ALT) is more specific for hepatocellular damage than **aspartate aminotransferase (AST)** in dogs, however, AST may be more sensitive (Chapter 9). Many canine tissues contain AST, limiting its usefulness as a liver-specific plasma test for hepatocellular necrosis.

7. Although **lipemia** can sometimes cause artifactually low serum electrolyte concentrations, it was probably not the cause of Sugar's hyponatremia and hypochloremia. **Hyponatremia** can sometimes reflect a free water excess rather than extracellular fluid (ECF) Na$^+$ depletion. In patients like Sugar that have marked hyperglycemia, the increased ECF osmolarity causes water to move out of cells, thus diluting the Na$^+$ concentration (it typically decreases by 1.5 mEq/L for each 100 mg/dl increase in the glucose concentration (above the upper end of the normal range)). Alterations in the Na$^+$ concentration are often accompanied by proportional movements in the **Cl$^-$ concentration**.

Intracellular **K$^+$** and **Mg^{++}** (the two most prevalent intracellular cations) concentrations increase because of cellular H$_2$O loss, and these two electrolytes begin to move down their concentrations gradients into ECF. Additionally, a decrease in ECF pH also causes further cellular K$^+$ and Mg^{++} loss (Chapter 88). **Insulin** causes K$^+$ and Mg^{++} to enter insulin-sensitive tissues, thus inadequate amounts of this hormone will impair this process. Renal elimination of these two electrolytes, however, is being maintained largely due to the presence of **KB$^-$ anions** in the renal filtrate. These anions hold on to Na$^+$, K$^+$ and Mg^{++}, and along with the **glucosuria** account for the osmotic diuresis that occurs with DM (Fig. 88-1). Urinary osmotic fluid loss leads to dehydration and hemoconcentration, which in turn can lead to **hypovolemic shock** if left untreated.

The **hypercalcemia** and **hyperphosphatemia** are secondary to the metabolic acidosis (increased mobilization of Ca^{++} and $HPO_4^=$ from bone, and increased release of Ca^{++} from plasma albumin), and **insulin** also promotes movement of $HPO_4^=$ into muscle cells (along with **glucose, amino acids, nucleosides, K^+, & Mg^{++}**). In its absence hyperphosphatemia develops, and when insulin is administered in excess, a serious hypophosphatemia can develop.

8. Sugar has **diabetic ketoacidosis (DKA)** secondary to hepatic overproduction of ketone bodies (**acetoacetic acid, β-OH-butyric acid**, and **acetone**; Chapters 70-72, 87 & 88). Acetone, which is not an acid, is formed by the spontaneous decarboxylation of acetoacetate, and is only detectable when the concentration of the latter is abnormally high. Unlike the other two ketone bodies, acetone is not further metabolized, but rather is excreted through the lungs and kidneys (where it accounts for the characteristic sweet or fruity smell on the breath and in the urine of severely diabetic animals).

Metabolic acidosis is usually a consequence of an increase in the amount of fixed acids in the body, from either ingestion or overproduction. The excess fixed acids (ketoacids) shift the bicarbonate buffer equation to the **left**, and as a result the **plasma HCO_3^- concentration decreases**:

$$CO_2 + H_2O \leftarrow H_2CO_2 \leftarrow H^+ + HCO_2^-$$

Acidemia also mildly **increases the respiratory rate**, so that during the initial acute phase of this disturbance, the **Pco_2** does not change (Chapter 87). However, as plasma buffer base (i.e., HCO_3^-) continues to decreases without a concomitant decrease in **Pco_2**, the **$HCO_3^-/(S \times Pco_2)$** ratio goes down, as does the plasma **pH** (Chapters 85 & 87):

$$pH = 6.1 + \log (HCO_3^-/(0.03 \times Pco_2))$$

Carotid chemoreceptors sense the continued decline in plasma pH, and signal the respiratory center to significantly increase the depth and rate of respiration (Fig. 87-2). As a result the Pco_2 decreases during this more chronic, respiratory compensatory phase, the $HCO_3^-/(S \times Pco_2)$ ratio moves back toward 20, and the plasma pH begins to normalize. Additionally, acid anions (ketone body (KB^-) anions, acetoacetate and β-OH-butyrate) plus Na^+ are filtered by the kidneys to retain electrical neutrality, and renal tubular epithelial cells secrete protons into both the proximal and distal tubular filtrate in exchange for Na^+ and HCO_3^-, which are added to peritubular blood (Fig. 87-3). This comprises the renal compensation to metabolic acidosis, but the extent to which the kidneys can secrete H^+ corresponds to a urine pH of about 4.5 (Sugar's is 5.0). Additionally, as stated above, KB^- anions that exceed the renal threshold for reabsorption tend to hold on to cations (particularly Na^+, K^+ & Mg^{++}), so the ketonuria begins to deplete these ECF electrolytes (Fig. 88-1).

9. The plasma **anion gap (AG)** is about the **"law" of electroneutrality**, which is an absolute requirement for all body fluid compartments (Chapter 86). That is, in each compartment (intra- and extracellular), the total concentration of cations must equal that of anions. In plasma it is customary to measure the **Na⁺, Cl⁻, HCO₃⁻** and **K⁺** concentrations. When the sum of the Na⁺ and K⁺ concentrations is subtracted from those of Cl⁻ and HCO₃⁻, there is normally a **"gap"** of about **17 mEq/L**. Since the plasma **K⁺** concentration is normally low, and doesn't deviate (much) in most pathophysiologic situations, it is often omitted from this calculation:

$$AG = ([Na^+] + [K^+]) - ([Cl^-] + [HCO_3^-]) \approx 17 \text{ mEq/L}$$

$$AG = [Na^+] - ([Cl^-] + [HCO_3^-]) \approx 12 \text{ mEq/L}$$

The **AG** is somewhat of a misnomer, because it is normally accounted for by the presence of anions such as plasma **protein⁻, HPO₄⁼, SO₄⁼, citrate⁻ lactate⁻**, and other **organic anions** not routinely measured by the diagnostic lab. In Sugar's case, the AG increased because of the presence of increasing amounts of KB⁻ anions, while the plasma HCO₃⁻ concentration decreased.

The decline in the plasma HCO₃⁻ concentration should be about equal to the increase in the fixed acid anion concentration (i.e., the KB⁻ anions) if there are no other contributing factors to the acidemia.

Case Study #6: Feline Lower Urinary Tract Disease (FLUTD)

Spot, a three year-old castrated male cat belonging to Dick and Jane, employees at the local Super Walmart, was brought to your veterinary clinic exhibiting signs of partial urinary tract obstruction (inappropriate urination and straining to urinate frequently). When questioned about Spot's diet, Jane conceded that she had been attempting to save money by feeding her cat more cereal-based plant material, and less meat-based material. She realized, however, that acid/base considerations of the feline diet were probably critical to Spot's well-being, but was unsure as to why, and what the consequences of an inappropriate diet might be. She had read cat food labels indicating that some diets were acidifying, while others were apparently alkalinizing.

A plasma profile shows a higher than normal pH and bicarbonate concentration, as well as an arterial $P_{CO_2} = 45$ mmHg. Most electrolyte concentrations were within the normal range. Urinalysis, however, indicated higher than normal amounts of potassium, magnesium, phosphate and bicarbonate to be present, with a pH of 7.1.

Questions

1. What type of acid/base disorder is indicated here, and what is the probable cause?
2. What are "fixed cations" and "fixed anions," and why are they important to acid/base physiology.
3. In terms of acid/base physiology, how does the high-meat protein diet of the cat compare to the plant-based protein diet?
4. What are the symptoms of FLUTD, and what causes them?
5. How are struvite uroliths formed?
6. How can formation of stuvite uroliths be prevented?
7. How are calcium oxalate uroliths formed?

Answers

1. Spot has **compensated metabolic alkalosis**, with a **right-shift** in the bicarbonate buffer equation (see Chapter 89).

During digestion and absorption of dietary constituents, animals generate variable amounts of organic and inorganic acids and bases which are nongaseous, and must eventually be excreted by the kidneys. Dietary acid load is determined by the acid/base nature of several strong and weak electrolytes ingested and absorbed (see Chapters 82 & 92), as well as by dietary sulfate and phosphate residues. For example, intestinal hydrolysis of ingested phosphoesters, and the breakdown of phosphoproteins, nucleoproteins and phosphatides result in phosphoric acid

Copyright © 2015 Elsevier Inc. All rights reserved.

(H_3PO_4) production (Chapters 82 & 85). When this strong acid, which significantly dissociates in solution, enters the circulation, it presents a major H^+ load to body buffers. In normal animals, the majority of phosphate will be eliminated in urine, and the more acid produced during metabolism, the more buffered the extracellular phosphate groups become ($\uparrow$**$H_2PO_4^-$:$HPO_4^=$ ratio**). Similarly, the amount of base produced, which is primarily excreted as urinary HCO_3^-, depends on the buffer base anions generated during metabolism, and the fixed cations present that tie up buffer base.

The **"alkaline tide"** that occurs after a meal refers to the **transient rise** in **plasma and urinary pH** resulting from efflux of **HCO_3^-** from parietal cells into blood in exchange for **Cl^-**, then secretion of **HCl** into the stomach lumen during gastric secretion. (Note: Although an "alkaline tide" (i.e., HCO_3^- tide) into blood is observed while food is in the stomach, a reciprocal and off-setting "acid tide" (i.e., H^+ tide) is normally observed later during pancreatic and biliary $NaHCO_3$ secretion into the intestinal lumen). Factors that delay gastric emptying and increase the **"gastric phase"** of digestion (e.g., dry, plant-based or high fat diets, or engorgement) accentuate the alkaline tide.

2. Key factors in acid/base physiology are the differential absorption of specific cations and anions by the intestine, the form or composition of dietary minerals, and the metabolism of nutrients to acidic or alkaline products. Physiologically significant **fixed cations** (namely **Na^+, K^+, Ca^{++}** and **Mg^{++}**) are those that cannot be altered by metabolism, but contribute to the whole-body alkaline load by promoting HCO_3^- retention. The monovalent fixed cations, Na^+ and K^+, are extensively absorbed across the intestinal wall, while the divalent cations, Ca^{++} and Mg^{++}, are more regulated (and therefore less extensively absorbed). The only physiologically significant **fixed anion** that readily substitutes for HCO_3^- is **Cl^-**, which, like Na^+ and K^+, is also extensively absorbed across the intestinal wall. By **increasing dietary Cl^-** (i.e., induced **hyperchloremic acidosis** -- addition of **Cl^-** without **Na^+**, ($\downarrow$**SID**; Chapter 92)), while manipulating the fixed cation absorbed and/or excreted by the kidneys (in practical terms via **$CaCl_2$** or **NH_4Cl** administration), it is possible to enhance **Cl^-** excretion (relative to the fixed cations), and thus induce urinary acidification.

3. The mineral load, as well as the amount and character of protein ingested, can substantially affect acid/base balance. The **high-meat protein diet** of the cat normally tends to be rich in sulfur-containing amino acids and phosphates, which are **acidifying**. By contrast, **plant proteins** are low in sulfur-containing amino acids and high in K^+, Mg^{++} and organic anions, thus favoring alkalosis (and presenting an inordinate load of phosphates, K^+ and HCO_3^- for excretion by the kidneys). As urinary HCO_3^- excretion increases, Cl^- excretion decreases, and the urine becomes **alkalinized**. These consequences, unfortunately, can be devastating for the cat.

4. **Feline Lower Urinary Tract Disease (FLUTD;** also referred to as **Feline Urologic Syndrome (FUS))**, can result in several signs, including **hematuria** (blood in the urine), **stranguria** (slow and painful discharge of urine), **pollakiuria** (unduly frequent passage of urine), and **urethral obstruction**. Causes include bacterial infections, **urolithiasis** (urinary calculi), or they may be unknown (**idiopathic**). Cats with a "first occurrence" of FLUTD usually have sterile urine, and those that have had prior catheterizations are more likely to have secondary bacterial infections. Thus far, attempts at securing a viral cause have been negative.

The morbidity rate (prevalence of the disease in a population) at veterinary hospitals is reportedly about **7%**. Although **idiopathic FLUTD** is diagnosed in about **65%** of cats with lower urinary tract signs, urolithiasis is diagnosed in about **25%** of cases. Approximately **50-60%** of these are **calcium oxalate** uroliths, and **40-50%** are **struvite uroliths**. Urolithiasis occurs most often in males, castrated or intact, older than two years of age, and is less likely to occur in females (presumably due to their anatomic advantage of a short, wide and straight urethra). Damage to the urethral mucosa or bladder wall, which can result from the initial episode, reportedly contributes to the likelihood of future episodes.

5. **Struvite crystals** are composed of **magnesium ammonium phosphate**:

$$Mg^{++}$$
$$O^- \cdots \cdots O^-$$
$$PO$$
$$O^-$$
$$NH_4^+$$

Excretion of struvite is linked directly to the concentration of its components in the nephron, and since **phosphate** and the **ammonium ion (NH_4^+)** are normally high in feline urine (as a consequence of the cat's high dietary protein diet), the primary manipulable variables become **H^+, H_2O** and **Mg^{++}**. Even when Mg^{++} is high in the diet and urine, struvite will not necessarily form or precipitate if the urine is sufficiently acidified or diluted for the amount of struvite present. In essence, struvite will only precipitate in concentrated urine as fine crystalline sand when the urinary pH rises **above 6.4** for a prolonged period of time (probably a few days initially, but more likely a matter of hours with each succeeding attack). Urinary pH is critical because trivalent phosphate (**PO_4^{-3}**) is required for conjugation with both divalent **Mg^{++}** and monovalent **NH_4^+**. Below pH 6.4, filtered phosphate acts as a buffer, becoming progressively enriched with hydrogen (**$HPO_4^=$** and then **$H_2PO_4^-$**), thereby removing the **PO_4^{-3}** necessary for struvite formation (Chapter 85).

6. If left to its ecological niche as a predator, the cat does not generally have a problem with urinary struvite formation, since its self-selected high-meat, low-carbohydrate diet is conducive to acid urine formation. The problem arises, as stated above, when large quantities of plant material are introduced into cat foods (cereal-based protein diets), because they tend to generate an alkaline urine, which may include a substantial Mg^{++} load. A diet with a **low ash** (low Mg^{++}) content can be fed to reduce urinary Mg^{++} excretion, thus lowering the divalent cation needed for struvite formation. This approach, however, may not be appropriate during pregnancy and/or lactation. The diet can also be formulated to maintain a relatively **low urinary pH of 5.5-6.5**, which favors H^+ buffering and urinary phosphate excretion as either $H_2PO_4^-$ or $HPO_4^=$. This rids the urine of the undesired trivalent PO_4^{-3} needed for struvite formation. Excessive acidification can, however, induce bone loss over time (Chapter 87), predisposing animals to **calcium oxalate urolith** formation (see below).

Patients predisposed to struvite formation can be managed by taking measures to **acidify the urine**, and **increase its volume**. This can be accomplished by feeding an acidified diet, perhaps with 1.5% NH_4Cl, coupled with a dietary program that maintains a moderately acid urinary pH (i.e., feeding a meat-based formula adequately balanced with vitamins and minerals). Although such diets may be more expensive, this approach could be cost-effective in terms of overall health and nutrition. Adding a small amount of salt to the diet can further stimulate H_2O intake, dilute the urine, and promote some urinary acidification. *Ad libitum* feeding of a canned food diet (or multiple, small-portion feedings) ensures a higher H_2O intake, and reduces the **"alkaline tide"** (because the gastric phase of digestion is shorter with high-moisture, premacerated diets). Meal feeding, particularly of dry food, accentuates the alkaline tide, which can sharply increase urinary pH and promote struvite formation.

7.   The incidence of **calcium oxalate uroliths** in cats has been increasing in recent years, probably due in part to the types of diets being fed in an attempt to reduce the incidence of struvite crystalluria and urolithiasis. These diets are generally restricted in Mg^{++}, contain supplemental NaCl to increase water consumption and decrease urine concentration, and, as indicated above, are formulated to produce an acid urine. The amount of acidifying agent in the diet may result in some degree of **metabolic acidosis**. Although excess H^+ is buffered by several different mechanisms, it has a tendency over time to **dissolve bone**, thus promoting urinary Ca^{++} excretion. When urine becomes supersaturated with Ca^{++} and oxalate, crystallization occurs. Calcium oxalate uroliths, as opposed to struvite uroliths, cannot be dissolved medically, and therefore can require surgical removal.

Appendix

Appendix Table I

Summary of Primary Digestive Processes

Source of Secretion	Enzyme (Inhibitor or Emulsifier)	Method of Activation & Optimal pH	Substrates	End Products
Salivary glands	α-Amylase (ptyalin)	pH 6.6-6.8	Starch & Glycogen	Maltose + 1:6 glucosides (oligosaccharides) + maltotriose
	Lipase	pH 4-4.5	Triglyceride	Diglyceride + fatty acid
Breast milk	Lipase	pH 4-4.5	Triglyceride	Diglycerides + fatty acids
(Colostrum)	Trypsin Inhibitor	pH 7	Trypsinogen	Prevents luminal trypsin acitvation
Stomach (abomasum)	Pepsinogen I (fundus) Pepsinogen II (pylorus)	Pepsinogens to pepsins by HCl (pH 1-2)	Protein	Oligopeptides
	Rennin (chymosin)	Ca^{++} (pH 4)	Casein of milk	Paracasein
	Lipase	pH 4-4.5	Triglyceride	Diglycerides + fatty acids
Pancreas	Trypsinogen (endopeptidase)	Trypsinogen to trypsin by enterokinase (pH 5.2-6), autocatalytic at pH 7.9	Protein & peptides	Polypeptides, dipeptides
	Tyrpsin Inhibitor	pH 7	Trypsinogen	Prevents luminal tyrpsin activation
	Chymotrypsinogen (endopeptidase)	Conversion to chymotrypsin by trypsin (pH 8)	Protein & peptides	Polypeptides, dipeptides

Appendix Table I Continued

Source of Secretion	Enzyme (Inhibitor or Emulsifier)	Method of Activation & Optimal pH	Substrates	End Products
	Proelastase (endopeptidase)	Conversion to elastase by trypsin (pH 8)	Protein & peptides	Polypeptides, dipeptides
	Procarboxypeptidase (exopeptidase)	Conversion to carboxypeptidase by tyrpsin (pH 8)	Polypeptides at the free carboxyl end of the chain	Small peptides, amino acids
	α-Amylase	pH 7.1	Starch & glycogen	Maltose + 1:6 glucosides (oligosaccharides) + maltotriose
	Lipase	Bile acids, lecithin, & pancreatic colipase (pH 8)	Triglyceride	2-Monoglycerides + fatty acids
	Cholesterol esterase (hydrolase)	Bile acids	Cholesterol esters	Cholesterol + fatty acids
	Phospholipase A_2	Activated by trypsin & Ca^{++}	Phospholipids	Lysophospholipids + fatty acids
	Ribonuclease (endonuclease)	pH 7	RNA	Oligonucleotides
	Deoxyribonuclease (endonuclease)	pH 7	DNA	Oligonucleotides
	Exonucleases	pH 7	Oligonucleotides	Mononucleotides
Liver	Bile Acids & Lecithin	pH 7	Insoluble fats	Mixed micelles
Duodenum (Brush border)	Enterokinase (enteropeptidase)	pH 7	Trypsinogen	Trypsin

Appendix Table I Continued

Source of Secretion	Enzyme (Inhibitor or Emulsifier)	Method of Activation & Optimal pH	Substrates	End Products
Small Intestine (Brush border)	Nucleotidases & Phosphatases	pH 7	Mononucleotides	Nucleosides + phosphate
	Nucleoside phosphorylases	pH 7	Nucleosides	Purines, pyrimidines & sugar phosphates
	b-Galactosidase (Lactase) (Galactocerebrosidase)	pH 7	Lactose Galactocerebrosides	Glucose + galactose + ceramides
	Trehalase	pH 7	Trehalose	Glucose
	Isomaltase (a-Dextrinase or a-1,6-Glucosidase)	pH 5-7	Dextrins, maltose, isomaltose, & maltotriose	Glucose
	Sucrase (Invertase) & maltotriose)	pH 5-7	Sucrose (maltose, fructose	Glucose +
	Maltase (a-Glucosidase)	pH 5.8-6.2	Dextrins, maltose & maltotriose	Glucose
	Glucoamylase (exo-a-1,4-glucosidase)	pH 5-7	Dextrins & Amylose	Glucose
	b-Glucosidase (Glucocerebrosidase)	pH 5-7	Glucosyl-ceramides (Glucocerebrosides)	Ceramides + glucose
	Peptidases (20%)	pH 7	Peptides	Smaller peptides & amino acids
Small Intestine (Cytoplasm of mucosal cells)	Peptidases (80%)	pH 7-8	Peptides	Smaller peptides & amino acids
	Phosphatases	pH 7-8	Organic phosphates	Organic molecule + free phosphate

Appendix Table II

Blood Chemistry Values for Domestic Animals

Constituent	Units	Dog	Cat	Horse	Cow	Pig	Sheep
AG (plasma)							
$(Na^+) - (Cl^- + HCO_3^-)$	mEq/L	10-16	12-19	9	11-25	20-21	14-18
Ammonia (NH_3)	μmol/L	0-40	0-40	0-40	--	--	--
ALP (SAP)	IU/L	12-127	10-79	109-352	29-99	26-362	68-387
ALT (SGPT)	IU/L	14-86	25-145	4-12	17-37	32-84	60-84
AST (SGOT)	IU/L	9-54	5-42	189-385	48-100	9-113	98-278
Amylase	IU/L	409-1250	496-1940	9-34	12-107	--	--
Bicarbonate	mEq/L	18-24	17-21	20-28	17-29	18-27	20-25
Bile acids – fast	μmol/L	<5	<2	<15	--	--	--
Postprandial	μmol/L	<15	<15	--	--	--	--
Bilirubin (total)	mg/dl	0.10-0.30	0.10-0.30	0.3-3.10	0.04-0.74	0-0.6	0.1-0.39
Direct	mg/dl	0.06-0.12	0.05-0.07	0.0-0.50	0-0.3	0-0.3	0-0.12
Indirect	mg/dl	0.04-0.18	0.05-0.23	0.2-3.00	0.04-0.44	0-0.3	0.1-0.27
BUN	mg/dl	8-30	15-33	11-27	10-26	8-24	18-31
Calcium	mg/dl	9.4-11.8	8.8-11.7	11.0-13.9	7.9-10.0	8-12	10.4-13
Chloride	mEq/L	106-116	110-125	99-105	94-104	100-105	98-115
Cholesterol	mg/dl	82-355	38-186	77-258	87-254	36-54	50-140
Cholinesterase	IU/L	1347-2269	1000-2000	--	--	--	--
CO_2 (content)	mEq/L	14-28	13-22	24-31	24-32	18-26	21-28
Pco_2	mmHg	38	36	42	40	--	41
Cortisol (basal)	μg/dl	1.0-6.8	0.3-2.6	--	--	--	--
CPK (CK)	IU/L	22-422	59-527	58-524	44-228	24-225	81-129
Creatinine	mg/dl	0.6-2.0	0.9-2.1	1.0-1.9	0.7-1.1	1.0-2.7	1.2-1.9
Fibrinogen	g/L	1-4	1-3	1-5	2-7	1-5	1-5
Folate	μg/L	7.5-17.5	13.4-38	--	--	--	--
GGT	IU/L	2-10	0-5	5-24	20-48	--	--
Glucose	mg/dl	67-135	70-120	60-128	37-71	65-95	50-80
Hemoglobin	g/L	130-190	90-150	110-170	80-150	100-180	80-160
Hct (PCV)	%	37-54	30-47	32-47	24-46	33-50	24-49
Iron	μg/dl	84-233	65-233	74-209	57-162	91-199	166-222
LDH	IU/L	10-36	16-69	41-104	178-365	96-150	60-111
Lipase	IU/L	13-200	0-83	--	--	--	--
Magnesium	mEq/L	1.8-2.6	2.0-2.7	1.8-2.6	1.4-2.3	2.7-3.7	2.2-2.8
Osmolarity	mosm/L	291-315	292-356	282-302	--	--	--

Appendix Table II Continued

Constituent	Units	Dog	Cat	Horse	Cow	Pig	Sheep
pH	pH units	7.31-7.42	7.24-7.40	7.32-7.44	7.31-7.53	--	7.32-7.54
Phosphorus (Pi)	mg/dl	2.6-7.2	3.0-6.3	1.9-6.0	4.6-9.0	5.3-9.6	5.0-7.3
Potassium	mEq/L	3.7-5.4	3.4-5.2	2.7-4.8	4.0-5.3	4.9-7.0	4.0-6.0
Protein (total)	g/dl	5.5-7.8	6.0-8.4	5.6-7.0	5.9-7.7	7.0-8.9	6.0-7.9
Albumin	g/dl	2.8-4.0	2.2-4.0	2.4-4.0	2.7-4.3	1.9-3.3	2.4-3.9
Globulin	g/dl	2.3-4.2	2.5-5.8	2.5-4.9	2.5-4.1	5.3-6.4	3.5-5.7
SDH	IU/L	2.9-8.2	3.9-7.7	1.9-5.8	4.3-15.3	1-6	6-28
SID	mEq/L	34	33-36	29-37	40-42	39-47	38-39
Sodium	mEq/L	140-150	146-158	128-142	136-144	139-152	136-154
T3	ng/dl	85-250	85-250	--	--	--	--
T4	µg/dl	1.2-3.0	1.2-3.0	--	--	--	--
T4 (free)	ng/dl	0.7-3.0	--	--	--	--	--
TG	mg/dl	29-40	25-191	9-52	0-14	--	--
Other Miscellaneous Variables							
Body Temperature	°F	99.5-102.5	100-102.5	99-100.5	100-102.5	100.5-104	102-104
Respiration	Per min	10-30	20-30	8-16	10-30	10-20	10-20
Pulse	Per min	60-120	110-130	28-40	40-80	60-80	70-80
Urine SG		1.025	1.030	1.040	1.032	1.012	1.030
Urine pH		Acidic	Acidic	Alkaline	Alkaline	Neutral	Alkaline
Puberty	Months	6-12	7-12	12-24	6-18	5-10	6-12
Estrus cycle length	Days	--	8-30 (non-mated)	19-26	21	21	16.5
Estrus duration	Days	3-12	4-20 (with & without male)	5-7	0.75	2-3	1.5
Ovulation time (in relation to estrus)		First 1/3	24-48 hr. (postcoitus)	Last 1/3	12-16 hrs. after	Last 1/2	Last 1/2
Gestation length	Days	58-63	63	330	280	114	150
Litter size	Ave.	1-8	4-5	1	1	4-14	1-3

Data from various sources, including the **Tufts Small** and **Large Animal Hospitals**, and the **Veterinary Laboratory Medicine (Interpretation and Diagnosis)** text by **Meyer DJ**, and **Harvey JW**. It should be noted, however, that most all of the standard ranges presented in this table will vary to some degree between and within diagnostic laboratories.

Appendix Table III

Basic Physiological Units

The relationship of concentration to mass and volume can be used in a number of physiological situations (e.g., fluid and electrolyte balance, xenobiotic administration, and renal function):

Concentration = Mass/Volume (or Amount/Volume)

Units: Amount = gm, mg, moles, osmoles, milliosmoles, etc.

Volume = liter (L), ml, 100 ml (deciliter, dl or dL), etc.

Concentration = gm/L, mg/ml, milliosmoles/L, etc.

% = gm% = gm/100 ml = gm/dl

mg% = mg/100 ml = mg/dl

Knowing these relationships and two of the three values above, the third can be calculated.

Rearranging:

Volume = Amount/Concentration

This relationship taken over time yields:

Flow = Volume/Time = (Amount/Time)/Concentration

Units (example): ml/min = (mg/min)/(mg/ml)

Solutions

Molar (M) = One gram-molecular wt. made up to 1L in solvent

Millimolar (mM) = M/1000

Molal (m) = One gram-molecular weight dissoved in 1000 gm solvent

mOsmolar = mOsm/L

1 mmole NaCl = 2 mOsm

1 mmole $CaCl_2$ = 3 mOsm

Appendix Table III Continued

Milligrams/deciliter (mg/dl or mg%) can be converted to milliequivalents/L (mEq/L) as follows:

$$mEq/L = (mg/dl \times 10 \times valence)/mg \ atomic \ mass$$

Example: If the plasma Na^+ concentration = 346 mg/dl, then the plasma contains 3460 mg Na^+/L. The equivalent mass of Na^+ is 23, and the valence is 1; therefore:

$$mEq/L = (346 \times 10 \times 1) / 23 = 150$$

Units for plasma electrolytes are sometimes given as millimoles/L (mmol/L or mM), and for Na^+ would be the same as mEq/L.

IU = International Unit (i.e., a unit of biological material (e.g., enzyme, hormone, vitamin, etc.), established by the **International Conference for the Unification of Formulas**).

Prefixes representing powers of ten

Powers of ten	Prefix	Symbol
10^{12}	tera-	T
10^9	giga-	G
10^6	mega-	M
10^3	kilo-	k
10^2	hekto-	h
10	deka-	da
10^{-1}	deci-	d
10^{-2}	centi-	c
10^{-3}	milli-	m
10^{-6}	micro-	μ
10^{-9}	nano-	n
10^{-12}	pico-	p
10^{-15}	femto-	f
10^{-18}	atto-	a

Abbreviations

A

A	Alanine
A	Apoprotein-A
A⁻	Any anion
A (vitamin)	Retinol
A₁ (vitamin)	Retinal (trans)
AA	Amino acid
AAA	Aromatic amino acid
AAV	Adeno-associated virus
AcAc	Acetoacetate
ACAT	Acyl-CoA:cholesterol acyltransferase
ACE	Angiotensin-converting enzyme
ACh	Acetylcholine
AChE	Acetylcholinesterase
ACP	Acyl carrier protein
ACTH	Adrenocorticotropic hormone
Acyl-CoA	An acyl derivative of coenzyme A
ADH	Antidiuretic hormone (vasopressin)
ADP	Adenosine diphosphate
AG	Anion gap (plasma)
AIDS	Acquired immunodeficiency syndrome
Ala	Alanine
ALA	Aminolevulinic acid
Alb⁻	Anionic albumin
ALS	Amyotrophic lateral sclerosis
ALT	Alanine aminotransferase (see SGPT)
ALP	Alkaline phosphatase (see SAP)
AMP	Adenosine monophosphate
AMPS	Adenylosuccinate
Apo-A, B₄₈, B₁₀₀, C, & E	Apoproteins
APRT	Adenine-specific PRT
AR	Aldose reductase
Arg	Arginine
As	Arsenic
Asn	Asparagine
Asp	Aspartate
AST	Aspartate aminotransferase (see SGOT)
AT	Active transport and acetyl transacetylase
ATCase	Aspartate transcarbamoylase
ATP	Adenosine triphosphate
AZT	Zidovudine

B

B	Boron
(B)	Brain isozyme
B₁ (vitamin)	Thiamin
B₂ (vitamin)	Riboflavin
B₃ (vitamin)	Niacin
B₅ (vitamin)	Pantothenic acid
B₆ (vitamin)	Pyridoxine
B₁₂ (vitamin)	Cobalamin
BA	Bile acid
BAL	Dimercaprol
BBB	Blood-brain-barrier
BC	Biliary cholesterol
BCAA	Branched-chain amino acid
BCP	Biotin carrier protein
BE	Base excess
-BE	Base deficit
BGP	Bone Gla protein (osteocalcin)
BMR	Basal metabolic rate
BNP	Brain natriuretic peptide
BP	Binding protein, or blood pressure
BPG	Biphosphoglycerate (diphosphoglycerate)
Br⁻	Bromine
BS	Bile salt
BSE	Bovine spongiform encephalopathy
BUN	Blood urea nitrogen
BW	Body weight

C

C	Carbon, catalase, apoprotein-C, or constriction
C	Cysteine
C (vitamin)	L-Ascorbate
C$_\alpha$	Alpha-carbon atom of an amino acid
^{14}C	Radiolabelled carbon
Ca⁺⁺	Calcium
CA	Carbonic anhydrase and cholic acid
CAA	Carbamoyl aspartic acid
CAD	Multifunctional cytoplasmic protein
CaHPO₄	Brushite of bone
CAM	Calmodulin
cAMP	3',5'-Cyclic adenosine monophosphate, cyclic-AMP
CAP	Carbamoyl phosphate
Ca₁₀(PO₄)₆(OH)₂	Hydroxyapatite of bone
CAT	Carnitine-acylcarnitine translocase
CB	Conjugated bilirubin
CBC	Complete blood count

c-BNP	Cardiac brain natriuretic peptide
CCK	Cholecystokinin
CCl$_4$	Carbon tetrachloride
CDC	Chenodeoxycholate
CDP	Cytidine diphosphate
CE	Cholesterol ester
Cer	Ceramide
CETP	Cholesterol ester transfer protein
cGMP	3',5'-Guanosine monophosphate, cyclic-GMP
CH	Cholesterol
CHE	Cholesterol ester
CK$_c$ (CPK$_c$)	Cytoplasmic creatine kinase (or phosphokinase)
CK$_g$ (CPK$_g$)	Cytoplasmic creatine kinase (or phosphokinase) associated with anaerobic glycolysis
CK$_m$ (or CPK$_m$)	Mitochondrial creatine kinase (or phosphokinase)
Cl$^-$	Chloride
ClO$_4^-$	Perchlorate
CM	Chylomicron
CMC	Critical micelle concentration
CMP	Cytidine monophosphate; 5'-phosphoribosyl cytosine
Cn$^-$	Cyanide
CN	Cyano(cobalamin)
CNZ	Condensing enzyme
cNMP	Cyclic nucleoside monophosphate (e.g., 3',5' cAMP)
CNS	Central nervous system
Co	Cobalt
^{60}Co	Radiolabelled cobalt
CO	Carbon monoxide or cardiac output
CO$_2$	Carbon dioxide
CO$_3^=$	Carbonate
CoA	Coenzyme A
CoA.SH	Free (uncombined) coenzyme A containing pantothenate (vitamin B$_5$)
CoQ	Coenzyme Q
COX	Cyclooxygenase
C~PO$_3$	Creatine phosphate
CPK	Creatine phosphokinase (see CK)
CPS-1	Carbamoyl phosphate synthetase-1 (mitochondrial)
CPS-2	Carbamoyl phosphate synthetase-2 (cytoplasmic)
CPT-1	Carnitine palmitoyltransferase I
CPT-2	Carnitine palmitoyltransferase II
Cr	Chromium
CSF	Cerebrospinal fluid
CT	Calcitonin
cTn-I	Cardiac troponin-I
CTP	Cytidine triphosphate

Cu^{++}	Copper
CV	Cardiovascular
Cys	Cysteine
Cyt c	Cytochrome c

D

D	Aspartate
D-	Dextrorotatory
D$_1$ (vitamin)	Impure preparation of D$_2$
D$_2$ (vitamin)	Ergocalciferol
D$_3$ (vitamin)	Cholecalciferol
1,25(OH)$_2$D$_3$ (vitamin)	Calcitriol (1,25-Dihydroxycholecalciferol; 1,25-DHC)
25(OH)D$_3$ (vitamin)	Calcifediol
dA	Deoxyadenosine
DAG	Diacylglycerol
DBP	Vitamin D binding protein
dC	Deoxycytidine
DC	Deoxycholate
DCA	Dicarboxylic acid
DCT	Distal convoluted tubule (kidney)
dFUMP	Deoxyfluorouracil monophosphate
dG	Deoxyguanosine
DG	Diglyceride
DHA	Dehydroascorbate
DHAP	Dihydroxyacetone phosphate
DHFA	Dihydrofolate (H$_2$ folate)
DHOA	Dihydroorotic acid
DHT	Dihydrotachysterol (Reduced D$_2$) or dihydrotestosterone
DI	Diabetes insipidus
DIPF	Diisopropylphosphofluoridate
DIT	Diiodotyrosine
DKA	Diabetic ketoacidosis
dl	Deciliter (100 ml)
DM	Diabetes mellitus
DMT-1	Apical iron transporter
DNA	Deoxyribonucleic acid
DNP	Dinitrophenol
dNDP	Deoxyribonucleoside diphosphate
dNMP	Deoxyribonucleoside monophosphate
DPG	Diphosphoglycerate (bisphosphoglycerate)
DSH	Domestic shorthair
dT	Deoxythymidine
dTMP	Deoxythymidine 5'-monophosphate
dUDP	Deoxyuridine diphosphate
dUMP	Deoxyribose uridine 5'-phosphate

E

E	Enzyme (also Enz), or apoprotein-E
E	Glutamate
e^-	Electron
E (vitamin)	Tocopherol
ECF	Extracellular fluid
ECG (EKG)	Electrocardiogram
ECW	Extracellular water
EGF	Epidermal growth factor
EHC	Enterohepatic circulation
eicosa	Twenty carbon
enoic	Containing double bonds
EMP	Embden-Meyerhoff pathway (glycolysis)
Enz	Enzyme (also E)
Epi	Epinephrine
EPO	Erythropoietin
Eq (or eq)	Equivalent
ER	Endoplasmic reticulum
ES	Enzyme-substrate complex
ETC	Electron transport chain

F

F	Fluorine (F^- fluoride)
F	Phenylalanine
FA	Folic acid (folate or pteroylglutamate)
FAs	Fatty acids
FAD	Flavin adenine dinucleotide (oxidized)
$FADH_2$	Flavin adenine dinucleotide (reduced)
FAR	Folic acid reductase
FAS	Fatty acid synthetase
Fe^{++}	Ferrous iron
Fe^{+++}	Ferric iron
FeS	Iron-sulfur complex
FFA	Free fatty acid
FH	Familial hypercholesterolemia
FLUTD	Feline lower urinary tract disease
FMN	Flavin mononucleotide (riboflavin 5′-monophosphate)
F-6-P	Fructose 6-phosphate
Fru	Fructose
FSH	Follicle stimulating hormone
5-FU	5′-Fluorouracil
Fuc	Fucose
FUdR	5′Fluoro 2′-deoxyuridine
FUS	Feline urologic syndrome
FXR	Farnesoid X receptor

G

G	Glycine
g, gm, or Gm	Gram
G$_i$	G-inhibitory protein
G$_s$	G-stimulatory protein
G$_{t1}$	Transducin
GABA	Gamma-aminobutyric acid (4-aminobutyrate)
GABA-AT	GABA-aminotransferase
Gal	Galactose
GalNAc	N-Acetylgalactosamine
GC	Glycocholate
GCDC	Glycochenodeoxycholate
GDP	Guanosine diphosphate
GFR	Glomerular filtration rate
GGT	Gamma-glutamyl transferase
GGT	γ-Glutamyl transpeptidase or γ-glutamyl transferase
GH	Growth hormone
GI	Gastrointestinal
GK	Glucokinase
Gla	γ-Carboxyglutamate
Glc	Glucose
Glc-1-P	Glucose 1-phosphate
Glc-6-P	Glucose 6-phosphate
Glc-6-Pase	Glucose 6-phosphatase
Glc-6-PD	Glucose 6-phosphate dehydrogenase
GlcNAc	N-Acetylglucosamine
GlcUA	Glucuronic acid
GLDH	Glutamate dehydrogenase
Gln	Glutamine
Gl-3-P	Glyceraldehyde 3-phosphate
Glu	Glutamic acid
GLUT	Glucose transporter
Gly	Glycine
G$_{m1}$-G$_{m4}$	Gangliosides
gm% (or %)	Gram percent (gm/100 ml)
GMP	Guanosine monophosphate
GP	Glutathione peroxidase
GR	Glutathione reductase
GSH	Glutathione (reduced)
GSSG	Glutathione (oxidized)
GTP	Guanosine triphosphate

H

(H)	Heart isozyme
H	Histidine

H⁺	Hydrogen ion (proton)
³H	Tritium
HA	Undissociated acid (inorganic or organic)
Hb⁻	Deoxygenated hemoglobin
HbA$_{1c}$	Hemoglobin-A$_{1c}$ (glycosylated adult Hb)
HbF	Fetal hemoglobin
HbO$_2$⁻ (or HHbO$_2$)	Oxygenated hemoglobin
HCl	Hydrochloric acid
HCO$_3$⁻	Bicarbonate
Hct	Hematocrit
HDL	High density lipoprotein
HE	Hepatoencephalopathy
HETE	Hydroxyeicosatetraenoic acid
H$_2$ Folate	Dihydrofolate (see DHFA)
H$_4$ Folate	Tetrahydrofolate (see THFA)
HGPRT	Hypoxanthine and guanine-specific PRT
HHb	Protonated hemoglobin
His	Histidine
HK	Hexokinase
HL	Hepatic lipase
HMG-CoA	3-Hydroxy-3-methylglutaryl-CoA
HMS	Hexose monophosphate shunt
H$_2$O	Water
H$_2$O$_2$	Hydrogen peroxide
HPETE	Hydroperoxyeicosatetraenoic acid
HPO$_4$⁼/H$_2$PO$_4$⁻	Dibasic/monobasic phosphate
HProt	Undissociated protein
HQ	Hydroquinone (vitamin K, active form)
H$_2$S	Hydrogen sulfide
HSL	Hormone sensitive lipase (adipolytic TG lipase)
HT	Hypertension
5-HT	Serotonin (5-hydroxytryptamine)
HU	Hydroxyurea
HX	Hypoxanthine
I	
I	Iodine (I⁻ iodide)
I	Isoleucine
¹³¹I	Radiolabelled iodine
IBD	Inflammatory bowel disease
ICD	Isocitrate dehydrogenase
ICF	Intracellular fluid
ICW	Intracellular water
IDL	Intermediate-density lipoprotein
IDP	Inosine diphosphate

IF	Intrinsic factor
IGF	Insulin like growth factor
IL	Interleukin
Ile	Isoleucine
IMP	Inosine monophosphate (hypoxanthine ribonucleotide)
IP$_3$	Inositol triphosphate
ITP	Inosine triphosphate
IU	International unit(s)

J
JG	Juxtaglomerular

K
K	Lysine
k$_1$	First rate constant
K$^+$	Potassium
K'	"Apparent" dissociation or ionization constant
K (vitamin)	Hydroquinone
K$_1$ (vitamin)	Phylloquinone
K$_2$ (vitamin)	Menaquinone
K$_3$ (vitamin)	Menadione
KB$^-$	Ketone body anion
kCal	Kilocalorie
kg	Kilogram
α-KG$^=$	Alpha-ketoglutarate
K$_m$	Substrate concentration producing half-maximal velocity (Michaelis constant)

L
L	Leucine
L	Liter
L-	Levorotatory
LBM	Lean body mass
LC	Lithocholate
LCAD	Long-chain acyl-CoA dehydrogenase
LCAT	Lecithin:cholesterol acyltransferase
LCFA	Long-chain fatty acid
LDH	Lactate dehydrogenase
LDL	Low-density lipoprotein
LES	Lower esophageal sphincter
Leu	Leucine
LH	Luteinizing hormone
Li	Lithium
log	Log to the base 10
LOH	Loop of Henle
LP	Lipoprotein
LPL	Lipoprotein lipase

LTA₄-LTE₄	Leukotrienes
Lys	Lysine
M	
M	Methionine
M	Molar
(M)	Muscle isozyme or methyl
MA	Megaloblastic anemia
Mal	Malate
MAO	Monoamine oxidase
Man	Mannose
MCAD	Medium-chain acyl-CoA dehydrogenase
MCFA	Medium-chain fatty acid
MD	Malate dehydrogenase
ME	Malic enzyme
mEq	Milliequivalent (10^{-3})
Met	Methionine
MetHb	Methemoglobin
Mg⁺⁺	Magnesium
mg%	Milligrams/100 ml
MG	Monoglyceride
MI	Myocardial infarction
MIT	Monoiodotyrosine
mM	Millimolar
mmHg	Millimeters of mercury
Mn⁺⁺	Manganese
Mo	Molybdenum
mol	Mole(s)
4-MP	4-Methylpyrazole
mph	Miles per hour
MR	Methemoglobin reductase
mRNA	Messenger RNA
MSH	Melanocyte stimulating hormone
MT	Malonyl transacetylase
MTX	Methotrexate
MW	Molecular weight
N	
N	Asparagine
N	Nitrogen
Na⁺	Sodium
NAG	N-acetyl-β-D-glucosaminidase
NaUA	Sodium urate
NAD⁺	Nicotinamide adenine dinucleotide (oxidized)
NADH	Nicotinamide adenine dinucleotide (reduced)
NADPH	Nicotinamide adenine dinucleotide phosphate (reduced)

NANA	N-Acetylneuraminic acid (NeuAc or sialic acid)
NDI	Nephrogenic diabetes insipidus
NDP	Nucleoside 5'-diphosphate
NE	Norepinephrine
NEFA	Non-esterified fatty acid
nEq	Nanoequivalents (10^{-9})
NeuAc	N-Acetylneuraminic acid (NANA or sialic acid)
NGF	Nerve growth factor
NH$_2$	Amine
NH$_3$	Ammonia
NH$_4^+$	Ammonium ion
Ni	Nickel
NMN	Nicotinate mononucleotide
NMP	Nucleoside 5'-monophosphate
NMR	Nuclear magnetic resonance
NO	Nitric oxide
NP	Nucleoside phosphate (nucleotide)
NPN	Non-protein nitrogen (e.g., urea)
NSAID	Nonsteroidal anti-inflammatory drug
NT	Neurotransmitter
NTP	Nucleoside 5'-triphosphate
NT-proBNP	N-Terminal-pro brain natriuretic peptide

O	
O$_2^-\cdot$	Superoxide anion (oxygen free radical)
OA	Orotic acid
OAA	Oxaloacetate
Obs	Observed
OH$^-$	Hydroxyl
OH·	Hydroxyl radical
Oligos	Oligonucleotides or oligosaccharides
OMP	Orotidine monophosphate

P	
P	Proline
P	Product or phosphorus
(P)	Propionyl
^{31}P	Isotope of phosphorus
PABA	Para-aminobenzoic acid
PAF	Platelet activating factor
PAPS	3'-Phosphoadenosine-5'-phosphosulfate
Pb$^-$	Lead
PBG	Porphobilinogen
Pco$_2$	CO_2 partial pressure
PCSK9	Proprotein convertase subtilisen/kexin type 9

PCT	Proximal convoluted tubule (kidney)
PCV	Packed cell volume
PD	Potential difference
PDH	Pyruvate dehydrogenase
peATP	Pre-existing ATP
Pi	Inorganic phosphate
peGTP	Pre-existing GTP
PEP	Phosphoenolpyruvate
PFK	Phosphofructokinase
15-PGDH	15-Hydroxyprostaglandin dehydrogenase
PGs (e.g., PGF$_2$)	Prostaglandins
PGI$_2$	Prostacyclin
3-PG	3-Phosphoglycerate
PGAT	PRPP glutamyl amidotransferase
PGM	Phosphoglucomutase
pH	-log [H$^+$]
Phe	Phenylalanine
pK'	pH at which the protonated (HA) and unprotonated (A$^-$) species of a weak acid are present at equal concentrations.
PKA	Protein kinase A
PKC	Protein kinase C
PL	Placental lactogen and phospholipid
PLA$_1$	Phospholipase A$_1$
PLA$_2$	Phospholipase A$_2$
PLases	Phospholipases
PLC	Phospholipase C
PLD	Phospholipase D
PNS	Parasympathetic nervous system
PO$_4^=$	Phosphate
pols	Polymerases (e.g., DNA polymerase)
PPi (or ppi)	Inorganic pyrophosphate
PRA	Phosphoribosylamine
PRL	Prolactin
Pro	Proline
Prot$^-$	Anionic protein
Prot$_{tot}$	Total plasma protein
PrP	Prion protein
PRPP	5'-Phosphoribosyl-1-pyrophosphate
PRS	PRPP synthetase
PrPSc	Infective prion isoform
PRT	Phosphoribosyl transferase
PSS	Portosystemic vascular shunt
PTC	Phosphatidylcholine
PTH	Parathyroid hormone
PTU	Propylthiouracil

PUFA	Polyunsaturated fatty acid
PU/PD	Polyuria/polydipsia
Q	
Q	Glutamine
R	
R	Arginine
R	Ribose, receptor, or relaxation
R-	Side chain
RBC	Red blood cell (erythrocyte)
RBP	Retinol binding protein
RDR	Ribonucleoside diphosphate reductase
RE cell	Reticuloendothelial cell (mononuclear phagocyte)
RER	Rough endoplasmic reticulum
RHO	Rhodopsin
RNA	Ribonucleic acid
rRNA	Ribosomal RNA
ROO·	Peroxyl free radical
RQ	Respiratory quotient
RTA	Renal tubular acidosis
S	
S	Serine
S	Substrate, sulfur, or solubility coefficient for CO_2 (i.e., 0.03 mmol/L/mmHg)
SAA	Serum amyloid A
SAH	S-Adenosylhomocysteine
SAM	S-Adenosylmethionine
SAP	Serum alkaline phosphatase (see ALP)
SCAD	Short-chain acyl-CoA dehydrogenase
SCFA	Short-chain fatty acid
SDH	Sorbitol dehydrogenase
Se	Selenium
Ser	Serine
SER	Smooth endoplasmic reticulum
SFA	Saturated fatty acid
SG	Specific gravity
SGLT	Sodium-dependent glucose transporter
SGOT	Serum glutamate oxaloacetate transaminase (see AST)
SGPT	Serum glutamate pyruvate transaminase (see ALT)
-SH	Sulfhydryl
Si	Silicon
SID	Strong ion difference ($[Na^+] - [Cl^-]$)
Sn	Tin
SNS	Sympathetic nervous system

SO$_4^=$	Sulfate
SOD	Superoxide dismutase
SRS-A	Slow-reacting substance of anaphylaxis
-S-S-	Disulfide
SVCT	Sodium-coupled vitamin C transporter

T

T	Threonine
T$_3$	Triiodothyronine
rT$_3$	Reverse T$_3$
T$_4$	Tetraiodothyronine
Tau	Taurine
TBG	Thyroid binding globulin
TBW	Total body water
TC	Transcobalamin
TC	Taurocholate or transcobalamin
TCA	Tricarboxylic acid
TCDC	Taurochenodeoxycholate
TcO$_4^-$	Pertechnetate
TG	Triglyceride (triacylglycerol)
THFA	Tetrahydrofolate (H$_4$ folate)
Thr	Threonine
TMAO	Trimethylamine-N-oxide
TML	Trimethyllysyl
TMP	Thymidine monophosphate
Tn-c	Troponin-C
TNF	Tumor necrosis factor
Tn-I	Troponin-I
Tn-T	Troponin-T
TP	Transepithelial potential
TPA	12-O-Tetradecanoylphorbol-13-acetate
TRH	Thyrotropin-releasing hormone
TR	Thioredoxin reductase
tRNA	Transfer RNA
Trp	Tryptophan
TS	Thymidylate synthase
TSE	Transmissible spongiform encephalopathy
TSH	Thyroid-stimulating hormone (thyrotropin)
TTFA	Fe-chelating agent
TTP	Thymidine triphosphate
TXA$_2$, TXB$_2$	Thromboxanes
Tyr	Tyrosine

U

UA	Uric acid
UA$^-$	Unidentified anion

UAG	Urinary anion gap
UCB	Unconjugatod bilirubin
UDP	Uridine diphosphate
UDPGal	UDP-galactose
UDPGlc	UDP-glucose
UDPGluc	UDP-glucuronic acid
UFA	Unsaturated fatty acid
UGT	UDP-glucuronosyltransferase
UMP	Uridine monophosphate
UTI	Urinary tract infection
UTP	Uridine triphosphate
UV	Ultraviolet
V	
V	Valine
V	Reaction velocity or vanadium
(V)	Vinyl
Val	Valine
VCO$_2$	Rate of CO_2 production
VFA	Volatile fatty acid
Vit E-O·	Phenoxy free radical
VLDL	Very low-density lipoprotein
V$_{max}$	Maximal velocity
V̇O$_{2(max)}$	Maximal rate of O_2 consumption
W	
W	Tryptophan
WBC	White blood cell
WHHL	Watanabe heritable-hyperlipidemic (rabbit)
WHWT	West Highland white terrier
wt	Weight
X	
X	Xanthine
x-Axis	Abscissa
XMP	Xanthosine monophosphate
XO	Xanthine oxidase
Xyl	Xylose
Y	
Y	Tyrosine
y-Axis	Ordinate
Z	
Zn^{++}	Zinc
ZP	Zona pellucida

References

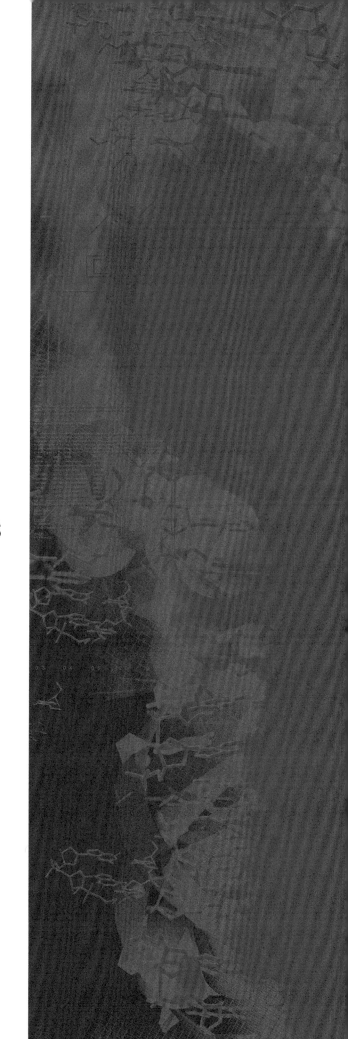

Abelow B: Understanding acid-base, 1st ed, Media, PA: Williams & Wilkins, 1998.

Adair GS, Barcroft J, Bock AV: The identity of haemoglobin in human beings. J Physiol (London) 55:332, 1921.

Alberts B, et al: Molecular biology of the cell, Garland, 1983.

Ahima RE, et al: Leptin regulation of neuroendocrine systems. Front Neuroendocrinol 21:51, 2000.

Anwer MS, Engelking LR: Intracellular pH regulation - Ch 24. In, Hepatic transport and bile secretion: physiology and pathophysiology, 1st ed, edited by PD Berk & N Tavoloni, Raven Press, Ltd, New York, NY, pgs 351-361, 1993.

Anwer MS, Engelking LR, Gronwall R, et al: Plasma bile acid elevation following CCl_4-induced liver damage in dogs, sheep, calves and ponies. Res Vet Sci 20:127, 1976.

Anwer MS, Gronwall R, Engelking LR, et al: Bile acid kinetics and bile secretion in the pony. Am J Physiol 229:592, 1975.

Ahlborg G, Felig P, Hagenfeldt L, et al: Substrate turnover during prolonged exercise in man. J Clin Invest 53:1080, 1974.

Allen EE, Bartlett DH: Structure and regulation of the omega-3 polyunsaturated fatty acid synthase genes from the deep-sea bacterium *Photobacterium profundum* strain SS9. Microbiology 148:1903, 2002.

Alonso FRA: Distribution of the cardiac output in the greyhound. Thesis, University of Pennsylvania Library, 1972.

Bartley W, Birt LM, Banks P: The biochemistry of the tissues, 1st ed, New York, NY: John Wiley & Sons, 1968.

Bartley W, Kornberg HL, Quayle JR: Essays in cell metabolism, 1st ed, New York, NY: John Wiley & Sons, 1970.

Bauer JE: Diet-induced alterations of lipoprotein metabolism. JAVMA 201:1691, 1992.

Beale EG, Hammer RE, Antoine B, et al: Glyceroneogenesis comes of age. FASEB J 16:1695, 2002.

Benesch R, Benesch RE. The effect of organic phosphates from the human erythrocyte on the allosteric properties of haemoglobin. Biochem Biophs Res Commun 26:162, 1967.

Bennion LJ, Grundy SM: Effects of diabetes mellitus on cholesterol metabolism in man. N Engl J Med 296:1365, 1977.

Berg JM, Tymoczko JL, Stryer L: Biochemistry, 5th ed, New York, NY: WH Freeman, 2001.

Bergman EN: Disorders of carbohydrate and fat metabolism. In, Dukes' physiology of domestic animals, 10th ed, Swenson MS, editor, Ithaca, NY: Cornell Univ Press, 1984.

Bohinski RC: Modern concepts in biochemistry, 5th ed, Englewood Cliffs, NJ: Prentice Hall, 1987.

Bohr C, Hasselback MA, Krogh AS: Scand Arch Physiol 16:402, 1904.

Botion LM, Brito MN, Brito NA, et al: Glucose contribution to *in vivo* synthesis of glyceride-glycerol and fatty acids in rats adapted to a high-protein, carbohydrate-free diet. Metabolism 47:1217, 1998.

Bretscher MS: The molecules of the cell membrane. Sci Am 253(4):100, 1985.

Briggs T, Chandler AM: Biochemistry, 1st ed, New York, NY: Springer-Verlag, 1987.

Brown MS, Goldstein JL: A receptor-mediated pathway for cholesterol homeostasis. Science 232:34, 1986.

Cahill GF: Starvation in man. J Engl J Med 282:668, 1970.

Cahill GF: Role of T_3 in fasted man. Life Sci 28:1721, 1981.

Cahill GF, Owen OE: Starvation and survival. Trans Am Clin Climat Assn 79:13, 1967.

Caldwell BV, Behrman HR: Prostaglandins in reproductive processes. Med Clin N Am 65:927, 1981.

Caldwell BV, Tillson SA, et al: Prostaglandins 1:217, 1972.

Carew TE, Covell JW: Left ventricular function in exercise-induced hypertrophy in dogs. Am J Cardiol 42:82, 1978.

Chanutin A, Curnish RR: Effect of organic and inorganic phosphates on the oxygen equilibrium of human erythrocytes. Arch Biochem Biophys 121:96, 1967.

Chiang JYL: Bile acid regulation of gene expression: Roles of nuclear hormone receptors. Endocr Rev 23:443, 2002.

Chiang JYL: Bile acid metabolism and signaling. Comp Physiol 3:1191, 2013.

Cipriani S, Mencarelli A, Palladino G, et al: FSR activation reverses insulin resistance and lipid abnormalities, and protects against liver steatosis in Zucker (fa/fa) obese rats. J Lipid Res 51:771, 2010.

Clarenburg R: Physiological chemistry of domestic animals, 1st ed, St. Louis, MO: Mosby-Year Book, 1992.

Coker RH, Simonsen L, Bulow J, et al: Stimulation of splanchnic glucose production during exercise in humans contains a glucagon-independent component. Am J Physiol 280:E918, 2001.

Coffee CF: Metabolism, 1st ed, Madison, CT: Fence Creek, 1998.

Cogan MC: Fluid and electrolytes, 1st ed, Norwalk, CN: Appleton & Lange, 1991.

Conn EE, Stumpf PK: Outlines of biochemistry, 1st ed, New York, NY: John Wiley & Sons, 1963.

Costanzo LS: Physiology, Cases and Problems, 4th ed, Lippincott Williams & Wilkins, 2012.

Costill DL, Daniels J, Evans W, et al: Skeletal muscle enzymes and fibre composition in male and female track athletes. J Applied physiol 40:149, 1976.

Cotter SM: Hematology, 1st ed, Jackson Hole, WY: Teton NewMedia, 2001.

Courtice FC. The blood volume of normal animals. J Physiol 102:290, 1943.

Cox RH, Peterson LH, Detweiler KD: Hemodynamics in the mongrel dog and the racing greyhound. Am J Physiol 230:211, 1976.

Cunningham JG, Klein BG: Textbook of veterinary physiology, 4th ed, Philadelphia, PA: WB Saunders/Elsevier Science, 2007.

Darnell J, Lodish H, Baltimore D: Molecular cell biology, 2nd ed, New York, NY: Scientific American Books, 1990.

Davenport HW: The ABC of acid-base chemistry, 6th ed, University of Chicago Press, 1974.

De Fabiani E, Mitro N, Gilardi F, et al: Coordinated control of cholesterol catabolism to bile acids and of gluconeogenesis via a novel mechanism of transcription regulation linked to the fasted-to-fed cycle. J Biol Chem 278:39124, 2003.

de Morais HSA: A nontraditional approach to acid-base disorders. In: DiBartola SP, ed: Fluid therapy in small animal proctice, Philadelphia, PA: WB Saunders Co, 1992.

Detweiler DK, Cox RH, Alonso FR, et al: Hemodynamic characteristics of the young adult greyhound. Fed Proc 33:360, 1974.

Devlin TM: Textbook of biochemistry with clinical correclations, 6th ed, Hoboken, NJ: John Wiley & Sons, 2005.

DiMauro S, Bresolin N, Hays AP: Disorders of glycogen metabolism of muscle. CRC Crit Rev Clin Neurol 1:85, 1984.

Dobson CM: Protein folding. Phil Trans R Soc Lond B348:1-119, 1995.

Dobson CM: Protein misfolding, evolution and disease. Trends Biochem Sci 24:329-332, 1999.

Donald DW, Ferguson DA: Response of heart rate, oxygen consumption, and arterial blood pressure to graded exercise in dogs. Proc Soc Exp Biol Med 121:626, 1966.

DuPraw E: Cell and molecular biology, 1st ed, New York, NY: Academic Press, 1968.

Engelking LR: Biochemical and physiological manifestations of acute insulin withdrawal, Sem Vet Med Surg 14(4):230, 1997.

Engelking LR: Disorders of bilirubin metabolism in small animal species. Comp Contin Educ Pract Vet 10:712, 1988.

Engelking LR: Equine fasting hyperbilirubinemia - Ch 9. In, Advances in veterinary science and comparative medicine; animal models in liver research, Vol 37, Cornelius CE, editor, Academic Press, Inc, 1993.

Engelking LR: Evaluation of equine bilirubin and bile acid metabolism. Comp Contin Educ Pract Vet 11:328, 1989.

Engelking LR: Metabolic and endocrine physiology, 3rd ed, Jackson Hole, WY: Teton NewMedia, 2012.

Engelking LR: Physiology of the endocrine pancreas, Sem Vet Med Surg 14(4):224, 1997.

Engelking LR: Review of veterinary physiology, 1st ed, Jackson Hole, WY: Teton NewMedia, 2002.

Engelking LR, Anwer MS: Liver and biliary tract - Ch 1 of section II - Pathophysiologic mechanisms and metabolic complications. In, Veterinary gastroenterology, 2nd ed, edited by Anderson NV, Lea & Febiger, 1992.

Engelking LR, Anwer MS, Hofmann AF: Basal and bile salt-stimulated bile flow and biliary lipid excretion in ponies. Am J Vet Res 50:578, 1982.

Engelking LR, Anwer MS, McConnel J, et al: Cocaine and lidocaine interfere with epinephrine-induced changes in intracellular calcium concentration and glucose efflux from rat hepatocytes. Int J Exp Clin Pharmacol 40:129, 1990.

Engelking LR, Barnes S, Dasher CA, et al: Radiolabeled bile acid clearance in control subjects and patients with liver disease. Clin Sci 57:499, 1979.

Engelking LR, Barnes S, Hirschowitz BI, et al: Determination of the pool size and synthesis rate of bile acids by measurement in blood of patients with liver diseases. Clin Sci 58:485, 1980.

Engelking LR, Dasher CA, Hirschowitz BI: Within-day fluctuations in serum bile-acid concentrations among normal subjects and patients with liver disease. Am J Clin Path 73:196, 1980.

Engelking LR, Dodman NH, Hartman G, et al: Effects of halothane anesthesia on equine liver function. Am J Vet Res 45:607, 1984.

Engelking LR, Gronwall R: Bile acid clearance in sheep with hereditary hyperbilirubinemia. Am J Vet Res 40:1277, 1979.

Engelking LR, Gronwall R: Effects of fasting on hepatic bile acid clearance. Proc Soc Exp Biol Med 161:123, 1979.

Engelking LR, Gronwall R, Anwer MS: Effect of bile acid on hepatic excretion and storage of bilirubin in ponies. Am J Vet Res 37:47, 1976.

Engelking LR, Gronwall R, Anwer MS: Effects of dehydrocholic, chenodeoxycholic, and taurocholic acids on the excretion of bilirubin. Am J Vet Res 41:355, 1980.

Engelking LR, Mariner JC: Enhanced biliary bilirubin excretion after heparin-induced erythrocyte mass depletion. Am J Vet Res 46:2175, 1985.

Engelking LR, Milliken GA, Smith JE: Studies on glucose-6-phosphate dehydrogenase: comparison of kinetic constants among different breeds of sheep. Am J Vet Res 35:1313, 1974.

Engelking LR, Paradis MR: Evaluation of hepatobiliary disorders in the horse - Ch 7 - In, Veterinary clinics of north america: equine practice, Ramanauskas D, Brobst DF, editors, WE Saunders, Vol 3(3):562, 1987.

Erickson HH: Exercise physiology - Ch 15 of part II - Respiration and exercise. In, Dukes' physiology of domestic animals, 11th ed, edited by Swenson MJ, and Reece WO, Ithaca, NY: Cornell Univ Press, 1993.

Ettinger SJ, Feldman EC: Textbook of veterinary internal medicine, 4th ed, Philadelphia, PA: WB Saunders Co, 1995.

Evans DL, Rose RJ: Cardiovascular and respiratory responses to submaximal exercise training in the thoroughbred horse. Pflügers Arch 411:316, 1988.

Fang Y, Studer E, Mitchell C, et al: Conjugated bile acids regulate hepatocyte glycogen synthase activity in vitro and in vivo via Galphai signaling. Mol Pharmacol 71:1122, 2007.

Febbraio MA, Pedersen BK: Muscle-derived interleukin-6 mechanisms for activation and possible biological roles. The FASEB J 16:1335, 2002.

Felig P, Wahren J: Fuel homeostasis in exercise. New Engl J Med 293:1078, 1975.

Fencl V, Gabel RA: Respiratory adaptations in acid-base disturbances. Contrib Nephrol 21:145, 1980.

Fencl V, Leith DE: Frontiers in respiratory physiology; Stewart's quantitative acid-base chemistry:applications in biology and medicine. Resp Physiol 91:1, 1993.

Fink WJ, Costill DL, Pollock ML: Submaximal and maximal working capacity of elite distance runners: muscle fiber composition and enzyme activities. Annals NY Acad Sci 301:323, 1977.

Fiorucci S, Baldelli F: Farnesoid X receptor agonists in biliary tract disease. Curr Opin Gastroenterol 25:252, 2010.

Francis GA, Fayard E, Picard F, et al: Nuclear receptors and the control of metabolism. Annu Rev Physiol 65:261, 2003.

Fredrickson DS, Gordon RS: Transport of fatty acids. Physiol Rev 38:585, 1958.

Friedman PJ: Biochemistry, 4th ed, Boston, MA: Little, Brown & Co, 1992.

Galbo H, Holst JJ, Christensen NJ: Glucagon and plasma catecholamine response to graded and prolonged exercise in man. J Applied Physiol 38:70, 1975.

Galbo H, Holst JJ, Christensen NJ: The effect of different diets and of insulin on the hormonal response to prolonged exercise. Acta Physiol Scand 107:19, 1979.

Ganong WF: The renin-angiotensin system and the central nervous system. Fed Proc 36:1771, 1977.

Ganong WF: Review of medical physiology, 20th ed, New York, NY: Lange Medical Books/McGraw-Hill, 2001.

Geissman TA: Principles of organic chemistry, 1st ed, San Francisco, CA: WH Freeman, 1959.

Giger U, Harvey JW: Hemolysis caused by phosphofructokinase deficiency in English Springer Spaniels: seven cases (1983-1986). JAVMA 191(4):453, 1987.

Gleeson M: Interleukins and exercise. J Physiol (London) 529:1, 2000.

Goldstein JL, Brown MS: Familial hypercholesterolemia. In, The metabolic basis of inherited disease, 5th edition. Stanbury JB, Wyngaarden DS, Fredrickson DS, Goldstein JL, Brown MS, editors, New York, NY: McGraw-Hill, 1983.

Gorin E, Tal-Or A, Shafrir E: Glyceroneogenesis in adipose tissue of fasted, diabetic and triamcinolone treated rats. Eur J Biochem 8:370, 1969.

Gow AJ, Luchsinger BP, Pawloski JR, et al: The oxyhemoglobin reaction of nitric oxide. Proc Natl Acad Sci USA 96:9027, 1999.

Gray GM: Carbohydrate digestion and absorption. N Eng J Med 292:1225, 1975.

Green A, Newsholme EA: Sensitivity of glucose uptake and lipolysis of white adipocytes of the rat to insulin and effects of some metabolites. Biochem J 180:365, 1979.

Greenstein B, Greenstein A: Medical biochemistry at a glance, 1st ed, Oxford, England: Blackwell Science, 1996.

Gronwall R, Engelking LR: Effect of glucose administration on equine fasting hyperbilirubinemia. Am J Vet Res 43:801, 1982.

Greville GD, Tubbs PK: The catabolism of long chain fatty acids in mammalian tissues. Essays in biochem 4:155, 1968.

Gruder WG: Renal and hepatic nitrogen metabolism in systemic acid-base regulation. J Clin Chem Clin Biochem 25:457, 1987.

Gurr MI, Harwood JL: Lipid biochemistry, 4th ed, New York, NY: Chapman & Hall, 1991.

Guy PS, Snow DH: A preliminary survey of skeletal muscle fibre types in equine and canine species. J Anat 124:499, 1977.

Hafez ESE, Dyer IA: Animal growth and nutrition, 1st ed, Philadelphia, PA: Lea & Febiger, 1969.

Hales EN, Luzio JP, Siddle K: Hormonal control of adipose tissue lipolysis. Biochem Soc Symp 43:97, 1978.

Hall GM, Lucke JN, Masheter K, et al: Metabolic and hormonal changes during prolonged exercise in the horse. In: Biochemistry of exercise IV; Poortmans J, and Niset G editors. Baltimore, University Park Press, 1981.

Hall JA: Potential adverse effects of long-term consumption of (n-3) fatty acids. Comp Contin Educ Pract Vet 18:879, 1996.

Harvey JW, Ling GV, Kaneko JJ: Methemoglobin reductase deficiency in a dog. J Am Vet Med Assoc 164:1030, 1974.

Hawk PB, Oser BL, Summerson WH: Practical physiological chemistry, 13th ed, New York, NY: McGraw-Hill, 1954.

Hediger MA: Gateway to a long life? Nature 417:393, 2002.

Hediger MA: Nature Med 8:445, 2002.

Henderson KM, McNatty KP: Prostaglandins 9:779, 1975.

Hess RS, Saunders M, VanWinkle TH, Ward CR: Concurrent disorders in dogs with diabetes mellitus: 221 cases (1993-1998). J Am Vet Med Assoc:1166, 2000.

Hopper MK: Ph.D. dissertation, Kansas State University Library, 1989.

Hopper MK, Pieschl RL, Pelletier, NG, et al: Cardiopulmonary effects of acute blood volume alteration prior to exercise. In, Equine exercise physiology 3, Persson SGB, Lindholm A, Jeffcott LB, eds, Cambridge, England: Granta Editions, 1983.

How KL, Hazelwinkel HAW, Mol JA: Dietary vitamin D dependence of cat and dog due to inadequate cutaneous synthesis of viatamin D. Gen Comp Endo 96:12, 1994.

Huang KT, et al: Modulation of nitric oxide bioavailability by erythrocytes. Proc Natl Acad Sci USA 98:11771, 2001.

Jones NL: Blood gases and acid-base physiology, 2nd ed, New York, NY: Thieme Medical Publishers, 1987.

Kanemaki T, Kitade H, Kaibori M, et al: Interleukin 1 beta and interleukin 6, but not tumor necrosis factor alpha, inhibit insulin-stimulated glycogen synthesis in rat hepatocytes. Hepatology 27:1296, 1998.

Kast HR, Nguyen CM, Sinal CJ, et al: Farnesoid X-activated receptor induces apolipoprotein C-II transcription: A molecular mechanism linking plasma triglyceride levels to bile acids. Mol Endocrinol 15:1720, 2001.

Katzung BG: Basic and clinical pharmacology, 3rd ed, Norwalk, CN: Appleton & Lange, 1987.

Koeth RA, Wang Z, Levison BS, et al: Intestinal microbiota metabolism of L-carnitine, a nutrient in red meat, promotes atherosclerosis. Nature Medicine doi:10.1038/nm.3145, Apr, 2013.

Krieger NM, Sherrard DJ: Practical fluids and electrolytes, 1st ed, Norwald, CN: Appleton & Lange, 1991.

Lane MD, Cha SH: Effect of glucose and fructose on food intake via malonyl-CoA signaling in the brain. Biochem & Biophys Res Communications 382:1, 2009.

Leaf A, Newburgh LH: Significance of the body fluids in clinical medicine, 2nd ed, Thomas, 1955.

Lefebvre P, Cariou B, Lien F, et al: Role of bile acids and bile acid receptors in metabolic regulation. Physiol Rev 89:147, 2009.

Lehninger AL: Biochemistry, 2nd ed, New York, NY: Worth Publishers, 1975.

Li T, Chanda D, Zhang Y, et al: Glucose stimulates cholesterol 7alpha-hydroxylase gene transcription in human hepatocytes. J Lipid Res 51:832, 2010.

Li T, Owsley E, Matozel J, et al: Transgenic expression of cholesterol 7alpha-hydroxylase in the liver prevents high-fat diet-induced obesity and insulin resistance in mice. Hepatology 52:678, 2010.

Liao JC: Blood feud: Keeping hemoglobin from nixing NO. Nature Med 8(12):1350, 2002.

Liao JC, Hein TW, Vaughn MW, et al: Intravascular flow decreases erythrocyte consumption of nitric oxide. Proc Natl Acad Sci USA 96:8757, 1999.

Linder MC: Nutritional biochemistry and metabolism, 2nd ed, Norwalk, CT: Appleton & Lange, 1991.

Liu X, et al: Diffusion-limited reaction of free nitric oxide with erythrocytes. J Biol Chem 273:18709, 1998.

Ma K, Saha PK, Chan L, et al: Farnesoid X receptor is essential for normal glucose homeostasis. J Clin Invest 116:1102, 2006.

MacKay EM: The significance of ketosis. J Clin Endocrinol 3:101, 1943.

Marsh MM, Walker VR, Curtiss LK, et al: Protection against atherosclerosis by estrogen is independent of plasma cholesterol levels in LDL receptor-deficient mice. J Lipid Res 40:893, 1999.

McAuliffe JJ, Lind LJ, Leith DE, et al: Hypoproteinemic alkalosis, Am J Med 81:86, 1986.

McDonald LE, Pineda MH: Veterinary endocrinology & reproduction, 4th ed, Lea & Febiger, 1989.

Meixner R, Hornicke H, Ehrlein H: Oxygen consumption, pulmonary ventilation, and heart rate of riding horses during walk, trot, and gallop. In, Biotelemetry VI, Sanson W, editor, Fayetteville, Arkansas: Univ of Arkansas Press, 1981.

Meyer DJ, Harvey JW: Veterinary laboratory medicine, interpretation and diagnosis, 3rd ed, St Louis, MO: Saunders, 2004.

Montgomery R, Conway TW, Spector AA: Biochemistry, a case-oriented approach, 5th ed, St Louis, MO: CV Mosby, 1990.

Mudaliar S, Henry RR, Sanyal AJ, et al: Efficacy and safety of the farnesoid X receptor agonist obeticholic acid in patients with type 2 diabetes and nonalcoholic fatty liver disease. Gastroenterology 145(3):574, 2013.

Murray RK, Granner DK, Mayes RA, et al: Harper's biochemistry, 25th ed, New York, NY: Lange Medical Books/ McGraw-Hill, 2000.

Murtaugh RJ: Critical care, 1st ed, Jackson Hole, WY: Teton NewMedia, 2002.

Newsholme EA, Leech AR: Biochemistry for the medical sciences, 1st ed, Chichester, England: John Wiley & Sons, 1983.

Ogilvie GK, Engelking LR, Anwer MS: Effects of storage on blood ammonia, bilirubin and urea nitrogen from cats and horses. Am J Vet Res 46:2278, 1985.

Ogilvie GK, Moore AS: Feline oncology, 1st ed, Trenton, NJ: Veterinary Learning Systems, 2001.

Paul P, Holmes WL: Free fatty acids and glucose metabolism during increased energy expenditure and after training. Med and Sci in Sports, 7:176, 1975.

Pols TW, Noriega LG, Nomura M, et al: The bile acid membrane receptor TGR5 as an emerging target in metabolism and inflammation. J Hepatol 54:1263, 2011.

Porez G, Prawill J, Gross B, et al: Bile acid receptors as targets for the treatment of dyslipidemia and cardiovascular disease: Thematic review series: New lipid and lipoprotein targets for the treatment of cardiometabolic diseases. J Lipid Res 53:1723, 2012.

Reaven G, Abbasi F, McLaughlin T: Obesity, insulin resistance, and cardiovascular disease. Recent Prog Horm Res 59:207, 2004.

Reiter, et al: Cell-free hemoglobin limits nitric oxide bioavailability in sickle-cell disease. Nature Med 8:1383, 2002.

Reshef L, Hanson RW, Ballard FJ: Glyceride-glycerol synthesis from pyruvate. Adaptive changes in phosphoenolpyruvate carboxykinase and pyruvate carboxylase in adipose tissue and liver. J Biol Chem 244:1994, 1969.

Reshef L, Hanson RW, Ballard FJ: A possible physiological role for glyceroneogenesis in rat adipose tissue. J Biol Chem 245:5979, 1970.

Rice ME: Ascorbate regulation and its neuroprotective role in the brain. Trends Neurosci 23:209, 2000.

Rice ME, Lee DJ, Choy Y: High levels of ascorbic acid, not glutathione, in the CNS of anoxia-tolerant reptiles contrasted with levels in anoxia-intolerant species. J Neurochem 64:1790, 1995.

Robinson SH, Lester R, Crigler JF Jr: Early-labeled peaks of bile pigment: Studies with glycine-^{14}C and Δ-ALA-^{3}H. New Engl J Med 27:1323, 1967.

Roch-Ramel F, Guisan B: News Physiol Sci 14:80, 1999.

Rose RJ: Exercise physiology, Vet Clin N Am Equine Prac 1(3):437, 1985.

Rowland LP, DiMauro S, Layzer RB: Phosphofructokinase deficiency. In: Engel AG, Bunker BQ, eds. Myology. New York: McGraw-Hill Book Co:1603-1617, 1986.

Salway JG: Metabolism at a glance, 1st ed, Oxford, England: Blackwell Science, 1994.

Schumm DE: Essentials of biochemistry, 2nd ed, Boston, MA: Little, Brown & Co, 1995.

Sharkey LC, Radin MJ: Manual of Veterinary Clinical Chemistry: A Case Study Approach, 1st ed, Jackson Hole, WY: Teton NewMedia, 2010.

Shin DJ, Campos JA, Gil G, et al: PGC-1a activates CYP7A1 and bile acid biosynthesis. J Biol Chem 278:50047, 2003.

Siggaard-Andersen O: The pH, log Pco_2 blood acid-base nomogram revised. Scand J Clin Lab Invest 14:598, 1962.

Siggaard-Andersen O: The acid-base status of the blood. Copenhagen:Munksgaard, 1963.

Simoyi MF, Van Dyke K, Klandorf H: Am J Physiol, Regul Integr Comp Physiol 282, R791, 2002.

Smith JE, Ryer K, Wallace L: Glucose-6-phosphate dehydrogenase deficiency in a dog. Enzyme 21:379,1976.

Snow DH, Guy PS: Muscle fibre type composition of a number of limb muscles in different types of horse. Res Vet Sci 28:137, 1980.

Snow DH, Harris RC, Stuttard E: Changes in haematology and plasma biochemistry during maximal exercise in greyhounds. Vet Rec 123:487, 1988.

Sotiriou S, et al: Ascorbid-acid transporter Slc23a1 is essential for vitamin C transport into the brain and for perinatal survival. Nature Med 8:514, 2002.

Spector R: Vitamin homeostasis in the central nervous system. N Engl J Med 296:1393, 1977.

Staaden RV: Exercise physiology. In, Electrocardiography and cardiology, the JD Steel memorial refresher course. Sydney, Australia: Univ of Sydney, Proceedings No 50:261, 1980.

Stewart PA: How to understand acid-base. A quantitative acid-base primer for biology and medicine, New York, NY: Elsevier North Holland Inc, 1981.

Stewart PA: Modern quantitative acid-base chemistry. Canad J Physiol Pharmacol 61:1444, 1983.

Subbiah MTR, Yunker RL: Cholesterol 7α-hydroxylase of rat liver: an insulin sensitive enzyme. Biochem Biophys Res Commun 124:896, 1984.

Swan RC, Pitts RF: Neutralization of infused acid by nephrectomized dogs. J Clin Invest 34:205, 1955.

Tepperman J, Tepperman HM: Metabolic and endocrine physiology, 5th ed, Yearbook Medical Publishers, 1987.

Thomas C, Auwerx J, Schoonjans K: Bile acids and the membrane bile acid receptor TGR5-connecting nutrition and metabolism. Thyroid 18:167, 2008.

Thomas C, Pellicciari R, Pruzanske M, et al: Targeting bile-acid signaling for metabolic diseases. Nat Rev Drug Discov 7:678, 2008.

Tsukaguchi H, et al: A family of mammalian Na^+-dependent L-ascorbic acid transporters. Nature 399:70, 1999.

Turley SD, Dietschy JM: The metabolism and excretion of cholesterol by the liver. In, The liver: biology and pathobiology, 2nd ed, Arias IM, Jakoby H, Popper D, Schachter D, Scafritz DA, editors, New York, NY: Raven Press, 1988.

Verma S, Fedak PWM, Weisel RD, et al: Fundamentals of reperfusion injury for the clinical cardiologist. Circulation 105:2332, 2002.

Vora S, Davidson M, Seaman C, et al: Heterogeneity of the molecular lesions in inherited phosphofructokinase deficiency. J Clin Invest 72:1995,1983.

von Engelhardt W: Cardiovascular effects of exercise and training in horses. Adv Vet Sci Comp Med 21:173, 1977.

Wang YX, Zhang CL, Yu RT, et al: Regulation of muscle fiber type and running endurance by PPARδ. PLos Biol 2(10):e294, 2004.

Watanabe Y: Serial inbreeding of rabbits with hereditary hyperlipidemia (WHHL-rabbit). Incidence and development of atherosclerosis and eanthoma. Atherosclerosis 36:261, 1980.

Watson JD: Molecular biology of the gene, 2nd ed, Philadelphia, PA: Saunders, 1972.

Webster CRL: Clinical pharmacology, 1st ed, Jackson Hole, WY: Teton NewMedia, 2001.

West JB: Best and Taylor's physiological basis of medical practice, Williams and Wilkins, 1990.

Whitby LG, Smith AF, Beckett GJ, et al.: Lecture notes on clinical biochemistry, 5th ed, Oxford, England: Blackwell Science, 1993.

White A, Handler P, Smith EL, et al.: Principles of biochemistry, 6th ed, New York, NY: McGraw-Hill, 1978.

Whitehair KJ, Haskins SC, Whitehar JG, et al: Clinical applications of quantitative acid-base chemistry. J Vet Intern Med 9:1, 1995.

Williamson DH: Ketone body metabolism during development. Fed Proc 44:2342, 1985.

Wilson JX: Antioxidant defense of the brain: A role for astrocytes. Can J Physiol Pharmacol 75:1149, 1997.

Wilson JX, Dixon SJ, Yu J, et al: Ascorbate uptake by microvascular endothelial cells of rat skeletal muscle. Microcirculation 3:211, 1996.

Winters RW, Dell R: Regulation of acid-base equilibrium. In: Physiological controls and regulations, Yamamoto WS, Brobeck JR, editors. Philadelphia, PA: WB Saunders:181, 1965.

Zhang JV, Ren PG, Avsian-Kretchmer O, et al: Obestatin, a peptide encoded by the ghrelin gene, opposes ghrelin's effects on food intake. Science 310 (5750):996, 2005.

Zhang Y, Lee FY, Barrera G, et al: Activation of the nuclear receptor TXR improves hyperglycemia and hyperlipidemia in diabetic mice. Proc Natl Acad Sci USA 103:1006, 2006.

Zimmerman HJ: Hepatotoxicity, New York, NY: Appleton-Century-Crofts, 1978.

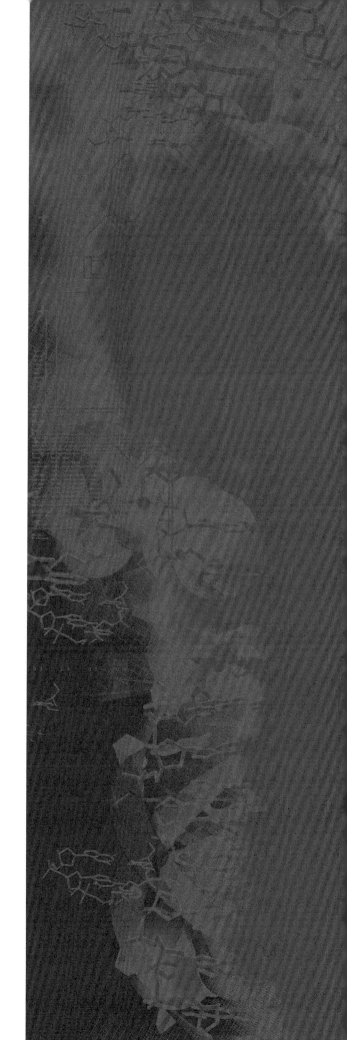

Index

Printed and bound by CPI Group (UK) Ltd, Croydon, CR0 4YY

03/10/2024

01040318-0020